List of the Elements

Name	Symbol	Atomic Number	Atomic Weight
Actinium	Ac	89	227.028
Aluminum	Al	13	26.982
Americium	Am	95	243
Antimony	Sb	51	121.76
Argon	Ar	18	39.948
Arsenic	As	33	74.922
Astatine	At	85	210
Barium	Ba	56	137.327
Berkelium	Bk	97	247
Beryllium	Be	4	9.012
Bismuth	Bi	83	208.980
Bohrium	Bh	107	262
Boron	B	5	10.811
Bromine	Br	35	79.904
Cadmium	Cd	48	112.411
Calcium	Ca	20	40.078
Californium	Cf	98	251
Carbon	C	6	12.011
Cerium	Ce	58	140.115
Cesium	Cs	55	132.905
Chlorine	Cl	17	35.453
Chromium	Cr	24	51.996
Cobalt	Co	27	58.933
Copernicium	Cn	112	277
Copper	Cu	29	63.546
Curium	Cm	96	247
Darmstadtium	Ds	110	269
Dubnium	Db	105	262
Dysprosium	Dy	66	162.5
Einsteinium	Es	99	252
Erbium	Er	68	167.26
Europium	Eu	63	151.964
Fermium	Fm	100	257
Fluorine	F	9	18.998
Francium	Fr	87	223
Gadolinium	Gd	64	157.25
Gallium	Ga	31	69.723
Germanium	Ge	32	72.61
Gold	Au	79	196.967
Hafnium	Hf	72	178.49
Hassium	Hs	108	265
Helium	He	2	4.003
Holmium	Ho	67	164.93
Hydrogen	H	1	1.0079
Indium	In	49	114.82
Iodine	I	53	126.905
Iridium	Ir	77	192.22
Iron	Fe	26	55.845
Krypton	Kr	36	83.8
Lanthanum	La	57	138.906
Lawrencium	Lr	103	262
Lead	Pb	82	207.2
Lithium	Li	3	6.941
Lutetium	Lu	71	174.967
Magnesium	Mg	12	24.305
Manganese	Mn	25	54.938
Meitnerium	Mt	109	266
Mendelevium	Md	101	258
Mercury	Hg	80	200.59
Molybdenum	Mo	42	95.94
Neodymium	Nd	60	144.24
Neon	Ne	10	20.180
Neptunium	Np	93	237.048
Nickel	Ni	28	58.69
Niobium	Nb	41	92.906
Nitrogen	N	7	14.007
Nobelium	No	102	259
Osmium	Os	76	190.23
Oxygen	O	8	15.999
Palladium	Pd	46	106.42
Phosphorus	P	15	30.974
Platinum	Pt	78	195.08
Plutonium	Pu	94	244
Polonium	Po	84	209
Potassium	K	19	39.098
Praseodymium	Pr	59	140.908
Promethium	Pm	61	145
Protactinium	Pa	91	231.036
Radium	Ra	88	226.025
Radon	Rn	86	222
Rhenium	Re	75	186.207
Rhodium	Rh	45	102.906
Roentgenium	Rg	111	272
Rubidium	Rb	37	85.468
Ruthenium	Ru	44	101.07
Rutherfordium	Rf	104	261
Samarium	Sm	62	150.36
Scandium	Sc	21	44.956
Seaborgium	Sg	106	263
Selenium	Se	34	78.96
Silicon	Si	14	28.086
Silver	Ag	47	107.868
Sodium	Na	11	22.990
Strontium	Sr	38	87.62
Sulfur	S	16	32.066
Tantalum	Ta	73	180.948
Technetium	Tc	43	98
Tellurium	Te	52	127.60
Terbium	Tb	65	158.925
Thallium	Tl	81	204.383
Thorium	Th	90	232.038
Thulium	Tm	69	168.934
Tin	Sn	50	118.71
Titanium	Ti	22	47.88
Tungsten	W	74	183.84
Uranium	U	92	238.029
Vanadium	V	23	50.942
Xenon	Xe	54	131.29
Ytterbium	Yb	70	173.04
Yttrium	Y	39	88.906
Zinc	Zn	30	65.39
Zirconium	Zr	40	91.224
Flerovium	Fl	114	—
Livermorium	Lv	116	—

PEARSON ALWAYS LEARNING

Steve Russo • Mike Silver

Preparatory Chemistry
CHEM 1301

Custom Edition for Texas Tech University

Taken from:
Introductory Chemistry: Atoms First, Fifth Edition
by Steve Russo and Mike Silver

Buy : Chemical principles
Peter Atkins, Loretta Jones, Leroy Laverman
7th Edition

Cover Art: Courtesy of Photodisc/Getty Images.

Taken from:

Introductory Chemistry: Atoms First, Fifth Edition
by Steve Russo and Mike Silver

Upper Saddle River, New Jersey 07458

This special edition published in cooperation with Pearson Learning Solutions.

Pearson Learning Solutions, 501 Boylston Street, Suite 900, Boston, MA 02116
A Pearson Education Company
www.pearsoned.com

Printed in the United States of America

3 4 5 6 7 8 9 10 V092 18 17 16 15 14

000200010271919437

DH

PEARSON

ISBN 10: 1-269-96756-8
ISBN 13: 978-1-269-96756-3

About the Authors

Steve Russo (right) has taken an early retirement as a Senior Lecturer in the Department of Chemistry at Cornell University and is now teaching chemistry part time at Ithaca College. Prior to that, he was an Assistant Professor at Indiana University. While there, he designed and implemented a state-of-the-art computer resource center for the undergraduate chemistry curriculum. He received his B.S. in chemistry from St. Francis College in New York City and his Ph.D. in physical organic chemistry from Cornell University. He is a member of the American Chemical Society and has been a recipient of the Dupont Teaching Award, Clark Teaching Awards while at Cornell University, and an Amoco Distinguished Teaching Award from Indiana University.

Mike Silver (left) is a Professor Emeritus of Chemistry at Hope College. He received his B.S. in chemistry from Farleigh Dickinson University in New Jersey and his Ph.D. in inorganic chemistry from Cornell University. He is also a member of the American Chemical Society (ACS) and past president of the ACS West Michigan Section, as well as a member of the Society of Cosmetic Chemists. He has received the Camille and Henry Dreyfus Teacher-Scholar Award for Excellence in Teaching and Research and the Provost's Award for Teaching Excellence. Currently, he teaches a course a semester at Hope College or Grand Rapids Community College, and he also serves as a consultant to the chemical and manufacturing industries in projects ranging from cancer drug design and synthesis to electrochromic automobile mirrors to adhesives for use in oral applications. Dr. Silver has designed and synthesized a variety of novel molecules, including an immune system stimulator to combat skin cancer and eczema, and an emulsifier that allows oil and water to coexist in what is called a multiple or triple emulsion. Both have been submitted for patent protection.

Brief Contents

Contents

Preface

Many instructors have told us that our text is the most readable introductory chemistry textbook their students have ever used. They also told us that their students can use as much help as possible when learning how to solve problems. With each edition we have included more worked examples and more end-of-chapter problems, including one-step-beyond problems, visually oriented and concept-based problems, and in-class group exercises, and stronger tools to help students learn the skills they need. To make the material more memorable, we have chapter-opening "teasers." These are stories and anecdotes utilizing materials from history to the present day that pertain to the material in the chapter and are fun and interesting to read. Also, every edition of our text, starting from the first, has employed an **"atoms first approach."** Why atoms first? The simple answer is that this is the way we have always taught this material since we began teaching decades ago. Both of us were influenced by the late Professor Michell J. Sienko of Cornell University, a textbook author himself and a fabulously talented lecturer who could hold the attention of 500 students at a time. His textbook and lectures always built up the material from the "bottom up," from atoms to molecules to properties of molecules to stoichiometry and reactions and beyond. It instantly made sense for us to teach this way. Following an atoms first approach, we never had to say things like "I'm teaching you this now because it's covered in lab this week," or "This will make more sense when we cover polarity later." When building a house, one doesn't usually pour half of the foundation, start work on the second floor, and return later to complete the foundation. In the same way, the atoms first approach lays a proper foundation and then continually builds atop it; while it gives some choice as to the order of topics, the path it builds is consistently straightforward and does not have leaps of faith or distracting U-turns.

Our goal with this text has always been to help students make sense of chemistry. As chemists, we know that chemistry is intrinsically interesting, that its principles do form a reasonably coherent whole, and that its problem-solving skills can be mastered by anyone. But, too often, students see the subject as incomprehensible and the course as a frightening labyrinth. All too frequently, they fall back on rote learning—memorizing algorithms, plugging numbers into formulas, and forgetting much of what they learned starting as soon as the pencil is dropped at the end of the final exam. As chemistry instructors, we hate to see that. Therefore, we designed this book to promote comprehension and problem solving as complementary skills. A student who understands the principles of chemistry does not have to rely strictly on memorization and is more likely to enjoy and retain the material. This understanding and sufficient practice at solving problems will help students master the principles. We hope that our book will help students come out of the course with a body of knowledge and skills that will serve them and that they will want to retain.

New to This Edition

What can authors possibly add to a fifth edition of a textbook? The answer turns out to be "Plenty!"

1. **Learning Outcomes:** Each end-of-chapter review section now begins with a table of learning outcomes for each section of the chapter, correlated to the end-of-chapter problems. The learning outcomes consist of goals and skills the student should learn and practice before moving on to the next chapter.
2. **One More Thing:** We have always been proud of the conceptual depth that this book plumbs, but some instructors have told us that they skip sections that go to great conceptual depth. Still, we wanted to add even more of this depth for this fifth edition that other reviewers and users have asked for. How to satisfy both? Our solution was to create a final section for every chapter titled "One More Thing." Being so titled, it gives the instructor an option to skip it and be guilt-free when time requirements are tight. We have made sure that it can be skipped without doing damage to the flow of the material that follows. In some cases, this section contains material that was previously incorporated in the main body of the chapter in the previous edition. In other cases it contains totally new material that we thought would be great to add for its conceptual and interest value. Now the instructor, and each individual student, has the option to jump into the deepest end of the conceptual pool if he or she wants.
3. **WorkPatches Hints Available in MasteringChemistry:** Judging from instructor and student feedback over the years, WorkPatches have always been a favorite feature of this book. They were always meant to be a way to break the reader out of the passive reading mode and make students more interactive by encouraging them to think about what was just read before going on. They accomplish this by posing a conceptual question, followed by a continuation of the main text that assumes the question was successfully answered without ever revealing the actual answer. Skipping a WorkPatch therefore is like joining in the middle of a conversation, something a reader would not want to do. We realize we are forcing the reader to do something here, and the answers are provided at the end of the chapter, but we don't want readers looking up these answers without first trying to come up with an answer on their own. So, in this edition, for the first time, we offer some help and something in step with the times. Today many, if not most, students have a smartphone. Included in the fifth edition is a QR code that will link the reader's phone to a site where he or she can get a hint to any WorkPatch. Given the love affair that students have with this technology, our hope is that this will be an incentive for them to interact more with the text. For readers who do not have a smartphone or a tablet, the hints can still be accessed by visiting MasteringChemistry for the fifth edition.
4. **Concept Questions:** In this edition we wanted to give readers more than just lists of topics that we call "Have You Learned This?" and summaries that we call "Skills to Know," which we have always included at the end of each chapter. Now, for the first time, each chapter ends with ten multiple-choice questions called "Concept Questions." These are meant to be done quickly and will give readers an opportunity to see if they have grasped the fundamental concepts presented in the chapter. Instructors know that if we throw a little technobabble at students, we can often shake students' confidence in what they know. The concept questions often use this approach, but if students have a true understanding, then smoke won't get in the way of their finding the correct response. The concept questions give them this opportunity.

5. **New End-of-Chapter Problems:** As always, we endeavor to refine and improve our end-of-chapter problems. We have gone through each chapter and, based on our own ideas and with input from users, replaced or improved more than 20% of the existing end-of-chapter questions. We receive many compliments on the number, variety, and quality of our end-of-chapter problems. Every instructor knows the value of time students spend working on such problems. As Professor Sienko would tell every one of his classes on the first day, "If you are not in some degree of pain during this course, you are not learning." Our book provides an enjoyable read due to its conversational style, atoms first approach, conceptual depth, and clean layout. The pain comes from struggling with well-thought-out end-of-chapter problems. Of course, the payback is good performance on exams and a lifelong retention of the major concepts of chemistry.

The conversational tone and the atoms first approach have always been and remain the principle hallmarks of our text. We are big believers in their power to help students learn. With each new edition, and with generous help from our users, we have hunted down errors and hopefully corrected them all. If you find any, please email us at silver@hope.edu (Mike Silver) and srusso@ithaca.edu (Steve Russo). Please send along any other comments you might have as well, good or bad. We love hearing from our users, and we accept constructive criticism reasonably well (although we prefer high praise).

Promoting Active Learning

How can we, as textbook authors, promote active learning? First, quite simply, we provide a book that makes sense to students. A student who understands the material is less likely to fall back on passive memorization. As one instructor told us, "This book allows me to spend class time doing hands-on learning versus spending time explaining the book."

Second, we incorporate devices to encourage active reading. A flip through the book will show many sets of practice problems. These practice problems are located so that students can immediately apply the skill the text has just presented. They are presented in sets of three or four; the first problem is solved step-by-step in the text, and the answers to the others are given at the back of the book. Each chapter contains an average of 25 practice problems.

However, we are not naïve. We know that many students routinely skip over these in-chapter practice problems. Therefore, we have included conceptual practice problems called WorkPatches. A WorkPatch is a "stealth" problem. It follows smoothly from the preceding text and is not boxed off. What is more, a student who tries to read through a WorkPatch without solving it will find that the subsequent text refers to, and often depends on, the answer—but does not say what it is. (The solutions to all WorkPatches are given at the end of the chapter.) In some cases, a WorkPatch serves as the springboard into the next topic, and all WorkPatches are now connected to hints, which can be accessed by scanning the QR code in the next few pages or back of the textbook. WorkPatches are denoted by a yellow stop sign/red light icon.

Helping Students Master Problem-Solving Skills

An instructor flipping through our text might be inclined to ask, "Where are all the worked examples?" In fact, this text has an abundance of worked examples and other problem-solving aids, but we have handled them in a way that preserves the text's coherence. As in the previous editions, many worked examples are presented in the text itself. We feel strongly that problem-solving techniques should be explained with the same care and continuity we use for concepts.

When students are working on problems, however, they also need access to compact summaries and examples. We have augmented these resources in the following ways:

- Important problem-solving methods are summarized in charts in the text.
- The same charts, accompanied by worked examples and additional material, appear in a special Skills to Know section at the end of the chapter immediately preceding the end-of-chapter problems. This section is intended as a "help center" to which students can refer while working problems. (A few of the more purely conceptual chapters do not have a Skills to Know section.)
- An abundance of additional step-by-step methods, worked examples, and practice problems are provided in the *Student Workbook and Selected Solutions* that accompanies the text, authored by our colleagues Saundra Yancy McGuire and Elzbieta Cook at Louisiana State University. For a weak or struggling student, nothing is more important than abundant, guided, confidence-building practice. This workbook enables us to offer a truly realistic amount of help while maintaining a clean, readable textbook. The *Workbook* also contains a generous mathematics review. We came to know Dr. McGuire when she directed the Center for Learning and Teaching at Cornell; she currently directs the Center for Academic Success at Louisiana State University. We are extremely glad that she chose to join her expertise with ours and are thankful to Elzbieta Cook for agreeing to revise the supplements for the fifth edition.
- As noted earlier, each chapter contains internal practice problems and WorkPatches in addition to end-of-chapter problems.

Room to Practice: Extensive Problem Sets

- Each problem set at the end of the chapter includes a section of additional problems that are not categorized by chapter section.
- There are now even more visually oriented problems.
- Each problem set contains a sufficient abundance of each of the types of problems an instructor might require.
- As in the previous editions, answers to selected problems are provided at the end of the book. (The full solutions for these selected problems, as well as for all the practice problems, are available in the *Student Workbook and Selected Solutions* manual.)

Making Chemistry Memorable

Instructors have told us that a surprising number of their students actually read our book rather than using it mainly as a resource while solving problems. We explain chemistry using everyday, conversational language, and we tie the concepts and calculations to stories and examples that help bring them to life. We use humor in places. We also ensure that the students know which points are fundamental and which represent additional detail. You will notice that most of the illustrations lack legends. That is because the text and the illustrations work hand-in-hand, and each illustration is placed exactly where it belongs.

We Want to Hear from You

One of the pleasures of revising a book is hearing from instructors who use it—learning what works and doesn't work, and gathering ideas. If you have any comments or suggestions, please feel free to contact us at the following email addresses: srusso@ithaca.edu (Steve Russo); silver@hope.edu (Mike Silver).

Acknowledgments

We very much wish to acknowledge the people who were instrumental in helping us with this fifth edition. We are indebted to our editors, Terry Haugen and Chris Hess, for believing enthusiastically in this book and championing a fifth edition. We very much want to thank Lisa Pierce, Jenna Gray, Wendy Perez, and Gina Cheselka, our project managers. These are the people who are the conductors of the orchestra. They overlook all the details, crack the whip when necessary, and make sure the final product is beautiful music rather than a cacophony of noise. Fran Falk, our editorial coordinator, was responsible for acquiring the prescriptive reviews, which greatly helped shape the direction of this edition. Divya Narayanan and Kerri Wilson did a wonderful job in photo research and finding new images that were needed. We are very grateful to Wynne Au Yeung and Mark Ong, who were in charge of the art and design program and are responsible for maintaining and improving what we believe to be one of the key features of our book, its clean, nonchoppy, and organized look that helps to make it so readable. Our genuine thanks also go to Wanda España who was in charge of the cover and interior design. We have had a tradition of some great and beautiful covers for the book, and that tradition has been maintained. We certainly also must acknowledge Jonathan Cottrell, our marketing manager, whose job is to get the message out about what makes this book so unique, and Jackie Jakob, our associate content producer, who has led the effort to tie our book to MasteringChemistry, the extremely useful online tutorial and assessment program.

We wish to extend our genuine thanks to the reviewers whose patient and thoughtful comments helped us shape this fifth edition. Their ability to provide excellent suggestions and find errors that we could not see no matter how hard we looked was incredible.

Jeff Allison, *Austin Community College*
Chris Bahn, *Montana State University*
Michael Hauser, *St. Louis Community College at Meramec*
Theresa Hill, *Rochester Community & Technical College*
Amy Johnson, *Eastern Michigan University*
Jesudoss Kingston, *Iowa State University*
Laurie LeBlanc, *Cuyamaca College*
Luigi Marzilli, *Louisiana State University*
Manish Mehta, *Oberlin College*
Mary Sohn, *Florida Institute of Technology*
Susan Thomas, *University of Texas at San Antonio*

We also want to thank Elzbieta Cook at Louisiana State University for her efforts in updating the *Student Workbook and Selected Solutions Manual* as well as the *Instructors Guide and Complete Solutions Manual*.

With regard to the previous editions that helped to shape the nature of this book, we want to thank all the people who have been involved over the years. In addition, for as long as new editions of this book are written, we will always want to thank the people who were on board and who believed in the vision and helped to create the earliest editions: Emiko Koike, Blakely Kim,

Irene Nunes, Jonathan Peck, Lisa Leung, Tony Asaro, and Saundra McGuire. And last, but certainly not least, Maureen Kennedy, Joan Marsh, Margot Otway, and Ben Roberts—the original four chambers who helped to inspire and create the heart of this book. We will always be in your debt.

Supplements

Name of Supplement	Available in Print	Available Online	Instructor or Student Supplement	Description
Student Workbook and Selected Solutions for Introductory Chemistry: Atoms First, 5th edition (0-321-95693-1)	✓		Supplement for Students	By Saundra Yancy McGuire of Louisiana State University and Elzbieta Cook of, Louisiana State University. Features examples from each chapter, learning objectives, review of key concepts from the text, and additional problems for student practice. Also provides comprehensive answers and explanations to selected end-of-chapter problems from the text. Provides over 200 worked examples and more than 550 practice problems and quiz questions to help students develop and practice their problem-solving skills.
Laboratory Manual for Introductory Chemistry (0-321-73025-9)		✓	Supplement for Students	By Wendy Gloffke and coauthored by Doris Kimbrough of the University of Colorado at Denver with contributions from Chris Bahn of Montana State University. Helps students develop data acquisition, organization, and analysis skills while teaching basic techniques. Written to accompany the text, this manual offers 25 experiments. This lab manual is available via Catalyst: The Pearson Custom Laboratory Program for Chemistry. This program allows you to custom build a chemistry lab manual that matches your content needs and course organization. You can either write your own labs using the Lab Authoring Kit tool or select from the hundreds of labs available at http://www.pearsonlearningsolutions.com/custom-library/catalyst. This program also allows you to add your own course notes, syllabi, or other materials.
MasteringChemistry® with Pearson eText for Introductory Chemistry: Atoms First, 5th edition (0-321-92456-8) www.masteringchemistry.com		✓	Supplement for Students and Instructors and Students	MasteringChemistry is the leading online homework and tutorial system for the sciences. Instructors can create online assignments for their students by choosing from a wide range of items, including end-of-chapter problems and research-enhanced tutorials. Assignments are automatically graded with up-to-date diagnostic information, helping instructors pinpoint where students struggle either individually or as a whole class.
Instructor's Guide with Complete Solutions for Introductory Chemistry: Atoms First, 5th edition (0-321-95692-3)	✓		Supplement for Instructors	By Saundra Yancy McGuire of Louisiana State University and Elzbieta Cook of Louisiana State University Includes chapter summaries, complete descriptions of appropriate chemical demonstrations for lecture, suggestions for addressing common student misconceptions, and examples of everyday applications of selected topics for lecture use, as well as the solutions for all the problems in the text.

Test Bank (Download Only) for Introductory Chemistry: Atoms First, 5th edition (0-321-95691-5)	✓	Supplement for Instructors	By Christine Hermann of Radford University This printed test bank includes over 1700 questions that correspond to the major topics in the text.
Instructor Manual for the Laboratory Manual (0-321-73026-7)	✓	Supplement for Instructors	By Wendy Gloffke and coauthored by Doris Kimbrough of the University of Colorado at Denver with contributions from Chris Bahn of Montana State University. This manual includes lists of equipment and chemicals needed to perform each lab.
Instructor's Resource Materials (Download Only) for Introductory Chemistry: Atoms First, 5th edition (0-321-95734-2)	✓	Supplement for Instructors	The online instructor resources include all the art, photos, and tables from the book in high-resolution format for use in classroom projection or for creating study materials and tests. In addition, the Instructor can access modifiable PowerPoint® lecture outlines to highlight key points in his or her lecture. Also available are downloadable files of the *Instructor Teaching Guide*, the Test Bank with more than 1000 questions, and a set of "clicker questions" suitable for use with classroom-response systems.

The atoms first approach—
Making chemistry accessible

Students in this course tend to be vastly diverse in terms of chemical and mathematical background. Written from an **atoms first approach**, the authors focus on the concepts behind chemical equations to help students become more proficient problem-solvers. Combined with a conversational tone, this text encourages mastery of conceptual understanding and quantitative skills students need to gain a deeper understanding of chemistry, rather than just memorization.

1.6 Learning Chemistry with This Book **19**

the Bohr Model of the atom, developed in the early 1900s. More sophisticated models of the atom exist, but many working chemists still use this model to make predictions about their chemistry. Therefore, we will go ahead and use the Bohr Model, keeping in mind that it predicts some chemical properties pretty well so long as you don't push it beyond its capabilities.

1.6 Learning Chemistry with This Book

A common misconception is that chemistry is all math, calculations, and numerical problem-solving. In all honesty, chemistry does have that side to it. But when a chemist is presented with a question about matter, the first thing that comes to mind is *not* a complicated mathematical formula. Instead, chemists use a basic set of fundamental concepts, often best represented with images instead of mathematical equations. For example, a chemist and a non-chemist picture the concept of melting differently:

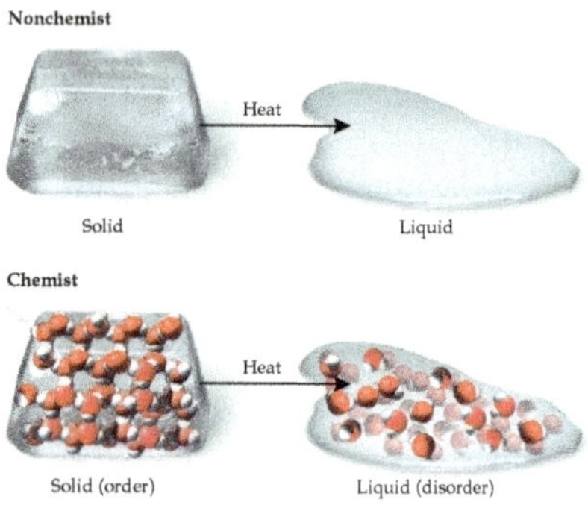

Advantages of an Atom's First Approach:

- starts at the very beginning, providing a solid foundation; building up material from the "bottom-up"
- first discusses atoms—history, their electronic structure, the modern model of atoms and then moves on to chemical bonding and molecules.

Learning Outcomes

2.1	Numbers in Chemistry—Precision and Accuracy	Explain what an exact number is and what a measured number is. [55, 56] Describe the difference between precision and accuracy. [57, 58, 59]
2.2	Numbers in Chemistry—Uncertainty and Significant Figures	Explain why measured numbers always have uncertainty associated with them. [61, 62, 63, 65]
2.3	Zeros and Significant Figures	Identify leading zeros and trailing zeros in a measured number. [64, 66, 67] Identify the amount of uncertainty in a written measured quantity. [63, 68, 69]
2.4	Scientific Notation	Convert a number from standard to scientific notation and vice versa. [71, 75, 173, 182] Identify the degree of uncertainty in a number written in scientific notation. [72, 74, 75fg]
2.5	How to Handle Significant Figures and Scientific Notation When Doing Math	Correctly add or subtract numbers according to the rules of significant figures. [77, 80acd, 161ad] Correctly multiply or divide numbers according to the rules of significant figures. [78, 80b, 161bc]
2.6	Number With a Name—Units of Measure	Identify the common SI units used by science. [81, 82, 83, 85, 88]
2.7	Density: A Useful Physical Property of Matter	Define density, and use it to do calculations. [95, 96, 97, 98]
2.8	Doing Calculations in Chemistry—Unit Analysis	Convert the units of a measured quantity to some other units using unit analysis. [102, 107, 110]

New! Learning Outcomes

Each chapter ends with a section of Learning Outcomes, correlated to the end-of-chapter problems. The Learning Outcomes consist of goals and skills the student should learn and practice before moving on to the next chapter.

Making chemistry manageable and relevant with practical study tools

Students often ask, "When will I use this?" and have difficulty relating chemistry to real topics or events in everyday life. Russo and Silver's text includes various real-world applications and learning tools, placed strategically to boost interest and help students develop essential study skills as they assess their own progress.

7.1 Why Does Matter Exist in Different Phases?

It's the middle of July, 34 °C, and you feel as if you're covered with a wet blanket. Sound familiar?

These are the sweltering, high-humidity days of summer when there is just too much water in the air. You can't see this water because it exists in the gas, or vapor, phase, but you can certainly feel it. It makes you feel miserable. Today's solution to this problem is, of course, the air-conditioner, which cools the air by drawing it in with a fan and passing it over cold refrigeration coils. If you were to look at the coils inside an air-conditioner, you would notice they are dripping wet. Cooling humid air causes the water vapor in the air to *condense* to a liquid on the cold coils, and in this way the air is dried as it is "conditioned." This is why air-conditioned air feels so good on a humid summer day—it's not just cooler, it's also drier. By setting an air-conditioner on its highest setting, you can sometimes cause the refrigeration coils to get so cold that the liquid water on them freezes. Thus, simply by cooling the air, an air-conditioner can drive atmospheric water through three phases—from the *gas phase* thro… and into the *solid phase*.

Section-Level Applications

Compelling applications and vignettes also appear as appropriate at the section level throughout the text.

Real-World Applications

Applications are woven directly into the text's conversational narrative, evoking powerful images in students' minds and helping them grasp and retain concepts. Examples from areas such as nutrition and the atmosphere and including specific examples related to DNA modification and corrosion show that chemistry is a fascinating science.

r such a model
elieve it or not,
rom examining
any light, but
when atoms of
gas state were
large amounts
u are looking
very time you
gn. Neon lights
es filled with
s that give off a
range-red light
with electrical
yellow light is
sodium-vapor
ven a tube filled
en gas glows
hen electrified.
rious nature,

Concept Questions

1. Which of the following statements is true regarding science and technology?
   (a) They are completely unrelated.
   (b) Technology precedes science.
   (c) Science precedes technology.
   (d) Science is the application of technology.
2. Matter can exist as which of the following?
   (a) As a homogeneous mixture of substances.
   (b) As a pure substance.
   (c) As a heterogeneous mixture of substances.
   (d) As all of the above.
3. To be a compound, a substance must
   (a) be made from more than one type of element.
   (b) not be a mixture.
   (c) be homogeneous.
   (d) be all of the above.
4. An elemental substance
   (a) can also be a compound.
   (b) can never be a compound.
   (c) cannot participate in a chemical reaction.
   (d) cannot change state since it is a pure element.
5. When a pure substance melts, it
   (a) undergoes a chemical change.
   (b) changes state.
   (c) changes from a liquid to a solid.
   (d) sublimes.
6. The melting point and the freezing point of a pure substance are
   (a) unrelated.
   (b) identical.
   (c) for the same change of state.
   (d) chemical properties of the substance.
7. A chemical property of hydrogen is that it
   (a) is a gas at room temperature.
   (b) is odorless.
   (c) combines with chlorine to make the compound hydrogen chloride (HCl).
   (d) combines with air to give a mixture of hydrogen and air.
8. In a chemical reaction
   (a) a physical change takes place.
   (b) products are converted into reactants.
   (c) one or more substances are converted into different substances.
   (d) elemental substances change into different elements.
9. A law is explained by
   (a) experimental data.
   (b) bias.
   (c) a theory.
   (d) an experiment.
10. The difference between a hypothesis and a theory is that
    (a) a hypothesis is a guess, whereas a theory is fact.
    (b) a theory is an untested hypothesis.
    (c) a theory is a well-tested hypothesis.
    (d) a theory is never wrong, whereas a hypothesis may be wrong.

Answers: 1c, 2d, 3a, 4b, 5b, 6b, 7c, 8c, 9c, 10c

New! Concept Questions

Asking students to think a bit further, these questions were written to gauge level of understanding and encourage students to review and apply the important concepts of a specific chapter.

Motivating students and providing personalized coaching and feedback

MasteringChemistry®

MasteringChemistry is the most effective, widely used online tutorial, homework and assessment system for chemistry. It helps teachers maximize class time with customizable, easy-to-assign, and automatically graded assessment that motivate students to learn outside of class and arrive prepared for lecture. These assessments can easily be customized and personalized by teachers to suit their individual teaching style. To learn more, visit www.masteringchemistry.com.

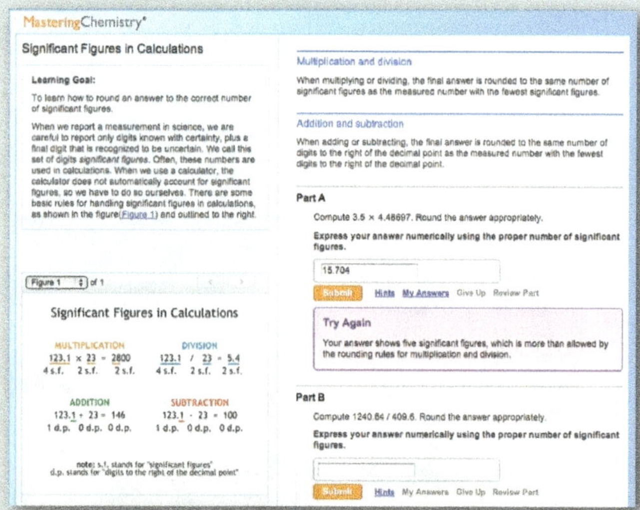

Student Tutorials

Tutorials have been adapted and authored by an advisory board of expert chemists who teach with the atoms first approach. Immediate feedback helps students develop the problem-solving skills and motivation needed to succeed in this course.

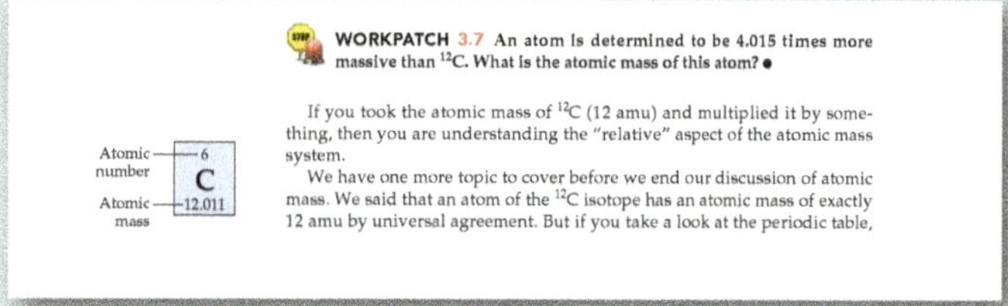

New! WorkPatch Hints

Conceptual practice problems called WorkPatches are now assignable in MasteringChemistry. Students who are having trouble completing a WorkPatch problem can scan the QR code on this page or visit the Study Area in MasteringChemistry for hints, and then go back and try the WorkPatch problem again.

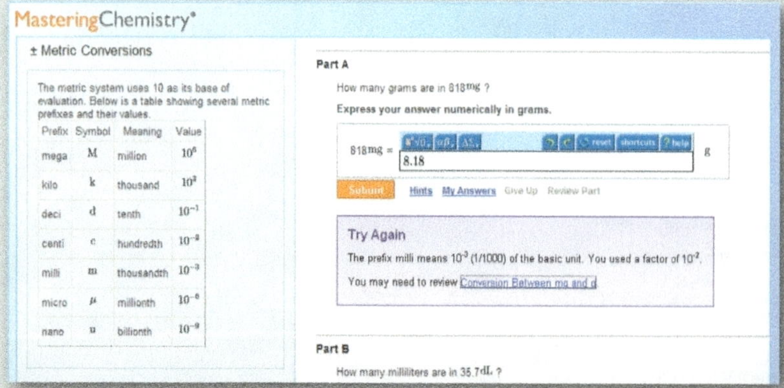

Math Review

MasteringChemistry offers a variety of math remediation options for students to brush up on required quantitative skills including both Math Review Tutorials that can be assigned as prerequisites before moving onto more difficult material and Math Remediation Tutorials that offer just in time help to specific students based on their individual answer inputs. These exercises include guided solutions, sample problems, and feedback when students answer incorrectly.

www.masteringchemistry.com

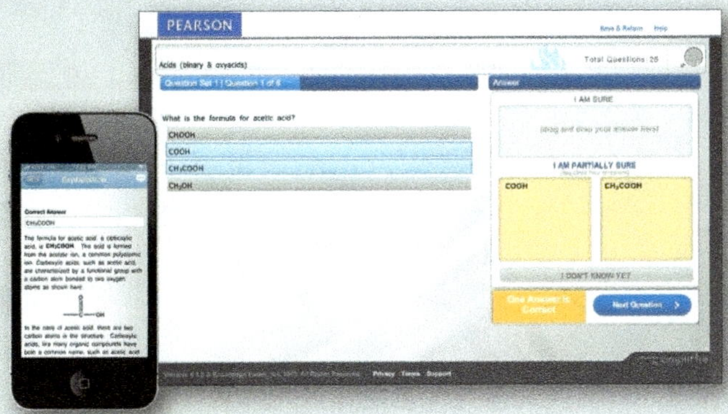

NEW! Dynamic Study Modules

These are designed to enable students to study effectively on their own as well as help students quickly access and learn the nomenclature they need to be more successful in chemistry.

These modules can be accessed on smartphones, tablets, and computers and results can be tracked in the MasteringChemistry® Gradebook. How it works:

1. Students receive an initial set of questions and benefit from the metacognition involved with asking them to indicate how confident they are with their answer.
2. After answering each set of questions, students review their answers.
3. Each question has explanation material that reinforces the correct answer response and addresses the misconceptions found in the wrong answer choices.
4. Once students review the explanations, they are presented with a new set of questions. Students cycle through this dynamic process of test-learn-retest until they achieve mastery of the material.

Gradebook and Student Performance Snapshot

The Gradebook feature captures the step-by-step work of each student in class, including the time taken on every step. With a single click, charts summarize the most difficult problems, vulnerable students, grade distribution, and even score improvement over the course.

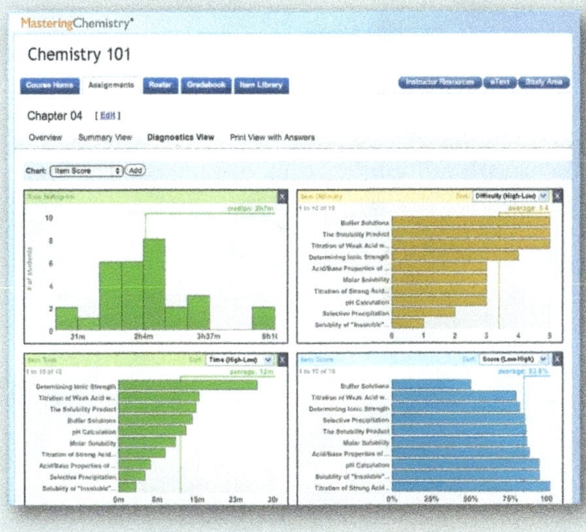

Calendar Features

The Course Home default page features a Calendar View displaying upcoming assignments and due dates.

- Instructors can schedule assignments by dragging and dropping the assignment onto a date in the calendar.
- The calendar view gives students a syllabus-style overview of due dates, making it easy to see all assignments due in a given month.

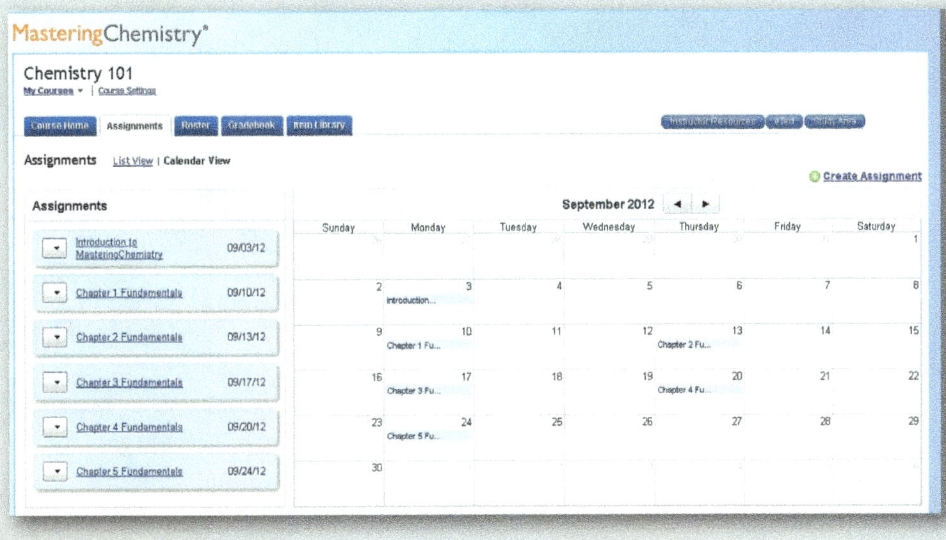

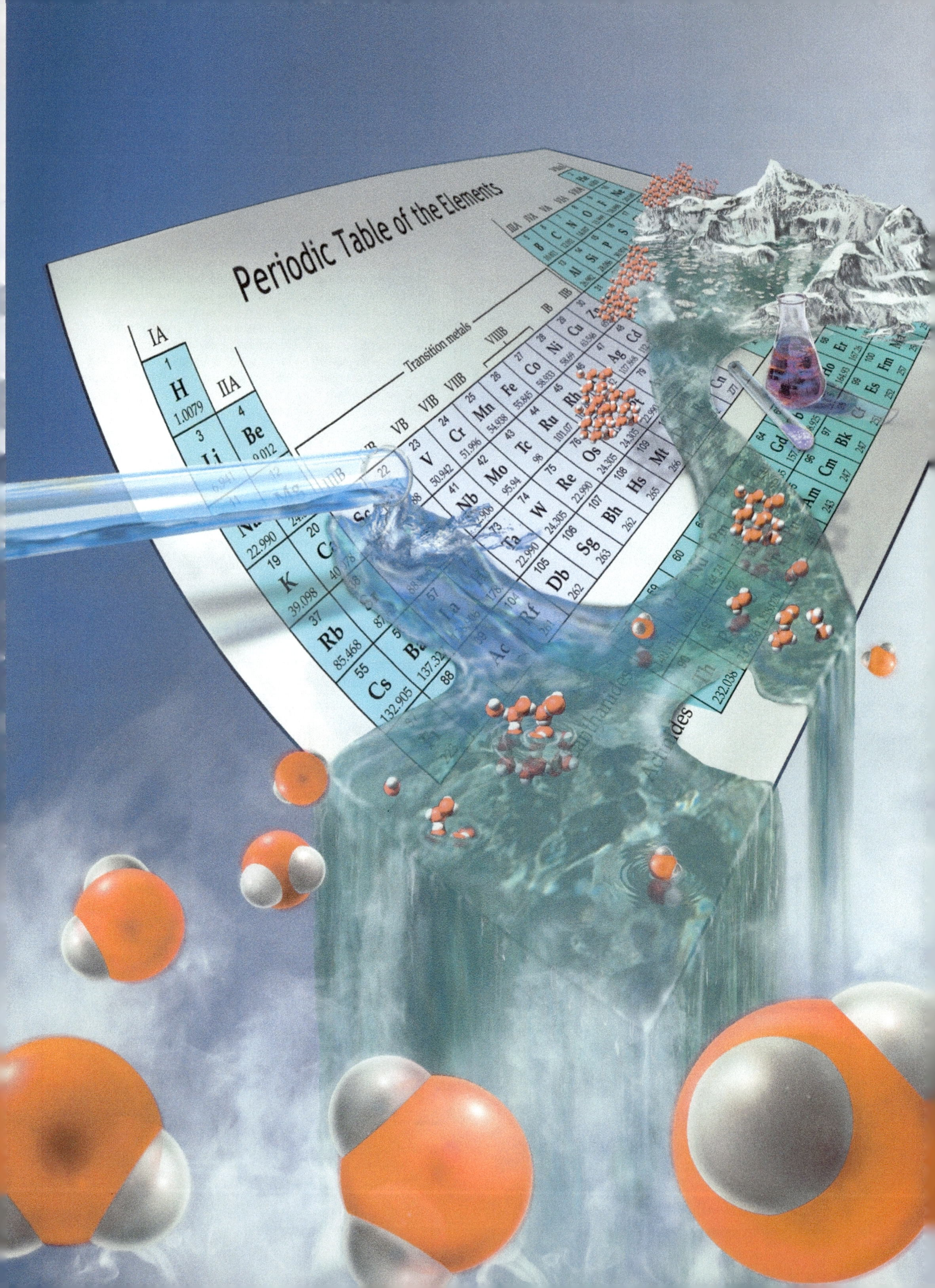

CHAPTER 1

What Is Chemistry?

Congratulations! You are taking a chemistry course. Chances are that you are taking it for one or more of the following reasons:

1. I have to take a science course, geology was closed, and physics is too hard.
2. All my friends are in the class.
3. It fits into my schedule.
4. It is required for my major.

There are much better reasons. Read on!

It is the early part of the twentieth century, 1914 to be exact. I am overcome with grief. My youngest child is burning with fever. The sickness has spread from her ear to her entire body. Her skin has a scarlet look, and she is in great pain. The doctor has applied some tincture of iodine, but he does not know how to cure her. He has told us to make arrangements. My beloved child will not see her fourth birthday.

It is the early part of the twenty-first century, 2014 to be exact. My daughter was ill yesterday with an earache. Our pediatrician diagnosed a streptococcus infection and administered the antibiotic amoxicillin. My daughter thought it tasted good, and she is back in preschool today, completely free of fever and pain.

It is the year 2036. We have chosen to have a daughter. Unlike most of today's parents, we will not preselect her IQ. However, we do agree with our genetic counselor that her system should be genetically engineered so that she will be immune to all known bacterial and viral infections.

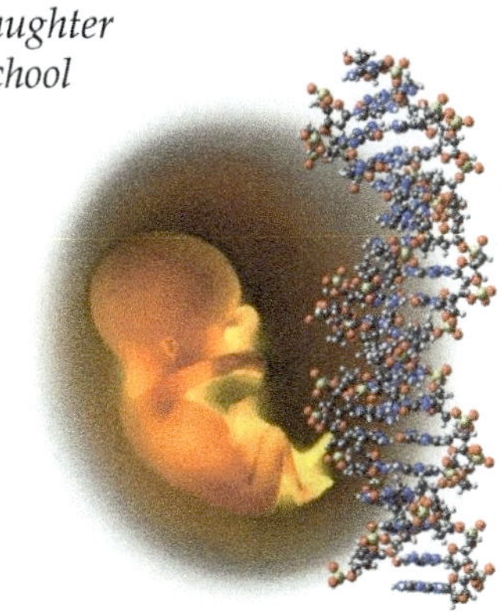

1.1 Science and Technology

The span from 1914 to 2014 was 100 years. The year 2034 is only 20 years away. The pace at which things are changing is accelerating at an unbelievable rate. One hundred years from now will be as different from today as today is from 500 years ago. Within the past decade, a significant portion of the genetic code for the human genome has been unraveled. One hundred years ago we had never even heard of DNA; today we are cloning it. One hundred years ago we burned coal; in less than half that time from now we'll generate energy in fusion reactors powered by hydrogen taken from seawater, duplicating the process that occurs within the cores of stars.

Change this rapid is new for humanity. Until recently, each new generation could expect to live pretty much like the one before it. Not anymore. We scarcely have time to get used to one change before ten more are upon us. We've barely had time to think about the ethics of birth control, a development of the 1970s, and now stem cell research, life prolongation, and cloning are knocking at our door. Your personal computer and its software are almost outdated the day you buy them. What is feeding all this change? The answer is *science*.

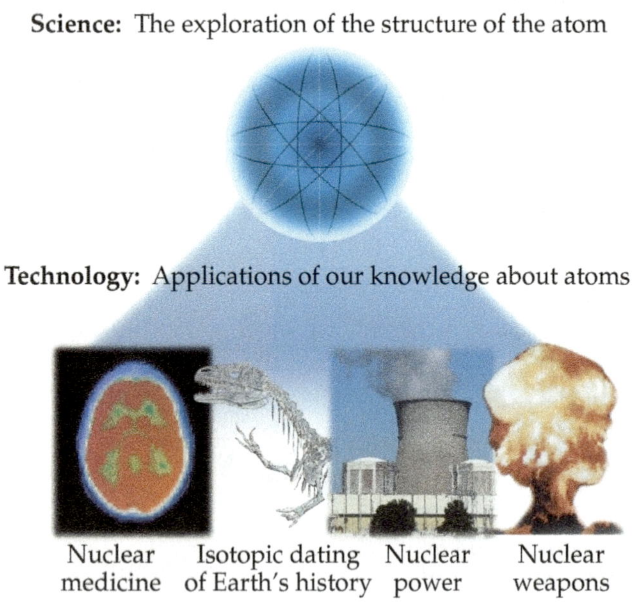

The dictionary defines **science** as "the experimental investigation and explanation of natural phenomena" or "knowledge from experience." Science begins with a simple question, like "how?" or "why?" How are atoms put together? Why do bats fly at night? Science is the pursuit of knowledge for its own sake, because we are curious. But science itself can't cause change unless something is done with the knowledge it uncovers. For that we need **technology**, the *application* of scientific knowledge. So science is also the pipeline for technology—feeding it and supplying it with ideas. For example, scientists asked "How are atoms put together?" and their experiments led them to an answer. Today, technologists can use that knowledge to change our lives by developing nuclear medical technologies to treat cancer and by building nuclear bombs. As is so often the case with scientific knowledge, it can be used to achieve very different ends.

Because science feeds technology, it's not a bad idea to ask the question "Is science always right?" Just consider these headlines:

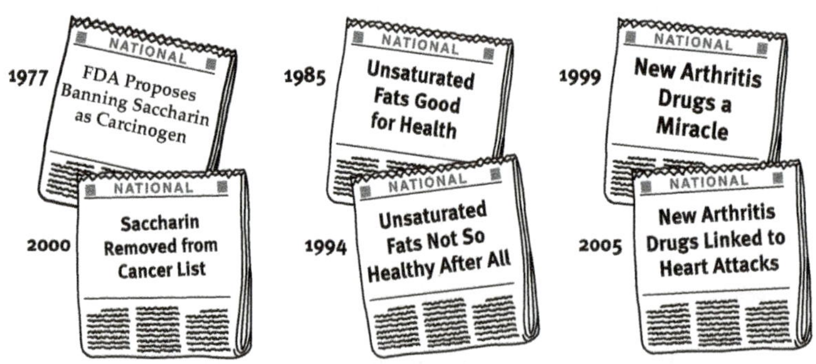

What is going on? Isn't science only about absolute, fundamental, provable truths? Unfortunately, no. Science, like literature, art, and music, is a human endeavor. And because it is a human endeavor, carried out by human scientists, you would not be wise to bet on science's infallibility. Ego, mistakes, and stubbornness can all get in the way of finding the truth. And what about technology? Is the technological application of scientific knowledge always good? Consider some of the forces that drive technology: the desire to benefit humankind, the desire to make a profit, the desire to be stronger than our enemies. Which motive do you think drives the cigarette industry in its application of knowledge concerning the effects of nicotine?

This brings us to some very important questions. Who should decide which areas of science are explored? Who should decide if a scientific result is real or a hoax? Who should decide what is done with scientific knowledge? In a free society, the answer is supposed to be the collective "you." After all, what scientists discover and technologists put into practice will dramatically affect how you live. In addition, much of the scientific research in this country is paid for by your tax dollars. Does this mean you have to earn a Ph.D. in chemistry, biology, or physics so that you can make informed decisions? For most people, that is not a practical or even a desirable solution. But total ignorance of science is not the answer either, because science and technology affect us directly in our everyday living.

This book is about the branch of science known as *chemistry*, often considered the "central science" since it forms a bridge between the principles of physics and the practice of biology. **Chemistry** is the study of matter and the transformations it undergoes (matter is being transformed within the space shuttle's rocket engines illustrated here). To understand this definition, we need to focus on two key concepts: *matter* and *transformation*.

Hydrogen and oxygen combine and *transform* into water vapor and heat in the space shuttle's liquid-fueled engine, helping to propel the shuttle into space.

1.2 Matter

Matter can be simply defined as "stuff"—anything that has mass and occupies space. Some matter you can feel and see, like this book. Other matter, like the air that surrounds you, is difficult to detect, but it still exists. It occupies space (the volume of the room), and it has mass (a tank filled with helium gas weighs more than the empty tank). Defining matter as "stuff" is adequate for most everyday situations. Since scientists often ask very specific questions, however, it almost always requires much more precise and detailed definitions. Chemists define matter by dividing it into two broad types, *pure substances* and *mixtures*. In **pure substances**, only a single type of matter is present. **Mixtures** occur when two or more pure substances are intermingled with each other. For example, table salt (chemical name, sodium chloride) is a pure substance. So is water. And so is table sugar (chemical name, sucrose). If you put salt and sugar in a jar together and shake, however, you have a mixture. Dissolve sugar in water, and you have another mixture. Some things that you might not think of as mixtures actually do fit the definition—a rock, for example. In most rocks, you'll see a mixture of different minerals, each a different pure substance.

Chemists further subdivide mixtures into two types, *homogeneous* and *heterogeneous*. **Homogeneous mixtures** (*homo-*, meaning "the same") are ones in which the composition of the mixture is identical throughout. A cup of tea with some sugar dissolved in it is a homogeneous mixture. Once well stirred, such a mixture is exactly the same no matter where you sample it.

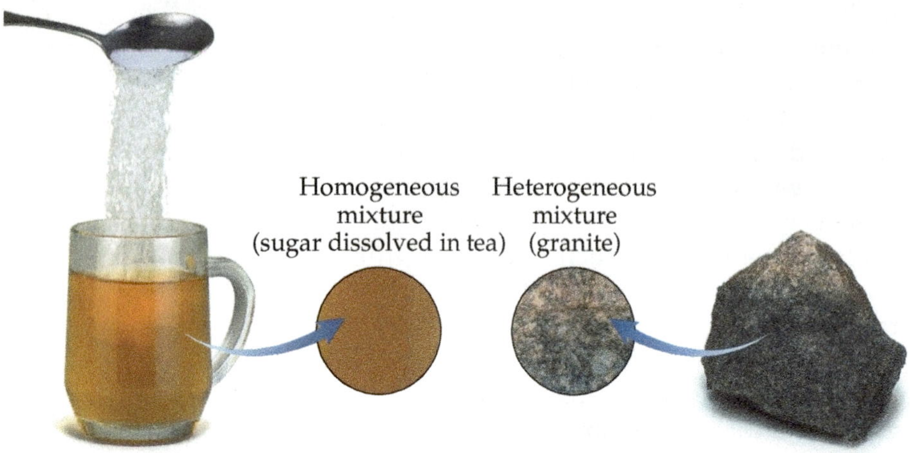

Another name for a homogeneous mixture is a **solution**. Our well-stirred cup of tea with sugar is a solution. The air you are breathing, consisting of a mixture made mostly of nitrogen gas and a smaller amount of oxygen gas, is also a homogeneous mixture, or a solution. Under normal conditions, no matter where you sample the air in the room, its composition (percent nitrogen and percent oxygen) is the same.

The rock pictured above is an example of a **heterogeneous mixture** (*hetero-*, meaning "different") because its composition is not the same throughout. Sometimes, heterogeneous mixtures can appear to be homogeneous even when they aren't. A mixture of table salt and table sugar is a heterogeneous mixture even when the two substances are ground together into a fine powder. This is true because a tiny sample taken at one place in the mixture might contain a different ratio of salt to sugar than a tiny sample taken at some other place. If the amount taken is small enough, it would even be possible to get a sample from this mixture that was pure sugar or pure salt, no matter how well you mechanically ground the two together. Only when such a tiny sample has the same composition wherever you sample it (as in the tea with sugar) can you call the mixture homogeneous.

A precise dividing line between a heterogeneous mixture and a homogeneous one is difficult to specify. Just how uniform does a mixture have to be in order to be considered homogeneous? Obviously, if the sample analyzed is of atomic size and includes only one atom, then no mixture can be considered homogeneous. In a larger sample containing billions and billions of atoms, the limiting factor is the analysis procedure used and how small a difference in composition is detectable by the experiment. In this text, we shall consider only obvious examples for discussion and classification and leave the more ambiguous determinations to the philosophers in the audience.

WORKPATCH 1.1 Classify the following examples of matter as pure substances, heterogeneous mixtures, or homogeneous mixtures (solutions).

(a) piece of wood
(b) iron nail
(c) rusty iron nail
(d) well-stirred mixture of food dye in water
(e) beeswax and candle wax mixed together by hand
(f) beeswax and candle wax melted together, stirred well, then allowed to solidify ●

Check your answers to WorkPatch 1.1 against the answers given at the end of the chapter. Did you get the right answers for the last two items? These items should make you think. Only one of the mixtures of beeswax and candle wax represents a solution because only one of the mixing processes (melting) can produce a homogeneous mixture.

Practice Problems

1.1 Which of the following are solutions?

(a) our atmosphere
(b) gold powder and silver powder ground together
(c) piece of 18-karat jewelry made from melting gold and silver metals together
(d) detergent dissolved in water
(e) oil droplets suspended in water

Answer: (a), (c), and (d)

1.2 Examination of "homogenized" milk under a microscope reveals suspended globules of fat. Is milk a heterogeneous mixture or a homogeneous mixture? Explain.

1.3 Fog is a suspension of tiny droplets of water in air. Is fog a heterogeneous mixture or a homogeneous mixture? Explain.

Before leaving the definition of matter, we want to say something about what matter is made of. All the matter that exists on our planet—from the air, to the dust in the air, to the ground we walk on, to the water that covers most of the planet, to the lifeforms that live on it—is made from **elements**, which are the basic building blocks of matter. At the time of writing, there are 118 known elements, 90 that occur naturally and 28 that can be synthetically prepared. All the known elements have been organized in a tabular form known as the *periodic table.*

The periodic table is so important to chemistry that we shall devote most of Chapter 3 to it. (A more complete periodic table and a list of the full names of all the elements appear inside the front cover of this book.)

Fe *(ferrum)* Iron
Au *(aurum)* Gold
Al Aluminum
C Carbon
S Sulfur

Periodic Table of the Elements

	1 IA	2 IIA	3 IIIB	4 IVB	5 VB	6 VIB	7 VIIB	8	9 VIIIB	10	11 IB	12 IIB	13 IIIA	14 IVA	15 VA	16 VIA	17 VIIA	18 VIIIA
1	1 H 1.0079																	2 He 4.003
2	3 Li 6.941	4 Be 9.012											5 B 10.811	6 C 12.011	7 N 14.007	8 O 15.999	9 F 18.998	10 Ne 20.180
3	11 Na 22.990	12 Mg 24.305											13 Al 26.982	14 Si 28.086	15 P 30.974	16 S 32.066	17 Cl 35.453	18 Ar 39.948
4	19 K 39.098	20 Ca 40.078	21 Sc 44.956	22 Ti 47.88	23 V 50.942	24 Cr 51.996	25 Mn 54.938	26 Fe 55.845	27 Co 58.933	28 Ni 58.69	29 Cu 63.546	30 Zn 65.39	31 Ga 69.723	32 Ge 72.61	33 As 74.922	34 Se 78.96	35 Br 79.904	36 Kr 83.8
5	37 Rb 85.468	38 Sr 87.62	39 Y 88.906	40 Zr 91.224	41 Nb 92.906	42 Mo 95.94	43 Tc 98	44 Ru 101.07	45 Rh 102.906	46 Pd 106.42	47 Ag 107.868	48 Cd 112.441	49 In 114.82	50 Sn 118.71	51 Sb 121.76	52 Te 127.60	53 I 126.905	54 Xe 131.29
6	55 Cs 132.905	56 Ba 137.327	57 La 138.906	72 Hf 178.49	73 Ta 180.948	74 W 183.84	75 Re 186.207	76 Os 190.23	77 Ir 192.22	78 Pt 195.08	79 Au 196.967	80 Hg 200.59	81 Tl 204.383	82 Pb 207.2	83 Bi 208.980	84 Po 209	85 At 210	86 Rn 222
7	87 Fr 223	88 Ra 226.025	89 Ac 227.028	104 Rf 261	105 Db 262	106 Sg 263	107 Bh 262	108 Hs 265	109 Mt 266	110 Ds 269	111 Rg 272	112 Cn 277		114 Fl		116 Lv		

58 Ce 140.115	59 Pr 140.908	60 Nd 144.24	61 Pm 145	62 Sm 150.36	63 Eu 151.964	64 Gd 157.25	65 Tb 158.925	66 Dy 162.5	67 Ho 164.93	68 Er 167.26	69 Tm 168.934	70 Yb 173.04	71 Lu 174.967
90 Th 232.038	91 Pa 231.036	92 U 238.029	93 Np 237.048	94 Pu 244	95 Am 243	96 Cm 247	97 Bk 247	98 Cf 251	99 Es 252	100 Fm 257	101 Md 258	102 No 259	89 Lr 262

Many elements have names that you are probably quite familiar with—carbon, silver, gold, iron, aluminum, uranium, hydrogen, helium, oxygen, and nitrogen are a few. Chemists represent the elements with one-, two-, or three-letter symbols (as shown in the periodic table), some taken from the English names (like C for carbon, H for hydrogen, and Al for aluminum) and others taken from the Latin names (like Fe for iron from the Latin word *ferrum;* Au for gold from the Latin word *aurum*) or other languages.

When we say that elements are the basic building blocks of matter, how basic is basic? For example, suppose you had a piece of pure elemental gold. What would happen if you cut the piece of gold in half, and then in half again, and then again, and then again? How many times could you divide the piece and still have elemental gold? People have been trying to answer this question for centuries. Around 400 B.C., the Greek philosopher Democritus suggested an atomic theory of the universe. Simply put, this theory said that all things are made up of minute, indivisible, indestructible particles called *atoms.* That this might be true was by no means self-evident. A piece of gold does not appear to be made of individual particles. Nevertheless, Democritus would have said that you can keep dividing your piece of gold only until you get down to a single atom of gold. In fact, Democritus was right, but his theory was not an easy sell. Aristotle, another Greek philosopher living around the same time, placed more trust in his senses and said that matter was continuous and not made up of discrete individual atoms. According to Aristotle, you could keep dividing your gold into smaller and smaller pieces forever; at no point would you reach an indivisible particle. Because Aristotle's proposition seemed more obviously correct, his theory carried the day for more than 2000 years.

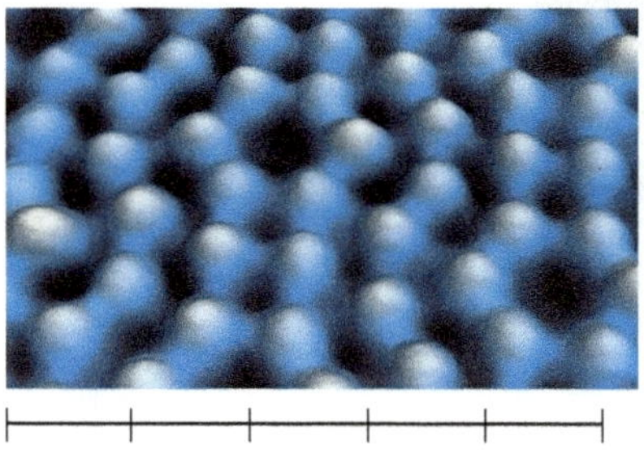

An image of silicon taken with a scanning tunneling microscope. Each purple sphere is a silicon atom.

As you will see in Chapter 3, scientists did eventually return to the atomic theory. We now know that all the elements exist as atoms, and an **atom** is the smallest possible piece of an element. Atoms are so tiny that, until recently, scientists thought we would never be able to see them. They spoke too soon. Though no one has seen an atom through an ordinary microscope, in the early 1980s, a device called a scanning tunneling microscope produced the first images of individual atoms—like the silicon atoms that appear as purple spheres on the surface of the silicon crystal shown in the photograph at left.

With the knowledge that matter is made up of atoms of the elements, we can now divide pure substances into two types, *elemental substances* and *compounds*. An **elemental substance** is one that is made from atoms of just one element. For example, our piece of pure gold is made from just gold atoms and nothing else. The same is true of a piece of pure iron (or any other pure metal), the oxygen you breathe, and the helium gas in a balloon.

Compounds, on the other hand, are pure substances made from atoms of two or more different elements. Water, for example, is a compound made from atoms of hydrogen and oxygen. A **chemical formula** for a compound indicates the number of atoms of each element that make up the smallest possible piece of that compound. The chemical formula for water, H_2O, tells us that the smallest possible piece of water is made from two hydrogen atoms and one oxygen atom. Because water is made from two different elements, it is classified as a compound. So is table sugar, whose formula is $C_{12}H_{22}O_{11}$. The oxygen in the air you breathe is a different case. The formula for oxygen, O_2, tells us that the smallest piece of life-sustaining oxygen gas has two oxygen atoms in it. However, though oxygen is made from two atoms, both are atoms of the same element, and so oxygen is not considered to be a compound; oxygen is an elemental substance. To summarize:

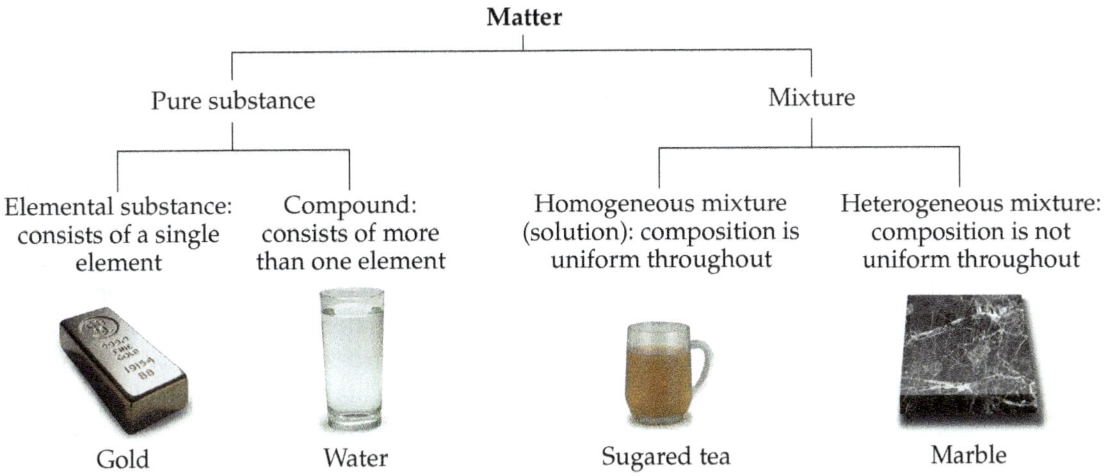

Classifying matter according to this scheme sometimes requires careful examination. Consider, for example, a glass of water, a glass of lemonade made from a powdered mix, and a glass of lemonade made from freshly squeezed lemons.

The water is a pure substance and a compound. The lemonade made from the mix is a homogeneous mixture (a solution). The lemonade made from lemons is a heterogeneous mixture because of the lemon pulp bits floating in it.

Pure water | Lemonade from powdered mix | Lemonade from freshly squeezed lemons

Practice Problems

1.4 Which of the following are compounds?

(a) iron oxide (Fe_2O_3)
(b) ozone (O_3)
(c) iron (Fe)
(d) carbon monoxide (CO)
(e) propane (C_3H_8)

Answer: (a), (d), and (e) because each is made from more than one kind of element; (b) and (c) are elemental substances, not compounds, because each is made of only one element.

1.5 Which of the following are compounds?

(a) sulfur (S_8)
(b) mixture of iron powder and aluminum powder
(c) mixture of O_2 gas and N_2 gas
(d) sulfur dioxide (SO_2)
(e) ammonia (NH_3)

1.6 True or false? A compound is a pure substance, but a pure substance need not be a compound. Give examples to prove your answer.

1.3 Matter and Its Physical Transformations

Let's get back to the definition of chemistry, which is why we started looking at matter in the first place. We said that chemistry is the study of matter and the transformations it undergoes. Since *transformation* means "change," chemistry is the study of changes in matter. And just as the word *matter* had to be subdivided into pure substances and mixtures, we must be very careful with the word *change*. In the science of chemistry, there are two kinds of changes matter can undergo: *physical* and *chemical*.

A **physical transformation** of a pure substance is one that leaves it as the same substance but in a different physical state. This leads us to another question: What do we mean by a *state of matter*? You are already familiar with the three most common **states of matter**: solid, liquid, and gas (or vapor). We shall examine these states in a detailed way later in this book, but for now your common knowledge of them will do just fine. Let's use water to demonstrate physical changes in matter. If you put a glass of liquid water in a freezer, the water will change to the solid state in the process we call **freezing**. This is an example of a physical change because only the state of the substance has changed. It's still water after the change. The reverse of freezing, called **melting**, is another example of a physical change. Naturally, the temperature below which liquid water freezes is also the temperature above which solid water melts, 32 °F or 0 °C at normal atmospheric pressure. (The Fahrenheit and Celsius temperature scales are described in Chapter 2.) Thus, this temperature is called either the freezing point or the melting point of water, depending on which way you are going. When heated either to or above its boiling point (212 °F or 100 °C at normal atmospheric pressure), water boils and undergoes **vaporization**, the change from the liquid to the vapor state. In the reverse of vaporization, **condensation**, the vapor returns to the liquid state. After either change, we still have water, and so vaporization and condensation are two more examples of physical changes.

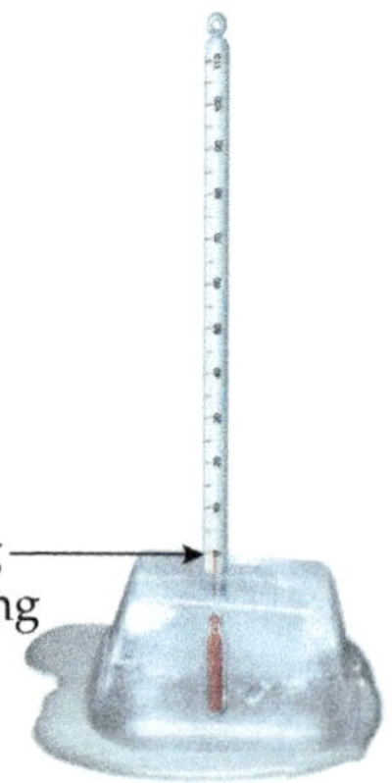

Water melts when its temperature goes above 0 °C; it freezes when its temperature goes below 0 °C.

Water boils when its temperature is 100 °C; it condenses when its temperature goes below 100 °C.

The melting point and boiling point of water are among its physical properties. The **physical properties** of a pure substance characterize its physical state and physical behavior. Other physical properties of water include its color (pure water is colorless), its odor (pure water is odorless), and its taste (pure water is tasteless). The following table summarizes some of the physical properties of water:

Physical Properties of Pure Water (at Normal Atmospheric Pressure)	
Melting point	0 °C
Boiling point	100 °C
Color	None
Odor	None
Taste	None

Each pure substance has its own unique set of physical properties. For example, pure ethanol, the component of beer and wine that gets you drunk, is a colorless liquid that looks just like water. However, it melts at −117 °C and boils at 78 °C. Thus, if you are presented with a colorless liquid and asked to determine whether it is water or ethanol, you can place a thermometer in it and heat it to boiling. If the thermometer reads 78 °C when it boils, then the liquid is ethanol.

Another important physical change is **sublimation**, which occurs when a pure substance goes directly from the solid state to the gas state without passing through the liquid state. For example, when you heat an ordinary ice cube, it first melts to liquid water before boiling to enter the vapor state. Carbon dioxide, however, another pure substance, behaves quite differently. Carbon dioxide exists as a solid below −78 °C, but if you warm it above this temperature, it slowly vanishes into thin air without ever forming a puddle or displaying any obvious wetness. (For this reason, solid carbon dioxide is called "dry ice.") It goes directly from the solid state to the gas state—it sublimes. Unlike water, carbon dioxide cannot exist in the liquid state unless a great deal of pressure is applied to it.

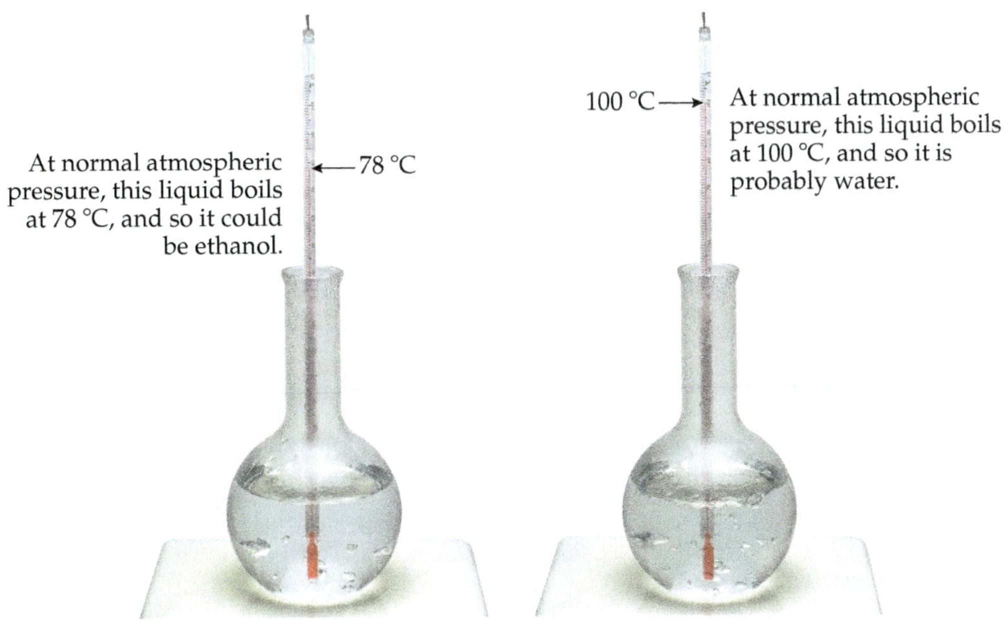

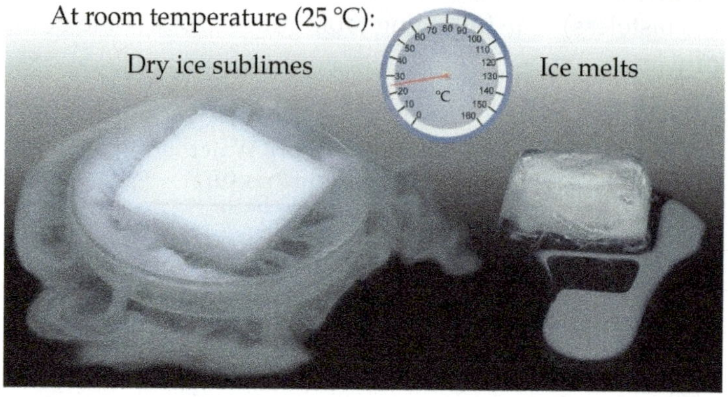

It is even possible for regular ice (solid water) to sublime if conditions are right. Have you ever noticed how snow disappears even when the temperature remains below freezing? The same thing happens inside the freezer compartment of a frost-free refrigerator.

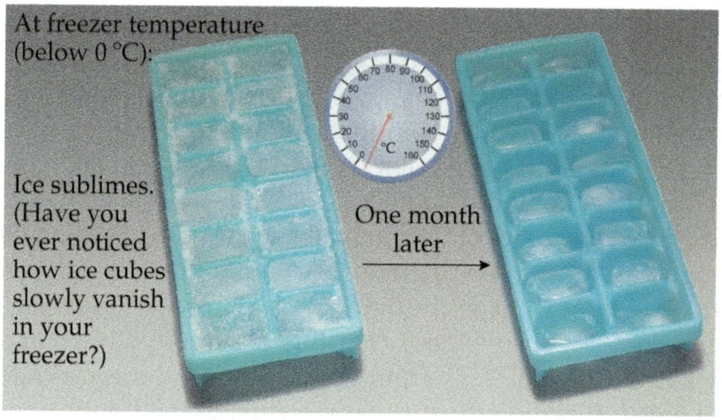

Practice Problems

1.7 Ice cubes slowly vanishing while stored in a freezer is an example of

(a) vaporization
(b) condensation
(c) melting
(d) sublimation

Answer: (d); the ice cubes "vanish" by going directly from the solid state to the gas state.

1.8 Molten iron cooling to solid iron is an example of

(a) sublimation
(b) condensation
(c) freezing
(d) melting

1.9 True or false? After being heated to above its boiling point, ethanol is no longer ethanol but something else. Explain.

1.10 You are presented with a block made of some pure metal and told the metal is gold, but you have your doubts. Using a thermometer, how can you determine whether the metal is gold?

1.4 Matter and Its Chemical Transformations

So far, we have been concentrating on the physical changes and physical properties of pure substances. It's now time to switch to chemical changes and chemical properties of pure substances. A pure substance is said to have undergone a **chemical change (chemical transformation)** if, after the change, it is a different substance or substances. For example, if you combine appropriate amounts of hydrogen gas, a pure substance, with oxygen gas, another pure substance, and then supply a spark, a new substance—water—will appear, and the hydrogen and oxygen will be gone. The hydrogen and oxygen have undergone a chemical change. The **chemical properties** of a pure substance are properties that can be described only when it undergoes a chemical change. For example, color isn't a chemical property because the color of a substance is apparent without any need to change it chemically in order to make this determination. Flammability, on the other hand, is a chemical property because burning a pure substance converts it into an entirely different substance or substances. For example, hydrogen is considered to be a flammable gas. Reactivity is a chemical property of a substance that describes the ability of a pure substance to undergo chemical change when combined with other pure substances. For example, helium gas is considered nonreactive because, when combined with other substances, no chemical change occurs. Chlorine (a greenish gas having the formula Cl_2) is considered to be very reactive because it combines with most substances it comes into contact with to create new substances. An exciting example of chemical change is the combination of chlorine gas with sodium (a soft, shiny metal having the elemental symbol Na, from the Latin *natrium*). If you mix sodium and chlorine together, a rather incredible thing happens. A tremendous amount of heat and light is given off. When things calm down enough for us to take a look, we see that the sodium metal and chlorine gas are both gone, replaced by an entirely new pure substance, the compound sodium chloride (formula NaCl):

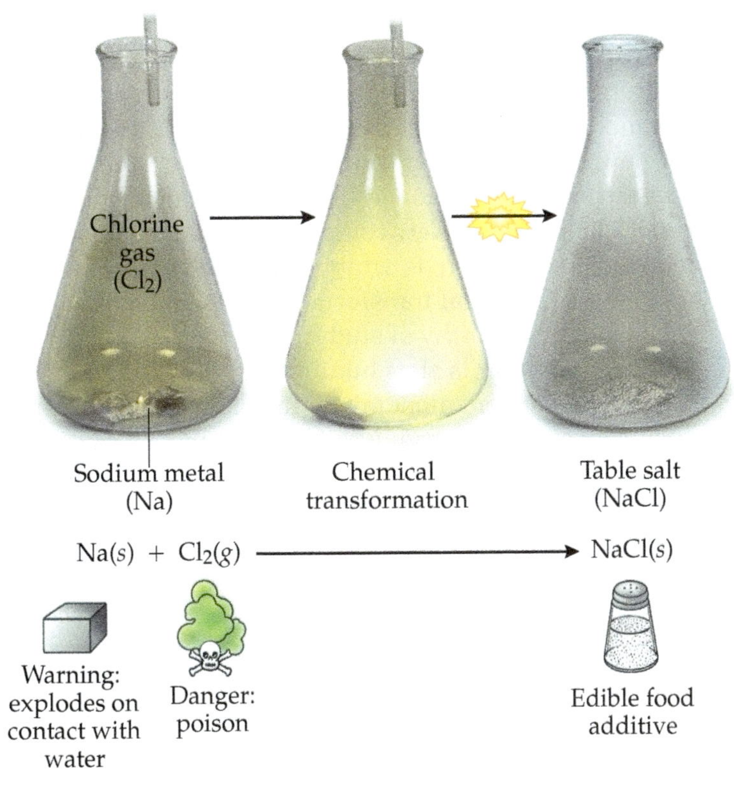

The new compound has chemical properties entirely different from those of either of the pure substances from which it was made. It no longer is metallic, like sodium, and it no longer burns things it comes into contact with like chlorine. It's ordinary table salt. What you have witnessed is a *chemical transformation*. As we said, a chemical transformation occurs when new substances, having new physical and chemical properties, are formed from some starting substance or substances. The starting substances are called **reactants**, and the new substances formed are called **products**. The process that goes on during the conversion of reactants to products is called a **chemical reaction**.

Chemists write chemical reactions using a symbolic representation, with the reactant substances to the left of a horizontal arrow and the product substances to the right. The arrow itself means "react to give." For instance, a chemist would write the chemical reaction between sodium and chlorine as

$$\underset{\text{Reactants}}{2\,\text{Na} + \text{Cl}_2} \longrightarrow \underset{\text{Products}}{2\,\text{NaCl}}$$

In English, this says "sodium and chlorine react to give sodium chloride." The matter has undergone a chemical transformation. Often, chemists include the notation (*g*) for gas, (*l*) for liquid, or (*s*) for solid to represent the physical state of each substance in a reaction. Thus, we could have written

$$2\,\text{Na}(s) + \text{Cl}_2(g) \rightarrow 2\,\text{NaCl}(s)$$

Most chemical transformations are carried out by nature, not by chemists. The rusting of iron, be it in a nail or on your car, is a chemical transformation in which iron reacts with oxygen in the presence of water to give the new compound, a hydrate of iron(III) oxide (formula $Fe_2O_3 \cdot H_2O$), commonly known as rust.

$$4\,\text{Fe} + 3\,\text{O}_2 + \text{H}_2\text{O} \longrightarrow 2\,\text{Fe}_2\text{O}_3 \cdot \text{H}_2\text{O}$$

Even more important to humans are the chemical transformations that occur during the digestion of food. Some of the food you eat gets converted to carbon dioxide and water, which you exhale. Your very thought processes involve many chemical reactions.

In spite of the stereotype of the crazed scientist madly mixing test tubes of chemicals for the joy of it, that's not the reason chemists mix substances together. Chemists are constantly searching for new compounds with new chemical and physical properties that can extend our knowledge about how matter behaves and can benefit humankind. New compounds to combat disease, to protect metal from corrosion, to whiten and brighten socks—the list goes on and on. Chemical transformation is what makes chemistry so interesting and worthy of study.

Chemistry got its start hundreds of years ago in the medieval period. Back then, alchemists, the forerunners of modern-day chemists, were busy trying to turn base metals, such as lead, into gold. They didn't succeed, but what they discovered while trying laid the foundation for the science of chemistry. Today we know that the alchemists could never have succeeded, for while turning lead into gold is an example of transforming matter from one substance into another, it is not an example of a *chemical* transformation. In a chemical transformation, none of the elements in the reactants change. If your reactants have carbon, hydrogen, and oxygen in them, then your products must have only carbon, hydrogen, and oxygen in them and no other elements. A chemical transformation can't convert one element to another element, and so turning lead into gold by chemical means is impossible. (Modern-day scientists can turn lead into gold via a *nuclear transformation*, which converts one element to another; you'll learn more about this in Chapter 16.)

Practice Problems

Refer to the periodic table inside the front cover for names of elements not yet familiar to you.

1.11 Which of the following represent a chemical transformation?

(a) $4\,P(s) + 5\,O_2(g) \rightarrow 2\,P_2O_5(s)$
(b) $H_2O(g) \rightarrow H_2O(l)$
(c) $3\,O_2 \rightarrow 2\,O_3$

Answer: (a) and (c) because the products are different from the reactants; (b) is a physical change because we have water on both sides of the arrow.

1.12 Water (H_2O) and carbon dioxide (CO_2) are produced in a chemical reaction when methane (CH_4) and oxygen (O_2) are combined and heated. Which are the reactants, and which are the products?

1.13 What, if anything, is wrong with the following chemical reaction?

$Cl_2(g) + Br_2(l) \rightarrow 2\,HCl(g) + 2\,HBr(g)$

(a) It is not a chemical reaction. It is only a physical transformation.
(b) The products include elements that are not in the reactants.
(c) There is nothing wrong with the chemical reaction.

1.5 How Science Is Done—The Scientific Method

Concepts such as what atoms are made of and how they react are essential to our understanding of chemistry. Many of the basic ideas that underlie chemistry were developed in the last 150 years, but the process that scientists used to develop these concepts has been around much longer. In fact, any curious person seeking answers to a question might have, and probably has, used the *scientific method*. For example, consider Albert. It is a commonly held belief that scientific types have little or no romantic life. This is, in fact, not true. In his younger days, Albert found himself newly arrived at a university, seeking dates.

Being of a curious nature, he decided to study the topic of dating. He went on many dates, making observations, collecting data, and recording it all in his diary (his "laboratory notebook"). These dates were his **experiments**, which are procedures scientists carry out to study some phenomenon. For Albert, some dates were successful and some were not. On each date, he would vary one aspect of his appearance (a racy new tie, a bold set of socks, an eye patch), observe its effect on the outcome, and diligently record the results.

After several months of collecting data from many dates, he came to the conclusion that changes in his appearance above his neckline had great influence on the outcome of his dates, but modifications below his neckline had little or no effect. This led him to postulate the following Law of Dating: "For a successful date, look as good as you can from the neck up." A **law** is a generalization that concisely summarizes the outcome of a series of experiments. At first, Albert thought this law a bit odd, but after much thought, he developed a *theory* that explained why it was true. A **theory** is an attempt to explain *why* a law exists, and a **model** is some kind of physical picture or mathematical expression of a theory. He theorized that in any stressful situation between two people, they watch each other's faces intensely for visual cues about how things are going. This Stress-Induced Intense Facial Watching Theory explained the Law of Dating.

The Scientific Method

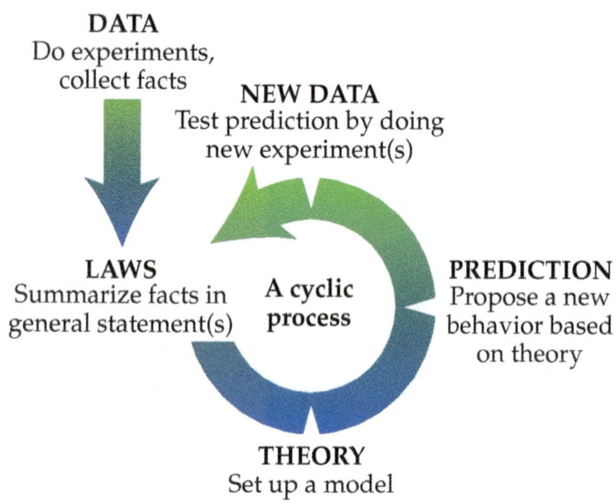

The process Albert used to satisfy his curiosity about dating—collecting data from experiments, using it to fashion a law, and then postulating a theory to account for the law—is called the **scientific method**.

Although the dating scenario we used to illustrate the scientific method is tongue-in-cheek, it does possess all the steps of the scientific method. However, even though Albert was pleased with his procedure, he probably should have discussed it with his advisor because he made some classic beginner's mistakes. Perhaps you noticed them. The most obvious is that he carried out each experiment (date) with a different woman. As a result, his observations of what dates respond to most strongly were based on the likes and dislikes of a different person each time. The only true way to make this study general would be for Albert to go on thousands of dates with thousands of different people in an attempt to average out their responses. This dilemma is one of sampling and is very much like the problem pollsters face when they take a pre-election poll. They must be careful to interview enough people so that they include Democrats, Republicans, Independents, and all other affiliations in roughly the percentages that these groups exist in the general population. Only then will the pollsters have a representative sample of the general population and a conclusion that has a chance of being correct.

Working with a large enough sample is rarely a problem in chemical studies because most involve samples that contain millions of billions of atoms or molecules, and we are therefore always measuring the average behavior of very large groups. Quite often, however, care must be taken to be sure the sample being analyzed is representative of the object being studied.

The scientific method is a cyclical process in which scientists continuously uncover new information (data), postulate new laws, and either modify or discard old ones. Based on these new or revised laws, theories are further tested and, if necessary, refined or completely revamped. In repeatedly going through this cycle, sometimes theories and laws are proved to be wrong, and sometimes they hold up. This is how science is supposed to work. Even the great Sherlock Holmes knew this (from *The Yellow Face* by Arthur Conan Doyle):

Actually, the first time a theory is postulated it is often called a *hypothesis* and not a theory. Only when the hypothesis holds up to further testing (further cycles of the scientific method) is it elevated to theory status. A theory, then, is a well-tested hypothesis, but it can be proven to be incorrect at any time with sufficient data. Albert's Intense Facial Watching Hypothesis may not have held up to the scrutiny of further cycles of the scientific method and made it all the way to being a theory. If so, then perhaps Albert's dating problems lie elsewhere.

Practice Problems

1.14 What is the relationship between a theory and a law?

Answer: A theory attempts to explain why *a law is correct.*

1.15 What is the relationship between a law and experimental data?

1.16 What is good evidence that a theory is correct?

It would seem at first glance that the scientific method should guarantee that science is always correct. If only it were so simple. Unfortunately (or sometimes fortunately), this logical process is carried out by sometimes illogical, emotional human scientists. Sometimes investigators believe so strongly in a particular theory that they enter the cycle from the wrong direction and attempt to interpret all they see in terms of their personal *bias*. **Bias** is a strong preference or inclination that inhibits impartial judgment. For example, today scientists believe that atoms are among the fundamental building blocks of matter. As we mentioned earlier, this is not a new idea. Democritus was battling Aristotle over the possibility of the existence of atoms more than 2000 years ago. Now we know that Aristotle, who rejected atoms, was wrong, but we can't be too hard on him—he had no ability to examine matter on a submicroscopic level. The beliefs of the ancient Greeks were founded not on verifiable experimental evidence but on their opinions and convictions about the natural world. These personal biases can really get in the way of scientific development. Indeed, Aristotle went to his grave convinced that an adult man has more teeth in his mouth than an adult woman. As the renowned philosopher Bertrand Russell once commented, "All Aristotle had to do was ask Mrs. Aristotle to open her mouth," and he would have discovered the truth.

The fact that Aristotle never did look into his wife's mouth to check his theory—that he would not even conceive of such a check being necessary—is an ideal illustration of why scientists must always be on guard not to let personal bias, ambition, politics, ideology, or theology divert them from the scientific method. Scientists must be neutral to the point that they are willing to throw out all they were taught and replace it with something else if the scientific method so demands. That is not always easy. It has been observed quite often that "a new theory in science is really accepted only when the last of its opponents dies off." Now, in case you are feeling smug that we, in modern times, would never succumb to such bias, just think "global warming." If ever there was a phenomenon that was susceptible to bias, this is it. We are not going to argue for or against global warming, except to say that some people on both sides of this issue have such a strong bias that they have essentially abandoned the scientific method and see only what they want to see. In addition to human bias, scientific findings can also be compromised by flaws in the design of an experiment. For example, how would you design an experiment to determine the effects of a vitamin C deficiency in humans? For ethical reasons, you cannot deprive a large group of humans of vitamin C. Instead, you might use animals, but can you be sure the results you see in your animals represent what happens in humans? If you used mice or rats, for instance, you'd see no ill effects because, unlike humans, mice and rats can synthesize their own vitamin C. If you used guinea pigs, on the other hand, the results would be similar to those for humans because guinea pigs, apes, and some fruit bats are the only other mammals besides humans that are unable to produce vitamin C. (We know what happens to humans deprived of vitamin C because the symptoms were once common during long sea voyages, before sailors learned to carry foods containing vitamin C.)

You might think chemical experiments would be less prone to flaws, but that's not always so. One source of trouble is that we usually can't observe atoms directly. To figure out what is happening on the atomic level, we measure changes we can observe, such as color change or the absorption or emission of heat, and then *infer* what must have happened to the atoms. Our inference is based on how we think the observed changes relate to changes occurring on the atomic level. If our understanding of this relationship is correct, our conclusions are likely to be correct. If our understanding is incomplete, however, our conclusions may be wrong.

Without the ability to gather data and do experiments, you can't employ the scientific method. For this reason, Democritus was unable to support his original atomic theory. In contrast, some of the most beautiful examples of the application of the scientific method and the replacement of old theories with new ones come from the development of modern atomic theory, beginning with John Dalton in the early 1800s. When you read about this development in Chapters 3 and 4, keep the scientific method in mind, and you will understand why theories came and went.

Finally we want to say a few words about models. We often use models to help us apply, understand, and visualize our theories, but you have to be careful with models. By their very nature, models are incomplete representations of the reality they are supposed to explain. This is fine as long as you are always aware of the limitations of the particular model you are using. You never want to push a model beyond its capabilities. For example, consider an airplane model glued together from the plastic pieces in a kit. If it is a true scale model, then it can be used to accurately determine the wingspan of the actual plane. But this type of model would never be useful in predicting the airplane's stall speed or fuel consumption because it is "incomplete" as far as these questions are concerned. For this task, you would be better served by a computer model that uses equations to calculate the thrust of the engines, the lift produced by the wings, and the drag produced by the plane's body. Of course, such a computer model is far more complicated than the plastic one and takes quite a bit more knowledge to use and correctly interpret its predictions. Does this mean the plastic model is useless? Absolutely not, just so long as you understand its limitations. In Chapter 4, you will be presented with the Bohr Model of the atom, developed in the early 1900s. More sophisticated models of the atom exist, but many working chemists still use this model to make predictions about their chemistry. Therefore, we will go ahead and use the Bohr Model, keeping in mind that it predicts some chemical properties pretty well so long as you don't push it beyond its capabilities.

1.6 Learning Chemistry with This Book

A common misconception is that chemistry is all math, calculations, and numerical problem-solving. In all honesty, chemistry does have that side to it. But when a chemist is presented with a question about matter, the first thing that comes to mind is *not* a complicated mathematical formula. Instead, chemists use a basic set of fundamental concepts, often best represented with images instead of mathematical equations. For example, a chemist and a nonchemist picture the concept of melting differently:

Nonchemist

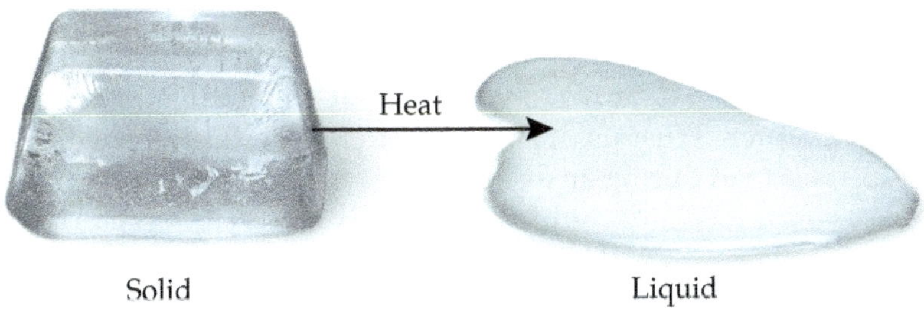

Chemist

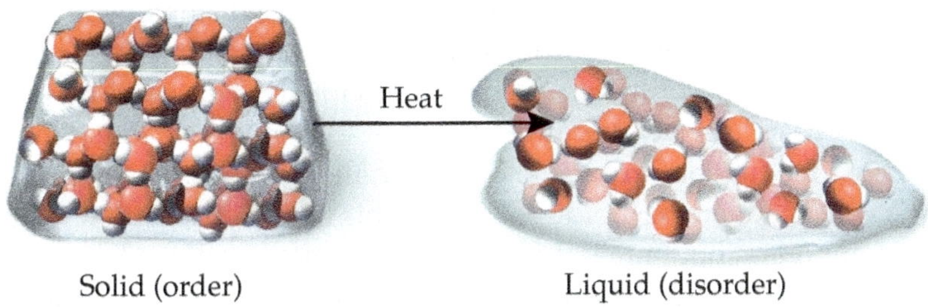

These fundamental concepts and images are the tools chemists use in answering questions about matter and the transformations it undergoes. Though this book tackles the numerical side of chemistry, it focuses on the fundamental concepts, illustrating them with pictures and everyday experiences. Even chemists forget mathematical expressions and memorized equations, but the fundamental concepts, stored as images in their minds, are with them all their lives.

Finally, we want to give you some advice about how to use this textbook. This advice comes from our own experience (after all, we were beginning chemistry students once). Read each chapter slowly, with paper and pencil in hand so you can take notes, draw pictures, and write down questions on points you are not sure about. We have done three things to encourage you to do this. First, each chapter includes a number of WorkPatches marked with a stop sign. When you reach one of these, *stop* reading and try to do the problem or sketch the concept. *Don't go on until you can answer the question.* Check your answer against the answer given at the end of the chapter.

Second, clusters of Practice Problems appear throughout each chapter. In each cluster, the worked-out solution is given for the first problem. The other problems in the cluster are similar to the first. In some cases, we provide simple answers to these problems, but it is up to you to actually work out the solution. Blue numbers indicate Practice Problems for which answers are given at the back of the book; solutions for these problems are provided in the *Student Workbook and Selected Solutions Manual* that accompanies this text (see Preface). Now, you could skip over the WorkPatches and Practice Problems, but you would be doing yourself a disservice. The best time to find out if you understand what you read is just after you've read it. In addition, working the Practice Problems and WorkPatches as you encounter them will help you retain the material. Keep paper and pencil handy, and do the WorkPatches and Practice Problems *as you encounter them.*

We think that doing the WorkPatches as you encounter them is so important that we have included something new in this edition: hints that are available to you via your smartphone or by accessing www.masteringchemistry.com for this textbook. If you are having trouble completing a WorkPatch, scan the QR code at the front of this book with your smartphone for a hint (go back and try it for the WorkPatch in this chapter).

Finally, each chapter ends with a section called Learning Outcomes that lists things you can do to prove to yourself that you understand the key concepts in the chapter. Trust us, it's much better to do this after you have just read the chapter than during an exam.

+1 One More Thing

Every chapter will have this final section called "One More Thing." Its job is to delve a bit further into the material, a next advanced step beyond the usual level of coverage in a typical Introductory Chemistry course. Your instructor will decide if this section is required, but obviously, you need no one's permission to read it, and we guarantee that this section will be interesting.

This being the first or introductory chapter, there really isn't too much to say yet in the way of advanced topics. So, we felt this would be a good time to tackle some contemporary, hot-button questions that science has played a part in. What questions? How about "Did humans actually evolve from simpler life forms?" "Is the cost of a modern industrial society to make the planet dangerously warmer?" "If so, is there anything we can actually do about it?" "How did the universe come into existence? Was it with a Big Bang or a tiny whimper?" Science has theories that address all of these questions. Evolution addresses where humans came from and is a theory first backed by scientific arguments put forth by Charles Darwin in 1859. Global warming, now called climate change, is a theory that explains how the burning of fossil fuels is warming the planet. The Big Bang theory explains how the universe came into being from a singularity where all matter in the universe was on top of itself. All of these theories are a result of the scientific method. All started as hypotheses that have been tested to some degree by predictions, experiments, and the collection of new data. So these theories are incontrovertible facts, right? Obviously, anyone who doubts these theories must be a full-time member of the Flat Earth Society. It might surprise you to know that there is one group that holds these and *all* theories of science in doubt. And that group is one made up of scientists themselves.

"Really?"

"Absolutely!"

This doesn't mean that scientists have no confidence in the theories of science. But to be true to the scientific method, we must be ready at all times to take a theory that we have tremendous confidence in and modify it or even toss it aside completely if the next cycle of the scientific method insists that we do so. The key word here is "confidence," which is different from "faith," the foundation of religion. The nineteenth-century English biologist, T. H. Huxley, stated it well when he wrote "The great tragedy of Science is the slaying of a beautiful hypothesis (a theory) by an ugly fact (contradictory data)!" Scientists must always be ready for such a slaying should new data so insist. In Chapter 3, you'll read about how a scientist by the name of Ernest Rutherford and his students went into a darkened room to do an experiment to confirm and further prove his mentor's theory of the structure of the atom. Instead, he was faced with data that said the theory put forth by his mentor, a world-renowned physicist, a man he admired and respected and intended to prove right, was, in fact, wrong.

As scientists, we understand that it is impossible to prove a theory. It is possible only to add support for it or to disprove it. If data from some new experiment challenge an established theory, then the theory must be modified, as Rutherford did. When you hear scientists argue in support of a theory, it is because those scientists are convinced that all the data collected to that point support the theory. Of course, data are open to interpretation. Some scientists are convinced that the available data support global warming; others are not. And even if there were consensus among all scientists that the theory of global warming was correct, did Superstorm Sandy, the storm that damaged the East Coast in October 2012, arise because of global warming? An opinion piece in the publication *USA Today* says "The answer is unknowable. No individual weather event can be conclusively linked to man-made climate change." It is right. Every theory has limitations as to how far it can be applied. Perhaps the biggest challenge faced by scientists is that they are not allowed to pick and chose which theories they will support based on their likes, dislikes, political views, or religious beliefs. But scientists are human, and bias does occur on occasion. If it does, then ultimately, the scientific method presses on, and we trust that, given enough cycles through the scientific method by others in the scientific community, this situation will eventually be corrected.

Have You Learned This?

Science (p. 4)
Technology (p. 4)
Chemistry (p. 5)
Matter (p. 5)
Pure substance (p. 5)
Mixture (p. 5)
Homogeneous mixture (p. 5)
Solution (p. 6)
Heterogeneous mixture (p. 6)
Element (p. 7)
Atom (p. 8)
Elemental substance (p. 8)
Compound (p. 8)
Chemical formula (p. 8)
Physical transformation (p. 10)
State of matter (p. 10)
Freezing (p. 10)
Melting (p. 10)
Vaporization (p. 10)
Condensation (p. 10)
Physical property (p. 10)
Sublimation (p. 11)
Chemical change (p. 13)
Chemical transformation (p. 13)
Chemical property (p. 13)
Reactant (p. 14)
Product (p. 14)
Chemical reaction (p. 14)
Experiment (p. 15)
Law (p. 15)
Theory (p. 16)
Model (p. 16)
Scientific method (p. 16)
Bias (p. 17)

Learning Outcomes

1.1	**Science and Technology**	Describe the difference between science and technology. [17, 19, 65]
1.2	**Matter**	Define the types of mixtures in terms of the types of matter. [23, 24, 25, 26] Describe how an elemental substance differs from a compound. [35, 37, 38] Explain what a chemical formula is (what it tells you). [40, 42, 43]
1.3	**Matter and Its Physical Transformations**	Name the states of matter and the types of physical transformation. [44, 45, 47, 48, 53]
1.4	**Matter and Its Chemical Transformations**	Explain what happens to a substance or substances after they undergo a chemical change. [56, 57, 55, 58, 88]
1.5	**How Science Is Done—The Scientific Method**	Describe the parts of the scientific method and how they are related to each other. [59, 60, 61]

Concept Questions

1. Which of the following statements is true regarding science and technology?
   (a) They are completely unrelated.
   (b) Technology precedes science.
   (c) Science precedes technology.
   (d) Science is the application of technology.

2. Matter can exist as which of the following?
   (a) As a homogeneous mixture of substances.
   (b) As a pure substance.
   (c) As a heterogeneous mixture of substances.
   (d) As all of the above.

3. To be a compound, a substance must
   (a) be made from more than one type of element.
   (b) not be a mixture.
   (c) be homogeneous.
   (d) be all of the above.

4. An elemental substance
   (a) can also be a compound.
   (b) can never be a compound.
   (c) cannot participate in a chemical reaction.
   (d) cannot change state since it is a pure element.

5. When a pure substance melts, it
   (a) undergoes a chemical change.
   (b) changes state.
   (c) changes from a liquid to a solid.
   (d) sublimes.

6. The melting point and the freezing point of a pure substance are
   (a) unrelated.
   (b) identical.
   (c) for the same change of state.
   (d) chemical properties of the substance.

7. A chemical property of hydrogen is that it
   (a) is a gas at room temperature.
   (b) is odorless.
   (c) combines with chlorine to make the compound hydrogen chloride (HCl).
   (d) combines with air to give a mixture of hydrogen and air.

8. In a chemical reaction
   (a) a physical change takes place.
   (b) products are converted into reactants.
   (c) one or more substances are converted into different substances.
   (d) elemental substances change into different elements.

9. A law is explained by
   (a) experimental data.
   (b) bias.
   (c) a theory.
   (d) an experiment.

10. The difference between a hypothesis and a theory is that
    (a) a hypothesis is a guess, whereas a theory is fact.
    (b) a theory is an untested hypothesis.
    (c) a theory is a well-tested hypothesis.
    (d) a theory is never wrong, whereas a hypothesis may be wrong.

Answers: 1c, 2d, 3a, 4b, 5b, 6b, 7c, 8c, 9c, 10c

Science and Technology

Answers to problems numbered in blue ink appear at the back of this book. Complete solutions to these selected problems appear in the Study Guide and Selected Solutions Manual *that accompanies this book.*

1.17 The United States started to undergo electrification in the late 1800s, with the first example of widespread electric lighting at the 1893 Chicago World's Fair by Nikola Tesla. In 1831, Michael Faraday discovered electromagnetic induction, which uses magnets to produce electricity in a wire. Explain which is an example of science and which an example of technology.

1.18 Define *chemistry*.

1.19 Give an example (not from this book) of a piece of scientific information and cite both a positive and a negative result of its technological application.

1.20 In the United States, most basic chemical research (science) is funded by the government, but most of the money spent on applying scientific knowledge (technology) comes from industry. Why do you suppose industry is reluctant to fund basic research?

1.21 Suppose study of the human brain (science) leads to the development of a device (technology) that allows a user to both download information from another person's brain ("mind-reading") and upload information into her or his own brain (instant learning). One extreme opinion is to make the device freely accessible to everyone. The other extreme opinion is to outlaw the device and destroy the plans for creating it. Where would you stand on this issue? Give your reasons, and then elaborate on who should decide.

Matter and Its Transformations

1.22 When is a mixture not a solution?

1.23 Is it possible to have a mixture of compounds? If yes, give an example.

1.24 How does a solution differ from a heterogeneous mixture?

1.25 Flour is mixed with powdered sugar, and then the combination is ground to a fine dust. Is the resulting product a solution or a heterogeneous mixture? Explain.

1.26 Brass is an *alloy* (a solid solution) of copper, Cu, and zinc, Zn. Given this fact, answer the following:

(a) Copper powder and zinc powder are ground together. Is the resulting product a solution or a heterogeneous mixture? Explain.

(b) The Cu and Zn powders are ground together, and then the mixture is heated until it is molten metal. Is the molten metal a solution or a heterogeneous mixture? Explain.

(c) The molten metal from part (b) is allowed to cool and solidify. Is the resulting solid a solution or a heterogeneous mixture? Explain.

1.27 The air you breathe is properly called a solution. Explain why.

1.28 What is an element? How many elements are known today?

1.29 If H is the symbol for the element *h*ydrogen and O is the symbol for the element *o*xygen, why is the symbol for sodium Na rather than S, and the symbol for iron Fe rather than I?

1.30 What is the smallest possible piece of an element called?

For Problems 1.31–34, refer to the list of elements that appears on the inside front cover.

1.31 What are the symbols for the elements lead, molybdenum, tungsten, chromium, and mercury?

1.32 What are the symbols for the elements sulfur, chlorine, phosphorus, magnesium, and manganese?

1.33 Give the element name that goes with each symbol: Ti, Zn, Sn, He, Xe, Li.

1.34 Give the element name that goes with each symbol: U, Pu, Cs, Ba, F, Si.

1.35 Sulfur is known in many forms, including yellow sulfur (formula S_8) and orange-red sulfur (formula S_6). Heating either of these forms to very high temperatures can yield S_2. Also known is a substance with the formula SO_2. Since all of these substances are made of elements, are they all elemental substances? Explain.

1.36 What does its chemical formula tell you about a pure substance?

1.37 Compounds have properties that are different from those of their constituent elements. Discuss a real example to support this statement.

1.38 Which of the following are elemental substances and which are compounds: F_2, $BrCl_3$, P_4, C_2H_2, HCl, Ar, Al_2O_3, Al?

1.39 Which of the following are elemental substances and which are compounds: chlorine, Cl_2; octane, C_8H_{18}; sulfur, S_8; neon, Ne?

1.40 The smallest possible piece of the compound hydrogen peroxide contains two hydrogen atoms and two oxygen atoms. Write the chemical formula for hydrogen peroxide, with hydrogen first.

1.41 The smallest possible piece of the compound nonane, a component of gasoline, contains 9 carbon atoms and 20 hydrogen atoms. Write the chemical formula for nonane, with carbon first.

1.42 The smallest possible piece of the sugar glucose has twice as many hydrogen atoms as oxygen atoms or carbon atoms. Given that glucose has six carbon atoms, write its chemical formula with carbon first, then hydrogen, then oxygen.

1.43 The air you breathe is made up mostly of nitrogen gas, N_2.
(a) What can you say about the smallest possible piece of nitrogen gas?
(b) Is nitrogen gas a compound? Explain.

1.44 What are the three most common states of matter?

1.45 What is the name of the process by which matter changes directly from the solid state to the gas state?

1.46 What is the name of the process that is the opposite of evaporation?

1.47 Propane has a boiling point below room temperature, and hexane has a boiling point above room temperature. In what physical state would you expect to find each substance at room temperature (which is usually about 20 °C)?

1.48 At normal atmospheric pressure, the freezing point of ethanol is −114 °C and its boiling point is 78.4 °C. What is the melting point of ethanol?

1.49 Solid mothballs work by filling a closet with a chemical vapor that is toxic to moths, and the balls shrink and disappear slowly with time. What is going on?

1.50 What are the melting point and boiling point of water in degrees Celsius and degrees Fahrenheit (at normal atmospheric pressure)?

1.51 Even though both are composed entirely of oxygen, the elemental substance O_2 is a life-sustaining gas but the elemental substance O_3 is a toxic gas. What does this say about the chemical properties of substances and their compositions?

1.52 Water, H_2O, is a harmless liquid, but hydrogen peroxide, H_2O_2, is a poison. How is this possible, given that both are compounds of hydrogen and oxygen?

1.53 When you eat sugar ($C_{12}H_{22}O_{11}$), it is eventually converted into CO_2 and H_2O, which you exhale. If you exhale onto a cold window, the window fogs up.
(a) What chemical change is being mentioned here?
(b) What physical change is being mentioned here?
(c) CO_2 is a solid that sublimes at −78 °C. Temperatures on planet Earth almost never get this cold. What does that tell you about the fog on the window?

1.54 When a candle burns, the solid wax melts and burns. In the process, the wax is combined with oxygen from the atmosphere and converted into carbon dioxide and water vapor. Explain what is going on in terms of chemical change and physical change.

1.55 Gasoline can be thought of as having the formula C_8H_{18}. When gasoline is used in a car's engine, it reacts with oxygen (O_2). If you look closely at the exhaust pipe of a running car, you will see water dripping out of it. Where did this water come form (what kind of change is involved)?

1.56 When exposed to oxygen in the air, white phosphorus, P_4, catches fire and produces the compound P_4O_{10}. Is this a description of a physical property or a chemical property of phosphorus? Explain.

1.57 Is the tarnishing of silver, Ag, a physical change or a chemical change? (The chemical formula for the tarnish on silver is Ag_2S.)

1.58 Nitrogen, N_2, combines with hydrogen, H_2, to give ammonia, NH_3. Nitrogen and hydrogen are odorless gases that are not very soluble in water. Ammonia is also a gas, but it has an extremely powerful odor and is very soluble in water. How can you explain this great difference in properties?

How Science Is Done—The Scientific Method

1.59 How does a law differ from a theory?

1.60 According to the scientific method, when does a theory reach the point of being unchallengeable?

1.61 Suppose you wanted to explore the role of sunlight on plant growth. Explain what you would do to go through one cycle of the scientific method, including a prediction you might make.

1.62 The scientific method is often considered foolproof and incapable of giving wrong answers, and yet history shows that scientists are sometimes wrong. What are some factors that cause the scientific method to give invalid results?

Additional Problems

1.63 Which of the following are elemental substances and which are compounds: baking soda, $NaHCO_3$; graphite, C; dry ice, solid CO_2; metallic sodium, Na; metallic mercury, Hg?

1.64 Which of the following are elemental substances and which are compounds: diamond, a solid form of carbon, $C(s)$; liquid water, $H_2O(l)$; nitric acid, $HNO_3(l)$; gaseous nitrogen, $N_2(g)$; liquid nitrogen, $N_2(l)$?

1.65 Classify each of the following as science or technology:
(a) Trying to understand where the Sun gets its energy
(b) The search for anticancer drugs
(c) The invention of the transistor
(d) The discovery of the electron
(e) Sending a person to the Moon

1.66 Which one of the following is a compound: ozone, O_3; stainless steel, a combination of iron and other metals; sulfuric acid, H_2SO_4; brewed coffee.

1.67 Residential ozonators work by passing an electric current through the air, converting the oxygen in the air to ozone. Is this process a chemical reaction or a physical transformation of oxygen? Explain.

1.68 Classify each of the following as elemental substance, compound, or mixture:
(a) Milk
(b) Chocolate cake
(c) Clean snow, $H_2O(s)$
(d) The simple sugar fructose, $C_6H_{12}O_6$
(e) The mercury, Hg, in a thermometer

1.69 What is the name of the process by which matter changes from the liquid state to the gas state?

1.70 Match each symbol with the correct element name:
(a) Na (b) Fe (c) Co (d) Sn (e) Mn

Possible names: neon, fermium, copper, mercury, cobalt, scandium, magnesium, phosphorus, manganese, iron, sodium, potassium, argon, tin, silver, gold, calcium.

1.71 Match each symbol with the correct element from the list of names in Problem 1.70:

(a) Au (b) Hg (c) K (d) P (e) Ag

1.72 The development of rust spots on an automobile body is an example of a __________ change. (physical or chemical)

1.73 The concise generalization of the results of a number of experiments is known as scientific __________. (experiment, law, or theory)

1.74 Which of the following can form a homogeneous mixture when mixed thoroughly with water: table salt, sand, table sugar, gold?

1.75 What is the difference between a spoonful of sugar dissolved in a glass of water and a spoonful of sugar melted over a candle flame?

1.76 Give three examples of a homogeneous mixture of two liquids.

1.77 Give an example of a homogeneous mixture of a gas in a liquid.

1.78 Brass is an example of a homogeneous mixture that involves two materials that are solids at room temperature, copper and zinc. How would you go about making such a mixture?

1.79 Give an example of a gas dissolved in another gas. Is this mixture homogeneous?

1.80 When dry ice (solid carbon dioxide, CO_2) goes directly from the solid state to the gas state, the process is called __________.

1.81 When nitrogen dioxide gas is cooled, it changes from brown to colorless. Is this more likely a chemical change or a physical change?

1.82 The explosion of a stick of dynamite is most likely a __________ change. (chemical or physical)

1.83 Mixtures can always be separated by physical means. Describe how you would separate the components in a mixture of sand and iron filings.

1.84 Mixtures can always be separated by physical means. Describe how you would separate the components in a mixture of table salt and sand.

1.85 Can you think of a physical means of separating hydrogen from oxygen in a sample of pure water, H_2O?

1.86 Which one of the following is an elemental substance: table salt, NaCl; vinegar, a solution of $C_2H_4O_2$ in water; table sugar, $C_{12}H_{22}O_{11}$; liquid nitrogen, N_2?

1.87 All the metals in the first column of the periodic table react with water because they all have a single electron in their outermost occupied electron shell. This statement is an example of a scientific __________. (law or theory)

1.88 In the chemical reaction for the "burning" of sugar, $C_6H_{12}O_6 + 6O_2 \rightarrow 6\,CO_2 + 6\,H_2O$, which substances are the reactants and which are the products? Does this equation represent a chemical change or a physical change?

1.89 Which one of the following is a homogeneous mixture: milk shake, 14-karat gold (made by melting gold and other metals together), fog, blood?

1.90 Which one of the following is a heterogeneous mixture: fog, gasoline, aluminum–magnesium alloy (made by melting these metals together), mixture of $N_2(g)$ and $Ar(g)$?

1.91 True or false? The melting point and freezing point of water are the same temperature at normal atmospheric pressure.

1.92 Which one of the following is an elemental substance: water, H_2O; bronze, made by melting copper and tin together; table salt, NaCl; ozone, O_3; sulfur dioxide, SO_2?

1.93 Which one of the following is a homogeneous mixture: orange juice with pulp, wood, fog, bronze (made by melting copper and tin together)?

1.94 True or false? A theory is an attempt to explain why a law is true.

1.95 Which one of the following is a compound: ozone, 18-karat gold (made by melting gold and other metals together), NaCl, liquid nitrogen, iced tea?

1.96 Consider the following two pieces of metal. Both consist of gold atoms (filled spheres) and silver atoms (open spheres). Which would be considered heterogeneous? Explain why.

(a) (b)

1.97 Beaker (a) contains the pure compound called hydrogen peroxide (formula H_2O_2). Beaker (b) contains the pure compound called water (H_2O). Both compounds are made to undergo a change.

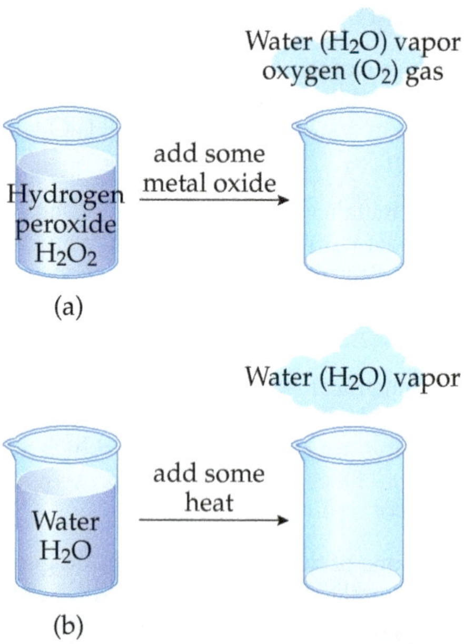

Which compound is undergoing a physical change, and which is undergoing a chemical change? Explain the reason for each.

One Step Beyond

1.98 What must Albert do to turn his "Intense Facial Watching Hypothesis" into a theory?"

1.99 Which is larger, a degree Celsius or a degree Fahrenheit, and by roughly how much is it larger? Justify your answer.

1.100 Based on your answer to Problem 1.99, which would be more noticeable to you, a 10 °C or a 10 °F increase in temperature?

1.101 The "splitting" of liquid water (H_2O) into oxygen gas (O_2) and hydrogen gas (H_2) is often proposed as a source of hydrogen for fuel. Since the state of matter is changing from liquid to gas, is it correct to call this an example of a physical change? Explain.

1.102 The freezing point of nitrogen is −210.0 °C. The boiling point of nitrogen is −195.8 °C.

Group Exercises (Work on these exercises with someone else in the class.)

1. Below are listed many different properties of octane (formula C_8H_{18}), one of the components of gasoline. Group these properties into chemical and physical, and justify each choice.
   (a) Octane is a liquid at room temperature.
   (b) Octane is a volatile liquid (meaning it evaporates rather quickly at room temperature).
   (c) Octane burns (combines with oxygen) to form carbon dioxide and water vapor.
   (d) Liquid octane is colorless.
   (e) Octane is less dense than water (it floats on water).
   (f) Octane is insoluble in water.
   (g) Octane is a poison.
   (h) Liquid octane dissolves grease well.
   (i) Octane forms octyl chloride ($C_8H_{17}Cl$) when combined with chlorine gas.
   (j) Octane has a bad taste and a slight odor.
   (k) Octane freezes below −57 °C.
2. Think of some common everyday observations for which you could propose a law. Write down the law and the observations that support it. Next, postulate a theory or model that explains why the law might be. Finally, propose an experiment that might test the theory.

WORKPATCH SOLUTIONS

1.1. (a) Heterogeneous mixture; different regions can easily be seen in the wood.

(b) Pure substance; made of a single element—iron.

(c) Heterogeneous mixture; the rust is a layer of iron(III) oxide over the elemental iron of the nail.

(d) Homogeneous mixture (solution); the dye is homogeneously distributed throughout the water.

(e) Heterogeneous mixture; no amount of hand-mixing can result in a homogeneous mixture.

(f) Homogeneous mixture (solution); melting allows the many compounds making up the two waxes to mingle and become a homogeneous mixture.

Density=m/V
101010100011
8:49
3:50

CHAPTER 2

The Numerical Side of Chemistry

"No human investigation can be called real science if it cannot be demonstrated mathematically."

Leonardo da Vinci

"I often say that when you can measure what you are speaking about, and express it in numbers, you know something about it; but when you cannot measure it, when you cannot express it in numbers, your knowledge is of a meagre and unsatisfactory kind."

Lord Kelvin (William Thomson, 1st Baron) (1824–1907)
English physicist and mathematician

"If it can't be expressed in figures, it is not science; it is opinion."

Lazarus Long (a character in the science fiction of Robert A. Heinlein)

"Measure twice, cut once."

Unknown master carpenter and the best advice ever given to novice fix-it-uppers.

2.1 Numbers in Chemistry—Precision and Accuracy

Although we promised in Chapter 1 to focus on fundamental concepts in this book, chemistry does have its numerical side. Numbers can be very helpful in expressing, understanding, and applying chemical concepts. For example, a question currently under debate is whether modern society's love for burning fossil fuels is causing global warming. Burning such fuels as gasoline in cars and natural gas in homes produces carbon dioxide. This carbon dioxide helps trap the Sun's warmth in our atmosphere. It is possible that we are producing so much carbon dioxide that we are affecting the planet's temperature, causing it to rise to the point where weather patterns are changing in undesirable ways. The question is, are we really doing this or is our output of carbon dioxide negligible compared with the amounts generated by natural producers of global-warming gases (decomposing vegetation, volcanic activity, etc.)? Only by a careful numerical analysis, at both the level of the chemical reactions involved and the global level, can we hope to find an answer to this question.

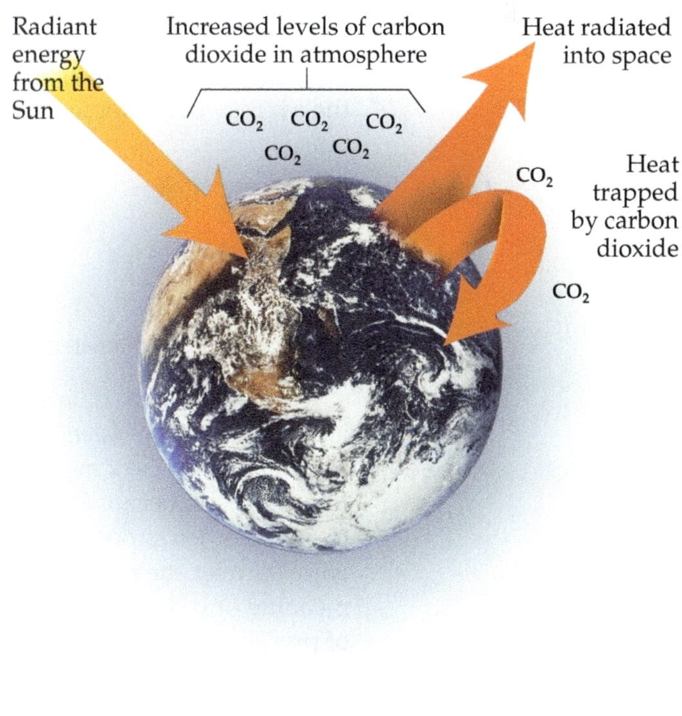

Let's tackle this numerical side of chemistry right now so we can use it when we need to. You may already have a feel for some of the fundamental concepts that govern chemistry's numerical side. If you think about it a bit, any numerical quantity is presented to you in one of two ways, either as an *exact number* or as the result of some *measurement*. For example, if someone tells you that there are seven days in a week, the numerical quantity 7 is an **exact number** because the number of days in a week is *universally defined* to be 7, and this number has no uncertainty. Also, the number of coins in your pocket is an exact number because you can't have a partial coin. That is, if you can *count* the number of items, the result is an exact number.

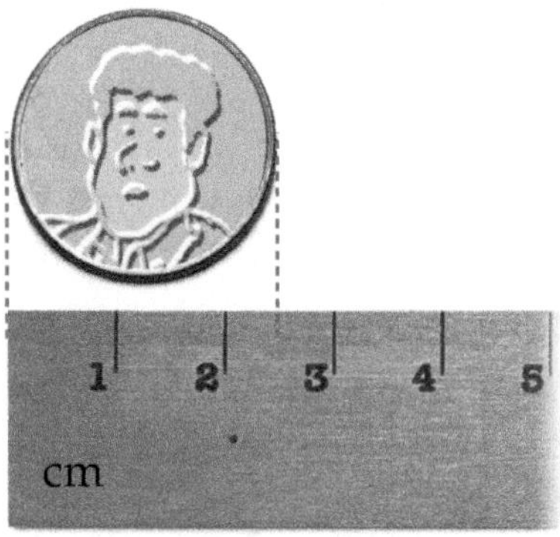

On the other hand, some numerical quantities are the result of a **measurement**. For example, we could use a ruler to measure the diameter of a coin. Let's agree that 2.5 cm is the true diameter of the coin. (The abbreviation "cm" stands for *centimeter*, a length that is a little shorter than 0.5 in. We'll study this and other units in Section 2.6.) Next, we divide six students into two teams and ask each student to measure the diameter once. Here are the results:

Team A	Team B
2.0 cm	2.2 cm
3.0 cm	2.2 cm
2.5 cm	2.2 cm

It looks like only the third student in team A measured the diameter correctly. How can we explain the other students' results? Well, maybe one of the students in team A was tired, and another had difficulty locating the diameter. In team B, perhaps two of the team members were swayed by the third, who was absolutely convinced that the diameter was 2.2 cm. Bad measurements or not, these are the data presented by the teams, so let's work with them.

One thing you can do with repetitive measurements is calculate an average by summing the measurements and then dividing by the total number of measurements. For example, (2.0 cm + 3.0 cm + 2.5 cm)/3 = 2.5 cm. This is the average diameter for team A. The average diameters for both teams are:

	Team A	Team B
	2.0 cm	2.2 cm
	3.0 cm	2.2 cm
	2.5 cm	2.2 cm
Average diameter:	2.5 cm	2.2 cm

We can use these data to discuss two important aspects of measured numerical quantities, *precision* and *accuracy*. **Precision** refers to the closeness to one another of a series of measurements made on the same object. In our case, team B was precise because its measurements were all the same, and thus as close to one another as possible. Team A was not precise because its measurements were all over the place, ranging from 2.0 cm to 3.0 cm. The term *precision* has meaning only when two or more measurements are made. We could not speak of precision if we measured the diameter just once.

If we look at the averages of the measurements, however, team A was more *accurate*. **Accuracy** refers to how close the measured result is to the true value, and team A's average diameter was closer to the true diameter (2.5 cm) than team B's. Unlike *precision*, the term *accuracy* can be applied if just a single measurement is made. For example, the third member of team A made the most accurate single measurement. We can summarize team B's results as being precise but not accurate, whereas team A's results are accurate but not precise. Thus, precision and accuracy do not have to go together. Of course, the goal in taking any measurement is to achieve both accuracy and precision.

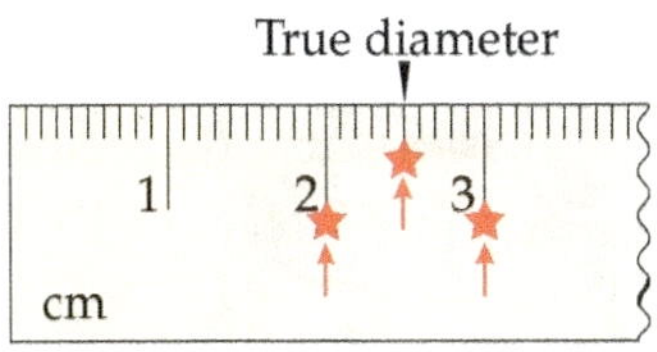

Team A Two students are quite inaccurate, although the *average* of the team's results is accurate.

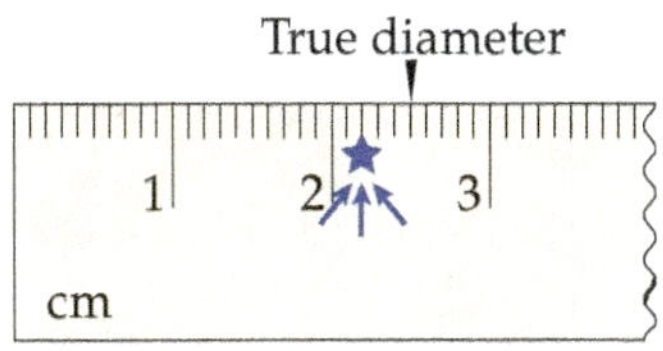

Team B Very precise; quite inaccurate.

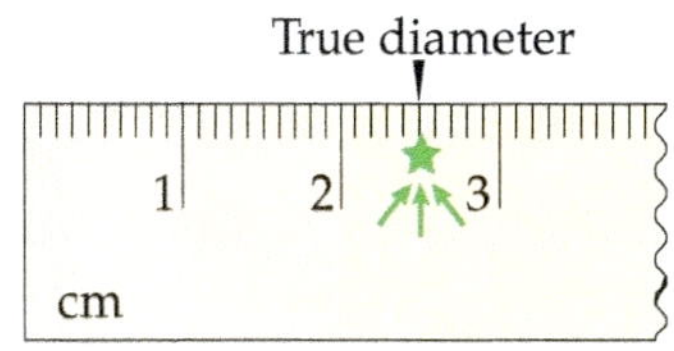

The kind of measurements a scientist likes: accurate *and* precise.

Practice Problems

2.1 Four people listening to a 2800-word speech counted the number of words in the speech as they were listening. They came up with these results:

Fred: 2736 words　　Wilma: 2810 words
Barney: 2792 words　　Betty: 2734 words

(a) Which person was most accurate?
(b) Which person was most precise?
(c) Would you judge the group's average word count as being accurate or inaccurate?
(d) Four other people heard the same speech and came up with the word counts 2722, 2724, 2719, and 2723. Is this group more or less accurate than the first? Is this group more or less precise than the first?

Answer:

(a) *Barney was most accurate because 2792 is closest to 2800, the actual number of words in the speech.*

(b) *Because each person made only one count, the term* precision *doesn't apply.*

(c) *The average word count for the group is (2736 + 2810 + 2792 + 2734)/4 = 2768. This is 32 words fewer than the actual word count of 2800. Because 32 is quite small relative to 2800, the group's determination is pretty accurate.*

(d) *The average word count for the second group is (2722 + 2724 + 2719 + 2723)/4 = 2722. This is 78 words fewer than the actual word count, and so the second group is less accurate than the first. As for precision, the first group ranges from a high of 2810 to a low of 2734 (a spread of 76), whereas the second group ranges from a high of 2724 to a low of 2719 (a spread of 5). The smaller spread makes the second group more precise.*

2.2 Two students count the grains of uncooked rice in a small cup. Both students repeat this measurement four times, with the following results:

Mike: 256, 263, 262, 266　　Ike: 250, 242, 270, 278

The actual number of grains is 260. Which student is more accurate? Which is more precise? Explain your answers.

2.3 Two students attempt to measure out a quart of water into a bucket. Jack has a half-quart container, and Jill has a 10-gallon container. Which student will probably be more accurate at putting a quart of water into the bucket? Explain.

2.2 Numbers in Chemistry—Uncertainty and Significant Figures

Now, let's take a closer look at numerical quantities that are the result of some measurement. For example, consider the following exchange between two people discussing the age of a dinosaur bone:

Even before you continue reading, we think you can sense something wrong with the logic used in this analysis. The problem is that dating techniques used to determine the age of fossils are accurate to within a few hundred thousand years at best. Therefore, the original age determination of 65 million years was approximate. When compared with this uncertainty, the additional 23 years that passed since the dating was carried out mean nothing. The 23 years are simply not significant relative to the error in the original measurement. In other words, while the age stated (65,000,023) has eight digits in it, not all of these digits have equal importance. Some, in fact, have absolutely no importance at all.

Like the fossil's age of 65 million years, most numbers in chemistry come from some kind of measurement. You should always be aware that no measuring device can measure without some uncertainty. If you spend a lot of money on the best measuring device, you may lower the uncertainty, but there will always be some uncertainty associated with any measurement. For example, earlier we measured the diameter of a coin with a ruler marked in centimeters:

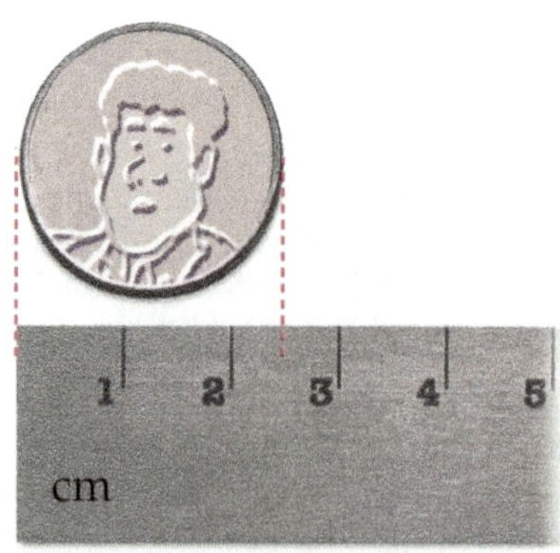

You can see that the diameter is somewhere between 2 cm and 3 cm. The dashed line on the right *seems* to fall halfway between 2 cm and 3 cm, and so you might record the diameter as 2.5 cm. The 0.5 is only an estimate, however, so someone else might estimate the diameter to be 2.4 cm. Yet another person might estimate 2.6 cm. In other words, there is uncertainty in the last digit.

What if you had not seen the ruler used and were presented with a measured diameter of 2.5 cm? How could you know where the uncertainty in this measured number is?

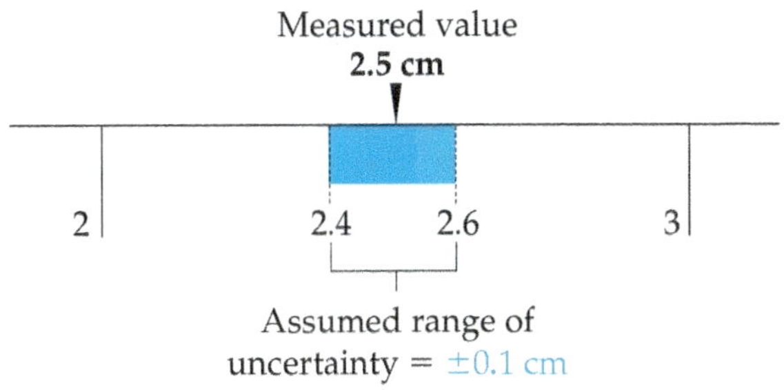

Without access to information about the measuring device (the ruler, in this case), you have no way of knowing where the uncertainty in a reported measurement lies unless some agreed-upon convention is used. Well, not surprisingly, there is just such a convention. The standard convention is to assume that *the uncertainty lies in the last digit written in the number*. Thus, a reported measurement of 2.5 cm is assumed to be uncertain in the tenths place. (In this book, when there is no access to information about the measuring device, we will *assume* an uncertainty of ±1 in the position of the uncertain digit. Hence, the number 2.5, which is uncertain in the tenths position, would be written as 2.5 ± 0.1. That is, the actual number could be anywhere from 2.4 cm to 2.6 cm.)

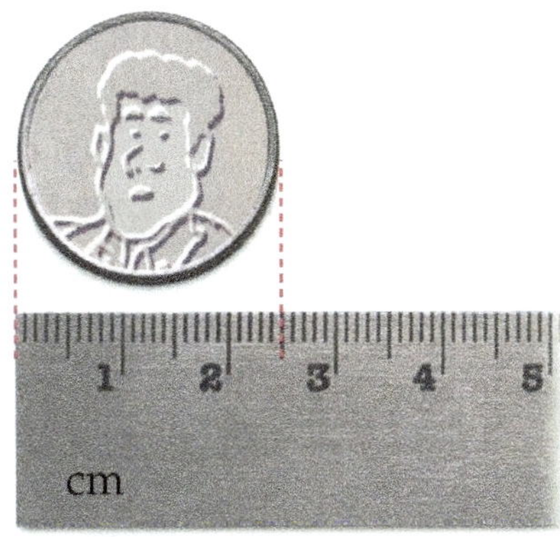

If this same coin were measured with a ruler calibrated in millimeters (as shown in the figure to the right), you would report the coin's diameter as 2.50 cm and not as 2.5 cm, because the uncertainty now lies one place farther to the right, in the hundredths position. By this same convention, if someone reported the result of a measurement to be 102 cm, it would be considered uncertain to ±1 cm because the last uncertain digit is in the ones position.

WORKPATCH 2.1 What is the uncertainty associated with the above ruler calibrated in millimeters? Explain your answer.

(a) ±1 cm **(b)** ±0.1 cm **(c)** ±0.01 cm •

Practice Problems

2.4 A desk is reported to weigh 185 pounds. What is the assumed uncertainty in this weight?

Answer: The uncertainty is ±1 pound because the last digit in the reported number is in the ones column.

2.5 A volume of liquid is measured as being 16.0 gallons. What is the assumed uncertainty in this measured volume?

2.6 The voltmeter shown here is used to measure the voltage of a battery. What is the uncertainty in the measured voltage if the digits shown represent volts?

(a) ±1 V (b) ±0.1 V (c) ±0.01 V

Now let's get back to the estimated coin diameter of 2.5 cm found by using the ruler calibrated in centimeters. Suppose you are asked to calculate the radius of the coin. To do this, you need to divide the diameter by 2:

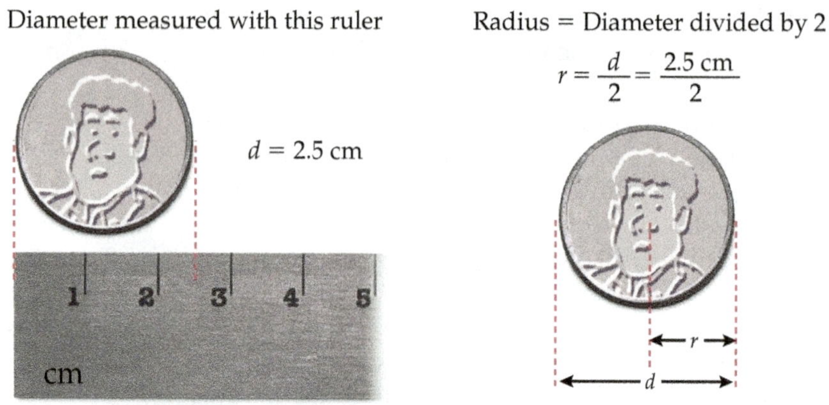

Your calculator gives 1.25 cm as the result. But wait! The last digit, 5, is in the hundredths column, implying that the radius is known to ± 0.01 cm. How can this be? The ruler used to measure the diameter was accurate only to ± 0.1 cm. Dividing the diameter by 2 didn't make the ruler more accurate. Even though your calculator gives a radius of 1.25 cm, the 5 is still beyond the ruler's ability to estimate. Instead, the correct value for the radius, rounding up, is 1.3 cm. (The rules for rounding numbers will be covered shortly.) This value is once again assumed to be uncertain to ± 0.1 cm.

The moral is that you can't always keep all the digits a calculator produces. You can keep only the *significant* ones—the ones that are not beyond the accuracy of the measuring device. These digits are called the **significant figures** (often shortened to **sig figs**) or **significant digits**.

2.3 Zeros and Significant Figures

The most important concept from Section 2.2 is that when you are presented with a measurement and you don't know what measuring device was used, you should assume that the uncertainty lies in the last written digit. For example, if you are told that a board is 103.6 in. long, you should assume that the uncertainty is ± 0.1 in. (because the 6 is in the tenths place) and all the digits are significant (in other words, there are four significant figures). But what about the number 0.06 in.? Are all three digits, the two 0s and the 6, significant? The answer is *no*. There is only one significant figure (the 6). The zeros are just placeholders, called *leading zeros.*

0.06 in.

↑↑ **Leading (placeholder) zeros are not significant**

Leading zeros are all the zeros that precede the first nonzero digit. They are generally found in numbers whose absolute value is less than 1. Our rule for the location of the uncertainty in the number has not changed. The last written digit, 6, is in the hundredths column, and so the uncertainty in the measurement 0.06 in. is ± 0.01 in. Notice that while the zeros in 0.06 in. are not significant, the zero in our first example, 103.6 in., *is* significant. That's because the zero in 103.6 in. is not a leading zero. Leading zeros come before the *first nonzero* digit, never after it. Consider the two numbers:

0.001 02 and 0.001 020

Both have three leading zeros (they are underlined). The first number, therefore, has only three significant figures, while the second number has four (the digits that are not underlined).

Rule 1: Leading zeros (all zeros before the first nonzero digit) are not significant.

The last zero in the number 0.001020 is given a special name, a *trailing zero*. **Trailing zeros** are those zeros that appear to the right of the last nonzero digit in a number:

0.001 02 No trailing zeros
0.001 020 One trailing zero

Trailing zeros that follow the decimal point are always significant and, as such, have meaning. The trailing zero in 0.001 020 is in the millionths column, and so the number 0.001 020 is known to an uncertainty of one-millionth ($\pm$0.000 001). The trailing zero gives the number 0.001 020 an uncertainty that is smaller than the uncertainty in the number 0.001 02 ($\pm$ 0.000 01).

Rule 2: Trailing zeros that follow the decimal point are significant.

Try now to determine the number of trailing zeros and significant figures in the following numbers. Don't go on until you get the correct answers.

WORKPATCH 2.2

	Number of trailing zeros	Number of significant figures
20.201	?	?
20.210	?	?
20.000 2	?	?
20.0	?	?
120.	?	?

The last number in this WorkPatch is an interesting case. It has one trailing zero, and its uncertainty is $\pm$1. Now, you might think it a bit ridiculous to write the number 120 with a decimal point at the end. Why not just write it without the decimal point? If you wanted to buy one hundred twenty eggs, you would write 120 eggs on your shopping list, not 120. eggs. In fact, writing 120 eggs on your list is OK because it is an exact (counted) number. However, if the number is a measurement, then leaving out the decimal point is the wrong thing to do. Without the decimal point, there are *two* ways to interpret the number. It could be interpreted as 120 $\pm$ 10 (meaning the zero is just a placeholder and not significant) or as 120 $\pm$ 1 (meaning the zero is a significant trailing zero). In other words, you can't be sure if zeros immediately to the left of the decimal point are trailing zeros unless the decimal point is actually written. With the decimal point included, there is no ambiguity. The decimal point indicates that any trailing zeros to the left of the decimal point *are significant*. Because chemistry is a quantitative science, chemists must make sure that the numbers they write communicate a single clear meaning.

Practice Problems

2.7 How many interpretations are there for the number 600?

Answer: With no decimal point at the end, 600 can be interpreted as having an uncertainty of ±100, ±10, or ±1. There is no way to know.

2.8 How would you express 600 in. ± 1 in.?

2.9 Fill out the following table:

	Number of significant figures	Uncertainty
10.0	?	?
0.00460	?	?
123	?	?

This brings us to an interesting question. Suppose you measured the length of a building as 120 ft using a ruler marked every 100 ft. This puts the uncertainty in the tens column, which means the measurement is uncertain to ±10 ft.

From the picture, it looks as if the building is about 120 ft long. Of course, this is an estimate. Because the ruler is marked in only hundreds of feet, it is safe to say that you could be off by about 10 ft. Another person might read the ruler as 110 ft, and yet another might read it as 130 ft. That is, the length of 120 ft is uncertain to ±10 ft. Now, how are you supposed to write the number 120 and at the same time communicate to the world that the uncertainty in this measured number is in the tens position (i.e., 120 ± 10)? You cannot write 120. because that implies an uncertainty of ±1 foot. You cannot write 120 because that is ambiguous. You can put a line under the middle digit ($1\underline{2}0$) and say "This is the last significant figure," but that is not how it is usually done. To express the uncertainty unambiguously, we write the number in a different way, using *scientific notation*.

2.4 Scientific Notation

The accepted way to unambiguously indicate the uncertainty in a measured number is to use **scientific notation**—writing the numerical quantity as a number (A) multiplied by 10 raised to an exponent (x).

The general form of a numerical quantity written in scientific notation is:

$A \times 10^x$ — A represents some number. x is an exponent (or power) of 10.

Example: 1.2×10^2 — $A = 1.2$, $x = 2$

The thing to do now is *not panic*. Using scientific notation is easier than it might first appear. First, you should always include a decimal point in the (A) part of the number. The 10^x part is simply an instruction that tells you what to do with the decimal point in order to determine the real value of the number A. This principle works so long as the exponent (x) is a whole number, which it will always be in this book. That is, x will be 1, 2, 3, . . . , or 0, or −1, −2, −3, . . . , but not something like 1.28. When the exponent x is a positive number, you move the decimal point in A to the right x places. When the exponent is negative, you move the decimal point x places to the left. If the exponent is zero, then leave the decimal point alone.

Scientific notation	Exponent means move decimal point	Standard notation
1.23×10^2	Two places to right	123.
1.23×10^{-1}	One place to left	0.123
1.23×10^0	Don't move it	1.23

How can you remember this? Look at what happened to the number 1.23. When the exponent was *positive* (10^2), the number 1.23 got *bigger* (it became 123.) because $10^2 = 100$. When the exponent was *negative* (10^{-1}), the number 1.23 got *smaller* (it became 0.123) because $10^{-1} = \frac{1}{10}$. When the exponent was zero (10^0), then 1.23 remained unchanged because $10^0 = 1$.

Practice Problems

2.10 Convert 4.68×10^{-1} to standard notation.

Answer: Because the exponent is negative, the number must get smaller. Move the decimal point one place to the left: 0.468

2.11 Convert 47.3×10^{-2} to standard notation.

2.12 Convert 47.325×10^3 to standard notation.

Sometimes in moving the decimal point to convert a number from scientific notation to standard notation, you will find that the decimal point moves beyond the beginning or end of the number. For example, consider the number 4.6×10^4. The exponent is positive, and so we are supposed to move the decimal point *four places to the right* to make the number larger. If the exponent were negative (4.6×10^{-4}), the decimal point would move *four places to the left*. But the starting number has only two digits. What are we to do in these cases? Simple—we let the decimal point move off the number. Each time it moves farther off the number, it creates an empty column into which we place a zero. Look closely at how we handle this for our two examples.

What to do when the decimal point moves off the number on the right side:

4.6×10^4 means move the decimal point four places to the right:

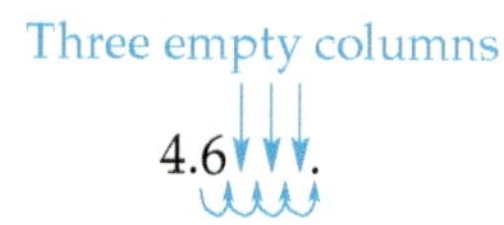

Moving the decimal point four places to the right creates three empty columns, and you should fill these empty columns with zeros:

46,000

What to do when the decimal point moves off the number on the left side:

4.6×10^{-4} means move the decimal point four places to the left:

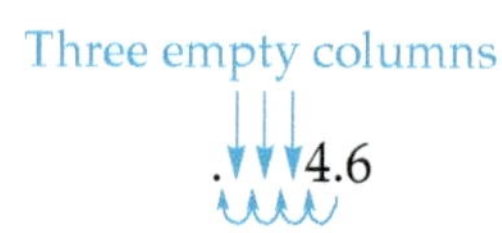

Moving the decimal point four places to the left creates three empty columns that must be filled with zeros:

.00046 or 0.00046 (the zero before the decimal is sometimes added to make the decimal point more noticeable)

Practice Problems

Write the following numbers in standard notation.

2.13 0.400×10^{-6}

Answer: 0.000 000400

2.14 2.35×10^{-3}

2.15 6.0×10^{3}

Whichever way the decimal point moves, the procedure is the same. Simply fill the newly created empty columns with zeros.

We have been converting numbers from scientific to standard notation. You should also be able to go the other way and convert standard notation to scientific notation. To do this, you first have to find the decimal point in the number. This is not a problem for a number like 62.8, where the decimal point is clearly written. When no decimal point is written, as in 125, assume there is a decimal point at the far right end.

Once you have located the decimal point, move it to the right of the first nonzero digit. In the number 125., the first nonzero digit is 1, so move the decimal point two places to the left, making the number 1.25. Next, you must multiply 1.25 by the value of 10^x that would undo what you just did (in order to maintain the same value). Because you moved the decimal point two places to the left, you must set $x = 2$ because multiplying by 10^2 will move the decimal point back two places to the right, giving the original number. So, in scientific notation, 125. is written 1.25×10^2. Likewise, the number 0.000 50 is written 5.0×10^{-4}. (We moved the decimal four places to the right to place it after the first nonzero digit and therefore must multiply by 10^{-4} to undo this.)

WORKPATCH 2.3 Convert the following numbers from standard to scientific notation.

(a) 123 (b) 0.000 06 (c) 0.000 060 (d) 1002.0 ●

Did you notice that (b) and (c) in the WorkPatch look very similar? There's an important difference, however: The number in (c) has a trailing zero. Because this zero is significant, it must show up when the number is written in scientific notation. We never throw away significant figures, even if they are zero.

Now, if you recall, we started this discussion of scientific notation seeking a way to write the number 120 ft ± 10 ft. We can do this using scientific notation, but first we need to mention one more critical thing. When converting from scientific to standard notation, *any zeros that are the result of filling in empty columns created as the decimal point was moved off the number are not significant.* For example, consider the measured quantity 1.2×10^2 ft. Let's convert it to standard notation. The 10^2 tells us to move the decimal point two places to the right, creating one empty column that we must fill with a zero to arrive at 120 ft:

Empty column, fill with a zero

$$1.2 \times 10^2 \text{ ft} = 1.2\_\ \text{ft} = 120 \text{ ft}$$

That added zero is not significant, however, as explained earlier. The last significant digit, which is the digit that always carries the uncertainty, is the 2, and it is in the tens column. Thus, by writing the number 120 as 1.2×10^2, we are indicating that we mean 120 ± 10.

Last significant digit

120 ± 10 is written in scientific notation as 1.2×10^2

What if, when measuring our building length, we had used a ruler calibrated in feet rather than in hundreds of feet? In that case, we would write 1.20×10^2 ft. The 10^2 still moves the decimal point two places to the right, and we still get 120. ft, but now the last significant digit is the zero. Why? Because we included the zero in 1.20×10^2 (a trailing zero) to begin with, to indicate the reduced level of uncertainty.

Last significant digit

120 ± 1 is written in scientific notation as 1.20×10^2

So, while 1.2×10^2 and 1.20×10^2 both say "one hundred and twenty," they indicate different degrees of uncertainty.

Scientific notation gives us a way to indicate unambiguously where the uncertainty in a measured number is. This is one reason we use this notation.

Try the next problem, in which the diameter of a coin is measured with a ruler marked in both centimeters and millimeters.

WORKPATCH 2.4 What is the uncertainty associated with this ruler where the numbered divisions are in centimeters?

(a) ±1 mm **(b)** ±0.1 mm **(c)** ±0.01 mm

Which choice for the diameter of the coin is correct?

(a) 5×10^1 **mm**
(b) 5.0×10^1 **mm**
(c) 5.00×10^1 **mm**
(d) 5.000×10^1 **mm** ●

Check your answers to the WorkPatch and make sure you understand before going on.

Another advantage of scientific notation is that it provides a convenient way to write numbers that are very large or very small. For example, suppose you measure the length of some tiny particle under a powerful microscope and find the length to be 0.000 000 120 m. All the zeros to the left of the 1 are placeholders and not significant (the last zero is a trailing zero and therefore is significant). To avoid having to write all those zeros, a scientist would write this measurement as 1.20×10^{-7}m.

WORKPATCH 2.5 Stop, use pencil and paper, and prove to yourself that 1.20×10^{-7} is the same as 0.000 000 120. ●

Did you put your pencil at the decimal point and then move it seven columns? Which way did you move your pencil, to the left or to the right? Why? Writing 1.20×10^{-7} does away with all the placeholder zeros in this very small number. This strategy also works for very large numbers. Suppose a particular asteroid is measured to be 600,000,000,000 (six hundred billion) miles from us, and the uncertainty in the measurement is ±10 million miles. The uncertainty is in the 10-millions column:

600,000,000,000
↑
10-millions column

Thus, we want to indicate that all the zeros that come after this column are not significant and are simply placeholders. To do this, we would write this number in scientific notation as 6.0000×10^{11}.

WORKPATCH 2.6 With pencil and paper, prove to yourself that if you were given the number 6.0000×10^{11}, you could (a) come up with the standard notation 600,000,000,000 and (b) explain why the uncertainty is ±10 million. ●

If you have trouble with this WorkPatch problem, examine the solution at the end of the chapter.

Practice Problems

Write the following numbers in scientific notation (to prove to yourself that it sometimes saves a lot of effort by not writing zeros). Be sure you are showing the correct number of significant figures.

2.16 0.000 000 000 020

Answer: 2.0×10^{-11}

2.17 47,100,000,000,000, assuming the uncertainty to be ±1 million

2.18 47,100,000,000,000, assuming the uncertainty to be ±10 million

2.5 How to Handle Significant Figures and Scientific Notation When Doing Math

When you do calculations with measurements, you must follow two rules to ensure that you don't end up with nonsignificant digits in your result. One rule applies to addition and subtraction. The other rule applies to multiplication and division. We'll cover the multiplication/division rule first.

Recall the earlier example of calculating the radius of a coin by dividing the measured diameter of 2.5 cm by 2. We couldn't keep the digit 5 in the calculated radius of 1.25 cm because it was not significant (it implied an uncertainty in the hundredths place, which was beyond the uncertainty of the ruler used to make the measurement). The rule determining the number of significant figures allowed in the result of multiplying or dividing two measured numbers is simple.

Rule 1: The result of multiplying or dividing two measurements cannot have more significant figures than there are in the measurement that has the smaller number of significant figures.

For example, consider these two multiplications:

2.0 cm	← Two sig figs
× 2 cm	← One sig fig
$4\ cm^2$	← Answer can have only one sig fig

BUT

2.0 cm	← Two sig figs
× 2.0 cm	← Two sig figs
$4.0\ cm^2$	← Answer can have two sig figs

Sometimes the only way to keep your answer from having too many significant figures is to round off. For example:

2.0 cm	← Two sig figs
× 8 cm	← One sig fig
$16\ cm^2$	← Answer can have only one sig fig

Because you must round to one significant figure, the correct answer is:

$20\ cm^2$ or, indicating one significant figure, $2 \times 10^1\ cm^2$

You must always round off your answer if that is the only way to get the right number of significant figures. You round up when the digit following the last digit to be kept is 5 or greater, and down when the digit following the last digit to be kept is 4 or less. Thus, rounding to three significant figures, 3625 becomes 3.63×10^3 and 3622 becomes 3.62×10^3. Rounding to two significant figures, 36.7 becomes 37 and 36.4 becomes 36.

Before you try the practice problems, we want to make something very clear. The first sentence in this section said, "When you do calculations with *measurements, ...*" All of the preceding multiplications involved numbers that came from measurements of length. There are times, however, when you will multiply or divide a measurement by an *exact number*. We did this when we took the measured coin diameter of 2.5 cm and divided it by 2 to find the coin's radius. The 2 did not come from some measurement. It is an exact number. The way to deal with an exact number is to treat it as if it has an infinite number of significant figures (i.e., 2 is really 2.000 000 00 ...). Therefore, when we divided 2.5 cm by 2 to get an answer of 1.25 cm on our calculator, we rounded it to 1.3 cm and not 1 cm. Exact numbers have no effect on the number of significant figures allowed in the result of a calculation. Now try your hand at the following practice problems.

Practice Problems

2.19 27.5 in./2.0 h = ?

Answer: 27.5 in./2.0 h = 13.75 in./h, but this is not correct. The answer can have only two significant figures, and so it is rounded to 14 in./h or 1.4×10^1 in./h.

2.20 22.0 miles × 2.0 miles = ?

2.21 220. h × 3 = ? (the 3 is an exact number)

Now try this one.

WORKPATCH 2.7 What is the result of 222 miles/2.0 h when written with the correct number of significant figures? ●

Obviously, 222 miles/2.0 h is 111 miles/h, but your result can have only two significant figures. To make this perfectly clear, you need to write the answer in scientific notation. The answer is 1.1×10^2 miles/h. It is rounded off (1.1×10^2 miles/h = 110 miles/h), and it uses only two significant figures. The moral here is, use scientific notation to get out of significant figure jams.

Practice Problems

Do the following calculations and give your answers to the correct number of significant figures using correct units.

2.22 1222 pounds/2 in. = ?

Answer: 6×10^2 pounds/in. (Your calculator gives 611, but your answer can have only one significant figure.)

2.23 Suppose you measure the length of a house to be 60.50 ft. How many yards is that? (1 yard = 3 ft)

Answer: There are exactly 3 ft in 1 yard because that is how a yard is defined. Therefore, the numbers 3 and 1 are exact and have no effect on the number of significant figures in the calculated result. Because the only measurement has four significant figures, your answer should also have four significant figures:

$$60.50 \text{ ft} \times \frac{1 \text{ yard}}{3 \text{ ft}} = 20.17 \text{ yards}$$

Measurement (60.50 ft) *Exact numbers* (1 yard / 3 ft)

2.24 (a) 1222 pounds/2.0 in. = ?
(b) 1222 pounds/2.00 in. = ?
(c) What do you get when you quadruple 21.72 cm?

The rule for how many significant figures are allowed in the result of an addition or subtraction is different from the multiplication/division rule.

Rule 2: When adding or subtracting a series of measurements, the result can be no more certain than the least certain measurement in the series.

For example, consider adding the following measurements:

	Position of uncertainty
123 cm	← Ones position (least certain of the three values being added)
+ 1.006 cm	← Thousandths position
+ 2.6 cm	← Tenths position
126.606 cm	← Answer calculator gives
127 cm	← CORRECT ANSWER because it can be no more certain than the ones position

The least certain measurement is 123 cm, which is uncertain in the ones position. The correct answer of 127 cm comes from rounding up the calculated result, 126.606, to the column where the least uncertainty is, the ones position.

Try the following practice problems now.

Practice Problems

Do the following problems and report your answers to the correct number of significant figures.

2.25 1555 in. + 0.001 in. + 0.2 in. = ?

Answer: 1555 in.

2.26 1555 cm + 0.001 cm + 0.8 cm = ?

2.27 142 cm − 0.48 cm = ?

If you follow these two rules, you will know how to round off the results your calculator gives you. Ignoring the rules of significant figures would put more certainty into your results than justified.

You will use a scientific calculator to do most of your calculations with numbers expressed in scientific notation. You can simply enter numbers directly in scientific notation and then add, subtract, multiply, or divide by pressing the appropriate keys. The calculator will handle the exponents for you. On most scientific calculators, to enter a number in scientific notation you first punch in the digits, then the [EE] or [EXP] key, then the value of the exponent (x), and then the [+/−] key if you want to make the exponent negative. For example, see the following:

Number	Key	How it looks on calculator
1.23×10^3	[1] [.] [2] [3] [EE] [3]	*1.23 E 03*
1.23×10^{-3}	[1] [.] [2] [3] [EE] [3] [+/−]	*1.23 E −03*

If you have trouble working with scientific notation on your calculator and the instruction manual doesn't help, see your instructor.

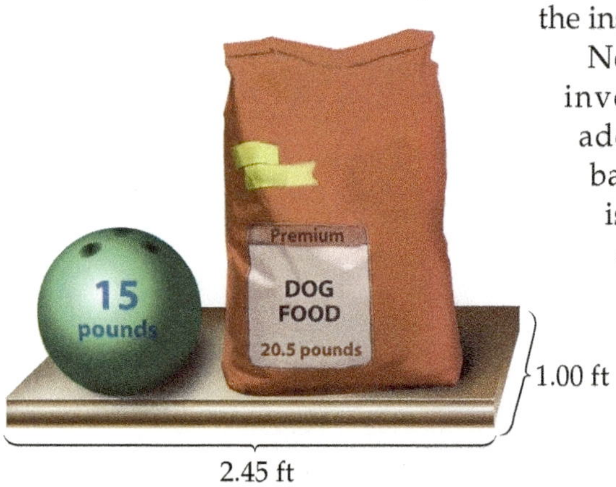

Now let's pull out all the stops and do a calculation that involves both the multiplication/division rule and the addition/subtraction rule. Suppose a bowling ball and a bag of dog food are placed on a rectangular shelf. The shelf is measured to be 1.00 ft wide and 2.45 ft long. The weight of the bowling ball is stamped on it, and the weight of the dog food is stamped on the bag. What is the weight pressing down per square foot of shelf surface area? To figure this out, all we have to do is sum up the weights of the ball and bag and divide by the surface area of the shelf (the area of a rectangle is its width times its length).

$$\text{Weight per square foot} = \frac{20.5 \text{ pounds} + 15 \text{ pounds}}{2.45 \text{ ft} \times 1.00 \text{ ft}}$$

You can do either the addition or the multiplication first—it does not matter. Either way you do it, however, the final answer is allowed to have only two significant figures. Do you see why? When we sum 20.5 pounds and 15 pounds on a calculator, we get 35.5 pounds, but by the addition/subtraction rule, we must round to 36 pounds.

After doing both the addition and the multiplication, we have:

$$\text{Weight per square foot} = \frac{36 \text{ pounds}}{2.45 \text{ ft}^2}$$

Now we have a division, and the multiplication/division rule clearly limits our result to two significant figures. A calculator gives 14.69387755 pounds/ft^2, but the answer after rounding is 15 pounds/ft^2. *We are not done with this problem yet, however!* Look back and you'll see we rounded numbers twice, once in the middle of our calculations and again at the very end. First we rounded the summed weight of 35.5 pounds to 36 pounds because the addition/subtraction rule limited this intermediate result to two significant figures. But suppose we had not done this intermediate rounding. Rather, suppose we used the result our calculator gave us and waited until all our calculations were done before applying the two-significant-figures restriction. This would change our final step to:

$$\text{Weight per square foot} = \frac{35.5 \text{ pounds}}{2.45 \text{ ft}^2}$$

Now a calculator gives 14.48979592 pounds/ft^2, which to two significant figures is 14 pounds/ft^2. Which result is more correct—15 pounds/ft^2 or 14 pounds/ft^2? The latter one is more correct because we did less rounding off in obtaining it. In any multistep calculation, it is always best not to do any rounding until the very end. Otherwise, you start to accumulate rounding error, with each rounding of an intermediate value putting more and more error into your final result.

2.6 Numbers With a Name—Units of Measure

A measurement makes no sense unless you know its units. For instance, it makes no sense to be told you can get to the airport in 3. Three minutes? Three hours? Three gallons of gas? You need the units.

Deciding which unit of measurement to use can be complicated because there are many possibilities for each type of measurement. Length can be measured in the system of inches, feet, yards, and miles, or it can be measured in the metric system of centimeters, meters, and kilometers. Volume can be measured in the system of quarts and gallons, or in the metric system of liters. Mass can be measured in a variety of units. So, there's potential for confusion in measurement, sometimes to a disastrous degree.

Consider, for example, the Mars Polar Lander launched by NASA. Instead of landing as planned on September 23, 2000, the $125 million probe burned up in the red planet's atmosphere. The company that built the retrorockets used U.S. customary units (inches and feet for length, pounds for force) to measure the rocket's thrust. It then communicated these data to NASA. Unfortunately, NASA assumed the data were in metric units (meters for length, newtons for force). Hence the headline "Metric Mars Orbiter Takes a Pounding." The joke is that Martians are currently taking bets on how deep a hole the next lander will drill into the surface of their planet.

To minimize this kind of confusion, scientists the world over and engineers in every nation except the United States (don't ask!) use a standardized system of units called **SI units**. (SI stands for *Système International*.) Of course, that they were dealing with two systems of units should have been made clear to the NASA scientists, but as one of NASA's directors commented afterwards, "People sometimes make errors."

Table 2.1 presents the base SI units used in chemistry.

Table 2.1 Base SI Units Used in Chemistry

Physical quantity	Name of unit	Abbreviation
Length	meter	m
Mass	kilogram	kg
Time	second	s
Temperature	kelvin	K
Amount of substance	mole	mol

The base SI unit of length is the meter (m). A meter is about $3\frac{1}{3}$ in. longer than a yard (1 yard equals 3 ft).

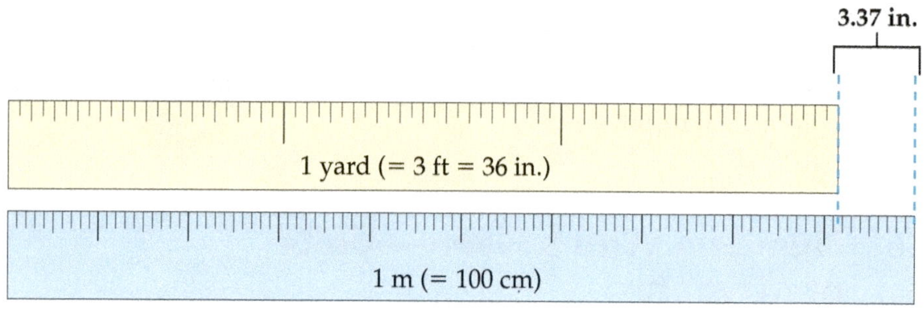

A kilogram (kg) equals a bit more than 2 pounds (2.2046 pounds, to 5 sig figs). Notice from Table 2.1 that we call the kilogram a unit of *mass* and not a unit of *weight*. **Mass** is a measure of the quantity of matter in an object and depends only on the quantity present. The **weight** of an object depends on the strength of the gravitational force exerted on the object. If you went to the moon, your mass would be the same as it is on Earth, but you would weigh only one-sixth of what you weigh on Earth because the moon has only one-sixth of our gravity. (Nevertheless, many people use the words *mass* and *weight* interchangeably.) The non-SI metric unit of mass, the gram, is used often in chemistry. One gram (g) is equal to 0.001 kg; that is, there are 1000 g in 1 kg. There are 453.6 g in 1 pounds.

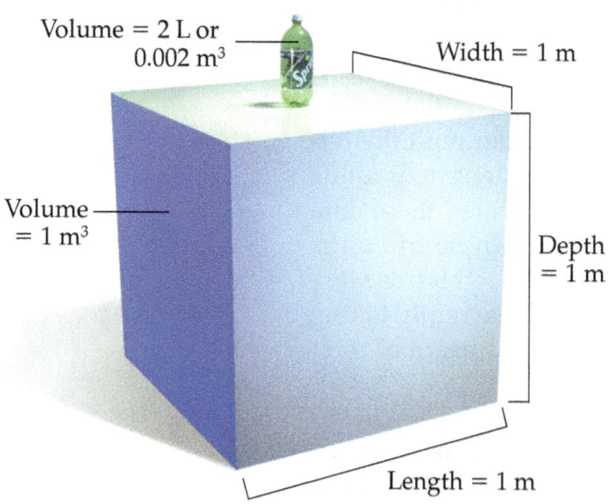

Derived SI units come from combinations of the base SI units. For instance, the SI unit for **volume** (the amount of space occupied by an object) is the cubic meter (m^3). A cubic meter of volume is the space enclosed by a box whose length, width, and depth are all 1 m. One cubic meter equals approximately 264 gallons. As we'll see in a moment, the liter is another volume unit.

Several derived SI units are listed in Table 2.2.

We often find the SI units in Tables 2.1 and 2.2 inconvenient because they are either too big or too small for the job at hand. For example, suppose you wanted to measure out a teaspoon of liquid and express this volume in cubic meters. Such a small volume of liquid is only a tiny fraction of a cubic meter

Table 2.2 Derived SI Units

Physical quantity	Name of unit	Abbreviation
Volume	cubic meter	m^3
Pressure	pascal	Pa
Energy	joule	J
Electrical charge	coulomb	C

(1 teaspoon = 0.000005 m^3). Would you want to have to ask for 0.000005 m^3 of cough syrup every time you needed a teaspoon of it? And suppose you wanted to express the thickness of a human hair in the SI unit of meters. A meter is so much larger than the thickness of a human hair that, again, it would be silly to report its thickness in meters. What we do, therefore, is modify the SI units using the Greek prefixes shown in Table 2.3.

Table 2.3 Greek Prefixes Used With SI Units

Greek prefix	Meaning
giga- (G)	one billion (10^9)
mega- (M)	one million (10^6)
kilo- (k)	one thousand (1000 or 10^3)
deci- (d)	one-tenth (.1 or 10^{-1})
centi- (c)	one-hundredth (.01 or 10^{-2})
milli- (m)	one-thousandth (.001 or 10^{-3})
micro- (μ)	one-millionth (10^{-6})
nano- (n)	one-billionth (10^{-9})
pico- (p)	one-trillionth (10^{-12})

Thus, instead of saying that the measured width of a human hair is 0.0002 m, we can say that it is 0.2 millimeter (0.2 mm). Because a millimeter is the same as 10^{-3} m, 0.2 mm is the same as 0.2×10^{-3} m, which is 0.0002 m. Instead of saying that the measured diameter of a blood cell is 0.000 010 0 m, we can say that it is 10.0 micrometers (10.0 μm). Because a micrometer is the same as 10^{-6} m, then 10.0 μm is the same as 10.0×10^{-6} m, which is 0.000 010 0 m. You try it now.

Practice Problems

2.28 Express the distance 24,000,000,000 m $\pm$100 million m using the appropriate Greek prefix.

Answer: The number is 24 billion, so let's use gigameters (Gm). Because 1 Gm is 1 billion meters, 24 billion meters is 24 Gm. However, to indicate an uncertainty of $\pm$100 million, we must write 24.0 Gm (the zero is in the 100 millions place).

2.29 Express the distance 4736 m in kilometers (1 km = 1000 m).

2.30 Express 0.025 m in millimeters (1 mm = 0.001 m).

The Greek prefixes combined with the SI units provide a considerable amount of flexibility for describing a wide range of measurements, from tiny to huge, but scientists needed still more flexibility. For example, the Greek prefixes can be applied to the liter (L), a non-SI metric unit that is used frequently.

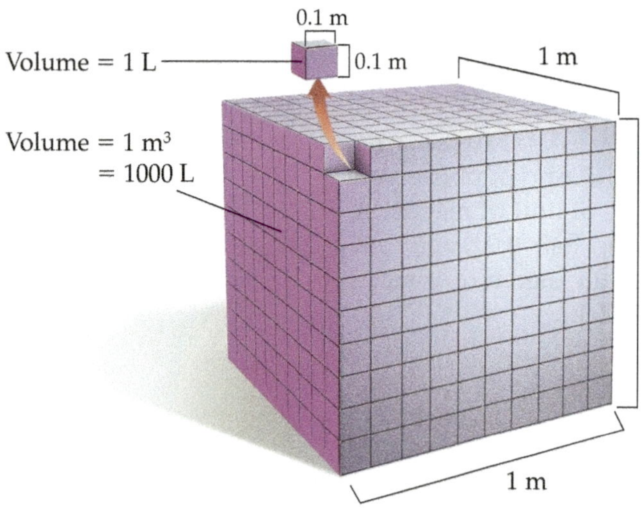

There are 1000 L in a cubic meter, the SI unit of volume. This means that a liter is the volume enclosed in a box that has a length, width, and height of 0.1 m. A liter is just a little bit larger than a quart; 1 L = 1.057 quart.

An extremely common unit for describing teaspoon-sized volumes is the milliliter (mL), which is equal to one-thousandth of a liter (1 mL = 0.001 L). Just as 1000 L fit inside 1 m^3, 1000 mL fit inside 1 L. A milliliter is the volume enclosed by a box whose length, width, and height are all 1 cm (0.01 m); thus, a milliliter is also a centimeter cubed (cm^3, commonly called a cubic centimeter or cc, pronounced "cee-cee"). You can always say either milliliter or cubic centimeter because the two mean the same thing.

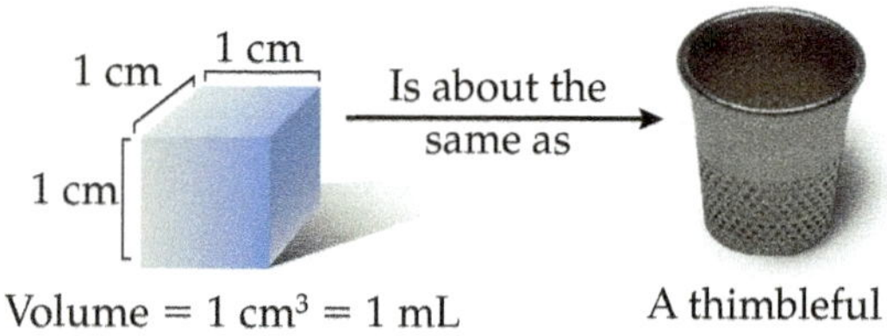

Non-SI units are also used for pressure and temperature. Rather than using SI units of pascals (Pa), chemists prefer to measure pressure either in millimeters of mercury or in atmospheres, as you will see in Chapter 11. And rather than using the SI unit of temperature, the Kelvin (represented as K, *not* °K), chemists typically describe temperatures using metric units of degrees Celsius (°C), which can be converted to kelvins when need be. The Kelvin temperature scale is also called the *absolute temperature scale* because nothing can be colder than 0 K (there is no such thing as a negative Kelvin temperature). You are probably more familiar with the Fahrenheit temperature scale. This unit has now been abandoned by most of the world except the United States. A comparison of the three temperature scales is shown at the bottom of the next page.

The Celsius scale sets the freezing point of water at 0 °C and the boiling point at 100 °C, making 100 divisions between these points. The Fahrenheit scale has 180 divisions between water's freezing point of 32 °F and its boiling point of 212 °F. This makes the Celsius divisions almost twice as large as the Fahrenheit divisions (180/100 = 9/5 as large, to be exact). The Fahrenheit-to-Celsius conversion formulas are:

$$°F = 32 + \frac{9}{5}°C \quad \text{or} \quad °C = \frac{5}{9}(°F - 32)$$

Note in the comparison chart that the Kelvin scale is divided the same way the Celsius scale is. Therefore, to convert from Celsius to Kelvin, all you have to do is add 273.15:

$$K = °C + 273.15$$

In the United States, people are a bit stubborn about using the Fahrenheit scale. Schools teach that water freezes at 32 °F and boils at 212 °F. Our normal body temperature is 98.6 °F, and a pleasant day is about 75 °F. Tell an average person in the United States that it is 35 degrees outside, and the reaction would probably be, "Thirty-five degrees! I'd better wear a warm coat." Unfortunately, if that were 35 °C, the person would be in for a shock, because 35 °C is the same as 95 °F!

You should get used to the Celsius scale. Remember, a Celsius degree is about twice as large as a Fahrenheit degree, and so if you go from a pleasant 24 °C (75 °F) to 30 °C, you are increasing the temperature by 6 °C, or almost 12 °F (it takes roughly twice as many °F to equal a change in °C). This takes you from 75 °F to almost 87 °F, a tad on the warm side. The exact conversion is:

$$
\begin{aligned}
°F &= 32 + \frac{9}{5} \times 30°C \\
&= 32 + \left(\frac{9 \times 30}{5}\right) \\
&= 32 + 54 \\
&= 86°F
\end{aligned}
$$

If you're used to units such as pounds, miles, quarts, gallons, and degrees Fahrenheit because you've worked with them all your life, it will take some practice to get used to the new units of measure. As you practice using SI and metric units, you will get more of a feel for them. Try the following WorkPatch to see if you are getting a feel for the metric volume units.

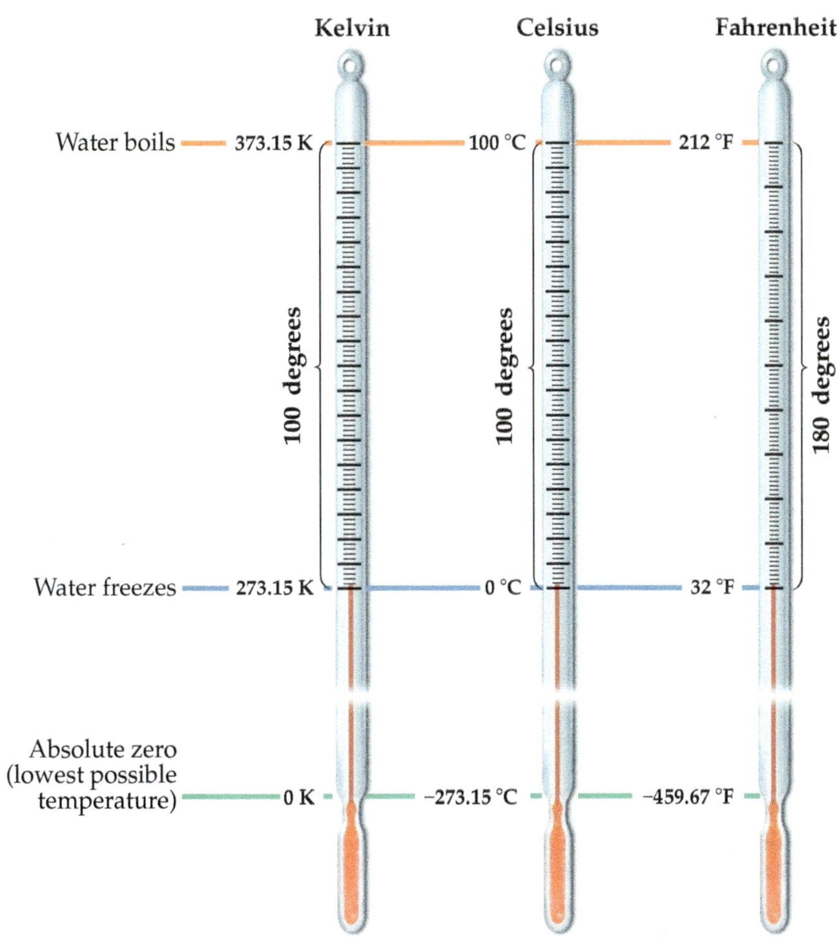

WORKPATCH 2.8 Suppose you want to use a 100-mL beaker to fill a 2.5-L bucket with water. How many times will you have to fill your beaker? ●

Could you do the WorkPatch in your head? The trick is to realize that 100 mL is one-tenth of a liter. Now try the following practice problems.

Practice Problems

2.31 How many milliliters are there in 1.000 L?

Answer: Because a milliliter is 1/1000 of a liter, there are 1000 mL in a liter.

2.32 How many milliliters are there in 2.500 L?

2.33 How many milliliters are there in 246.7 cm^3?

2.34 The temperature outside is 263.5 K. What is the temperature in degrees Celsius and in degrees Fahrenheit?

2.7 Density: A Useful Physical Property of Matter

Density is a measure of the "compactness of matter." It is a property that tells you how much matter there is in a given volume. The *how much* is expressed as a mass, usually in grams. Mathematically, density is defined as the mass per unit volume of the material in question:

$$\text{Density} = \frac{\text{Mass}}{\text{Volume}} = \frac{m}{V}$$

For example, there is much less mass per unit volume of a cloud than per unit volume of lead. Therefore, a cloud has a much lower density than a piece of lead.

Chemists usually express densities in units of g/mL or g/cm^3. Table 2.4 lists the densities of some common substances. Notice how the last three entries in the table, all gases, have much lower densities than the other entries, all liquids and solids.

Table 2.4 Densities of Some Common Substances*

Substance	Density (g/mL)	Substance	Density (g/mL)
Gold (solid)	19.3	Butter (solid)	0.9
Mercury (liquid)	13.6	Gasoline (liquid)	0.7
Lead (solid)	11.4	Oxygen (gas)	0.001 43
Water (liquid)	0.997	Air (gas)	0.001 30
Water (solid ice at 0 °C)	0.917	Helium (gas)	0.000 18

*Densities of all liquids and solids except ice measured at 25 °C; densities of gases measured at 0 °C and 1 atm pressure.

Try the following WorkPatch to see if you have a good feel for density.

WORKPATCH 2.9 Some metals can exist in more than one form, differing in how their atoms are arranged in the solid. Shown here are two forms of the same metal. One form is denser than the other. Which is denser and why?

Form A **Form B**

Your answer explains that along with the mass of the atoms that make up a substance, something else is also important in determining its density.

The density of liquid water is 0.997 g/mL, or essentially 1.00 g/mL. This number simply means that 1.00 mL of water has a mass of 1.00 g. Mercury, a liquid metal used in thermometers, has a density of 13.6 g/mL. This means that 1.00 mL of mercury has a mass of 13.6 g. Thus, 1.00 mL of mercury is 13.6 times more massive than the same volume of water.

Density can be very useful in identifying an unknown substance. For example, suppose you were given a small cube of golden yellow metal that measured 2.00 cm × 2.00 cm × 2.00 cm, and you had to determine whether it was gold or painted lead. One way to do this would be to determine the mass of the cube and then divide this mass by the volume of the cube to get its density. You could then compare this density with the entries in Table 2.4.

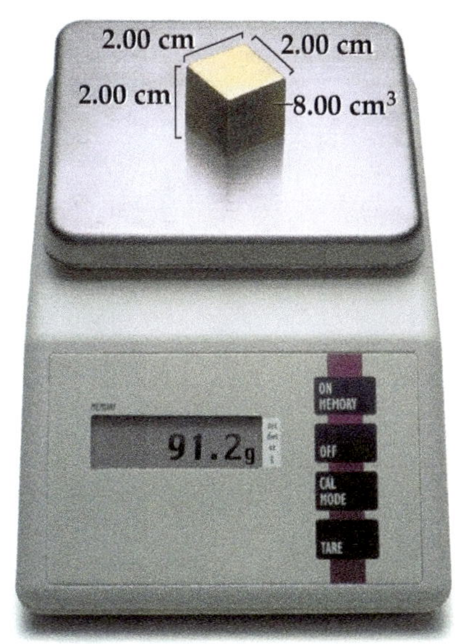

$$\text{Volume of cube} = 2.00\text{ cm} \times 2.00\text{ cm} \times 2.00\text{ cm} = 8.00\text{ cm}^3 = 8.00\text{ mL}$$

$$\text{Mass of cube} = 91.2\text{ g}$$

$$\text{Density} = \frac{91.2\text{ g}}{8.00\text{ mL}} = \frac{11.4\text{ g}}{\text{mL}}$$

Because lead has a density of 11.4 g/mL and gold has a density of 19.3 g/mL, the yellow metal block must be lead painted to look like gold. Now you have a very accurate way to determine if the gold someone is selling you is really gold, with the density value being totally independent of the size of the sample. This is true because density is an **intensive property**, a property that does not depend on the amount of material. Had you been given a larger or smaller cube of the same metal we just analyzed, the density would have been the same. The opposite of an intensive property is an **extensive property**, one that does depend on the amount present. Examples of extensive properties of matter are mass and volume. Because density equals mass/volume, we see that an intensive property—density—can be derived from two extensive properties.

Measuring the volume of an irregularly shaped object by displacement

Measuring the density of an object requires determining its mass and its volume. Determining mass is usually the easy part. Determining volume, however, can be difficult if the object is irregularly shaped. For example, how could you easily measure your own volume? You could not simply multiply your length times your width times your depth as we do for a box because your width and depth change as you go from feet to head. An easy way to measure the volume of an irregularly shaped object is to immerse it completely in a container filled to the very top with water. Doing so will cause a volume of water equal to the volume of the object to be displaced to spill from the container. Collecting this overflow water and measuring its volume will give you the volume of the object. This is called *determining the volume of an object by displacement.* It was a concept first observed by Archimedes (a Greek mathematician, 287 BC–212 BC), who ran naked through the streets yelling "Eureka" after making the mental leap that the rising level of water occurring after he got into the bathtub could be used to determine volume.

Practice Problems

2.35 Which is denser, a 200 pound block of lead or a 0.1-g piece of gold?

Answer: The piece of gold is more dense (see Table 2.4). The size of the piece does not matter because density is an intensive property.

2.36 A cube that is 10.0 mm × 10.0 mm × 10.0 mm has a mass of 4.70 g. What is its density in grams per milliliter?

2.37 A small statue that has a mass of 500.0 g displaces 150.5 mL of water. What is its density in grams per milliliter?

2.8 Doing Calculations in Chemistry—Unit Analysis

Now we want to propose a general strategy for solving many of the numerical problems that come up in chemistry. Many times, the biggest problem for students is knowing where to start. It can often appear that a blizzard of numbers has blown your way. For example, consider the following problem: If you have 2.00 g of the liquid metal mercury (Hg), what volume in milliliters will it occupy, given that the density of mercury is 13.6 g/mL?

We have a measured mass of mercury (2.00 g) and the density of mercury (13.6 g/mL), and we need to find a volume in milliliters. Where should we begin? The answer comes from looking very closely at the units associated with the numerical quantities in the problem. The density of mercury has two units, grams and milliliters. The mass of mercury, 2.00 g, has just one unit. The general strategy for solving a numerical problem makes use of this information in a three-step approach called **unit analysis**.

The first step of unit analysis is to write down the given measurement.

Step 1: Write down the given measurement and its units.

2.00 g mercury

The second step is to use the density of mercury, but we first want you to look at density in a new way—as a *conversion factor*. A **conversion factor** can be recognized because *it always has two different units, one over the other*. For example, you know that there are 60 seconds in a minute,

$$60 \text{ s} = 1 \text{ min}$$

This is a conversion factor. It is often written in ratio form as:

$$\frac{60 \text{ s}}{1 \text{ min}} \quad \text{or} \quad \frac{1 \text{ min}}{60 \text{ s}}$$

60 s per 1 min   1 min per 60 s

You can write it either way. The left ratio says there are 60 s per 1 min (60 s/min), and the right ratio says there is 1 min per 60 s (min/60 s). Both are true. Both ratios have two units—seconds and minutes—one over the other. *This is the hallmark of a conversion factor.*

Because the density of mercury is 13.6 g/mL, we can write

$$1 \text{ mL mercury} = 13.6 \text{ g}$$

Note: The "1" in "1 mL mercury" is treated as an exact number in all calculations.

and we can write this in ratio form as:

$$\frac{13.6 \text{ g mercury}}{1 \text{ mL mercury}} \quad \text{or} \quad \frac{1 \text{ mL mercury}}{13.6 \text{ g mercury}}$$

Either form is correct, although traditionally, density is written as mass over volume.

So, step 1 was to write down the measurement and units (2.00 g mercury). Step 2 is to multiply the measurement by the appropriate conversion factor in such a way that the units of the measurement cancel out. For our example, this means we do the following:

Step 2: Multiply the measurement by a conversion factor so that the original units of the measurement cancel out.

$$2.00 \text{ g mercury} \times \frac{1 \text{ mL mercury}}{13.6 \text{ g mercury}}$$

The conversion factor is written as 1 mL/13.6 g because this way, the units "grams" cancel out. Identical units cancel out when they appear in both the top (numerator) and the bottom (denominator) of a ratio because anything over itself equals 1.

Identical units on top and bottom cancel:

$$2.00 \text{ g mercury} \times \frac{1 \text{ mL mercury}}{13.6 \text{ g mercury}} \longleftarrow \text{The unit that survives}$$

Had we written the density conversion factor the other way (with grams on top and milliliters on the bottom), the units "grams" would appear twice on the top and would not cancel. Step 2 says to multiply by the conversion factor *so as to make the original units of the measurement cancel out!* The unit that survives is "milliliter," a unit of volume. This is what the question asked us for (the volume occupied by the mercury). So, we're ready for the final step, calculating our answer:

Step 3: Perform the calculation.

$$2.00 \text{ g mercury} \times \frac{1 \text{ mL mercury}}{13.6 \text{ g mercury}} = 0.147 \text{ mL mercury}$$

Practice Problems

2.38 Which of the following can be thought of as conversion factors?

(a) 49.3 kg
(b) 4.184 J/°C
(c) 350 miles/h
(d) 12 eggs per dozen
(e) 1 dozen grams

Answer: (b), (c), and (d).

2.39 Write two conversion factors (two different ratios) to express the fact that there are 24 h in a day.

2.40 Given a speed of 600.0 miles/h and a measured distance of 50.0 miles, multiply the conversion factor by the measured distance to make identical units cancel. Calculate the result and include its units.

2.41 Given a speed of 600.0 miles/h and a measured time of 50.0 h, multiply the conversion factor by the measured time to make identical units cancel. Calculate the result and include its units.

You might be wondering why we can multiply by conversion factors the way we do. To see why this multiplication is allowed, let's again use the density of mercury as our example. That the density is 13.6 g/mL tells us that 1 mL of mercury = 13.6 g of mercury. This means that the conversion factors

$$\frac{13.6 \text{ g mercury}}{1 \text{ mL mercury}} \quad \text{or} \quad \frac{1 \text{ mL mercury}}{13.6 \text{ g mercury}}$$

both equal 1 since anything divided by something equal to it yields 1. In essence, therefore, when we multiply a number by an appropriate conversion factor, we are not changing the value of the number (because we are multiplying it by 1, and multiplying anything by 1 does not change its value). All we are doing is changing its units.

You must always make sure that when you use conversion factors, the units you want to "disappear" are arranged so that they cancel. Not doing so is one of the most common mistakes students make when doing chemistry calculations.

One of the benefits of unit analysis is that you can use step 2 over and over again, multiplying by conversion factor after conversion factor, to put the final answer into any units you desire. For example, imagine you are told that the density of a liquid is 2.50 g/mL and asked to convert this to pounds per gallon. You have to make *both* g and mL "disappear" and replace them with pounds and gallons. Here is how it's done:

If you are not at the units you want, keep multiplying by conversion factors:

$$\frac{2.50\,\cancel{g}}{\cancel{mL}} \times \underbrace{\frac{1\text{ lb}}{453.6\,\cancel{g}} \times \frac{1000\,\cancel{mL}}{1\,\cancel{L}} \times \frac{3.7854\,\cancel{L}}{1\text{ gal}}}_{\text{Conversion factors}} = \frac{20.9\text{ lb}}{\text{gal}}$$

The conversion factors come from the fact that there are 453.6 g per pound, 1000 mL per liter, and 3.7854 L per gallon. Note that each conversion factor is written in such a way as to cancel out some unit so that you are left with lb/gal.

You can string out as many conversion factors as necessary to come up with the unit the problem is asking for. To do the math when many conversion factors are strung together, simply multiply everything on top and then divide by everything on the bottom. For the problem we just did, you would key the following into your calculator: $2.50 \times 1000 \times 3.7854 \div 453.6 =$.

You must also be careful to multiply by each of the numbers in the numerator and divide by each of the numbers in the denominator. It is better to do all of the multiplications first, then all of the divisions, hitting the "divide" key before each entry.

The next example shows you the power of stringing out conversion factors. Suppose it took you a total of 3.5 weeks to read this book. How many seconds did it take you to read it? You could use the following conversion factors:

$$\begin{aligned} 60\text{ s} &= 1\text{ min} \\ 60\text{ min} &= 1\text{ h} \\ 24\text{ h} &= 1\text{ day} \\ 7\text{ days} &= 1\text{ week} \end{aligned}$$

Once again, we use the three-step approach of unit analysis: Step 1, write down the measured time it took you to read the book; step 2, multiply it by conversion factors to cancel out its units and arrive at the units you want; step 3, do the calculation. With respect to significant figures, you can treat all of the earlier conversion factors as exact because they derive not from some measurement but from a definition (e.g., a minute is defined as having 60 s in it). Remember, when you are doing math, exact quantities have no effect on the number of significant figures in the answer.

How many seconds did it take to read this book if it took you 3.5 weeks to read it?

Surviving unit

$$3.5\ \cancel{\text{weeks}} \times \frac{7\ \cancel{\text{days}}}{1\ \cancel{\text{week}}} \times \frac{24\ \cancel{\text{h}}}{1\ \cancel{\text{day}}} \times \frac{60\ \cancel{\text{min}}}{1\ \cancel{\text{h}}} \times \frac{60\ \text{s}}{1\ \cancel{\text{min}}} = 2,116,800\ \text{s}$$

The measurement and its units — Conversion factors — Calculated answer

Because the starting number has only two significant figures, the answer can have only two significant figures and so is 2.1×10^6 s.

The method of unit analysis, starting with a measurement and multiplying by conversion factors to make units cancel, is very simple and yet very powerful. Now let's use it in one more problem. Suppose you want to build a driveway 10.0 ft wide and 60.0 ft long. The Acme Driveway Construction Company quotes you a price of \$2.25 per square meter (m^2) of driveway. (The president of the company is into metric units.) How much will it cost to build your driveway? Unit analysis will give you your answer. Area is just length times width, and so it's no problem to figure out the area of your proposed driveway:

$$10.0\ \text{ft} \times 60.0\ \text{ft} = 6.00 \times 10^2\ \text{ft}^2$$

We have a problem, though. The company quoted you a cost per square meter, not per square foot. Suppose you know the following:

$$\begin{aligned} 12\ \text{in.} &= 1\ \text{ft} \\ 2.54\ \text{cm} &= 1\ \text{in.} \\ 100\ \text{cm} &= 1\ \text{m} \end{aligned}$$

With this information and unit analysis, you are off to the races. Follow it through:

$$6.00 \times 10^2\text{ft}^2 \times \frac{12\ \text{in.}}{1\ \text{ft}} \times \frac{2.54\ \text{cm}}{1\ \text{in.}} \times \frac{1\ \text{m}}{100\ \text{cm}} \times \frac{\$2.25}{\text{m}^2}$$

The conversion factors are arranged in such a way that all but one of the units cancel and we are left with units of dollars, but there is a problem! We started with our measurement in ft^2, but the first conversion factor has ft in the denominator rather than ft^2. The unit ft^2 is not the same as the unit ft, and so the two don't cancel! To fix this, all we need to do is square the first conversion factor:

Square it

$$6.00 \times 10^2\text{ft}^2 \times \left(\frac{12\ \text{in.}}{1\ \text{ft}}\right)^2 \times \frac{2.54\ \text{cm}}{1\ \text{in.}} \times \frac{1\ \text{m}}{100\ \text{cm}} \times \frac{\$2.25}{\text{m}^2}$$

This squaring changes the first conversion factor and allows us to cancel ft^2:

$$6.00 \times 10^2\cancel{\text{ft}^2} \times \frac{144\ \text{in.}^2}{1\ \cancel{\text{ft}^2}} \times \frac{2.54\ \text{cm}}{1\ \text{in.}} \times \frac{1\ \text{m}}{100\ \text{cm}} \times \frac{\$2.25}{\text{m}^2}$$

You can always square or cube a conversion factor if necessary, which is a good thing because we are going to have to do it again. When we changed ft to ft^2, we also changed in. to in.^{2}, and so we must somehow cancel out in.^{2}, cm^2, and m^2. Here is what we actually need to do:

$$6.00\times10^2\,\text{ft}^2\times\left(\frac{12\text{ in.}}{1\text{ ft}}\right)^2\times\left(\frac{2.54\text{ cm}}{1\text{ in.}}\right)^2\times\left(\frac{1\text{ m}}{100\text{ cm}}\right)^2\times\frac{\$2.25}{\text{m}^2}$$

On your calculator you simply key in

$$600\times12\times12\times2.54\times2.54\times2.25\div100\div100=\$125$$

That's the answer. Whether or not you can afford it, only you know.

The chart below sums up the method of unit analysis.

Using unit analysis to convert a measurement to another set of units

Step 1:
- Write the measurement with its units.
- Identify the units to which you need to convert.

Step 2:
- Multiply the measurement by a conversion factor that lets you cancel the units you don't want.
- Cancel units. You can cancel identical units on the top and bottom of the expression.

If one conversion factor won't yield the units you need, multiply by additional conversion factors.

Step 3:
- Perform the calculation.

Many conversion factors useful for doing numerical calculations in chemistry are given inside the back cover of this book. Try your hand now at the following practice problems.

Practice Problems

2.42 The density of gold is 19.3 g/cm^3. What volume in milliliters will 20.0 g of gold occupy? (*Hint*: Don't be fooled. Remember that 1 cm^3 = 1 mL.)

Answer:

Step 1: The measurement and its units are 20.0 g. The unit conversion needed is grams to milliliters.

Step 2: The conversion factors we have to work with are

$$\frac{19.3\text{ g}}{1\text{ cm}^3}\quad\text{and}\quad\frac{1\text{ cm}^3}{19.3\text{ g}}$$

from the density information and

$$\frac{1\text{ cm}^3}{1\text{ mL}}\quad\text{and}\quad\frac{1\text{ mL}}{1\text{ cm}^3}$$

from what we learned in Section 2.6. Choosing those conversion factors that let us cancel units correctly, we get

$$20.0\text{ g} \times \frac{1\text{ cm}^3}{19.3\text{ g}} \times \frac{1\text{ mL}}{1\text{ cm}^3}$$

Step 3: Cancel and do the math:

$$20.0\text{ g} \times \frac{1\text{ cm}^3}{19.3\text{ g}} \times \frac{1\text{ mL}}{1\text{ cm}^3} = 1.04\text{ mL}$$

2.43 The density of air is 0.001 30 g/mL. What is the mass in grams of 500.0 L of air? What is this mass in kilograms?

2.44 The measured density of lead is 11.4 g/mL. What volume in milliliters will 1.50 pounds of lead occupy? [1 pound = 453.6 g]

2.45 It takes six cups of flour to bake one cake; exactly one cup of flour has a mass of 120.0 g. If you have 6955 g of flour, how many cakes can you bake? (*Hint*: Two conversion factors are given in this problem. Find them and write them down first in ratio form. Then use them in the correct form with the measured quantity, which is 6955 g of flour.)

2.46 A floor installer can cover 250.0 m^2 of floor area per hour with floor tiles. How many square feet per minute can he cover?

Finally, we want you to realize that the use of conversion factors via the method of unit analysis is very general and is not just for chemistry. With the appropriate conversion factors (if they exist), you can convert just about anything to anything else. For example, suppose statistical analysis shows that the average mass of a student in a college class is 165 pounds and that the average percentage of A's received in college chemistry classes throughout the nation is 7% (i.e., 7 A's for every 100 students). You have just been given two conversion factors. Now suppose you are asked, "How many A's can we expect from a class of college chemistry students where the total mass of students is 2,455 pounds?" No problem:

$$2{,}455\text{ pounds} \times \frac{1\text{ student}}{165\text{ pounds}} \times \frac{7\text{ A's}}{100\text{ students}} = 1.042\text{ A's}$$

Only one A? Ouch! Hopefully, this is not your class.

2.9 Rearranging Equations—Algebraic Manipulations With Density

When you work with mathematical equations, it is often desirable to rearrange them (rewrite them) to make it easier to arrive at the answer. This rearranging of an equation is called **algebraic manipulation** (or **algebra**, for short). You can often avoid having to do this by using unit analysis; however, sometimes it will be to your advantage to do some algebra. The fundamental rule when rearranging equations algebraically is simple: *When rearranging an equation, you are not allowed to do anything that changes the meaning of the equation; you can change only its appearance.*

A good example for demonstrating algebraic manipulation is density. We have seen that density is defined as mass over volume ($d = m/V$). Like any other equation, $d = m/V$ can be algebraically rearranged at will. Let's return to an earlier problem that we already solved by unit analysis. If you have

2.00 g of liquid mercury, what volume will it occupy if its density is 13.6 g/mL? In this problem, we are being asked to find a volume. This means we want to algebraically rearrange the density equation $d = m/V$ so that it ends up looking like $V =$ something. This is called *solving the equation for volume*. How can we do this? By following the fundamental rule: Do anything you want to this equation so long as what you do doesn't change its meaning. This means you can add anything to the equation, subtract anything from it, multiply it by anything, or divide it by anything, so long as you don't change its meaning. The way to ensure that the meaning isn't changed is *to do the same thing to both sides of the equation*. If we add something on the left side of the equal sign, we must add the same thing on the right side of the equal sign. If we multiply by something on the left side, we must multiply by the same thing on the right side. Let's try this now for our density equation. Suppose we multiply both sides of the equation by volume (V), the quantity we are being asked to solve for:

$$d \times V = \frac{m}{V} \times V$$

In a mathematical equation, anything over itself is equal to 1. On the right side of the equation, we now have V on the top and V on the bottom. This is V/V, which equals 1. Follow the diagram given here to see how this changes the appearance (but not the meaning) of our equation.

$$d \times V = \frac{m}{V} \times V$$

V on top

V on bottom

$$d \times V = m \times \frac{V}{V}$$

$V/V = 1$, and $m \times 1 = m$

$$d \times V = m$$

We now have $d \times V = m$. We are almost at our target of $V =$ something. We just have to remove d from the left side. To do this, we divide both sides of the equation by d. Notice how this changes the appearance of our equation:

$$\frac{d \times V}{d} = \frac{m}{d}$$

Dividing both sides by d gives us d/d on the left side, which equals 1, and $V \times 1 = V$

$$V = \frac{m}{d}$$

We have done it! By algebraic manipulation, we have taken the equation for density $d = m/V$ and rewritten it as $V = m/d$ without changing its meaning. We have solved it for V. Now, to calculate the volume asked for, we simply divide the mass by the density:

$$V = \frac{m}{d} = \frac{2.00\ \text{g}}{13.6\ \text{g/mL}} = 0.147\ \text{mL}$$

This is the same answer we arrived at by the method of unit analysis.

Another way to look at algebraic manipulations is to consider yourself an eliminator of variables. The variables are just the quantities in the equation. In our last example, for instance, mass (m), volume (V), and density (d) were the variables. Your job is to eliminate the ones you don't want from one side of an equation, leaving behind only the one you want. For example, given the equation for density $d = m/V$, how could you quickly solve it for m? The answer is to eliminate the V from the right side and get m all by itself. The question then becomes, what do you have to do to eliminate V from the right side? The answer is to multiply both sides by V. Doing so gives $Vd = Vm/V$, and $Vm/V = m$ (the V's cancel). You end up with $Vd = m$, and you have solved for m.

Sometimes you have to add or subtract instead of multiply or divide. For example, if you are given the equation $p = q + r$ and asked to solve for q, you need to get rid of the r from the right side. Because r stands alone (i.e., it is not multiplied or divided by anything else), simply subtract r from both sides (*always do the same thing to both sides*). This gives $p - r = q + r - r$. On the right side, $r - r = 0$ (anything minus itself is zero), and so the r's are eliminated from the right side. You end up with $p - r = q$. You have solved the equation for q.

Try your hand at algebraic manipulations in the following problems now. Do whatever it takes to eliminate unwanted variables from one side of an equation. Remember that whatever you do to one side of the equation you must also do to the other side. The first two problems involve multiplying or dividing. The last problem involves adding or subtracting.

Practice Problems

Carry out the following algebraic manipulations. Show how you solved each problem, as is done in the answer to Problem 2.47.

2.47 Given $p/q = r$, solve for p.

Answer: To solve $\frac{p}{q} = r$ *for* p*, you must eliminate* q *from the denominator of the left side.*

To do this, multiply both sides by q*:*

$$\frac{\not{q}p}{\not{q}} = rq$$

The q *cancels from the left side because it is in both the numerator and the denominator. The result is* $p = rq$.

2.48 Given $p/q = r$, solve for q.

Answer: $q = \frac{p}{r}$

2.49 Given $P + Q = z$, solve for P.

Answer: $P = z - Q$

+1 One More Thing

2.10 Quantifying Energy

Another thing chemists quantify is *energy*. We do that because every change that matter undergoes (whether physical or chemical) involves energy. For instance, consider the melting of ice (a physical change). A piece of ice can't melt unless you add energy in the form of heat. Likewise, many chemical reactions, such as the reaction of gasoline and oxygen to form carbon dioxide and water, release lots of energy (heat and light—fire). So, to understand the changes matter undergoes, we must quantify energy.

What exactly is energy? You no doubt have an idea. A bolt of lightning, the light from our Sun, the explosion of an atomic bomb: These are all things you intuitively know have energy. In science, **energy** is defined as the capacity for doing work. The more energy something has, the more work it can do. Energy can also be thought of as having many forms. Two forms we shall dwell on in this book are heat and light. We shall have more to say about these forms later. What we want to do now is to *quantify* energy (which simply means to give it units), just as we did with mass and volume.

A well-known unit of energy is the **calorie** (cal). This is not an SI unit, but it is so well known (who hasn't tried to "cut calories") that we can't ignore it. Actually, there are two types of calories—the "little c" calorie and the "big C" Calorie (Cal), where 1 Cal = 1000 cal = 1 kilocalorie. The Calorie is the one used by dietitians and the food industry. When you go to the gym and lift weights, that 250 Cal (=250,000 cal) you "burned" is the amount of energy you expended. Some of that energy went into the work you performed when you lifted the weights, and some of it went into heating you up (which is why you broke into a sweat).

To give you a feel for how much energy one calorie is, consider 1 g of water at 25 °C. A mass of 1 g of water is about 20 drops from an eyedropper, an amount you could easily hold in the palm of your cupped hand. It takes 1 cal of heat energy to warm 1 g of water from 25 °C to 26 °C. (That is how a calorie is defined, in fact.) An increase in temperature of 1 °C would barely be noticeable by your hand, and so on an everyday scale, 1 cal is not much energy. On the other hand, 1 Cal would increase the temperature of that same 1 g of water by 1000 °C (assuming the water stayed a liquid with properties identical to those of 25 °C water). You would have little trouble feeling *that* amount of energy!

1 cal 1 Cal

The SI unit of energy is the **joule** (J). It is named in honor of James Prescott Joule (1818–1889), a British physicist who participated in discovering the **law of conservation of energy**. (This law states that energy is neither created nor destroyed; it just goes from one form to another, such as from light to heat.) There are 4.184 J per calorie (and therefore 4.184 kJ per Calorie). This means that 1 J of heat energy is roughly one-fourth the amount of heat energy in 1 cal. From this information, we can write the following conversion factors:

$$\frac{4.184\ \text{J}}{\text{cal}} \quad \text{or} \quad \frac{\text{cal}}{4.184\ \text{J}} \quad \text{and} \quad \frac{4.184\ \text{kJ}}{\text{Cal}} \quad \text{or} \quad \frac{\text{Cal}}{4.184\ \text{kJ}}$$

To convert from one energy unit to another, all you do is perform unit analysis using the appropriate conversion factor. For example, suppose that Twinkie you just couldn't resist contained 200.0 Calories. We could convert this to joules as follows:

$$200.0\ \cancel{\text{Cal}} \times \frac{4.184\ \cancel{\text{kJ}}}{\cancel{\text{Cal}}} \times \frac{1000\ \text{J}}{\cancel{\text{kJ}}} = 8.368 \times 10^5\ \text{J}$$

That's almost 837,000 J! Maybe you want to think twice about snacking on Twinkies.

This brings us to a final question about energy—how do you measure it? For example, how did someone determine that a delicious Twinkie has 200.0 Cal? Believe it or not, one way is to use water. Recall that 1 cal is the amount of heat energy needed to warm 1 g of water by 1 °C. With this knowledge, we can define the *specific heat* of water. The **specific heat** of a substance is the amount of heat energy needed to warm 1 g of that substance by 1 °C. Thus, the specific heat of water is 1 cal/g · °C. *Notice the units.* You read them "1 calorie *per* gram *per* Celsius degree." That is, it takes 1 cal to warm 1 g of water by 1 °C. Table 2.5 lists the specific heats of some common substances.

Table 2.5 Specific Heats of Some Common Substances

Substance	Specific heat cal/g·°C	Specific heat J/g·°C
Water	1.000	4.184
Iron	0.107	0.449
Aluminum	0.215	0.901
Ethanol	0.581	2.43

Notice two things: (1) The values differ quite a lot from one another and (2) the specific heats expressed in joules are 4.184 times larger than those expressed in calories. Now it's time to see if you understand what these specific heats are telling you.

WORKPATCH 2.10 Suppose you have 1.0 g of water and 1.0 g of aluminum, both at 25.0 °C. You want to warm them to 26.0 °C. Would you need to add the same amount of heat to each? How much heat in joules would you need in each case? ●

Just from looking at the specific heats in Table 2.5, you can see that it takes almost five times as many calories to warm water by 1 °C as to warm the same mass of aluminum by 1 °C. Make sure you understand this before going on, or what follows won't make much sense.

Once we know a substance's specific heat, we can use that substance to measure energy. For example, let's measure the calorie count for a Twinkie using water. First, a little background. When you eat a Twinkie, your body essentially "burns" it. It does this *slowly* and in many steps, but the result is the same as if the Twinkie had been set on fire in the air. When you burn a Twinkie, you convert its stored chemical energy to heat energy, which is released to the surroundings. Therefore, if we can measure the heat released when we burn a Twinkie, we will have measured its calorie content.

So to measure the calorie content of a Twinkie, we simply burn it in such a way that we can capture and measure the heat released. The picture below shows how. We put the Twinkie in a small metal box that contains enough oxygen to let the Twinkie burn completely. This metal box is called a **bomb**. Then we immerse the bomb in a larger, insulated vessel (called a **calorimeter**) that is filled with a known amount of water. For our Twinkie measurement, we shall use a calorimeter that contains 1.00×10^4 g of water (10,000 g; about 2 gallons). When we burn the Twinkie, the heat flows out into the water, warming it. We carefully measure this change in temperature.

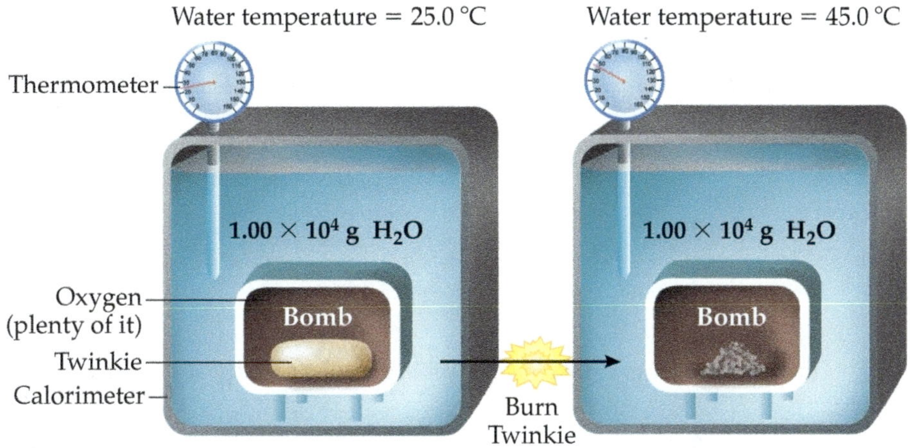

Because we know the specific heat of water and how much water we have, we can use the temperature change to determine the amount of heat the Twinkie released as it burned. (A real calorimeter is also calibrated to allow for the slight amount of heat absorbed by the bomb and the insulated calorimeter walls, but we will ignore that.)

Our calorimeter contains 1.00×10^4 g of water, and the temperature of this water increased by 20.0 °C (from 25.0 °C to 45.0 °C). We know that the specific heat of water is 1 cal/g·°C. To calculate the heat energy released by burning the Twinkie, all we do is multiply these three things together:

$$\text{Heat} = \begin{array}{c}\text{Specific heat}\\ \text{of water}\end{array} \times \begin{array}{c}\text{Mass of water}\\ \text{in calorimeter}\end{array} \times \begin{array}{c}\text{Change}\\ \text{in water}\\ \text{temperature}\end{array}$$

$$= \frac{1.000\ \text{cal}}{\cancel{\text{g}} \cdot \cancel{°\text{C}}} \times 1.00 \times 10^4\ \cancel{\text{g}} \times 20.0\ \cancel{°\text{C}} = 2.00 \times 10^5\ \text{cal}$$

Notice that the unit that remains after we cancel units is calories. Thus, a Twinkie contains 2.00×10^5 cal of energy. Let's convert this to Calories:

$$2.00 \times 10^5\ \text{cal} \times \frac{\text{Cal}}{1000\ \text{cal}} = 2.00 \times 10^2\ \text{Cal} = 200\ \text{Cal}$$

And there it is, the Calorie content of a Twinkie. Of course, if you were discussing this with a scientist instead of a dieter, you would convert the result to joules or kilojoules so as to be speaking in SI units.

Do you need to memorize the equation for calculating the amount of heat released in a calorimeter? The answer is *yes*, but if you can visualize the calorimeter (the bomb, the water, and the thermometer), then you can reason it out. We want to give you a word of caution, though. Because specific heat has grams as one of its units, the mass of water you enter into the equation must also be in grams (not kilograms, not pounds, not anything else but grams). If you violate this rule, the mass units will not cancel, and your calculated amount of heat will be wrong. Finally, the equation for calculating the heat released in a calorimeter is also useful for calculating heat in other situations. Follow through the solutions for the next two practice problems, and then try the last three on your own.

Practice Problems

2.50 Suppose you have a 2.000-pound block of iron at 50.0 °C. How much heat in joules would it take to warm this block to 75.0 °C?

Answer: To calculate the amount of heat it takes to warm up any substance, use the following equation—first multiply the specific heat by the mass of your sample and then multiply that product by the temperature change:

$$\text{Heat} = \text{Specific heat of iron} \times \text{Mass of iron} \times \text{Change in iron temperature}$$

$$= \frac{0.449\ \text{J}}{\cancel{\text{g}} \cdot \cancel{^\circ\text{C}}} \times 907.2\ \cancel{\text{g}} \times 25.0\ \cancel{^\circ\text{C}} = 1.02 \times 10^4\ \text{J}$$

Notice that we converted the mass from pounds to grams before plugging it into the heat equation:

$$2.000\ \cancel{\text{pounds}} \times \frac{453.6\ \text{g}}{\cancel{\text{pounds}}} = 907.2\ \text{g}$$

2.51 A 20.0-g block of iron initially at 25.0 °C has 100.0 J of heat energy added to it. What is its temperature after the heat energy has been added?

Answer: This problem involves doing algebra calisthenics with the equation used in Problem 2.50. All we need to do is rearrange the equation to solve it for "change in temperature" and then plug in the values for everything we know.

Heat = Specific heat of substance × mass of substance × change in substance temperature

If we divide both sides by specific heat of substance × mass of substance, we get

$$\frac{\overset{100.0\ \text{J}}{\text{Heat}}}{\underset{0.449\ \text{J/g}\cdot^\circ\text{C}}{\text{Specific heat of substance}} \times \underset{20.0\ \text{g}}{\text{mass of substance}}} = \text{Change in substance temperature} = 11.1\ ^\circ\text{C}$$

Because the initial temperature is 25.0 °C, the final temperature is 25.0 °C + 11.1 °C = 36.1 °C.

2.52 Suppose you have a 50.0-mL sample of ethanol at 22.0 °C and want to warm it to 60.0 °C. How many kilojoules of heat will this take? The density of ethanol is 0.785 g/mL. (*Hint*: You need to use density to convert the given volume to grams.)

2.53 Convert the answer you calculated in Problem 2.52 to joules, Calories, and calories.

2.54 A 0.100-g sample of your favorite candy is burned in a calorimeter that contains 1.00 kg of water initially at 22.0 °C. After the candy is burned, the water temperature is 35.5 °C. How many Calories are there per gram of your candy?

As we said earlier, you can't avoid the numerical side of chemistry if you hope to study it in any detail. This chapter has prepared you for dealing with this numerical side, and it is important. However, the fundamental concepts of chemistry and the pictures, models, and stories that are discussed in the remainder of this book are just as important, if not more so. After all, you can't use the mathematical tools you just learned about to solve a problem until you understand what the problem is about.

Have You Learned This?

Exact number (p. 30)
Measurement (p. 30)
Precision (p. 30)
Accuracy (p. 30)
Significant figure or sig fig or significant digit (p. 34)
Leading zero (p. 34)
Trailing zero (p. 35)
Scientific notation (p. 37)
SI unit (p. 45)
Mass (p. 46)
Weight (p. 46)
Derived SI unit (p. 46)
Volume (p. 46)
Density (p. 50)
Intensive property (p. 52)
Extensive property (p. 52)
Unit analysis (p. 53)
Conversion factor (p. 53)
Algebraic manipulation (p. 58)
Energy (p. 61)
Calorie (p. 61)
Joule (p. 61)
Law of conservation of energy (p. 61)
Specific heat (p. 62)
Bomb (p. 63)
Calorimeter (p. 63)

Learning Outcomes for Chapter 2

2.1	**Numbers in Chemistry—Precision and Accuracy**	Explain what an exact number is and what a measured number is. [55, 56] Describe the difference between precision and accuracy. [57, 58, 59]
2.2	**Numbers in Chemistry—Uncertainty and Significant Figures**	Explain why measured numbers always have uncertainty associated with them. [61, 62, 63, 65]
2.3	**Zeros and Significant Figures**	Identify leading zeros and trailing zeros in a measured number. [64, 66, 67] Identify the amount of uncertainty in a written measured quantity. [63, 68, 69]
2.4	**Scientific Notation**	Convert a number from standard to scientific notation and vice versa. [71, 75, 173, 182] Identify the degree of uncertainty in a number written in scientific notation. [72, 74, 75fg]
2.5	**How to Handle Significant Figures and Scientific Notation When Doing Math**	Correctly add or subtract numbers according to the rules of significant figures. [77, 80acd, 161ad] Correctly multiply or divide numbers according to the rules of significant figures. [78, 80b, 161bc]
2.6	**Number With a Name—Units of Measure**	Identify the common SI units used by science. [81, 82, 83, 85, 88]
2.7	**Density: A Useful Physical Property of Matter**	Define density, and use it to do calculations. [95, 96, 97, 98]
2.8	**Doing Calculations in Chemistry—Unit Analysis**	Convert the units of a measured quantity to some other units using unit analysis. [102, 107, 110]

2.9	**Rearranging Equations—Algebraic Manipulations With Density**	Algebraically rearrange a mathematical equation to solve it for a desired variable. [113, 114, 115]
2.10	**Quantifying Energy**	Calculate the amount of heat energy using specific heat. [122, 123]

Skills to Know

Uncertainty

1. All numbers derived from measurements have uncertainty. By convention, this uncertainty is assumed to be in the number's final significant digit.	2.46 — Certain (2.4) Uncertain (estimated) (6)
2. To calculate the exact amount of uncertainty, identify the column of the uncertain digit and put a 1 in that column. For instance, if the uncertain digit is in the hundredths column, the uncertainty is ±0.01.	2.46 ← Uncertain digit in hundredths column Uncertainty = ±0.01

Scientific notation

A number written in scientific notation has the form $A \times 10^x$, where A is a number between 1 and 9.9999 ... and x is a power of 10. To convert the number to standard notation, move the decimal point in A by x places. If x is positive, move the decimal point to the right; if x is negative, move the decimal point to the left.

Scientific notation		Standard notation
2.46×10^4	=	24600.
2.46×10^0	=	2.46
2.46×10^{-4}	=	0.000246

Rules of significant figures

1. Leading zeros are just placeholders; they are not significant.

Trailing zeros that follow the decimal point are always significant. **Trailing zeros** that precede the decimal point may or may not be significant; use a decimal point to avoid ambiguity.

Not significant (0.0) — Significant (600)
0.0600 cm

This number has three significant figures.

1200. — Decimal point tells you these trailing zeros are significant.

1200 — No way to know whether these two trailing zeros are significant.

2. When you **multiply** or **divide** numbers that have uncertainty, your answer can have no more significant figures than the starting number with the fewest significant figures.

2.22	← 3 sig figs
× 2.0	← 2 sig figs
4.4	← Answer can have only two significant figures.

3. When you **add** or **subtract** numbers that have uncertainty, your answer can be no more certain than the least certain number.	2.22 ← Uncertainty in hundredths position + 2.0 ← Uncertainty in tenths position 4.2 ← Answer as certain as least certain number (uncertainty in tenths position).
4. An exact number has no effect on the number of significant figures in the result of a calculation.	4 sig figs, Exact number, 4 sig figs $6.525\text{ ft} \times \dfrac{12\text{ in.}}{\text{ft}} = 78.30\text{ in.}$

Using unit analysis to solve numerical problems in chemistry

Example: What is the volume in milliliters of 4.00 g of gold, Au? The density of gold is 19.3 g/mL.

Step 1 • Write the measurement with its units. • Identify the units to which you need to convert.	Measurement: 4.00 g Au Convert from grams to milliliters.
Step 2 • Multiply the measurement by a conversion factor that lets you cancel the units you don't want. • Cancel identical units on the top and bottom of the expression.	$4.00\text{ g Au} \times \dfrac{1\text{ mL Au}}{19.3\text{ g Au}}$ $4.00\ \cancel{\text{g Au}} \times \dfrac{1\text{ mL Au}}{19.3\ \cancel{\text{g Au}}}$

If one conversion factor is not enough to give the units you want, multiply by additional conversion factors.

Step 3 Perform the calculation.	$4.00\ \cancel{\text{g Au}} \times \dfrac{1\text{ mL Au}}{19.3\ \cancel{\text{g Au}}} = 0.207\text{ mL Au}$

Heat

1. The common units of heat energy are calories (cal), Calories (Cal), and joules (J). The joule is the SI unit of energy (including heat energy).	1 Cal = 1000 cal 1 cal = 4.184 J 1 Cal = 4184 J = 4.184 kJ

2. The **specific heat** of a substance is the amount of heat energy you must add to warm 1 g of the substance by 1 °C. For instance, it takes more heat energy to warm an aluminum pan than an equal-mass iron one because aluminum has a larger specific heat than iron.

Units of specific heat:

Any of these heat energy units may be used . . .

$$\frac{\text{J, kJ, cal, Cal}}{\text{g}\cdot{}^{\circ}\text{C}}$$

. . . but mass is always in grams and temperature change is always in °C.

3. To calculate how much heat energy you must add to a sample of a substance to raise its temperature by a certain amount, you must know the specific heat of that substance and the mass of the sample:

$$\text{Heat energy added} = \text{Specific heat} \times \text{Mass in grams} \times \text{Change in temperature in }{}^{\circ}\text{C}$$

Example: How much heat energy does a 140.-g aluminum pan absorb as it heats from 25.0 °C to 175 °C?

Specific heat of aluminum = 0.901 J/g·°C

$$\text{Temperature change} = 175\ {}^{\circ}\text{C} - 25.0\ {}^{\circ}\text{C} = 150.\ {}^{\circ}\text{C}$$

$$\text{Heat energy added} = \frac{0.901\ \text{J}}{\text{g}\cdot{}^{\circ}\text{C}} \times 140.\ \text{g} \times 150.\ {}^{\circ}\text{C} = 1.89 \times 10^4\ \text{J}$$

Concept Questions

1. A measured quantity is exact if
   (a) it is close to the true value.
   (b) it is precise.
   (c) it is measured many times.
   (d) All of the above.

2. A measured value will have uncertainty
   (a) only if it is measured incorrectly.
   (b) if it is not measured more than once.
   (c) always.
   (d) if the measuring device is inexpensive.

3. The difference between 0.1 cm and 0.10 cm is which of the following:
   (a) There is no difference. They are both one tenth of a centimeter.
   (b) There is more uncertainty in the 0.10 cm measurement.
   (c) There is more uncertainty in the 0.1 cm measurement.
   (d) 0.1 cm has uncertainty whereas 0.10 cm does not.

4. The measured quantity 0.00100200 seconds has
   (a) 9 significant figures.
   (b) 8 significant figures.
   (c) 6 significant figures.
   (d) 2 significant figures.

5. The measured quantity 0.00100200 s can be written as
   (a) 1.002×10^3 s.
   (b) 1.002×10^{-3} s.
   (c) 1.00200×10^{-3} s.
   (d) 1.00200×10^3 s.

6. The difference between 2.3×10^{-2} m and 2.30×10^{-2} m is
   (a) precision (2.30×10^{-2} m is more precise).
   (b) uncertainty (2.30×10^{-2} has a small uncertainty).
   (c) accuracy (2.3×10^{-2} has less uncertainty).
   (d) None of the answers are correct.

7. The sum of 1.1 cm and 0.04 cm is
   (a) 1.14 cm.
   (b) 1 cm.
   (c) 1.1 cm.
   (d) 1.2 cm.

8. Unit analysis
   (a) says that units are optional when writing measured quantities.
   (b) says that you should analyze the units of a measurement.
   (c) is a method for solving numerical problems.
   (d) is a study of the number of elements in a compound.

9. Density
   (a) is an extensive property that depends on the amount of mass of a sample.
   (b) is an intensive property that does not depend on the mass or volume of the sample.
   (c) can never serve as a conversion factor because it consists of two different units.
   (d) has units of grams times centimeters.

10. The following is true about energy:
   (a) It cannot be accurately quantified.
   (b) It can be measured by directing it into a known mass of water.
   (c) Chemical change does not depend on energy; only physical change does.
   (d) It has units of joules per calorie.

Answers: 1a, 2c, 3c, 4c, 5c, 6b, 7c, 8c, 9b, 10b

Numbers in Chemistry—Precision and Accuracy

2.55 There are 3 ft in a yard. A certain piece of wood is 3 ft long. What is the fundamental difference between the value of 3 *ft* in these two statements?

2.56 Jack is asked to determine the number of pennies in \$1.00. Jill is asked to determine the number of liters in 1 gallon. Whose answer will be an exact number, and whose will be measured? Explain why.

2.57 Someone is going to measure your height three times and then take an average. This height will then be fed into an automated gun that will aim 1 in. higher than your height to shoot an apple off of it. Suppose the height measurer could only be precise or accurate, but not both. Which would you choose, and why?

2.58 Three people measure the distance from Main Street to Market Avenue using their best estimates. Their data are 2.5 miles, 2.5 miles, and 2.6 miles. Survey charts show the actual distance to be 1.8 miles. Characterize the collected data in terms of accuracy and precision.

2.59 Two people attempt to measure the length in feet of a parking lot they know to be about 100 ft long. One person uses a 6-in. ruler; the other uses a 120-ft tape measure. If both measuring devices are graduated in 1/16 in., which person is likely to make the more accurate measurement? Explain.

Numbers in Chemistry—Uncertainty and Significant Figures

2.60 A ruler is marked in intervals of 1/8 in. To what fraction of an inch can you estimate a measurement?

2.61 Given a written numerical value for some measurement, how do you know what the uncertainty in that measurement is?

2.62 Is it possible for a measured quantity to have no uncertainty associated with it? Explain.

2.63 What is the uncertainty in each measured number:
(a) 12.60 cm (b) 12.6 cm
(c) 0.000 000 03 in. (d) 125 ft

2.64 Indicate the number of significant figures in each measurement in Problem 2.63.

2.65 Using radioactive elements, scientists determined the age of a dinosaur skeleton to be 78.5 million years. In years, what is the uncertainty in this measurement?

2.66 Underline any trailing zeros in these measurements:
(a) 12.202 km (b) 0.01 mL
(c) 205 °C (d) 0.010 g

2.67 For the measurements in Problem 2.66, underline all the zeros that are significant.

2.68 What is the uncertainty associated with each measurement in Problem 2.66?

2.69 The measurement 30 ft is ambiguous, but the measurement 30. ft is not. Explain what the ambiguity is, and how adding the decimal point eliminates the ambiguity.

2.70 Give all interpretations possible for the measurement 2200 ft.

Scientific Notation

2.71 Convert the following measured values from scientific notation to standard notation. For each one, indicate the number of significant figures.
(a) 5.60×10^1 kg (b) 2.5×10^{-4} m
(c) 5.600×10^6 miles (d) 0.02×10^2 ft

2.72 What is the uncertainty for each measured value in Problem 2.71?

2.73 Using scientific notation, write the measurement 30 ft as having an uncertainty of:
(a) ± 1 ft (b) ± 0.1 ft (c) ± 0.01 ft

2.74 Using scientific notation, write the measurement 2200 ft as having an uncertainty of ± 100 ft.

2.75 Convert the following numbers from standard notation to scientific notation:
(a) 226 (b) 226.0 (c) 0.000 000 000 50
(d) 0.3 (e) 0.30
(f) 900,000,574 with an uncertainty of ± 1 million
(g) 900,000,574 with an uncertainty of ± 100

2.76 How many significant figures are there in the following measured values, and what is the uncertainty in each measurement?
(a) 0.001 kg (b) 0.000 10 m
(c) 102 L (d) 2.600×10^{-3} m
(e) 1.1×10^6 km

How to Handle Significant Figures and Scientific Notation When Doing Math

2.77 To the correct number of significant figures, what is the result of adding the measured values 100 in. + 2 in. + 0.001 in.? What is the uncertainty in the result?

2.78 To the correct number of significant figures, what is the product of each mathematical operation? Use scientific notation when necessary. (No units shown means a number is exact.)
(a) 2.30 cm × 2 (b) 2 m × 2.000 m
(c) 1001 J × 10 (d) 124 mm ÷ 0.1 mm

2.79 A student walks 20,450.2 ft to school every day. A mile is defined as 5280 ft. Doing the division 20,450.2 ft ÷ 5280 ft per mile on a calculator gives 3.873143939 miles. What, if anything, is wrong with this answer?

2.80 Do these calculations using a scientific calculator and report your answers in scientific notation:
(a) $(3.33 \times 10^4 \text{ km}) + (2.22 \times 10^5 \text{ km})$
(b) $(2.444 \times 10^9 \text{ J}) \div (2.444 \times 10^{-9} \text{ J})$
(c) $(2.34 \times 10^2 \text{ m}) - (2.34 \times 10^1 \text{ m})$
(d) $(4.00 \times 10^4 \text{ L}) + (6.00 \times 10^{-1} \text{ L})$

Numbers With a Name—Units of Measure

2.81 What is the base SI unit of length? What is the SI unit of volume?

2.82 What are two metric but non-SI units of volume, and why are they more often used than the SI unit of volume?

2.83 When is it correct to use cm^3 instead of mL? Explain.

2.84 Why was the SI unit system developed by scientists?

2.85 Convert each length to meters. Report your answers in scientific notation and watch your significant figures.
(a) 2.31 gigameters (Gm)
(b) 5.00 micrometers (μm)
(c) 1004 millimeters (mm)
(d) 5.00 picometers (pm)
(e) 0.25 kilometer (km)

2.86 Which is larger, a Celsius degree or a Fahrenheit degree? Explain.

2.87 Of the three temperature scales, which can have negative temperatures? For the one(s) that can't, explain why not.

2.88 Convert:
(a) 22.5 °C to Fahrenheit and Kelvin
(b) −3.00 °F to Celsius and Kelvin
(c) 100.0 °C to Kelvin and Fahrenheit
(d) 65.1 °C to Fahrenheit and Kelvin

2.89 How cold does it have to be for water to freeze in °F? In °C? In kelvins?

2.90 What is the coldest temperature possible in Celsius and Fahrenheit? Give your answers to an uncertainty of plus or minus one-hundredth of a degree.

2.91 You fill a bottle with water by pouring into it the contents of these two graduated cylinders:

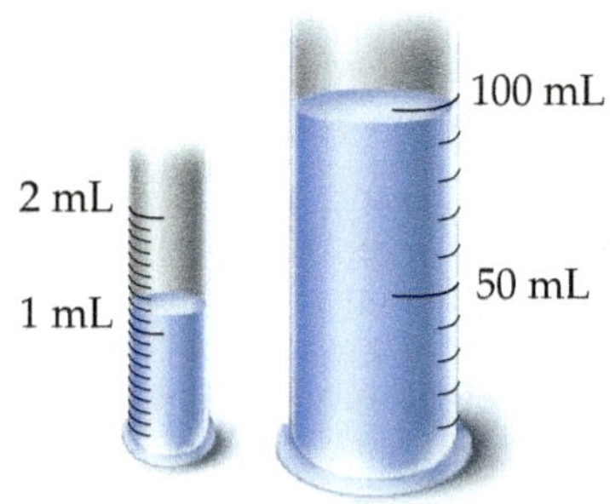

(a) What is the uncertainty in each graduated cylinder?
(b) To the correct number of significant figures, what is the final volume of water in the bottle after you pour the contents of both cylinders into it? What is the uncertainty in this volume? Explain.

2.92 A student measures the length of one side of a perfect cube with the ruler shown here, which is marked in centimeters. She then calculates the volume V by cubing the length l ($V = l^3 = l \times l \times l$). What volume should she report? Give the answer to the proper number of significant figures and with the proper units.

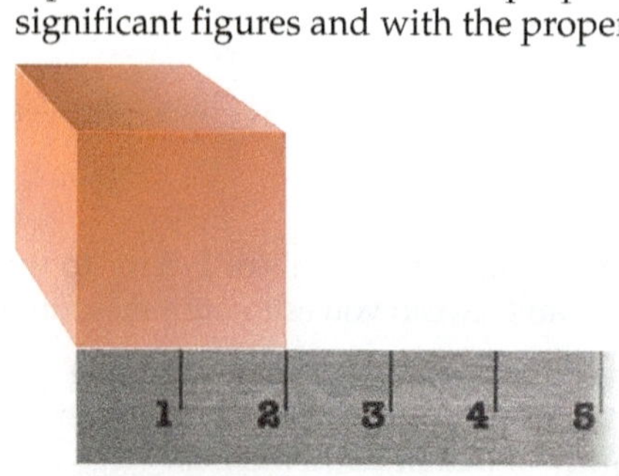

2.93 Using a ruler marked in centimeters and millimeters, a student measures the diameter of a ball to be 1.5 cm. His partner measures the same ball with the same ruler and comes up with 1.50 cm. Which student used the ruler incorrectly? How did that student use the ruler incorrectly?

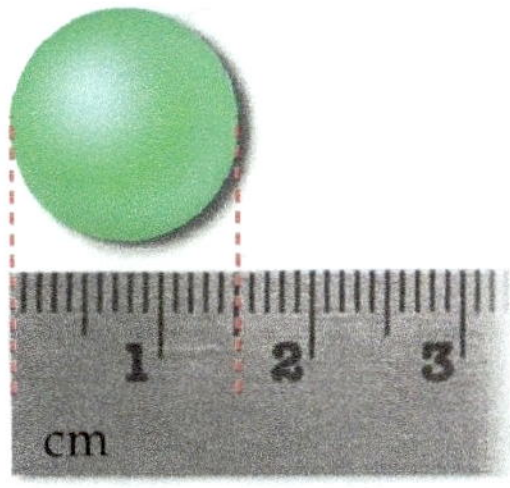

2.94 The students measure another ball with the ruler in Problem 2.93 and determine that its diameter is 2.55 cm. What is the radius of the ball to the correct number of significant figures?

Density

2.95 Define density, and explain why the unit for density is called a derived SI unit.

2.96 At 25 °C, what is the mass in grams of 1000.0 mL of water? (*Hint*: Use data from Table 2.4.)

2.97 At 25 °C, what is the mass in grams of 2.0 L of mercury? (*Hint*: Use data from Table 2.4.)

2.98 At 25 °C, the dimensions of a stick of butter are 10.0 cm × 10.0 cm × 10.0 cm. What is the mass of this butter in grams? (*Hint*: Use data from Table 2.4.)

2.99 Using a bathroom scale, a tub, a sponge, and a measuring cup, explain how you would measure your own density.

2.100 Two students measure the density of gold. One works with a 100-g bar of pure gold. The other works with a 200-g bar of pure gold. Which student measures the larger density? Explain your answer.

2.101 Pyrite is a mineral having the formula FeS_2 and a density of 5.01 g/mL. It looks like gold and is often called fool's gold as it is sometimes found in association with gold in gold deposits. Any object sinks in a liquid if the object is denser than the liquid, and it floats if the object is less dense than the liquid. Using Table 2.4, how could you separate gold from fool's gold?

Unit Analysis

2.102 Suppose it takes you 0.850 weeks to reach the moon in a space ship. How many seconds does it take for you to get there? Use unit analysis to calculate your answer, and show your work.

2.103 A train traveling at 45.0 miles/h has to make a trip of 100.0 miles. How many minutes will the trip take? Use unit analysis to calculate your answer, and show your work.

2.104 You have a great job in which you earn $25.50 per hour. How many dollars do you earn per second? Use unit analysis to calculate your answer, and show your work.

2.105 Gold has a density of 19.3 g/mL. Suppose you have 100.0 glonkins of gold. What volume in liters will the gold occupy? Here are some conversion factors to help you: 0.911 ounce per glonkin and 28.35 g per ounce. Use unit analysis to calculate your answer, and show your work. Treat both conversion factors as exact.

2.106 One liter is equal to 0.264 gallon. Suppose you have 1.000×10^3 cm^3 of water. How many gallons do you have? Use unit analysis to calculate your answer, and show your work. Treat all conversion factors as exact.

2.107 The dimensions of a rectangular box are given to be 10.2 cm × 43.7 cm × 9.56×10^2 mm. What is its volume in liters? Be careful! The units are not all the same.

2.108 The block of Problem 2.107 has a mass of 2.43×10^2 kg. What is the density of the block in grams per milliliter?

2.109 You measure one edge of a cube using a meterstick marked in centimeters. Unfortunately, the edge is longer than 1 m. You mark the 1-m point on the cube edge with a pen and then, using a 15-cm ruler marked in millimeters, measure the remaining distance to be 1.40 cm.
(a) What is the length of the edge in centimeters?
(b) What is the volume of the cube in cubic centimeters? (Remember, the lengths of all edges of a cube are equal.) Watch your significant figures. Use scientific notation if you have to.
(c) The cube has a mass of 111 kg. What is its density in grams per milliliter? Watch your significant figures.

2.110 A rectangular box measures 6.00 in. in length, 7.00 in. in width, and 8.00 in. in height. What is the volume of the box in liters? [2.54 cm = 1 in.]

2.111 An object travels 80.0 m/s. How fast is it traveling in miles per hour? [1 m = 3.28 ft, 1 mile = 5280 ft]

Algebraic Manipulations

2.112 Why can't you multiply just one side of an equation by something when algebraically rearranging the equation?

2.113 (a) Solve the equation $y = z/x$ for x. (b) Solve the equation $y = z/2x$ for x.

2.114 Solve the equation $y = z - x$ for x.

2.115 Solve the equation $5x - 6 = 3x - 8$ (find the value of x that makes this equation true).

2.116 The density of a certain liquid is 1.15 g/mL. What mass in grams of the liquid is needed to fill a 50.00-mL container? Do this problem by the method of algebraic manipulation, beginning with the equation density = mass/volume and showing all steps.

2.117 Repeat Problem 2.116 using unit analysis. Show all steps. Do you get the same answer you got in Problem 2.116?

Quantifying Energy

2.118 Define *energy*.

2.119 How much heat energy is 1 cal? Give your answer in terms of changing the temperature of water.

2.120 Convert:
(a) 4.50 Cal to calories
(b) 600.0 Cal to kilojoules
(c) 1.000 J to calories
(d) 50.0 Cal to joules

2.121 Define *specific heat*.

2.122 A 100.0-g block of iron and a 100.0-g block of aluminum are both initially at 25.0 °C. Both are then warmed to 100.0 °C. Does one block require more heat energy than the other to reach 100.0 °C? If so, how much more?

2.123 How many "big C" Calories does it take to raise the temperature of 2.00 L of water from 22.0 °C to 40.0 °C? How many kilojoules? Take the density of water to be 1.00 g/mL.

2.124 Why is it necessary for a calorimeter to be insulated?

2.125 A 2.50-g piece of wood is burned in a calorimeter that contains 0.200 kg of water. The burning causes the water temperature to increase from 22.1 °C to 28.7 °C. How much heat energy is released in joules? What is the energy content of the wood in joules per gram of wood?

2.126 How many joules of heat energy would it take to raise the temperature of 2.00 pounds of iron from 30.0 °C to 90.0 °C? [453.6 g = 1 pound]

2.127 How many grams of wood from Problem 2.125 would you have to burn to accomplish the task of Problem 2.126?

2.128 A 2.000-ton block of aluminum at room temperature (22.0 °C) is struck by lightning. The lightning transfers 8.000×10^6 kJ of energy to the block. How hot does the block get? [1 ton = 2000 pounds, 1 pound = 453.6 g]

Additional Problems

2.129 Which one of the following expresses the measured value 1230.0 m with the correct number of significant figures and appropriate Greek prefix?
(a) 1.23 km
(b) 1.230 cm
(c) 1.2300 km
(d) 1.2300 mm
(e) 12.3 km

2.130 Convert:
(a) 7.98×10^{23} μL to liters
(b) 3.00×10^{-3} mg to grams
(c) 4.21×10^8 mL to gallons [1 m^3 = 264 gallons]

2.131 A metal sphere has a radius r of 4.00 cm. What is the volume V of this sphere in cubic centimeters? The formula for the volume of a sphere is $V = (4/3)\pi r^3$, where $\pi = 3.14159$.

2.132 What is the density, in g/cm³, of the sphere in Problem 2.131 if its mass is 2.5 kg?

2.133 Which one of the following expresses the measured value 0.000 003 L with the correct number of significant figures?
(a) 3 mL
(b) 3 μL
(c) 3.00×10^{-6} L
(d) 3.00×10^{-3} mL

2.134 The SI unit of speed is meters per second. In the United States, speed is often expressed in miles per hour. If a car is traveling 60.0 miles/h, what is its speed in meters per second? [1 mile = 1.61 km]

2.135 In the United States, car fuel efficiency is expressed in miles per gallon of gasoline. However, fuel efficiency can also be expressed in kilometers per liter of gasoline. If the fuel efficiency of a car is 11.0 km/L, what is its fuel efficiency in miles per gallon? [1 mile = 1.61 km, 1 gallon = 3.79 L]

2.136 Convert:
(a) 72 °F to °C and K
(b) −12 °C to °F and K
(c) 178 K to °F and °C

2.137 Use Greek prefixes to express the relationship between the calorie and the Calorie.

2.138 When a 16.74-g rubber stopper is placed in a graduated cylinder containing 25.46 mL of water, the water level rises to 37.42 mL. What is the density of the stopper in grams per milliliter?

2.139 Solve each equation for the underlined quantity:
(a) $°F = 9/5\,\underline{°C} + 32$
(b) $PV = n\underline{RT}$
(c) $E = hc/\underline{\lambda}$

2.140 Convert:
(a) 2.37×10^2 L to milliliter
(b) 800 kg to grams
(c) 0.592 mm to meters
(d) 8.31 g to kilograms
(e) 9.62×10^{-6} L to microliters
(f) 8000 m to kilometers
(g) 19.3 mg to grams
(h) 0.00345 mL to liters

2.141 For solids, the amount of material per unit volume is often expressed in grams per milliliter, whereas for gases the amount of material per unit volume is usually expressed in grams per liter. If the amount of matter in air is 1.34 g/L, what is this value in:
(a) g/mL (b) kg/L (c) kg/mL?

2.142 The specific heat of methane gas is 2.20 J/g · °C. If the temperature of a sample of methane gas rises by 15 °C when 8.8 kJ of heat energy is added to the sample, what is the mass of the sample?

2.143 Explain how determining the number of significant figures allowed in an answer when measured values are multiplied or divided is different from determining the number of significant figures allowed in an answer when measured values are added or subtracted.

2.144 Convert:
(a) 23.0 °C to K (b) 98.6 °F to °C
(c) 296 K to °F (d) 32 °F to K
(e) 523 K to °C (f) 38 °C to °F

2.145 A student takes three measurements of the mass of an object. If the actual mass is 8.54 g, indicate whether each set of measurements is precise but not accurate, accurate but not precise, both accurate and precise, or neither accurate nor precise:
(a) 6.38 g, 9.23 g, 4.36 g
(b) 8.53 g, 8.59 g, 8.55 g
(c) 9.53 g, 8.54 g, 7.54 g
(d) 6.25 g, 6.27 g, 6.26 g

2.146 A slice of bread is burned in a calorimeter containing 1.000 kg of water. The water temperature rises from 25.0 °C to 33.0 °C. What is the calorie content of the bread? What is the Calorie content?

2.147 A block of metal measuring 3.0 cm × 4.0 cm × 5.0 cm has a mass of 470.0 g. What is the density of the metal in grams per cubic centimeter?

2.148 A student measures the mass of an object three times and reports the numeric average of her measurements. If her three measurements are 212 g, 260 g, and 233 g and the actual mass is 235 g, which of the following statements is true:
(a) The student is accurate but not precise.
(b) The student is precise but not accurate.
(c) The student is both accurate and precise.
(d) It is impossible to tell whether the student is accurate and/or precise without knowing how she determined the mass.

2.149 The density of an irregularly shaped object is determined by immersing the object in water. If the mass of the object is 8.34 g and the water level rises from 25.00 mL to 28.10 mL, what is the density of the object in grams per milliliter?

2.150 The specific heat of copper is 0.385 J/g · °C. How much heat energy in kilojoules is required to raise the temperature of 454 g of copper from 40.0 °C to 75.0 °C?

2.151 A friend tells you a quick way to convert °C to °F: double the °C, subtract one-tenth of the doubled value, and add 30 to the result. Use the Celsius–Fahrenheit conversion equation to explain why this is true.

2.152 Express 23,000,000 in scientific notation having:
(a) Two significant figures
(b) Three significant figures
(c) Five significant figures
(d) Six significant figures
(e) Eight significant figures

2.153 Dieters are often told that drinking ice-cold water burns more energy than drinking room-temperature water. Why is this true?

2.154 If a graduated cylinder has markings that indicate 0.01 mL, what is the uncertainty in any volumes determined using this graduated cylinder?

2.155 The density of gold is 19.3 g/mL, that of lead is 11.4 g/mL, that of iron 7.8 g/mL, and that of aluminum 2.7 g/mL. A student is given separate samples of three substances A, B, and C, along with a graduated cylinder containing 50.0 mL of water. Each sample has a mass of 200.0 g, and the student finds that the volumes of the samples are A, 25.64 mL; B, 10.36 mL; and C, 17.54 mL. What is the identity of each substance?

2.156 When 10 kJ of heat energy is added to a beaker containing 250 g of water initially at 23.0 °C, what is the final temperature of the water?

2.157 Indicate the uncertainty in:
(a) 74.8 m (b) 0.0026 g
(c) 1.250×10^3 L (d) 18 cm
(e) 18 pennies

2.158 Round each number to three significant digits and express the answer in scientific notation:
(a) 0.592861 (b) 438932
(c) 0.000073978 (d) 0.235469
(e) 82.550 (f) 529.8

2.159 Explain the relationship between a calorie and a Calorie.

2.160 Explain what is wrong with the statement "I took one measurement of the mass, and it was a very precise measurement."

2.161 Do the following calculations and express each answer in scientific notation:
(a) $(5.03 \times 10^2) + (8.1 \times 10^1)$
(b) $(8.32 \times 10^{-5}) \times (0.53 \times 10^4)$
(c) $\frac{6.02 \times 10^{23}}{3}$, where the 3 is an exact number
(d) $(3.960 \times 10^3) - (4.62 \times 10^2)$

2.162 A block measures 6.0 cm on each side. What is the volume of the block in cubic meters?

2.163 The mass of an average neon atom is 20.2 atomic mass units (amu), where 1 amu $= 1.66 \times 10^{-24}$ g.
(a) What is the mass in atomic mass units of 20 neon atoms?
(b) What is the mass in grams of 20 neon atoms?
(c) What is the mass in grams of 6.022×10^{23} neon atoms?

2.164 If the same amount of heat energy is added to a beaker containing 100 mL of ethanol and a beaker containing 100 mL of water, which liquid experiences the greater rise in temperature?

2.165 What is the volume in milliliters of 15.0 g of (a) liquid water at 25 °C, (b) ice at 0 °C, (c) gasoline at 25 °C, (d) lead at 25 °C, (e) mercury at 25 °C, (f) helium gas at 0 °C and 1 atm pressure? (Use data from Table 2.4.)

2.166 Indicate whether the trailing zero in each value is significant, not significant, or possibly significant:
(a) 540 ± 0.5 (b) 540 ± 5
(c) 0.540 (d) 0.000 540
(e) 540

2.167 Indicate the number of significant figures in:
(a) 0.503 200 mL (b) 2000 ± 5
(c) 2000 ± 50 (d) 2000 ± 0.5
(e) 200.1

2.168 As you will learn later in this course, some liquids do not mix with each other, oil and water being an example. When such nonmixing liquids are combined, they form separate layers, with the densest substance at the bottom. Suppose the nonmixing liquids gasoline, liquid mercury, and liquid water are added to an empty graduated cylinder. Using the data in Table 2.4, draw the graduated cylinder and show the positions of the three liquids.

2.169 True or false? If any statement is false, rewrite it to make it true.
(a) When multiplying or dividing a series of measured values, the number of significant figures in the answer is determined by the measured value having the fewest significant figures.
(b) When adding or subtracting a series of measured values, the number of significant figures in the answer is determined by the measured value having the fewest significant figures.

2.170 Convert to cal/g·°C:
(a) 1.04 J/g·°C (the specific heat of nitrogen gas)
(b) 0.84 J/g·°C (the specific heat of carbon dioxide gas)

2.171 A one-semester chemistry course meets for 1-h sessions three times a week for 15 weeks. How many milliseconds will a student with perfect attendance spend in class during the semester?

2.172 Describe how the uncertainty in a measured value is determined.

2.173 Write each number in scientific notation, taking all trailing zeros to be nonsignificant:
(a) 502,000 (b) 0.000 038 402
(c) 436,000,000 (d) 8470
(e) 0.00591 (f) 0.658

2.174 A company wants 800 square feet of carpet, but the carpet store sells only by the square meter. How many square meters does the company need to buy? (1 m = 39.37 in.)

2.175 If 1 U.S. dollar is worth 1.54 Canadian dollars, how many U.S. dollars are needed to purchase an item that costs 350 Canadian dollars?

2.176 Which contains more matter: a 50-g block of lead or a 50-g block of gold? Explain.

2.177 Indicate the number of significant zeros in each value:
(a) 2.300 (b) 2.3003
(c) 0.0023 (d) 2300
(e) 23.000

2.178 Fill in the blanks:
(a) $1.89 \times 10^3 =$ ________ $\times 10^4$
(b) $7.932 \times 10^{-5} =$ ________ $\times 10^{-7}$
(c) $4.68 \times 10^{-12} = 468 \times$ ________
(d) $3.46 \times 10^{-1} =$ ________ $\times 10^0$

2.179 Is reporting a measured mass as 580. g the same as reporting it as 580 g? Explain.

2.180 Which is larger, a Kelvin or a Celsius degree? Explain.

2.181 A student reports a series of five length measurements that are accurate but not precise. Is it more likely that his laboratory technique is very good but the measuring instrument is bad, or that his laboratory technique is bad? Explain.

2.182 Write each number in standard notation:
(a) 1.79×10^{-2} (b) 8.76×10^{-9}
(c) 4.88×10^{10} (d) 7.52×10^1
(e) 8.37×10^0 (f) 4.184×10^4

2.183 (a) If 25.0 cm^3 of an unknown substance has a mass of 195 g, what is the density of the substance in grams per cubic centimeter?
(b) How many cubic centimeters does 500.0 g of the substance occupy?
(c) Does this substance sink or float in mercury, which has a density of 13.6 g/mL?

2.184 Express each mass in grams, both in standard notation and in scientific notation:
(a) 536 mg (b) 8.26 dg
(c) 0.0057 μg (d) 139 kg
(e) 836 ng (f) 0.073 Mg

2.185 At 25 °C, air has a density of 1.3×10^{-3} g/mL. What is this density in (a) kilograms per liter and (b) pounds per gallon?

2.186 On a hot summer day, you want to cool two glasses of warm lemonade, but have no ice. Not wanting to wait until you can make some, you place two small metal blocks in the freezer. One block is pure iron, the other pure aluminum, and each has a mass of exactly 50 g. After both have cooled to −10 °C, you put them into separate glasses and add 200 mL of warm lemonade to each. After a few minutes, both blocks have warmed up to +10 °C. At this point, is the lemonade in one glass cooler than the lemonade in the other glass? If so, which is cooler and why? (Despite all the numerical information, you should be able to use specific heat values from Table 2.5 to answer without doing any calculations.)

2.187 Use a scientific calculator to do the following calculations. Express each answer in scientific notation and to the correct number of significant figures.

(a) $9.865 \times 10^3 + 8.61 \times 10^2$

(b) $\dfrac{(6.626 \times 10^{23}) \times (3.00 \times 10^8)}{4.5 \times 10^{-7}}$

(c) $\dfrac{5.6200 \times 10^{-9}}{3.821 \times 10^9}$

(d) $\dfrac{4.5600 \times 10^3 - 2.91 \times 10^1}{5}$, where the 5 is an exact number

2.188 What mass of each substance occupies a volume of 50.0 mL? (Densities are shown in parentheses.)

(a) Lead (11.4 g/mL)
(b) Ethanol (0.785 g/mL)
(c) Oxygen gas (1.4×10^{-3} g/mL)
(d) Hydrogen gas (8.4×10^{-5} g/mL)
(e) Mercury (13.6 g/mL)
(f) Gold (19.3 g/mL)

2.189 Write two conversion factors that express the relationship between:

(a) Grams and kilograms, using 1 and 1000
(b) Kilograms and grams, using 1 and 0.001
(c) Yards and feet
(d) Meters and centimeters, using 1 and 100
(e) Meters and centimeters, using 1 and 0.01

2.190 The recommended tire pressure in a bicycle is 125 pounds/in^2. What is this tire pressure in atmospheres? (1 atm = 14.70 pounds/in.^{2})

2.191 Do the following calculations and express each answer to the correct number of significant figures. (All values are measurements.)

(a) $\dfrac{5.03 + 7.2}{0.003}$

(b) $\dfrac{8.93 \times 0.054}{1.32}$

(c) $(6.23 \times 0.042) + 9.86$

2.192 You overhear a classmate telling another student that 1 ft equals 12 ± 0.5 in. Is this statement correct or incorrect? Why?

2.193 Indicate the correct number of significant figures for each answer, given that all values are measurements:

(a) $(6.350 \times 10^{-8}) \times (0.0080)$
(b) $(5.30 \times 10^2) + (22.1 \times 10^2)$
(c) $(5.830 \times 10^2) + (22.100 \times 10^2)$

(d) $\dfrac{100.0 \times 0.1500}{58.443}$

(e) $\dfrac{100.0 \times 0.15}{58.4}$

2.194 If 1 U.S. dollar is worth 0.690 English pounds, how many U.S. dollars are needed to purchase an item that costs 350 pounds?

2.195 A graduated cylinder was marked incorrectly at the factory and indicates a volume that is 5 mL more than the actual volume of liquid in the cylinder. If an experienced laboratory technician who does not know about the error is using the cylinder, are the volumes she measures more likely to be accurate, precise, or neither? Explain.

2.196 The density of water at 4.00 °C is 1.00 g/mL. The density of ice at 0 °C is 0.917 g/mL. Water is different from most other substances in that the solid phase (ice) is less dense than the liquid phase. Explain why this characteristic makes ice fishing possible.

2.197 How many significant figures are there in each number:

(a) 5.300×10^{-2}
(b) 3.2×10^5
(c) 0.00890×10^{-4}
(d) $7.960\,000\,0 \times 10^{10}$
(e) 8.030×10^{21}

2.198 The speed limit on some interstate highways is 70 miles/h. What is this speed limit in (a) kilometers per hour, (b) kilometers per second, (c) meters per hour, (d) meters per second? (1 mile = 1.61 km)

2.199 Explain what effect compressing a gas has on the density of the gas.

2.200 Calculate the amount of heat energy in joules required to heat 50.0 g of each substance from 25.0 °C to 37.0 °C. (specific heats shown in parentheses):

(a) Iron (0.449 J/g·°C)
(b) Aluminum (0.901 J/g·°C)
(c) Mercury (0.14 J/g·°C)
(d) Water (4.18 J/g·°C)

2.201 Which substance in Problem 2.200 undergoes the most gradual change in temperature when heated? Which undergoes the fastest change in temperature when heated? Explain.

2.202 Shooters A and B each shoot three arrows at their target, aiming for a bull's-eye. Which shooter is more precise? Which shooter is more accurate? Explain.

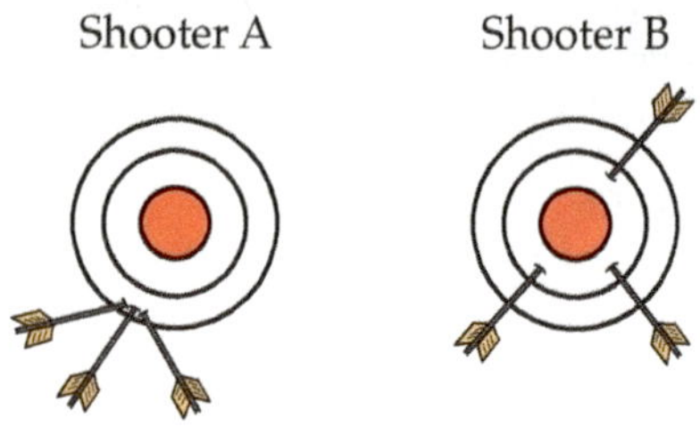

2.203 One of the bars represents a joule of energy and the other represents a calorie of energy. Which is which? Explain how you knew.

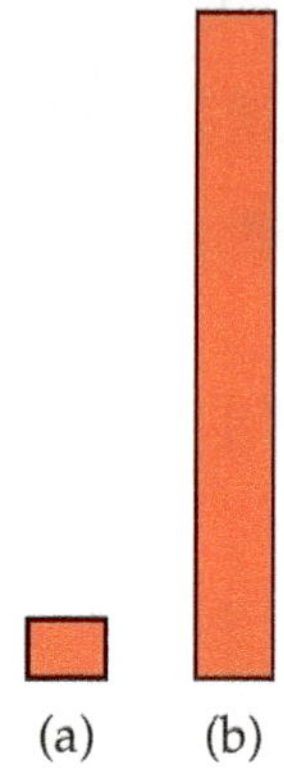

One Step Beyond

2.204 A rock is experimentally determined to be 3.5 billion years old, with an uncertainty of plus or minus 100,000 years. Write the age of this rock in scientific notation.

2.205 A microscopic particle is measured and found to have a 40.2 micron diameter. What is its diameter in meters? What is its diameter in nanometers?

2.206 The mass of an ordinary Twinkie is 36.4 g, and its length is 36 cm. How many tons would a 6.4 ft Twinkie weigh?

2.207 10.6 g of a substance was burned in an insulated calorimeter that contained 1.00×10^4 g of water and a bomb made of 9.00×10^1 g of aluminum. The initial temperature inside the calorimeter was 25.0 °C. After the burning, the temperature inside the calorimeter was 45.0 °C. How many kilojoules per gram of energy is released upon burning this substance?

2.208 A student uses a digital balance to determine the mass of an object. Its digital display reads 2.635 g. He keeps looking at the balance and it stays fixed at 2.635 g for half an hour. He then states, "The mass of the object is 2.635 g exactly. There is no uncertainty in this measurement because the last digit '5' stayed constant and did not change over time." Is he right? Is the mass of the object exactly known with no uncertainty? Explain.

2.209 Without doing any numerical calculations, determine which would have the smallest volume:
(a) 50 g of water (density = 1.0 g/mL)
(b) 50 g of salt water (density = 2.3 g/mL)
(c) 50 g of mercury (density = 13.6 g/mL)
(d) 50 g of alcohol (density = 0.89 g/mL)

Explain your reasoning.

2.210 A physician has ordered 325 mg of atropine to be given to a patient intramuscularly. If atropine were available as a solution that contained 5000 μg/mL, how many milliliters would you need to give the patient?

Group Exercises (Work on these exercises together with someone else in the class.)

1. Suppose you needed someone to repeatedly make a measurement, the value of which could determine if you live or die. Suppose also that you could choose Experimenter A or Experimenter B to do the measuring. Experimenter A has the reputation of making extremely precise but woefully inaccurate measurements. Experimenter B has the reputation of making extremely accurate but woefully imprecise measurements. Who should you choose to do the measuring and why?

2. Suppose you are teaching someone how to multiply a measured quantity such as 12.5 in. by a conversion factor that converts it to feet. Devise a general rule to tell the student how to write the conversion factor.

3. You work for a company whose job is to take inexpensive rulers, marked in centimeters, and do all that is possible to lower the uncertainty associated with measuring the diameter of a coin. List all the things that you might do. What would this do to the expense of your measuring system?

4. Suppose you needed a material that could absorb heat without having its temperature increase very much. Would you choose aluminum or water? Make sure your explanation uses the concept of specific heat (somebody consult Table 2.5).

WORKPATCH SOLUTIONS

2.1 (c) ±0.01 cm. Because of the millimeter divisions (remember, 1 mm = 0.1 cm), you can read a value exactly to 0.1 cm. Then you estimate between two adjacent lines to get a digit for your hundredths place because the estimating (uncertainty) is in the hundredths place.

2.2

	Number of trailing zeros	Number of significant figures
20.201	None	5
20.210	1	5
20.0002	None	6
20.0	2	3
120.	1	3

2.3 (a) 1.23×10^2 (b) 6×10^{-5}
(c) 6.0×10^{-5} (d) 1.0020×10^3

2.4 Each small division represents 1 mm. You can estimate between any two adjacent lines, which means you are estimating to 0.1 mm. This gives an uncertainty of (b) ± 0.1 mm. The diameter of the coin is (c) 5.00×10^1 mm.

2.5 In 1.20×10^{-7}, the exponent -7 instructs you to move the decimal point seven places to the left, filling in with zeros the empty columns you generate:

Fill with zeros

$$.\,\underline{}\,1.20 = .000000120$$

(The zero shown to the left of the decimal point in the WorkPatch—0.000 000 120—is there merely to make it easier to spot the decimal point. This zero is usually shown: 0.5, 0.003, etc.)

2.6 (a) In 6.0000×10^{11}, the exponent instructs you to move the decimal point 11 places to the right, filling with zeros any empty columns generated:

Last significant digit — Fill with zeros — Last significant digit

$$6.0000\,\underline{} = 600{,}000{,}000{,}000$$

(b) The last significant digit is the fourth zero from the 6. This zero is in the 10-millions column, and so the uncertainty is ± 10 million.

2.7 1.1×10^2 miles/h.

2.8 Because it takes 1000 mL to get 1 L, you have to fill your 100-mL beaker 10 times to get 1 L. To get 2.5 L, you have to fill your beaker 25 times.

2.9 Form B is denser because the atoms are packed more closely together (notice that the gaps between atoms in B are smaller than the gaps between atoms in A).

2.10 The specific heat of water is much larger than the specific heat of aluminum. Therefore, it takes more heat to warm up a given mass of water than it does to warm up the same mass of aluminum. For the water, you need to add

$$1.0\,\text{g} \times \frac{4.184\,\text{J}}{\cancel{\text{g}}\cdot\cancel{^\circ\text{C}}} \times 1.0\,\cancel{^\circ\text{C}} = 4.2\,\text{J}$$

of heat energy to raise the temperature from 25.0 °C to 26.0 °C. To increase the temperature of the aluminum from 25.0 °C to 26.0 °C, you must add

$$1.0\,\cancel{\text{g}} \times \frac{0.901\,\text{J}}{\cancel{\text{g}}\cdot\cancel{^\circ\text{C}}} \times 1.0\,\cancel{^\circ\text{C}} = 0.90\,\text{J}$$

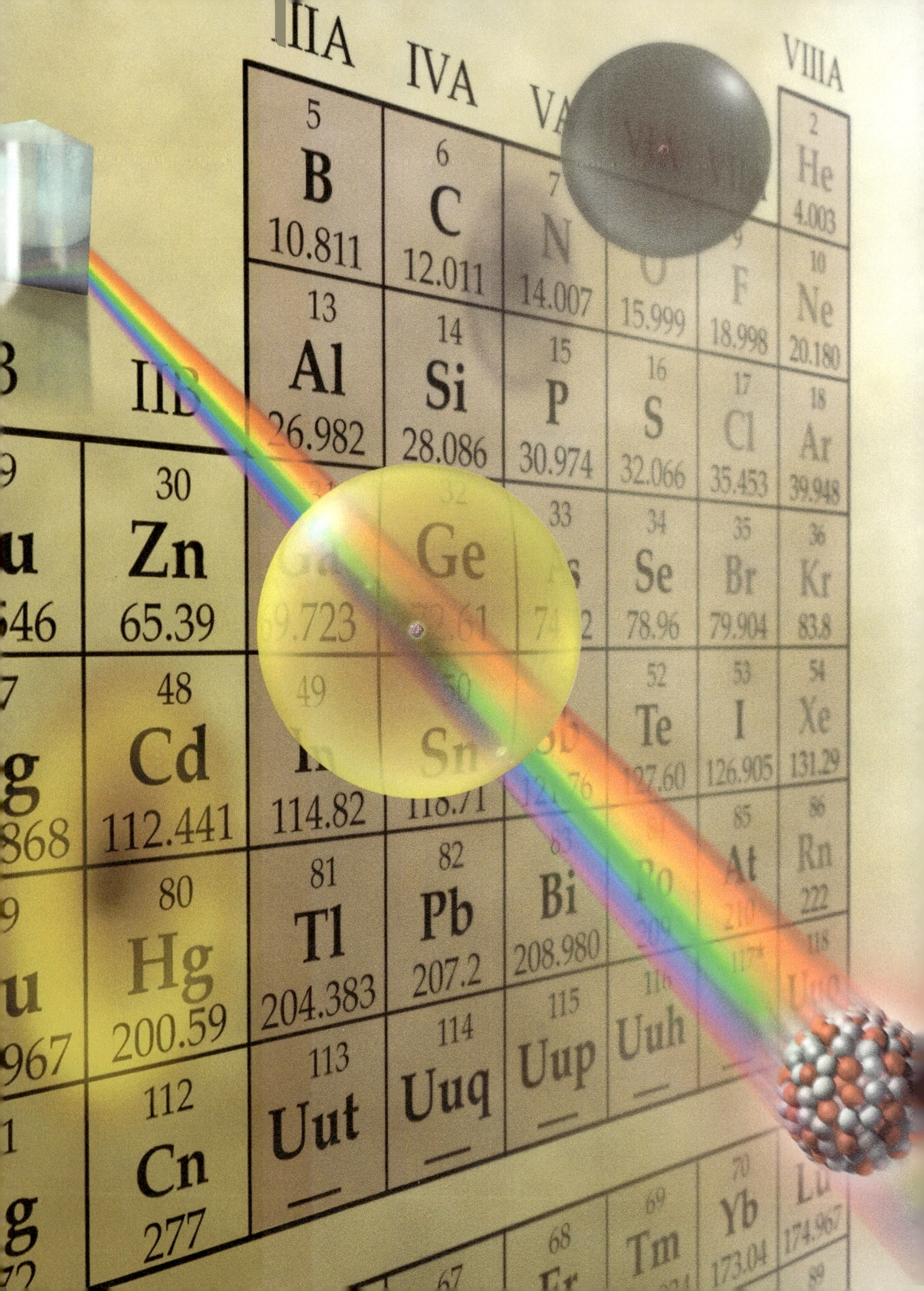
IIIA
IVA
VIIIA
IIB
5
B
10.811
6
C
12.011
7
N
14.007
15.999
F
18.998
2
He
4.003
10
Ne
20.180
13
Al
26.982
14
Si
28.086
15
P
30.974
16
S
32.066
17
Cl
35.453
18
Ar
39.948
30
Zn
65.39
Ge
33
34
Se
78.96
35
Br
79.904
36
Kr
83.8
48
Cd
112.441
114.82
52
Te
53
I
126.905
54
Xe
131.29
80
Hg
200.59
81
Tl
204.383
82
Pb
207.2
Bi
208.980
85
At
86
Rn
222
112
Cn
277
113
Uut
114
Uuq
115
Uup
Uuh
68
69
Tm
70
Yb
173.04

CHAPTER 3

The Evolution of Atomic Theory

Water Is Water (or Is It)?

The following is a hoax from a website titled "Ban dihydrogen monoxide": "In 2005, there were 3,482 unintentional dihydrogen monoxide–related deaths in the United States, an average of nine people per day. Dihydrogen monoxide is killing our children. Dihydrogen monoxide is the second-leading cause of death among children under the age of 15. Children under five and adolescents between the ages of 15–24 have the highest dihydrogen monoxide–related death rates. An estimated 5,000 children ages 14 and under are hospitalized due to dihydrogen monoxide–related incidents each year; 15 percent die in the hospital and as many as 20 percent suffer severe, permanent neurological disability. The U.S. government has known about the dangers of dihydrogen monoxide for many years. However, lawmakers refuse to address this desperate situation. Dihydrogen monoxide needs to be outlawed now. Dihydrogen monoxide can be found in many food and beverage products that Americans consume on a daily basis. And as the above statistics indicate, dihydrogen monoxide can be deadly. Please sign this petition, and help us force the hand of our government to ban dihydrogen monoxide now."

Yes, dihydrogen monoxide, a name which indicates the atoms from which it is made, is water. We use water in this chapter as an example to illustrate some of the concepts that led to the acceptance of atoms.

3.1 Dalton's Atomic Theory

Recall from Chapter 1 the battle between the philosophies of Democritus and Aristotle that took place around 400 B.C. Aristotle proposed that matter was continuous and could be infinitely subdivided into smaller and smaller particles. Democritus postulated that matter was ultimately made up of tiny, indivisible particles that he called *atoms*. Aristotle's view won the day, and it took more than 2000 years before the atomic theory of matter made a comeback. Historians generally credit the beginning of modern atomic theory to the work of a mild-mannered English schoolteacher named John Dalton in the early 1800s, and it is here that we shall begin.

John Dalton's atomic model was just that, a model in the true spirit of the scientific method. It was formulated as an explanation for a number of "laws" about how matter behaves in a chemical reaction. These laws were the result of careful experimental work done in the latter part of the 1700s and were based on observations of a great number of chemical substances and their reactions. These laws are the foundation on which modern atomic theory is based.

The first of these is the **law of conservation of matter**:

Law of conservation of matter: When a chemical reaction takes place, matter is neither created nor destroyed.

Today we accept this law as fact (with minor modifications). However, that matter is conserved in any chemical reaction was not at all obvious at first because many reactions *seem* to either create or destroy matter. Two examples are shown in the following illustration:

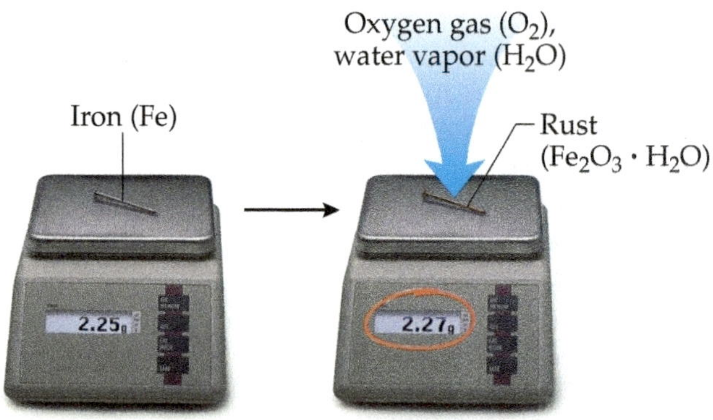

Matter seems not to be conserved:
A nail gains mass as it rusts.

Actually, matter is conserved:
Rusting is a reaction in which oxygen gas and water combine with iron to form rust ($Fe_2O_3 \cdot H_2O$).

Matter seems not to be conserved:
A glass of water with an Alka-Seltzer tablet loses mass as the tablet dissolves.

Actually, matter is conserved:
The Alka-Seltzer reacts with water to produce carbon dioxide gas, which escapes from the beaker.

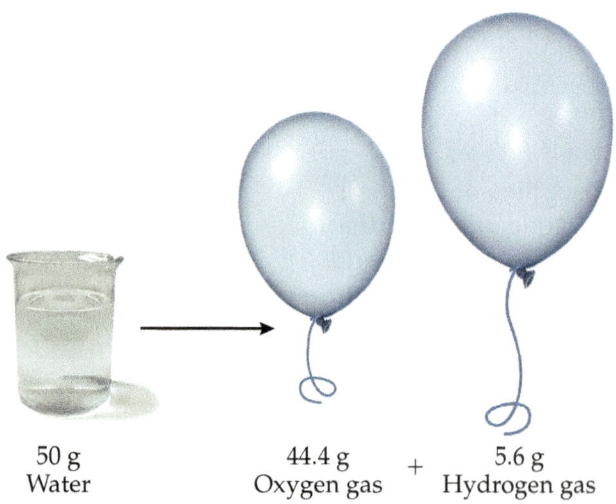

The issue was set straight in 1775 by the French chemist Antoine Lavoisier, along with his wife, Marie, who worked with him in his laboratory. They carried out reactions in sealed vessels that allowed no matter—including gases—to enter or leave during the reactions. Under these conditions, they found that matter is always conserved.

The second law that led Dalton to postulate his atomic theory is known as the **law of constant composition** (also sometimes called the **law of definite proportions**):

Law of constant composition: Multiple samples of any pure chemical compound always contain the same percent by mass of each element making up the compound.

For example, if a 50.0 g sample of pure water (H_2O) is decomposed into its component elements, you will obtain 5.6 g of hydrogen gas and 44.4 g of oxygen gas. The **percent by mass** of these elements is therefore

$$\frac{\text{Mass of hydrogen in compound}}{\text{Total mass of compound}} = \frac{5.6\text{ g}}{50.0\text{ g}} \times 100 = 11\%$$

$$\frac{\text{Mass of oxygen in compound}}{\text{Total mass of compound}} = \frac{44.4\text{ g}}{50.0\text{ g}} \times 100 = 88.8\%$$

These same percentages are found in any sample of pure water, no matter where it comes from or what the size of the sample is.

The law of constant composition also holds when a chemical compound is synthesized from its elements. For example, no matter how water is made, no matter how much hydrogen and oxygen are combined with each other, the water produced always has the same percentage of oxygen and hydrogen. In the following WorkPatch, water is made first by combining just the right amounts of hydrogen and oxygen (so that they are both used up) and then made again with only half the amount of hydrogen.

Law of Constant Composition

All samples of pure water have the same composition by mass: 11.2% hydrogen, 88.8% oxygen.

WORKPATCH 3.1 Use these data to calculate the percentage of oxygen and hydrogen in these two samples of water:

Hydrogen +	Oxygen →	Water
100.0 g	**793.6 g**	**893.6 g**

(a) Is the law of conservation of matter obeyed?
(b) What is the percent hydrogen in water?
(c) What is the percent oxygen in water?

Hydrogen +	Oxygen →	Water	leftover oxygen
50.0 g	**793.6 g**	**446.8 g +**	**396.8 g**

(d) Is the law of conservation of matter obeyed?
(e) What is the percent hydrogen in water?
(f) What is the percent oxygen in water? ●

Did you get the same percentages for both samples? If you did (and you should have), you have demonstrated that the law of constant composition is obeyed.

To Dalton, constant composition was very good evidence that the elements did indeed exist in the form of atoms. When you formed a chemical compound, he reasoned, you were combining atoms of the constituent elements. An atom of one element would always combine with a fixed number of atoms of another element, and that fixed number depends on the identity of the elements and on the identity of the compound being made.

Practice Problems

3.1 A student makes a compound of sulfur and oxygen. She uses 5.00 g of sulfur and 4.99 g of oxygen, and all of the elemental substances are completely used up. What is the percent sulfur in the compound?

Answer: $(5.00\ g\ S \div 9.99\ g\ compound\ 1) \times 100 = 50.1\%\ S$

3.2 A student makes two different compounds of nitrogen and oxygen using the masses shown below. In both cases, all of the elemental substances are completely used up.

Compound 1 10.0 g nitrogen + 11.42 g oxygen
Compound 2 10.0 g nitrogen + 22.84 g oxygen

What is the percent oxygen in each compound?

Answer: Compound 1, 53.3% oxygen; compound 2, 69.5% oxygen

3.3 A 101.96-g sample of a compound of aluminum and oxygen is 47.1% by mass oxygen.

(a) What is the percent by mass aluminum in this compound?
(b) Of the 101.96 g of this compound, how many grams are aluminum?

Answer: (a) 52.9% aluminum (b) 53.9 g aluminum

Based on the law of conservation of matter and the law of constant composition, Dalton formulated his atomic theory, which can be summarized in five short statements:

Dalton's atomic theory

1. All matter is made up of atoms (small, indivisible, indestructible, fundamental particles).
2. Atoms can neither be created nor destroyed (they persist unchanged for all eternity).
3. Atoms of a particular element are alike (in size, mass, and properties).
4. Atoms of different elements are different from one another (different sizes, masses, and properties).
5. A chemical reaction involves either the union or the separation of individual atoms.

Molecules of the chemical compound we call water are always formed from one atom of oxygen and two atoms of hydrogen, H_2O. Why do atoms of one element combine with only certain numbers of atoms of other elements in forming chemical compounds? Dalton wanted a physical model—a picture of the atom—that fit his theory and helped explain it. Chemists of the time ultimately envisioned a ball-and-hook model that allowed atoms to combine with one another in particular ways. This model described atoms of different elements as balls of different sizes (and masses) with a characteristic number of hooks embedded in them.

$$2\,H + O \longrightarrow H_2O$$

The number of hooks in each type of atom would account for its chemical reactivity and explain why elements of certain atoms characteristically combined with only certain numbers of other atoms. The atoms "reacted" with one another until their hooks were filled.

Today, we know that atoms don't have hooks, but for the early 1800s, this was a good model. It made visual sense of the behaviors summarized by the two conservation laws Dalton knew about. If you accept the five statements of Dalton's theory as a viable model of reality, the laws of conservation of matter and constant composition can be easily explained and accounted for.

Many scientists at the time did not accept Dalton's theory or his model and continued to argue that atoms did not exist. Needless to say, however, no other proposed theory was able to account for the observed laws.

Law of Conservation of Matter

$$H \;\; H \qquad O \longrightarrow H_2O$$

Matter is conserved in a chemical reaction because the reaction consists simply of atoms either hooking together or unhooking.

Nearly 200 years after Dalton proposed his atomic model, we now know that none of the five statements of his theory is *entirely* true. Atoms are not the most fundamental of particles—they are composed of the even smaller particles we call electrons, protons, and neutrons. Atoms can be created and destroyed, but a nuclear process is required to do so. We shall encounter these and other exceptions to Dalton's five statements as we proceed through this book. None of these later findings should be taken as diminishing Dalton's accomplishments, however. For the time, his was a superb model. It adequately explained experimental observations, and it laid the foundation for further developments in atomic theory. Of course, even the inability of Dalton's detractors to come up with any plausible alternatives did not make the acceptance of the atom an overnight phenomenon. It took nearly 100 years for the last of the holdouts to become convinced (or to die off!), but by the beginning of the twentieth century, the concept of the atom was firmly established.

Practice Problems

3.4 You visit France, where a local salesperson tries to sell you special water from a mountain spring. This water is supposed to be special because it has the formula H_3O. What law does this claim violate?

Answer: The law of constant composition, which tells us that a compound has the same composition no matter where it comes from. Water is always H_2O.

3.5 When coal (essentially pure carbon) is burned, it combines with atmospheric oxygen to produce carbon dioxide gas. Nevertheless, when the coal burns, it seems to disappear. Why is this not a violation of the law of conservation of matter?

3.6 Using Dalton's ball-and-hook atomic model, sketch an explanation of how it is possible for hydrogen to combine with oxygen to form two different compounds—water (H_2O) and hydrogen peroxide (H_2O_2).

3.2 Development of a Model for Atomic Structure

Once the existence of atoms was accepted, the obvious next question was, "What do atoms look like?" A series of extremely clever experiments done by physicists J. J. Thomson, James Chadwick, and others decisively showed that atoms are not the fundamental particles of matter but are themselves composed of even smaller particles.

In 1897, Thomson discovered the first *subatomic* particle to be identified, the **electron**. He found that all atoms contained electrons and that the electrons from all atoms were identical. These electrons were very small, lightweight particles indeed, having only 1/1836 the mass of a hydrogen atom, the smallest and lightest of all atoms. In addition, the electron had a negative electrical charge, which for convenience was assigned a numerical value of −1.

Just 10 years later, in 1907, Thomson and E. Goldstein found another subatomic particle that was present in all atoms, the **proton**. This particle was much heavier than the electron, with a mass almost equal to that of a hydrogen atom. The proton was found to have a positive electrical charge equal in magnitude but opposite in sign to that of the electron. This charge was assigned a value of +1.

Twenty-five years later, Chadwick demonstrated the existence of a third subatomic particle, the **neutron**. It had about the same mass as a proton but lacked any charge (it was electrically neutral, hence the name neutron). The lack of a charge made the neutron much more difficult to study and helps to explain why its discovery came so much later.

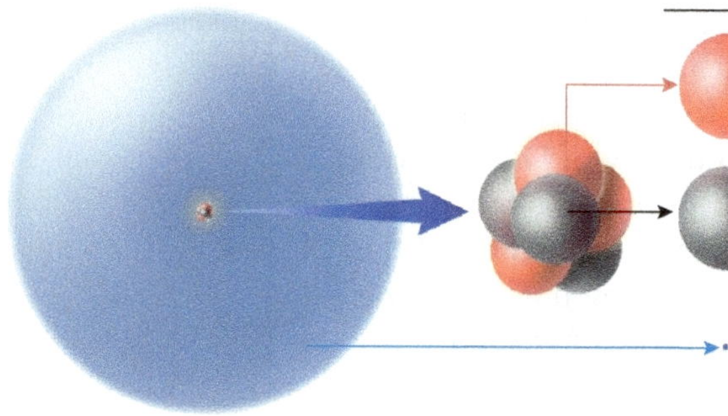

Particle	Charge	Mass and occurrence
Proton	+1	Mass of 1.007 amu*; found in all nuclei
Neutron	0	Mass about the same as that of proton (1.009 amu); found in almost all nuclei (exception: most H nuclei)
Electron	−1	Mass about 1/1836 that of proton (5.486×10^{-4} amu); found outside nucleus

*This unit will be explained in Section 3.4.

Once it was established that all atoms are composed of electrons and protons, the path of investigation turned to the exploration of the internal structure of the atom. How could only three subatomic particles be put together to make all the different atoms in the periodic table?

Thomson's Plum-Pudding Model of the Atom

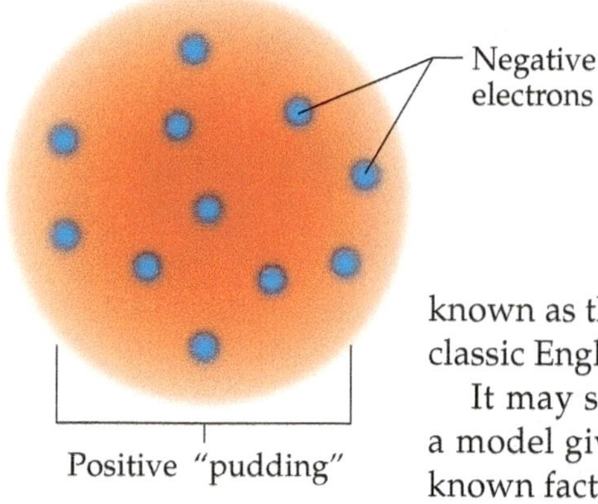

The first model was proposed by Thomson shortly after his discovery of the electron. He knew two basic facts about the atom: (*1*) Atoms contain small, negatively charged particles called electrons, and (*2*) the atoms of an element behave as if they have no electrical charge at all—they are electrically neutral. Thomson reasoned that there must be something in the atom that carries a positive charge to neutralize the electrons (protons had yet to be discovered). Thomson's model for the atom consisted of a "cloud of positive electricity" in which the negatively charged electrons were embedded. This became known as the plum-pudding model because it reminded many scientists of a classic English pudding.

It may seem a silly model today, but it fulfilled all the requirements of a model given the data available at that time. It was consistent with all the known facts about atoms and successfully accounted for their neutral behavior. What ultimately banished this model to the trash heap was an experiment performed just a few years later, in 1909, by the British physicist Ernest Rutherford.

3.3 The Nucleus

Rutherford's experiment, like many of the other great experiments that have shaped modern chemistry and physics, was beautifully simple in its concept. Rutherford had been studying **alpha** (α) **particles**, small chunks of positively charged matter that are spontaneously and randomly given off by many naturally occurring radioactive elements, such as radium, polonium, and radon. We now know that any alpha particle consists of two protons and two neutrons, giving it a +2 charge. Rutherford did not know this precisely, but he did know that alpha particles were about 7000 times more massive than electrons,

had a positive charge, and were ejected from certain types of matter at very high speeds—about 9300 miles/s (1.5×10^4 km/s), or approximately one-twentieth the speed of light. Rutherford thought of alpha particles as positively charged "bullets" that could be fired at various targets. The targets he chose were thin foils of pure metals. One was made of pure gold, a metal so soft and malleable that it could be hammered into a thin film only a couple of thousand atoms thick. He decided to "fire" the alpha particles at the gold foil and "see" what happened.

Of course, he could not actually see either the alpha particles or the individual atoms of gold in the foil target. However, a screen made of glass coated with zinc sulfide was known to glow (fluoresce) with a tiny point of green light when struck by an alpha particle. Rutherford simply positioned a zinc sulfide screen behind his gold target so that any alpha particle that passed through the gold target would strike the screen and cause it to glow, thus revealing the particle's path. Because the plum-pudding model of the atom was in vogue at the time, Rutherford expected the particles to cruise right through the gold atoms pretty much undisturbed. After all, the electrons in the gold atoms were much too small and lightweight to cause any resistance, and the positive charge in each gold atom was spread out in a thin cloud throughout the entire volume of the atom. If this model of the atom were correct, the fast-moving alpha particles should go straight through the gold foil and hit the zinc sulfide screen in a straight-line trajectory from the particle source. To make sure that he and his students missed nothing, they sat in a pitch-black room for an hour before each experiment so that their eyes would be ultrasensitive to any green flashes on the fluorescent screen. What they saw was *almost* what they expected.

Rutherford's Alpha-Particle Experiment

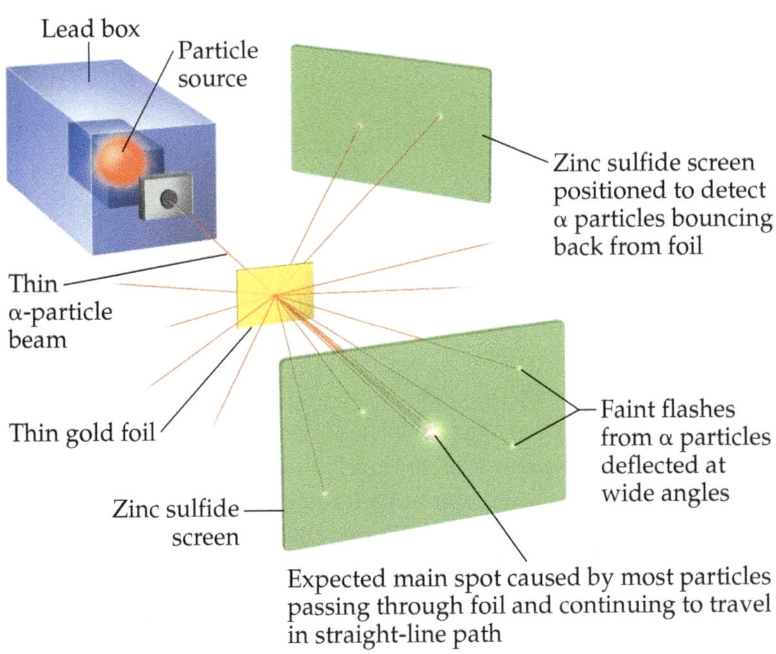

Most of the alpha particles did indeed go straight through the foil as predicted. But a few, barely noticeable as green flashes on the zinc sulfide screen, were deflected at wide angles. Some alpha particles even bounced right back toward the source, causing a glow on a screen positioned alongside the source! This was incredible. In Rutherford's own words, "It was . . . as if you fired a 15-inch shell at a piece of tissue paper and it came back and hit you." If the results were to be believed, the plum-pudding model of how the atom was put together could not possibly be correct. A new model for the internal structure of the atom would have to be created to account for the results of Rutherford's experiment. In particular, two observations needed to be explained: (*1*) Most of the alpha particles went straight through the foil, and (*2*) a few (about 1 in 20,000) were deflected from a straight-line path, some by very extreme angles.

The fact that *any* of the fast-moving alpha particles were deflected by the gold atoms in the foil led Rutherford to conclude that there must be something small and massive inside each gold atom. This conclusion was inescapable. Except for this tiny massive something inside each atom, the atom must be mostly empty space. (It was through this empty space that all the undeflected alpha particles had passed.) It took Rutherford more than a year to come up with this model of the atom, and close analysis of his results told him that this tiny massive something must be positively charged.

WORKPATCH 3.2 What led Rutherford to the conclusion that this massive something inside the atom was very small? ●

In his attempt to verify the plum-pudding model of the atom, Rutherford ended up having to discard it and replace it with a model that showed the atom as being mostly empty space but with something tiny, massive, and positively charged inside it. This something he called the atom's **nucleus**, which we now know holds an atom's protons and neutrons. The "empty space" that makes up most of the volume of an atom is occupied by the much less massive electrons that are also present in the atom.

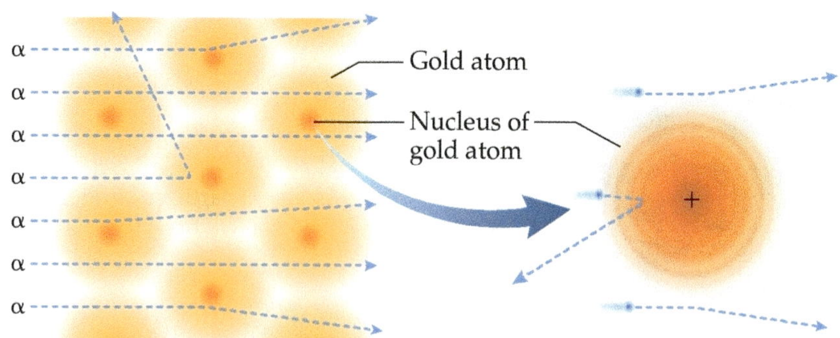

Based on the number of alpha particles deflected and their angles of deflection, Rutherford was able to calculate the relative sizes of the "electron space" and the nucleus. He calculated the nucleus to be about 10^{-13} cm in diameter, while the diameter of the entire atom was about 10^{-8} cm. Thus, the nucleus and the atom differ in size by a factor of 10^5—meaning that an atom is 100,000 times larger than its nucleus.

Rutherford's atom was mostly empty space. At the center of the spherical atom was a tiny, dense nucleus; the rest of the volume was occupied by the electrons, which moved around the nucleus in some fashion. A modern-day comparison to Rutherford's model might be a Ping-Pong ball suspended in the middle of the Dallas Cowboys Stadium. The Dallas Stadium represents the total volume of the atom, and the Ping-Pong ball represents its nucleus. The electrons would be equivalent to a few mosquitoes flying around inside the stadium.

It certainly was not easy for Rutherford to discard an accepted model. Another, less thorough scientist might have been tempted to ignore the few faint flashes of green light that showed up at large deflection angles or might even have neglected to design the experiment to look for them because "obviously" all the action would be in a straight-line path. In this case, however, Rutherford and the scientific method triumphed, replacing an old model with a new one and inviting new sets of experiments to challenge and further refine it.

The Dallas Cowboys Stadium

Rutherford realized that more work needed to be done. For instance, the electrons presented a problem. Rutherford assumed they had to be continuously moving because, if they stopped, they would be sucked into the nucleus by its strong positive charge (opposite charges attract each other, even on the atomic level). But *how* were they moving? According to the physics known in Rutherford's day, an electron moving around a nucleus must radiate energy but only for a finite amount of time until its energy is gone. Thus, the electrons moving about inside an atom should quickly lose energy, slow down, and spiral into the nucleus. In other words, the principles of physics demanded that a Rutherford type of atom should collapse in on itself within a fraction of a second. Because it was pretty clear that all matter in the universe was in no immediate danger of collapsing, Rutherford knew that either something was wrong with his model or something was wrong with physics as it was then understood. He had difficulty accepting either proposition, and so, as you might imagine, he put forth his model of the atom with a great deal of trepidation.

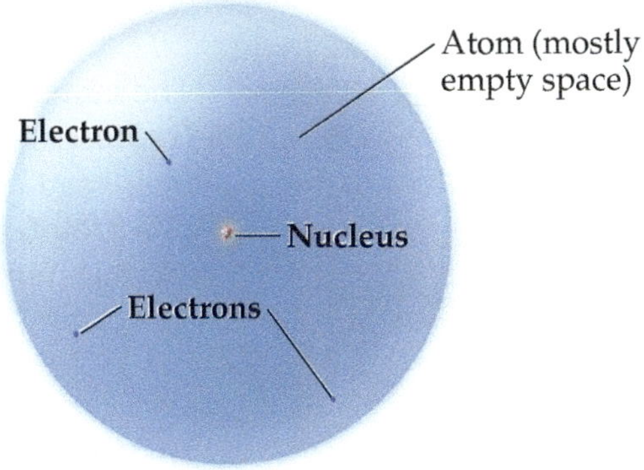

Rutherford knew that his model was in need of refinement and modification. What he did not know was that the solution to the problem was going to be so revolutionary, so inconceivable, that even the scientists who developed it found it almost impossible to accept. Albert Einstein went to his grave not fully believing in it and working feverishly to find flaws in the analysis. The laws of classical physics, used for centuries to calculate the force of gravity, the motions of the planets, and the trajectory of missiles, just did not work when applied to particles the size of an atom. To solve the problem of the atom, classical physics itself—the foundation of modern science—would have to be tossed away and replaced with something else. Over the next 20 years (1910–1930), a revolution in quantum physics (the physics of atomic and subatomic particles) was born.

3.4 The Structure of the Atom

Let's take stock of where we are with regard to the subatomic pieces that go into the making of an atom. There are the relatively massive, positively charged (+1) protons (p) that reside at the center of the atom in a tiny nucleus. There are the much less massive, negatively charged (−1) electrons (e^-) that move about outside the positive nucleus. A fundamental law of physics (quantum as well as classical) is that oppositely charged particles attract each other. The significance of this is that the electrons can't leave the atom, at least not without the input of a good deal of energy. They are attracted to the nucleus and so are kept from wandering away.

The third subatomic particle is the neutron (n). This particle has no charge; it is electrically neutral. Neutrons have approximately the same mass as protons.

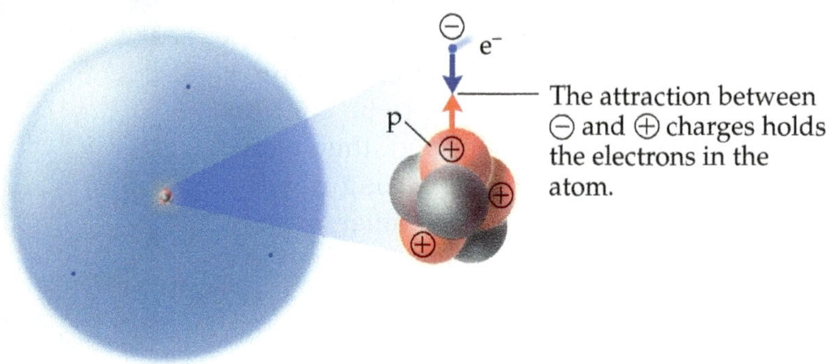

Atomic Number

The number of protons in the nucleus of an atom is called the **atomic number** (represented by the letter Z) of that atom. The atomic number is extremely important because it alone determines the identity of the atom. An atom with one proton (atomic number 1) is always hydrogen, an atom with six protons (atomic number 6) is always carbon, and so on.

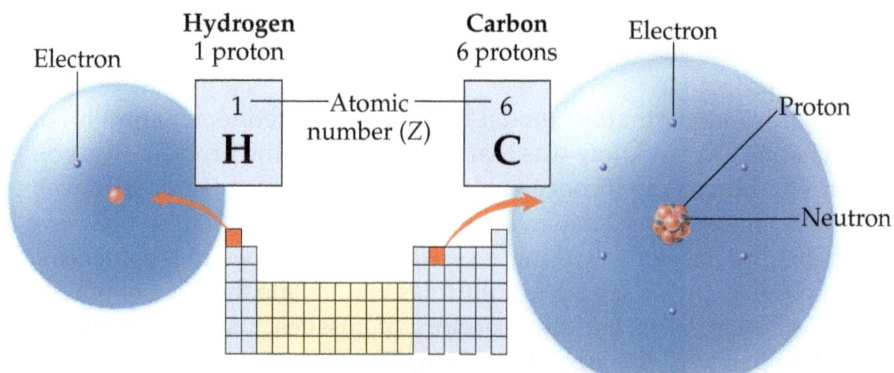

The atomic number is so important that the periodic table of the elements is arranged according to increasing atomic number. Examine the atoms in the figure above and then answer the following WorkPatch:

WORKPATCH 3.3 From the figure, what do you think is the rule for determining the number of electrons a neutral atom contains? ●

Your answer to the WorkPatch should help you understand how the overall neutrality of an atom depends on the number of protons and electrons it contains. (If you are not sure about this, check the answer at the end of the chapter before going on.)

Now, what about neutrons? These are fairly massive subatomic particles that reside in the nucleus. Most hydrogen atoms have no neutrons; all other elements have at least one. Carbon atoms, with six protons, usually have six neutrons, but there is no simple relationship between the number of neutrons and protons in the nucleus of an atom. This is especially true for the heavier atoms in the periodic table. For instance, a typical uranium nucleus, with 92 protons (atomic number 92), contains 146 neutrons.

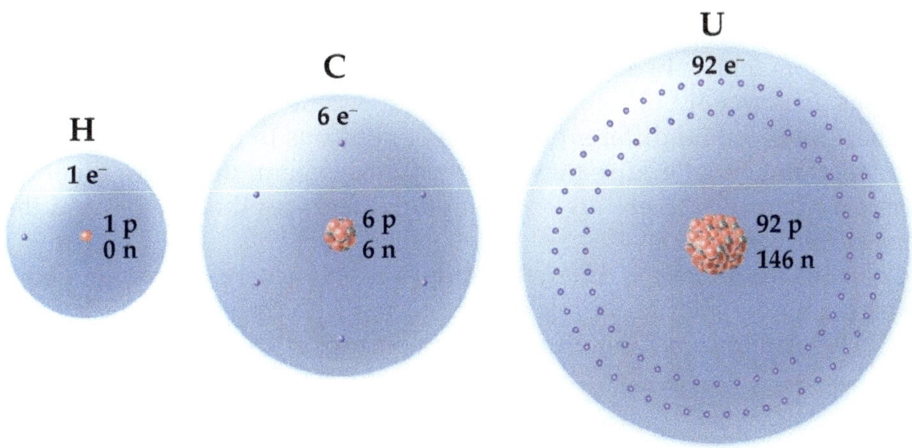

Mass Number

Along with the atomic number (Z), chemists have defined another atomic quantity, called *mass number* (sometimes represented by the letter A). An atom's **mass number** is simply the number of protons plus the number of neutrons in its nucleus. Thus, the mass numbers for H, C, and U are as follows:

H		C		U	
	1 p		6 p		92 p
Mass	+ 0 n	Mass	+ 6 n	Mass	+ 146 n
number ⟶	1	number ⟶	12	number ⟶	238

Let's see if you're catching on to this. Complete the following WorkPatch, and then we'll discuss it:

WORKPATCH 3.4 Sketch the carbon atom in the following three ways, showing the correct numbers of protons, neutrons, and electrons. All three of your carbon atoms should be electrically neutral.

(a) Make the mass number equal to 12.
(b) Make the mass number equal to 13.
(c) Make the mass number equal to 14. ●

If you did things correctly, all three sketches should have six protons in the nucleus (because these atoms are all carbon) and six electrons outside the

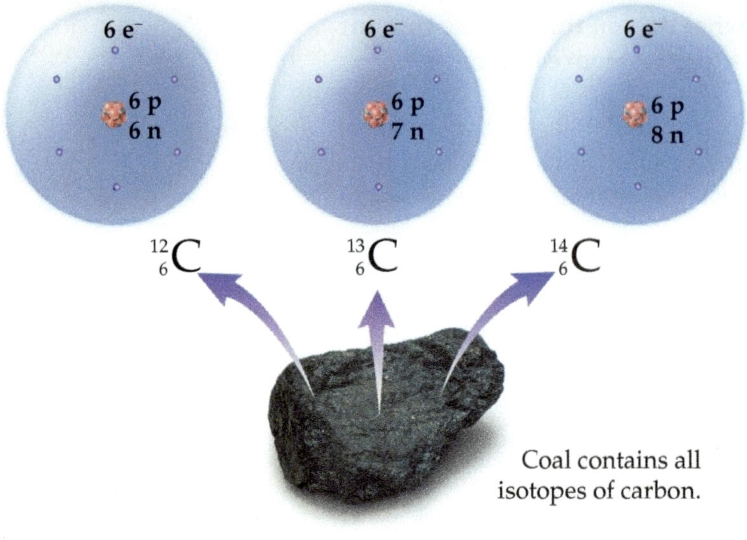

Coal contains all isotopes of carbon.

nucleus (to keep each atom electrically neutral). The only difference, therefore, must be in the number of neutrons in the nucleus: Sketch (a) should have six neutrons, sketch (b) should have seven neutrons, and sketch (c) should have eight neutrons. The question is, what should we call these three slightly different versions of the same element? The answer—in all three cases—is carbon. They must all be carbon because they all have an atomic number of 6. It is the atomic number and *only the atomic number* that determines the elemental identity of an atom. And yet these three carbon atoms are not identical because atom (b) has an extra neutron and atom (c) has two extra neutrons. These three atoms are examples of **isotopes**. Isotopes are different versions of the same element (same atomic number) that contain different numbers of neutrons in their nuclei (different mass numbers). Because they are the same element, they have identical chemical properties (there are some subtle differences that are beyond the scope of this book).

Chemists represent the different isotopes of an element by showing both the mass number and the atomic number along with the symbol for the element. The mass number is written as a superscript to the left of the symbol, and the atomic number is written as a subscript:

Mass number → $^{12}_{6}C$ ← Atomic number

Very often, the atomic number is omitted from this notation. Thus, the three isotopes of carbon can be written as ^{12}C, ^{13}C, and ^{14}C. Because these are all carbon atoms, the "missing" atomic number is, by definition, known to be 6. These isotopes can also be written as carbon-12, carbon-13, and carbon-14. When you are talking about an isotope, say either the letter or the name of the element plus the mass number. The isotope ^{12}C, for example, is pronounced either "c-twelve" or "carbon-twelve."

Relative Abundances of Carbon Isotopes on Earth

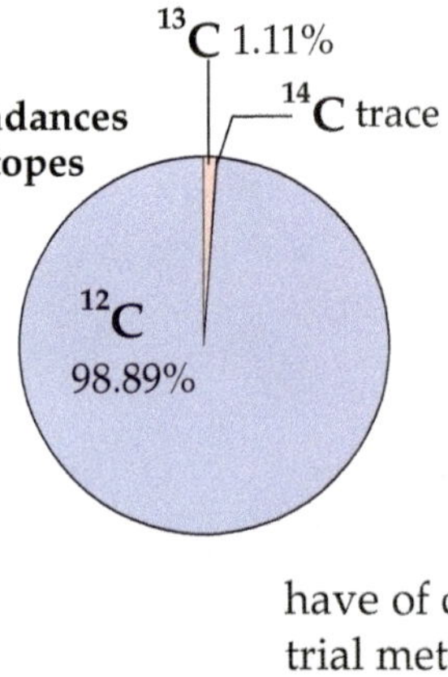

All three of these carbon isotopes exist in nature, although not in equal abundance. On this planet, 98.89% of all the carbon is ^{12}C, 1.11% is ^{13}C, and only a trace is ^{14}C. The relative abundances of these isotopes are most likely different on other planets. Analysis of the percentage of each isotope in a rock is one way chemists have of determining whether the rock is terrestrial or from some extraterrestrial meteorite.

Nearly all the elements exist as two or more naturally occurring isotopes. Even the simplest of the elements, hydrogen, comes in three isotopic forms. Although they are really all hydrogen atoms (because they all have one proton in their nucleus), they are given different names, as shown here. (This multiple naming is not done for isotopes of any other element.)

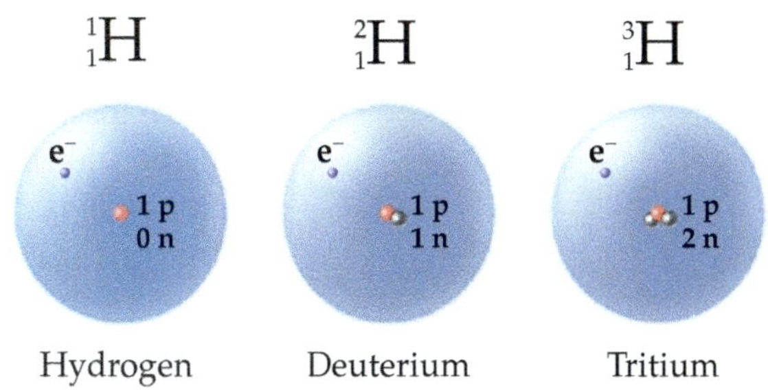

While isotopes of an element have essentially identical chemical properties, there is a single important difference between them. Recall that neutrons and protons carry most of an atom's mass. Because isotopes differ in the numbers of neutrons they contain, isotopes have different masses. Deuterium (^{2}H) is about twice as heavy as ^{1}H, for instance, and tritium (^{3}H) is heavier than deuterium. For this reason, ^{2}H is sometimes referred to as *heavy hydrogen,* and D_2O (the symbol D is used for a deuterium atom) is called *heavy water.*

Together, the mass number and atomic number tell you quite a bit about an atom, but what if you are given only one of these numbers? Try the following WorkPatch:

WORKPATCH 3.5 Suppose you are told that an atom has a mass number of 16. Is this enough information to identify the element? Why? ●

If you are not sure of the answer, think about what you need to know in order to determine an atom's identity. Now suppose you are given additional information, as in the next WorkPatch.

WORKPATCH 3.6 Suppose you are told that an atom with a mass number of 16 contains 9 neutrons. Can the atom be identified now? Which element is this? ●

This time you should be able to determine the atom's elemental identity (check the answer at the end of the chapter). The moral of this story is that to identify an atom, you need to know its atomic number, and you can determine this number in a variety of ways provided sufficient information is given.

Practice Problems

3.7 Give the full atomic symbol for an atom that has 16 neutrons and an atomic number of 15.

Answer: The atomic number 15 identifies the element as phosphorus (P). The mass number is 15 + 16 = 31. The full atomic symbol is $^{31}_{15}P$.

3.8 Bromine (Br) has two abundant isotopes, one with 44 neutrons and the other with 46 neutrons. Give the full atomic symbols for both isotopes.

3.9 How many electrons does a neutral atom of each isotope in Practice Problem 3.8 have?

3.10 Fill in this table:

	$^{14}_{7}N$	$^{24}_{12}Mg$	$^{23}_{11}Na$	$^{59}_{26}Fe$
Mass number	?	?	?	?
Atomic number	?	?	?	?
Number of protons	?	?	?	?
Number of neutrons	?	?	?	?
Number of electrons	?	?	?	?

Atomic Mass

We have looked closely at atomic number and mass number, but neither tells us the actual mass of an atom. For example, although the mass number of tritium ($^{3}_{1}H$) is 3, the actual mass of this atom is not 3 (not 3 g, not 3 lb, not 3 anything). So, what is the actual mass of a tritium atom? The actual mass of any atom is formally called its **atomic mass.**

To begin our discussion, let's look at the atomic masses for some isotopes of hydrogen, carbon, and magnesium. The atomic masses are given in *atomic mass units* (abbreviation amu).

Isotope	Atomic mass (amu)	Isotope	Atomic mass (amu)
^{1}H	1.007 83	^{12}C	12
^{2}H	2.014 10	^{13}C	13.003 35
^{3}H	3.016 05	^{24}Mg	23.985 04

An **atomic mass unit** (also known as a *dalton,* Da) is defined as exactly one-twelfth the mass of a $^{12}_{6}C$ atom and is equal to $1.660\ 54 \times 10^{-24}$ g:

$$1 \text{ amu} = \frac{1}{12} \text{ the mass of } ^{12}_{6}\text{C atom} = 1.660\ 54 \times 10^{-24} \text{ g}$$

Do you notice something about the atomic masses? Look closely. For all isotopes except ^{12}C, the atomic mass does not equal the superscript mass number, although they are close. Only ^{12}C has a mass in amu exactly equal to its mass number (this is due to a universal agreement by all chemists worldwide).

Because the atomic mass unit is defined this way, atomic masses are often called *relative atomic masses* (relative as in "compared to"). In other words, you can think of atomic masses as telling you how massive an atom is compared to a ^{12}C atom. For example, one ^{1}H atom has an atomic mass of 1.007 83 amu. This means that an ^{1}H atom is 1.007 83 ÷ 12, or roughly one-twelfth, as massive as a ^{12}C atom. One ^{24}Mg atom has an atomic mass of 23.985 04 amu, which means that a ^{24}Mg atom is 23.985 04 ÷ 12, or roughly twice, as massive as a ^{12}C atom.

Of course, because you know how many grams 1 amu equals, you can calculate the mass of an atom in grams. To see this, let's calculate the mass in grams of one ^{12}C atom, *the only atom* whose mass number and atomic mass are equal:

$$12 \text{ amu} \times \frac{1.660\ 54 \times 10^{-24} \text{ g}}{\text{amu}} = 1.992\ 65 \times 10^{-23} \text{ g}$$

WORKPATCH 3.7 An atom is determined to be 4.015 times more massive than ^{12}C. What is the atomic mass of this atom? ●

If you took the atomic mass of ^{12}C (12 amu) and multiplied it by something, then you are understanding the "relative" aspect of the atomic mass system.

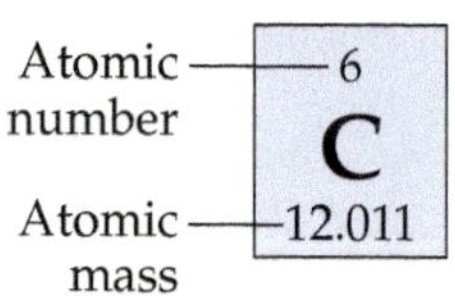

We have one more topic to cover before we end our discussion of atomic mass. We said that an atom of the ^{12}C isotope has an atomic mass of exactly 12 amu by universal agreement. But if you take a look at the periodic table, you will see that carbon's atomic mass is listed as 12.011, not 12. This atomic mass takes into account the fact that not all carbon atoms in nature are ^{12}C atoms. As we have seen, carbon consists predominantly of two isotopes, ^{12}C (98.89% natural abundance) and ^{13}C (1.11% natural abundance).

Any naturally occurring sample of carbon atoms will contain these two isotopes in these abundances. The atomic mass of carbon reported in the periodic table is called a *weighted average* of the atomic masses of these two isotopes. Each isotope's atomic mass is multiplied by its percent abundance (in decimal form, meaning the percent divided by 100), and then the results are summed:

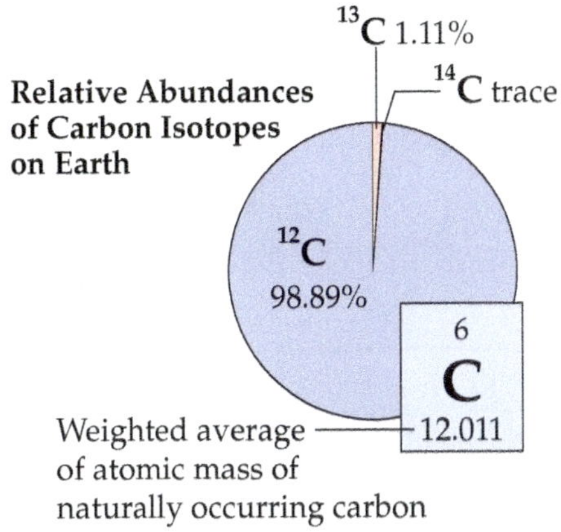

$$\text{Weighted average atomic mass of naturally occurring carbon} = \overbrace{(12\ \text{amu})(0.9889)}^{^{12}_{6}\text{C portion}} + \overbrace{(13.003\,35\ \text{amu})(0.0111)}^{^{13}_{6}\text{C portion}} = 12.01\ \text{amu}$$

(The ^{14}C isotope is so scarce that it can be ignored in this calculation.)

All the atomic masses listed in the periodic table of the elements are calculated in this way and are thus average masses for the naturally occurring mixture of isotopes. If you want the atomic mass of an individual isotope (other than ^{12}C), you need to consult a reference book such as the *Handbook of Chemistry and Physics* (CRC Press).

Practice Problems

3.11 Iron (Fe) and silicon (Si) are common elements in Earth's crust. How much more massive is iron than silicon?

Answer: Just take a ratio of the atomic masses from the periodic table:

$$\frac{\text{Fe}\ \ 55.845}{\text{Si}\ \ 28.086} = 1.9884$$

So, iron is roughly twice as massive as silicon.

3.12 What is the mass in grams of a billion billion (10^{18}) atoms of ^{24}Mg? The atomic mass of ^{24}Mg is 23.9850 amu.

3.13 Chlorine exists as two isotopes in nature, $^{35}_{17}Cl$ (atomic mass 34.969 amu, abundance 75.77%) and $^{37}_{17}Cl$ (atomic mass 36.966 amu).

(a) What is the percent abundance of the $^{37}_{17}Cl$ isotope?
(b) Calculate the weighted average of the atomic mass of naturally occurring chlorine.
(c) How many times more massive is $^{37}_{17}Cl$ than $^{35}_{17}Cl$?

3.5 The Law of Mendeleev—Chemical Periodicity

Chemists kept very busy during the nineteenth century. A major effort was under way to isolate and characterize all of the elements. Chemists throughout the world set out to decompose various chemical compounds into their component elements in order to determine the properties of these fundamental chemical building blocks. By 1860, as a result of these efforts, nearly 70 of the 118 elements we know about today had been isolated and studied. In the thousands of different chemical compounds and mixtures that chemists decomposed, each with its own unique physical and chemical properties, only these 70 elements were found. This represented a great simplification for the science of chemistry. Any object in the universe could now, at least in principle, be understood in terms of the relatively few elements that made it up.

As elements were discovered and their properties examined, it became necessary to organize the data in some useful way that would help to make sense of them all. One of the greatest advances toward this goal was made in the mid-1800s by a Russian chemist named Dmitri Mendeleev. He listed the elements and their properties on individual cards and then experimented with different arrangements of these cards to look for any patterns. The breakthrough came when he arranged the elements in order of increasing atomic mass (using the atomic mass values known at the time):

First 20 Elements Arranged from Least Massive to Most Massive

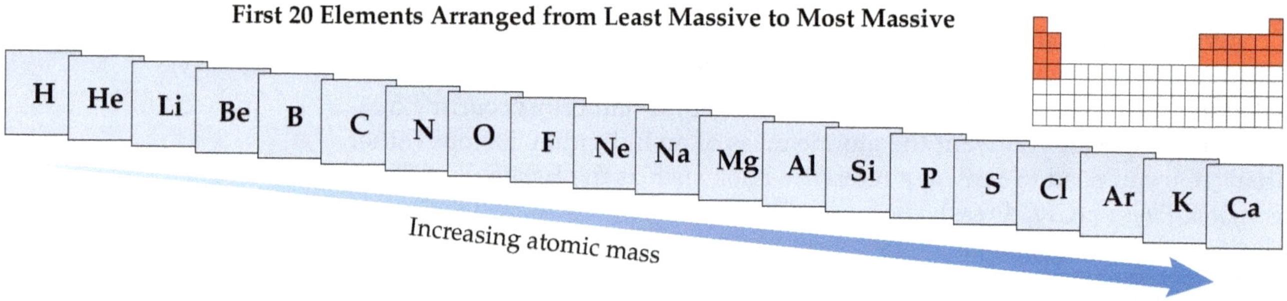

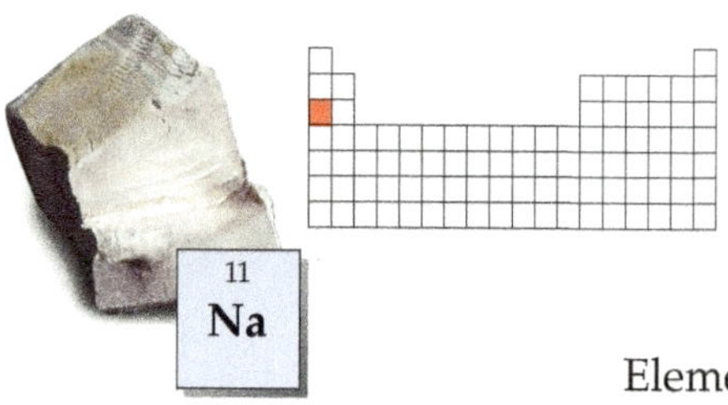

[We have inserted the elements helium (He), neon (Ne), and argon (Ar), which were completely unknown in Mendeleev's time. Mendeleev had other elements in these positions.] With the elements arranged this way, Mendeleev noted that their chemical properties repeated in a regular way. For example, consider the properties of the element sodium (Na) shown at right.

Elemental sodium is too reactive to be found in nature. However, chemists managed to isolate pure sodium from its compounds. It is a soft, silver-colored metal with a low density and a low melting point (for a metal). Like most other metals, it is a good conductor of electricity. It is also highly reactive, which can be readily demonstrated by dropping a piece of sodium metal in water. Sodium reacts violently with the water, producing a flammable hydrogen gas, which often ignites. In addition, if a plant dye called litmus is added to the water after the reaction has taken place, the water turns dark blue. An analysis of the reaction products shows that the sodium metal reacted with water to form NaOH, a compound known to make litmus turn blue.

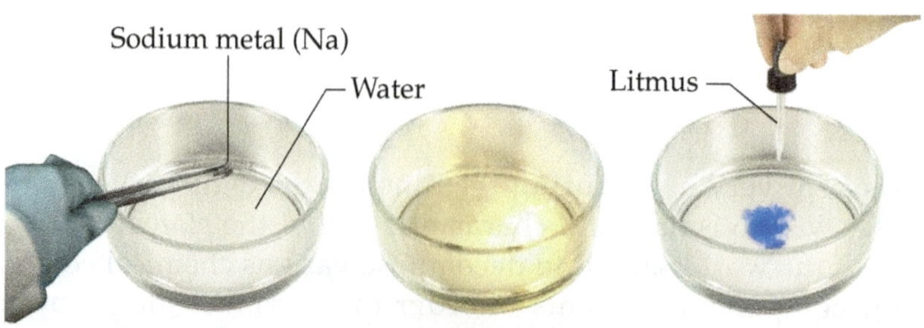

$$2\,Na + 2\,H_2O \longrightarrow 2\,NaOH + H_2$$

Like most other chemists of the time, Mendeleev knew of all this about sodium, and by itself this knowledge was nothing to get excited about. But when he examined his arrangement of cards, looking for elements with chemical properties similar to those of sodium, he noticed something interesting. Eight elements to the left and right of sodium in his list were elements with chemical and physical properties almost identical to those of sodium. Eight elements to the right was potassium (K), and eight elements to the left was lithium (Li):

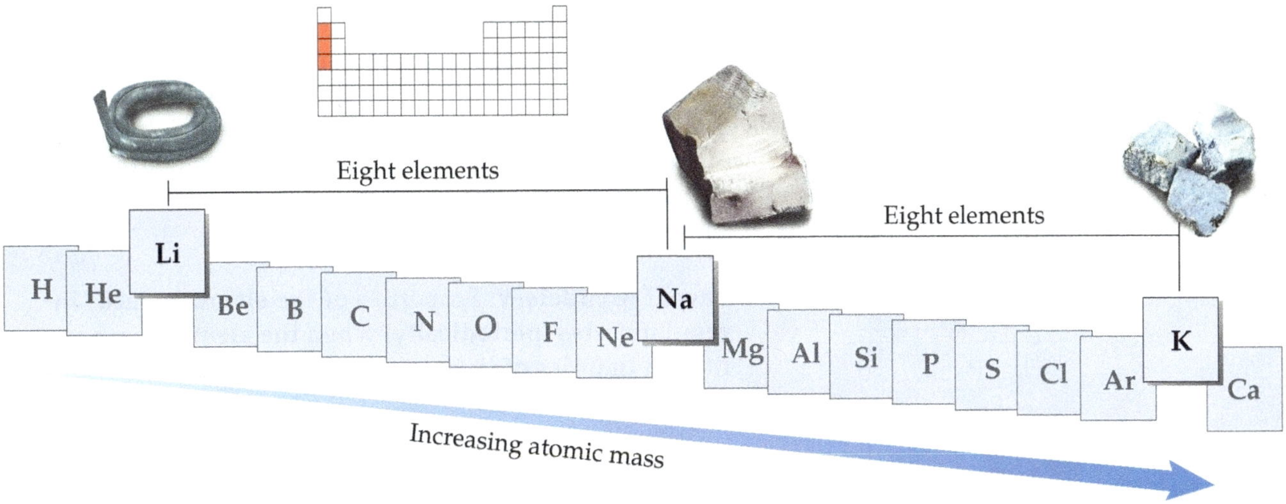

Both of these elements react with oxygen to form oxides (Li_2O and K_2O) that have chemical formulas similar to that of sodium oxide (Na_2O). Both react with water to form hydroxides (LiOH and KOH), just as sodium does (NaOH). All three are silver-colored metals that are too reactive to exist in their elemental form in nature. All are good electrical conductors; all three react vigorously with water, releasing flammable hydrogen gas; and all three produce a blue solution with litmus.

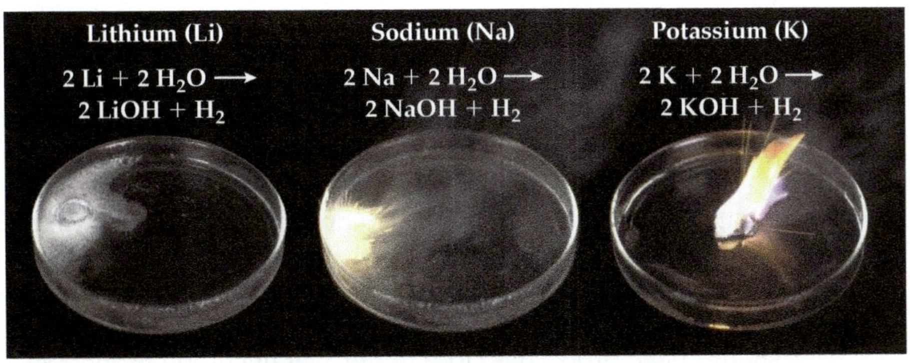

Was this just coincidence? Hardly! Mendeleev observed this same pattern of 8 for other elements in his uniquely arranged list. Magnesium (Mg) reacts with oxygen to give a 1:1 oxide having the chemical formula MgO. Eight elements away in either direction was another element that gave an oxide having a similar formula (beryllium oxide, BeO, and calcium oxide, CaO).

Also, as we now know, neon (Ne) refuses to react with oxygen at all. Sure enough, eight elements away in either direction was an element—helium (He) and argon (Ar)—that also refused to react with oxygen. There seems to be something almost magical about the number 8. This pattern was sometimes referred to as the *law of octaves*.

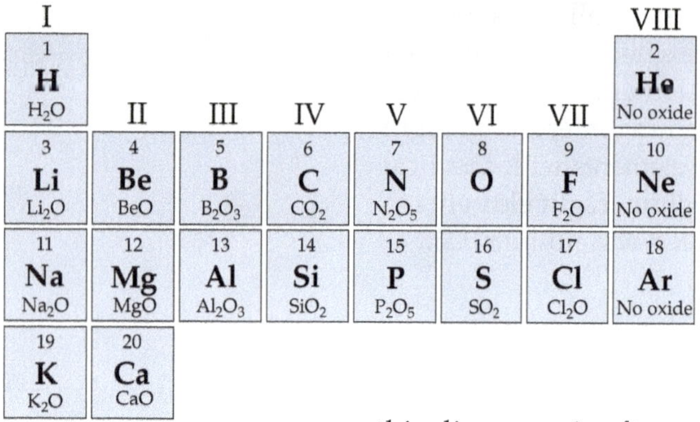

When Mendeleev rearranged his cards to stack all elements having similar chemical properties in one column, he got eight columns. No doubt about it. There had to be something special about this number 8. (While the elements of column 8 were unknown at the time, Mendeleev's table did have eight columns.)

This repeating behavior in the chemical properties of the elements is called either **chemical periodicity** or **periodic behavior** (thus the name *periodic table*). In recognition of his efforts, this discovery is often referred to as the **law of Mendeleev**. This same relationship was discovered independently, and at about the same time, by the German chemist Lothar Meyer, but Mendeleev received most of the credit for the periodic table.

Law of Mendeleev: Properties of the elements recur in regular cycles (periodically) when the elements are arranged in order of increasing atomic mass.

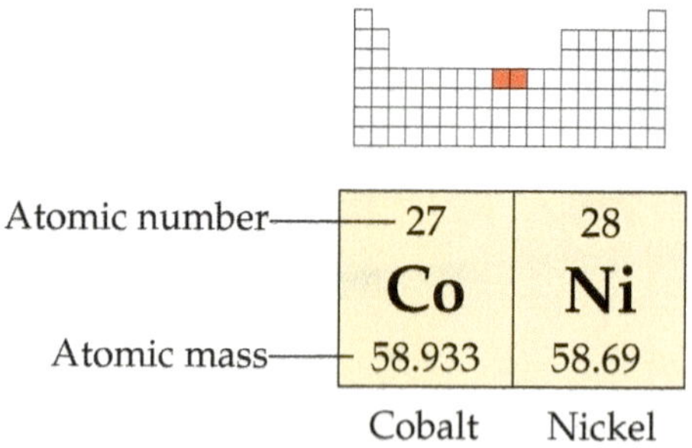

Cobalt comes before nickel in the modern periodic table. Mendeleev would have incorrectly put Co after Ni because the atomic mass of Co is greater than that of Ni.

Mendeleev was fortunate that his periodic table worked as well as it did. The modern periodic table is arranged according to increasing atomic number, not atomic mass. Fortunately, there are only a few places where Mendeleev's table puts elements in an order different from the order found in the modern table.

It was later discovered that 8 was not the only magic number with respect to periodic behavior. In portions of the periodic table, properties repeat every 18 or 32 elements, as we shall soon see.

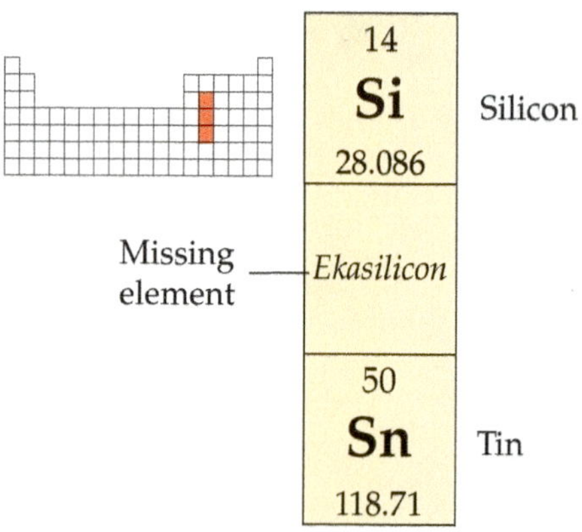

When Mendeleev began his table, only 70 elements were known. Indeed, when he arranged the elements into columns having similar chemical properties, there were some holes in his table. For example, the column containing the element carbon (C) also contained the elements silicon (Si) and tin (Sn), but there was a hole where an element with a mass somewhere between that of Si and Sn should be.

With his newly constructed table, Mendeleev was able to do what no chemist before him could do—predict the physical and chemical properties of elements that had yet to be discovered! This was quite a trick. Mendeleev reasoned that the missing element, which he called *ekasilicon* (*eka* meaning "next in order"), should have properties close to the average of the properties of the elements above and below it. This told chemists what properties to look for and led to the rapid discovery of the missing element. Today we call ekasilicon (represented below by X) germanium, Ge. Look how close Mendeleev's predictions about germanium were:

	Predicted properties (average of Si and Sn)	**Observed properties (measured after discovery)**
Atomic mass	72 amu	72.61 amu
Density	5.5 g/cm^3	5.32 g/cm^3

(*continued*)

	Predicted properties (average of Si and Sn)	Observed properties (measured after discovery)
Melting point	825 °C	938 °C
Oxide formula	XO_2	GeO_2
Density of oxide	$4.7\ g/cm^3$	$4.70\ g/cm^3$
Chloride formula	XCl_4	$GeCl_4$
Boiling point of chloride	100 °C	86 °C

WORKPATCH 3.8 Life on our planet is based on the element carbon, but science fiction writers often like to conjure up silicon-based life-forms. Why have these writers picked on the element Si to create alien life? ●

Mendeleev's insights greatly advanced our understanding of the chemistry of the elements, but they also presented a new challenge. Any chemically useful model of the atom would have to account for the existence of periodic chemical properties in the elements.

Rutherford's model did not even attempt to address the chemical behavior of the elements, let alone their periodic behavior, and it did nothing to explain that nasty classical physics problem of why electrons don't spiral into the nucleus. Modifying Rutherford's model was not going to be an easy task. The solution was to come with the new quantum physics and the resulting modern (quantum) model of the atom, which we'll see in the next chapter. In the rest of this chapter, we'll take a closer look at the modern periodic table, and then examine some of the clues that pointed the way to the quantum model of the atom.

3.6 The Modern Periodic Table

The **periodic table** is elegant in its simplicity, and yet it says so much. Its neat rows and columns underlie a detailed understanding of the subatomic structure and chemical properties of the elements. Each vertical column is called a **group**, and each horizontal row is called a **period**. Both the groups and the periods are numbered. There is no controversy on how to number the periods; they are numbered 1 through 7. There are, however, a few different ways both to number and to name the groups, as shown here. They can be numbered either with roman numerals or with Arabic numbers, and the groups colored violet and pale green have alternate names.

When the groups are numbered with roman numerals, each numeral is followed by either A or B. The large blocks of color in the table on the next page indicate that the periodic table separates the elements into three broad classes: Violet groups IA–VIIIA are called either the **representative elements** or the **main-group elements**. Tan groups IB–VIIIB are known as **transition metals**. Finally, the elements in the upper pale-green row at the bottom are called the **lanthanides** (or **rare earths**), and the elements in the lower pale-green row are called the **actinides**. The symbol of each element is accompanied by its atomic number above and its average atomic mass below.

The representative, or main-group, elements are the ones on which much of early chemistry was based. They also show the strongest periodic

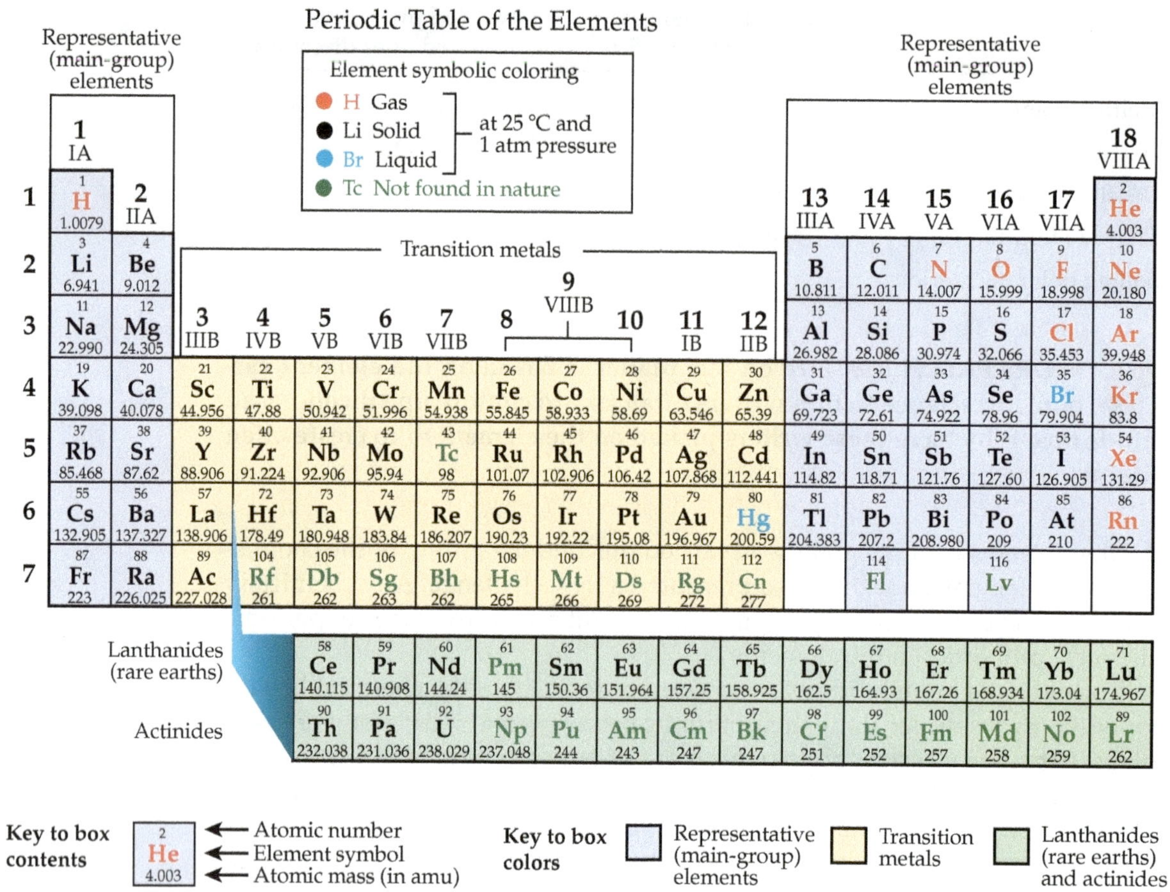

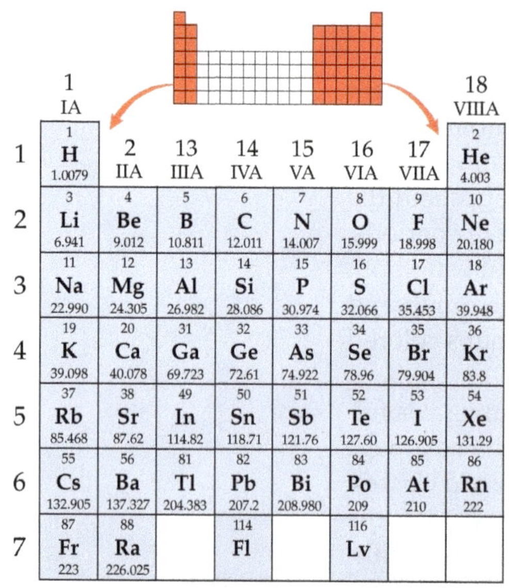

A periodic table showing only the representative elements is eight columns wide. Chemical periodicity occurs every eight elements.

relationships in the table. When applied only to the representative elements, the law of octaves works magnificently. Picking any element and then moving away eight elements in either direction brings you to another element whose properties are similar to those of the element you started with.

Look again at the entire periodic table, and pay close attention to how the groups are numbered. At first glance, the roman-numeral numbering system looks a bit bizarre. There is a big block of B groups jammed in between the A groups. Even worse, the B groups start with number III, go to VIII (three columns are assigned this number), and then end with I and II. There are good historical reasons for this strange numbering, but its awkwardness is what led IUPAC (the International Union of Pure and Applied Chemistry) to adopt the more straightforward 1-through-18 sequence which is also shown.

Our periodic table includes useful information about the phase (solid, liquid, or gas) in which the pure element exists under standard conditions (25 °C and 1 atm pressure).

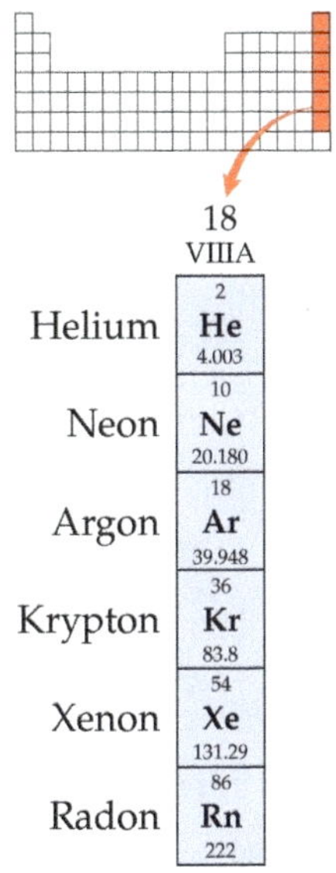

Now we shall take a brief tour of the periodic table, starting with the representative elements (groups IA–VIIIA). These groups contain solid, liquid, and gaseous elements. In fact, all the elements that exist as gases at standard temperature (25 °C) and standard pressure (1 atm) are representative elements, including the two that make up most of the air we breathe, nitrogen and oxygen.

The first representative elements we shall examine are those of group VIIIA (18). This is a very special column of the periodic table. It is the only group whose members are all gases at standard temperature and pressure. In addition, all the elements in this group are extremely unreactive. For this reason, the group VIIIA (18) elements are sometimes called the *inert gases*, and for some time chemists believed that these elements were completely chemically unreactive. In 1962, however, the British chemist Neil Bartlett challenged this assumption and prepared the first compound of the gas xenon (Xe). Today we refer to this group as either the **noble gases** or the **rare gases**. With the exception of helium, they are all found in trace amounts in the atmosphere. Earth's gravitational pull is not strong enough to keep any helium that gets into the atmosphere from escaping into space, and as a result, all naturally occurring helium comes from deposits trapped underground.

Radon (Rn), the only noble gas that is radioactive, has received some bad press lately. It has been discovered seeping into basements from naturally occurring underground sources. Sustained exposure is thought to be responsible for an increase in the incidence of lung cancer, and many people have purchased radon detectors for their basements.

As we continue to examine the representative elements, we find yet another way to divide them: They can be classified as *metals*, *nonmetals*, or *metalloids*, which have properties intermediate between those of metals and those of nonmetals. (*Semimetals* is an alternate name for metalloids.) This division is indicated in the following illustration. The red stair-step line that starts at boron, B, separates the metals from the nonmetals. The elements bordering this line (except for aluminum, Al, and polonium, Po) are the metalloids. Of the 44 naturally occurring representative elements, 7 are metalloids, 17 are nonmetals (including the noble gases), and the remaining 20 are metals.

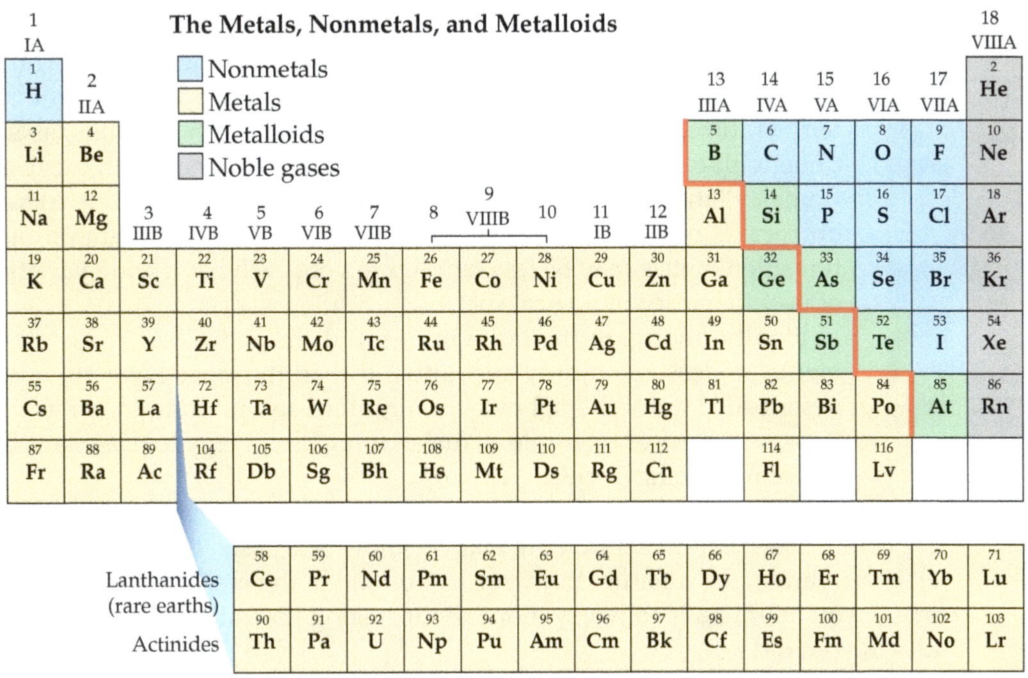

While we have not yet formally defined the terms *metal* and *nonmetal*, you probably have a pretty good feel for them based on everyday experience. **Metals** tend to be shiny solids that are bendable and malleable (which means they can be pounded into very thin sheets) and conduct heat and electricity well. **Nonmetals** tend to be brittle and do not conduct heat or electricity well. (They are *insulators*, the opposite of conductors.) Lead (Pb), a representative element in group IVA (14) once used for plumbing, is an obvious metal. Diamond, a pure form of carbon (C), is a nonmetal. For now, these everyday "definitions" of metal and nonmetal will suffice. Later, we'll give more chemically accurate definitions that relate to the tendency of these elements to gain or lose electrons when forming chemical compounds.

Representative metals such as aluminum (Al) and tin (Sn) are familiar to most people. You've certainly come across aluminum foil and tin cans (which are actually made from steel coated with tin). The metals of groups IA (1) and IIA (2) are less familiar because they are much too reactive to be found in nature in their elemental forms. However, as parts of compounds, some of them are also quite familiar. The compound sodium chloride (table salt) has the group IA (1) metal sodium (Na) in it. Many common antacids, such as Milk of Magnesia and Tums, contain compounds of the group IIA (2) metals magnesium (Mg) and calcium (Ca).

The **metalloids**, or **semimetals**, can act, depending on circumstances, like either a metal or a nonmetal. The elements silicon (Si), germanium (Ge), and arsenic (As) are good examples of elements that demonstrate classic metalloid character. They neither conduct electricity as well as metals nor insulate electrically as well as nonmetals. They are somewhere in between and are thus called *semiconductors*. These elements are used to make the sophisticated electronic chips that are the "brains" of computers. Indeed, the entire electronics industry owes its existence to the metalloid properties of these elements.

That completes our tour of the representative elements, and so let's move on to the transition metals. These elements form a bridge between the two parts of the representative elements. Perhaps, the most familiar transition metal is iron (Fe), the major component of steel, used in everything from bicycles to skyscrapers. Others, such as nickel (Ni), silver (Ag), gold (Au), copper (Cu), and zinc (Zn), are also familiar to most people. Mercury (Hg), commonly used in thermometers, is the only transition metal that is a liquid at room temperature. Chromium (Cr) is used for plating other metals, as is done with car bumpers. Titanium (Ti) is used to strengthen bicycle frames. Platinum (Pt), used in pollution-control catalytic converters on cars, is even more expensive than gold. While the remaining transition metals may be less familiar to you, each has a wide variety of uses in our technological society.

Finally, there are the lanthanides and the actinides. These elements are less commonly encountered than those we have already described. Two of them, however, uranium and plutonium, are well known because of their role as fuels for nuclear energy and as materials for producing nuclear weapons.

Of the 118 known elements, approximately 75% are metals and only 16% are nonmetals! Nevertheless, all life that we know of is based on a nonmetallic element, carbon. Had nature built you out of one of the majority elements, a metal, you might have to worry more about rusting than about getting wrinkles in your old age.

Practice Problems

3.14 How many groups constitute the representative (main-group) elements?
Answer: Eight

3.15 Does the stair-step boundary line that separates metals from nonmetals in the periodic table cross into the transition-metal portion of the table?

3.16 What do the elements in a group of the periodic table have in common with one another?

Mendeleev's arrangement of elements by mass resulted in the elements in a particular group (column) having similar chemical properties. Exactly why it worked out this way was not understood at the time. Nevertheless, these similarities invited giving each group a special name. Of the eight representative groups, five have commonly used names, the first two and the last three. When metals from either group IA (1) or group IIA (2) are placed in water, they react with it to make the water alkaline (basic), hence the names **alkali metals** for group IA (1) and **alkaline earth metals** for IIA (2). The group VIA (16) elements, which include oxygen and sulfur, are called the **chalcogens**. The name comes from the Greek words *chalkos* for "copper" and *genes* for "born." Most copper-containing minerals contain oxygen or sulfur. The group VIIA (17) elements, which include chlorine, bromine, and iodine, are called the **halogens** (*halos* is Greek for "salt"). Many salts have a halogen in them, such as sodium chloride. Group VIIIA (18) elements are called either noble gases or rare gases, as mentioned earlier. You should commit these group names to memory.

1 IA (Alkali metals)	2 IIA (Alkaline earth metals)	3 IIIB	4 IVB	5 VB	6 VIB	7 VIIB	8 VIIIB	9 VIIIB	10 VIIIB	11 IB	12 IIB	13 IIIA	14 IVA	15 VA	16 VIA (Chalcogens)	17 VIIA (Halogens)	18 VIIIA (Noble gases)
1 H 1.0079																	2 He 4.003
3 Li 6.941	4 Be 9.012											5 B 10.811	6 C 12.011	7 N 14.007	8 O 15.999	9 F 18.998	10 Ne 20.180
11 Na 22.990	12 Mg 24.305											13 Al 26.982	14 Si 28.086	15 P 30.974	16 S 32.066	17 Cl 35.453	18 Ar 39.948
19 K 39.098	20 Ca 40.078	21 Sc 44.956	22 Ti 47.88	23 V 50.942	24 Cr 51.996	25 Mn 54.938	26 Fe 55.847	27 Co 58.933	28 Ni 58.69	29 Cu 63.546	30 Zn 65.39	31 Ga 69.723	32 Ge 72.61	33 As 74.922	34 Se 78.96	35 Br 79.904	36 Kr 83.8
37 Rb 85.468	38 Sr 87.62	39 Y 88.906	40 Zr 91.224	41 Nb 92.906	42 Mo 95.94	43 Tc 98	44 Ru 101.07	45 Rh 102.906	46 Pd 106.42	47 Ag 107.868	48 Cd 112.411	49 In 114.82	50 Sn 118.71	51 Sb 121.76	52 Te 127.60	53 I 126.905	54 Xe 131.29
55 Cs 132.905	56 Ba 137.327	57 La 138.906	72 Hf 178.49	73 Ta 180.948	74 W 183.85	75 Re 186.207	76 Os 190.2	77 Ir 192.22	78 Pt 195.08	79 Au 196.967	80 Hg 200.59	81 Tl 204.383	82 Pb 207.2	83 Bi 208.980	84 Po 209	85 At 210	86 Rn 222
87 Fr 223	88 Ra 226.025	89 Ac 227.028	104 Rf 261	105 Db 262	106 Sg 263	107 Bh 262	108 Hs 265	109 Mt 266	110 Ds 269	111 Rg 272	112 Cn 277		114 Fl		116 Lv		

Given all we have said so far, you might wonder why hydrogen is included in group IA (1), the alkali metals. After all, hydrogen is not a metal. In fact, Mendeleev would not have put it in this group. We shall wait until the next chapter to explain why it is there. Suffice it to say that while hydrogen is shown in group IA (1), nobody thinks of it as an alkali metal.

As we have seen, the structure of the periodic table reflects what we have learned about the chemistry of the various elements. The law of octaves applies satisfactorily when you are dealing only with the representative elements. The chemical properties of these elements repeat every eight elements, and this part of the table contains eight columns to allow for this. But the law needs some modification if we are to include the rest of the elements in the table. If we continue to place elements into columns based primarily on their similar chemical behaviors, we end up with the form of the periodic table we know today, which has many more than eight columns.

The introduction of the transition metals into the table begins with row (period) 4 and causes the number of columns to swell to 18 (because there are 8 representative elements +10 transition metals per row). In this part of the table, therefore, chemical properties repeat every 18 elements and not every 8.

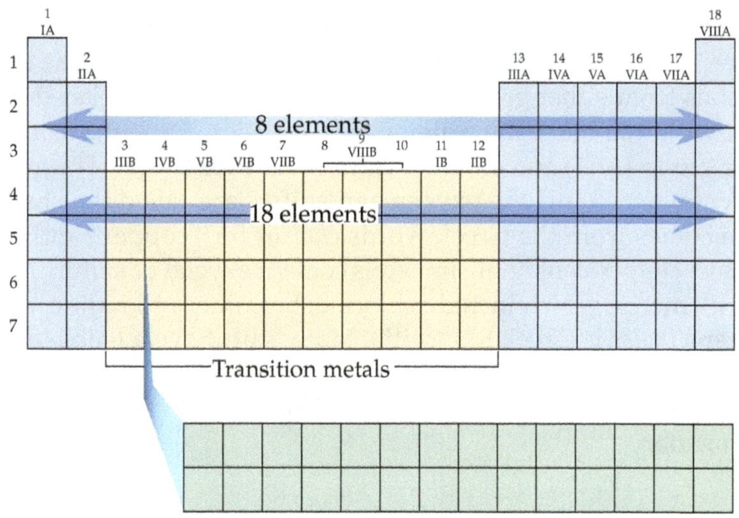

As we continue down the rows of the table, we come to periods 6 and 7, where we must now accommodate the lanthanide and actinide elements. So far, we have shown these elements as a separate block below the main part of the table. That's just a convenience to allow the table to fit better on the printed page. Really, the lanthanide and actinide elements should fit right into periods 6 and 7, like this:

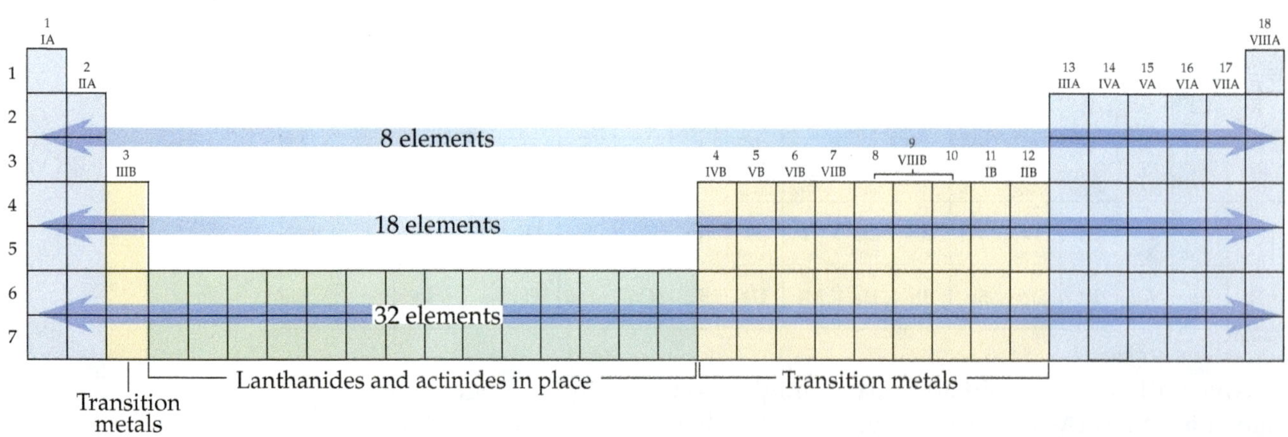

Thus, in periods 6 and 7, the table becomes 32 columns wide, and the chemical properties of the elements in these periods repeat with a cycle of 32 rather than 8 or 18. It's much easier to see how the periodicity changes from 18 to 32 at period 6 when you are looking at the periodic table with the lanthanide and actinide elements inserted.

Practice Problems

3.17 What are the period number and group number of the element that has atomic number 15?

Answer: The element that has atomic number 15 is phosphorus (P). It is in period 3 and group VA (15).

3.18 Which elements with atomic numbers greater than 40 can be expected to have chemical properties similar to those of the element bromine (Br)? What is the periodicity of these two elements?

3.19 What would the periodicity be if the transition metals, lanthanides, and actinides were removed from the periodic table?

3.7 An Introduction to Ions and the First Ionization Energy

An atom is electrically neutral because the number of negatively charged electrons outside the nucleus equals the number of positively charged protons inside the nucleus. An atom in which this electron–proton balance is not maintained has a net charge and is called an **ion**. The only practical way to create an ion is to add or remove electrons from a neutral atom. If you take a neutral silicon (Si) atom, for instance, and add one electron to it, the resulting ion has one more electron than it has protons and therefore has a net charge of −1. Such negatively charged ions are called **anions**. Likewise, if you remove one electron from a neutral silicon atom, the resulting ion will have one fewer electron than it has protons and a net positive charge of +1. Ions with a positive charge are called **cations**:

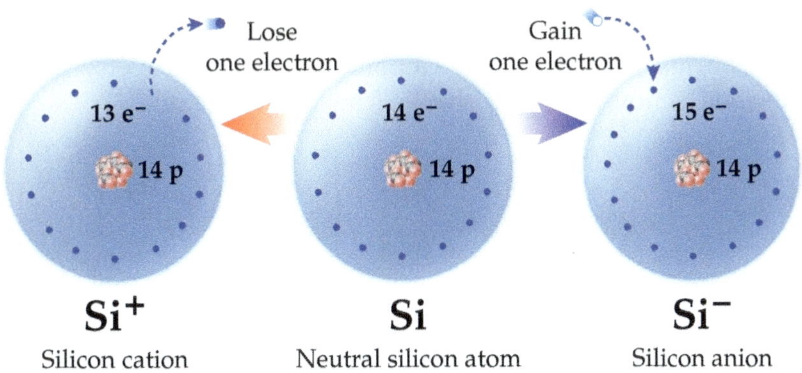

Now, what if we take another atom, say magnesium (Mg), and remove two electrons from the neutral atom? There will now be an excess of two protons, and we end up with a cation with a +2 charge. This ion is written Mg^{2+}. If we instead added two electrons to a magnesium atom (which for magnesium is actually very difficult to do), we would get an ion with a −2 charge, which is written Mg^{2-}. In both cases, the usual convention in writing ions is to give the numerical value of the charge before the + or − sign. In the case of ions with a single charge, the 1 is usually omitted and the ion is written simply, for example, Li^+ or Cl^-. An element symbol written without any charge (or with a zero, for example, Fe^0 or Cu^0) represents a neutral atom of that element.

S^{2-} Cl^-	C	Na^+ Mg^{2+}
Anions	Neutral atom (no charge indicated)	Cations

WORKPATCH 3.9 **Fill in the table:**

	$^{14}_{7}N^{3-}$	$^{24}_{12}Mg^{2+}$	$^{23}_{11}Na^{+}$	$^{56}_{26}Fe^{3+}$
Mass number	?	?	?	?
Atomic number	?	?	?	?
Number of protons	?	?	?	?
Number of neutrons	?	?	?	?
Number of electrons	?	?	?	?

This WorkPatch is meant to drive home an important point. The only way you can turn a (neutral) atom of an element into an ion of that element is by adding or removing electrons. You do not adjust the number of protons in the nucleus. If the number of protons were changed, the atom would not be the same element anymore.

We will have more to say about ions in the next chapter, but we want to finish this chapter by taking a closer look at the formation of a cation via the removal of an electron from a neutral atom. The **first ionization energy** is the minimum amount of energy it takes to completely remove an electron from a neutral atom, forming a cation with a +1 charge. Does it always take energy to remove an electron from a neutral atom? The answer is ***YES***! Remember, the protons in an atom's nucleus are positively charged, whereas electrons possess negative charge, and opposite electrical charges attract. An atom's electrons are "trapped" within the atom by this electrical (also called coulombic) attraction, and it will always take energy to overcome this attraction to remove them, just like it takes large amounts of energy to overcome gravity's attraction and remove the space shuttle from planet Earth.

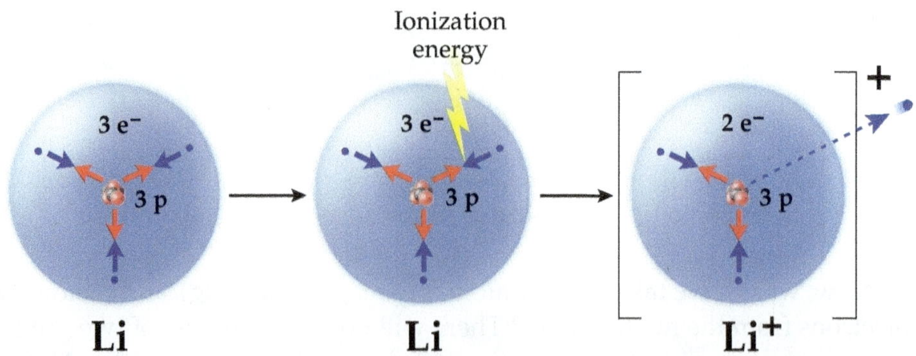

Electrons cannot leave the atom unless energy is supplied to overcome their attraction to the nucleus.

Ionization energy "kicks" an electron out of the atom.

The resulting cation has a +1 charge because it has one more proton than it has electrons.

We have seen that the chemical properties of the elements vary in a regular or periodic manner. Are there any other properties of the elements that vary in some repetitive way? Yes! One of them is the first ionization energy. Take a close look at the first ionization energies of the elements printed below each elemental symbol in the following periodic table:

First Ionization Energies

Representative (main-group) elements: groups 1–2 and 13–18. Transition metals: groups 3–12. First ionization energy (eV) is printed below each symbol.

Period	1 IA	2 IIA	3 IIIB	4 IVB	5 VB	6 VIB	7 VIIB	8	9 VIIIB	10	11 IB	12 IIB	13 IIIA	14 IVA	15 VA	16 VIA	17 VIIA	18 VIIIA
1	1 H 13.6																	2 He 24.6
2	3 Li 5.4	4 Be 9.3											5 B 8.3	6 C 11.3	7 N 14.5	8 O 13.6	9 F 17.4	10 Ne 21.6
3	11 Na 5.1	12 Mg 7.6											13 Al 6.0	14 Si 8.2	15 P 10.5	16 S 10.4	17 Cl 13.0	18 Ar 15.8
4	19 K 4.3	20 Ca 6.1	21 Sc 6.5	22 Ti 6.8	23 V 6.7	24 Cr 6.8	25 Mn 7.4	26 Fe 7.9	27 Co 7.9	28 Ni 7.6	29 Cu 7.7	30 Zn 9.4	31 Ga 6.0	32 Ge 7.9	33 As 9.8	34 Se 9.8	35 Br 11.8	36 Kr 14.0
5	37 Rb 4.2	38 Sr 5.7	39 Y 6.4	40 Zr 6.8	41 Nb 6.9	42 Mo 7.1	43 Tc 7.3	44 Ru 7.4	45 Rh 7.5	46 Pd 8.3	47 Ag 7.6	48 Cd 9.0	49 In 5.8	50 Sn 7.3	51 Sb 8.6	52 Te 9.0	53 I 10.5	54 Xe 12.1
6	55 Cs 3.9	56 Ba 5.2	57 La 5.6	72 Hf 6.7	73 Ta 7.9	74 W 8.0	75 Re 7.9	76 Os 8.7	77 Ir 9.1	78 Pt 9.0	79 Au 9.2	80 Hg 10.4	81 Tl 6.1	82 Pb 7.4	83 Bi 7.3	84 Po 8.4	85 At -	86 Rn 10.7
7	87 Fr -	88 Ra 5.3	89 Ac 5.2	104 Rf -	105 Db -	106 Sg -	107 Bh -	108 Hs -	109 Mt -	110 Ds -	111 Rg -	112 Cn -		114 Fl		116 Lv		

Lanthanides (rare earths)	58 Ce 5.5	59 Pr 5.4	60 Nd 5.5	61 Pm 5.6	62 Sm 5.6	63 Eu 5.7	64 Gd 6.2	65 Tb 5.9	66 Dy 5.9	67 Ho 6.0	68 Er 6.1	69 Tm 6.2	70 Yb 6.3	71 Lu 5.4
Actinides	90 Th 6.1	91 Pa 5.9	92 U 6.1	93 Np 6.2	94 Pu 6.1	95 Am 6.0	96 Cm 6.0	97 Bk 6.2	98 Cf 6.3	99 Es 6.4	100 Fm 6.5	101 Md 6.6	102 No 6.7	89 Lr -

The table lists the first ionization energies of the elements in units of electronvolts (eV), a unit of energy commonly used by physicists. Look at the table closely. Can you see any trends in the first ionization energies from left to right across any period? (*Hint*: Take a look at the figure below, which plots the first ionization energies of the elements versus increasing atomic number.)

Ionization Energy as a Function of Atomic Number

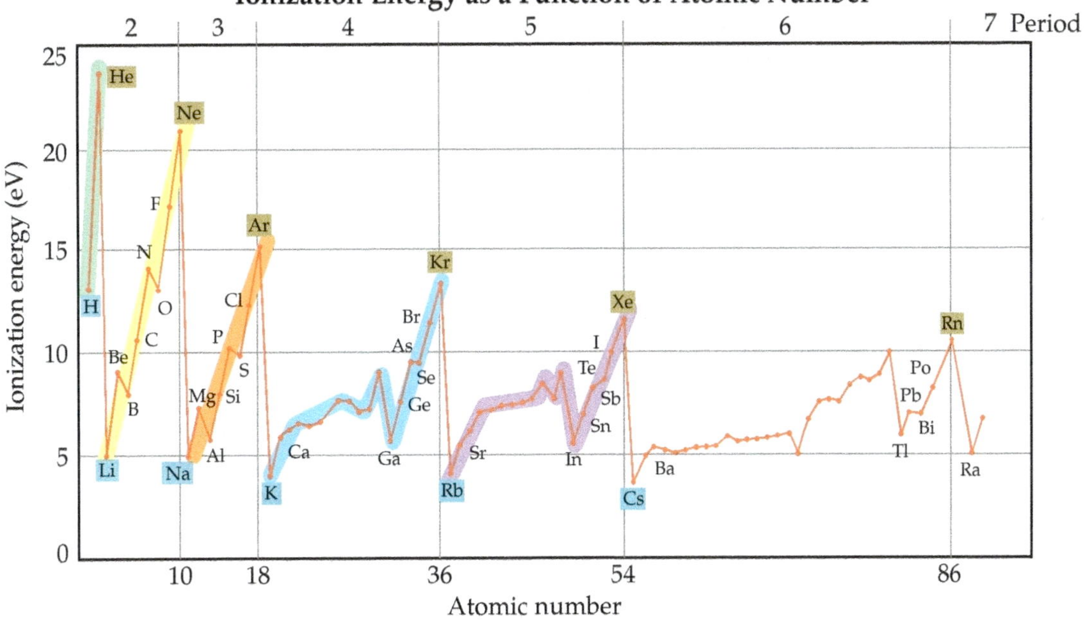

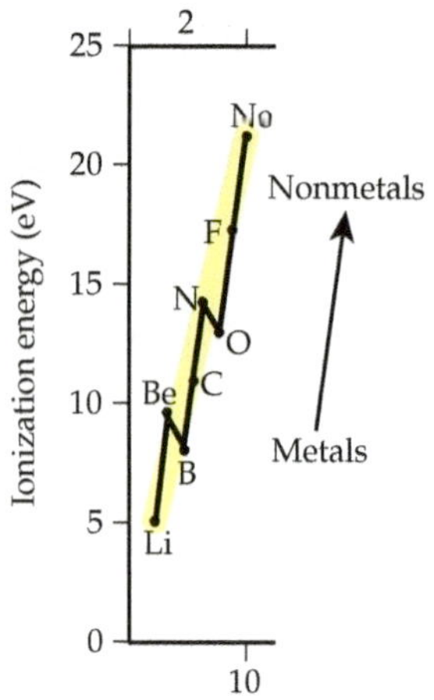

Each color represents a different period in the periodic table (green is the first period, yellow the second period, orange the third, and so on). As you climb up along each colored line, you are traversing the elements from left to right across that period, which means you are moving from metallic elements to nonmetallic elements. For example, consider climbing up along the yellow line (the second-period elements).

The plot makes obvious the periodic trend involving first ionization energy. As you move from left to right across a period, from metals to nonmetals, the first ionization energy increases (it becomes harder and harder to remove an electron from the neutral atom). When you go to the next period, this resets (the first ionization energy drops), but then the rise begins anew. What this teaches you is that no matter what period you are in, it doesn't take much energy to remove an electron from a metal atom, but it takes a whole lot of energy to remove an electron from a nonmetal. Indeed, it generally takes so much energy to remove an electron from a nonmetal that nonmetals typically gain electrons to become anions rather than lose electrons to become cations. This has led chemists to state the following generalization:

Metals like to lose electrons.
Nonmetals like to gain electrons.

Why this is true and what its implications are will be revealed to you in the next chapter, but keep it in mind. It is an important generalization.

Practice Problems

3.20 The alkali metals all react with water by giving up an electron to it. Based on their relative positions in the periodic table, which do you think would be more reactive with water—lithium or sodium?

Answer: Sodium. It has a lower first ionization energy, meaning it is easier for sodium to give up an electron. Because giving up an electron is how it reacts with water, this should make sodium the more reactive metal.

3.21 Arrange the atoms I, Zr, Ag, and Rb from smallest to largest first ionization energy. What does this say about metals versus nonmetals?

3.22 Give the full symbol for the atom or ion that has 26 protons and 30 neutrons in its nucleus and 23 electrons outside its nucleus. Also give the number of the group this element is in.

3.23 Give the full symbol for the atom or ion that has eight protons and eight neutrons in its nucleus and ten electrons outside its nucleus. Also give the name of the group this element is in.

Let's take stock of where chemistry was after Mendeleev's and Rutherford's contributions. Thanks to Mendeleev, chemists had a periodic table that summarized the periodic chemical behavior of the elements. Thanks to Rutherford, chemists knew the basics about atomic structure (a tiny nucleus that contained protons and accounted for most of the atom's mass; electrons somewhere outside the nucleus doing something). But Rutherford's model of the atom could not explain periodicity! An improved model that could explain periodicity was needed. Where

was the hint for such a model to be found? Believe it or not, the hint came from examining light. Not just any light, but light given off when atoms of elements in the gas state were provided with large amounts of energy. You are looking at such light every time you read a neon sign. Neon lights are glass tubes filled with neon gas atoms that give off a characteristic orange-red light when supplied with electrical energy. Bright yellow light is given off by sodium-vapor street lamps. Even a tube filled with hydrogen gas glows bluish-pink when electrified. Being of a curious nature, physicists passed this light through a prism. Passing light through a prism was nothing new. Sunlight (so-called white light) passed through a prism is separated, or *dispersed,* into its component colors. Drops of water in the atmosphere acting like prisms have given all of us a chance to see this dispersion in the form of a rainbow. A rainbow created from white light is called a *continuous spectrum* because the colors smoothly blend into one another without any breaks.

But look at what happened when physicists passed the light given off by a hydrogen or neon lamp through a prism. They saw separate lines of color (first detected for hydrogen by scientists in the mid-1800s). We must tell you that this was astounding! Never before had such a thing been observed. These are called either *discontinuous spectra* or *line spectra* because the colored lines do not blend smoothly into one another. Instead, there are regions of blackness between lines.

Line Spectra of Hydrogen and Neon

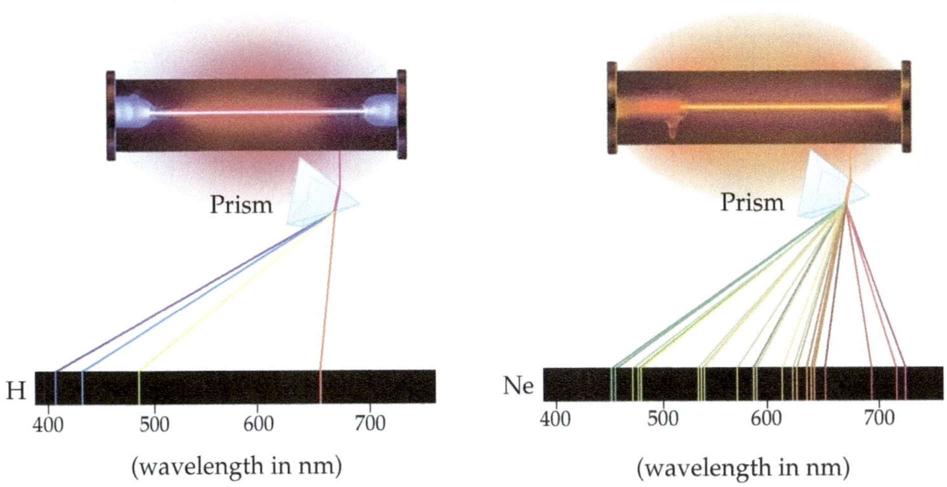

Line spectra were the hint for an improved model of the atom, one that could link periodicity to atomic structure. To understand this hint, we are going to have to take a much closer look at what light is, and so we'll begin the next chapter with light. To keep you interested, however, we shall give a little of the story away now. As it turns out, line spectra would lead to a new model of the atom so unusual, so bizarre, that even those who invented it would have trouble accepting it. Arguments would erupt, scientists would shake their heads in disbelief, and a lot of physics learned over the previous two centuries would be tossed out the window. The areas of chemistry and atomic physics were about to get very exciting.

+1 One More Thing

3.8 Isotopes, Mass Spectrometers, and Extraterrestrial Origins

As part of his atomic model, John Dalton stated that all atoms of a single element are identical to one another. Today we know that this is not true. Actually, we have known of exceptions to that aspect of Dalton's theory since 1932, when the neutron was discovered. Atoms can exist in slightly modified forms called isotopes. Let's take a closer look at isotopes. All of the elements in the periodic table exhibit isotopes. That is, every element can exist as a collection of atoms that have the same atomic number (number of protons in the nucleus) but contain a different number of neutrons (and thus have different mass numbers). In Section 3.4 we discussed the masses of the two principal carbon isotopes in atomic mass units (amu), where 1 amu equals the extremely tiny mass of $1.660\,54 \times 10^{-24}$ g.

	Mass	Abundance in nature
^{12}C	12 amu	98.89%
^{13}C	13.00335 amu	1.11%

Now let's face it. Even with the world's most sensitive balance, the mass of one ^{12}C atom or one ^{13}C atom would register as zero. So, how can anyone measure such a small mass? The answer is that a balance is not used. Instead, an instrument known as a mass spectrometer is used.

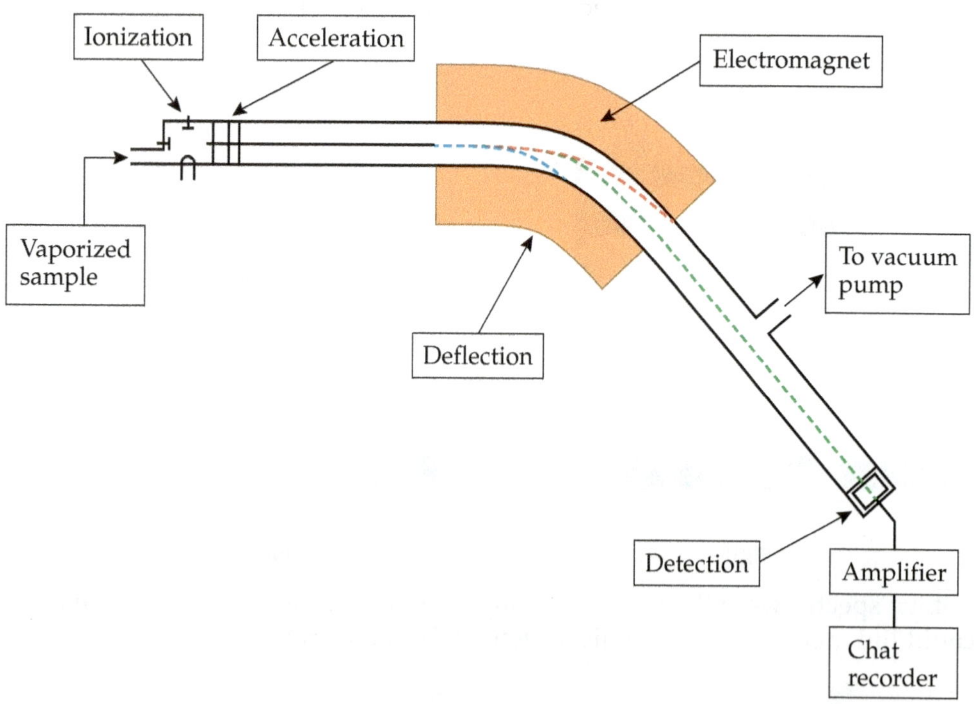

Rather than weigh an atom, a mass spectrometer imparts a charge on an atom (turns it into an ion) and then accelerates it toward a detector that can detect charged particles. However, the mass spectrometer is constructed such that, in order to reach the detector, the ion must be deflected from a straight flight path. An electromagnet is used to deflect the ion so it can reach the detector. By knowing how much voltage this process took, the mass of the ion can be calculated.

When a material is atomized and put through a mass spectrometer, the instrument can report not only the mass of each isotope of every element present but also its relative abundance. Interestingly, when zinc-bearing rocks from Earth are compared to those from the moon, it is found that the moon rocks are disproportionately rich in the lighter isotopes of zinc compared to terrestrial rocks. In other words, the percent abundances for the isotopes of zinc are different on the moon than they are on Earth. One explanation for this difference supports the theory that the moon formed from a collision with Earth billions of years ago that vaporized some of Earth, which then condensed to form the moon (lighter isotopes volatilize more easily than heavy ones, so vapor that condensed to form the moon was richer in the lighter isotopes of zinc). The *Curiosity* rover currently probing Mars has a mass spectrometer on board that can be used to analyze soil samples.

Have You Learned This?

Law of conservation of matter (p. 79)
Law of constant composition (p. 80)
Percent by mass (p. 80)
Dalton's atomic theory (p. 82)
Electron (p. 83)
Proton (p. 83)
Neutron (p. 83)
Alpha particle (p. 84)
Nucleus (p. 86)
Atomic number Z (p. 88)
Mass number A (p. 89)
Isotope (p. 90)
Atomic mass (p. 92)
Atomic mass unit (p. 92)
Chemical periodicity (p. 96)
Periodic behavior (p. 96)
Law of Mendeleev (p. 96)
Periodic table (p. 97)
Group (p. 97)
Period (p. 97)
Representative (main-group) element (p. 97)
Transition metal (p. 97)
Lanthanide (rare earth element) (p. 97)
Actinide (p. 97)
Noble gas (rare gas) (p. 99)
Metal (p. 100)
Nonmetal (p. 100)
Metalloid (semimetal) (p. 100)
Alkali metal (p. 101)
Alkaline earth metal (p. 101)
Chalcogen (p. 101)
Halogen (p. 101)
Ion (p. 103)
Anion (p. 103)
Cation (p. 103)
First ionization energy (p. 104)

Learning Outcomes

3.1	**Dalton's Atomic Theory**	Calculate the percent composition of each element in a compound. (25, 27, 28) Describe how Dalton explained formulas of compounds. (26, 29)
3.2	**Development of a Model for Atomic Structure**	Name and describe the subatomic particles in an atom. (31, 32)
3.3	**The Nucleus**	Describe Rutherford's alpha-particle experiment and what it proved. (33, 34, 37)

3.4	**The Structure of the Atom**	Explain what the atomic number and the mass number tell you about an atom. (35, 36, 38, 40, 41, 42) Explain the meaning of an atom's atomic mass and mass number. (44, 45, 46, 47, 50) Define *isotopes*. (38, 39, 52)
3.5	**The Law of Mendeleev—Chemical Periodicity**	Explain Mendeleev's discovery. (55, 56, 57)
3.6	**The Modern Periodic Table**	Identify groups, periods, metals, and nonmetals in the periodic table. (58, 59, 60, 61, 62, 63, 65, 66, 68)
3.7	**An Introduction to Ions and the First Ionization Energy**	Explain what cations and anions are and how to form them. (70, 71, 72, 73, 74, 75, 76, 77)
3.8	**Mass Spectrometry**	Understand what a mass spectrometer is and how it works. (144–147)

Skills to Know

Calculating a percentage

Often you must calculate what percentage one quantity is of another. What percent of your day is spent commuting? What percent yield can you expect from an investment? These questions all use the basic formula:

$$\text{Percent} = \frac{x}{y} \times 100$$

What percent of y is x?

Example: A sample of water consists of 4.435 g of hydrogen and 35.20 g of oxygen. What percent by mass of the sample is hydrogen?

$$\%\ \text{Hydrogen} = \frac{\text{Mass of hydrogen}}{\text{Mass of sample}} \times 100$$

$$= \frac{4.435\ \text{g H}}{4.435\ \text{g H} + 35.20\ \text{g O}} \times 100$$

$$= \frac{4.435\ \text{g H}}{39.64\ \text{g H}_2\text{O}} \times 100 = 11.19\%$$

Atomic number, mass number

1. The **atomic number Z** is the number of protons in an atom's nucleus. All atoms of a given element have the same atomic number.

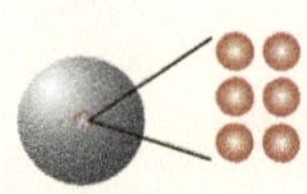

Six protons
Atomic number Z = 6

Carbon atom

2. The **mass number A** of an atom is the total number of protons plus neutrons in the nucleus. Because different isotopes of an element have different numbers of neutrons, each isotope has its own mass number.

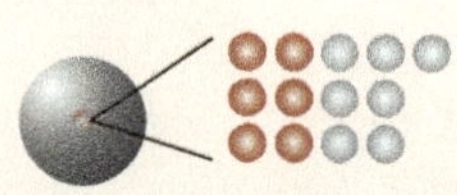

Six protons
Seven neutrons
Mass number = 13

Atom of carbon-13

3. An atom's mass number *is not* its atomic mass (except in the case of ^{12}C). However, the mass number is numerically close to the atomic mass because protons and neutrons account for most of an atom's mass, and have a mass of about 1 amu each.	Mass number $\neq$ Mass of atom! (except in the case of ^{12}C)

Atomic mass

1. Atomic mass is the mass of an atom. The atomic mass *of a specific isotope* is the exact mass of an atom of that isotope.

2. The atomic mass *of an element* reported in the periodic table takes into account the fact that each element consists of isotopes having slightly different masses. The atomic mass reported for an element is a weighted average of the masses of the element's naturally occurring isotopes. It is calculated by multiplying the mass of one atom of each isotope of the element by the relative natural abundance of that isotope.

Example: What is the atomic mass of carbon?

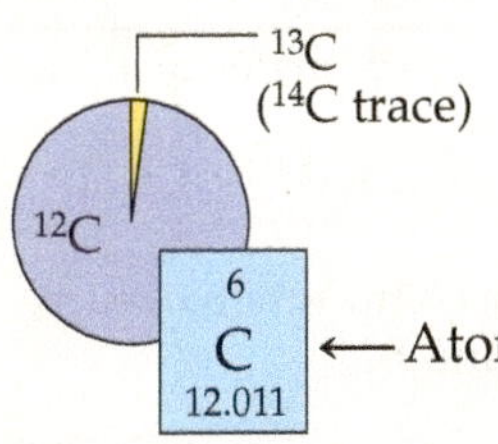

Carbon consists almost entirely of ^{12}C and ^{13}C (the isotope ^{14}C can be neglected). Here is how you calculate the atomic mass of carbon:

$$\longleftarrow \text{Atomic mass of C} = \left(\begin{array}{c}\text{Mass of}\\ ^{12}\text{C atom}\\ \text{in amu}\end{array} \times \frac{\%\text{ abundance of } ^{12}\text{C}}{100}\right) + \left(\begin{array}{c}\text{Mass of}\\ ^{13}\text{C atom}\\ \text{in amu}\end{array} \times \frac{\%\text{ abundance of } ^{13}\text{C}}{100}\right)$$

3. Atomic masses are reported in atomic mass units (amu), where 1 amu is exactly 1/12 the mass of an atom of ^{12}C, by definition. Thus, ^{12}C has an atomic mass of 12 amu. (Because atomic mass is defined relative to the mass of a ^{12}C atom, atomic mass is sometimes given without units, as *relative atomic mass*.)

Concept Questions

1. The law of conservation of matter says which of the following?
   (a) Matter is made of atoms.
   (b) In a chemical reaction, matter changes from reactants to products.
   (c) In a chemical reaction, matter is always created or destroyed.
   (d) In a chemical reaction, matter is neither created nor destroyed.

2. Rutherford's alpha-particle experiment proved which of the following?
   (a) Atoms are neutral.
   (b) Electrons have a negative charge.
   (c) An atom's nucleus is tiny and contains most of the atom's mass.
   (d) An atom's nucleus is large and contains most of the atom's mass.

3. An atom's mass number tells you
   (a) the mass of an atom.
   (b) how many protons are in the nucleus.
   (c) how many neutrons are in the nucleus.
   (d) how many (protons + neutrons) are in the nucleus.

4. An atom's atomic number tells you
   (a) how many electrons are in the neutral atom.
   (b) how many protons are in the neutral atom.
   (c) how many neutrons are in the nucleus when subtracted from the atom's mass number.
   (d) all of the above.

5. Isotopes of an element are
   (a) charged atoms, also called ions for short.
   (b) actually different elements.
   (c) the same element but with different number of neutrons in the nucleus.
   (d) the same elements but with different number of protons in the nucleus.

6. In the periodic table
   (a) elements in the same period have similar chemistry.
   (b) groups run left to right (they are rows).
   (c) periods run top to bottom (they are columns).
   (d) elements in the same group have similar chemistry.

7. The periodic table of elements is called periodic because
   (a) the chemical properties always repeat themselves in repeating intervals of atomic number.
   (b) the elements in a group have similar chemistry.
   (c) it was invented by the Russian Dmitri Mendeleev, and *periodisky* is the Russian word for table.
   (d) it was periodically updated with new elements over the years.

8. The difference between a cation and an anion is
   (a) a cation has more electrons than an anion.
   (b) an anion has more electrons than a cation.
   (c) a cation is negative whereas an anion is positive.
   (d) they have the same mass number but different atomic numbers.

9. Creating a cation from a neutral atom
   (a) always takes energy, called the ionization energy.
   (b) always requires the removal of an electron from the atom.
   (c) never requires the addition of an electron to the atom.
   (d) requires all of the above.

10. The following is true regarding metals and nonmetals:
   (a) The majority of elements in the periodic table are metals.
   (b) Metals like to lose electrons, whereas nonmetals like to gain electrons.
   (c) Only nonmetallic elements exist as gases at 25 °C and 1 atm.
   (d) All of the above are true.

Answers: 1d, 2c, 3d, 4d, 5c, 6d, 7a, 8b, 9d, 10d

Dalton's Atomic Theory

3.24 When wood burns, the mass of the ash is less than the mass of the original wood. Yet the law of conservation of matter says that mass cannot be created or destroyed in a chemical reaction. How do you reconcile the result of burning with this law?

3.25 Volcanoes spew off hydrogen sulfide, a poisonous, bad-smelling gas. A 100.00-g sample of hydrogen sulfide gas was obtained from some strange volcanic planet. Analysis showed that 94.08 g of it is the element sulfur. The rest of the sample consists of the element hydrogen.
   (a) What are the percentages by mass of sulfur and hydrogen in this sample?
   (b) Hydrogen sulfide from Earth has exactly the same percent compositions for S and H as hydrogen sulfide from the strange volcanic planet. Which law accounts for this?

3.26 If a scientist during Dalton's time found that hydrogen sulfide always had the formula H_2S, how would this have been explained?

3.27 Two compounds of iodine (I) and chlorine (Cl) are analyzed. Compound A consists of 126.9 g of I and 35.45 g of Cl. Compound B consists of 126.9 g of I and 106.4 g of Cl.
   (a) What is the percent composition of each element in compound A?
   (b) What is the percent composition of each element in compound B?

3.28 Suppose 12.0 g of carbon (C) reacts with 70.0 g of sulfur (S) to give 76.0 g of the compound carbon disulfide (CS_2). In the process, all the carbon gets used up, but some elemental sulfur is left over.
   (a) For the law of conservation of matter to be obeyed, how much sulfur is unused?
   (b) What is the percent C in CS_2?
   (c) What is the percent S in CS_2?
   (d) What is the sum of the %C and %S in CS_2?

3.29 Ammonia has the formula NH_3, and methane has the formula CH_4. According to Dalton, which atom has more hooks on it, N or C, and how many more hooks does it have?

3.30 Two different compounds, both consisting of sodium (Na) and oxygen (O), were analyzed. The data are given below:

Compound	Mass of sample analyzed	Mass of O present	Mass of Na present
A	19.50 g	8.00 g	?
B	61.98 g	16.00 g	?

   (a) Fill in the last column of the table.
   (b) Calculate the %Na and %O for both compounds.

Atomic Structure

3.31 Name the three subatomic particles found in an atom and give their relative masses and charges.

3.32 Given that the mass of an electron is 5.486×10^{-4} amu, and the mass of a proton is 1.007 amu, calculate how many times heavier a proton is than an electron.

3.33 In Rutherford's alpha-particle scattering experiment:
(a) How did he interpret the observation that only a very few of the α particles were scattered by the gold foil and most of them went straight through?
(b) What was wrong with his model of the atom as far as the physics of the day was concerned?

3.34 Suppose all the alpha particles in Rutherford's experiment went straight through the gold foil with absolutely no deflections. What would this imply about the structure of the atom?

3.35 What else besides an atom's mass number would you need to determine its elemental identity? Explain how you would use it along with the mass number.

3.36 Does knowing how many electrons a neutral atom has tell you its elemental identity? Explain.

3.37 Physicists are fond of saying that an atom is mostly empty space. What justifies this statement?

3.38 Fill in the following table for four neutral atoms:

	$^{15}_{8}O$	?	?	?
Mass number	?	16	37	?
Atomic number	?	8	?	?
Number of protons	?	?	?	?
Number of neutrons	?	?	?	12
Number of electrons	?	?	17	11

3.39 Uranium exists mainly as two isotopes in nature, possessing mass numbers 235 and 238. Write the full atomic symbols for both isotopes.

3.40 Consider the symbol $^{14}_{7}C$.
(a) What is wrong with this symbol?
(b) Adjust the atomic number so that the symbol is correct.
(c) Adjust the symbol letter so that the symbol is correct.

3.41 Give the full atomic symbol for this atom:

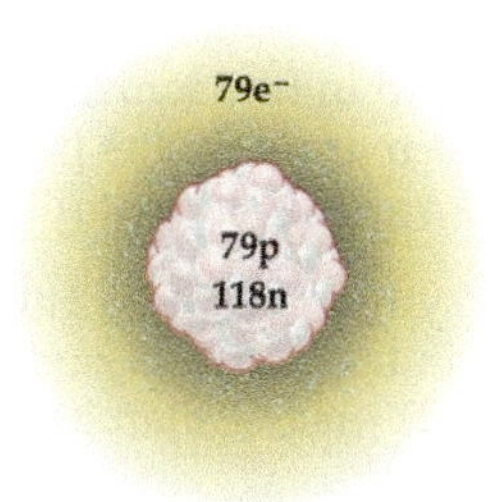

3.42 Write the full atomic symbol for the only atom that has its mass number equal to its atomic number.

3.43 Suppose you wanted to turn atoms of lead into atoms of gold. What would you have to do to the nucleus of the lead atoms?

3.44 What is the difference between an atom's atomic mass and its mass number?

3.45 Why is the atomic mass scale also called the relative atomic mass scale?

3.46 What is so special about the mass number and atomic mass of $^{12}_{6}C$?

3.47 How much more massive is an "average" oxygen atom than a $^{12}_{6}C$ atom? (Use the periodic table for the atomic mass of oxygen.)

3.48 How much more massive is an "average" titanium (Ti) atom than a $^{12}_{6}C$ atom? (Use the periodic table for the atomic mass of titanium.)

3.49 How much more massive is an "average" oxygen atom than a He atom? (Use the atomic masses from the periodic table.)

3.50 Why is the atomic mass of carbon not listed as 12 on the periodic table if the universal standard is that $^{12}_{6}C$ has an atomic mass of exactly 12 amu?

3.51 Suppose we wanted to use one of the more massive atoms, like $^{235}_{92}U$, as the standard reference for the atomic mass scale.
(a) What would be the universally agreed-upon atomic mass of the $^{235}_{92}U$ isotope?
(b) Why don't we use $^{235}_{92}U$ instead of $^{12}_{6}C$?

3.52 Bromine exists as only two isotopes in nature, $^{79}_{35}Br$ (atomic mass 78.918 336 amu, % natural abundance = 50.69%) and $^{81}_{35}Br$ (atomic mass 80.916 289 amu).
(a) What is the percent natural abundance for $^{81}_{35}Br$?
(b) Calculate the atomic mass of the naturally occurring mixture of isotopes.

3.53 Naturally occurring hydrogen on Earth has an atomic mass of 1.0079 amu. Suppose you were on another planet and found the atomic mass of hydrogen to be 1.2000 amu. How would you explain this? (The atomic mass of $^{1}_{1}H$ is 1.007 825 2 amu; the atomic mass of $^{2}_{1}H$ is 2.104 102 2 amu.)

3.54 While an atom's mass number and its atomic mass are not the same, they are often quite close to one another. Consider the following: Uranium-235 has an atomic mass of 235.043 93 amu and a percent abundance of 0.73%. Uranium-238 has an atomic mass of 238.0508 amu and a percent abundance of 99.27%. Without doing any calculations or consulting any other sources, which of the following do you think represents the atomic mass of naturally occurring uranium?
(a) 234.04 amu (b) 236.03 amu (c) 237.03 amu
(d) 238.03 amu (e) 238.07 amu

Explain how you made your choice.

The Modern Periodic Table

3.55 (a) In what unique way did Mendeleev order the elements to make his discovery?
(b) What was his discovery?
(c) How does his ordering differ from the modern ordering?

3.56 How did Mendeleev's discovery aid in discovering additional elements?

3.57 Define *chemical periodicity* with respect to chemical properties.

3.58 If one considers just the representative elements, how many groups would the periodic table have?

3.59 Considering just the representative elements, what percentage of them are metals, nonmetals, and metalloids? How do your results compare to the same percentages derived using the entire periodic table?

3.60 Give the names associated with groups IA (1), IIA (2), VIA (16), VIIA (17), and VIIIA (18).

3.61 What element seems poorly placed in group IA (1), the alkali metals? Why?

3.62 How does a group differ from a period in the periodic table?

3.63 Suppose you added one electron and one proton to an atom somewhere in the middle of a period. Which way would this move you on the periodic table and by how many places?

3.64 Magnesium (Mg) reacts with chlorine (Cl_2) to form the compound $MgCl_2$.
(a) Predict the formulas for the compounds formed when all the other alkaline earth metals react with chlorine.
(b) Predict the formulas for the compounds formed when all the alkaline earth metals react with bromine.
(c) Explain what principle you used to make your predictions.

3.65 State two things that are unique about the noble gases.

3.66 Name and give symbols for five of the transition metals.

3.67 How many elements are in each transition metal period of the periodic table? How many are in each lanthanide (rare earth)/actinide period?

3.68 Which elements are responsible for changing the chemical periodicity in the periodic table from 8 to 18? From 18 to 32?

3.69 Metals are excellent conductors of electricity. Nonmetals are poor conductors of electricity and are called insulators. Some elements have the ability to conduct electricity roughly between that of metals and nonmetals. Give some examples of these elements, and name the class they belong to.

Other Regular Variations in the Properties of Elements

3.70 It always takes energy to remove an electron from a neutral atom. Explain why, and name the minimum amount of energy it takes to do this.

3.71 A student claims that the way to convert a neutral atom of Cl into a Cl^- anion is to decrease its atomic number by one while leaving the number of electrons unchanged.
(a) Will this give the atom a −1 charge and convert it into an anion? Explain.
(b) Will the atom still be an atom of the element chlorine? Explain.
(c) What should the student have done?

3.72 What would the charge on a ion be if, starting from the neutral atom: (a) You add three electrons. (b) You remove four electrons. (c) You add two neutrons to the nucleus.

3.73 Two students are studying using a plastic model of the atom to which they can easily add or remove electrons, protons, and neutrons. They build a model for the neutral $^{14}_{6}C$ atom. Their next job is to make the +2 ion of this carbon isotope. Student X removes two electrons from the atom. Student Y adds two protons to the atom's nucleus. Did both students successfully construct a +2 carbon-14 cation? Explain.

3.74 Fill in the following table:

	$^{15}_{8}O^+$	?	?	?
Mass number	?	27	?	58
Atomic number	?	?	15	?
Number of protons	?	13	?	?
Number of neutrons	?	?	16	30
Number of electrons	?	?	?	27
Charge on ion	?	+3	−3	+1

3.75 Something is wrong with the following statement. Fix it. "Adding an electron to a +1 cation yields a neutral atom, but adding an electron to a neutral atom yields a cation."

3.76 A student says that it is possible for an atom's second ionization energy to be zero because all the work was done creating a cation with the first ionization energy. Is this student right or wrong? Explain.

3.77 What is the trend in first ionization energy as you go down a given group in the periodic table? As you go across a period from left to right?

3.78 Give the full symbol for the +2 alkaline earth cation that is in the fourth period and has a mass number 21 greater than its atomic number. Explain.

3.79 Of the atoms Na, Mg, and Al, which should be the most difficult to ionize? Which has the smallest first ionization energy?

3.80 Of the atoms Na, Mg, and K, which should be the most difficult to ionize? Which has the smallest first ionization energy?

3.81 Refer to the graph shown at the bottom of page 29. The trend is for first ionization energy to increase from left to right across a period. Describe any exceptions to this overall trend. Use the portion of the plot that is for the third period atoms to answer this question.

3.82 Sodium, Na, a metal, rapidly reacts with chlorine, Cl_2, a nonmetal.
(a) What reason is given for this high reactivity in this chapter?
(b) In what ways would you expect the reaction between lithium, Li, and bromine, Br_2, to be similar to the reaction between sodium and chlorine?

3.83 In the reaction of lithium, Li, with nitrogen, N_2, three lithium atoms react with one nitrogen atom to give the compound Li_3N. Atoms of one of these elements lose one electron; atoms of the other element gain three electrons.
(a) Which element gains the three electrons? Explain your choice.
(b) Does the element that gained three electrons become a cation or an anion?
(c) Give the full atomic symbol for the ion of part (b), assuming it has seven neutrons in its nucleus.

3.84 White light from an incandescent light bulb gives a continuous visible spectrum when the light is passed through a prism. What is meant by the word *continuous* in this context?

3.85 A heated gas made up of individual atoms gives off a line spectrum when the emitted light is passed through a prism. How does a line spectrum differ from a continuous spectrum?

Additional Problems

3.86 A 34.01-g sample of pure hydrogen peroxide produced at a chemical company contains 32.00 g of oxygen. The rest of the mass is due to hydrogen.
(a) What is the percent by mass of oxygen and hydrogen in this sample?
(b) Another company also produces pure hydrogen peroxide. What is the percent by mass hydrogen and oxygen in its hydrogen peroxide? What chemical law allowed you to answer this question?
(c) A 91.83-g sample of pure hydrogen peroxide is obtained. How many grams of it are oxygen?

3.87 Tin, Sn, and oxygen can combine to form two different compounds called oxides. Oxide A contains 78.77% by mass tin. Oxide B contains 88.12% by mass tin. What is the percent by mass of oxygen in each oxide?

3.88 When wood burns, it combines with oxygen from the atmosphere to produce carbon dioxide, water vapor, and ash. What would you have to do to prove that the law of conservation of matter is not disobeyed when wood burns?

3.89 What evidence exists to support the theory that an atom contains a massive nucleus that is very small relative to the size of the whole atom?

3.90 When 5 g of compound A reacts with an unlimited amount of compound B, 7 g of compound C is formed. How many grams of compound B must have reacted, and what law allows you to answer the question?

3.91 A student claims that isotopes of the same element have the same number of electrons and protons. Is she correct? Justify your answer.

3.92 A student claims that isotopes of the same element have the same number of neutrons but different numbers of protons. Is he correct? Justify your answer.

3.93 How do you convert a neutral sulfur atom to an anion carrying a charge of 2−?

3.94 A student says that the easiest way to convert a neutral sodium atom to a Na^+ cation is to add a proton to its nucleus. She reasons that this is so because protons have a 1+ charge. What is wrong with her argument? What does a neutral Na atom become when a proton is added to its nucleus?

3.95 The mass number of a neutral atom is 19, and its atomic number is 9. The atom contains how many protons, electrons, and neutrons?

3.96 The mass number of a 2+ cation is 56, and its atomic number is 26. The ion contains how many protons, electrons, and neutrons?

3.97 How many protons, electrons, and neutrons are there in:
(a) ^{79}Br (b) $^{81}Br^-$ (c) $^{23}Na^+$ (d) $^{3}H^+$

3.98 The element nitrogen has only two naturally occurring isotopes: ^{14}N with a mass of 14.00308 amu and an abundance of 99.635% and ^{15}N with a mass of 15.00011 amu and an abundance of 0.3650%. Calculate the atomic mass of nitrogen.

3.99 Silver has only two naturally occurring isotopes: ^{107}Ag with a mass of 106.90509 amu and an abundance of 51.84% and ^{109}Ag with a mass of 108.9047 amu. Calculate the atomic mass of silver.

3.100 Mendeleev arranged the elements by atomic mass and then separated them into various groups. What was his criterion for which elements were grouped together?

3.101 According to the periodic table on the inside front cover, how many main-group elements are known? How many of these are representative elements?

3.102 Arrange Na, Cs, S, Cl in order of increasing first ionization energy.

3.103 Scandium (Sc) is a metal that tends to lose three electrons. Copper (Cu) and zinc (Zn) are also metals that tend to lose one and two electrons, respectively. Knowing this, explain why the first group in the transition metals is numbered IIIB, whereas the last two groups in the transition series are numbered IB and IIB, respectively.

3.104 How many electrons are there in Zr^{4+}? What 2+ cation has the same number of electrons as Zr^{4+}?

3.105 How many electrons are there in Br^-? What 1+ cation has the same number of electrons as Br^-?

3.106 What is the identity of element X if its 2+ cation contains ten electrons?

3.107 Magnesium has a higher first ionization energy than sodium, consistent with the periodic trend for first ionization energies. Postulate a reason why Mg has the higher first ionization energy.

3.108 Use the periodic tables on pages 23 and 25 to complete this table:

Name of group or classification	Period	Group	Elemental symbol	Atomic number	Atomic mass (amu)	Metal metalloid, or, nonmetal?
				26		
Noble gas	1					
	5	VIIA (17)				
				19		
Halogen	3					
			Ne			
				8		
—	3					Metalloid
		—		92		

3.109 A 110.99-g sample of a compound containing only calcium and chlorine is found to contain 40.98 g of calcium.
(a) How many grams of chlorine are in the compound?
(b) What is the percent by mass of calcium in the compound?
(c) What is the percent by mass of chlorine in the compound?

3.110 Explain the major difference between Thomson's model of the atom and Rutherford's model. Draw a picture of each model.

3.111 Examine the periodic table that lists the first ionization energies of the elements, and state the trend for first ionization energies when you go down any group.

3.112 How does the arrangement of elements in the modern periodic table differ from that of Mendeleev's periodic table?

3.113 Write the elemental symbols for nitrogen, silicon, phosphorus, potassium, and gold.

3.114 Indicate atomic number, mass number, and number of protons, neutrons, and electrons in a neutral atom of each isotope:

$^{27}_{13}Al$ $^{60}_{27}Co$ $^{200}_{79}Au$ $^{238}_{92}U$ $^{127}_{53}I$

3.115 List three properties of metals and two properties of nonmetals. Describe the behavior of metalloids relative to the behaviors of metals and nonmetals.

3.116 Explain why Rutherford expected all alpha particles to go through the gold foil undisturbed.

3.117 (a) The elements in group IA (1) are called ______________.
(b) The elements in group IIA (2) are called ______________.
(c) The elements in group VIIA (17) are called ______________.
(d) The elements in group VIIIA (18) are called ______________.

3.118 Use Dalton's hook atomic models to sketch a molecule of H_2S.

3.119 Write the full atomic symbol for each isotope:
(a) 8 protons, 9 neutrons
(b) Atomic number 50, mass number 119
(c) 12 neutrons, atomic number 11
(d) 28 protons, mass number 58
(e) 81 neutrons, symbol Ba

3.120 Complete the following statements:
(a) The nucleus contains the ______________ charged particles in the atom.
(b) The particles in the nucleus are the __________ and ______________.
(c) Almost all of the mass of an atom is contained in the ______________.
(d) The two particles whose charges cancel to make an atom neutral are ______________ and ______________.

3.121 How do metals and nonmetals differ with regard to their tendency to gain or lose electrons? With their tendency to form cations or anions?

3.122 Calcium reacts with fluorine to form the compound CaF_2. In the reaction, each atom of one of the elements loses two electrons, and each atom of the other element gains one electron.
(a) Atoms of which element gain electrons?
(b) Atoms of which element lose electrons?
(c) What is the charge on the atoms that gain one electron?
(d) What is the charge on the atoms that lose two electrons?
(e) Write the full atomic symbol, including charges, for each element. Give Ca 20 neutrons and F 10 neutrons.

3.123 Use the elements lithium, potassium, and sodium to explain what is meant by chemical periodicity.

3.124 There are three isotopes of hydrogen—protium, deuterium, and tritium. Protium contains one proton and no neutrons. Deuterium contains one more neutron than protium, and tritium contains one more neutron than deuterium. For each isotope, give the atomic number, mass number, number of protons/neutrons/electrons, and full atomic symbol.

3.125 Given that one measure of metallic character is the tendency to lose electrons, arrange Cs, Be, Li, Ne, Na in order of increasing metallic character.

3.126 For each pair, indicate which atom loses an electron more easily:
(a) Na or K
(b) Na or Al
(c) Rb or Ca
(d) S or Cl
(e) Br or Kr

3.127 Explain the difference between mass number and atomic mass.

3.128 Naturally occurring copper, Cu, is composed of two isotopes. The isotope copper-63 is 69.17% of naturally occurring copper and has an atomic mass of 62.94 amu. The isotope copper-65 has an atomic mass of 64.93 amu. (a) What percent of naturally occurring copper is copper-65? (b) Calculate the atomic mass of naturally occurring copper.

3.129 Explain why an alternative name for the vertical columns in the periodic table is *families*.

3.130 Explain why the first and second statements of Dalton's atomic theory are not exactly true.

3.131 Write the name of each element:

(a) Be (b) Mg (c) Fe (d) S (e) Ar (f) Cu

3.132 In the pie-chart graphs, the percent natural abundance for Cl-35 (atomic mass 34.969 amu) is represented as green, and the percent natural abundance for Cl-37 (atomic mass 36.966 amu) is represented as yellow. Given that the atomic mass listed for Cl in the periodic table 35.353 amu, and without looking up the percent natural abundances for either isotope, which pie chart is most correct? Explain your reasoning.

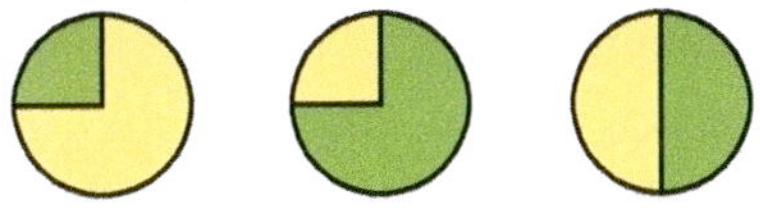

(a) (b) (c)

3.133 Consider the following process (e^- represents an electron):

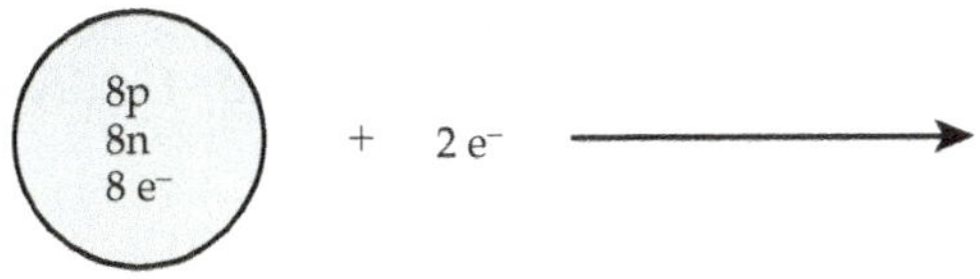

(a) Give the full symbol for the starting atom.
(b) Give the full symbol for the resulting ion.
(c) Is the product a cation or an anion? Explain.

3.134 Fill in the blanks for each of the atoms below (protons are red, neutrons are blue, and electrons are black):

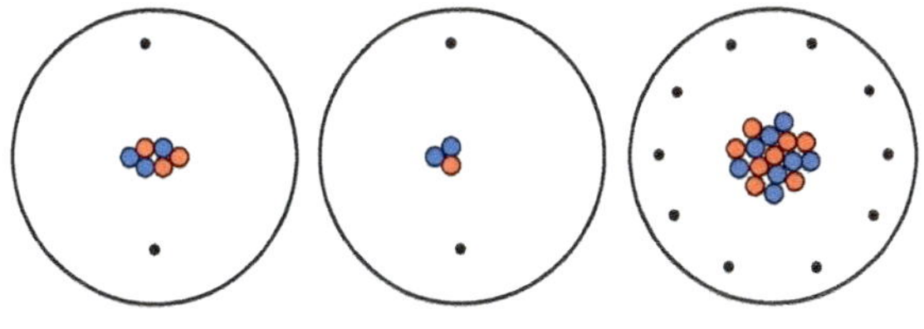

Element	______	______	______
Ion (yes or no)	______	______	______
Charge of atom	______	______	______
Atomic number	______	______	______
Mass number	______	______	______

3.135 The U.S. quarter, pictured below, has a diameter of 24.25 mm. If the nucleus inside an atom had the diameter of a U.S. quarter, then what would be the diameter of the atom in miles?

3.136 The following is a periodic table consisting only of the representative elements. Indicate where the transition metals get inserted. How many groups are in the smaller section of representative elements after the transition metals are inserted?

	IA	IIA	IIIA	IVA	VA	VIA	VIIA	VIIIA
1	1 H 1.0079							2 He 4.003
2	3 Li 6.941	4 Be 9.012	5 B 10.811	6 C 12.011	7 N 14.007	8 O 15.999	9 F 18.998	10 Ne 20.180
3	11 Na 22.990	12 Mg 24.305	13 Al 26.982	14 Si 28.086	15 P 30.974	16 S 32.066	17 Cl 35.453	18 Ar 39.948
4	19 K 39.098	20 Ca 40.078	31 Ga 69.723	32 Ge 72.61	33 As 74.922	34 Se 78.96	35 Br 79.904	36 Kr 83.8
5	37 Rb 85.468	38 Sr 87.62	49 In 114.82	50 Sn 118.71	51 Sb 121.76	52 Te 127.60	53 I 126.905	54 Xe 131.29
6	55 Cs 132.905	56 Ba 137.327	81 Tl 204.383	82 Pb 207.2	83 Bi 208.980	84 Po 209	85 At 210	86 Rn 222
7	87 Fr 223	88 Ra 226.025		114 Fl		116 Lv		

A periodic table showing only the representative elements is eight columns wide. Chemical periodicity occurs every eight elements.

One Step Beyond

3.137 Although not discussed until the next chapter, what process do you think chemists mean by an atom's second ionization energy?

3.138 Regarding problem 3.137, do you think an atom's second ionization energy would be smaller, the same, or larger than its first ionization energy? Do you think the second ionization energy for an atom would ever be less than the first? Explain.

3.139 Postulate a reason why the first ionization energy increases as we progress from left to right across a period.

3.140 A nitrogen atom can combine with three chlorine atoms to yield NCl_3. A phosphorus atom can combine with three chlorines to yield PCl_3 and also with five chlorine atoms to yield PCl_5. How would chemists have explained this in John Dalton's day?

3.141 Atom A has 35 protons and 44 neutrons in its nucleus. Atom B has 35 protons and 46 neutrons in its nucleus. Should we expect similar or different chemical properties for these atoms? Explain.

3.142 The first ionization energy is the minimum energy required to remove an electron from a neutral atom. Upon escape from the atom, how would the now free electron differ if you had used more energy than the minimum first ionization energy as compared to using the exact first ionization energy?

3.143 Suppose 12.0 g of silicon (Si) reacts with 70.0 g of oxygen (O_2) to give 25.6 g of the compound silicon dioxide (SiO_2). In the process, all the silicon gets used up, but some oxygen is left over.
(a) For the law of conservation of matter to be obeyed, how much oxygen is unused?
(b) What is the percent Si in SiO_2?
(c) What is the percent O in SiO_2?

3.144 In a mass spectrometer, atoms are ionized and then accelerated through the instrument by a nearby metal plate with a hole in it and an electrical charge on it. Suppose during the ionization process, atoms are turned into cations. What charge, positive or negative, must be put onto the metal plate to accelerate the ions? Explain.

3.145 An electromagnet bends the path of the fast moving ions in a mass spectrometer so they can reach the detector. The strength of an electromagnets magnetic field depends on the how much voltage is being fed to the magnet. How does a mass spectrometer measure the mass of a single atom?

3.146 Suppose two ions have been accelerated down a mass spectrometer, one of light mass and one of heavy mass. Which one do you think will take more voltage to the electromagnet to make it reach the detector? Explain. (*Hint*: Think about moving a chair versus a piano.)

3.147 How can a mass spectrometer determine if a rack sample is from Earth or from some extraterrestrial source?

Group Exercises (Work on these exercises together with someone else in the class.)

1. Rutherford's alpha-particle scattering experiment showed that the nucleus of an atom is tiny. Decide how the experimental results would differ from what Rutherford observed if:
(a) The nucleus was much larger than it actually is.
(b) The electron was much larger than it actually is.

2. An amu is defined as 1/12 the mass of a ^{12}C atom. Discuss why you think 1/12 was used, and explain why 1 amu is neither the mass of a proton or a neutron. Then, in terms of subatomic particles, define what 1 amu equals.

3. Other things in nature besides chemical properties are periodic. List as many things in nature as you can that are periodic, and include the period for each.

4. Devise a rule which tells how to turn an atom into an anion. Devise a rule which tells how to turn an atom into a cation. Both rules must include subrules which determine what the charge of the ion will be.

WORKPATCH SOLUTIONS

3.1 (a) The law of conservation of matter is obeyed because 100.0 g + 793.6 g = 893.6 g.
(b) The percent hydrogen is $(100.0\text{ g}/893.6\text{ g}) \times 100 = 11.19\%$ H.
(c) The percent oxygen is $(793.6\text{ g}/893.6\text{ g}) \times 100 = 88.81\%$ O.
(d) The law of conservation of matter is obeyed because 50.0 g + 793.6 g = 446.8 g + 396.8 g.
(e) The percent hydrogen is $(50.0\text{ g}/446.8\text{ g}) \times 100 = 11.2\%$ H.
(f) The percent oxygen is $(396.8\text{ g}/446.8\text{ g}) \times 100 = 88.81\%$ O. (Notice that the amount of oxygen (396.8 g) in the calculation is the amount actually used, which is the starting amount minus the leftover amount.)

3.2 Because most of the alpha particles passed through the gold atoms undeflected.

3.3 From the figure, it is obvious that an atom has as many electrons outside the nucleus as it has protons inside the nucleus. Because each electron has a −1 charge and each proton has a +1 charge, the overall atom must be neutral because equal numbers of +1 and −1 charges cancel each other.

3.4 (a)

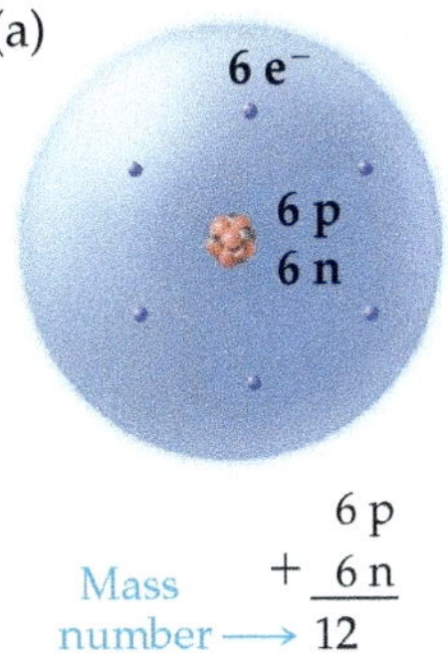

(b)

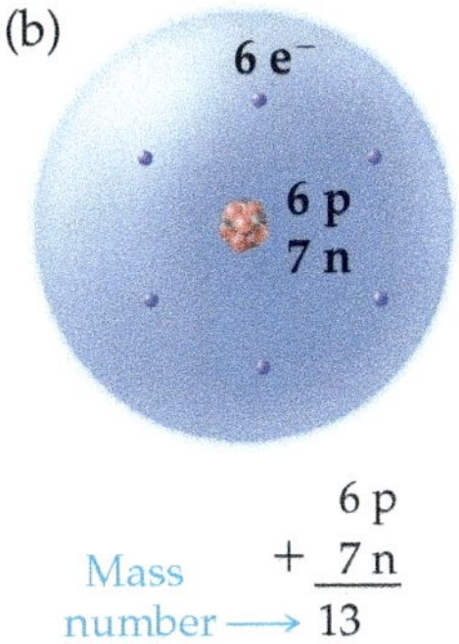

(c)

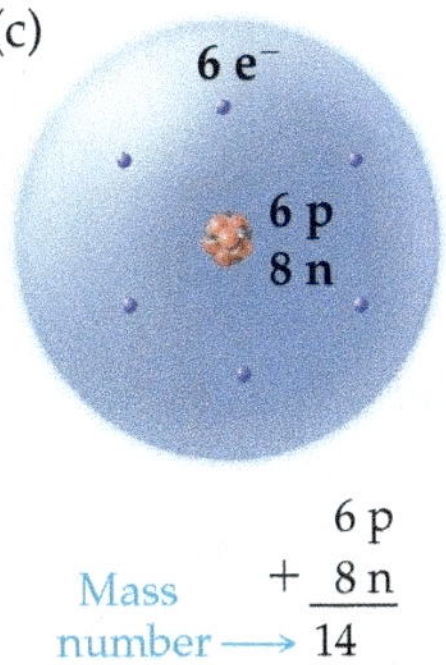

3.5 No. The mass number tells you the number of protons + neutrons. To identify the element, you need to know the number of protons (atomic number) in the nucleus.

3.6 Now you can figure out the atomic number. If the atom has a mass number of 16 and contains 9 neutrons, the number of protons is $16 - 9 = 7$. The element with atomic number 7 is nitrogen (N).

3.7 If the mass of an atom is 4.015 times the mass of a $^{12}_{6}C$ atom, the atom's relative atomic mass is $4.015 \times 12 = 48.18$. (Remember that 12 is an exact number, and so we don't need to round off.)

3.8 Because silicon (Si) is in the same group as carbon, its chemical properties should be similar to those of carbon (C).

3.9

	$^{14}_{7}N^{3-}$	$^{24}_{12}Mg^{2+}$	$^{23}_{11}Na^{+}$	$^{56}_{26}Fe^{3+}$
Mass number	14	24	23	56
Atomic number	7	12	11	26
Number of protons	7	12	11	26
Number of neutrons	7	12	12	30
Number of electrons	10	10	10	23

The Modern Model of the Atom

Just in case you thought chemistry was boring, here are quotes from some of the people who contributed to the theories you are about to learn.

"I don't like it, and I'm sorry I ever had anything to do with it."

Erwin Schrödinger commenting on the new physics he helped develop called Quantum Mechanics.

"How wonderful that we have met with a paradox. Now we have some hope of making progress."

Niels Bohr

"I am convinced that He (God) does not play dice with the universe."

Albert Einstein commenting on what he thought about the quantum mechanical model of the atom.

4.1 Seeing the Light—A New Model of the Atom

We ended Chapter 3 with a first look at the line spectrum given off by an electrified tube of hydrogen gas. To explain such a spectrum, we said, science would postulate a model of the atom so strange that even those who proposed it had trouble accepting it. Erwin Schrödinger, the physicist most credited with developing this model of the atom, would be quoted as saying, "I don't like it, and I'm sorry I ever had anything to do with it." The great physicist Albert Einstein refused to believe it could be correct. This is pretty exciting stuff for science, but to understand how science explained the hydrogen spectrum, we must first know more about the nature of light itself.

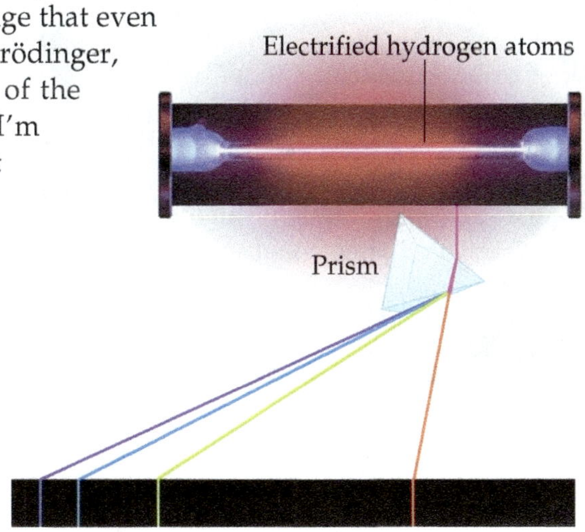

Line spectrum of hydrogen

We are all familiar with light. Flick a switch in a darkened room, and an electric current begins to flow through a tungsten filament until the filament glows white hot inside a glass bulb. The resulting white light illuminates the room.

Light moves through space at the incredible speed of 3.00×10^8 m/s. Called c, or the speed of light, this speed equals 186,000 miles/s or 671 million miles/h! If you were traveling at the speed of light, you could travel around the world at its equator seven and a half times in a single second. But what exactly *is* light? Modern science tells us that light is two different things at once. It consists of particles, or packets, of energy called *photons*, and yet at the same time it acts like a *wave of energy*. Here we want to focus on the view of light as a wave.

As a wave, light can be characterized by its *wavelength*, symbolized by the Greek letter λ (lambda). **Wavelength** is simply the distance between identical adjacent points on a wave—that is, the distance between one crest (or trough) and the next (see diagram on next page).

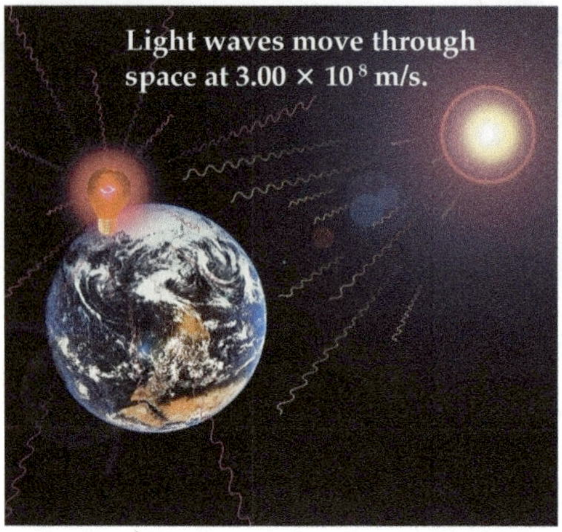

Our brains interpret light of different wavelengths as different colors. Light with a wavelength between 650 and 750 nm (nanometer, 10^{-9} m) is seen as red light. Violet light has shorter wavelengths, around 380–430 nm. The wavelengths between these extremes are interpreted by our brains as the different colors of the rainbow. The wavelengths of light between 380 and 750 nm are collectively called *visible light* because these are the wavelengths our eyes are designed to detect. When all these wavelengths impinge on our eyes at once, we see white light. Either a prism or a diffraction grating will separate, or *disperse,* this white light into its different wavelengths, giving rise to a continuous spectrum, the familiar rainbow of colors. As we said at the end of Chapter 3, this spectrum is called continuous because the colors blend smoothly into one another without any gaps or breaks.

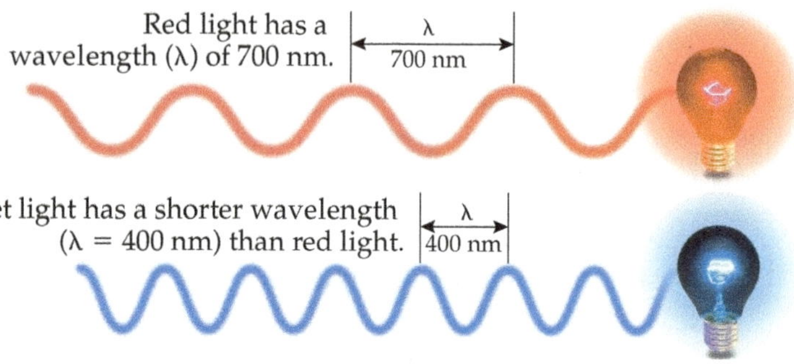

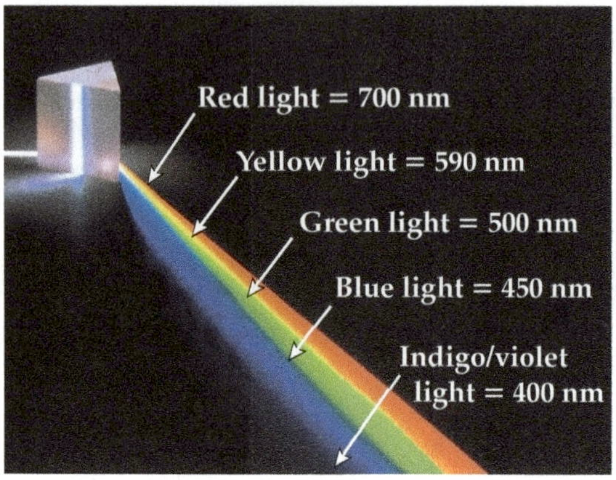

Though our world seems very bright and colorful, the fact is that we do not see most of the light that hits our eyes. Outside the visible range of 380 to 750 nm, there are many other wavelengths that are invisible to our eyes. All the visible and invisible wavelengths are collectively called **electromagnetic radiation** because they consist of both an electrical component and a magnetic component. The entire range of electromagnetic radiation is known as the **electromagnetic spectrum**.

The Electromagnetic Spectrum

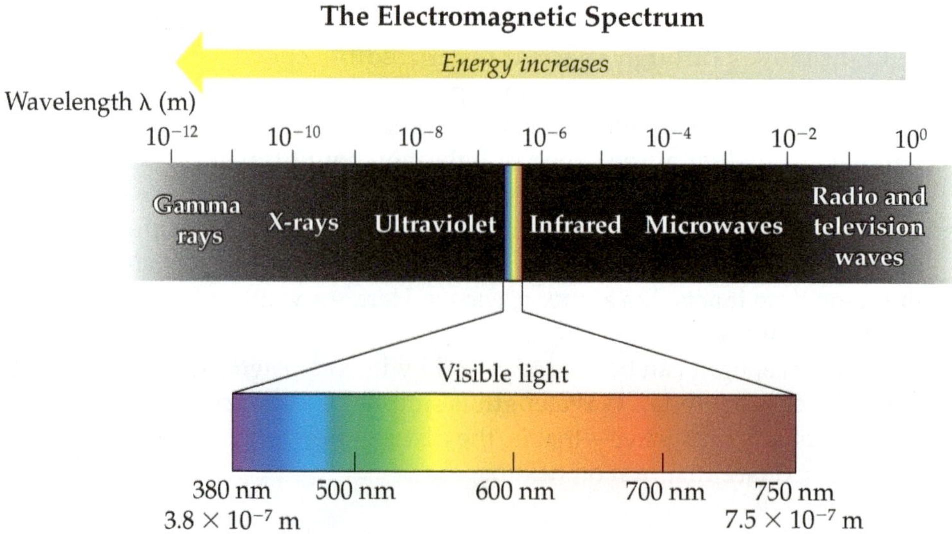

Even though we cannot see any electromagnetic radiation outside the narrow visible region, it can affect us. This is because all electromagnetic radiation has energy, and the amount of energy depends on the wavelength of the radiation. The equation relating the wavelength of a particular beam of electromagnetic radiation to its energy content is:

Equation to calculate the energy of electromagnetic radiation

$E = \frac{hc}{\lambda}$ where $E =$ energy of electromagnetic radiation

$c = 3.00 \times 10^8$ m/s (speed of light)

$h = 6.626 \times 10^{-34}$ J·s (Planck's constant)

$\lambda =$ wavelength of electromagnetic radiation

To use this equation, you simply insert the values for h and c (both constants) and the wavelength λ of the light. Watch the units, though! Because the speed of light c has units of meters per second, your wavelength must be expressed in meters. Because Planck's constant h has units of joules × seconds, your calculated energy will be in joules.

Practice Problems

4.1 What is the energy of blue light that has a wavelength of 450.0 nm?

Answer: $E = \frac{hc}{\lambda} = \frac{(6.626 \times 10^{-34}\ \text{J} \cdot \text{s})(3.00 \times 10^{8}\ \text{m/s})}{(450.0\ \text{nm})(1 \times 10^{-9}\ \text{m/nm})} = 4.42 \times 10^{-19}\ \text{J}$

4.2 What is the energy of red light with a wavelength of 660.5 nm?

Answer: 3.01×10^{-19} *J*

4.3 What are the wavelength and color of light that has an energy of 3.50×10^{-19} J? (*Hint:* Use algebraic manipulation to solve the energy equation for λ. You will get an answer in meters. Convert it to nanometers (10^{-9} nm = 1 m) and consult the electromagnetic spectrum shown on page 122.)

Answer: 568 nm, yellow

4.4 Which type of electromagnetic radiation has a wavelength roughly on the order of the height of a person? What can you say about the energy of this radiation?

The equation $E = hc/\lambda$ tells us that the energy of electromagnetic radiation is inversely proportional to its wavelength. As the wavelength of electromagnetic radiation gets smaller, the energy of the radiation increases. Compare the answers to Practice Problems 4.1 and 4.2 to convince yourself of this. Because the wavelength (660.5 nm) of the red light is larger (longer) than the wavelength (450.0 nm) of the blue light, the red light has less energy.

All the forms of electromagnetic radiation having wavelengths shorter than that of violet light (ultraviolet, X-rays, and gamma rays) are invisible to humans. Because they contain so much energy, ultraviolet rays can cause sunburn, X-rays can cause cancer, and gamma rays can kill. Also invisible to us are all the forms of electromagnetic radiation having wavelengths longer than that of red light (infrared, microwaves, radio, and television waves), although we can detect the presence of infrared radiation as heat.

Four Lines of the Hydrogen Spectrum in the Visible Region

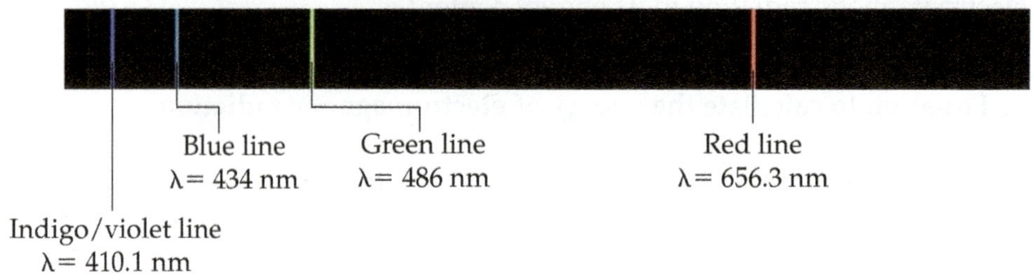

So, what should you get from this discussion? Simply that each color (wavelength λ) of light has a particular amount of energy. Thus, the fact that only four distinct lines are seen in the visible region of the spectrum of electrified hydrogen atoms means that these atoms can only emit the four particular amounts of energy associated with these four wavelengths, and not any other intermediate amounts of energy. This is why the spectrum is black between the lines, but why should atoms of hydrogen give off only these certain energies? What is so special about these energies and why does every atom of hydrogen behave the same way?

4.2 A New Kind of Physics—Energy Is Quantized

In 1913, the Danish physicist Niels Bohr had an answer that was revolutionary. He proposed that the energies of the electrons inside atoms are *quantized*. By **quantized**, he meant that the electrons inside an atom could have only specific, discrete amounts of energy. Such an assertion went against all the teachings of classical physics. Both classical physics and everyday observation tell us that an object can have any amount of energy. Consider a tennis ball, for example. It can have any amount of kinetic energy, depending on how hard you hit it. It can have low energy (a gentle lob over the net), high energy (a fierce slam shot), or any energy in between these two extremes.

But imagine a tennis ball that could only have certain energies and therefore can travel at only certain speeds. It could never travel at any speed other than the allowed ones, no matter how hard or soft it was hit. That would be quite a tennis ball! Actually, tennis balls *do* have quantized energies; we just can't detect this fact (read on). However, we *can* detect that electrons in atoms have quantized energies.

Before we go on, we want to remind you of the scientific method discussed in Chapter 1. The scientific method can result in a well-respected theory being set aside and replaced with another one. Most often, a major change of this kind is not the result of one person's work. For example, when Bohr proposed that the energies of the electrons in an atom are quantized, he was drawing on the work of the German physicist Max Planck. Planck had earlier proposed energy quantization in his efforts to explain the energy characteristics of the light emitted by heated objects. Planck saw this only as a mathematical trick, something that allowed him to arrive at the correct answer in his calculations. He did not believe that energy was actually quantized in any real physical system. Bohr was aware of Planck's work, and he took the concept one step further by applying it to the electrons in an atom. That was a bold thing to do, particularly since Bohr had no explanation for *why* electrons should exhibit quantization. He just knew that quantization explained certain experimental observations. The explanation of why quantization occurs would come about ten years later from yet another scientist, Louis de Broglie. This is the scientific method in action.

As noted at the end of Section 3.3, scientists today recognize two kinds of physics, classical and quantum. **Classical physics** says that objects can have any energy. It works extremely well for large objects, such as tennis balls, cannon balls, rockets, and planets. **Quantum physics** says that objects can have only certain particular energies. It is the physics that correctly explains the behavior of very small particles the size of atoms and electrons. In fact, quantum physics is the more general of the two. Earlier we said that the motion of a tennis ball is quantized, but we just can't detect the quantization. That is because the differences between the allowed energies for a large object like a tennis ball are so small that the energies seemingly blend into one another to yield a smooth continuum.

For example, consider heating a large amount of water on a stove. As you add heat energy to the water, the water's temperature (a measure of its energy content) appears to increase in a smooth, continuous fashion. Even with the world's most sensitive and accurate thermometer, you would not observe any temperature jumps. That is not because there are no temperature jumps but rather because the space between the jumps is too small to be measured. The jump size is on the order of Planck's constant, which means about 10^{-34} degree! No thermometer could ever detect such small changes. However, the differences between the allowed energies for very small objects, like the tiny electrons inside an atom, are much larger and therefore become detectable. This is why we can get away with using classical physics for the behavior of large "objects," but we must use quantum physics if we want to understand the behavior of electrons inside an atom.

WORKPATCH 4.1 Imagine quantum physics applied to objects as large as a car. What would traffic on a highway look like? •

4.3 The Bohr Theory of Atomic Structure

Bohr proposed that the electrons in an atom travel around the nucleus in circular orbits much as the planets orbit the Sun. The revolutionary part of his model was that these electrons could possess only particular, fixed energies, which in turn meant that only specific orbits, orbits at certain fixed distances from the nucleus, were allowed.

To understand the relationship between the size of an electron's orbit and the energy of an electron in that orbit, you need to only remember that the nucleus is positively charged and an electron is negatively charged, and so they are attracted to one another. Think of this attraction as a spring that connects the electron to the nucleus. When the electron is closest to the nucleus, the spring is relaxed and the electron is at its lowest possible energy. To pull the electron away from the nucleus into a larger orbit requires putting energy into the electron to overcome the electron–nuclear attraction. In our spring model, this is equivalent to stretching the spring. The energy we put in is stored in the stretched spring. The stretched spring (and thus the atom itself) now contains more energy than before and therefore is in a higher-energy, less-stable situation. Unless there is something preventing it, the electron will spontaneously snap back into the lower orbit and release the energy that was stored in the stretched spring.

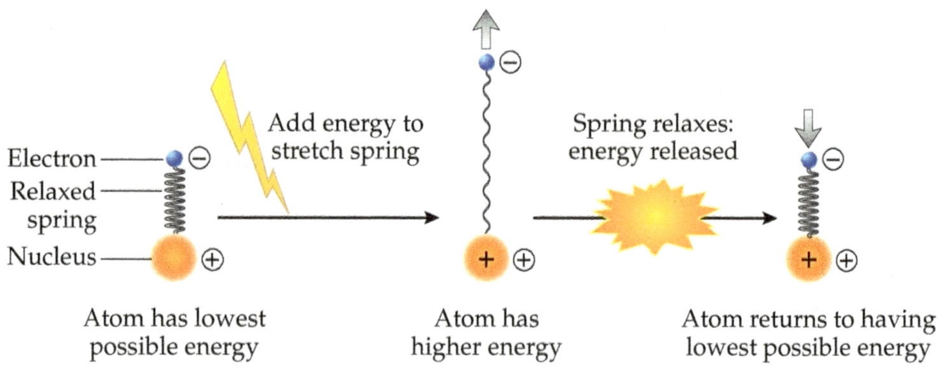

Therefore, the closer an electron is to the nucleus, the lower the electron's energy and the more stable the atom:

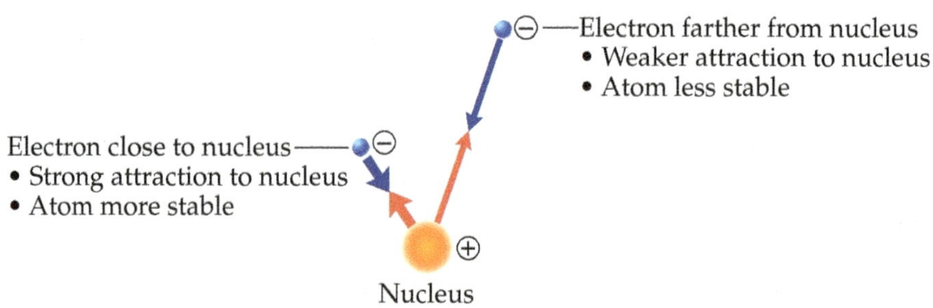

So in proposing that an electron could have only certain (quantized) energy values, Bohr was also saying that the electron in an atom could be found at only certain fixed distances from the nucleus, distances where the energy required to stretch the spring corresponds to the allowed energies.

Before we go on, it is important to tell you that the scientific world quickly moved beyond Bohr's circular orbits. No one today believes that electrons orbit an atom's nucleus on fixed paths. The modern model of the atom has a very different view of how electrons behave in an atom. (We'll introduce you to the modern model at the end of the chapter.) This current modern model does not mean that Bohr's model is not worth learning about. Far from it. Bohr's introduction of energy quantization and the relationship between an electron's energy and its distance from the nucleus are all correct and still a part of the modern model. In addition, Bohr's model is much easier to visualize than the modern model, and it works well in explaining certain aspects of chemistry, such as the periodic behavior of the elements and why certain compounds

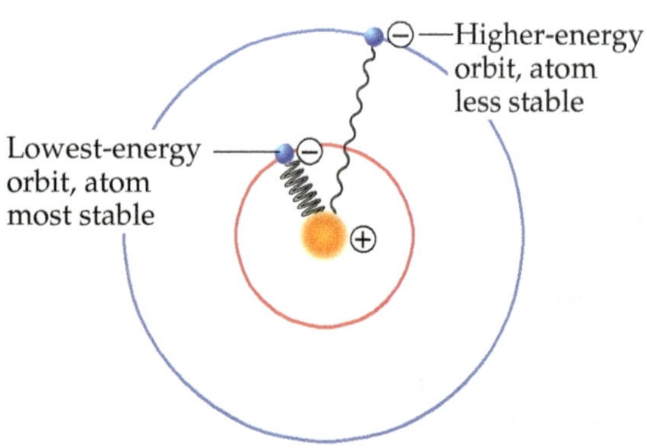

have the chemical formulas they do. Also, an acquaintance with Bohr's model is probably the best way to prepare for understanding the modern one.

In Bohr's model of the atom, electrons can jump from one allowed energy level to another or from one orbit to another (these movements are called *quantum jumps*), but they can never possess any energy between the allowed ones. In Bohr's view, electrons are positioned around the nucleus in an atom much like a person standing on a ladder. A person on a ladder can stand on the first rung, the second rung, and so on, but will never be found standing *between* any two rungs. The same is true for the electrons in an atom according to Bohr.

Bohr assigned each of his allowed electron orbits, which were called **shells**, a **principal quantum number (*n*)**, starting with $n = 1$ for the shell closest to the nucleus, $n = 2$ for the next shell out from the nucleus, and so on. The shells continue with $n = 6, 7, 8$, and so on up to infinity. As the orbits get larger, they take the electron farther away from the nucleus and thus require the electron to possess a greater amount of energy to exist in that orbit.

Ideally, all the electrons in an atom would like to occupy the $n = 1$ shell, because it is the shell of lowest energy (maximum electron–nucleus attraction). But electrons also repel one another, and crowding them into a single shell would quickly raise the energy of the atom dramatically. In order to deal with this problem, and also to explain line spectra and periodic chemical behavior, Bohr maintained that each shell can hold a fixed **maximum** number of electrons. The larger the shell, the more electrons it can hold, because a larger shell provides more room in which the electrons can spread out. Bohr maintained that each shell can hold a maximum of $2n^2$ electrons, where n is the principal quantum number of the shell.

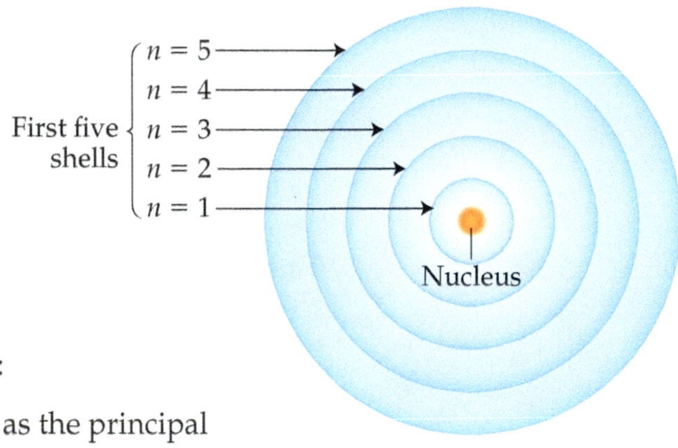

Here, then, are the main features of the Bohr model:

1. Orbits (shells) get larger (their radius increases) as the principal quantum number, n, increases.
2. The energy of an electron in an atom increases with n. Electrons in the $n = 1$ orbit have the lowest energy. Electrons found in the $n = 2$ orbit have more energy. Those in the $n = 3$ orbit have even more energy, and so on.
3. Each shell can hold a maximum of $2n^2$ electrons. Therefore, the electron capacities of the first four shells are:

Maximum Electron Capacities of the First Four Shells

$n = 4$	$2n^2 = 2 \times 4^2 = 32$ electrons
$n = 3$	$2n^2 = 2 \times 3^2 = 18$ electrons
$n = 2$	$2n^2 = 2 \times 2^2 = 8$ electrons
$n = 1$	$2n^2 = 2 \times 1^2 = 2$ electrons

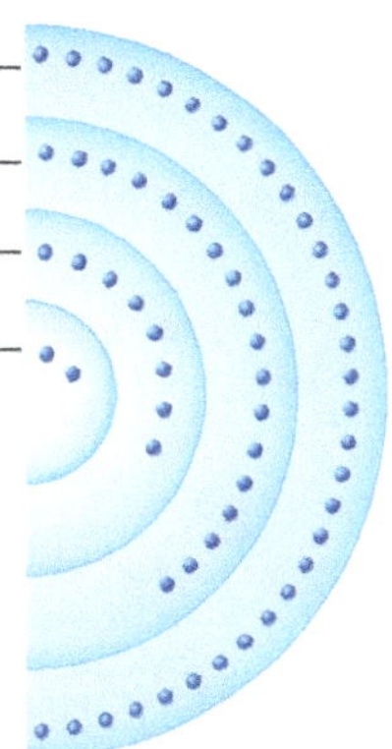

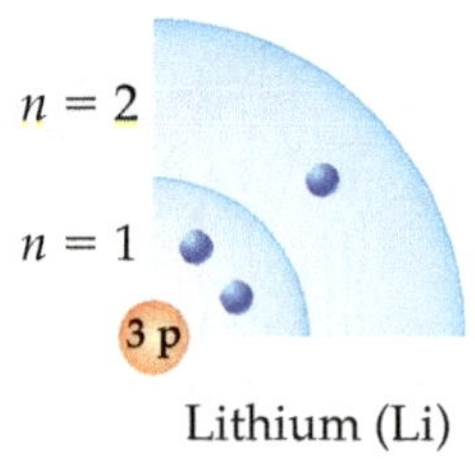

4. When "building" an atom, the innermost shell (the one lowest in energy) is filled first. Electrons are then added to each subsequent shell until it is filled to capacity ($2n^2$ electrons). For example, the lithium (Li) atom, with atomic number 3, has three protons in its nucleus and three electrons in its electron shells. Two of the electrons go into the most stable $n = 1$ shell, and the third electron must go into the next most stable $n = 2$ shell.

This is the essence of the Bohr atomic model. Now we can use it to show you why atoms exhibit line spectra and why the elements possess properties that vary in a periodic fashion.

Practice Problems

4.5 What is wrong with this Bohr model of the beryllium (Be) atom (atomic number 4)?

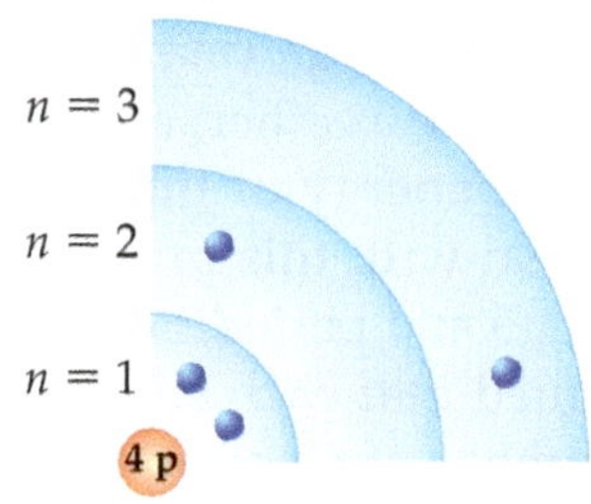

Answer: Don't begin filling the $n = 3$ shell until the $n = 2$ shell is full. Don't skip shells!

4.6 Draw a Bohr model for an atom of sulfur (S). How many additional electrons can fit into the $n = 3$ shell?

4.7 Why is an electron in a shell that has a low value of n in a more stable arrangement than one in a shell that has a higher value of n?

4.8 How many electrons can the $n = 5$ shell in an atom hold?

4.4 Periodicity and Line Spectra Explained

Explaining the periodic nature of the chemical properties of the elements was one of the major triumphs of Bohr's model. Consider the elements lithium (Li) and sodium (Na), both in group IA (1) and possessing similar chemical properties. According to Bohr, lithium, with 3 electrons, and sodium, with 11 electrons, have the structures shown at left. If you look closely at these diagrams, you'll see a similarity between lithium and sodium. They both have completely filled inner shells and only a single electron in the outermost occupied shell. This outermost occupied shell of an atom is called the **valence shell**. Perhaps these elements have similar chemical properties because they have identical valence-shell configurations. Does this pattern hold true throughout the periodic table?

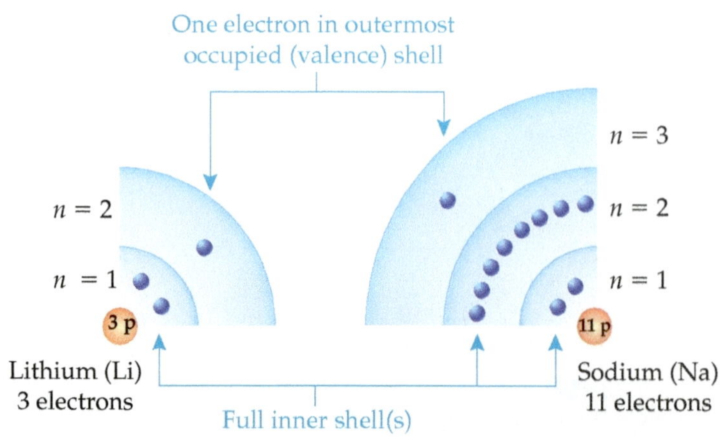

Let's look at another pair of elements from the other side of the table, fluorine (F) and chlorine (Cl). They are both in group VIIA (17), so we know they have similar chemical properties. What about their valence shells? When the electrons are put into the shells following Bohr's prescription, both elements end up with seven electrons in their valence shell. This, according to the Bohr model, is why F and Cl have similar chemical properties and belong in the same group of the periodic table.

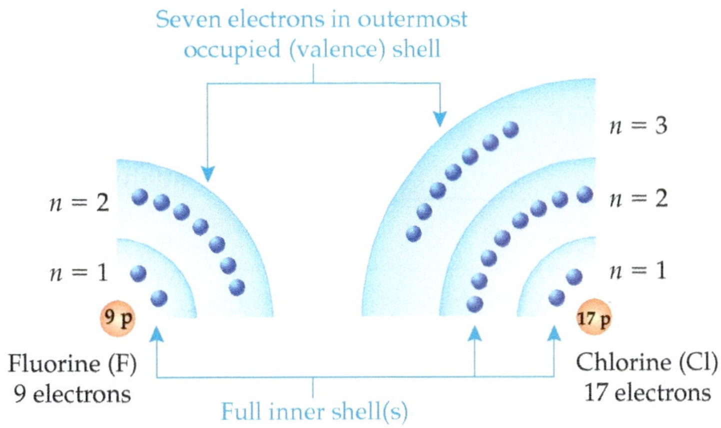

The electron-shell configurations for the first 18 elements in the periodic table, filled according to Bohr's rules, are shown in the following chart. Look at the outermost occupied shell in each group—the valence shell—and convince yourself that elements with similar chemical properties have identical valence-shell configurations. Only helium (He) seems to not follow this rule. Helium has only two valence electrons, whereas all the other members of group VIII (18) have eight. In a way, though, it does. Both helium and the atom below it, neon (Ne), have completely filled valence shells. The difference is that the valence shell in helium is the $n = 1$ shell and can accommodate only two electrons. All the atoms after helium in this group contain eight electrons in their valence shell.

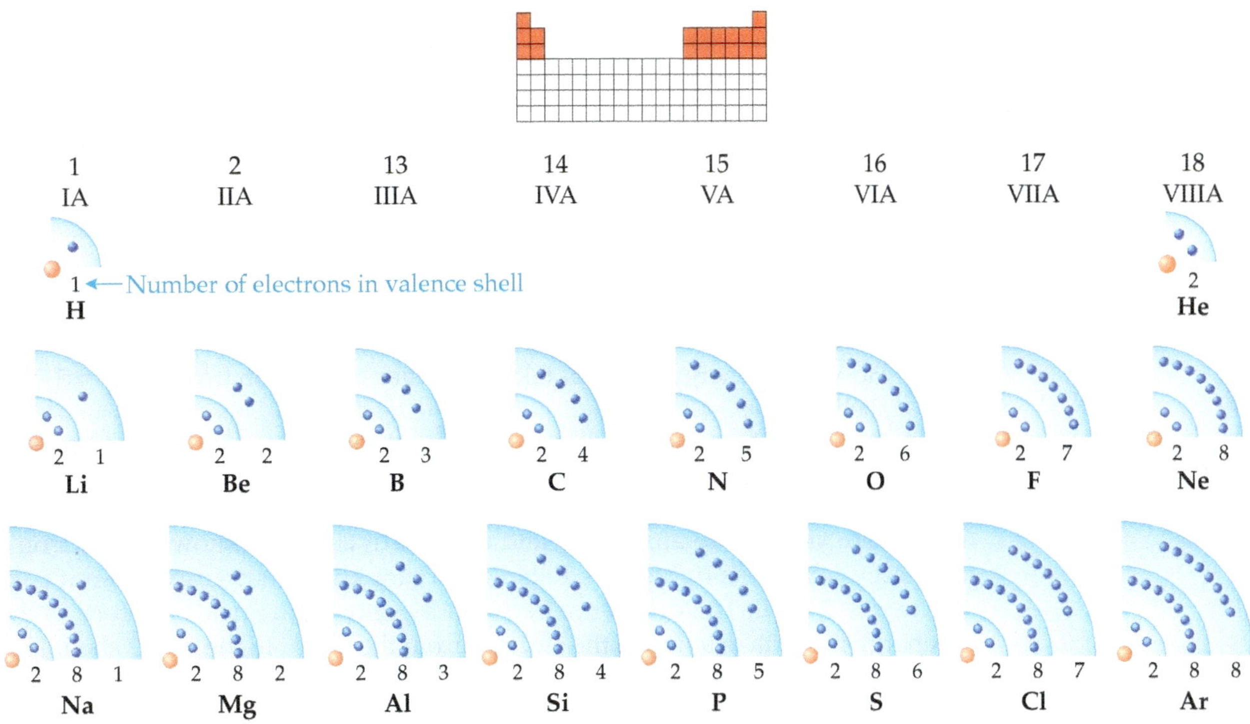

Careful examination of this chart reveals a shortcut for determining the number of electrons in the valence shell of an atom. Hydrogen, lithium, and sodium each have one valence-shell electron in the Bohr model, and they are also in group IA of the periodic table. Fluorine and chlorine both have seven valence-shell electrons and are found in group VIIA. For the representative elements, the roman-numeral group number directly gives the number of valence-shell electrons in the atoms of that group.

The number of electrons in the valence shell of an atom is equal to the roman-numeral group number for the representative (A group) elements.

So, according to Bohr's model, the chemical properties of the elements are periodic because, as successive shells of electrons get filled, the valence-shell configuration repeats.

The other great success of the Bohr model was its ability to explain the line spectrum of hydrogen. Not only did the model explain why line spectra exist, but the mathematical calculations that Bohr carried out based on the model's assumptions reproduced exactly the experimental line spectrum of hydrogen. From a physicist's perspective, this was the real success of the Bohr model. Bohr's equations could be used to calculate the energy of an electron in any allowed shell of the hydrogen atom. A scaled version of the energy values Bohr calculated for the first six shells of the hydrogen atom is shown at the left in the form of an **energy-level diagram**. As we shall see, the absolute energies of the shells are not important. It is only the differences between them that are important. Thus, we have arbitrarily rescaled Bohr's calculated energies so that the energy of the $n = 1$ shell is 1.0 eV.

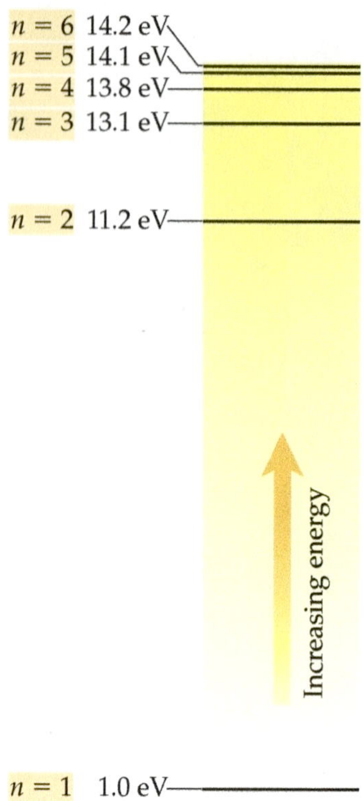

The calculated energies of the Bohr shells in the hydrogen atom scaled so that the energy of the $n = 1$ shell is 1.0 eV.

Notice how the energy differences between the orbits get smaller as n increases. This fact is a direct consequence of Bohr's calculations and will become important later on.

WORKPATCH 4.2 Examine the energy-level diagram for hydrogen shown above, and answer the following questions:

(a) If an electron is in the $n = 1$ shell, what is the electron's scaled energy in electronvolts?

(b) If an electron is in the $n = 2$ shell, what is the electron's scaled energy in electronvolts?

(c) How much energy would it take to transfer an electron from the $n = 1$ shell to the $n = 2$ shell?

(d) Would it take the same, more, or less energy than your answer to (c) to take an electron from the $n = 2$ shell to the $n = 3$ shell? Why? ●

Were you able to answer (c) and (d) intuitively? You know the electron's energy when it is in the $n = 1$ shell, and you know its energy when it is in the $n = 2$ shell. The difference between these two energies, therefore, is the energy it will take to move the electron from the $n = 1$ to the $n = 2$ shell. This is a very important concept, so make sure you understand it. As for part (d), shell energies get closer together with increasing n. Clearly then, the energy cost will be less to go from $n = 2$ to $n = 3$ than to go from $n = 1$ to $n = 2$.

To predict the line spectrum of hydrogen, Bohr assumed that the atom starts out in its *ground state*. The **ground state** of an atom is defined as the arrangement of electrons that has the lowest total energy. Thus, it is the configuration where all the electrons are placed in the shells having the lowest energy values. For a hydrogen atom, which has only one electron, the ground state is as shown at the left below (we use a Bohr diagram to show you where the electron is and an energy-level diagram to indicate the energy of the electron).

Under normal conditions, hydrogen atoms spend most of their time in the ground state. However, when enough energy is added to the atom (for instance, by heating it or passing an electric current through it), its electron can move ("jump") into a higher-energy shell. When this occurs, we say energy was absorbed by the atom and the atom is now in an *excited state*. In an **excited state**, one or more of an atom's electrons are located in higher-energy shells even though there is room for those electrons in a shell of lower energy.

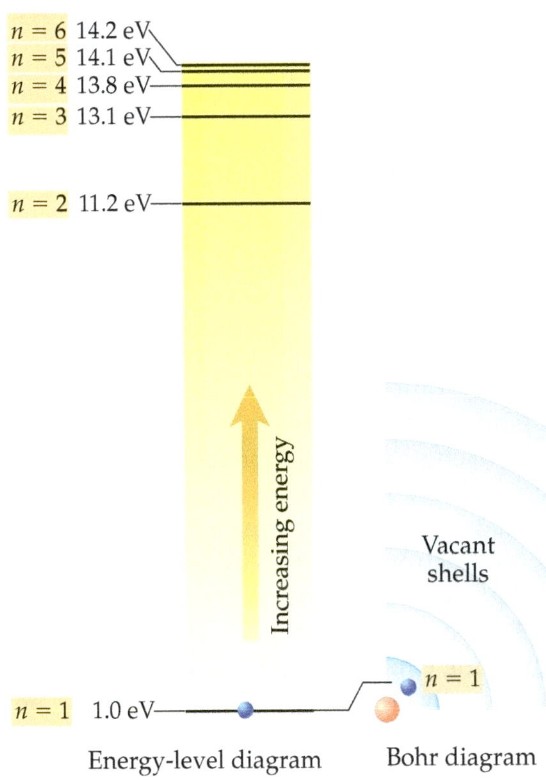

Ground state of hydrogen: electron in $n = 1$ shell

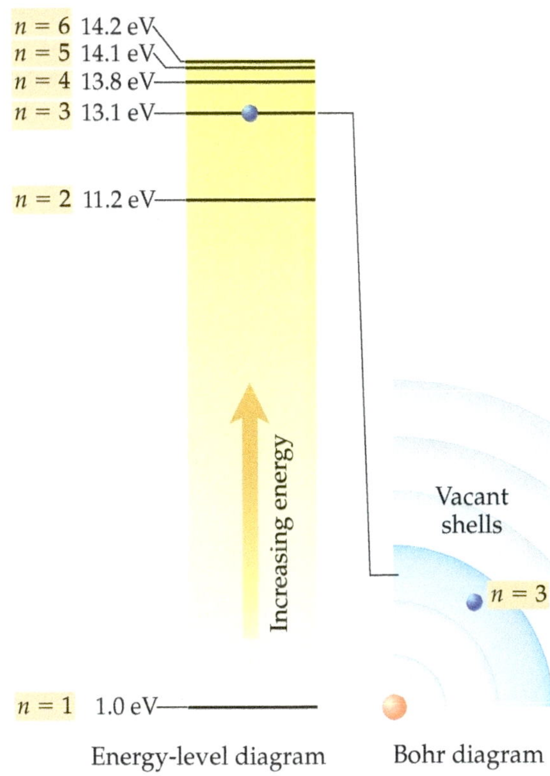

An excited state of hydrogen

One possible excited state for the hydrogen atom is shown on the right above. This energy-level diagram represents an excited state because there are vacant lower-energy shells below the electron (levels have been skipped before they are full). Another possible excited state could place the electron in the $n = 2$ shell, and yet another in the $n = 4$ shell, and so on. We can create any of these excited states by starting with the ground-state hydrogen atom and adding enough energy to allow the electron to jump from the $n = 1$ shell to any upper shell.

WORKPATCH 4.3 What is the lowest-energy excited state of the hydrogen atom? ●

The WorkPatch points out that all excited states are not created equal. For instance, in the hydrogen excited state with the electron in the $n = 2$ shell, the electron has less energy than in the excited state with the electron in the $n = 3$ level, because it takes more energy to raise the electron to $n = 3$ than to $n = 2$.

WORKPATCH 4.4 Starting from the ground state, how much more energy does it take to create the $n = 3$ excited state of hydrogen than the $n = 2$ excited state? ●

Check your WorkPatch answer against ours. When you can do the WorkPatch, then you are becoming skilled at understanding and using an energy-level diagram.

When a tube of hydrogen gas is electrified, the electricity supplies the energy to produce various excited states. To account for the line spectrum exhibited by such a gas, we must consider what happens *after* these excited states are created. As it turns out, atoms do not usually remain in an excited state for long. Because there are vacancies at lower energy levels, the electrons can drop back down into these levels (a process called *relaxation*). Relaxation is desirable from an energy point of view because, in general, things tend to settle in the lowest energy state possible (as, for instance, when a ball let go on the side of a hill rolls to the bottom of the hill). To create an excited state, an atom absorbs a certain amount of energy. The big news is when the atom returns to the ground state (relaxes), it loses exactly the same amount of energy.

One way an atom can lose energy is to emit it as light. The amount of energy possessed by the light wave during relaxation is the same as the amount of energy the electron loses in dropping to a lower energy level. This emitted energy determines the particular wavelength (color) of the light emitted, and this is where line spectra come from. When an excited hydrogen atom relaxes from the $n = 2$ state to the ground state, the energy of the electron goes from 11.2 eV down to 1.0 eV, meaning the electron loses 10.2 eV of energy in the process. This energy is released as light with a wavelength corresponding to an energy of 10.2 eV. Using Table 4.1, we can see that this particular electron transition results in ultraviolet light being emitted. Because ultraviolet light is outside the visible region, we can't visually observe this particular line in the hydrogen line spectrum.

Table 4.1 Energy and Wavelength Ranges for Ultraviolet and Visible Light

Radiation	Energy (eV)	Wavelength λ (nm)
Ultraviolet	> 3.74	< 380
Indigo/violet	3.34–2.95	380–430
Blue	2.95–2.64	430–480
Green	2.64–2.24	480–565
Yellow	2.24–2.10	565–620
Orange	2.10–1.95	620–650
Red	1.95–1.69	650–750

The observed line spectrum of hydrogen does show four lines that are located in the visible region of the electromagnetic spectrum—one red line, one green line, one blue line, and one indigo/violet line. According to Table 4.1, the red line corresponds to light with an energy in the range from 1.69 to 1.95 eV. A close inspection of the energy-level diagram for hydrogen shows that there is one, and only one, jump that corresponds to energy in this range. An electron in an excited hydrogen atom relaxing from the $n = 3$ level to the $n = 2$ level produces light with exactly 1.9 eV of energy, as you saw in the preceding WorkPatch.

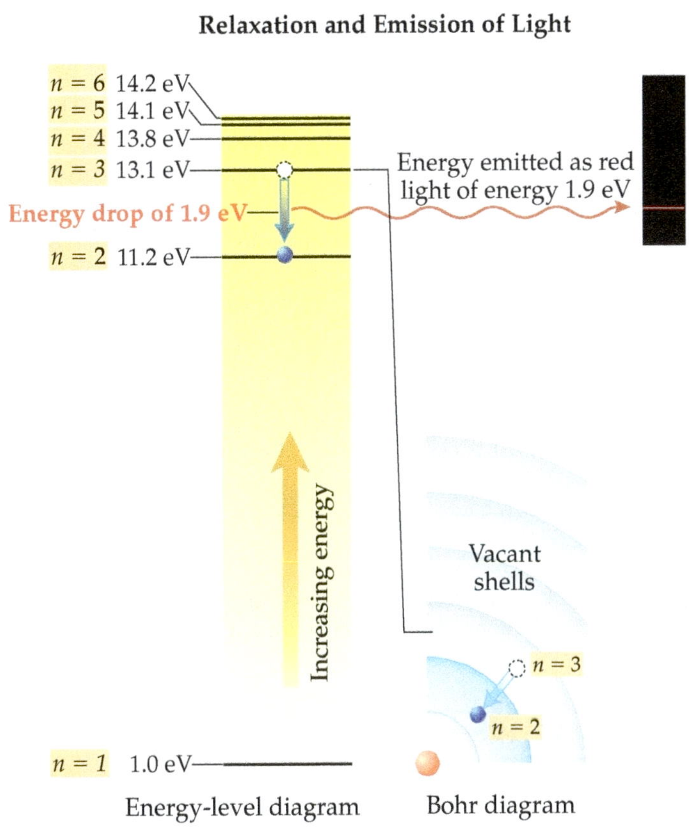

With the electron in the $n = 2$ shell, we are not yet to the ground state. Thus, we can conclude that the electron in an excited atom does not necessarily return to its ground state in a single jump. It can also reach the ground state by making a series of smaller downward jumps.

WORKPATCH 4.5 The hydrogen atom has three other visible lines in its line spectrum: green, blue, and indigo/violet. Use the energy-level diagram for hydrogen and Table 4.1 to determine the electron jumps responsible for these lines. ●

Were you able to do the WorkPatch? Because indigo/violet has the highest energy of these four colors of visible light, the jump corresponding to the indigo/violet line should be the largest. Check your answer to see that it is.

Bohr's model accounts for the existence of line spectra because the electrons in an atom are allowed to possess only certain energies and never anything in between. Because the energy levels are fixed, the differences between them are also fixed, and only certain wavelengths (energies) of light will be produced. The energy levels for the atoms of other elements have different allowed energy values and hence will exhibit different line spectra.

The rapid acceptance of Bohr's model was due mainly to its success in exactly reproducing the line spectrum of the hydrogen atom. Not only could Bohr correctly calculate the exact position and energy of the four lines in the visible part of the spectrum, but he could also match exactly the lines detected in other regions of the spectrum (ultraviolet and infrared). Nevertheless, Bohr's model of the atom did have some problems. The lines Bohr calculated should be present in the line spectrum of the helium atom did not match up at all with the actual spectrum. In fact, his calculations failed to predict the correct spectrum for any atom with more than one electron. In less than 20 years, Bohr's model would be replaced with the current model of the atom. However, his postulate that the energy levels of electrons in atoms are quantized, and his rules for the maximum number of electrons that each shell can accommodate, are still correct. Therefore, we can continue to use these aspects of the Bohr model to explain many things in chemistry. Before we do, however, we must look at an important modification to the model.

Practice Problems

4.9 The ground state for the lithium (Li) atom and the scaled energies of its shells are shown below. Draw a Bohr diagram for the lowest-energy excited state of lithium.

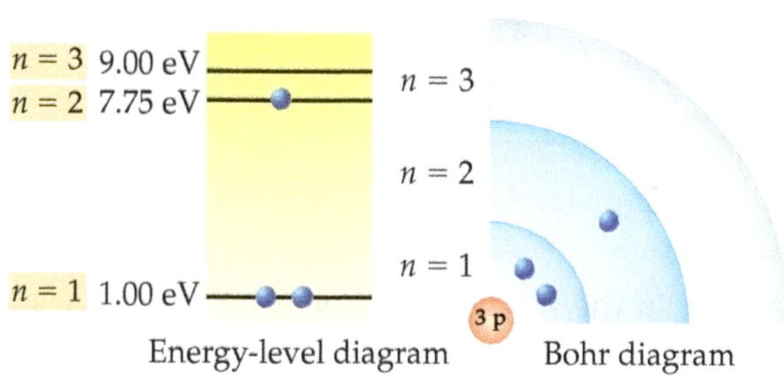

Answer: Two excited states are drawn below, but only (a) correctly answers the question. Remember, both excited states are created from the ground state. To produce (a), an electron must be lifted from the $n = 2$ shell to the $n = 3$ shell. This takes 1.25 eV (1.25 eV of energy must be put into the atom). To produce (b), an electron is lifted from the $n = 1$ shell to the $n = 2$ shell. This takes 6.75 eV. Because it took more energy to create (b), (b) is a higher-energy excited state than (a). Thus, (a) is the lowest-energy excited state that can be prepared from the ground-state lithium atom.

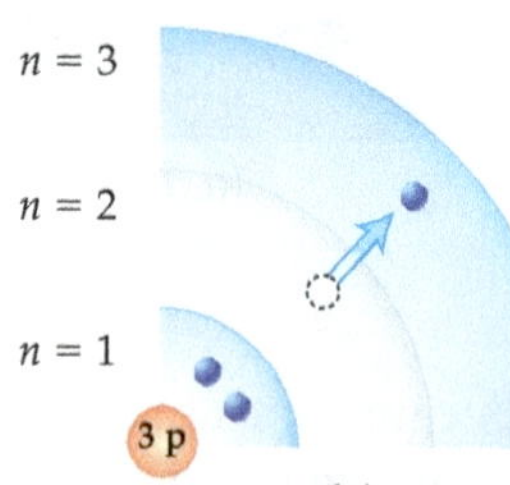

(a)

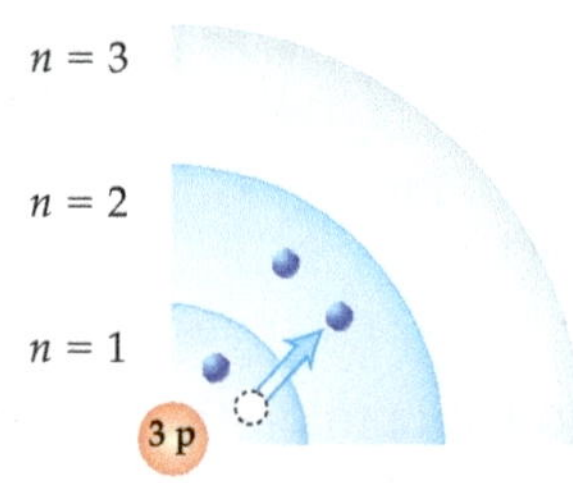

(b)

4.10 Draw a Bohr diagram for the highest-energy excited state you can make using the three shells shown in Practice Problem 4.9 and exciting only a single electron.

4.11 Draw a Bohr diagram for a Li^+ cation in its ground state.

4.12 An atom has atomic number 6 and has eight electrons.

(a) Which element is this?
(b) Is this a neutral atom, a cation, or an anion? If it is an ion, what is its charge?
(c) Draw a Bohr diagram for this atom in its ground state.

4.13 An F^- anion has ten electrons. A student draws the following ground-state Bohr diagram for this anion. There are two things wrong with the diagram. What are they?

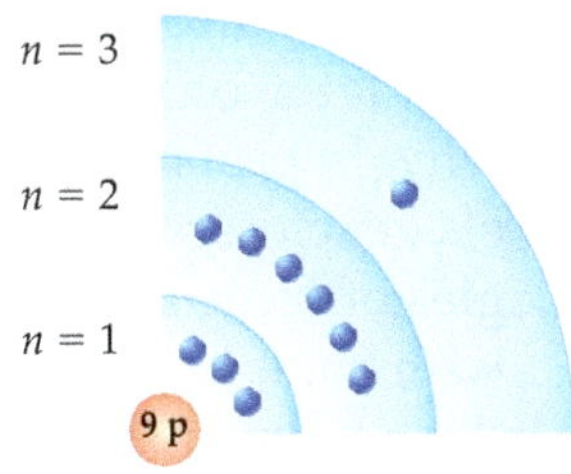

4.5 Subshells and Electron Configuration

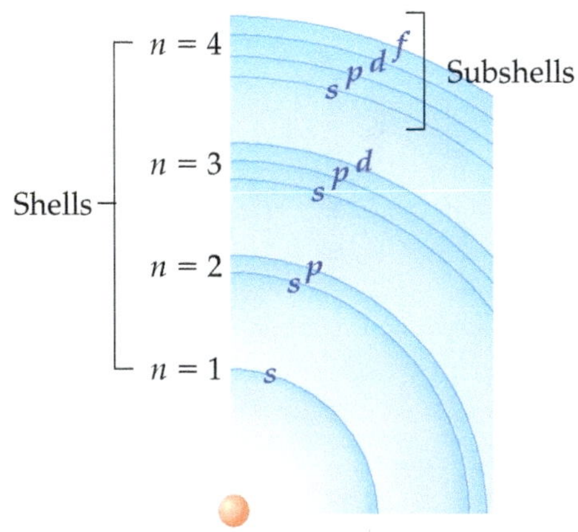

The principal quantum number n tells you the number of subshells in a shell. The $n = 1$ shell has one subshell, the $n = 2$ shell has two subshells, and so on. All subshells in a given shell are close to one another both in size and in energy.

Now we need to look at an important refinement to the Bohr model, the essence of which was important in the development of the modern model. During Bohr's time, instruments improved enough to show that the simple line spectra generated by excited atoms were not so simple after all. In many cases, what was originally thought to be a single line turned out to be a number of very closely spaced lines of nearly identical color (energy). Bohr struggled with these findings for several years and tried various modifications to his basic model in an attempt to account for them. What he and others eventually proposed was, in essence, that some of the electron shells in an atom actually consist of a set of **subshells** that are very closely spaced in energy and size.

As shown at the right, these subshells are given single-letter designations (s for the first subshell in a shell, p for the second, d for the third, and f for the fourth). Not every shell contains all of these subshells. The first shell ($n = 1$) contains just an s subshell. As we move outward from the $n = 1$ shell, each subsequent shell possesses an additional subshell, so the $n = 2$ shell consists of two subshells (s and p), the $n = 3$ shell has three subshells (s, p, and d), and so on. In addition, as shown at the top of the next page, each subshell has a different electron capacity. An s subshell can hold a maximum of 2 electrons, a p subshell can hold a maximum of 6 electrons, a d subshell can hold up to 10 electrons, and an f subshell can hold 14 electrons. These subshells also increase in energy and size, with the s subshell in any given shell being the smallest and having the lowest energy. Notice that the electron capacity of each shell

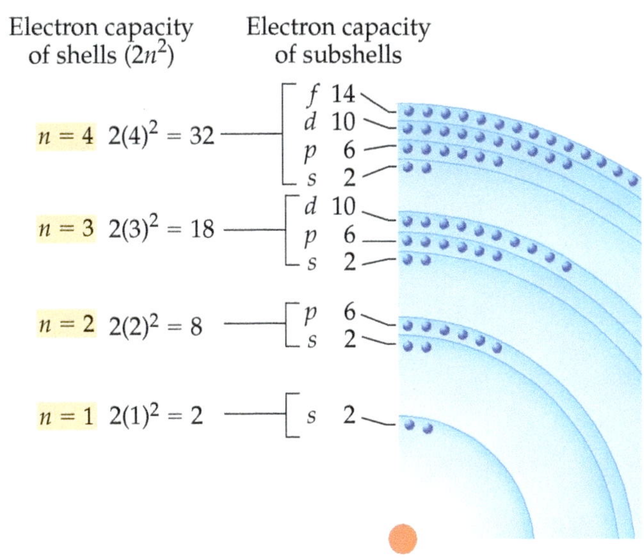

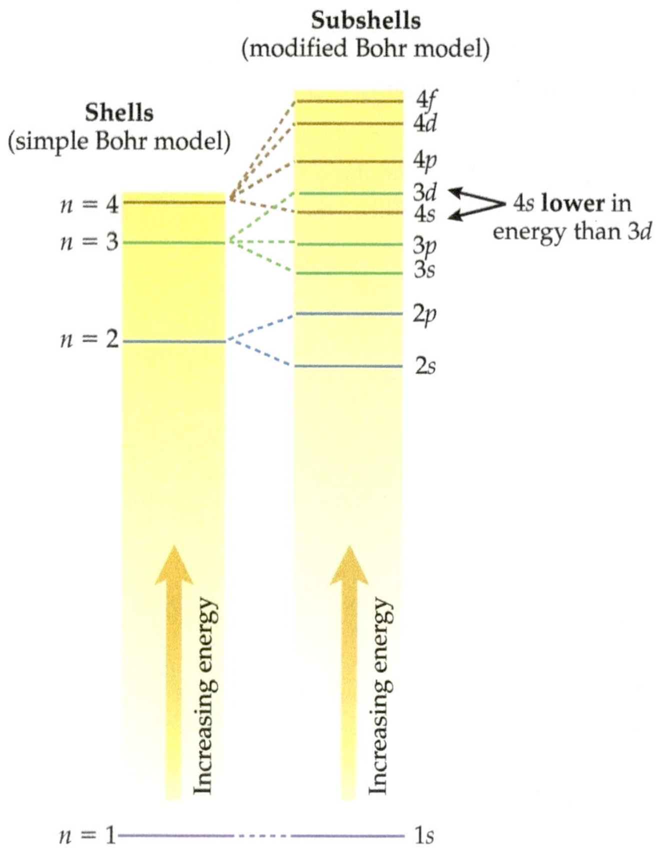

is unchanged from the original Bohr model. For instance, the $n = 3$ shell still holds a maximum of $2n^2$ or 18 electrons.

When we introduced the energy-level diagram for the hydrogen atom, we pointed out that the energy values of the principal shells get closer and closer together as n increases. We also said that this would become important later on. Now is the time to show you why. In part because the principal energy levels get closer together as n increases, the energies of some subshells can actually cross over with those in an adjacent shell. The diagram to the left illustrates this.

As we've said, the subshells in a shell increase in energy in the order $s \Rightarrow p \Rightarrow d \Rightarrow f$, and all of the $n = 3$ subshells are higher in energy than the $n = 2$ subshells. But look carefully at the 4*s* subshell. It is actually *lower* in energy than the 3*d* subshell, even though the 4*s* subshell has a larger value of n. Additional crossings occur as we proceed beyond $n = 4$. The consequences of these crossings are extremely important! Remember, Bohr stated that the ground state of an atom is achieved by filling the shells in order of increasing energy, starting with the $n = 1$ shell and working up. The same rule holds when we take subshells into account. This solves a problem that initially puzzled Bohr—what was called the potassium problem.

Potassium (K) is a group IA (1) atom possessing 19 electrons. Group IA atoms are supposed to have one valence electron, but that's not what Bohr's initial model produced. Look at what happens when we position potassium's 19 electrons using the simple version (principal shells only) of Bohr's model:

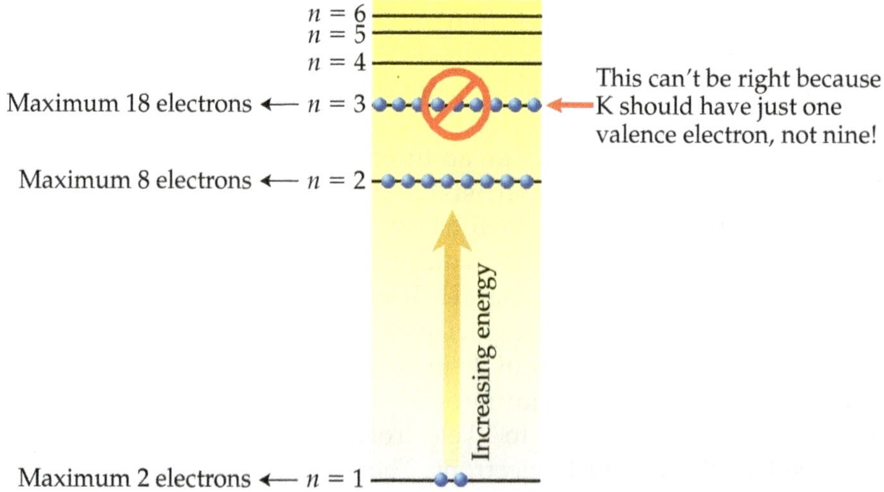

The ground state of K according to the simple Bohr model (shells only)

The $n = 1$ and $n = 2$ shells are filled, as expected. But the valence $n = 3$ shell has 9 electrons in it instead of the expected 1 electron for this group IA atom. The modified subshell model fixed this problem. Demonstrate this for yourself now.

WORKPATCH 4.6 Using dots to represent potassium's 19 electrons, fill the subshells in the diagram below to arrive at the ground-state energy-level diagram for the potassium atom. Select the diagram that gives an answer consistent with the periodic table. ●

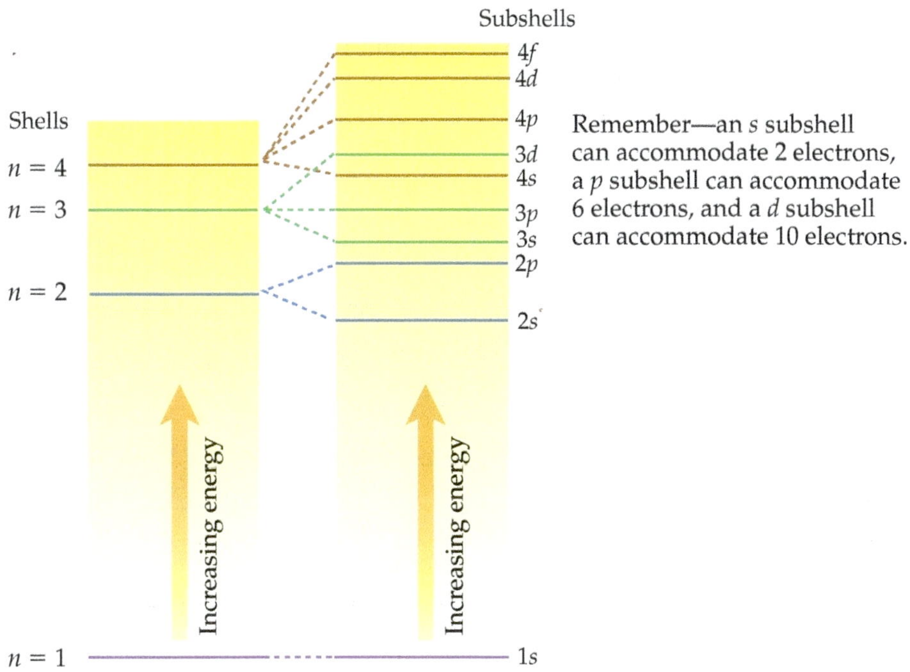

In doing the WorkPatch, you should have arrived at an atom with one electron in the valence 4*s* subshell. You have just solved the potassium problem! We are about to discuss this energy-level diagram in detail, so check your answer against the one at the end of this chapter.

Electron Configuration

The way the electrons are arranged in the shells and subshells of an atom is called the atom's **electron configuration**. An energy-level diagram is one way to depict the electron configuration of an atom; however, it is a tedious and elaborate representation. Chemists have developed an alternative notation to specify an atom's electron configuration. Each occupied subshell, starting with the lowest-energy subshell, is indicated with a lowercase italic letter (*s*, *p*, *d*, *f*) preceded by a number indicating the principal quantum number of the shell (n). The number of electrons occupying the subshell is indicated by a superscript. Thus, we can replace the modified Bohr diagram for potassium with the notation:

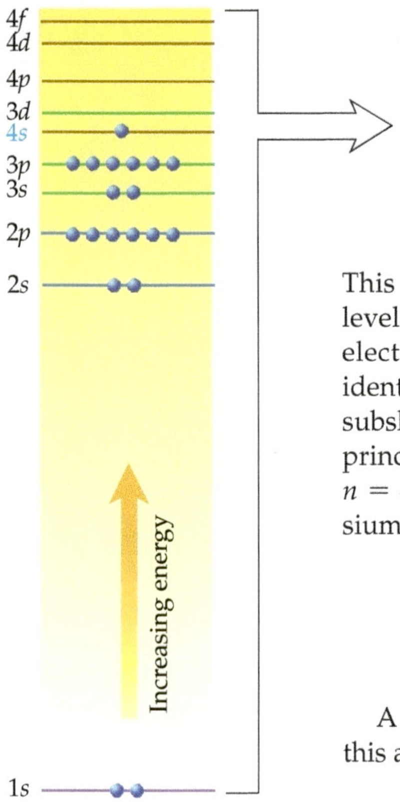

Number of electrons in subshell

$1s^2 2s^2 2p^6 3s^2 3p^6 4s^1$

n

Subshell

This electron-configuration notation is a lot easier to write than an energy-level diagram, and it communicates the same information. It tells us how the electrons are arranged in the subshells. In addition, it is an easy matter to identify the valence electrons. In general, the valence electrons are those in the subshells of highest principal quantum number n. For potassium, the highest principal quantum number is 4, and if we sum up the electrons present in the $n = 4$ subshells, we get 1, the correct number of valence electrons for a potassium atom.

Valence electron

$1s^2 2s^2 2p^6 3s^2 3p^6 \mathbf{4s^1}$

A slightly more difficult example is bromine (atomic number 35). When this atom is in the ground state, its electron configuration is:

$1s^2 2s^2 2p^6 3s^2 3p^6 \mathbf{4s^2} 3d^{10} \mathbf{4p^5}$

Valence electrons

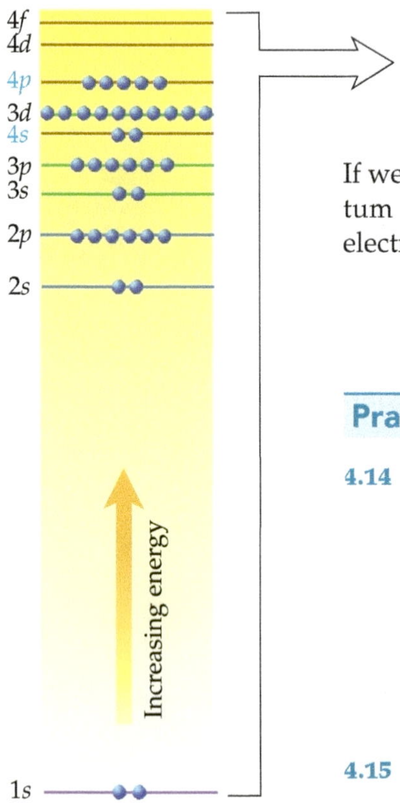

If we sum up the electrons in the subshells having the highest principal quantum number (4), we get seven valence electrons. To emphasize the valence electrons, chemists often write them last:

$1s^2 2s^2 2p^6 3s^2 3p^6 3d^{10} \mathbf{4s^2 4p^5}$

Practice Problems

4.14 Use electron-configuration notation to show how electrons are distributed in a ground-state silicon (Si) atom. How many valence electrons does this atom have?

Answer: Silicon has atomic number 14, so it has 14 electrons. We start by putting the first two electrons in the 1s subshell, the next two in the 2s subshell, the next six in the 2p subshell, the next two in the 3s subshell, and the remaining two in the 3p subshell:

$1s^2 2s^2 2p^6 \mathbf{3s^2 3p^2}$

The highest principal quantum number is 3, and summing up the electrons in the two $n = 3$ subshells tells us this atom has four valence electrons.

4.15 Without looking at a periodic table, use electron-configuration notation to show how electrons are distributed in a ground-state atom with 33 electrons. How many valence electrons does this atom have? What periodic table group would you expect to find it in? Now look at a periodic table and determine the atom's identity.

4.16 Use electron-configuration notation to show how electrons are distributed in a ground-state scandium (Sc) atom.

All the subshell energy-level diagrams we have presented so far go only to $n = 4$. As we proceed beyond this value of n, additional subshell energy crossings occur. You will be pleased to know that you do not need to memorize any of these crossings. That's because the periodic table can be used as a quick and easy guide for deducing the electron configuration of any element. To do this, we think of the periodic table as being constructed from four blocks—the *s* block, the *p* block, the *d* block, and the *f* block—with the designation (s, p, d, f) of each block representing the subshell being filled.

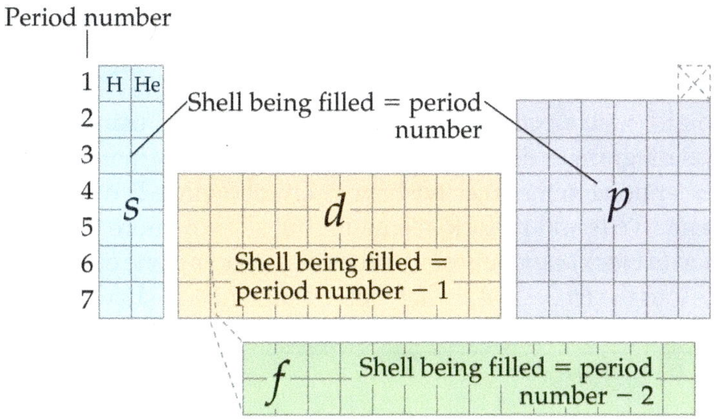

Notice that the *s* block is 2 elements wide. That is because an *s* subshell can hold a maximum of two electrons. The *p* block is 6 elements wide because a *p* subshell can hold up to six electrons. Likewise, the *d* and *f* blocks are 10 and 14 elements wide, respectively, indicating the maximum number of electrons those subshells can hold. Also, notice that we moved helium (He) from its usual position at the extreme right of the table to a position in the *s* block, adjacent to hydrogen.

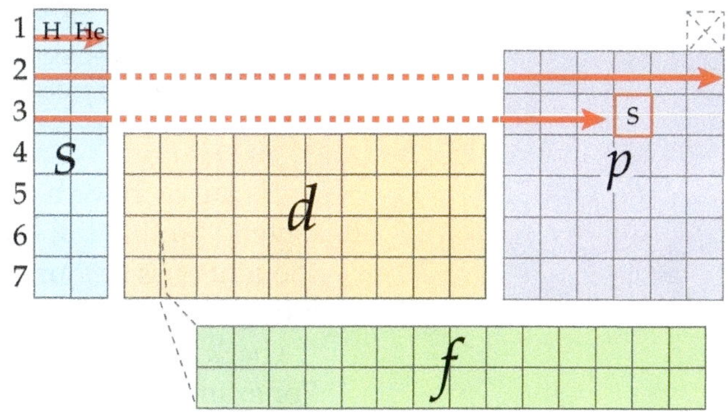

The period (row) number tells you which shell (n) is currently filling, *unless you are in the d or f block.* In the *d* block, the value of n for the *d* subshell being filled is one less than the period number. If you are in the *f* block, the value of n for the *f* subshell being filled is two less than the period number. This takes into account the subshell crossings. A few examples will help make this clear.

Suppose we want to determine the electron configuration of sulfur (S). Keep glancing at the periodic table below and follow the red arrows as we discuss how to do this. We start at hydrogen (as we always will) and move across the first period from left to right until we reach the end of the period at helium. Each element we encounter on our way adds one more electron to the atom. Helium is the second element, so we have two electrons so far. They are both in the 1*s* subshell because we have moved along a path in the first period and we are in the *s* block. So far we have $1s^2$. Next, we move to the beginning of period 2. Here we move through both the *s* and *p* blocks of the period, gaining $2s^2$ and $2p^6$. Having reached the end of period 2, we move on to period 3, going first through the *s* block ($3s^2$) and then four elements into the *p* block to reach

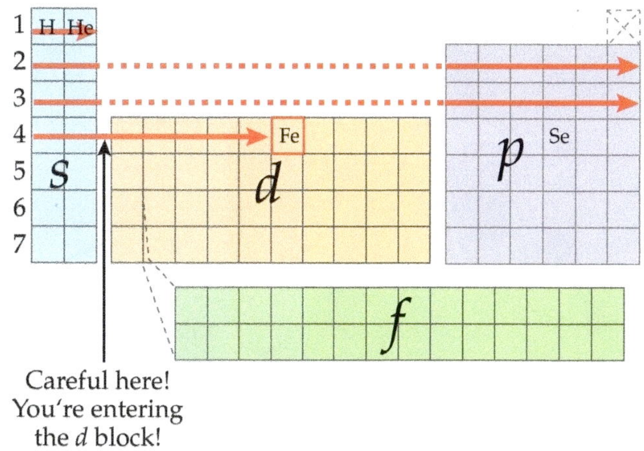

our target, sulfur. Four deep into the p block of period 3 gives us four electrons, or $3p^4$. The complete notation for sulfur (S) is, therefore, $1s^22s^22p^63s^23p^4$. As expected, there are six valence electrons in the $n = 3$ shell for this group VIA element.

Now let's try iron (Fe), a d-block element. Again we start at H. The only catch is to remember that when you plunge into the d block in period 4, you must subtract one from the period to get the correct n (to account for the subshell crossings). This means that the final six electrons should be listed as $3d^6$, not as $4d^6$. Thus, the electron configuration for iron is $1s^22s^22p^63s^23p^64s^2\,3d^6$.

If we were looking for the electron configuration of selenium (Se), our arrow would continue past iron, out of the d block, and into the p block. Once we leave the d block and enter the p block, we must bump n back up by one. The electron configuration of selenium (Se), therefore, is $1s^22s^2\,2p^63s^23p^64s^2\,3d^{10}4p^4$.

The ground-state electron configuration obtained by using either a subshell energy-level diagram or a four-block periodic table is not always correct because there are exceptions to the rules we have been following. The rules work well for the representative (s and p block) elements, but slightly more than one-quarter of the transition (d block) elements violate the rules, sending one or both of the valence ns electrons into the $(n - 1)\,d$ shell. Practice Problem 4.17 shows you one of them.

Practice Problems

4.17 Using a four-block periodic table as a guide, determine the ground-state electron configuration of copper (Cu).

Answer: You should have arrived at $1s^22s^22p^63s^23p^64s^23d^9$*, because copper is nine deep into the d block of period 4. However, copper is one of those exceptions, and its true electron configuration is* $1s^22s^22p^63s^23p^64s^13d^{10}$.

4.18 Using a four-block periodic table as a guide, determine the ground-state electron configuration of krypton (Kr). Explain why it is proper for this element to be in group VIIIA.

4.19 Using a four-block periodic table as a guide, determine the ground-state electron configuration of palladium (Pd).

So far we have been neglecting the f-block elements—that is, the elements beyond lanthanum (La) and actinium (Ac). Lanthanum and actinium can be thought of as detour signs to the f block. Remember that the value of n in the f block is two less than the period you are in. Once you get through the f block, it's back into the d block (where the value of n is only one less than the period). For example, let's determine the electron configuration of gadolinium (Gd).

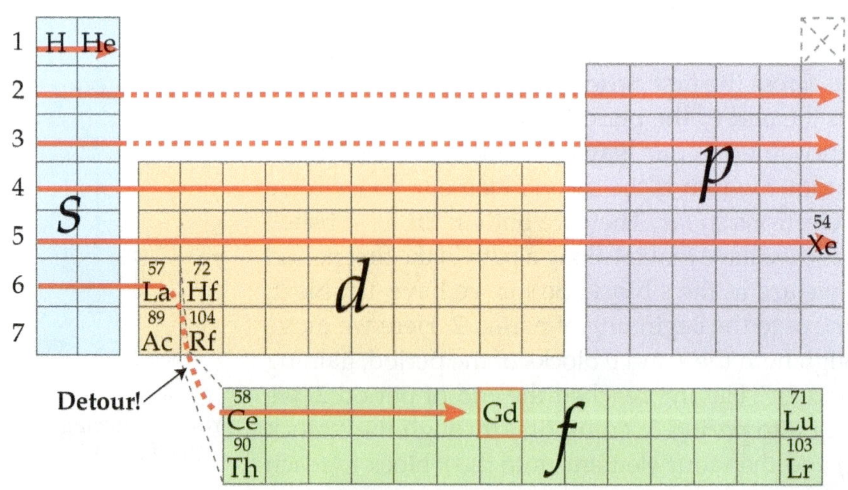

Follow the arrows. We get $1s^2$ from the first row, $2s^22p^6$ from the second row, $3s^23p^6$ from the third row, $4s^23d^{10}4p^6$ from the fourth row, $5s^24d^{10}5p^6$ from the fifth row, and finally $6s^25d^14f^7$ from the sixth row (we get only one deep into the *d* block and then must detour and go seven deep into the *f* block). The result is $1s^22s^22p^63s^23p^64s^23d^{10}4p^65s^24d^{10}5p^66s^25d^14f^7$. We must point out that for the *f*-block elements, there are more exceptions to this method of filling the subshells than there are elements that obey it. For most, you must transfer the single *d* electron into the *f* subshell that is being filled. This pattern is shown in the abbreviated electron configurations listed here:

Electron Configurations of Highest-Energy Subshells (the Ones Being Filled)

58	59	60	61	62	63	64	65	66	67	68	69	70	71
Ce	**Pr**	**Nd**	**Pm**	**Sm**	**Eu**	**Gd**	**Tb**	**Dy**	**Ho**	**Er**	**Tm**	**Yb**	**Lu**
$6s^25d^14f^1$	$6s^25d^14f^2$	$6s^25d^14f^3$	$6s^25d^14f^4$	$6s^25d^14f^5$	$6s^25d^14f^6$	$6s^25d^14f^7$	$6s^25d^14f^8$	$6s^25d^14f^9$	$6s^25d^14f^{10}$	$6s^25d^14f^{11}$	$6s^25d^14f^{12}$	$6s^25d^14f^{13}$	$6s^25d^14f^{14}$
	$6s^24f^3$	$6s^24f^4$	$6s^24f^5$	$6s^24f^6$	$6s^24f^7$		$6s^24f^9$	$6s^24f^{10}$	$6s^24f^{11}$	$6s^24f^{12}$	$6s^24f^{13}$	$6s^24f^{14}$	
90	91	92	93	94	95	96	97	98	99	100	101	102	103
Th	**Pa**	**U**	**Np**	**Pu**	**Am**	**Cm**	**Bk**	**Cf**	**Es**	**Fm**	**Md**	**No**	**Lr**
$7s^26d^15f^1$	$7s^26d^15f^2$	$7s^26d^15f^3$	$7s^26d^15f^4$	$7s^26d^15f^5$	$7s^26d^15f^6$	$7s^26d^15f^7$	$7s^26d^15f^8$	$7s^26d^15f^9$	$7s^26d^15f^{10}$	$7s^26d^15f^{11}$	$7s^26d^15f^{12}$	$7s^26d^15f^{13}$	$7s^26d^15f^{14}$
$7s^26d^2$				$7s^25f^6$	$7s^25f^7$		$7s^25f^9$	$7s^25f^{10}$	$7s^25f^{11}$	$7s^25f^{12}$	$7s^25f^{13}$	$7s^25f^{14}$	

red = predicted electron configuration
black = actual configuration when different from predicted

Writing electron configurations for atoms that possess many electrons, such as the *f*-block elements, can get pretty tedious. Chemists have developed an additional shortcut which recognizes that the electron configuration of the noble gas atom that directly precedes any element is identical to the early part of the electron configuration for that element. For example, the noble gas that directly precedes gadolinium (Gd, atomic number 64) is xenon (Xe, atomic number 54, look at the periodic table above). The first 54 electrons in Gd can thus be replaced with the Xe symbol, and the electron configuration of Gd can be written $[Xe]6s^25d^14f^7$. In this notation, the symbol for the noble gas that immediately proceeds the element is used and is always enclosed in square brackets.

Practice Problems

4.20 Write the ground-state electron configuration of chlorine (Cl) and lutetium (Lu), using both the full notation and the noble gas abbreviated notation.

Answer: Cl $1s^22s^22p^63s^23p^5$; $[Ne]3s^23p^5$

Lu $1s^22s^22p^63s^23p^64s^23d^{10}4p^65s^24d^{10}5p^66s^25d^14f^{14}$; $[Xe]6s^25d^14f^{14}$

4.21 Write the ground-state electron configuration of radium (Ra), using both the full notation and the noble gas abbreviated notation.

4.22 Write the ground-state electron configuration of uranium (U), using both the full notation and the noble gas abbreviated notation.

Finally, we can turn this around and ask questions about an atom once we've determined its electron configuration. For example, arsenic (As) has the ground-state electron configuration $1s^22s^22p^63s^23p^64s^23d^{10}4p^3$ (or $[Ar]4s^23d^{10}4p^3$). We could now ask, "Where does this atom belong in the periodic table? Which period is it in? Which group is it in?" You should be able to answer these questions. Because the outermost occupied (valence) shell is $n = 4$, we expect to find As in the fourth period. In addition, As has five valence electrons, which we can emphasize by rewriting the electron configuration as $1s^22s^22p^63s^23p^63d^{10}4s^24p^3$. Five valence electrons puts it in group VA.

Here are a few additional hints. If an element has a partially filled *d* subshell, it is a transition metal (a *d*-block element). If an element has a partially filled *f* subshell, it is an *f*-block element. A partially filled *s* or *p* subshell usually identifies the element as an *s*-block element or a *p*-block element, respectively, and thus as a representative element (there are exceptions, of course; however, we will not worry about these here).

Practice Problems

4.23 In which period and which group is the element that has the ground-state electron configuration $[Ar]4s^23d^{10}4p^5$? Which element is this?

Answer: Rewriting the electron configuration as $[Ar]3d^{10}4s^24p^5$ to emphasize the valence electrons, we see that the highest value of n is 4, so this atom is in the fourth period. There are a total of seven valence electrons, and the p subshell is partially filled, meaning this is a group VIIA element. It is bromine (Br).

4.24 In which period and which group is the element that has the ground-state electron configuration $1s^22s^22p^63s^23p^64s^23d^3$? Which element is this?

4.25 In which period and which group is the element that has the ground-state electron configuration $[Xe]6s^1$? Which element is this?

4.6 Regular Variations in the Properties of Elements

Chemical behavior is not the only property of the elements that varies in a periodic, repetitive way. Two other properties are *atomic size* and *ionization energy*. In the previous chapter, we examined how the ionization energy of the elements increases as you go from left to right across any period. That is, it requires increasingly larger amounts of energy to remove an electron from an atom as you go from left to right across a period. Since metallic elements are located on the left side of the periodic table, and nonmetallic elements are on the right side, this led to an important generalization:

Metals prefer to lose valence electrons.
Nonmetals prefer to gain valence electrons.

What we did not do in the previous chapter is explain what is behind the ionization energy trend. We'll do that now, and electron configuration will be the key. We will begin by looking at the periodic trend in atomic size first.

Atomic Size

If we think of atoms as spheres, we can characterize each by its radius, called the **atomic radius**. Atomic radii can be measured by experimental techniques such as X-ray diffraction. They are often reported in angstroms (Å, 1 Å=10^{-10} m), nanometers (10^{-9} m), or picometers (10^{-12} m). The chart on the next page illustrates the relative sizes of the representative elements. Notice that the atoms vary in size in a regular way as you go down a group or across a period. In general,

Atomic radius increases as you go down a group.
Atomic radius decreases as you go from left to right in a period.

There are local exceptions to these trends, but they hold overall.

Relative Atomic Sizes of the Representative Elements

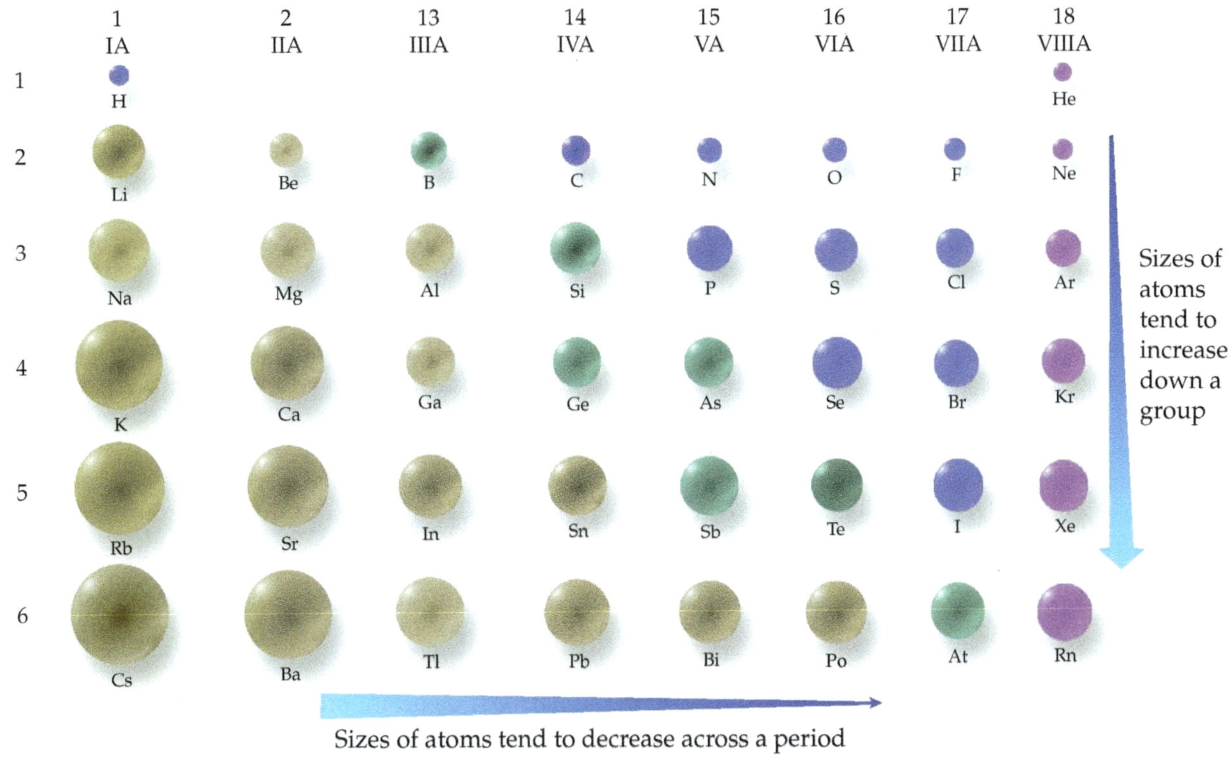

We are now in a position to understand why size varies in this way. Let's start with the trend within a group. The electron configurations for the first three elements in group IA are shown in the diagram to the right.

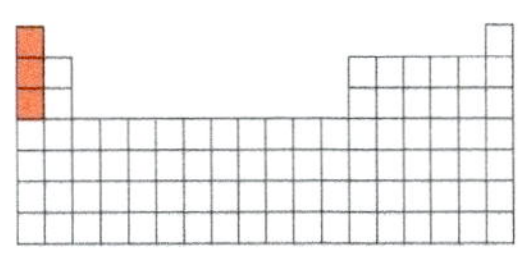

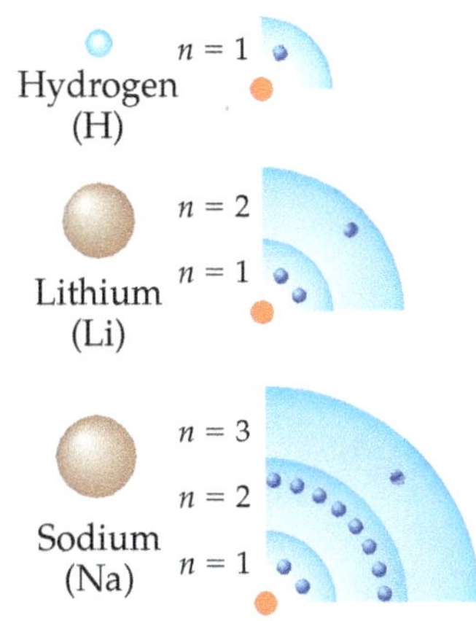

One of the consequences of Bohr's model is that shells get larger as n increases. Therefore, a sodium atom, with its last added electron in the third shell, is larger than a lithium atom, which has only two occupied shells. Each step down in a group adds an additional occupied shell, thus increasing the size of the atom.

As you can see from the chart below, the size changes within a row (period) of the periodic table are somewhat more subtle, but they still follow a definite trend. To examine the size trend across a period, let's look at the electron configurations for the first four elements in period 2:

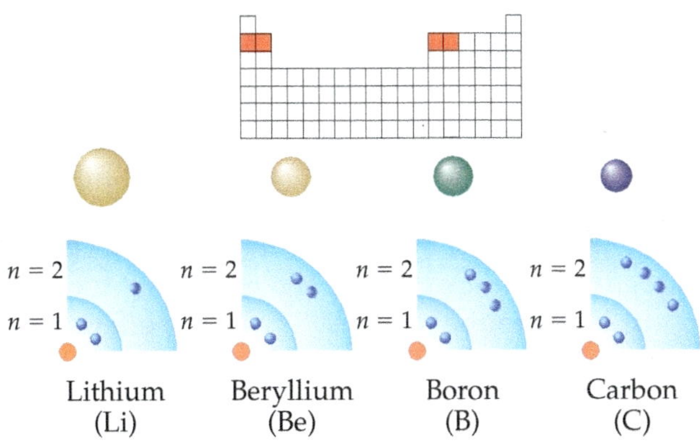

All of these atoms have their outermost electrons in the second shell of the atom, and as a result you might expect that the sizes of all the atoms would be roughly the same. But remember, as we go across a period from left to right, each step adds an additional electron to the valence shell and an additional proton to the nucleus. The added proton makes the nucleus more positive, which causes an increased attraction for all of the electrons, including the valence electrons. This increased positive nuclear charge results in all the electrons being pulled in more tightly and shrinks the shells, thus decreasing the size of the atom. This combination of the same valence shell and an increased nuclear charge is responsible for the observed size trend within a period.

WORKPATCH 4.7 Redraw the shells of Li, Be, B, and C shown in the above drawing to make the overall size of each atom fit the size trend. In each nucleus, write the total number of protons present. Which atom would be most difficult to ionize? ●

What happens to the relative sizes of the shells in your WorkPatch drawings as you go from left to right? Check your drawings carefully against the answers at the end of the chapter. Good drawings make the answer to the ionization question obvious.

Ionization Energy

Now we are set to explain the periodic trend in ionization energy. On the next page is the diagram you first saw in the previous chapter. Each color represents a different period. As you climb up each colored line, you are moving from left to right across a period, and the ionization energy generally increases (it becomes more difficult to remove an electron from the atom). The plot also shows that as you move down any group, the ionization energy decreases (it becomes easier to remove an electron from the atom). This is shown below for groups IA (the elements boxed in blue) and VIIIA (the elements boxed in green). Look at the elements outlined in blue. As you move across the plot from Li to Na to K,... and so on you are moving down group IA, and the ionization energy is decreasing. The same holds true for the green boxed group VIIIA elements, and for any group, for that matter.

WORKPATCH 4.8 Locate the elements in group VIIA in the ionization energy plot. Do any of these elements violate the expected trend in ionization energies? ●

Astantine (At), not labeled on the plot and the last element in group VIIA, is the dot to the immediate left of Radon (Rn).

We can now understand these period and group trends in ionization energy by looking again at the trends in atomic size. As you go down a group, atoms get bigger, and, at the same time, it becomes easier (requires less energy) to remove an electron. In a larger atom, the outermost electrons are, in general, farther from the nucleus. Also, the inner electrons partly cancel or shield some of the positive charge of the nucleus, with the result that the outer electrons "feel" a reduced attraction to the nucleus. These two effects mean that in a larger atom, the outer electrons are less strongly attracted to the nucleus than in a smaller atom. Therefore, the outer electrons are easier to remove from the larger atom—their ionization energy is lower as you progress down a group.

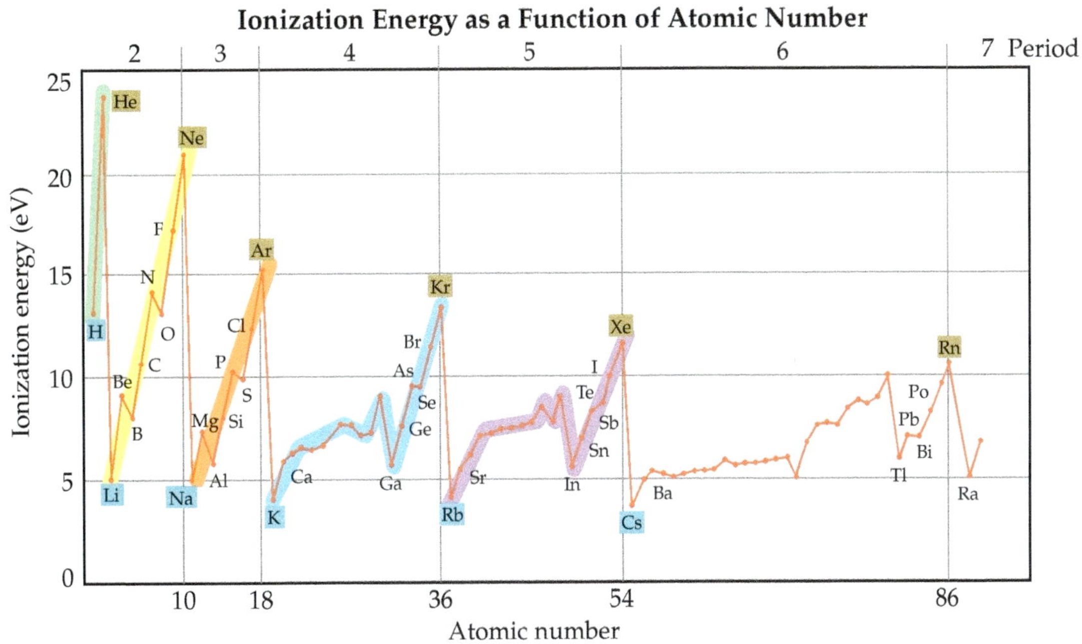

Of course, the opposite is also true. As atoms get smaller (either going up a group or left to right across a period), it becomes more difficult to remove an electron because they are getting closer and therefore more attracted to the nucleus, and so the ionization energy increases. This explains why ionization increases from left to right across a period.

These regular variations (periodic trends) in atomic size and ionization energy can be a great help to chemists in explaining the chemical behavior of the elements. For example, the alkali metals in group IA (1) have relatively low ionization energies. This means that an element like sodium easily loses an electron. Now, if you start at sodium and travel across period 3, you come to a much smaller atom that has a relatively large ionization energy, chlorine (Cl). It takes a large amount of energy for chlorine to lose an electron, and this element generally does not give up electrons when it reacts. In fact, just the opposite tends to occur: Cl tends to *gain* an electron when it reacts. Bring sodium and chlorine together, and this is exactly what happens. Each sodium atom loses an electron, becoming a Na^+ cation, and each chlorine atom picks up an electron and becomes a Cl^- anion. The result is the spontaneous and rather spectacular chemical transformation of sodium and chlorine into the compound sodium chloride (NaCl), table salt.

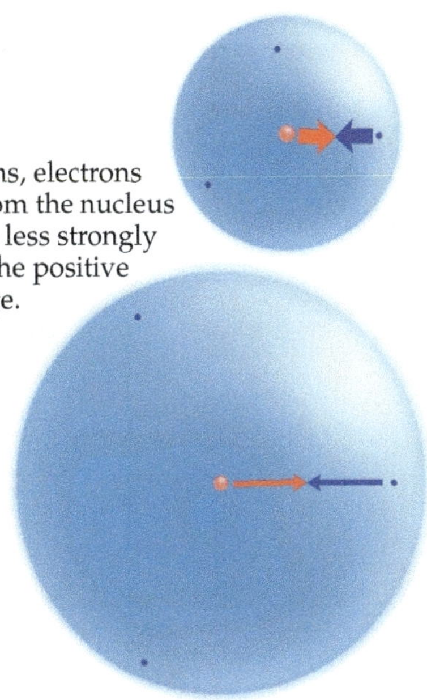

In larger atoms, electrons are farther from the nucleus and therefore less strongly attracted by the positive nuclear charge.

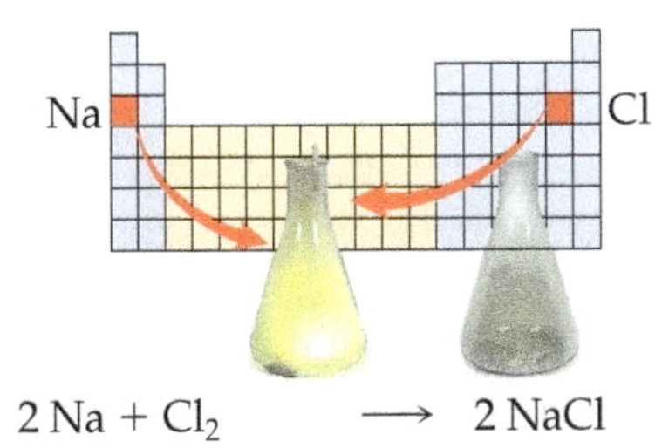

Practice Problems

4.26 The alkali metals all react with water by giving up an electron to it. Based on their relative positions in the periodic table, which do you think would be more reactive with water, lithium or sodium?

Answer: Sodium. It is farther down the group and therefore has a lower ionization energy, meaning it is easier for sodium to give up an electron. Because giving up an electron is how it reacts with water, this should make sodium the more reactive metal.

4.27 Arrange the atoms Mg, Sr, S, O, and Rb from smallest to largest.

4.28 Give the full symbol for the atom or ion that has 26 protons and 30 neutrons in its nucleus, and 23 electrons outside its nucleus. Also give the number of the group this element is in.

4.29 Give the full symbol for the atom or ion that has eight protons and eight neutrons in its nucleus, and ten electrons outside the nucleus. Also give the name of the group this element is located in.

Now that you understand why metals easily lose valence electrons whereas nonmetals lose them with difficulty (and in fact, would rather gain valence electrons), you are ready to use the Bohr model to understand why atoms of elements combine to form compounds the way they do.

4.7 Compound Formation and the Octet Rule

The Bohr model can be very useful in predicting and understanding the chemical composition of many compounds. For example, it can be used to explain why the compound sodium chloride has the formula NaCl and not Na_2Cl, $NaCl_2$, or any other variation. The one-to-one ratio of sodium to chlorine in this compound can be understood in terms of an early chemical principle known as the *octet rule*. This rule arose from the observation of the unique nature of the elements in group VIIIA (18), the noble gases. As we saw in Chapter 3, the elements in this group are exceptionally unreactive. This lack of "desire" to combine with other elements was interpreted as exceptional stability for the elements in group VIIIA. The Bohr model provided another unifying feature for these elements—they all have eight electrons in their valence shell (with the exception of helium, which has two). The obvious implication is that an element is or becomes unreactive (unusually stable) if it has eight valence electrons. Thus was born the octet rule.

> **The octet rule: Elements react to form compounds in such a way as to put eight electrons in their outermost valence shell, giving them a valence-shell configuration identical to that of a noble gas and making them exceptionally stable.**

According to the **octet rule** (from the Latin *octo*, meaning "eight"), the reason the other elements in the periodic table (groups IA–VIIA) participate in chemical reactions is to ultimately end up with eight electrons in their valence shell, thus gaining a stability similar to that of the noble gases.

The octet rule works well for many of the atoms found in biological and organic molecules (in particular, C, H, O, and N), and it is used to predict how the atoms in these molecules bind together. We can also use this rule with sodium and chlorine to understand why the chemical formula for sodium chloride is NaCl and not something else. Such applications make the octet rule very useful. However, the rule has many exceptions. For example, hydrogen's outermost occupied shell is the $n = 1$ shell, and so the hydrogen atom achieves stability (which means a filled outermost occupied shell) with only two electrons (a duet), so that hydrogen's electron configuration becomes like that of helium. The elements beryllium (Be) and boron (B) often settle for only four and six electrons, respectively, in many compounds instead of the usual eight. Many atoms in the third period and beyond can end up with more than an octet in some compounds, and the transition metals usually don't obey the octet rule at all. Despite all these exceptions, chemists still use this simple rule because so much of chemistry involves organic and biological molecules to which it usually applies—but remember that it has its limitations.

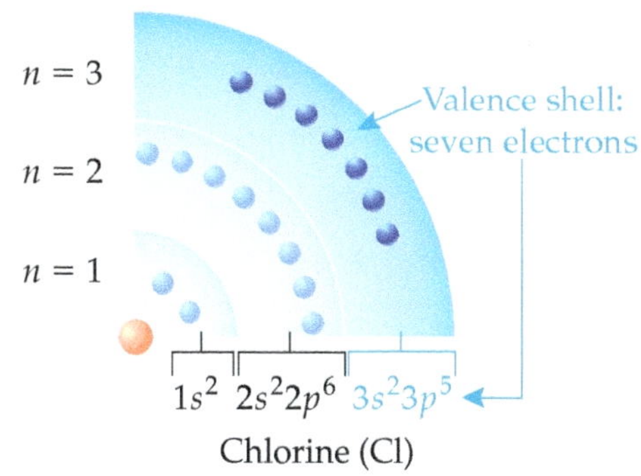

Let's consider the formation of sodium chloride from sodium and chlorine atoms in terms of the octet rule. We begin with a chlorine atom, which has 17 electrons and the electron configuration shown on the right. Because chlorine is a member of group VIIA, chlorine atoms have seven electrons in their outermost occupied shell before the reaction. To understand how chlorine atoms react, focus on the valence shell and apply the octet rule by asking, "How can a chlorine atom get eight electrons in this shell?" Because the valence shell is only a single electron short of an octet, the obvious answer is for each chlorine atom to pick up one electron, which would produce a chloride ion with a −1 charge. But where will this electron come from? To answer this, let's look at a sodium atom, which has 11 electrons and the electron configuration shown on the right.

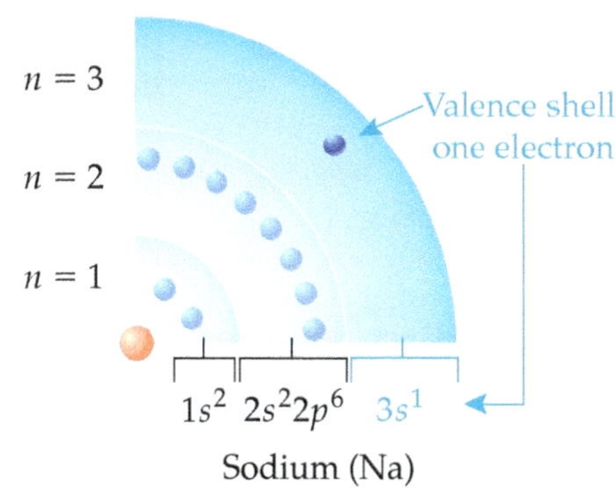

Sodium atoms have only a single electron in their valence shell. Again we apply the octet rule and ask, "What is the best way for a sodium atom to end up with eight electrons in its valence shell?" Like chlorine, sodium could do this by picking up more electrons, but each atom would have to gain seven electrons. This would result in a sodium ion with a −7 charge, and that's quite a bit of excess negative charge for the sodium nucleus to hold on to. An easier way for sodium to achieve its octet is for it to lose its single valence electron from the $n = 3$ shell, leaving this shell empty. The $n = 2$ shell then becomes the valence shell, and this shell already has eight electrons in it! In addition, if sodium loses one electron, then this electron is now available for a chlorine atom.

Recall that Cl needs to gain one electron to satisfy the octet rule. Now we know where it might come from. In reacting with chlorine, each sodium atom needs to lose one electron to satisfy the octet rule and become a cation with a

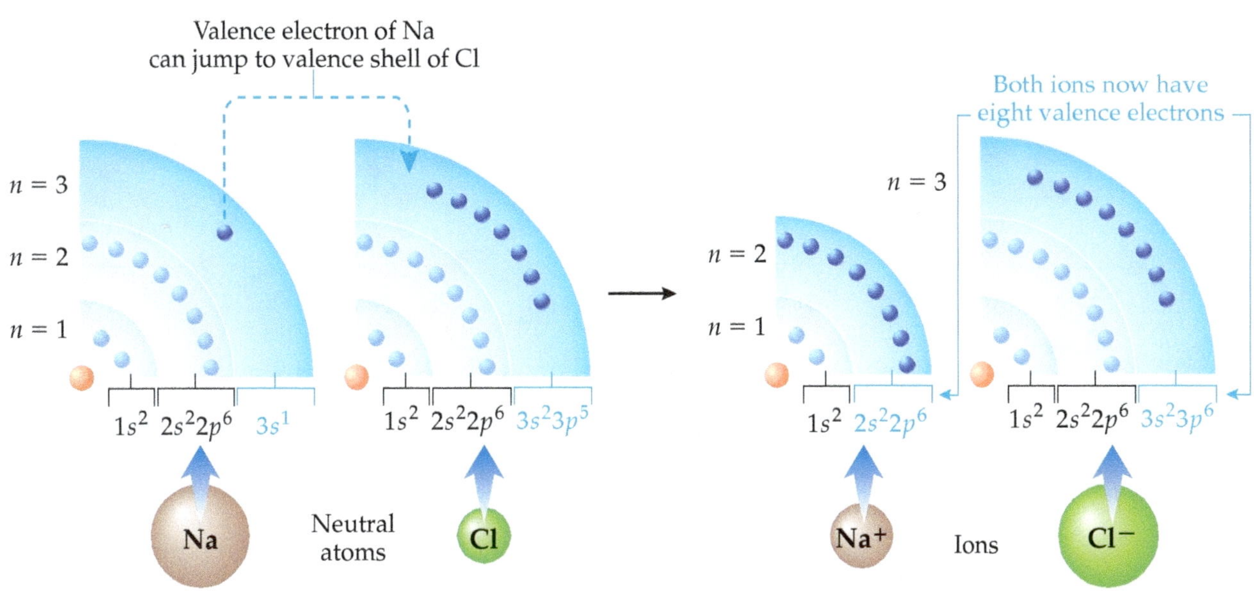

+1 charge. Each chlorine atom gains one electron and becomes an anion with a −1 charge. In doing so, both the newly formed ions satisfy the octet rule (eight electrons in the valence shell).

In essence, one electron is exchanged between the atoms. Note that in this exchange, the atom that loses the electron, sodium, is a metal. The atom that gains the electron, chlorine, is a nonmetal. This is consistent with our rule of thumb regarding valence electrons and the typical behavior of metals and nonmetals.

We can also now explain why the composition of sodium chloride is NaCl and not, say, $NaCl_2$. Each sodium atom loses one electron, and each chlorine atom picks up one electron. In order to balance this electron exchange and not end up with a deficit or excess of electrons, these elements must combine in a one-to-one ratio.

That is not the case for all elements. Let's predict the formula for the compound magnesium fluoride. We start by considering the electron configurations of Mg, a group IIA metal, and F, a group VIIA nonmetal:

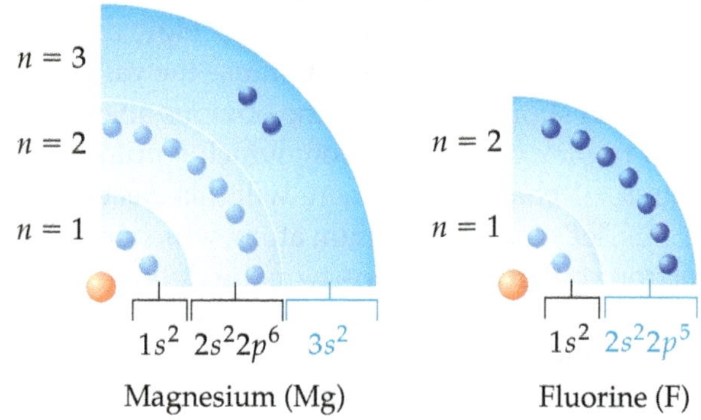

Now consider the best way for each atom to achieve an octet in its valence shell. Like chlorine, each nonmetal fluorine atom needs to gain only one electron and, in doing so, will form the anion F^-. Magnesium, a metal atom, can most easily arrive at an octet by losing electrons, but it must lose two electrons and will therefore

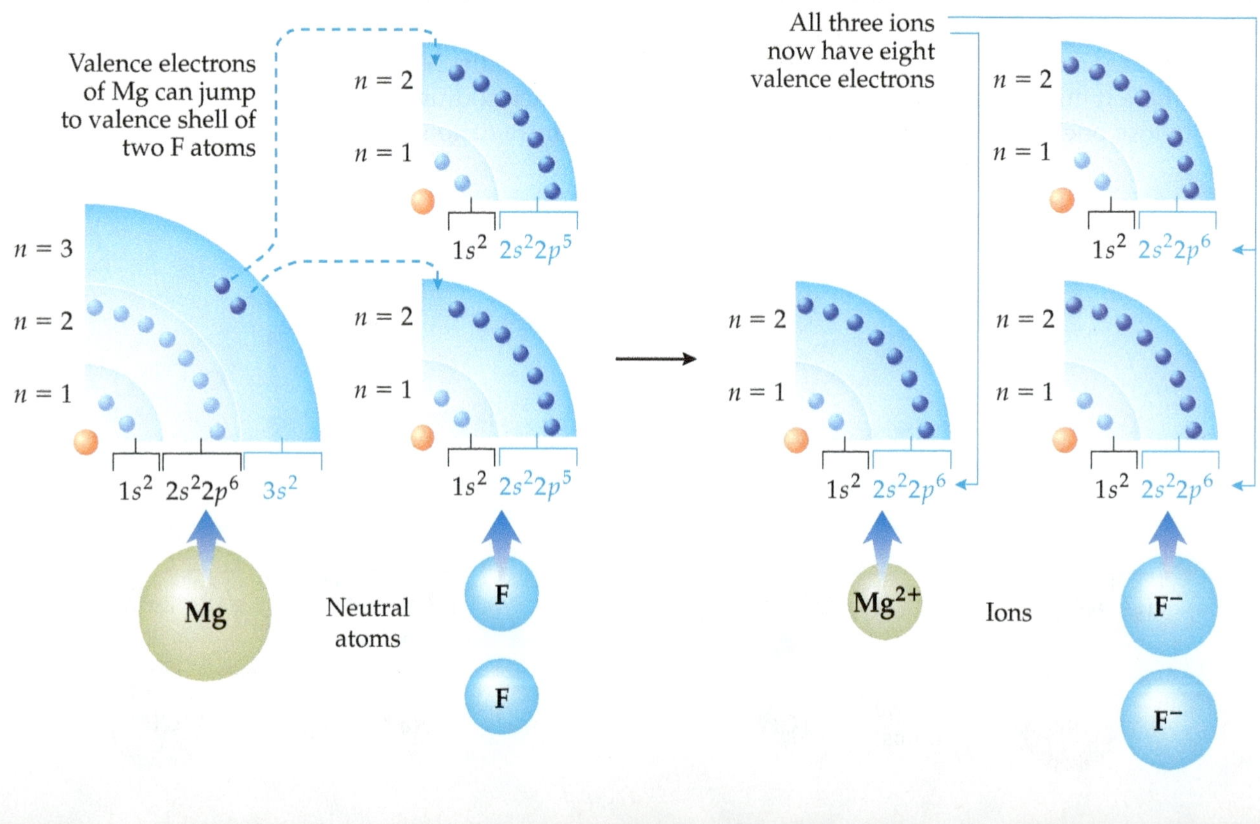

form a cation with a +2 charge (Mg^{2+}). This has a profound effect on the formula of the compound that can be formed. Because each Mg atom must lose two electrons but each F atom will pick up only one electron, it takes two F atoms reacting with one Mg atom to balance the electron exchange. The +2 charge and the two −1 charges of the resulting ions add up to zero, making the compound electrically neutral. The chemical formula of magnesium fluoride is thus MgF_2. The diagram on the bottom of the previous page summarizes all this. Keep your eyes on the blue valence electrons and you'll see that the octet rule is obeyed.

You might realize by now that it is not really necessary to draw the electron configurations of the atoms in order to determine how each achieves its octet. There is a shortcut, at least for the representative (group A) elements. Reconsider the magnesium fluoride example. Magnesium is in group IIA, so we know that its valence shell contains two electrons without resorting to an electron-configuration diagram. We also know that Mg is a metal and that metals tend to lose electrons when they react, so it's a good bet that Mg will lose both its valence electrons to satisfy the octet rule. Fluorine is a group VIIA element and thus has seven electrons in its valence shell. Fluorine is also a nonmetal, and nonmetals tend to gain electrons when they react, so it is likely that F will gain one electron to satisfy the octet rule. Thus, instead of having to draw electron configurations to determine formulas, we just need to consider the Roman numeral of the group the atoms are in and whether the elements are metals or nonmetals. Table 4.2 sums up the behavior of the representative elements for the metals in groups IA–IIIA and the nonmetals in groups VA–VIIA. You can use this table to predict the most likely formula for the compound that results from the reaction between any representative metal and nonmetal.

Table 4.2 Summary of the Chemistry of the Representative Elements

Group	Valence-shell electron configuration	To achieve octet in valence shell, atom	To become
IA	ns^1	Loses one electron	→ +1 cation
IIA	ns^2	Loses two electrons	→ +2 cation
IIIA	ns^2np^1	Loses three electrons	→ +3 cation
VA	ns^2np^3	Gains three electrons	→ −3 anion
VIA	ns^2np^4	Gains two electrons	→ −2 anion
VIIA	ns^2np^5	Gains one electron	→ −1 anion

Notice that the elements of group IVA are not included in this table. We purposely left them out because they can be a bit schizophrenic. Each of these elements has four valence electrons and therefore would need to either gain or lose four electrons to achieve an octet. Neither alternative is very satisfactory, and thus these elements tend to react in an entirely different way. We shall discuss this point in more detail in the next chapter.

Practice Problems

4.30 Predict the chemical formula Na_xO_y of the compound that results from the reaction between the elements Na and O.

Answer: Na is a group IA metal and thus loses one electron to become Na^+. *O is a group VIA nonmetal and so gains two electrons to become* O^{2-}. *In order for the compound formed to be electrically neutral, the formula must be* Na_2O.

4.31 Draw the reaction above in terms of electron configurations, as was done for MgF_2 in the text.

4.32 Predict the chemical formula of the compound that results from the reaction between the elements Ba and F.

4.33 Predict the chemical formula of the compound that results from the reaction between the elements Al and O.

You can see from these practice problems that the octet rule is a powerful predictive tool. To a chemist, the tendency to lose or gain valence electrons in a reaction is the defining property that makes an element a metal or a nonmetal. It is a rule of thumb to remember. Of course, like any rule of thumb, it is a generalization. There are certainly examples of reactions where metal atoms gain electrons or nonmetal atoms lose them. Even the octet rule is a generalization, a rule of thumb with plenty of exceptions. Nevertheless, it is extremely useful and often yields correct predictions.

+1 One More Thing

4.8 The Modern Quantum Mechanical Model of the Atom

The Bohr model began with the assumption that the electrons in an atom can possess only certain amounts of energy (in other words, that energies are quantized). It is an understatement to say that many scientists of the day, having been trained in classical physics, had difficulty adjusting to this concept. But there was little time to complain. In less than ten years, quantum physics went even further, with a new description of the electrons in an atom so unusual that even the scientists who gave birth to the field weren't very happy about this new description. Bohr himself said, "Anyone who is not shocked by quantum theory has not understood it." And, as noted at the beginning of this chapter, Erwin Schrödinger, the Austrian physicist who developed the equation on which much of quantum physics is based, has been quoted as saying, "I don't like it, and I'm sorry I ever had anything to do with it." Much of this discomfort was generated not so much by the concept of quantized energy levels but by the other bizarre behaviors that quantum physics demanded of electrons. For example, the equations of quantum physics predict that electrons can exhibit a form of behavior known as *tunneling*. To best understand this behavior, let's assume that a baseball follows the same laws of physics as an electron. The tunneling theory says that if you place a baseball in a solid steel box and weld a cover onto the box, you can come back sometime later and find the baseball sitting outside the still-sealed box, even though the baseball lacks sufficient energy to penetrate the box. No magic or trickery is involved here. It's just possible according to the laws of quantum physics. Baseballs are large objects and therefore follow the laws of classical physics, but electrons are small enough that quantum effects are important, and electron tunneling is experimentally observed. In fact, a number of electronic devices on the market today (for example, the scanning tunneling electron microscope and

Josephson junctions, used to make superfast switches for supercomputers) make use of this strictly quantum mechanical behavior. But this was not the worst of it.

This bizarre tunneling behavior where electrons appear in "forbidden" places (on the opposite side of barriers meant to constrain them) is a direct manifestation of what is called the **uncertainty principle**, put forth by Werner Heisenberg in the 1930s. The essence of this principle is that in dealing with particles the size of an electron, it is impossible to know, with any real accuracy, exactly where the electron is or where it is going. In addition, the more carefully and precisely you measure the location of an electron, the less well you will be able to predict where it will be in the next instant. This is exactly the opposite of classical physics, where, for example, careful determination of the present position and velocity of the Moon allows us to know exactly where it will be days or weeks (even centuries) later so that we can aim a rocket and land on it.

The uncertainty principle has some direct consequences when applied to an electron. Because we can never know an electron's precise location, we cannot describe it as a particle traveling on a well-defined orbit or path because such a particle could be precisely located (in other words, you could point to it). Couple this fact with the experimental observation made in the early 1900s that the electron displays wavelike behavior (*it has a measurable wavelength, just like light!*), and we really have a mess. Exactly how are we supposed to think of an electron in an atom? What picture should come to mind? The answer came from Schrödinger in 1925. He accepted the uncertainty principle and said that, instead of viewing the electron as a particle, we should think of it as a nebulous cloud of negatively charged matter. He went on to show that the density of this electron cloud varies. That is, in some places the cloud is like a thick fog, and in other places it is very thin. Places where the cloud density is high are places where the probability of finding the electron is high, and places where the density is low are places where the probability of finding the electron is low. The cloud might look like this:

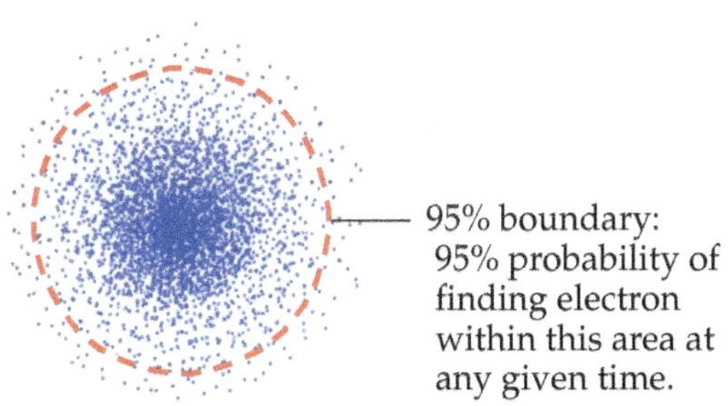

Electron cloud corresponding to 95% boundary.

The left side of the illustration shows a cross section of the cloud for a 1*s* electron (the actual cloud is a three-dimensional sphere). The red dotted line encompasses the volume within which the electron resides 95% of the time. The right side of the illustration shows the spherical shape of the cloud by representing the 95% boundary as a sphere.

Schrödinger called this electron cloud an **orbital**. For some physicists, this was too much to accept. Einstein, for instance, refused to believe in this newest quantum mechanical model and its probability-based view of the electron, insisting that "God does not play dice with the universe."

Today, however, most physicists and chemists have been forced by the weight of many experimental observations to subscribe to Schrödinger's description of the atom, which is the current, modern, quantum mechanical model of the atom. Of course, some "believers" have had to be dragged, kicking and screaming, to acceptance.

Schrödinger did not stop here. He treated the electron cloud (the orbital) almost like a mound of Jello. Have you ever shaken Jello? If you have, you know that it wiggles back and forth with periodic (wavelike) motion. Well, Schrödinger said that the electron cloud also exhibits wavelike motion and that this motion gives rise to orbitals (electron clouds) of various shapes. These shapes are generated by a mathematical equation called a wave equation. The three shapes for orbitals that you should know about are:

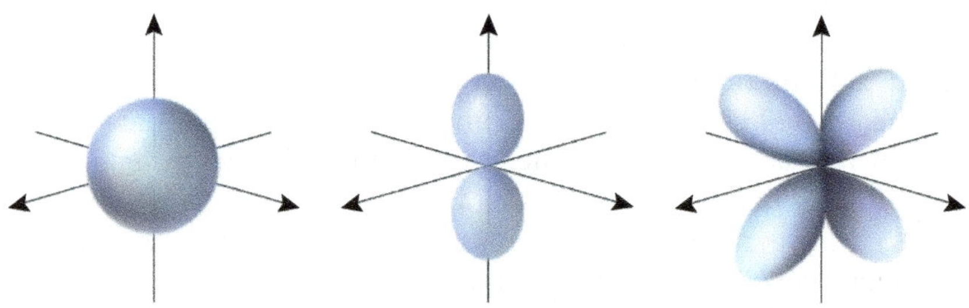

An *s* orbital is spherical, a *p* orbital is dumbbell-shaped, and a *d* orbital looks a bit like a four-leafed clover (there are also *f* orbitals). Notice the *s*, *p*, and *d* designations. These are the same designations we saw in Bohr's subshell model and when we were obtaining electron configurations from the periodic table. What is a good way to think of all this? One way is to imagine you could look inside of an atom and see its electrons, and then you would see clouds of various sizes and shapes made up of negatively charged matter.

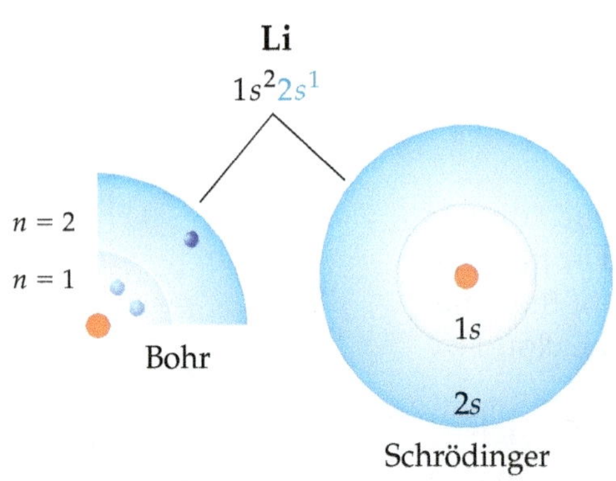

So here is how the modern model works. Let's consider lithium, with its $1s^2 2s^1$ ground-state electron configuration. Bohr would say that the two 1*s* electrons are in a 1*s* shell and the one 2*s* electron is in a larger 2*s* shell. Schrödinger would say that the two 1*s* electrons spend their time looking like a 1*s* orbital, and the single 2*s* electron looks like the larger 2*s* orbital. (You would see a small and very dense 1*s* cloud and a less dense but larger 2*s* cloud, both centered about the nucleus.)

According to Schrödinger, the orbitals get larger as the principal quantum number *n* increases, as shown here with *s* orbitals:

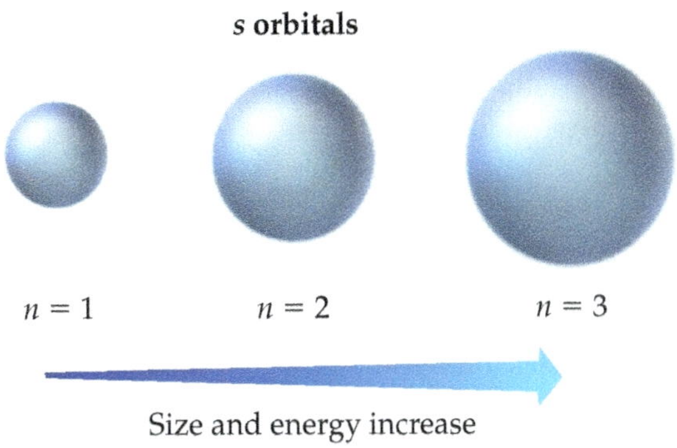

A larger cloud means that the electron spends more time, on average, farther from the nucleus. This means that as n increases, the electron spends more time farther away from the nucleus (just as it does in the Bohr model), increasing the electron's energy. Indeed, if we gave sufficient energy to the $1s^2 2s^1$ ground-state lithium atom so as to produce the excited-state $1s^2 2p^1$ configuration, Schrödinger's view of what's happening would be:

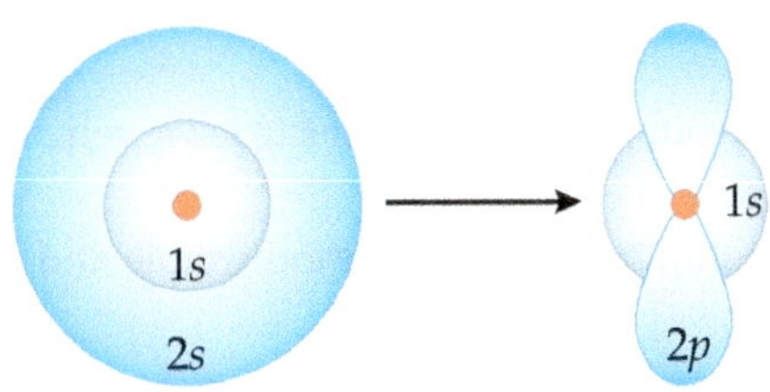

$1s^2 2s^1$
Ground state for Li

$1s^2 2p^1$
An excited state for Li

The valence electron has changed from a spherical $2s$ cloud to a dumbbell-shaped $2p$ cloud.

So let's sum up the current model of the atom. It retains Rutherford's tiny, massive, positively charged nucleus and Bohr's proposal about the quantized energy of the electrons, but it replaces the orbiting-particle version of an electron with a nebulous electron cloud (an *orbital*) that has (a) a particular shape resulting from the electron's wavelike motion and (b) a particular size and energy determined by its principal quantum number n. The orbital defines the volume of space about the nucleus where the electron is most likely to be found.

If you continue on with advanced courses in chemistry, you will study the modern quantum mechanical model of the atom in much greater depth. However, what you have learned in this chapter will serve you well for understanding chemical bonding, which comes next.

Have You Learned This?

Wavelength (p. 121)
Electromagnetic radiation (p. 122)
Electromagnetic spectrum (p. 122)
Quantized energy (p. 124)
Classical physics (p. 125)
Quantum physics (p. 125)
Shell (p. 127)
Principal quantum number (n) (p. 127)
Maximum (p. 127)
Valence shell (p. 128)
Energy-level diagram (p. 130)
Ground state (p. 130)
Excited state (p. 131)
Subshell (p. 135)
Electron configuration (p. 137)
Metal (p. 142)
Nonmetal (p. 142)
Atomic radius (p. 142)
Octet rule (p. 146)
Uncertainty principle (p. 151)
Orbital (p. 151)

Learning Outcomes

4.1	**Seeing the Light—A New Model of the Atom**	Describe electromagnetic radiation in terms of its wavelength, frequency, and energy. [34, 35, 39] Calculate the energy of light. [36, 44]
4.2	**A New Kind of Physics—Energy Is Quantized**	Explain what is meant by "quantized energy." [45, 46, 47, 48, 49]
4.3	**The Bohr Theory of Atomic Structure**	Describe Bohr's shells (orbits) and how they relate to quantized energy and *n*. [56, 57, 58] Explain the difference between a ground-state and an excited-state electron configuration. [59, 60, 61]
4.4	**Periodicity and Line Spectra Explained**	Explain how the Bohr model accounts for atomic line spectra. [70, 75, 76, 77] Explain how the Bohr model accounts for periodicity of the elements. [66, 69, 138]
4.5	**Subshells and Electron Configuration**	Describe Bohr's subshells and why they are needed. [78, 79, 80, 81, 82, 83] Write the ground-state electron configuration for an atom and an ion. [84, 85, 86, 87, 89, 92] Identify the period, group, and valence electrons given an electron configuration. [88, 90] Describe the relationship between the periodic table and an atom's electron configuration. [91, 93, 94]
4.6	**Regular Variations in the Properties of Elements**	Describe the periodic trends in atomic size and ionization energy. [95, 96, 97, 98, 99, 105]
4.7	**Compound Formation and the Octet Rule**	Describe the octet rule and how it applies to metals and nonmetals. [106, 107, 108, 109, 110] Predict the formula of a compound formed from a metal and a nonmetal using the octet rule. [111, 112, 114, 122]
4.8	**The Modern Quantum Mechanical Model of the Atom**	Describe what an orbital is and the types of orbitals that exist in an atom. [125, 126, 127, 128]

Skills to Know

Wavelength

The wavelength of light (symbolized by the Greek letter λ, lambda) is the distance from one wave crest to the next.

The shorter the wavelength of light, the higher the energy of the light. Visible light ranges from low-energy red light to high-energy indigo/violet light.

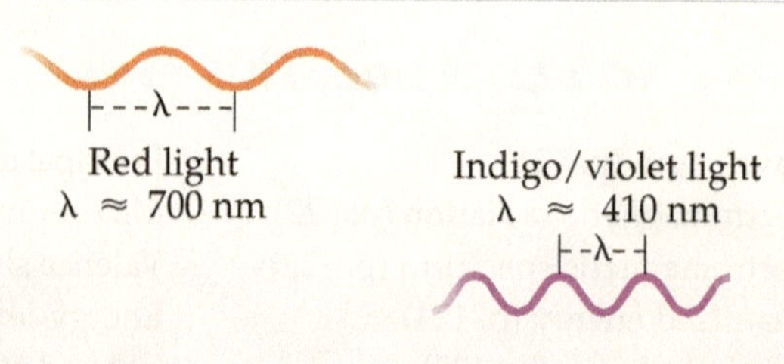

Converting between energy and wavelength

1. To find the energy in joules of light that has a particular wavelength, use the equation at right. But be careful! The wavelength must be in meters. [1 nm = 10^{-9} m]

Planck's constant $h = 6.626 \times 10^{-34}$ J · s

Speed of light $c = 3.00 \times 10^{8}$ m/s

$$E = \frac{hc}{\lambda}$$

Wavelength—*must be in meters*

2. Rearranging the above equation lets you find the wavelength of light that has a given energy. The energy must be in joules, and the equation gives you wavelength in meters.

$$\lambda = \frac{hc}{E}$$

Wavelength—*must be in meters*

Energy—*must be in joules*

Example: What is the energy of red light of wavelength 720 nm?

First convert wavelength to meters (as the equation demands): $720\ \text{nm} \times \frac{1 \times 10^{-9}\ \text{m}}{1\ \text{nm}} = 7.20 \times 10^{-7}\ \text{m}$

Then plug in the numerical values and solve: $E = \frac{hc}{\lambda} = \frac{(6.626 \times 10^{-34}\ \text{J} \cdot \text{s}) \times (3.00 \times 10^{8}\ \text{m/s})}{7.20 \times 10^{-7}\ \text{m}} = 2.76 \times 10^{-19}\ \text{J}$

Energy transitions between shells

In order for an electron to move to a higher shell, an atom must absorb an amount of energy equal to the energy difference between the shells.

When an electron relaxes to a lower shell, the atom emits an amount of energy equal to the energy difference between the shells.

The absorbed/emitted energy may be electromagnetic radiation having a wavelength corresponding to that energy.

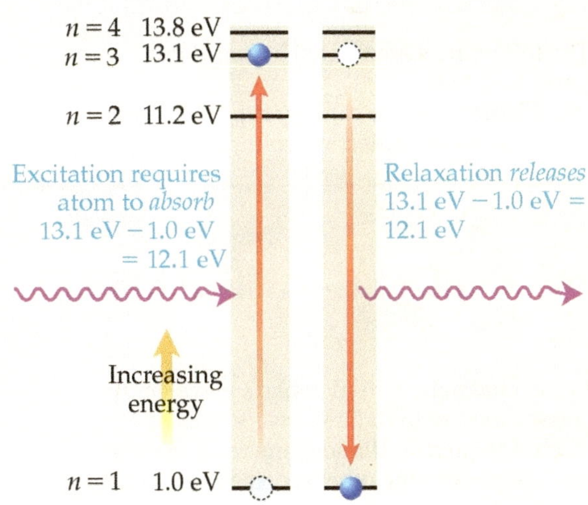

Concept Questions

1. The following is true about light:
   (a) It is also called electronic radiation.
   (b) Its energy depends on its speed.
   (c) The larger its wavelength, the smaller its frequency.
   (d) All of the above.

2. To say that something's energy is quantized means
   (a) its charge can be positive and negative.
   (b) it can have only very tiny amounts of energy.
   (c) it has energies that cannot be calculated.
   (d) it can have only certain allowed energies.

3. According to Bohr's theory of the atom
   (a) the energy of an atom's electrons are quantized.
   (b) there can be a maximum of $2n^2$ electrons at a given shell n.
   (c) both the energy and the size of an orbit (shell) increase as n increases.
   (d) all of the above.

4. Bohr explained that periodicity was due to
   (a) chemical properties repeating themselves.
   (b) similar valence-shell electron configurations.
   (c) ground-state electron configurations that were similar to excited-state configurations.
   (d) the existence of only eight main groups.

5. According to Bohr, the light emitted by atoms
   (a) is due to electron excitation between quantized energy levels.
   (b) is due to electron relaxation between quantized energy levels.
   (c) yields a continuous spectrum due to energy quantization of their electrons.
   (d) is always invisible to the human eye.

6. Why did Bohr split shells into subshells called *s*, *p*, *d*, and *f*?
   (a) To make chemistry complicated and harder for students to master.
   (b) To abide by the Uncertainty Principle.
   (c) To explain that "single" lines in a line spectrum were actually often made up of very closely spaced lines.
   (d) To explain why an atom's energy levels get closer together as n increases.

7. The following statement(s) is true about the periodic table:
   (a) Elements in the same group have similar valence-shell electron configurations.
   (b) Elements in the same period have the same value of n for their valence electrons.
   (c) When traversing an nth period in the d block, it is the $(n - 1)d$ orbitals that are being filled.
   (d) All of the above.

8. When you move from left to right across a period, atoms
   (a) shrink because the quantum number n increases.
   (b) grow because the number of protons in the nucleus increases.
   (c) shrink because the number of protons in the nucleus increases.
   (d) shrink because quantization becomes smaller.

9. Ionization energy increases left to right across a period because
   (a) it does not. Ionization energy decreases left to right across a period.
   (b) as you go down a group, atoms increase in size due to increasing values of n.
   (c) atoms shrink in size, resulting in a more positive nucleus and a tighter hold on electrons.
   (d) atoms shrink in size, resulting in increasing attraction between the electrons and the nucleus.

10. The octet rule says which of the following?
   (a) Main group metals will typically lose all their valence electrons in a chemical reaction.
   (b) Main group nonmetals will typically gain a negative charge equal to 8 – (group number).
   (c) Atoms will react to achieve eight electrons in their valence shells.
   (d) All of the above.

Answers: 1c, 2d, 3d, 4b, 5b, 6b, 7d, 8c, 9d, 10d

Seeing the Light—A New Model of the Atom

4.34 Electromagnetic radiation has a wavelength λ associated with it. Draw two waves of electromagnetic radiation, one representing green light and the other representing X-rays, showing in a relative way how the two differ in wavelength.

4.35 By how much will the value of $E\lambda/h$ differ for blue light of wavelength 460.0 nm from red light of wavelength 730.0 nm?

4.36 Consider X-rays ($\lambda \sim 10^{-10}$ m) and green light ($\lambda \sim 10^{-7}$ m).
   (a) Do a calculation to show how much more energetic X-rays are than green light.
   (b) Why are X-rays dangerous whereas green light is not?

4.37 Light travels extremely rapidly ($c = 3.00 \times 10^8$ m/s). Suppose you had to travel 30 miles to work every day. If you traveled at the speed of light for the entire trip, how long would it take you to get to work (in seconds)? [1 mile = 1.61 km]

4.38 The Earth is 9.3×10^7 miles from the Sun. How long does it take the Sun's visible light to reach us (in minutes)? The Sun also emits high-energy gamma rays. How long does it take the gamma rays to reach us (in minutes)? [$c = 3.00 \times 10^8$ m/s; 1 mile = 1.61 km]

4.39 According to the equation for the energy of light, which statement is true? (*1*) The energy of light increases as its wavelength increases. (*2*) The energy of light decreases as its wavelength increases. Explain how the equation for the energy of light tells you which is true.

4.40 The unit of nanometers (nm) is commonly used for the wavelength of visible light. What does 1.00 nm equal in meters? What does it equal in inches? [1 inch = 2.54 cm exactly]

4.41 Exposure to gamma rays can kill you, whereas exposure to radio waves is not harmful. Why is this so?

4.42 Suppose a radio wave has a wavelength of 10 m. What is the energy of this radiation (in joules)?

4.43 Suppose an X-ray has a wavelength of 10 pm. What is the energy of this radiation (in joules)? How many times more energetic is this radiation than the radio wave in the previous problem? [1 pm = 10^{-12} m]

4.44 The emission spectrum shown below is for neon.

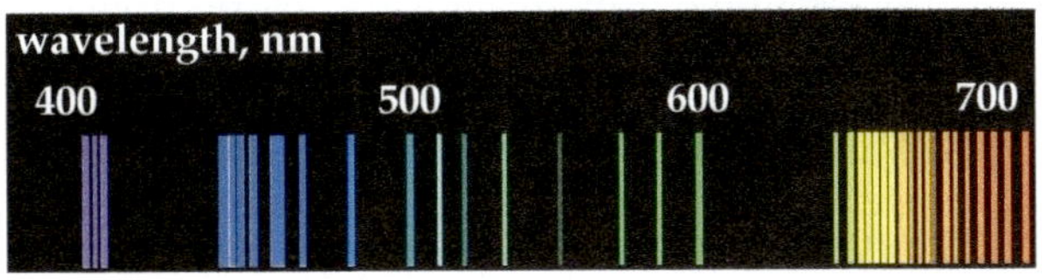

(a) Compare it to the emission spectrum of hydrogen. How is it different?
(b) Is it a continuous or discontinuous spectrum? Explain.
(c) Calculate the energy in joules of the light-blue line at 5.00×10^2 nm.

A New Kind of Physics—Energy Is Quantized

4.45 If you exhibited quantum behavior while running in a footrace, describe how you would appear to the people watching the race.

4.46 When we say the energy of something is quantized, what does this mean? For what type of objects is energy quantization easily observed?

4.47 What is meant by the term *quantized energy*?

4.48 Which is more general, classical physics or quantum physics? Explain your answer.

4.49 The fact that electrified atoms emit only certain colors of light in sharp lines, with not all colors blended together, tells us what about an atom?

The Bohr Theory of Atomic Structure

4.50 How does the energy of an object relate to its stability?

4.51 The following drawings represent charged particles. Which situation is more stable? Explain your answer.

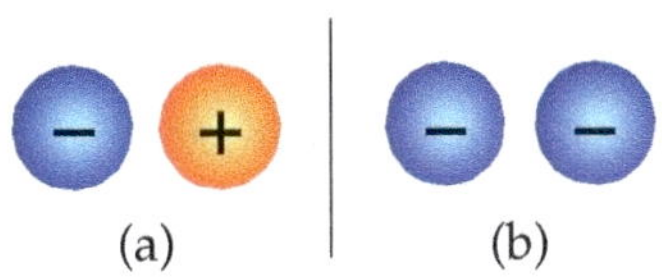

4.52 The following drawings represent charged particles. Which situation is most stable? Which situation is least stable? Explain your answers.

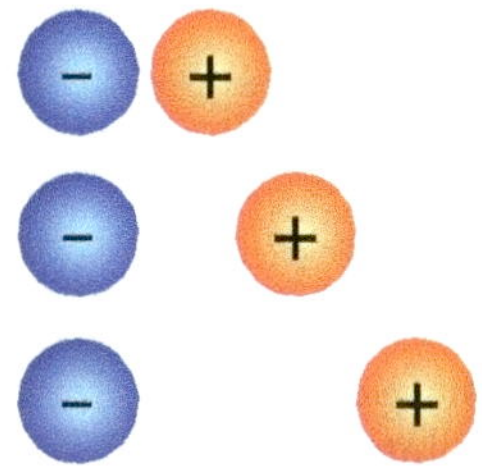

4.53 What happens to both the energy of an electron in an atom and its distance from the nucleus as n increases?

4.54 What type of physics could be used to describe an electron in an atom if the electron could have *any* energy?

4.55 What is another name for a Bohr orbit?

4.56 It always takes energy to remove an electron from an atom, no matter what n shell the electron is in. Also, the higher the n, the more energy an electron starts out with.
(a) Explain why it takes energy to remove an electron from an atom no matter what n shell it is in.
(b) Why does it take less energy to remove an electron from an atom the higher its n value?

4.57 What are two things that happen to an electron in an atom as the n value of the electron increases?

4.58 Of what significance is the quantity $2n^2$?

4.59 Explain why we construct a Bohr model of the atom by first filling a lower shell to capacity before going to an upper shell.

4.60 Does the following Bohr model represent a ground or excited state? Explain your answer.

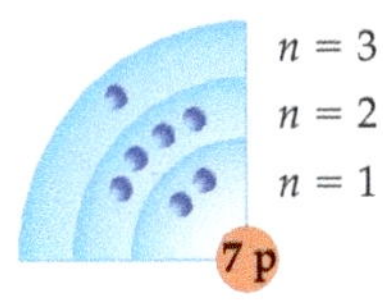

4.61 Using Bohr's rules, draw the lowest-energy configuration for the atom represented in Problem 4.56.

4.62 Why can't an atom's electrons ever be located between orbits?

4.63 Use the Bohr model of the atom to explain why saying that an electron can be only at certain distances from the nucleus is the same thing as saying that the electron can have only certain energies.

Periodicity and Line Spectra Explained

4.64 What is meant by an atom's valence shell?

4.65 According to Bohr, why do atoms in the same group in the periodic table have similar chemical properties?

4.66 Explain how the Bohr model of the atom accounts for the existence of atomic line spectra.

4.67 What is a ground-state electron configuration? What is an excited-state electron configuration? Explain in words and also by drawing a Bohr model of the Mg atom in both the ground state and an excited state.

4.68 True or false? It is impossible for the H^+ cation to exist in an excited state. Justify your answer.

4.69 Draw Bohr models and use them to explain why phosphorus (P) and arsenic (As) have similar chemical properties.

4.70 (a) How much energy would be necessary to take a hydrogen atom from its ground state to an excited state in which the $n = 3$ orbit is occupied?
(b) What would be the energy of light emitted when the excited atom relaxed back to the ground state?
(c) What kind of electromagnetic radiation would be emitted? (Consult Table 4.1.)

4.71 What would happen to the electron in a ground-state hydrogen atom if the atom were given 5.1 eV of energy?

4.72 In order for hydrogen atoms to give off continuous spectra, what would have to be true?

4.73 For each of the diagrams below do the following:
(a) Determine whether anything is wrong with the diagram. Explain.
(b) Give a full atomic symbol for the diagram if there is nothing wrong with it.
(c) For those diagrams that have nothing wrong with them, indicate whether they represent a ground state or an excited state.

1
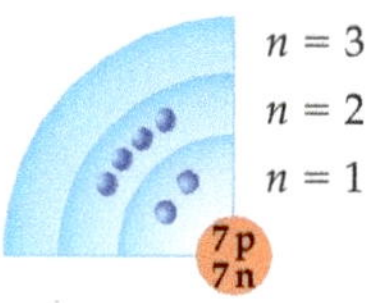

2
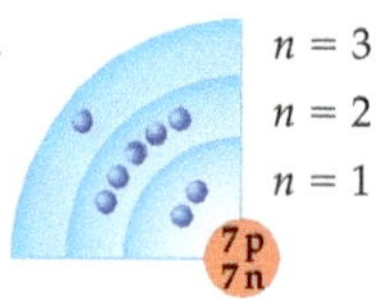

3
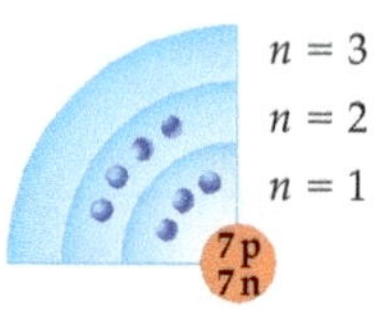

4.74 Regarding the representative elements:
(a) What does knowing the group number tell you about an element's valence electrons?
(b) What does knowing the period number tell you about an element's valence electrons?

4.75 Referring to Table 4.1, draw a wave of red light and a wave of blue light that show (relatively) how the wavelengths compare. Which light has the higher energy? Without doing any numerical calculations, which light would you expect to be associated with relaxation from the $n = 3$ energy level to the $n = 2$ energy level? Which light would you expect to be associated with relaxation from the $n = 4$ level to the $n = 2$ level?

4.76 Using the energy-level diagram on page 133, calculate the wavelength of the light emitted when the electron in a hydrogen atom falls from the $n = 5$ shell to the $n = 3$ shell. Report your answer in meters and in nanometers. [$1 \text{ eV} = 1.602 \times 10^{-19} \text{ J}$]

4.77 The electron in a hydrogen atom relaxes from the $n = 4$ shell to some lower-energy shell. The light emitted during the relaxation has a wavelength of 1772.6 nm. By calculating the energy of this light, determine the shell to which the electron relaxed. [$1 \text{ eV} = 1.602 \times 10^{-19} \text{ J}$]

Subshells and Electron Configuration

4.78 The simple Bohr model (the model without subshells) works well up to the 19th element, potassium (K). Explain how it fails at K.

4.79 What was the experimental evidence that supported the existence of subshells? Explain how this evidence suggested subshells.

4.80 (a) What is the numbering system used to label shells?
(b) What is the lettering system used to label subshells?
(c) How many subshells are there in a given shell?

4.81 How many electrons can each subshell hold before it is considered full?

4.82 Bohr solved the potassium problem by putting its last electron where? How did he justify this?

4.83 You have seen that the 4*s* subshell fills before the 3*d* subshell. This being true, what would be wrong with drawing the Bohr diagram as follows? (*Hint*: What does the diagram drawn this way suggest about relative subshell size that is, in fact, not true?)

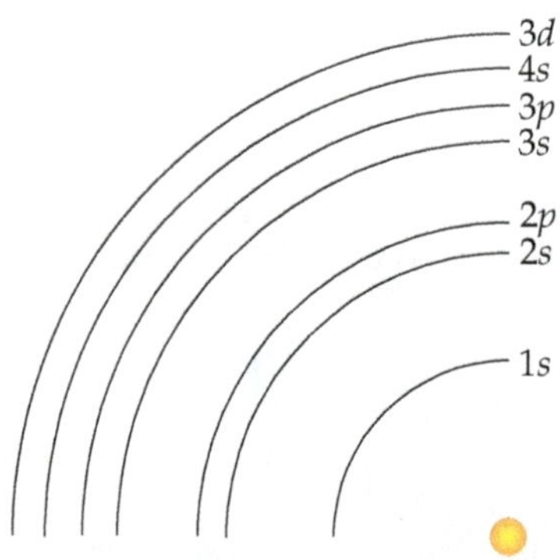

4.84 Write the electron configuration for the following elements without using the noble gas abbreviated form (use the periodic table to assist you).
(a) B (b) Sc (c) Co (d) Se (e) Ru

4.85 Repeat Problem 4.84, but use the noble gas abbreviated form this time.

4.86 Write the electron configuration for the following elements without using the noble gas abbreviated form.
(a) Ba (b) W (c) Pb (d) Pr (e) Pa

4.87 Repeat Problem 4.86, but use the noble gas abbreviated form this time.

4.88 In which period and group in the periodic table are these atoms found?
(a) $1s^22s^22p^3$
(b) $1s^22s^22p^63s^1$
(c) $1s^22s^22p^63s^23p^64s^23d^{10}4p^5$

4.89 A student has written what he thinks are some ground-state electron configurations. Which ones have something wrong with them? What is wrong?
(a) $1s^22p^63s^1$ (b) $1s^22s^63s^23p^64s^24p^6$
(c) $2s^22p^63s^23p^64s^23d^7$ (d) $1s^22s^22p^73s^33p^6$
(e) $1s^22s^22p^53s^1$
(f) $1s^22s^22p^63s^23p^64s^23d^{10}4p^65s^14d^4$ [This is the transition metal niobium, which violates the four-block electron-assignment rules, so be careful.]

4.90 How many valence electrons does each of these atoms have?
(a) $1s^22s^22p^3$ (b) $1s^22s^1$ (c) $1s^22s^22p^63s^23p^64s^23d^7$
(d) $1s^22s^22p^63s^23p^6$ (e) $1s^2$

4.91 Regarding primary quantum number n, which representative elements have valence electrons with an n value that is three times the n value of lithium's valence electrons? What period are they in?

4.92 Write electron configurations for O, O^{2+}, and O^{2-}. Which form would you expect to find in most compounds of oxygen? Why?

4.93 When using the periodic table to assign electron configurations, what is the rule for the n quantum number when you are in the d block? When you are in the f block?

4.94 Why are the s, p, d, and f blocks in the periodic table 2, 6, 10, and 14 blocks wide, respectively?

Regular Variations in the Properties of Elements

4.95 (a) According to the Bohr model, why might someone expect that atoms would not change in size as you go from left to right across a period?
(b) In fact, atoms do change in size across a period. What is the trend and what is the explanation for it?

4.96 According to the Bohr model, why do atoms get larger as you proceed down a group in the periodic table?

4.97 Which atom has a smaller 1s subshell, lithium (Li) or beryllium (Be)? Justify your answer.

4.98 Which atom has a smaller valence shell, lithium (Li) or sodium (Na)? Explain your answer.

4.99 Which atom is larger, lithium (Li) or beryllium (Be)? Explain your answer.

4.100 Which atom is larger, lithium (Li) or sodium (Na)? Explain your answer.

4.101 Rank the following atoms from smallest to largest: Si, Mg, Rb, Na.

4.102 As you move from left to right across a period you add more subatomic particles to the atoms, and yet the atoms become smaller.
(a) What subatomic particles are you adding?
(b) If you are adding particles, why don't the atoms become larger?

4.103 Order the following atoms from smallest to largest, judging from their relative positions in the periodic table: Cs, Fe, Ti, Hf.

4.104 Order the following atoms from smallest to largest, judging from their relative positions in the periodic table: Ca, Se, F, S.

4.105 Explain the trends in ionization energy in terms of the trends in atomic aize.

Compound Formation and the Octet Rule

4.106 What is the octet rule, and what is the justification behind it?

4.107 How do metal atoms usually attain an octet in chemical reactions?

4.108 How do nonmetal atoms usually attain an octet in chemical reactions?

4.109 Draw the Bohr model for a Cl atom and for a Cl^- ion. How many electrons are there in the valence shell in each drawing?

4.110 Draw a Bohr model for an Al atom and for an Al^{3+} ion. How many electrons are there in the valence shell in each drawing?

4.111 Predict the formula of the compound that forms when sodium atoms react with sulfur atoms. Completely explain your reasoning.

4.112 Predict the formula of the compound that forms when lithium atoms react with nitrogen atoms. Completely explain your reasoning.

4.113 Why are roman-numeral group numbers for the representative elements useful in predicting how many electrons an atom will gain or lose in a chemical reaction?

4.114 Consider the following Bohr diagrams for two reactants:

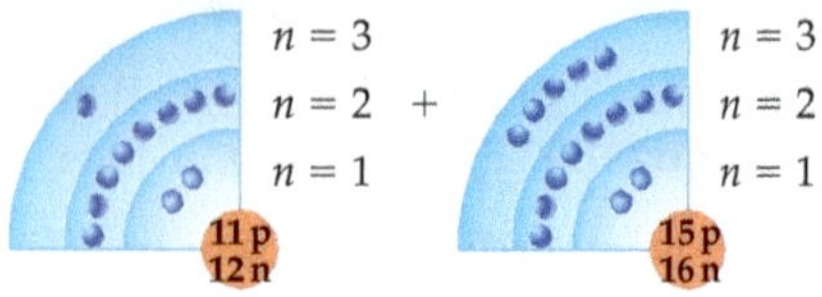

(a) What are the identities of the two elements reacting with each other? Give full atomic symbols.
(b) Which is the metal, and which is the nonmetal?
(c) What is the formula of the resulting compound?
(d) Draw Bohr diagrams for the ions formed.

4.115 When a group IIA element X reacts with a group VIIA element Y, what will be the formula? Why?

4.116 Explain what is meant by this statement: The element hydrogen is an exception to the octet rule and yet it obeys it in principle.

4.117 Aluminum forms compounds with both sulfur and oxygen. Why are the formulas of the resulting compounds Al_2S_3 and Al_2O_3, respectively? Why are the formulas similar?

4.118 True or false? The O^{2-} and F^- anions have identical electron configurations. Justify your answer with Bohr diagrams and electron configuration notation.

4.119 Which part of the following statement is true and which part is false? Mg^{2+} and Na^+ have identical electron configurations, and they also have similar properties. Explain your answer fully.

4.120 How can you tell how many electrons a representative metal is likely to lose? What, in general, will be the charge of the cation it forms?

4.121 How can you tell how many electrons a representative nonmetal is likely to gain? What, in general, will be the charge of the anion it forms?

4.122 Consider an anion and a cation joining to form a compound.
(a) How does knowing the charge of the ions help to determine the formula?
(b) What would be the formula of the compound formed from Ca^{2+} cations and N^{3-} anions?

The Modern Quantum Mechanical Model of the Atom

4.123 What is wrong with Bohr's planetary model of atomic electrons according to modern quantum mechanical theory? (*Hint:* Use Heisenberg's uncertainty principle in your answer.)

4.124 Quantum mechanical tunneling is one consequence of quantum theory. What is quantum mechanical tunneling?

4.125 What is an orbital?

4.126 Draw an *s* orbital, a *p* orbital, and a *d* orbital.

4.127 Suppose you could see orbitals (electron clouds) inside an atom. According to Schrödinger, describe what you would see as a 1*s* electron was excited to become a 2*p* electron.

4.128 What gave Schrödinger justification to think of an electron in an atom as a nebulous cloud?

Additional Problems

4.129 How many seconds does it take light to travel 3.00×10^3 miles from New York to California? [1 mile = 1.61 km]

4.130 A laser emits a beam of green light that has a wavelength of 5.00×10^{-5} cm. What is the wavelength of this light in nanometers?

4.131 What is the wavelength in nanometers of infrared light for which $\lambda = 2.50 \times 10^{-5}$ m? How many times longer is this wavelength than red light that has a wavelength of 750 nm?

4.132 Electromagnetic radiation emitted by magnesium has a wavelength of 285.2 nm. (a) Is this radiation visible to the eye? (b) What is the energy of this radiation?

4.133 Fill in the table:

Wavelength (m)	Wavelength (nm)	Energy (J)
		5.65×10^{-18} J
	2602.0 nm	
7.85×10^{-12} m		

4.134 Would moving an electron farther from an atom's nucleus give off light energy or require the absorption of light energy? Explain your answer.

4.135 According to Bohr's model, is energy absorbed or released when an electron moves to a shell of lower *n*?

4.136 According to Bohr's model, energy must be put into an atom to move an electron from a low-energy shell to a higher-energy shell. How do you calculate the amount of energy needed for the move?

4.137 Convert this excited boron, B, atom to (a) an excited state of lower energy and (b) the ground state:

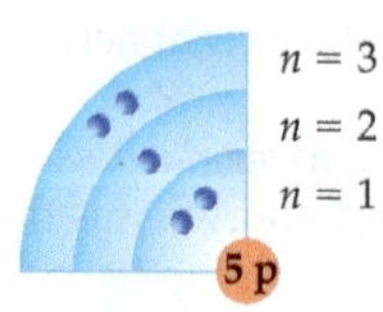

4.138 According to Bohr, what is so special about the valence shell of an atom?

4.139 What are the wavelength in nanometers and energy in joules of the light emitted when a hydrogen electron originally in the $n = 6$ shell relaxes to the ground state? $[1\ \text{eV} = 1.602 \times 10^{-19}\ \text{J}]$

4.140 Use electron configuration notation to explain why aluminum, Al, and gallium, Ga, have similar chemical properties even though gallium has *d* electrons but aluminum does not.

4.141 These energy-level diagrams are for iron, Fe. Indicate whether each represents the ground state, an excited state, or is incorrect. Explain your answers.

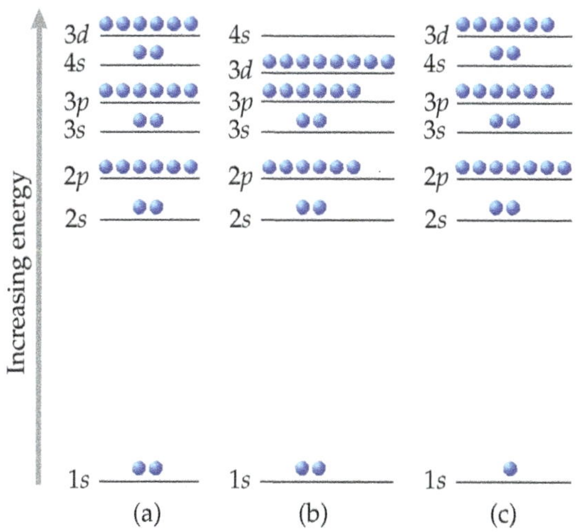

4.142 Name each element and tell how many valence electrons it has:
(a) $1s^22s^22p^63s^23p^2$
(b) $[Ne]3s^1$
(c) $[Ar]4s^23d^{10}4p^3$

4.143 Which subshell is filled in transition metals? What is unique about its order of filling?

4.144 Consider the anion whose charge is 2^- and whose electron configuration is identical to that of argon, Ar. What is the symbol for this anion?

4.145 What do F^-, O^{2-}, Na^+, and Mg^{2+} all have in common?

4.146 Why is the formula for calcium sulfide CaS and not Ca_2S?

4.147 Select the element from each pair expected to have the lower first ionization energy. Explain.
(a) N and F (b) Mg and Ba (c) N and Ca

4.148 How does the first ionization energy of the alkali metal in a given period compare with the first ionization energy of the halogen in the same period? What is the result in terms of how these elements tend to react with each other?

4.149 Arrange calcium, strontium, arsenic, bromine, and chlorine in order of (a) increasing atomic size and (b) increasing first ionization energy.

4.150 Of the atoms Na, Cl, K, Br, which has the largest atomic radius? Which has the largest first ionization energy?

4.151 Arrange Ca, Se, F, S, and Rb in order of increasing ionization energy.

4.152 Predict the formula for lithium nitride, made from lithium, Li, and nitrogen, N. Explain your reasoning.

4.153 Suppose the 3*s* valence electron in a ground-state sodium atom is excited to the 4*s* orbital. Why does the electron now have higher energy?

4.154 What is the wavelength in nanometers of electromagnetic waves that have an energy of 1.00×10^{-15} J?

4.155 Are the waves in Problem 4.144 more or less energetic than visible light ($\lambda = 400\text{–}750$ nm)?

4.156 Which electromagnetic radiation is most dangerous to humans?
(a) X-rays (b) Ultraviolet light
(c) Gamma rays (d) Infrared light
(e) Radio waves

4.157 Write the ground-state electron configuration for each of the following atoms or ions. Which have a valence-shell octet?
(a) Ca^+ (b) Li (c) P^{3-} (d) Ar^+ (e) Si^{2+}

4.158 Halogens are very reactive because (choose the correct answer):
(a) They need to gain only one electron to satisfy the octet rule.
(b) They have seven electrons in their valence shell, and the more electrons an atom has, the more reactive it is.
(c) They are nonmetals, and all nonmetals are reactive.
(d) They can easily lose their seven valence electrons to satisfy the octet rule.

4.159 Write the ground-state electron configuration for each of the following atoms or ions. Which have a valence-shell octet?
(a) Ar (b) Na^+ (c) C^{2-} (d) O^{2-} (e) Ca^{2+}

4.160 Which subshell has the lowest energy?
(a) 4*s* (b) 3*p* (c) 2*p* (d) 3*s* (e) 2*s*

4.161 Which is the correct ground-state electron configuration for antimony, Sb?
(a) $1s^22s^22p^63s^23p^63d^{10}4s^24p^64d^{10}5s^25d^3$
(b) $1s^22s^22p^63s^23p^63d^{10}4s^24p^64d^{10}5s^25p^3$
(c) $1s^22s^22p^63s^23p^64s^23d^{10}4p^65s^24d^{10}5p^4$
(d) $1s^22s^22p^63s^23p^64s^23d^{10}4p^65s^25p^3$
(e) $1s^22s^22p^63s^23p^64s^23d^{10}4p^65s^24f^3$

4.162 What is the energy in joules of green light that has a wavelength of 500 nm?

4.163 Indicate whether or not the following volumes are quantized:
(a) The volume of water available in 16-oz bottles
(b) The volume of water available from a drinking fountain
(c) The volume of soft drink available from a soda-fountain dispenser
(d) The volume of soft drink available in 12-oz cans

4.164 Predict the formula of the compound formed by the reaction between:
(a) Ca and Br (b) K and N (c) Al and S
(d) Na and I (e) Mg and O

4.165 Arrange in order of increasing valence-shell size: Sr, Mg, Ba, Ca, Be.

4.166 The order of the colors in the visible spectrum can be remembered by the acronym ROY G. BIV, which stands for red, orange, yellow, green, blue, indigo, violet. In the acronym, are the colors arranged in order of increasing or decreasing wavelength? In order of increasing or decreasing energy?

4.167 (a) Bohr is the scientist who proposed that the energy of electrons is ________.
(b) The two kinds of physics are ________ physics and ________ physics.

4.168 Write the ground-state electron configuration, both full notation and noble gas abbreviation, and indicate the number of valence electrons for:
(a) Al (b) I (c) Rb (d) Ar (e) Mg

4.169 Metals tend to ________ valence electrons, whereas nonmetals tend to ________ valence electrons.

4.170 What is the valence shell in Mg, Ge, W, Cl, Cs?

4.171 If gamma radiation has a wavelength of 1.00×10^{-12} m, what is the energy of gamma radiation in joules?

4.172 What is the formula for the maximum number of electrons in each shell of the Bohr atom? How many electrons are allowed in the $n = 2$ shell? The $n = 6$ shell?

4.173 What are the group number, period number, and name of the element whose electron configuration is $1s^22s^22p^63s^23p^64s^23d^{10}4p^2$?

4.174 Circle the correct choice to indicate how many electrons each element must gain or lose to form an octet:

(a) Mg	gains, loses	1, 2, 3	electrons
(b) Se	gains, loses	1, 2, 3	electrons
(c) Al	gains, loses	1, 2, 3	electrons
(d) Sr	gains, loses	1, 2, 3	electrons
(e) Br	gains, loses	1, 2, 3	electrons
(f) P	gains, loses	1, 2, 3	electrons

4.175 State the Heisenberg uncertainty principle and what it implies about the structure of an atom.

4.176 Conversion factor fiesta. If a car were able to travel at the speed of light, how many seconds would it take the car to travel 1000 miles? [1 mile = 5280 ft; 1 inch = 2.54 cm exactly]

4.177 In the Bohr model of the atom, are the electrons in shells closer to the nucleus higher or lower in energy than electrons in shells farther from the nucleus? Explain.

4.178 What is the total number of *p*-subshell electrons for each of the following atoms: P, Mg, Se, Zn?

4.179 Write the full ground-state electron configuration for Ca^{2+}, S^{2-}, Ar, K^+.

4.180 Rank visible light, gamma rays, X-rays, radio/television waves, infrared radiation, and ultraviolet light in order of (a) increasing wavelength and (b) increasing energy.

4.181 Draw a simple Bohr model (no subshells) for an oxygen atom. How many electrons are in the valence shell? How many more electrons can be put into the valence shell?

4.182 Taking all configurations to be for neutral atoms, identify the elements having the following electron configurations:
(a) $1s^22s^22p^63s^23p^63d^{10}4s^24p^5$
(b) $1s^22s^22p^1$
(c) $1s^22s^22p^63s^23p^64s^23d^{10}4p^65s^1$
(d) $1s^22s^22p^63s^23p^64s^23d^{10}4p^65s^24d^{10}5p^66s^24f^{14}5d^{10}6p^6 7s^26d^15f^3$
(e) $1s^22s^22p^63s^23p^64s^2$

4.183 Explain the major difference between the orbits in the Bohr model of the atom and the orbitals in the quantum mechanical model of the atom.

4.184 What is the highest value of n for the electrons in these atoms?
(a) Co (b) As (c) Sr (d) Po

4.185 The diagram below shows a ground-state hydrogen atom undergoing excitation to an excited state followed by relaxation.

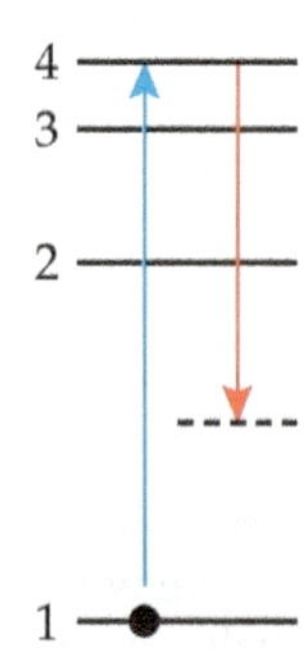

(a) What is fundamentally wrong with the diagram?

(b) Using the figure on page 133, calculate the amount of energy required to cause the above excitation.

(c) Illustrate at least two allowed relaxations from the excited state shown in the diagram.

4.186 (a) The following diagram depicts the formation of LiCl. According to the diagram, what are the charges of the lithium cation and the chlorine anion?

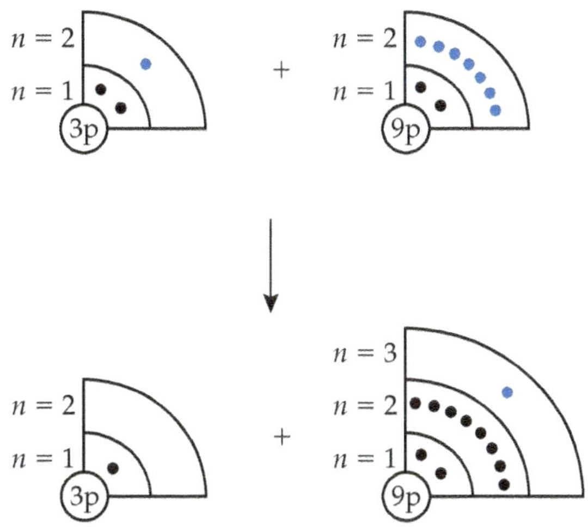

(b) The diagram yields charges for the ions that would be considered incorrect. Explain why the diagram leads to incorrect charges.

4.187 Regarding the *n*th period element whose position is indicated below with an "X", which statement(s) are correct?

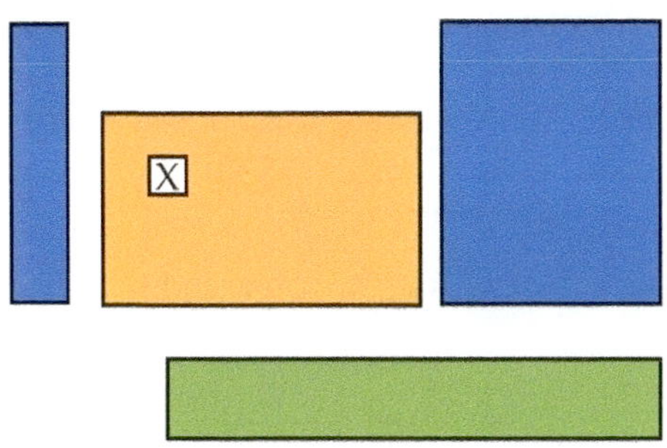

(a) The element is a representative nonmetal.
(b) The element is a transition metal.
(c) The element is in the *p* block.
(d) The element is in the *d* block.
(e) The valence electron configuration is $ns^2(n-1)d^3$.
(f) The valence electron configuration is nd^5.
(g) The valence electron configuration is $(n-1)d^4$.

4.188 Consider the ground-state beryllium (Be) and carbon (C) atoms, shown below.

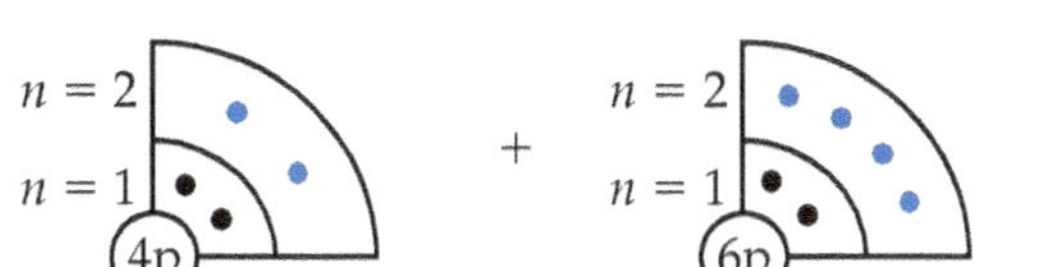

(a) Indicate which atom is smaller and explain why.
(b) Which ground-state atom has valence electrons that look like dumbell-shaped clouds? Explain how you knew this.

4.189 Explain what the following diagram has to say about where one might find lithium's valence electron.

One Step Beyond

4.190 The second ionization energy of an atom is the minimum energy required to remove an electron from the +1 cation of the atom, and it is always larger than the first ionization energy. Why is this so? (*Hint:* Think of atomic electrons as clouds, and each electron must "look" through every other electron cloud in the atom to "see" (feel) the nucleus.)

4.191 The electron configuration of a helium atom is found to be $1s^1$. What can you say about this helium atom?

4.192 What is the implication of thinking about atomic electrons as clouds that got Einstein so upset? What did Schrödinger call these clouds?

4.193 Predict the formula for the compound aluminum nitride made from the elements aluminum and nitrogen, and explain how you made your prediction.

4.194 Consider a $1s \rightarrow 2s$ electron transition. For which atom would this require shorter wavelength light, H or He? Justify your answer.

4.195 Which of the following electron transitions in hydrogen would absorb the largest amount of energy?

A. $n = 2$ to $n = 3$
B. $n = 2$ to $n = 4$
C. $n = 1$ to $n = 4$
D. $n = 3$ to $n = 1$
E. $n = 7$ to $n = 1$

4.196 True or false: Photons of green light have a higher frequency than photons of blue light.

True or false: Photons of blue light have a longer wavelength than photons of orange light.

4.197 Arrange the following sets of atoms in order of increasing atomic size:

(a) F, Cl, Br, I ______ ______ ______ ______
(b) Mg, Na, S, Al ______ ______ ______ ______
(c) Al, Si, Ge, Te ______ ______ ______ ______

4.198 Of the following atoms, which has the largest *first* ionization energy?
(a) Br (b) O (c) C (d) P (e) I

4.199 Identify the period 2 element that is described by the ionization data below.

$M(g) \rightarrow M^{+}\ 1\ e^{-}$	$IE(1) = 1.40 \times 10^3\ J/mol$
$M^{+}(g) \rightarrow M^{2+}\ 1\ e^{-}$	$IE(2) = 2.86 \times 10^3\ J/mol$
$M^{2+}(g) \rightarrow M^{3+}\ 1\ e^{-}$	$IE(3) = 4.58 \times 10^3\ J/mol$
$M^{3+}(g) \rightarrow M^{4+}\ 1\ e^{-}$	$IE(4) = 7.48 \times 10^3\ J/mol$
$M^{4+}(g) \rightarrow M^{5+}\ 1\ e^{-}$	$IE(5) = 9.44 \times 10^3\ J/mol$
$M^{5+}(g) \rightarrow M^{6+}\ 1\ e^{-}$	$IE(6) = 5.33 \times 10^4\ J/mol$
$M^{6+}(g) \rightarrow M^{7+}\ 1\ e^{-}$	$IE(7) = 6.44 \times 10^4\ J/mol$

4.200 Which emission spectrum corresponds to the energy level diagram shown? (*Hint*: Consider every possible relaxation among these levels.)

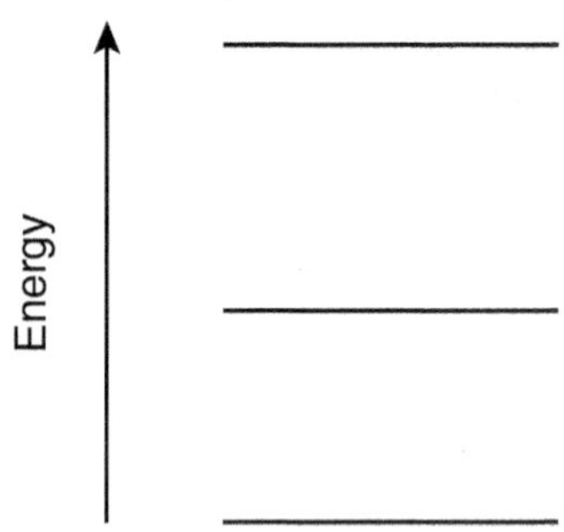

λ increases $\rightarrow$

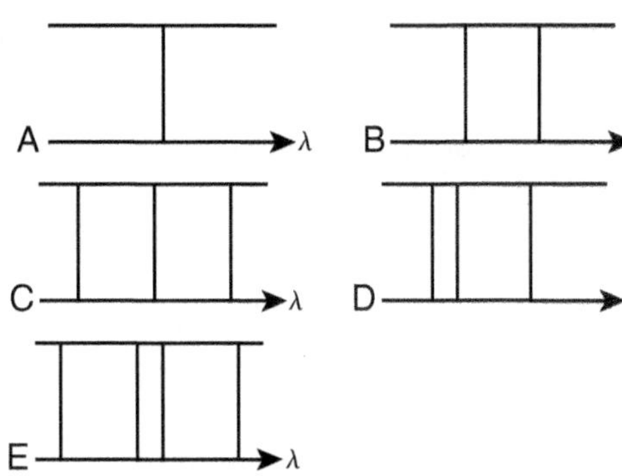

4.201 Below are data on the first four ionization energies for a fictitious element X.

First ionization energy = 500 kJ/mol

Second ionization energy = 2000 kJ/mol

Third ionization energy = 3500 kJ/mol

Fourth ionization energy = 25,000 kJ/mol

From the data, which of the following statements is(are) *incorrect*?
(a) X could belong to group IIIA.
(b) The fourth ionization energy is so much greater than the third ionization energy because X^{3+} consists of a noble-gas core or a pseudo-noble-gas core.
(c) The third ionization energy is greater than the second ionization energy because X^{2+} has a bigger charge than X^{+}.
(d) X could belong to group IIIB.
(e) X could belong to group VA.

4.202 In which group and period would you expect to find the MOST metallic element?

4.203 In which group and period would you expect to find the MOST nonmetallic element?

4.204 When a hydrogen atom is excited in a flame, a line of blue-violet light is emitted. This happens when an electron makes a transition between the $n = 5$ and the $n = 2$ orbit of the atom. If the energy of this light is 2.9 eVs and the energy of an electron in the $n = 2$ hydrogen orbit is 11.2 eVs, what must be the energy of the $n = 5$ orbit?

4.205 Of the following transitions in the Bohr hydrogen atom, which transition results in the emission of photons with the shortest wavelength?
(a) $n = 1 \rightarrow n = 6$
(b) $n = 4 \rightarrow n = 1$
(c) $n = 9 \rightarrow n = 5$
(d) $n = 3 \rightarrow n = 6$
(e) $n = 5 \rightarrow n = 9$

4.206 Arrange the following atoms in order of increasing atomic size: Ne, As, Se, K, S, Cl

4.207 Shown below is an energy level diagram and the emission line spectrum that arises from these energy levels. What electron transition (drop) gave rise to the line labeled B in the emission spectrum? (*Hint*: Consider every possible relaxation among these levels.)

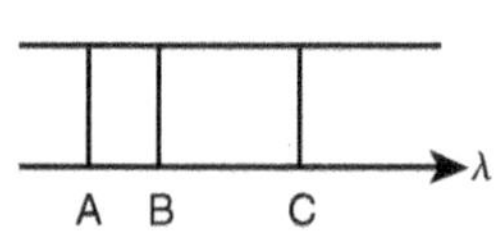

Group Exercises (Work on these exercises together with someone else in the class.)

1. Discuss the type of light that would be given off by electrified atoms if the electrons within these atoms could have any amount of energy. Then

work together to draw an energy-level diagram that would go along with such an atom.

2. Take turns picking an *s*-, *p*-, *d*-, and *f*-block element and asking your partner to write its ground-state electron configuration.

WORKPATCH SOLUTIONS

4.1 Atom-sized cars on the highway could travel only at certain allowed speeds. Thus, instead of accelerating smoothly from zero to 60 miles/h as it entered traffic, a car stopped at an entrance ramp would have to be at 0 miles/h one instant and, say, 45 miles/h the next instant, and 85 miles/h the next instant. If all the other cars on the highway were not traveling at one of these allowed speeds, think of the major pileup that would occur!

4.2 (a) 1.0 eV
(b) 11.2 eV
(c) The difference between them, 11.2 eV − 1.0 eV = 10.2 eV.
(d) Less energy, 13.1 eV − 11.2 eV = 1.9 eV.

4.3 The hydrogen atom with the electron in the $n = 2$ shell.

4.4 12.1 eV − 10.2 eV = 1.9 eV more energy.

4.5 Green: $n = 4$ to $n = 2$; blue: $n = 5$ to $n = 2$; indigo/violet: $n = 6$ to $n = 2$.

For example, the green line corresponds to light of energy between 2.24 and 2.64 eV (Table 4.1). The only relaxation that produces this energy is from the $n = 4$ level (13.8 eV) to the $n = 2$ level (11.2 eV): 13.8 eV − 11.2 eV = 2.6 eV.

4.6

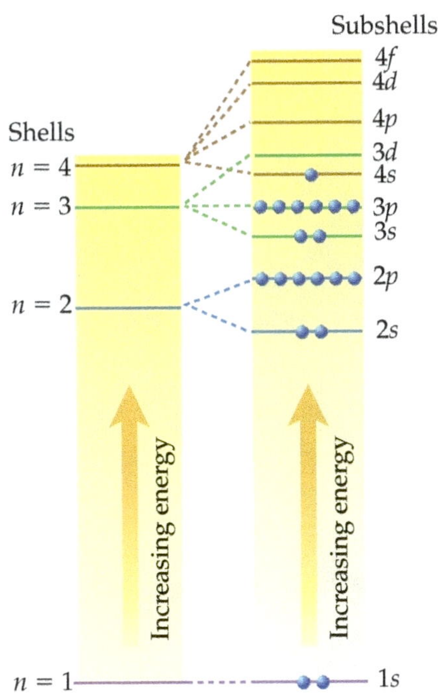

4.7

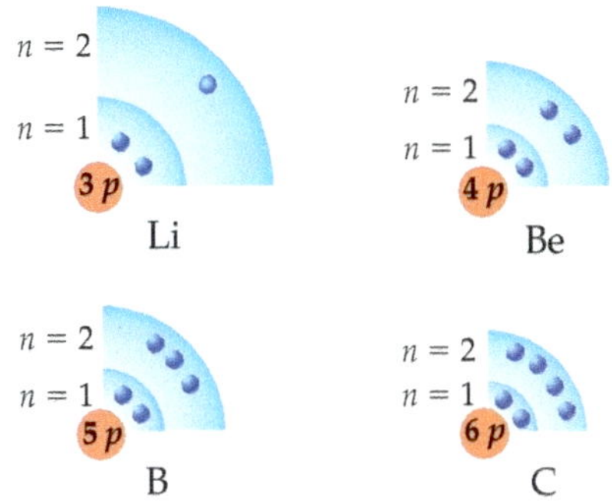

Carbon would be the hardest to ionize because its shells are closest to the nucleus, increasing the force of attraction between the negative electrons and the positive nucleus. The shells are smallest for C because it has the most positive nucleus of the atoms in this period.

4.8 The elements F, Cl, Br, and I obey the expected trend. You should find that At (Astatine) has a slightly lower ionization energy than I, so it also obeys the trend.

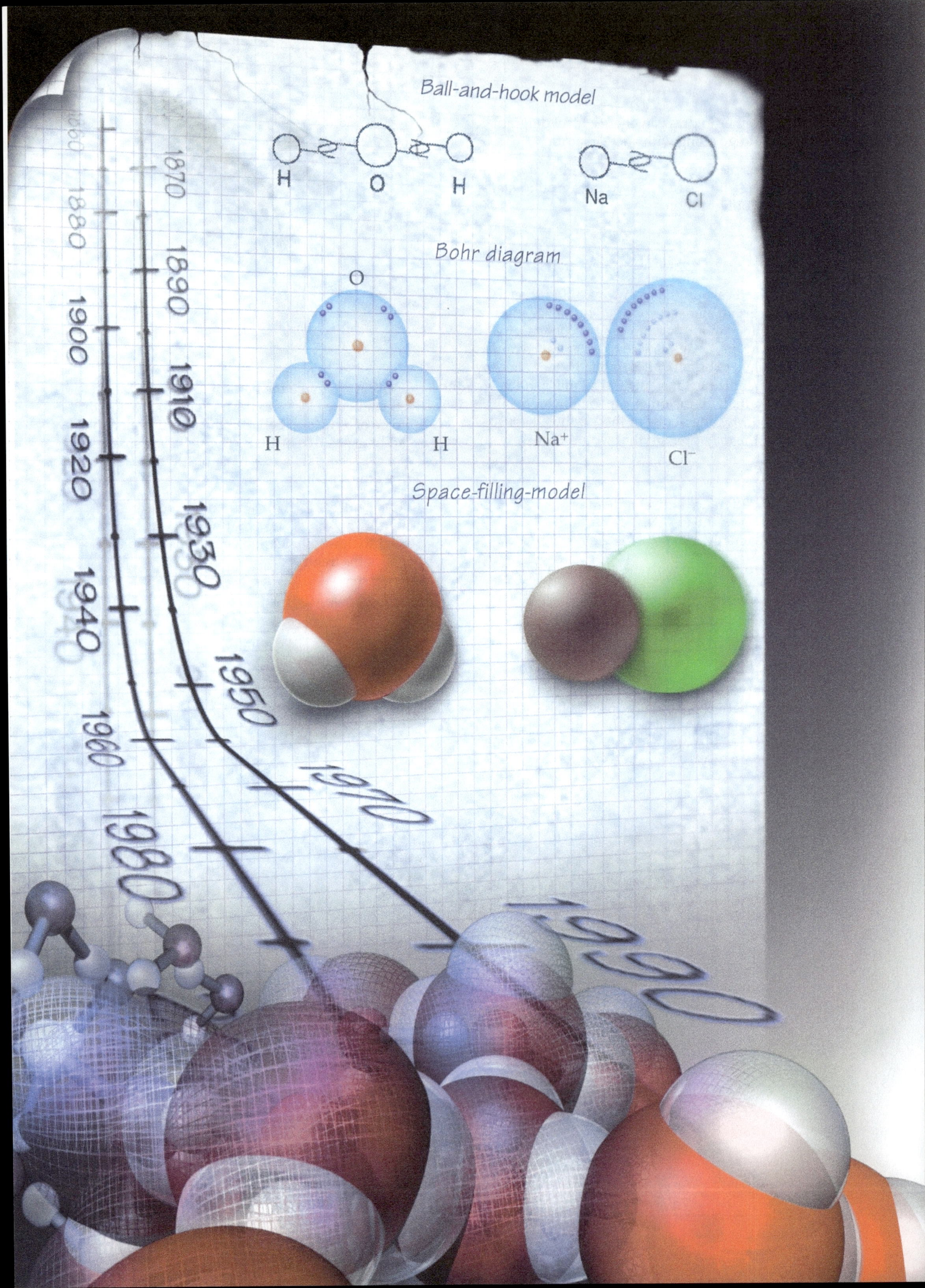
Ball-and-hook model
H O H
Na Cl
Bohr diagram
O
H H
Na+
Cl−
Space-filling-model
1870
1880
1890
1900
1910
1920
1930
1940
1950
1960
1970
1980
1990

CHAPTER 5

Chemical Bonding and Nomenclature

In the early 1800s, an English school teacher named John Dalton proposed one of the first models for a "chemical bond." He envisioned atoms as hard spheres with hooks embedded in them. The number of hooks in a particular atom determined the way in which it would combine with other atoms. More than 100 years later, chemists G.N. Lewis and Linus Pauling developed the first modern theories of chemical bonding. How excited were chemists about these theories? Consider the following note that Lewis wrote to Pauling: "I have just returned from a short vacation for which the only books I took were a half-dozen detective stories and your 'Chemical Bond'. I found yours the most exciting of the lot." Another chemist wrote to Pauling, "I have been very much interested by your new book....... As evidence of my interest in it, I can cite the fact that it is the first scientific book which I can remember reading during the course of a fishing trip..." Now it is your turn to be excited by chemical bonding.

5.1 Ionic Bonding

In the previous chapters, we saw that a characteristic of metals is that they tend to lose valence electrons, whereas nonmetals do not (in fact, nonmetals prefer to gain valence electrons). How many electrons a metal atom loses or a nonmetal atom gains could be predicted by the octet rule. For example, we considered what would happen when an atom of sodium (a group IA metal) reacts with an atom of chlorine (a group VIIA nonmetal). Each sodium atom loses an electron, and each chlorine atom gains one. In other words, there has been an electron transfer from the metal to the nonmetal. Stripped of its single valence electron, the Na atom is now a positively charged sodium ion, Na^+, a cation. Having gained an electron, the chlorine atom is now a negatively charged ion, Cl^-, an anion. This transfer is typically illustrated by a curved arrow, called an "electron pushing" arrow, and can be used to show how a valence electron or electrons move.

The creation of ions with opposite charges gives rise to an attractive force between the ions, and the result is the formation of a new compound, sodium chloride, (NaCl, or table salt). Since the compound is made from ions, it is called an **ionic compound**. That is, if you could look inside a piece of solid NaCl and "see" at the atomic level, you would see ions actually touching one another. On the top of the next page is a huge rock of halite, the name for the mineral that is sodium chloride, taken from a salt mine (the off-white colors are due to impurities).

The first thing you would notice if you could pick it up (this rock weighs hundreds of pounds) is that it is a hard material. If you tried to break it with

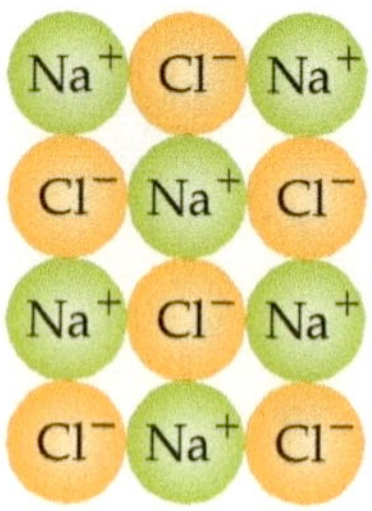

Halite

only your bare hands by pulling it apart or twisting it, you could not. Evidently, the Na^+ and Cl^- ions that are inside this chunk of NaCl are stuck to each other so fiercely that even using all your strength, you cannot pull them apart. The oppositely charged ions exhibit an extremely strong attraction for one another much like two strong magnets held together by a similar magnetic force. In the case of our oppositely charged ions, this force is known as an electrostatic force.

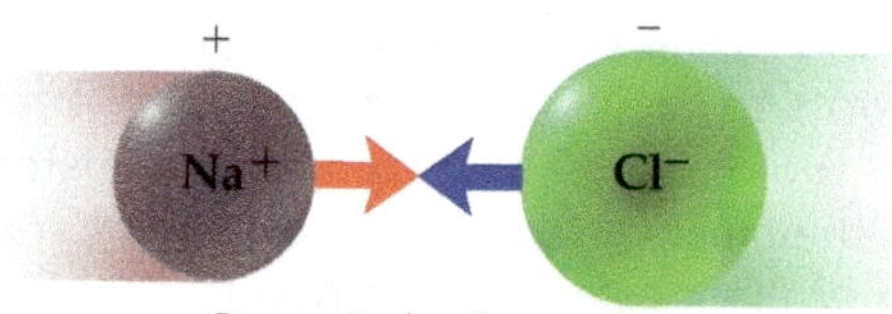

Oppositely charged ions attract each other.

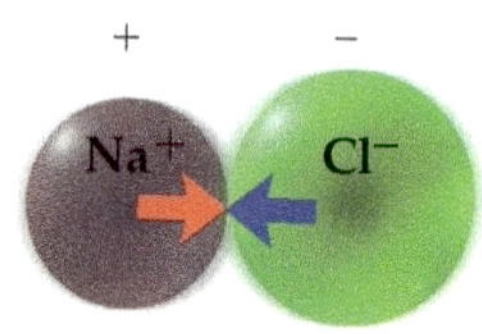

Ions stick together.

Coulomb's Law $F = k\left(\frac{q_1 q_2}{r^2}\right)$

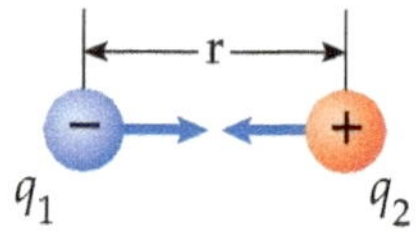

Since q_1 is negative and q_2 is positive, the electrostatic force F is negative (attractive)

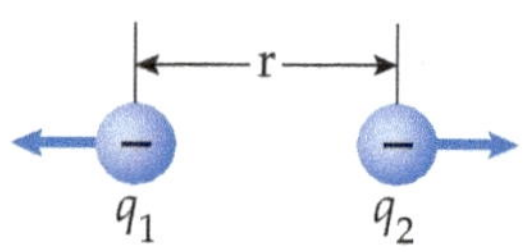

Since q_1 and q_2 are both negative, the electrostatic force F is positive (repulsive)

The electrostatic force is governed by an equation from physics known as **Coulomb's Law.** Using this equation, we see the factors that determine the strength of this force.

Here, k is a constant used to make the units work out right. The variable r is the distance between the centers of a Na^+ cation and a neighboring Cl^- anion (this can be determined by X-ray diffraction). q_1 is the charge of the cation (+1 for Na^+) and q_2 the charge of the anion (−1 for Cl^-). Since k and r are always positive numbers, the value of F will be positive whenever the product of q_1 and q_2 have the same sign (a repulsive force), and negative when they have different signs (an attractive force). Now try the following WorkPatch.

WORKPATCH 5.1 Suppose, instead of NaCl, we considered the ionic compound magnesium oxide, MgO, made from Mg^{2+} cations and O^{2-} anions. Judging just from the charges, would the force of attraction between the ions in MgO be greater or smaller than in NaCl? By how much? ●

Using Coulomb's Law, you have just learned about an important factor that determines how strongly oppositely charged ions "stick" or are attracted to one another (check your answer). But now we need a name for this attractive force between oppositely charged ions. Since it is an attractive force between ions, it is called an *ionic bond*. An **ionic bond** is the attractive force that arises between oppositely charged ions that were formed because of a transfer of electrons. An ionic bond is a type of *chemical bond*. A **chemical bond** is a generic name given to the attractive force that exists between atoms in a chemical compound. If the compound is an ionic substance, then the chemical bonds are called ionic bonds.

Finally, we want to consider the structure of ionic solids such as NaCl. By this we mean, how are the billions upon billions of Na^+ and Cl^- ions arranged inside a piece of solid sodium chloride? Ionic solids such as NaCl exist as a large, ordered, three-dimensional network of positively and negatively charged ions called an **ionic lattice**. In the ionic lattice, the ions are arranged in such a way that positive ions are adjacent to and in contact with the negative ions (resulting in ionic bonds) but ions with like charges (both positive or both negative), which repel one another, are kept farther apart and do not touch one another. A portion of the NaCl lattice is shown here. This entire lattice is held together by ionic bonds between oppositely charged ions.

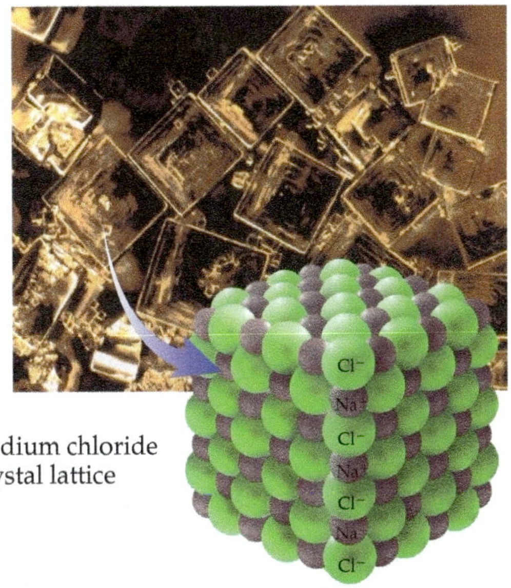

The sodium chloride crystal lattice

Such a lattice is often described as *crystalline*. A **crystalline** solid is one in which the particles that make it up, be they atoms, ions, or molecules, are arranged in a highly ordered fashion. The opposite of a crystalline solid is an **amorphous** solid, one in which there is no overall order among the particles that make it up.

Chemical bonds such as ionic bonds are typically very strong (indeed, chemical bonds are among the strongest attractive forces that exist in chemistry). This is why NaCl has such a high melting point (801 °C). It takes a tremendous amount of thermal energy to overcome or weaken the ionic bonds enough that the ions can be free to move about in the molten liquid. The sum of all the ionic bonds that are formed within the lattice give rise to the *lattice energy*. The **lattice energy** can be thought of as the energy it would take to break the ionic lattice up into individual, isolated ions (i.e., to completely break all of the the ionic bonds in the crystal). Coulomb's Law gives us some insight into lattice energy as well. Since the charges of the ions are in the numerator of the Coulomb's Law equation, we can expect the lattice energy to be higher for lattices made up of more highly charged ions. Indeed, as strong as the NaCl lattice is, the MgO lattice, made from Mg^{2+} and O^{2-} ions, has more than four times the lattice energy of NaCl, making its lattice more than four times stronger. Interestingly though, if you take a solid crystal of NaCl or MgO and drop it, it will often break along a perfectly straight plane (called **cleaving**). This is because the ions can momentarily shift due to the force of the impact, causing layers of like-charged ions to come into contact with one another, whereupon the layers repel and the lattice cleaves along a plane.

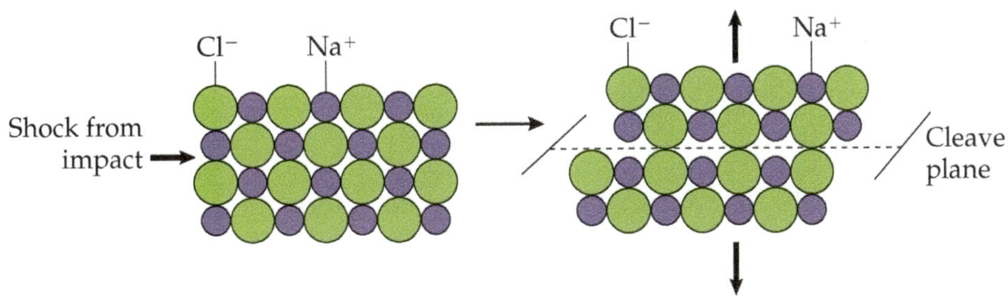

5.2 Molecules—What Are They? Why Are They?

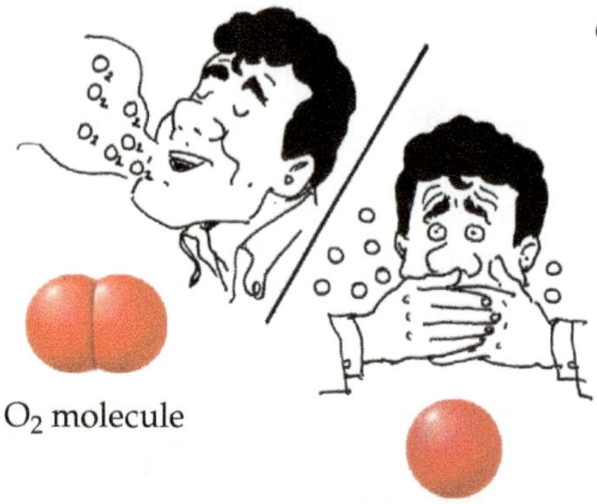

O_2 molecule

O atom

Not all substances are ionic in nature. Take a deep breath. As everyone knows, the air filling your lungs contains oxygen gas, but exactly what is it that you are breathing? The gas you are inhaling is not made up of oxygen ions or even of individual neutral (uncharged) oxygen (O) atoms. Neutral oxygen atoms are extremely reactive, and your lungs would be instantly destroyed by inhaling them. Instead, you are breathing oxygen *molecules* (O_2).

A **molecule** is defined as a collection of atoms bound together in some way. In the case of oxygen, two oxygen atoms are all that is needed to form a molecule. Thus, the molecular formula for the oxygen molecule is O_2.

Any substance that is made of discrete (distinct) molecules is called a **molecular substance**. An example of a molecule that is made from more than two atoms is methane, the main component of natural gas, the fuel burned in home furnaces, kitchen stoves, and the bunsen burners in your chemistry laboratory. The methane molecule, which has the molecular formula CH_4, is a collection of five atoms—a single carbon atom and four hydrogen atoms, as shown to the right of the burner. Methane gas is a collection of individual separate CH_4 molecules.

Methane

Note that while methane and oxygen both exist as molecules, only CH_4 is classified as a molecular *compound* because it consists of more than one type of element. O_2 is considered a molecular *elemental substance*. We should point out at this time that ionic compounds like NaCl are not molecular substances. That is because there are no discrete NaCl pairs in an NaCl lattice. (There is nothing that you could point to in an ionic lattice that stands out as a separate molecule). This is quite different from water, a molecular compound, where discrete H_2O molecules make up the lattice in solid ice.

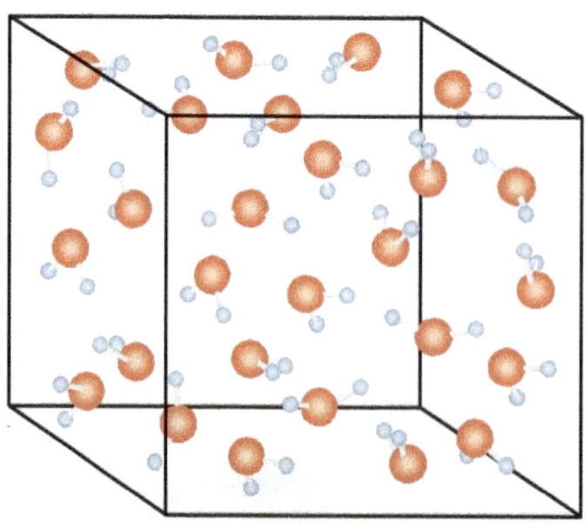

Individual water molecules are identifiable in the lattice of ice. Water is therefore a molecular compound.

One of the things that make molecules so interesting is that they can have chemical and physical properties that are very different from those of the atoms from which they are made. For example, a simple change from individual O atoms to O_2 molecules changes the substance from a lethal to a life-sustaining gas. Another example is the molecule glucose, one of the simplest kinds of sugar. Much of the food you eat either contains glucose or gets converted to glucose during digestion. The glucose molecule is assembled from 24 atoms (6 carbon atoms, 12 hydrogen atoms, and 6 oxygen atoms). Whereas individual carbon, hydrogen, and oxygen atoms have no particular nutritional value, the glucose molecule, $C_6H_{12}O_6$, is a major source of food for the cells in your body.

Natural sugars in honey are broken down to glucose in the body.

Huge corporations invest large amounts of money to design and produce new molecules having specific, tailor-made properties. Two examples you are probably familiar with are the compounds aspartame (NutraSweet), which has the property of being sweet without providing significant calories, and Viagra, which has been immensely profitable for its manufacturer. There are literally millions upon millions of different molecules, ranging from simple two- or three-atom combinations, such as oxygen, O_2, and water, H_2O, to huge supermolecules containing thousands or even hundreds of thousands of atoms, such as DNA.

WORKPATCH 5.2 Would you expect these two molecules to have nearly identical properties?

H_2O $\quad$ H_2O_2

Did you answer the WorkPatch correctly? The answer (given at the end of the chapter) emphasizes what we have been discussing about the chemical properties of molecules.

5.3 Holding Molecules Together—The Covalent Bond

Unlike ionic compounds, which contain charged atoms (ions), molecular compounds consist of molecules made from neutral atoms. One of the most fundamental and important questions in chemistry is, "What holds a molecule together?" Atoms aren't sticky, and they don't come with hooks attached. What is it then that prevents molecules from coming apart into separate atoms?

To answer this question, let's consider the element hydrogen. In the periodic table, an atom of hydrogen is represented by the letter H. However, a tank of hydrogen gas contains no individual H atoms. It is filled with H_2 molecules. Like oxygen, hydrogen is more stable as a *diatomic* (two-atom) molecule. Something is making the two H atoms in H_2 attract one another even though the individual H atoms in the molecule are neutral. Indeed, if you could get your hands on a tank full of H atoms, you would not have it for very long. The gas in the tank would rapidly transform into a collection of H_2 molecules. To discover why hydrogen prefers to exist this way, let's begin by doing a thought experiment, sometimes called by the German name Gedanken experiment. Imagine that you have been shrunk down to a height of 10^{-10} m so that you are about the size of a hydrogen atom. (This length 10^{-10} m is equivalent to 1 angstrom, Å, as noted in Section 3.7). In front of you are two H atoms. Each atom consists of a positively charged nucleus and a single negatively charged valence electron. The negative electron is attracted to the positive nucleus, and this force is what firmly holds the electron in place. Let's begin with the two H atoms far from each other (say, about 10 Å apart). At this distance, the atoms would just sit there, each unaware that the other one exists.

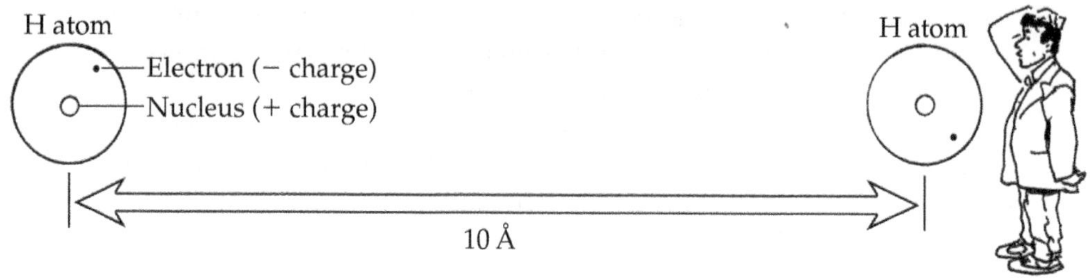

Now imagine moving the atoms closer to each other, a little bit at a time. At 9 Å, they still just sit there. At 8 Å, still no change.

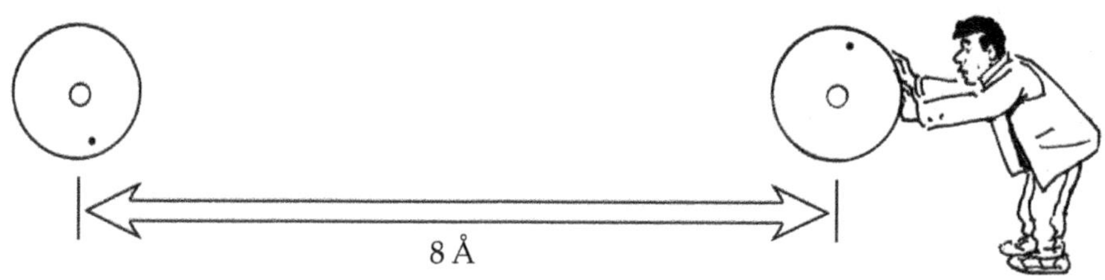

But when the atoms get close enough to each other, say to within 3 Å, something interesting happens. Suddenly, the atoms spontaneously start to move toward each other! Their movement is slow at first, but as they get closer, they begin to move faster. Soon they're racing toward each other!

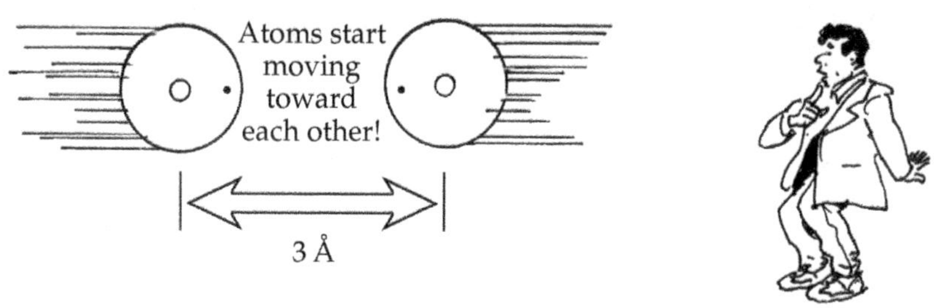

As the two atoms get closer and closer together, you turn away in anticipation of a messy crash.

When you turn back, however, you discover that the two atoms are still intact, with their nuclei separated by a distance of 0.74 Å. At the same time, the temperature in the vicinity of the atoms has increased. An H_2 molecule has been created.

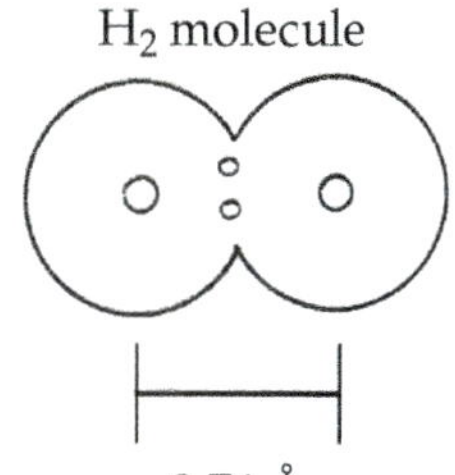

What happened? Why did the H atoms start moving toward each other when they got close, and why did they stop when the two nuclei were 0.74 Å apart? And where did the increase in temperature come from?

There is much to explain. Let's start with the first observation—when the atoms were brought to within 3 Å of each other, they began to move toward each other. This spontaneous movement was caused by the same force that holds each electron close to its own nucleus: opposite charges attract. When the two atoms got close enough, the negative electron of one atom began to be attracted to the positive nucleus of the other atom, and vice versa. These attractive forces *between atoms* began to pull the atoms toward each other.

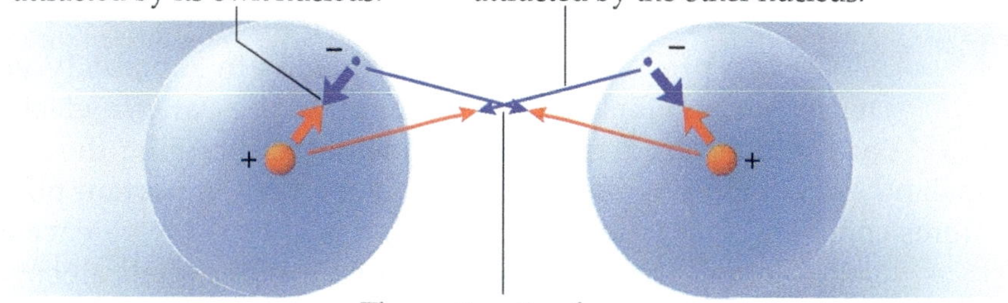

As the atoms got closer, these attractive forces became stronger, and so the atoms picked up speed. Upon formation of the H_2 molecule, each electron found itself strongly attracted to *two* nuclei instead of one. Thus, there is a greater total amount of positive–negative attraction in the H_2 molecule than in two isolated H atoms.

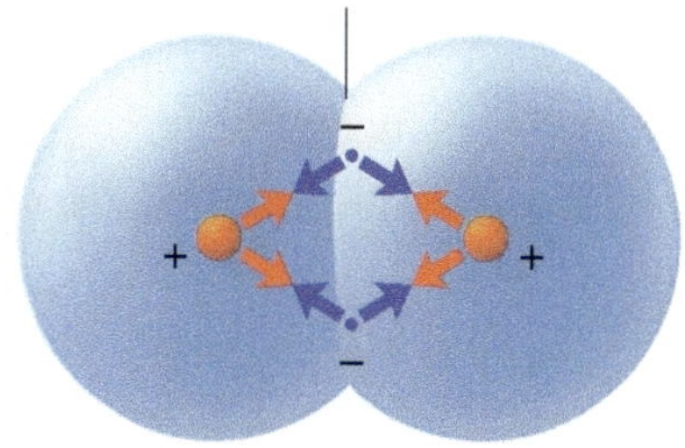

Even though the isolated H atoms started out neutral and remain neutral in the molecule, there is extra positive–negative attraction in the molecule relative to the two isolated atoms! It is this additional attractive force that holds the molecule together. We call this additional attractive force a *covalent bond*. To separate the H_2 molecule into isolated hydrogen atoms, you would have to provide enough energy to overcome this additional attractive force and break the bond.

Why did the atoms stop when the nuclei were 0.74 Å apart? Once again, the reason is a force between charges, but this time it's between like charges. Like charges repel each other, and the nuclei of both atoms are positive. When the nuclei are 0.74 Å apart, the repulsion between the two nuclei becomes great enough to keep the H atoms from getting any closer. A covalent bond is therefore a delicate balance between attractive and repulsive forces.

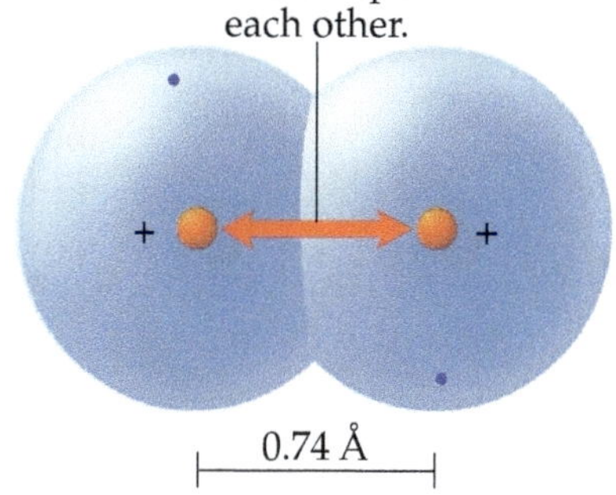

Finally, as the two atoms formed a molecule, the temperature increased. Why? Since attractive interactions stabilize particles, the extra positive–negative attraction that is the covalent bond results in the H_2 molecule being more stable than the two isolated H atoms. This

The electrons spend most of their time between the nuclei:

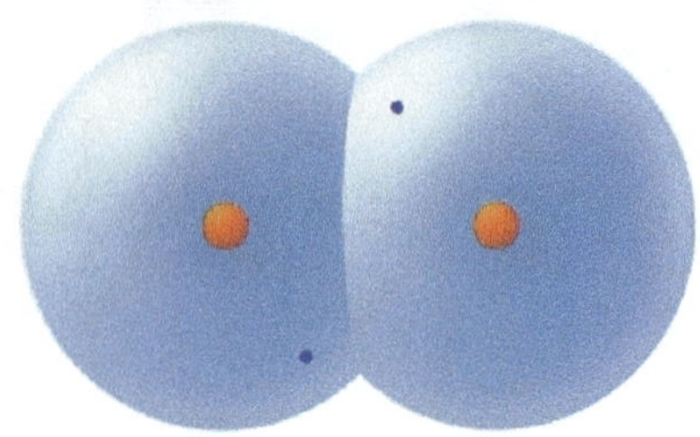

although they spend some time in other places:

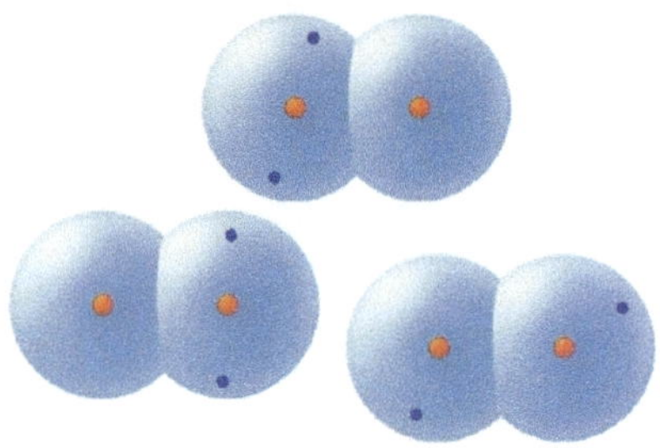

gain in stability is the source of the temperature increase because one way things can become more stable is to release energy. (The opposite is also true—feeding energy into something makes it less stable.) The energy released is manifested as heat. You would have to put this same amount of heat (energy) back into the molecule to separate it into isolated atoms—that is, to break the covalent bond.

In the H_2 molecule, both electrons spend their time "buzzing around" both nuclei, but when we draw a representation of the molecule, we usually show the electrons occupying the region between the nuclei. This is a reasonable place to put them because they are attracted equally to both nuclei, and this is in fact where they are located most of the time. Drawn this way, it looks as if the nuclei are sharing the two electrons. In fact, that is exactly what a chemist would say is happening. We can therefore define a covalent bond as follows:

A covalent bond is the additional force of attraction that results from valence electrons being *shared* between two nuclei. The bond holds molecules together.

Co- is a prefix that means partner, as in somebody you share with; *valent* refers to the fact that it is the valence electrons that are being shared and are responsible for bonding the H atoms to each other. So the term *covalent* emphasizes a very important point: When atoms come together to form molecules, it is always the valence electrons that are involved. An atom's inner-shell electrons (also called *core electrons*) are almost never involved in bonding because they are too close to their own nucleus and therefore too strongly attracted to it to be shared with another nucleus. We didn't have to worry about this for hydrogen because a hydrogen atom has only one electron. However, as we consider bonding between atoms of other elements, we shall concern ourselves only with the valence electrons.

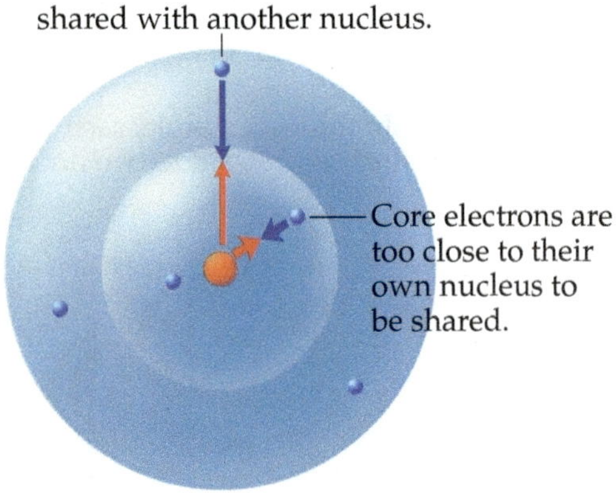

Only valence electrons participate in covalent bonding.

There are several ways to represent atoms joined by covalent bonds. Three of these ways are:

H:H H··H H—H

The line drawn between the H atoms stands for the two shared electrons that give rise to the covalent bond. A line is easier to draw than two dots, and it is the representation most often used. Every time you see a line drawn between atoms in a molecule, remember that the line stands for a pair of electrons shared between two nuclei and the covalent bond that results. Take another look at the water and methane molecules we saw earlier. In water, there are two covalent bonds (two lines); in methane, there are four:

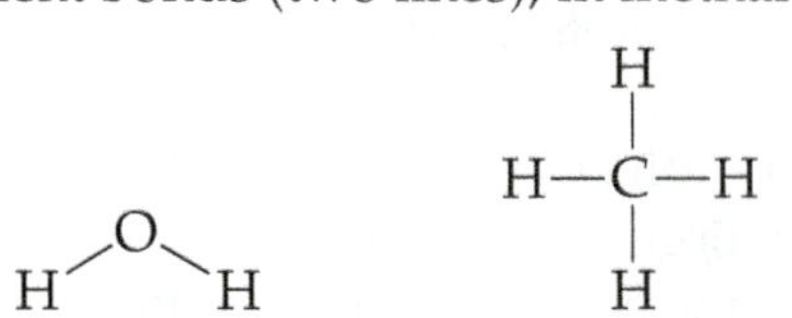

Water, H_2O Methane, CH_4

Although we shall use lines to represent covalent bonds throughout this book, you should remember that they are just a chemist's device. If you could see an individual water molecule, you would not see any lines. You would see only one oxygen atom and two hydrogen atoms that always stay close to one another.

Covalent bonds are quite strong; in fact, energies involved in this type of bonding are among the largest that chemists encounter (they are of the same order of strength as ionic bonds). The O—H covalent bonds in water molecules are the reason H_2O molecules don't fly apart into atoms even when you boil the water. The steam that rises from a pot of boiling water is made up of intact H_2O molecules. To break the O—H bonds in water would take an amount of energy far greater than the amount required to boil it.

To summarize, covalent bonds are strong, just as strong as ionic bonds (both are types of chemical bonds). It always takes energy to break a chemical bond. There are no exceptions to this statement. The converse is also true: Energy is always released upon forming a chemical bond, be it ionic or covalent.

To further explore covalent bonding, let's return to hydrogen atoms. Recall that when two hydrogen atoms are isolated and far apart, each has only one valence electron. What about after they come together to form a molecule of H_2? Does each hydrogen atom in the molecule still have only one valence electron? The answer depends on how we choose to do our electron counting. In the case of a covalent bond between atoms, chemists have decided to double-count the shared electrons. That is, in H_2, each H atom in the molecule is considered to have two electrons.

Covalent bonds are very strong. The amount of energy it takes to boil water is far too small to break them. Thus, steam and water are both made of intact H_2O molecules.

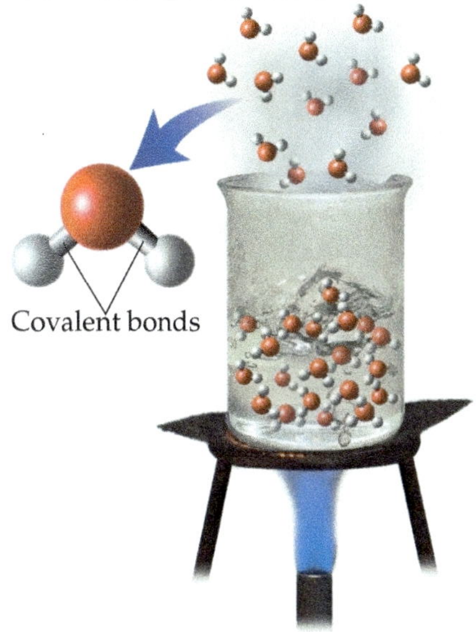

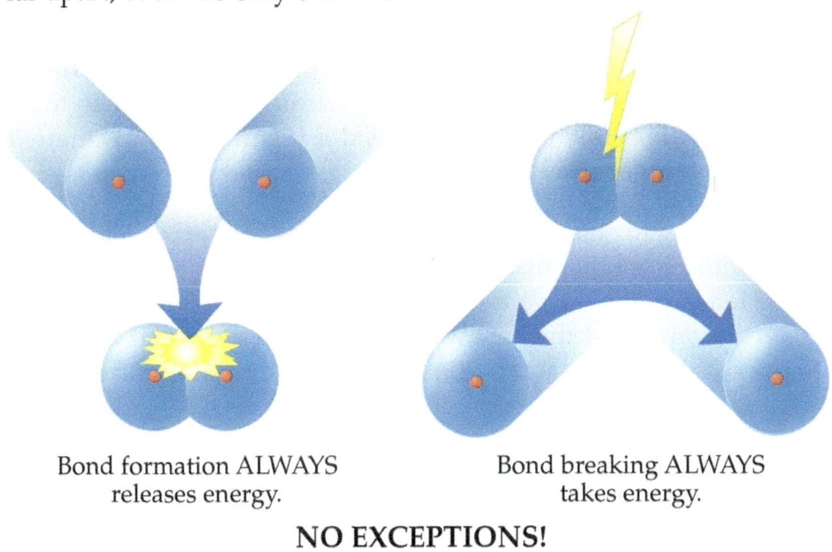

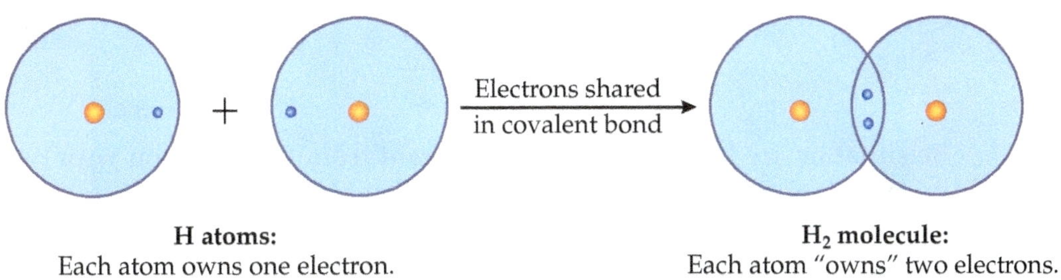

If you had a joint bank account with a friend and you did your accounting this way, you could quickly get into some financial trouble. Obviously, a shared \$100 bank account does not give each of you \$100 to spend. Nevertheless, this is how we count shared valence electrons. This means that sharing electrons is one way for atoms to get more electrons. The covalent bond in the H–H molecule effectively gives each hydrogen atom one more electron than it started with. Every covalent bond an atom forms increases its number of valence electrons by one. One justification for this counting method is that it helps us to predict how many atoms will combine when they come together, as we shall see shortly.

WORKPATCH 5.3 How many valence electrons does each atom in the methane molecule have? Double-count shared electrons.

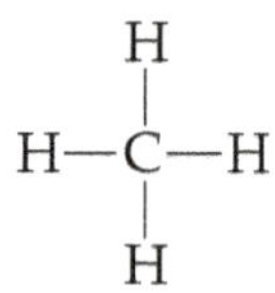
●

You should have found an octet on one of the atoms in the WorkPatch if you did your counting correctly.

WORKPATCH 5.4 How many valence electrons does a carbon atom normally have? How many valence electrons does it have in the methane molecule? How many electrons did it gain, and how many covalent bonds to hydrogen did it form? ●

Are you starting to get an insight into why methane is CH_4 and not CH_5 or CH_6, or anything else other than CH_4?

We are leading up to something important here regarding bonding between atoms, and the best way to introduce you to it is to consider bringing two helium (He) atoms together and attempting to form a helium molecule:

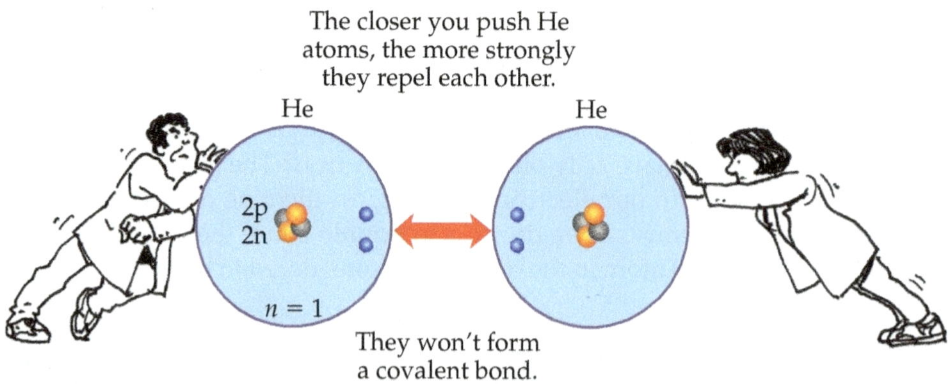

The situation we have now is very different from the situation with two hydrogen atoms. The closer two He atoms get to each other, the more they repel each other. Indeed, they refuse to form a covalent bond. Why the difference between this case and hydrogen? A hint comes from examining the behavior of the other elements in helium's group in the periodic table, group VIIIA (18), the noble gases. All the other elements in this group have eight electrons in their valence shell (a filled valence shell), and all of them strongly resist forming covalent bonds. Does the octet rule, based on a drive to achieve a filled valence shell, play a role in determining when covalent bonds form and how many of them form? The answer is an emphatic yes.

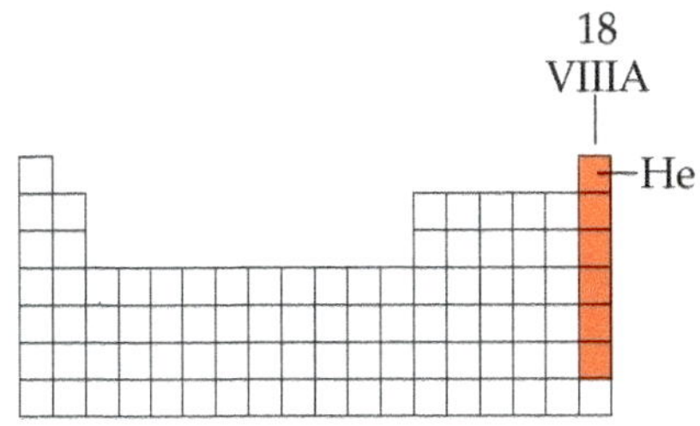

5.4 Molecules, Dot Structures, and the Octet Rule

What determines whether an element forms covalent bonds or not? How many covalent bonds is an element likely to form? Why is a water molecule always H_2O? Why not H_3O or H_4O or HO_2? The answers to these questions come from applying the octet rule, which says that many of the elements "want" to be like group VIIIA atoms and have eight electrons in their valence shell (or two electrons if their valence shell is the $n = 1$ shell). This rule often holds true when atoms are part of a molecule, where they achieve an octet by sharing electrons with other atoms.

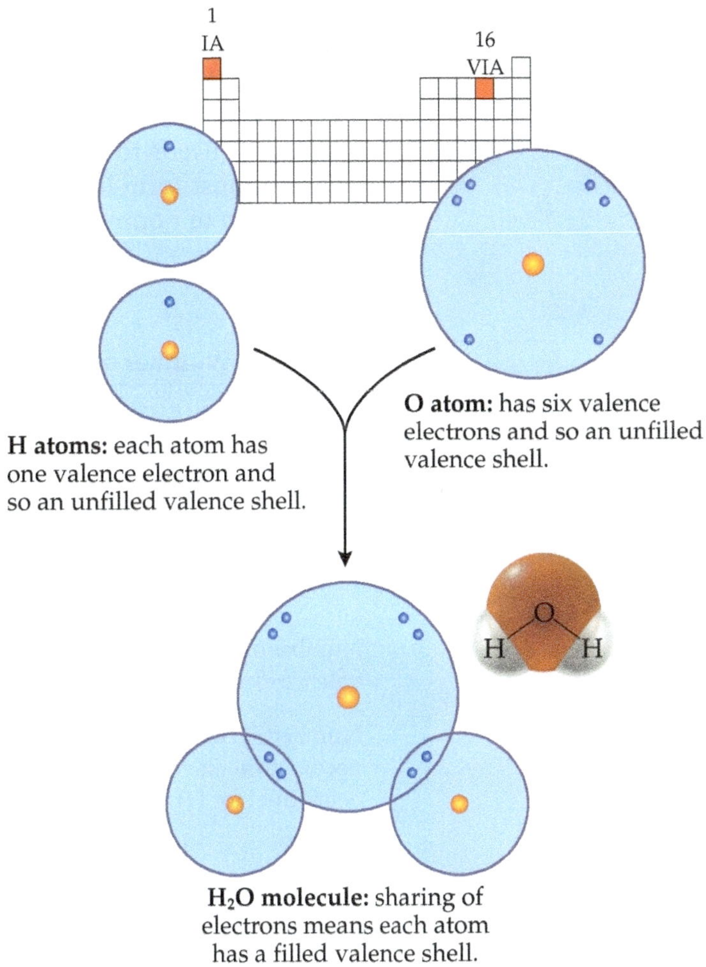

By understanding how many electrons an atom wants to share, we can better understand why water has the molecular formula H_2O and not something else, as illustrated in the adjacent figure. The best way to begin is to determine the number of valence electrons a neutral, isolated atom has before it becomes part of a molecule. This is easy. Recall that for a representative element, the Roman-numeral group number tells you the number of valence electrons. While the noble gases in group VIIIA have eight valence electrons (He, with two, is the only exception), all of the other representative elements have fewer than eight valence electrons—that is, less than an octet. For example, every group IVA element has four valence electrons and is therefore four electrons short of an octet.

One way group IVA elements can get eight electrons is to share all four of their valence electrons with other atoms. For example, remember the methane molecule, CH_4, we looked at earlier? Carbon starts with only four valence electrons and needs four more to complete an octet. Therefore, it forms four covalent bonds with other atoms:

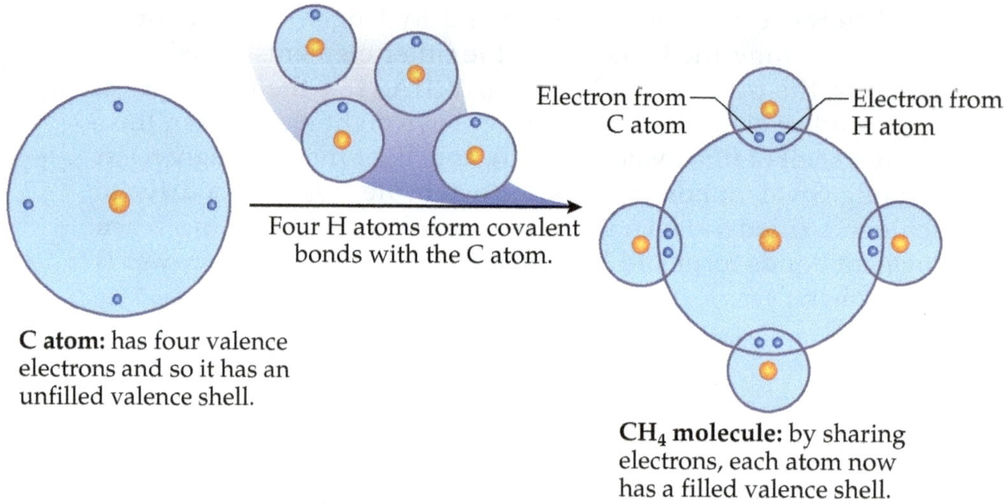

Because the roman-numeral group number tells you the number of valence electrons an atom has, it is a simple matter to figure out how many additional electrons it needs to reach eight. This number is also the number of covalent bonds it must form to get there. Table 5.1 summarizes the requirements for each group of nonmetal atoms (we'll save the metals for later). Examine it closely; then we'll see how it can be used to determine the molecular formula of a molecule.

Table 5.1 Number of Covalent Bonds Nonmetal Atoms Usually Form

Group number	**IA**	**IVA**	**VA**	**VIA**	**VIIA**
Nonmetal element	H	C Si Ge	N P As Sb	O S Se Te	F Cl Br I At
Number of valence electrons	1	4	5	6	7
Number of electrons needed to achieve octet (duet for H)	1	4	3	2	1 ← This is the number of bonds an atom of the element usually forms.

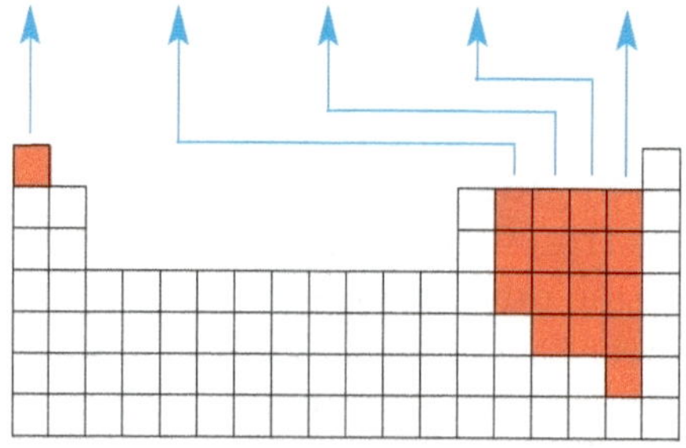

The blue number in the last row of each column in Table 5.1 is the number of covalent bonds the atoms in that column must form to achieve an octet. For example, let's look at ammonia, a molecule made from nitrogen and hydrogen. Find nitrogen in the table. It is in group VA and, therefore, has five valence electrons. It needs three more electrons to achieve an octet. In ammonia, these three electrons come from forming three covalent bonds to H atoms. That is why the formula for ammonia is NH_3—not NH_2 or NH_4.

Practice Problems

5.1 Predict the formula of the compound that forms between carbon and chlorine.

Answer: Table 5.1 tells you that carbon needs to form four bonds and chlorine needs to form one bond. The formula is therefore CCl_4.

5.2 Predict the formula of the compound that forms between phosphorus (P) and hydrogen.

5.3 Predict the formula of the compound that forms between silicon (Si) and bromine (Br).

Table 5.1 is fine, but to build more complex molecules, such as glucose and DNA, we need a system for drawing atoms that can tell us at a glance how many covalent bonds each atom forms. A chemist named G. N. Lewis invented just such a system for molecules back around 1916. His drawings are called **dot diagrams**, or sometimes **Lewis dot diagrams**. Dot diagrams are used to show an atom's valence electrons. In a dot diagram, the symbol of an element is surrounded by dots, one dot for each valence electron. For the period 2 elements, for instance, the dot diagrams are:

In every case, the number of dots is equal to the group number for these representative elements. Notice how the electrons are added. The first four electrons (Li through C) are positioned on the four sides of the symbol. It is not until we have a fifth electron, in N, that we begin to pair the electrons. Thus, carbon is surrounded by four single electrons, whereas nitrogen has three single electrons and one pair. We refer to the pairs as either **electron pairs** or **paired electrons**, and we call the single electrons **unpaired electrons**. Thus, nitrogen has three unpaired electrons and one electron pair. As we shall see, the difference between paired and unpaired electrons is important.

There is some flexibility with respect to how you arrange the unpaired and paired electrons. For example, all three of these dot diagrams for oxygen are correct:

All these forms are equivalent.

Because the number of valence electrons determines what the dot diagram for an atom looks like, all atoms with the same number of valence electrons have identical dot configurations. Thus, the dot diagrams for periods 3–7 of the representative elements are identical to those shown above for period 2. For example, sulfur (S) has a dot diagram that is identical to oxygen since it is immediately under oxygen in the periodic table. We won't concern ourselves with dot diagrams for the transition metals, lanthanides, or actinides. This leaves just two more elements to describe, hydrogen and helium, both of which are unique. Their dot diagrams are:

H·   He:

Note that helium's two valence electrons are always paired.

We now have simple diagrams for all the representative elements. These dot diagrams can help us determine the bonding that can occur between these elements. If you examine all the dot diagrams that we have considered, only the group VIIIA atoms (He, Ne, Ar, Kr, Xe, Rn) have all their valence electrons paired up and possess a complete octet (duet for He). All the other representative elements have unpaired valence electrons and thus less than an octet. The octet rule states that they must have eight valence electrons (duet for H) to achieve stability. One way to get eight valence electrons is for these elements to share their unpaired electrons with other atoms.

Let's see how this works by using fluorine as an example. Fluorine is in group VIIA, so each atom has seven valence electrons—three pairs and one unpaired electron:

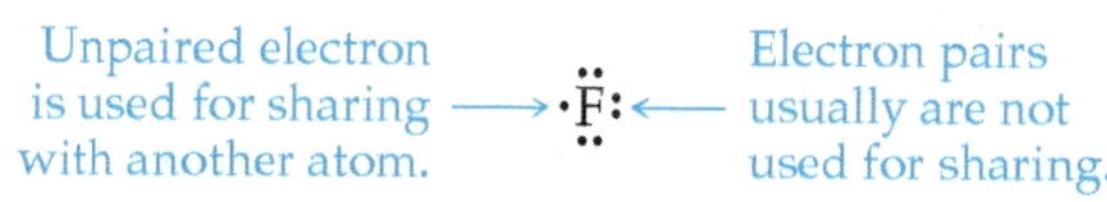

The fluorine atom needs one more valence electron to achieve an octet. If two F atoms come together and share their unpaired electrons, then both atoms can have eight valence electrons. When the atoms share in this way, a single covalent bond forms between them. In this case, no more bonds will form because both atoms now have octets. This is why fluorine exists as the diatomic molecule F_2 and not as F, F_3, or any other combination.

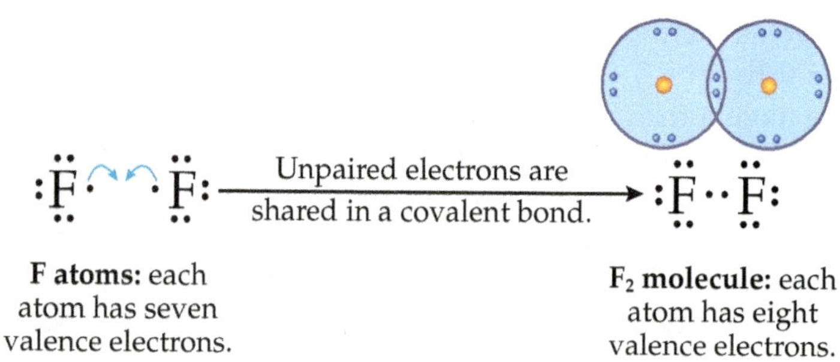

WORKPATCH 5.5 Why do two He atoms refuse to form the molecule He_2 when brought close to each other? ●

How did you answer the WorkPatch? You could have said that there were no unpaired electrons to pair up. Or you could have answered in terms of the octet rule. Either way is correct.

Practice Problems

5.4 In forming molecules, atoms can share unpaired electrons in order to achieve an octet in their valence shell. Hydrogen is an exception. What number of electrons does hydrogen "want," and why is this number not eight?

Answer: Hydrogen wants two electrons (a duet instead of an octet) because it is a period 1 atom and so its valence shell is the $n = 1$ shell. This shell is filled when there are two electrons in it.

5.5 What is the molecular formula of the compound that forms between H atoms and F atoms? Justify your answer with dot diagrams.

5.6 Why is water H_2O and not H_3O or something else? Justify your answer with dot diagrams.

The strategy of sharing unpaired electrons and aiming for octets works for all sorts of molecules. For example, the molecules NF_3 and CH_4 are formed from atoms as follows:

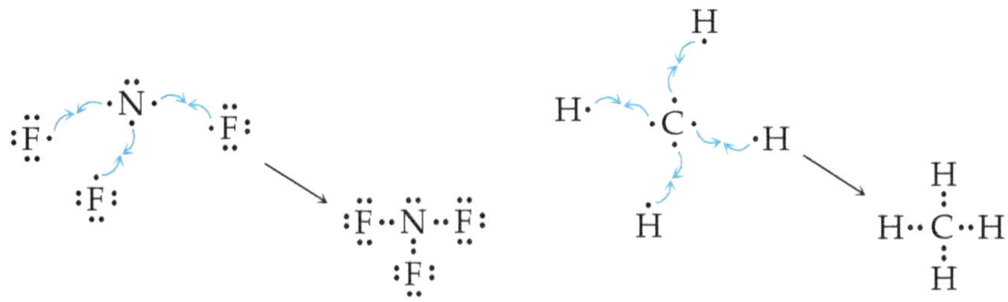

Drawing all these dots gets tedious, so we usually just draw a line for each shared pair of electrons:

As we said previously, a line between two atoms is understood to represent two shared electrons, which constitute a covalent bond. Therefore, shared pairs of electrons are often called **bonding pairs**. Any unshared electron pairs, such as the three pairs on the F atoms and the one pair on the N atom in NF_3, are called **lone pairs** because they belong to a single (lone) atom.

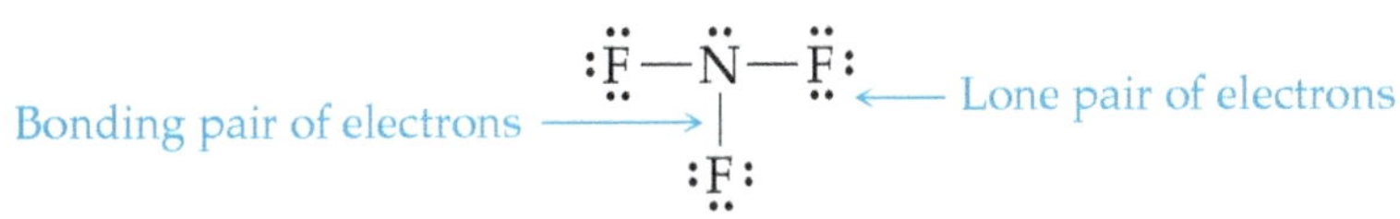

Note that not all atoms in a molecular dot diagram necessarily have lone pairs (two examples are the carbon atom in methane and hydrogen atoms in any molecule). Because they are not shared, lone pairs are not directly involved in bonding. Nevertheless, they are important and should never be left out of a molecular dot diagram.

WORKPATCH 5.6 How many valence electrons are present on each atom in the NF_3 molecule? ●

Did you count the lone pairs once but double-count the shared pairs when doing your electron counting for the WorkPatch? This exercise demonstrates the importance of lone pairs with respect to the octet rule.

With dot diagrams in hand, you are now ready to build more complicated molecules.

WORKPATCH 5.7 Consider ethane, C_2H_6, a minor component of natural gas. What does this molecule look like? Start with the atoms as dot diagrams and try putting them together so that each C ends up with eight valence electrons, and each H ends up with two valence electrons. Then replace every pair of shared electrons with a line. When you are done, compare your attempt with the answer at the end of the chapter. ●

There should be no lone pairs in your dot diagram for this WorkPatch. Remember, hydrogen never has a lone pair in a molecule, because two electrons are all it needs to fill its valence shell, and these will come from the covalent bond. As for carbon, it starts out with four unpaired electrons, but it will end up with no lone pairs when it forms bonds to four other atoms in order to gain four more electrons, which is the case in ethane. It begins to look like carbon will never have lone pairs in a molecule. As we shall see in the next section, lone pairs on carbon are possible but rare.

Practice Problems

5.7 Draw a dot diagram for hydrazine, N_2H_4, sometimes used as a rocket fuel.

Answer: Start with the pieces—individual dot diagrams for all the atoms—and then put them together in a way that satisfies the octet rule (duet for H):

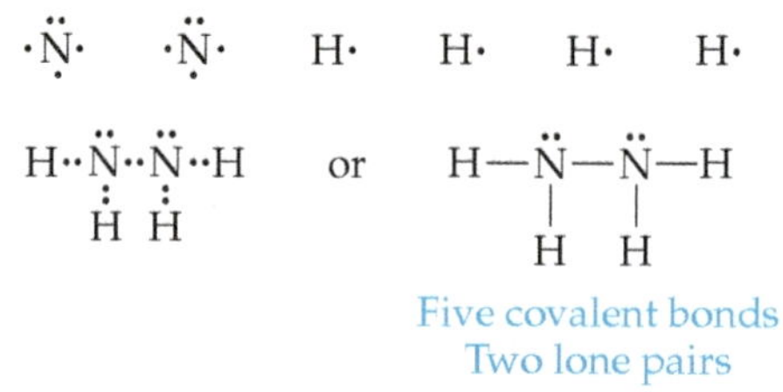

5.8 Draw a dot diagram for hypochlorous acid, HClO. (*Hint:* The O atom forms a bond with H and Cl.)

5.9 Draw a dot diagram for propane, C_3H_8, used as a fuel for heating.

5.5 Multiple Bonds

Sometimes two atoms share more than one pair of electrons with each other. Such a situation is called either a **multiple covalent bond** or simply a **multiple bond**. For example, let's reconsider those diatomic O_2 molecules you are breathing. To understand the bonding in this molecule, we start as usual by drawing a dot diagram for one oxygen atom:

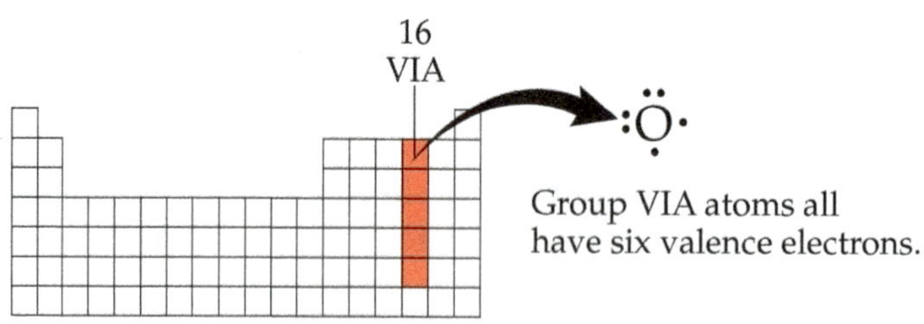

Being a group VIA atom, oxygen has two unpaired valence electrons. As two oxygen atoms get close, they can form a covalent bond, with each atom sharing one of its unpaired electrons:

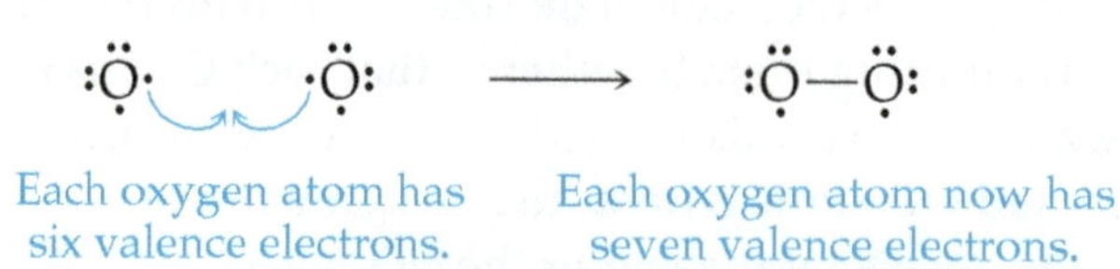

Where do we go from here? Both oxygen atoms now have seven valence electrons, not yet an octet. Each atom still has one unpaired electron left, though, and unpaired electrons are for sharing. So don't stop. Share the remaining unpaired electrons:

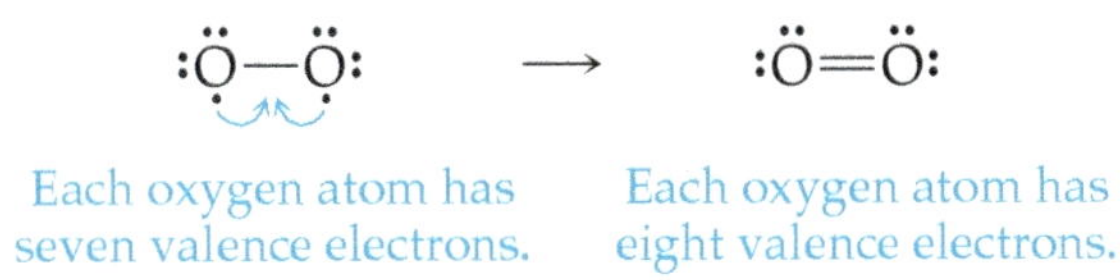

Look what this did for each oxygen atom. Each oxygen atom is now surrounded by eight valence electrons. The octet rule has been satisfied for both oxygen atoms by the formation of a multiple bond. In this case, there are two pairs of electrons shared between the atoms—in other words, two covalent bonds. We refer to this as a **double bond**. When there is just one covalent bond between two atoms, it is called a **single bond**. As you might expect, a double bond is stronger than a single bond.

The air you breathe contains another diatomic gas, nitrogen, N_2. To draw this molecule, start with dot diagrams for the isolated nitrogen atoms, then bring them together to share electrons until both atoms end up with eight:

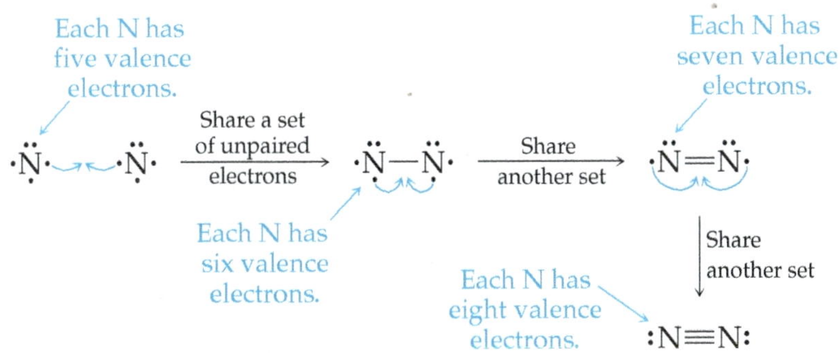

This time we end up with a **triple bond**, three shared pairs between the N atoms, a very strong bond indeed.

Related to the C_2H_6 (ethane) molecule of WorkPatch 5.7 is the acetylene molecule, C_2H_2, used in oxyacetylene torches that can cut through steel. This molecule also contains a multiple bond. Determine which kind in the following WorkPatch.

WORKPATCH 5.8 Draw a dot diagram for the acetylene molecule, C_2H_2. ●

You should have found that there is only one way to put the acetylene molecule together. If you put both hydrogens on the same carbon, you would end up violating the octet rule and leaving unpaired electrons on the other carbon atom.

Practice Problems

5.10 Draw a dot diagram for carbon dioxide, CO_2.

Answer: 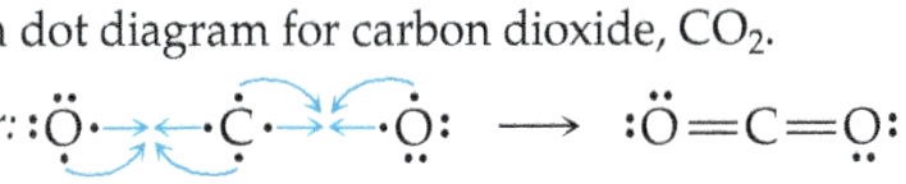

5.11 Draw a dot diagram for propyne, C_3H_4. One carbon has three hydrogens bound to it, one has one hydrogen bound to it, and one has no hydrogens bound to it.

5.12 Draw a dot diagram for hydrogen cyanide, HCN. The hydrogen is attached to the carbon.

5.13 Draw a dot diagram for acetone, C_3H_6O, the active ingredient in nail polish remover. The oxygen is bound to only one carbon, and that carbon has no hydrogens bound to it.

Drawing dot diagrams for molecules by starting with dot diagrams for each individual atom is a great way to learn about covalent bond formation, but it is a tedious process and doesn't always work. There is a simpler, four-step method that we want to show you now. We shall demonstrate this method using two examples—carbon monoxide, CO, and carbon dioxide, CO_2.

Step 1. Determine the *total* number of valence electrons (dots) that will be in the final diagram. To do this, simply add up the group numbers of all the atoms in the molecular formula.

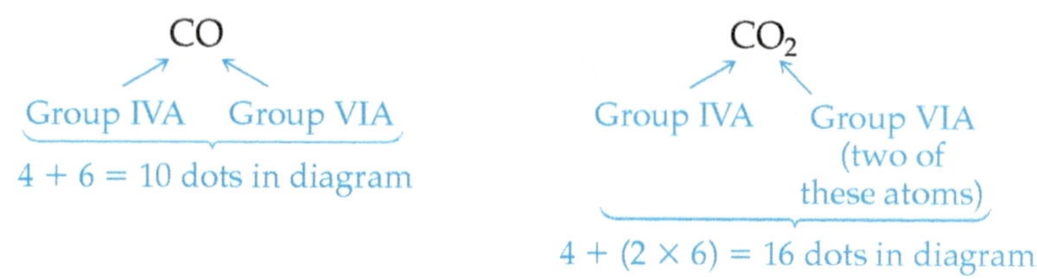

Step 2. Connect the atoms with single bonds. If you are not told how the atoms are connected to one another, it often works to assume that the first nonhydrogen atom in the molecular formula is the central atom. For example, the central atom in phosphorus trichloride, PCl_3, is P, and the central atom in sulfuric acid, H_2SO_4, is S. Also remember that every time you connect two atoms with a single bond, you are using two electrons.

C—O   O—C—O

2 dots used   4 dots used

Step 3. Put in the remaining dots two at a time as lone pairs, first on the terminal atoms of your structure and then, if there are any dots left, on the central atoms. Keep adding electrons to an atom until it has an octet before moving on to the next atom. You must stop putting in lone pairs when you run out of electrons. This is very important. Do not use more electrons than you calculated in step 1.

All 10 dots now in diagram   All 16 dots now in diagram

If at the end of this step, every atom except hydrogen has eight valence electrons around it and every hydrogen has two electrons, you are done. The molecule will have only single bonds. In our two examples, we are not done because each carbon has only four valence electrons. When this happens, we must go on to step 4.

Step 4. If there are atoms that do not yet have an octet, send lone pairs to the rescue by moving them into a bonding position between the atoms. The lone pairs must come from an *adjacent* atom (one that is attached to the octet-deficient atom).

CO

CO_2

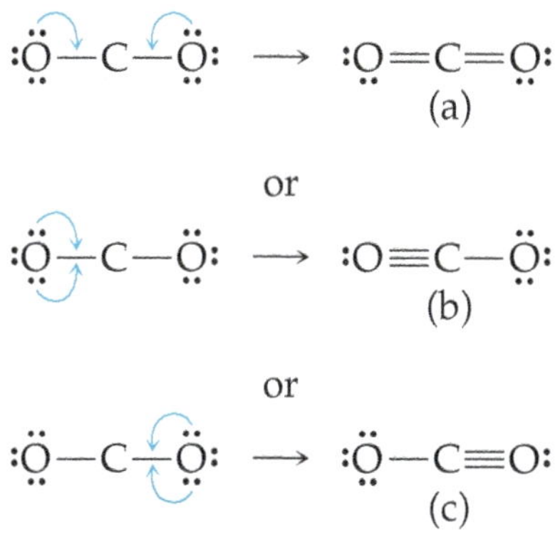

This solution reveals that the carbon monoxide dot diagram contains a triple bond between the carbon and oxygen atoms. Each atom also has one lone pair, and each atom has a complete octet. This was accomplished using only the ten valence electrons we started with. By following steps 1–4, we came up with a valid molecular dot diagram. For carbon dioxide, we drew three dot diagrams, all equally valid with respect to the octet rule. Convince yourself of this by examining the diagrams and answering the following WorkPatch.

WORKPATCH 5.9 How did we get three different dot diagrams for carbon dioxide? ●

Do you understand why we arrived at more than one valid dot diagram for CO_2? If you do, you will know when to expect this situation. But all three dot diagrams for CO_2 can't be correct, can they? Well, actually they are all correct. They are identical in their arrangement of atoms (carbon in the middle with oxygens attached to it). The only difference is the arrangement of electrons. Valid dot diagrams that differ from one another only in the arrangement of electrons are called **resonance forms**. The actual molecule is considered by chemists to be some combination of these three forms, with all the forms not necessarily equally weighted. Experiments and theoretical considerations show that carbon dioxide is most like resonance form (a), with perhaps just a tad of (b) and (c) mixed in.

Up to now, we have been generating dot diagrams for electrically neutral (uncharged) molecules. Before we present some practice problems, we want to modify the first step in the above procedure so that it can be extended to cover *polyatomic ions*. **Polyatomic ions** are just what they sound like—ions made from many atoms. Some examples are nitrate, NO_3^-, used to preserve meats; phosphate, PO_4^{3-}, found in soda pop; and ammonium ions, NH_4^+, often found in soaps and detergents. These are all ions because they have an overall electrical charge.

CO_2

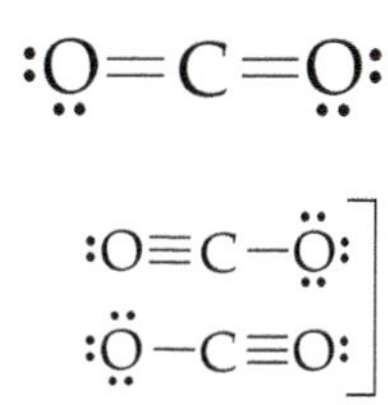

Main resonance form

Minor resonance forms

When drawing dot diagrams for ions, we follow steps 1–4 with one small change. The charge affects the total number of dots in the diagram. Step 1 has to be modified to account for this. To do this, we must consider how to convert a neutral molecule to a cation or an anion. An anion is negative because it has a surplus of electrons. The surplus is equal to the charge of the anion. For example, consider the nitrate anion, NO_3^-. Neutral NO_3 has 23 valence electrons (one group VA atom and three group VIA atoms). The anion NO_3^-, with its -1 charge, must therefore have 24 valence electrons. As a second example, consider the phosphate anion, PO_4^{3-}. Neutral PO_4 has 29 valence electrons (one group VA atom and four group VIA atoms). The anion PO_4^{3-}, with its -3 charge, must therefore have 32 valence electrons.

A cation is positive because it has a deficiency of electrons. For example, consider the ammonium ion, NH_4^+. Neutral NH_4 has nine valence electrons (one group VA atom and four group IA atoms). The NH_4^+ cation, with its $+1$ charge, therefore has eight valence electrons.

Drawing dot diagrams

Step 1: Determine the total number of valence electrons (dots) that will be in the diagram. For a polyatomic ion, adjust the electron count to reflect the charge.

Step 2: Connect the atoms using single covalent bonds. If you do not know how the atoms are arranged, try using the first nonhydrogen atom in the formula as the central atom, and arranging the other atoms around it symmetrically.

Step 3: Add the remaining electrons as lone pairs, completing octets as you go.

Step 4: If any atom lacks an octet, move lone pair(s) from adjacent atom(s) into bonding positions on the octet-deficient atom.

Modified Step 1: Determine the total number of valence electrons (dots) that will be in the final diagram. To do this, add up the group numbers of all the atoms in the molecular formula. *If you are dealing with an ion, adjust the dot count according to the charge (subtract electrons for cations, add them for anions).*

The method for drawing dot diagrams is summarized at right.

You are now ready for some practice problems. Some will have resonance forms, some will not—it's up to you to decide. When resonance forms exist, draw them all.

Practice Problems

5.14 Draw a dot diagram for the carbonate anion, CO_3^{2-}.

Answer:

Step 1: CO_3^{2-} has a total of 24 dots:

C is group IVA = 4 dots
O is group VIA (there are three of them) = $6 \times 3 = 18$ dots
-2 charge gives an additional two electrons = 2 dots
24 dots

Step 2: Assume the first atom written is the central atom and all other atoms are attached to it. Connect them with single bonds.

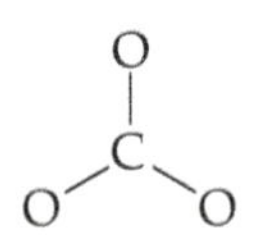

The three bonds consume 6 electrons, leaving 18 to go.

Step 3: Put in the remaining electrons as lone pairs, completing octets as you go.

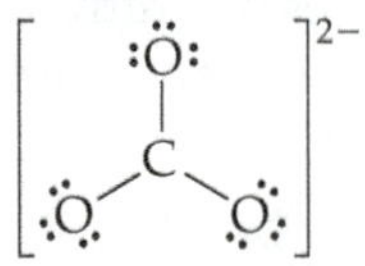

All 24 electrons are in. Are we done? No. Carbon does not have an octet. We must proceed to step 4.

Step 4: If all octets are not completed, move lone pairs from adjacent atoms to the atom needing electrons.

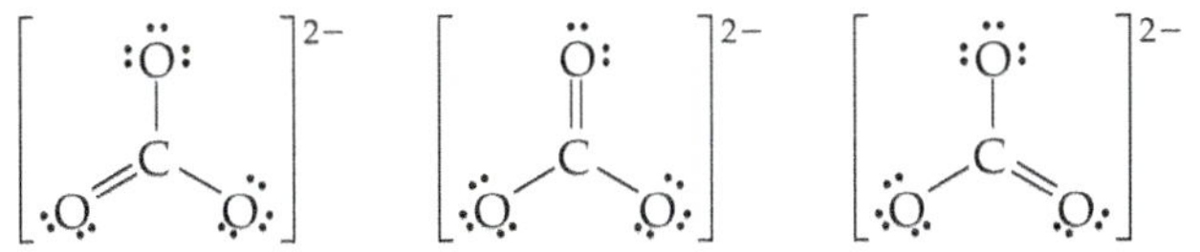

We get three resonance forms, depending on which lone pair we use to create the double bond. Note that if we are dealing with a polyatomic ion, it is traditional to put the ion in square brackets and indicate the overall charge as a superscript.

5.15 Draw a dot diagram for neutral SO_3.

5.16 Draw a dot diagram for the sulfite ion, SO_3^{2-}.

5.17 Draw a dot diagram for NO^+.

5.18 Are the dot diagrams shown below resonance forms? Explain.

5.6 Equal versus Unequal Sharing of Electrons—Electronegativity and the Polar Covalent Bond

When two atoms come close enough to each other, a bond may form between them. The bond occurs because the valence electrons of the two atoms redistribute themselves in such a way as to give rise to an attractive force between the atoms. So far, we have examined redistribution involving either a sharing (covalent bond) or a complete transfer (ionic bond) of electrons. We can represent this pictorially. We begin with two atoms, each having a single valence electron as illustrated at left. Two things can happen when these two atoms get close to each other. Either the valence electrons are shared, or they are transferred. Examine the pictures closely:

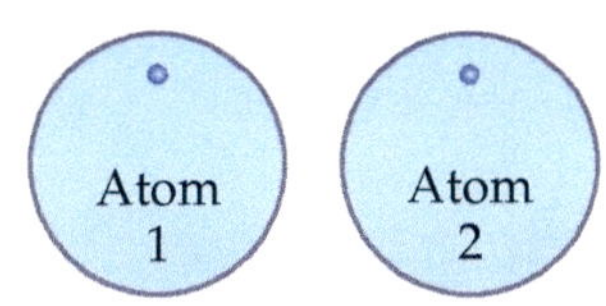

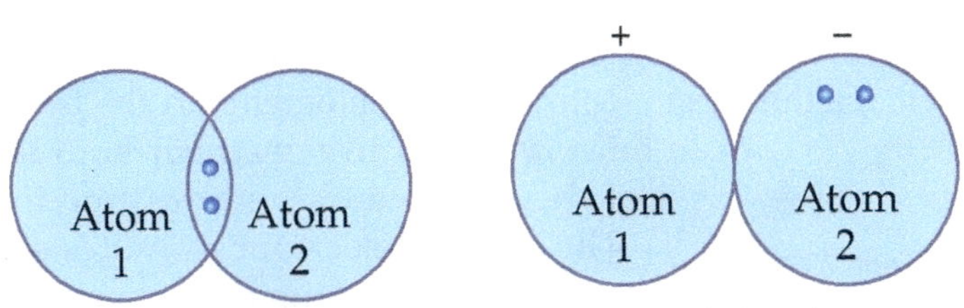

Covalent bond: electrons shared equally

Ionic bond: electron transferred

Whether the electrons are shared or transferred, the result is the same in that a strong attractive force—a chemical bond—arises. But are equal sharing or complete transfer the only possibilities? To gain further insight into this question, we must turn to a defined property of atoms called *electronegativity*. **Electronegativity (EN)** is a numerical rating of an atom's ability to attract to itself the shared electrons in a covalent bond. The chart below lists the electronegativity values of the representative elements.

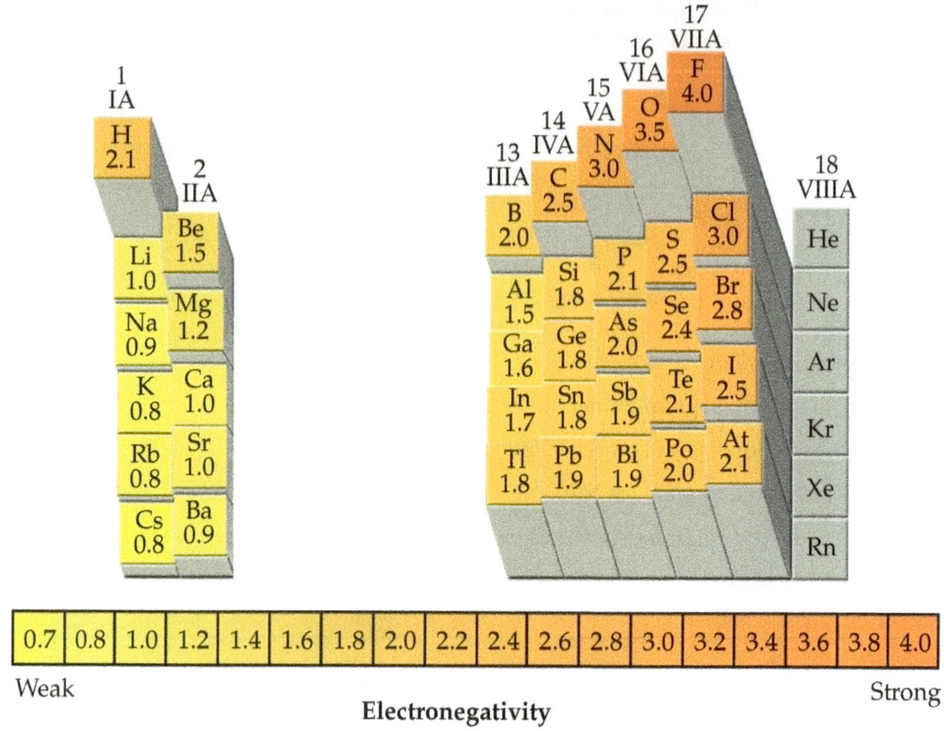

The electronegativity scale ranges from 0.7 to 4.0. The higher the value, the greater an atom's ability to attract shared electrons to itself. Fluorine is the best at attracting shared electrons to itself and is assigned the highest electronegativity value, 4. All other electronegativity values are scaled relative to fluorine $= 4.0$. Francium (not shown, occurs under Cs), at 0.7, is the least electronegative atom. In general, metal atoms tend to have low electronegativity values, and nonmetal atoms tend to have high electronegativity values.

There is a noticeable periodic trend in electronegativity values, which is emphasized by the color shading in the chart on this page. The highest values are in the upper right-hand corner (orange), and the lowest values are in the lower left-hand corner (yellow).

Nitrogen, N, oxygen, O, fluorine, F, and chlorine, Cl, are the most electronegative elements in the periodic table. These are the electron "hogs." Because they are the most electronegative elements in the periodic table, it can be very useful to remember both the identity of these hogs and their position in the table. The relative electronegativities of other elements can be estimated by considering how close in the table they are to this group. The closer an element is to these hogs, the more electronegative the element is. This means that as you go either up a group or left to right across a period, electronegativity generally increases.

Notice that the noble gas atoms of group VIIIA are not assigned any electronegativities. This is because they don't typically form bonds to other atoms, so their ability to attract shared electrons is a moot point.

One place where electronegativity is useful is in drawing Lewis dot diagrams. Earlier, we said that you can often assume that the first nonhydrogen atom in the chemical formula of a simple molecular compound is the central atom. Thus, for instance, C is the central atom in $COCl_2$:

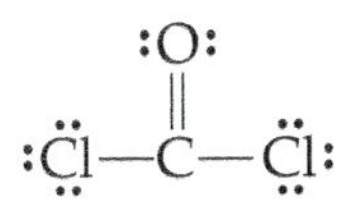

However, if someone chooses to write the formula $OCCl_2$ (which is a perfectly correct way to do it), blindly following this rule would lead to an incorrect dot diagram:

Electronegativity can help here. For simple molecules, it is generally true that the least electronegative atom (except hydrogen) is the central atom. Thus, we would place C in the middle of the molecule even if the formula were written $OCCl_2$ or Cl_2OC.

Another important use of electronegativity is to help understand what type of bonding occurs between two atoms. When two nonmetal atoms of identical electronegativity bond, the result is an equal sharing of electrons, giving rise to a pure covalent bond. This is the case for the bond in such molecules as H_2, F_2, Cl_2, Br_2, I_2, O_2, and N_2. If the atoms have different electronegativities, however, the bonding electrons will spend more time near the atom of greater electronegativity. In other words, the more electronegative atom will hog the bonding electrons. This results in a pileup of the bonding electrons around the more electronegative atom, causing it to develop a partial negative charge (indicated by the symbol $\delta-$, where the Greek lowercase letter delta, δ, means "partial"). The opposite happens around the less electronegative atom—it develops a partial positive charge, $\delta+$, that is equal in magnitude to the $\delta-$ charge on the more electronegative atom. The result is a **polar covalent bond**, a covalent bond in which the electrons are shared *unequally*. This is the case for the bond or bonds in molecules like HCl, H_2O, and HF.

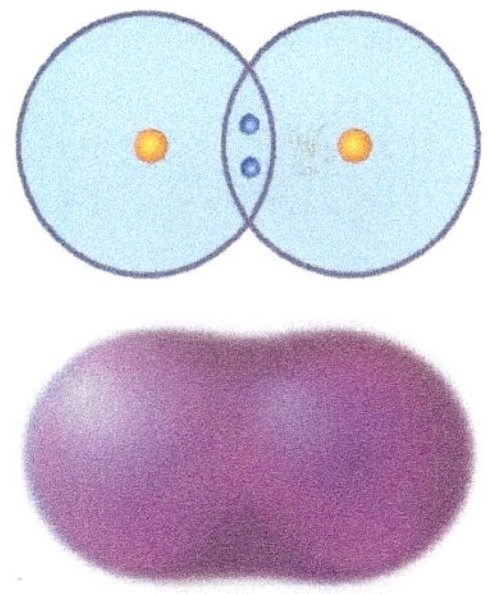

Covalent bond:
electrons shared equally

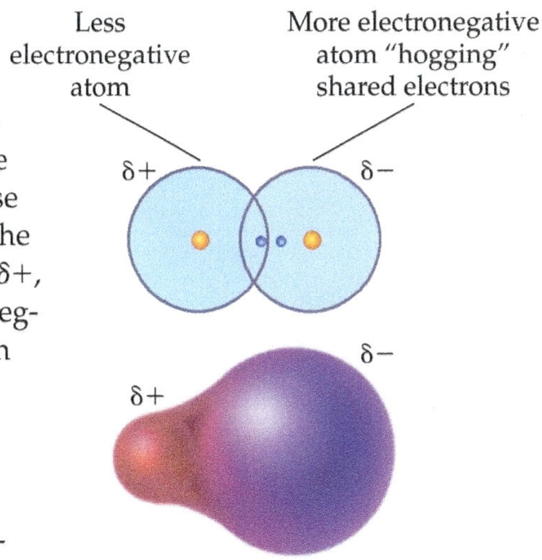

Polar covalent bond:
electrons shared unequally

The greater the electronegativity difference, ΔEN, between the atoms, the greater the partial charges that develop. (The symbol Δ is the Greek uppercase letter delta. In science, this symbol usually stands for a difference in two quantities.) Increasing ΔEN values between atoms involved in a chemical bond are said to increase the polarity of the bond.

Do you notice something here? A polar covalent bond with its $\delta+$ and $\delta-$ atoms is starting to look like an ionic bond. A pure ionic bond, resulting from the complete transfer of electrons, has full positive and negative charges on the ions:

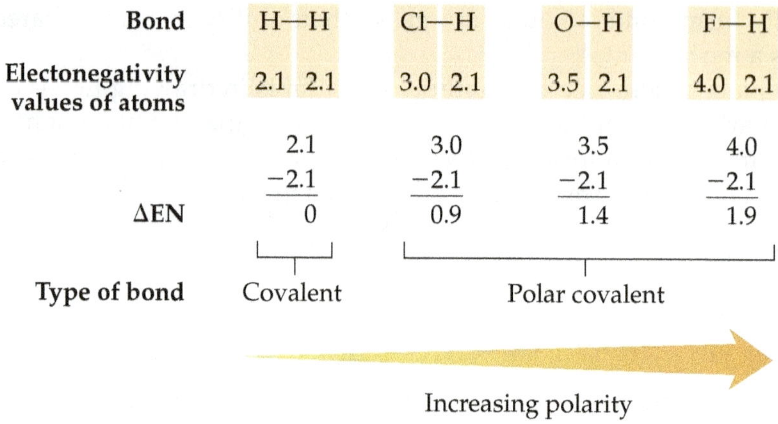

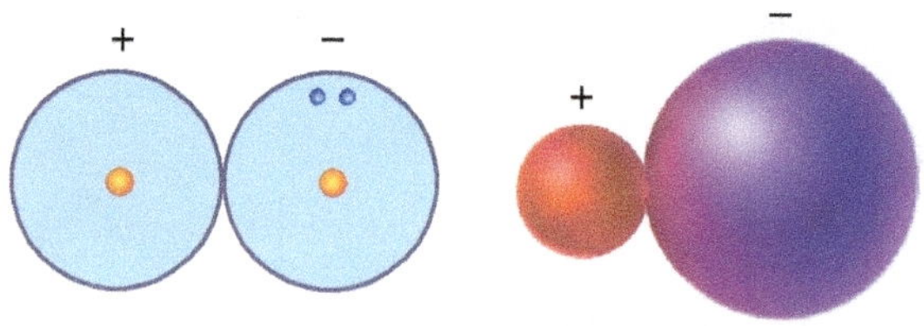

Ionic bond: electron transferred

A polar covalent bond can therefore be thought of as a covalent bond that has some fraction of ionic character. Indeed, the brilliant chemist Linus Pauling, who originated the concept of electronegativity and determined the values listed in this book, developed a simple relationship between electronegativity difference and the percent ionic character in a bond. This relationship is nicely shown in the graph below. The graph shows that when ΔEN equals 1.7, a bond can be considered to be 50% ionic (and therefore 50% covalent).

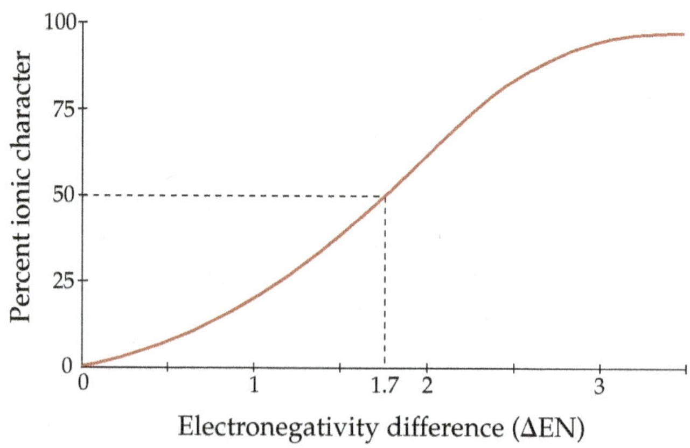

What should you take away from this discussion? Simply that it is often an oversimplification to refer to a bond between two elements as being either ionic or covalent. Except when ΔEN = 0, the bond will have some of the characteristics of both bond types. Ionic and covalent are therefore extremes at the ends of a continuum of bonding types:

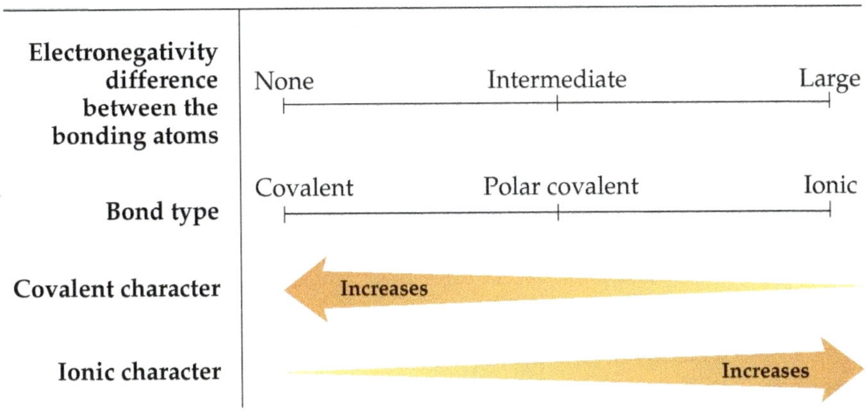

So how do we classify bonds? To answer this question, consider these cases:

F_2	$\Delta EN = 0$
CO_2	$\Delta EN = 1.0$
HF	$\Delta EN = 1.9$
NaBr	$\Delta EN = 1.9$
NaF	$\Delta EN = 3.1$

The bond in the fluorine molecule, F_2, is purely covalent and nonpolar. The bond in NaF is, according to the graph we just looked at, about 90% ionic. What do we call a bond that is 90% ionic? When a metal is bound to a nonmetal (as is the case with NaF) and the percent ionic character is greater than 50%, it is traditional to refer to the compound and the bonds in it as being ionic (even though at 90% ionic, the electron transfer is not actually complete).

What about HF? Here, the electronegativity difference is quite large and leads to a bond that is roughly 60% ionic. Nevertheless, this bond is classified as polar covalent and not ionic.

F_2	$\Delta EN = 0$	0% ionic	Covalent
CO_2	$\Delta EN = 1.0$	23% ionic	Polar covalent
HF	$\Delta EN = 1.9$	60% ionic	Polar covalent
NaBr	$\Delta EN = 1.9$	60% ionic	Ionic
NaF	$\Delta EN = 3.1$	90% ionic	Ionic

The HF bond is classified as polar covalent because the term *ionic* is generally reserved for molecules in which a metal bonds to a nonmetal, as in NaBr. For NaBr, we have a case where the ΔEN value is again 1.9, just as with polar covalent HF, but we call this bond ionic because a metal, Na, is involved.

There are exceptions to the metal–nonmetal rule. For example, the compound ammonium chloride, NH_4Cl, is best thought of as an ionic compound made up of NH_4^+ cations and Cl^- anions, even though it contains no metals.

Practice Problems

For Practice Problems 5.19–5.21, use the electronegativity values in the chart on page 189 to calculate ΔEN *and predict whether the bonds* (Ba—Cl, O—O, *and* Si—O) *are covalent, polar covalent, or ionic.*

5.19 $BaCl_2$

Answer: EN values: Ba 0.9, Cl 3.0; $\Delta EN = 2.1$.

The bond between Ba and Cl is ionic because the bonding is between a metal (Ba) and a nonmetal (Cl), and this ΔEN value corresponds to about 62% ionic character according to our graph.

5.20 O_3

5.21 SiO_2

5.22 Without knowing electronegativity values, a student claims $BaCl_2$ is more ionic than $BeCl_2$. All she has access to is a periodic table. How does she know she is right?

In summary, substances such as Cl_2 and HCl and NaCl do not fly apart because valence electrons are equally shared, unequally shared, or transferred between one atom and the other, respectively. All three situations give rise to an attractive electrostatic force between the atoms, a force we call a chemical bond (covalent, polar covalent, or ionic). Valence electrons truly are the "glue" that bonds atoms to each other.

5.7 Nomenclature—Naming Chemical Compounds

In the early days of chemistry, as compounds were discovered and characterized, they were often given names that had little to do with their molecular formula, if the formula was even known. For example, the compounds H_2O, NH_3, and Hg_2Cl_2 have the *common names* water, ammonia, and calomel, respectively. The problem with these common names is that they give no clue to the chemical formula of the compound. For example, the only way to know that *ammonia* means NH_3 is memorization. Such a naming system is doomed to fail when you consider that today there are literally millions of known compounds. Just imagine trying to memorize a few hundred thousand different names and their associated formulas and you see the problem.

For this reason, it quickly became obvious that chemistry needed a naming system (a *nomenclature* system) that would generate a unique name for each compound based on its chemical formula. Through international agreement, just such a nomenclature system has been developed. Once you understand the rules of this system, just hearing the name of a compound will allow you to write its molecular formula. We shall show you these rules for naming the simplest of compounds—**binary compounds**, which are those made up of only two elements. This category can be further subdivided into *binary ionic compounds* and *binary covalent compounds*. **Binary ionic compounds** are composed of a metal and a nonmetal and are usually ionic. **Binary covalent compounds** are composed of two nonmetals or metalloids and are usually covalent. We'll start with the binary ionic compounds.

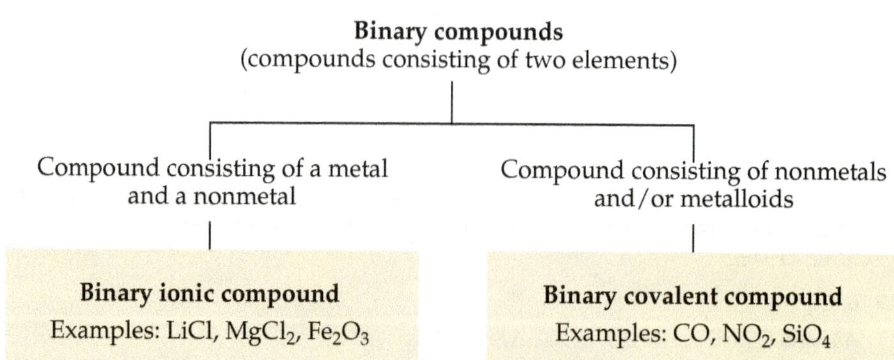

Naming Binary Ionic Compounds

The molecular formula for a binary ionic compound is always written with the metal preceding the nonmetal. For example, the chemical formula for sodium chloride is always written NaCl and never ClNa. When naming the compound, the metal is considered the cation and the nonmetal the anion. Like the formula, the name identifies the cation first, then the anion. The cation is given the name of its element, and the anion has the suffix *-ide* added to its elemental name (*-ide* means "negative"). For example, the ionic compound made from fluorine and lithium has the formula LiF and is called lithium fluor*ide*. Here are a few more examples of binary ionic compounds:

Cation — Anion — *-ide* suffix

	Cation	**Anion**	
Na_2S	Na^+	S^{2-}	Sodium sulf*ide*
$MgCl_2$	Mg^{2+}	Cl^-	Magnesium chlor*ide*
Al_2O_3	Al^{3+}	O^{2-}	Aluminum ox*ide*
Li_3N	Li^+	N^{3-}	Lithium nitr*ide*

Notice that each name tells you which elements are present in the compound and even identifies the anion (the one with the *-ide* ending), but does not seem to tell you how many of each ion are in the chemical formula. In fact, the name does tell you how many. Just by looking at the name *magnesium chloride*, you should be able to determine that the formula is $MgCl_2$. You already have the knowledge necessary to do this. We know that the metal in an ionic compound has lost electrons. Because magnesium is in group IIA, it has lost two electrons to become Mg^{2+}. We also know that nonmetals gain electrons. Because chlorine is in group VIIA, it has gained one electron and become Cl^-. Finally, in order for magnesium chloride to be electrically neutral, there must be two Cl^- ions for every Mg^{2+} ion, making the chemical formula $MgCl_2$. So, just from seeing the name *magnesium chloride*, you can deduce the formula. This is the beauty of this naming system.

WORKPATCH 5.10 Gallium, Ga, forms a compound with oxygen called gallium oxide. Based on the location of these elements in the periodic table, what is the expected chemical formula of this compound? ●

Check your answer to the WorkPatch against that given at the end of the chapter. You should be able to go from chemical formula to name or from name to chemical formula for any binary ionic compound.

Practice Problems

5.23 What is wrong with the chemical formula MgCl?

Answer: To have this formula, the ions would need to have charges that were equal in magnitude but opposite in charge, such as +1 and −1, but Mg forms Mg^{2+} cations and Cl forms Cl^- anions.

5.24 Fill in the blanks:

Formula	Cation	Anion	Name
CaF_2	?	?	?
?	?	?	Cesium bromide
?	Al^{3+}	S^{2-}	?
K_2O	?	?	?

5.25 What do all the chemical formulas of the group IIA bromides have in common?

The transition metals and some of the heavier representative metals throw a monkey wrench into this simple nomenclature system because these elements can form more than one stable cation. For example, unlike sodium, which always exists in compounds as Na^+, iron commonly exists as either Fe^{2+} or Fe^{3+}. This means that *iron chloride* could be the name for either $FeCl_2$ or $FeCl_3$. This is not good. We can't allow ambiguity in our naming system. Fortunately, a simple addition to the system fixes this problem. When dealing with a transition metal or one of the heavier representative metals, a roman numeral in parentheses is included in the name after the cation to indicate the magnitude of the positive charge. For example, iron(II) chloride (pronounced "iron two chloride") implies Fe^{2+} and has the chemical formula $FeCl_2$, whereas iron(III) chloride is $FeCl_3$ (Fe^{3+}). This Roman-numeral notation takes care of the problem and allows us to go from name to formula or from formula to name. Table 5.2 below shows some of the transition metals along with their most common ionic charges.

Table 5.2 The Most Common Ionic Charges of Some Transition Metal Atoms

Sc 3+	Ti 4+ 3+	V 5+ 4+ 3+	Cr 6+ 3+	Mn 7+ 4+ 3+ 2+	Fe 3+ 2+	Co 3+ 2+	Ni 2+	Cu 2+ 1+	Zn 2+
Y 3+	Zr 4+	Nb 5+	Mo 6+					Ag 1+	Cd 2+
La 3+	Hf 4+	Ta 5+	W 6+				Pt 4+ 2+	Au 3+ 1+	Hg 2+ 1+

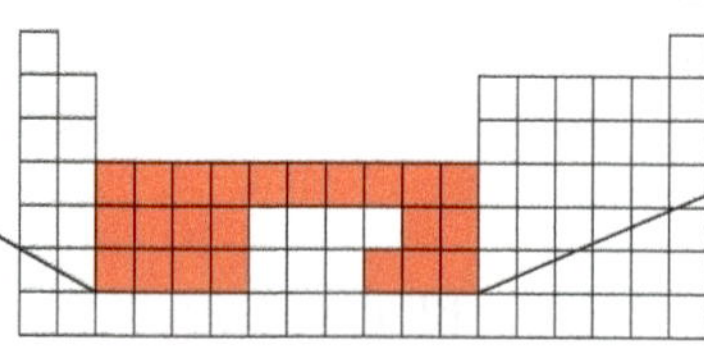

Practice Problems

5.26 Give the proper name for the compound $TiCl_4$.

Answer: Cl is in group VIIA and so has gained one electron to become a Cl^- anion. Because there are four of these anions, titanium, Ti, must have a +4 charge to maintain electrical neutrality. The name is therefore titanium(IV) chloride.

5.27 Give the proper name for the compound $TiCl_3$.

5.28 Give the molecular formula for tin(II) fluoride.

You should be aware of an older but still commonly used system for naming these types of compounds, a system that does not use Roman numerals to indicate ionic charge. Instead, the name of the cation is changed by adding a suffix to indicate charge. For the element iron, the suffix *-ous* is used for Fe^{2+} (ferrous) and the suffix *-ic* is used for Fe^{3+} (ferric). However, the suffixes *-ous* and *-ic* do not necessarily mean +2 and +3, respectively. Rather, *-ous* is used for the ion with lower positive charge and *-ic* is used for the ion with higher charge. For example, tin, Sn, commonly exists in two cationic forms, +2 and +4. The compounds formed between tin and fluorine could be either SnF_2 or SnF_4. In this case, SnF_2, tin(II) fluoride, is stannous fluoride and SnF_4, tin(IV) fluoride, is stannic fluoride. Think about this the next time you are brushing your teeth with fluoride toothpaste, whose active ingredient is stannous fluoride. Table 5.3 contrasts the old versus the modern systematic names for some binary ionic compounds.

Table 5.3 Old Versus Systematic Names for Some Binary Ionic Compounds

Ion	Compound formed Ion with Cl	Old name	Systematic name
Fe^{2+}	$FeCl_2$	Ferrous chloride	Iron(II) chloride
Fe^{3+}	$FeCl_3$	Ferric chloride	Iron(III) chloride
Cu^{+}	$CuCl$	Cuprous chloride	Copper(I) chloride
Cu^{2+}	$CuCl_2$	Cupric chloride	Copper(II) chloride
Co^{2+}	$CoCl_2$	Cobaltous chloride	Cobalt(II) chloride
Co^{3+}	$CoCl_3$	Cobaltic chloride	Cobalt(III) chloride
Hg_2^{2+} (each Hg is +1)	Hg_2Cl_2	Mercurous chloride	Mercury(I) chloride
Hg^{2+}	$HgCl_2$	Mercuric chloride	Mercury(II) chloride
Sn^{2+}	$SnCl_2$	Stannous chloride	Tin(II) chloride
Sn^{4+}	$SnCl_4$	Stannic chloride	Tin(IV) chloride

Naming binary ionic compounds

The name of the compound consists of the names of the two ions, with the metal ion named first.

- *Metal ion name:* based on element name.
- *Nonmetal ion name:* element name with suffix *-ide* (Cl^-, chloride; S^{2-}, sulfide; O^{2-}, oxide).

When the metal has the same ionic charge in all compounds (most representative metals),

the name of the metal ion is simply the name of the metal:

$LiCl$	Lithium chloride
$MgCl_2$	Magnesium chloride
CaO	Calcium oxide

Only one chemical formula is possible in each case, and so the name is unambiguous.

When the metal has different ionic charges in different compounds (most transition metals and some heavier representative metals),

Systematic names
the charge of the metal ion is indicated by a Roman numeral in parentheses after the metal name:

$FeCl_2$	Iron(II) chloride
$FeCl_3$	Iron(III) chloride

or

Older system names
the charge of the metal ion is indicated by a suffix. When a metal commonly occurs as either of two ions, *-ous* is used for the ion of lower charge and *-ic* for the ion of higher charge:

$FeCl_2$	Ferrous chloride
$FeCl_3$	Ferric chloride

Naming Binary Covalent Compounds

Binary covalent compounds generally consist of two different nonmetals or metalloids and are usually covalent. They are named in much the same way as binary ionic compounds with one slight difference. In the case of ionic compounds, the name *magnesium chloride* was enough to tell us that the molecular formula was $MgCl_2$ because we could count on the magnesium always being +2 and the chlorine always being −1. With binary covalent compounds, this is no longer the case. When two nonmetallic elements combine, a range of possible charges exists. This means that we need to insert prefixes into the name to tell us how many atoms of each element are present in the molecular formula. The prefixes we use are shown in Table 5.4 and are derived from early Greek names for numbers.

Table 5.4 Greek Prefixes

1	mono-	5	penta-	9	nona-
2	di-	6	hexa-	10	deca-
3	tri-	7	hepta-		
4	tetra-	8	octa-		

Also, because both elements in these compounds are nonmetals, we treat the less electronegative element as if it were a metal, list it first, and use the unmodified name of the element. The more electronegative element is listed second and has the suffix *-ide* appended to its name. For example, the compound HCl is called hydrogen chloride.

WORKPATCH 5.11 Why does it make sense to give the more electronegative element in a binary covalent compound the suffix *-ide*? ●

Let's consider a few examples. Carbon and oxygen can combine to form two binary compounds, CO and CO_2. These compounds are called carbon *mon*oxide and carbon *di*oxide. We use the prefixes *mon(o)-* and *di-* to indicate that either one or two atoms of oxygen are present in the compound. These prefixes can also be used with the first element. For example, the compounds NO_2 and N_2O_3 are named nitrogen dioxide and dinitrogen trioxide, respectively. Notice that NO_2 is *not* called mononitrogen dioxide. If there is only one atom of the less electronegative element, the prefix *mono-* is generally left out. Some additional examples are

PCl_3	Phosphorus trichloride
P_4O_6	Tetraphosphorus hexoxide*
P_4O_{10}	Tetraphosphorus decoxide*
SF_6	Sulfur hexafluoride

SO_2	Sulfur dioxide
SO_3	Sulfur trioxide
NO_2	Nitrogen dioxide
N_2O_4	Dinitrogen tetroxide*
N_2O_5	Dinitrogen pentoxide*

*To avoid two vowels together, the final *a* on a prefix is dropped when the element name begins with a vowel.

Naming binary covalent compounds

The name of the compound is constructed from the names of the two elements.

- The *less electronegative element* is named first (just as with the metal in a binary ionic compound).
- The *more electronegative element* is named second and takes the suffix *-ide* (just as with the nonmetal in a binary ionic compound).
- *For each element,* a prefix indicates the number of atoms of that element in the formula. If only one atom of the first element is present, the prefix is generally omitted:

SF_6	Sulfur hexafluoride
N_2O_4	Dinitrogen tetroxide

Practice Problems

5.29 Give the proper name for the compound N_2O.
Answer: Dinitrogen monoxide

5.30 Give the proper name for the compound NO.

5.31 Give the proper name for the compound PCl_5.

5.32 Give the chemical formula for tetraphosphorus decasulfide.

Nomenclature and Polyatomic Ions

Because polyatomic ions show up in many of the compounds chemists deal with, it pays to become familiar with the names and formulas of these ions. Many of the important (common) ones are listed in Table 5.5. You should commit these names, chemical formulas, and charges to memory. Examine the ions now, and then we'll show you how to name compounds that contain them.

You may be familiar with some of these polyatomic ions. Aqueous solutions of the hypochlorite ion are sold as chlorine bleach. Nitrites are used to preserve meats. Aqueous solutions of peroxide ions are found in many medicine cabinets and used as an antiseptic. The cyanide ion is extremely toxic and is the favorite way for Hollywood movie spies to do themselves in. Only one

Table 5.5 Some Common Polyatomic Ions

Ion formula	Ion name	Ion formula	Ion name
NH_4^+	Ammonium		
$CH_3CO_2^-$	Acetate	PO_4^{3-}	Phosphate
CN^-	Cyanide	HPO_4^{2-}	Hydrogen phosphate
NO_2^-	Nitrite	$H_2PO_4^-$	Dihydrogen phosphate
NO_3^-	Nitrate	ClO^-	Hypochlorite
CO_3^{2-}	Carbonate	ClO_2^-	Chlorite
HCO_3^-	Hydrogen carbonate (or bicarbonate)	ClO_3^-	Chlorate
SO_3^{2-}	Sulfite	ClO_4^-	Perchlorate
SO_4^{2-}	Sulfate	CrO_4^{2-}	Chromate
HSO_4^-	Hydrogen sulfate (or bisulfate)	$Cr_2O_7^{2-}$	Dichromate
MnO_4^-	Permanganate	O_2^{2-}	Peroxide

of the ions in Table 5.5 is a cation (ammonium, NH_4^+). Several of the anions consist of a nonmetal plus one or more oxygen atoms, such as sulfate, SO_4^{2-}. These ions are called oxyanions, and they often exist in a series in which the individual members differ only in the number of oxygen atoms that accompany the nonmetal atom. When there are two members in such a series, the ion with the fewer oxygens gets the *-ite* ending, and the ion with the greater number of oxygens gets the *-ate* ending:

SO_3^{2-}	Sulfite
SO_4^{2-}	Sulfate

Naming polyatomic ionic compounds

The compound is named the way a binary ionic compound is, except that the anion name is the name of the polyatomic ion.

- The metal ion is named first, using the rules for binary ionic compounds.
- The polyatomic ion is named second. The names of the common polyatomic ions are given in Table 5.5.

$NaNO_2$	Sodium nitrite
$NaNO_3$	Sodium nitrate
$Cu(NO_3)_2$	Copper(II) nitrate or cupric nitrate

When there are more than two oxyanions in a series, the prefix *hypo-* is used for the ion with the fewest oxygens, and the prefix *per-* is used for the oxyanion with the most oxygens:

ClO^-	Hypochlorite
ClO_2^-	Chlorite
ClO_3^-	Chlorate
ClO_4^-	Perchlorate

Naming compounds that contain polyatomic ions is easy. You simply use the names from Table 5.5 and then name the compound as you would a binary ionic compound. For example, you call NaCl sodium chloride. Therefore, you call NaClO sodium hypochlorite and NH_4Cl ammonium chloride.

And just as for binary ionic compounds, knowledge of the charges of the ions allows you to go from the name of a compound to its chemical formula. As an example, let's consider the compound sodium phosphate. From Table 5.5, we know that the phosphate ion has the composition PO_4 and a −3 charge. Because sodium, a group IA element, forms ions with a charge of +1, the molecular formula must be Na_3PO_4 to achieve charge neutrality:

$$\underbrace{\begin{matrix} Na^+ \\ Na^+ \\ Na^+ \end{matrix}}_{+3} + \underbrace{PO_4^{3-}}_{-3} \longrightarrow \text{Neutral compound, } Na_3PO_4$$

What is the formula for magnesium phosphate? The phosphate ion still has a −3 charge, but now the magnesium cation, a group IIA element, has a +2 charge. To arrive at an electrically neutral compound, we need to combine two PO_4^{3-} ions with three Mg^{2+} ions:

$$\underbrace{\begin{matrix} Mg^{2+} \\ Mg^{2+} \\ Mg^{2+} \end{matrix}}_{+6} + \underbrace{\begin{matrix} PO_4^{3-} \\ PO_4^{3-} \end{matrix}}_{-6} \longrightarrow \text{Neutral compound, } Mg_3(PO_4)_2$$

The formula for magnesium phosphate is therefore $Mg_3(PO_4)_2$. Notice that we put parentheses around the phosphate; otherwise, the formula would look like this: Mg_3PO_{42}, which says there are 42 oxygen atoms in the compound. Whenever there is more than one polyatomic ion in the formula for a compound, the formula for the polyatomic ion must be surrounded by parentheses: $CuNO_3$ for copper(I) nitrate but $Cu(NO_3)_2$ for copper(II) nitrate.

Practice Problems

5.33 Write the formula for ammonium phosphate.

Answer: Because ammonium is NH_4^+ and phosphate is PO_4^{3-}, the formula must be $(NH_4)_3PO_4$ to obtain charge neutrality.

5.34 Write the formula for sodium carbonate.

5.35 Write the proper name for $Ca(ClO)_2$.

5.36 Write the formula for magnesium bicarbonate.

Nomenclature of Acids

Although the subject of acids (and bases) is covered in later chapters, we can say a bit about the nomenclature of acids now. For the purposes of this discussion, any compound that dissolves in water and falls apart (*dissociates*) to produce H^+ ions in the water is an **acid**.

Acids are named by one of two systems, depending on whether or not the acid contains oxygen. Let's first consider acids that do not contain oxygen. When gaseous HBr is dissolved in water, the HBr dissociates to produce H^+ ions. Therefore, HBr is an acid. Because this acid does not contain oxygen, it is named by adding the prefix *hydro-* and the suffix *-ic acid* to the root name of the nonhydrogen element (or elements). So HBr is hydrobromic acid. Some other examples are:

HF	Hydrofluoric acid
HCl	Hydrochloric acid
HI	Hydroiodic acid
HCN	Hydrocyanic acid
H_2S	Hydrosulfuric acid

These compounds produce acidic solutions when dissolved in water, and thus their aqueous solutions are named as indicated above. However, when they are not dissolved in water, they are named according to the rules for covalent compounds. Thus, an aqueous solution of HCl gas is called hydrochloric acid, and pure HCl gas is called hydrogen chloride.

If the acid contains one or more oxygens, it is called an *oxyacid*, and the rules for naming it are somewhat different. Now the focus is on the name of the polyatomic anion in the compound. If the anion name ends in *-ate*, we name the acid by changing the suffix from *-ate* to *-ic*:

Formula of acid	Formula of ion	Name of ion	Name of acid
H_2SO_4	SO_4^{2-}	Sulf*ate*	Sulfur*ic* acid
HNO_3	NO_3^-	Nitr*ate*	Nitr*ic* acid
$HClO_4$	ClO_4^-	Perchlor*ate*	Perchlor*ic* acid
$HC_2H_3O_2$	$C_2H_3O_2^-$	Acet*ate*	Acet*ic* acid

If the anion name ends in *-ite*, we name the acid by changing the suffix from *-ite* to *-ous*:

Formula of acid	Formula of ion	Name of ion	Name of acid
H_2SO_3	SO_3^{2-}	Sulf*ite*	Sulfur*ous* acid
HNO_2	NO_2^-	Nitr*ite*	Nitr*ous* acid
HClO	ClO^-	Hypochlor*ite*	Hypochlor*ous* acid

We shall learn much more about the nomenclature and properties of acids in Chapter 15.

Naming acids

- For now, an acid is any substance that dissociates in water to yield an H^+ ion.
- The naming rules given here are used when the acid is dissolved in water.
- When not dissolved in water, acids of this type are either binary compounds or polyatomic ionic compounds and are named by the appropriate rules.

When the formula does not contain oxygen,

add the prefix *hydro-* and the suffix *-ic acid* to the name of the nonhydrogen part or polyatomic ion:

HF	Hydrofluoric acid
HCN	Hydrocyanic acid
H_2S	Hydrosulfuric acid

When the formula contains oxygen (these acids are all polyatomic ionic compounds)

and the name of the anion ends in *-ite,*

replace the suffix *-ite* with *-ous acid:*

H_2SO_3 (sulfite ion)	Sulfurous acid
HNO_2 (nitrite ion)	Nitrous acid

and the name of the anion ends in *-ate,*

replace the suffix *-ate* with *-ic acid:*

H_2SO_4 (sulfate ion)	Sulfuric acid
HNO_3 (nitrate ion)	Nitric acid

Finally, some compounds have been around for so long that the common names are the only ones used. For example, NH_3 is always referred to as ammonia and not nitrogen trihydride. And absolutely nobody calls H_2O dihydrogen monoxide! To prove this to yourself, the next time you go to a restaurant, see how clever your waiter thinks you are when you ask for a glass of dihydrogen monoxide.

+1 One More Thing

5.8 Exceptions to the Octet Rule

In Chapter 4, we defined the octet rule. The rule stated that atoms share electrons so as to achieve eight electrons in their outermost valence shells. We also said that this was a *rule of thumb* with plenty of exceptions. It's time now to face those exceptions, and they go in both directions. That is, there exist molecules that contain atoms that possess less than an octet of valence electrons (such atoms are often called **octet deficient** or **electron deficient**), and there are also molecules with atoms that possess more than an octet of valence electrons (these atoms are said to possess an **expanded octet**). Let's look at some *octet-deficient* examples first. Beryllium, (Be), a group IIA atom, reacts with two chlorine atoms to form only two single covalent bonds, and the group IIIA atom boron (B) reacts with three chlorine atoms to form only three single covalent bonds. In both cases, the central atom is left with less than an octet of electrons.

There are compounds where Be and B do obey the octet rule, but the above exceptions are noteworthy and remind us that the octet rule is indeed a *rule of thumb*. However, the next three atoms, carbon (C), nitrogen (N), and oxygen (O), can be counted on to obey the octet rule with very few exceptions.

Octet Deficient Atoms

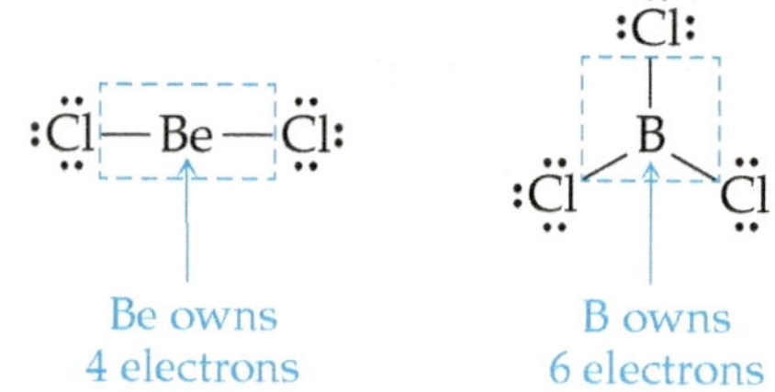

On the other side are the *expanded-octet* atoms, which take on more than eight valence electrons. These can only be atoms in periods 3 and beyond. Being in period 2, carbon, nitrogen, and oxygen can never exceed an octet. This is because the principal quantum number n for the valence shell of period 2 atoms is 2, and such a shell can accommodate no more than $2n^2 = 2(2)^2 = 8$ electrons (see Chapter 4.5). This expands to 18 when $n = 3$, so exceeding an octet of electrons becomes possible for period 3 atoms and beyond. Thus, while nitrogen, a period 2 atom, bonds to three chlorine atoms to yield NCl_3, in which the central N atom obeys the octet rule, period 3 phosphorus (P) forms not only PCl_3 but also PCl_5, in which the central P atom has a total of ten valence electrons (an expanded octet). Likewise, period 3 sulfur (S) can react with chorine to form SCl_6, with the sulfur atom possessing a total of 12 valence electrons.

Expanded Octet Atoms

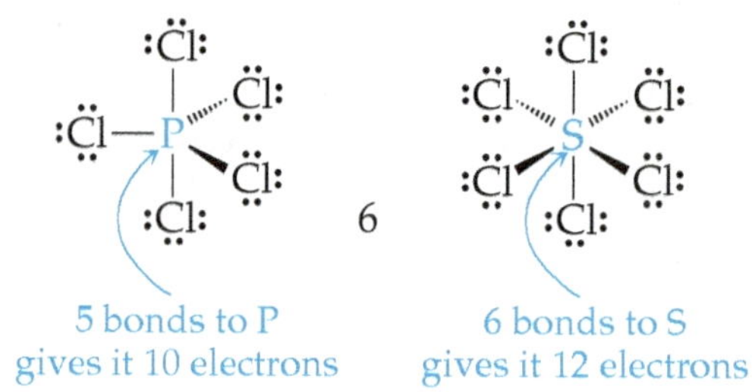

Expanded-octet atoms rear their head even when unexpected. For example, consider the sulfate anion, SO_4^{2-}. Two dot diagrams for sulfate are shown below, both with the same total number (32) of valence electrons. The one on the left has all of the oxygen atoms and the sulfur obeying the octet rule. The one on the right moves two lone pairs from two oxygen atoms into the bonding position (see the red arrows) to make two multiple (double) bonds, giving sulfur an expanded octet (12 electrons around S) while maintaining octets about the oxygen atoms.

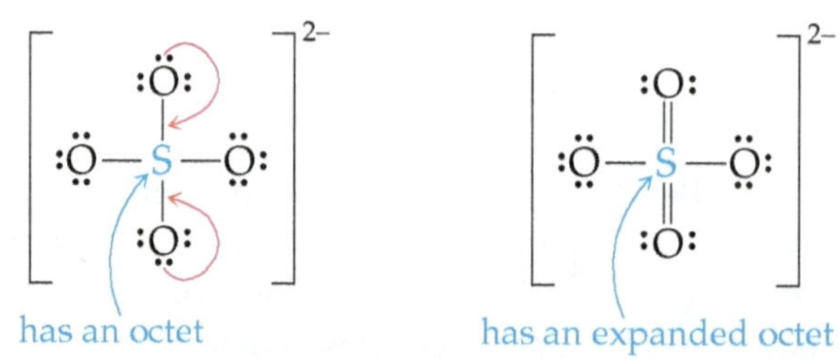

Measurements of bond energies and bond lengths indicate that the dot diagram on the right is a more accurate representation of this polyatomic anion if you also include all the resonance forms in which the two double bonds are distributed among all the possible positions. Therefore, don't be surprised when you encounter atoms in molecules that disobey the octet rule when those atoms are other than C, N, and O.

Have You Learned This?

Ionic compound (p. 167)
Coulomb's Law (p. 168)
Ionic bond (p. 168)
Chemical bond (p. 168)
Ionic lattice (p. 169)
Crystalline (p. 169)
Amorphous (p. 169)
Lattice energy (p. 169)
Cleaving (p. 169)
Molecule (p. 170)
Molecular substance (p. 170)
Covalent bond (p. 174)
Dot diagrams (p. 179)
Lewis dot diagrams (p. 179)
Electron pairs (p. 179)
Paired electrons (p. 179)
Unpaired electrons (p. 179)
Bonding pairs (p. 181)
Lone pairs (p. 181)
Multiple covalent bond (p. 182)
Multiple bond (p. 182)
Double bond (p. 183)
Single bond (p. 183)
Triple bond (p. 183)
Resonance forms (p. 185)
Polyatomic ions (p. 185)
Electronegativity (EN) (p. 187)
Polar covalent bond (p. 189)
Binary compounds (p. 192)
Binary ionic compounds (p. 192)
Binary covalent compounds (p. 192)
Acid (p. 200)
octet deficient or electron deficient (p. 201)
expanded octet (p. 201)

Learning Outcomes

5.1	Ionic Bonding	Define an ionic bond and predict when it is likely to occur. [38, 39, 40] Predict the formula of an ionic compound. [41, 42, 43] Describe what an ionic lattice and lattice energy are. [43, 45, 46, 47, 48]
5.2	Molecules—What Are they? Why Are They?	Describe what a molecule is and give some examples. [49, 50, 51, 52, 53, 54, 55]
5.3	Holding Molecules Together—The Covalent Bond	Describe what a covalent bond is and what gives rise to it in terms of attractions. [56, 57, 58, 60, 61, 62] Draw a covalent bond and count the valence electrons owned by each atom. [66, 67, 68]
5.4	Molecules, Dot Structures, and the Octet Rule	Draw dot diagrams for atoms. [69, 70, 71, 73] Draw dot diagrams for molecules. [72, 74, 75, 76, 77, 78, 79, 80, 81]
5.5	Multiple Bonds	Draw dot diagrams for molecules that have multiple bonds. [84, 87, 92, 93, 95, 96, 97, 99, 101, 102, 104] Draw resonance forms. [86, 88, 98]
5.6	Equal versus Unequal Sharing of Electrons—Electronegativity and the Polar Covalent Bond	Define *electronegativity* and describe its periodic trends. [105, 106, 107, 108, 109, 119] Describe bond polarity using electronegativity. [115, 116, 117, 120] Predict if a bond will be covalent, polar covalent, or ionic. [110, 114, 115, 118, 121]
5.7	Nomenclature—Naming Chemical Compounds	Name and give formulas for binary ionic compounds. [127, 128, 138] Name and give formulas for compounds possessing transition metals. [129, 130] Name and give formulas for binary covalent compounds. [126, 133, 139] Name and give formulas for compounds possessing polyatomic ions. [131, 132, 136] Name and give formulas for acids. [142, 143, 145, 146]
5.8	Exceptions to the Octet Rule	Draw dot diagrams for molecules that possess electron deficient atoms. [201, 202] Draw dot diagrams for molecules that possess expanded octet atoms. [193, 203–206]

Skills to Know

Method for drawing dot diagrams

Example: Draw a dot diagram for HCO_3^-.

1. Determine the total number of valence electrons (dots) that will be in the diagram. For a polyatomic ion, adjust the electron count to reflect the charge.

C	$1 \times 4 =$	4 electrons
O	$3 \times 6 =$	18 electrons
H	$1 \times 1 =$	1 electron
Ionic charge of −1		1 electron
Total		24 electrons

2. Connect the atoms with single bonds. If you do not know how the atoms are arranged, use the least electronegative atom in the formula (excluding hydrogen) as the central atom.

Choose C as central atom

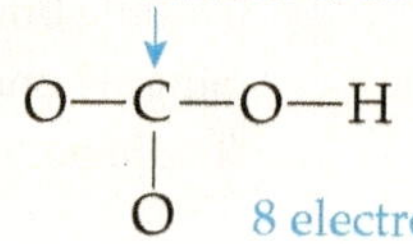

8 electrons assigned

3. Add the remaining electrons as lone pairs, satisfying octets as you go. Do terminal atoms first (except H, which needs only two electrons).

Lacks an octet

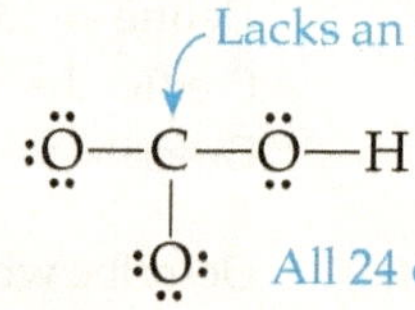

All 24 electrons assigned

4. If any atom lacks an octet, move lone pair(s) from adjacent atom(s) into bonding positions on the octet-deficient atom.

Lone pair to the rescue

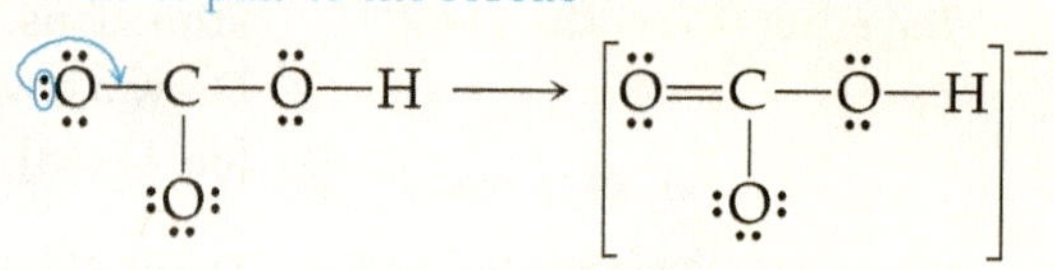

Final dot structure (one of a resonance set)

Electronegativity and bond polarity

1. The electronegativity, EN, of an element is a number that tells how strongly an atom of the element attracts bonding electrons to itself (how much of an electron "hog" it is). EN values range from 0.7 (the metal francium) to 4.0 (the nonmetal fluorine).

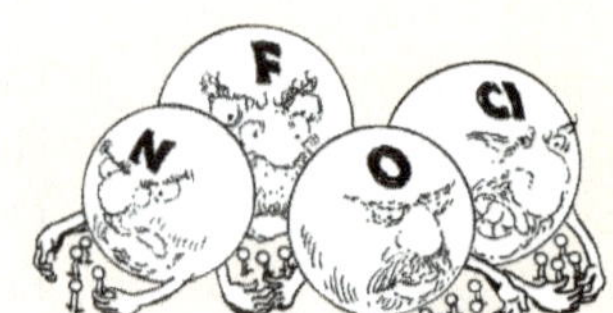

The biggest electron hogs of all

2. The polarity of a bond depends on the difference in electronegativity, ΔEN, between the bonded atoms. ΔEN is calculated by subtracting the smaller from the larger EN.

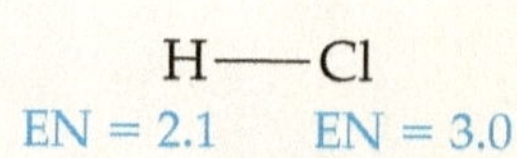

$\Delta EN = 3.0 - 2.1 = 0.9$

3. The greater the ΔEN of a bond, the stronger are the $\delta+$ and $\delta-$ charges on the bonded atoms and hence the more polar is the bond. When $\Delta EN = 0$, the bonding is pure covalent. When ΔEN is very large, the bonding is ionic. In between, the bonding is polar covalent.

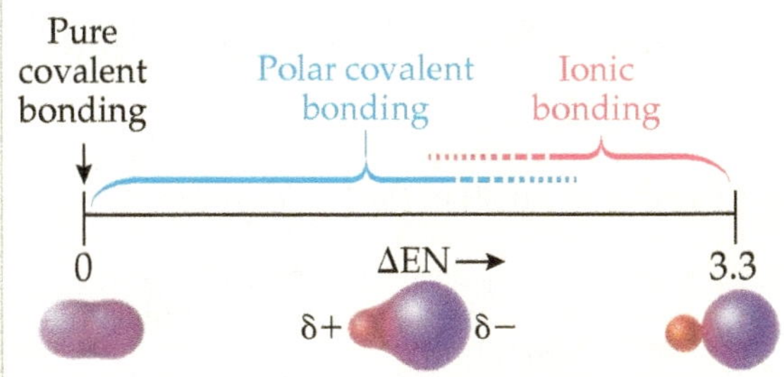

Binary compounds
(compounds consisting of two elements)

- Compound consisting of a metal and a nonmetal
  - **Binary ionic compound**
    Examples: LiCl, $MgCl_2$, Fe_2O_3
- Compound consisting of nonmetals and/or metalloids
  - **Binary covalent compound**
    Examples: CO, NO_2, SiO_4

Naming binary ionic compounds

The name of the compound consists of the names of the two ions, with the metal ion named first.

- *Metal ion name:* based on element name.
- *Nonmetal ion name:* element name with suffix *-ide* (Cl^-, chloride; S^{2-}, sulfide; O^{2-}, oxide).

When the metal has the same ionic charge in all compounds (most representative metals),

the name of the metal ion is simply the name of the metal:

LiCl	Lithium chloride
$MgCl_2$	Magnesium chloride
CaO	Calcium oxide

Only one chemical formula is possible in each case, and so the name is unambiguous.

When the metal has different ionic charges in different compounds (most transition metals and some heavier representative metals),

Systematic names
the charge of the metal ion is indicated by a Roman numeral in parentheses after the metal name:

$FeCl_2$	Iron(II) chloride
$FeCl_3$	Iron(III) chloride

or

Older system names
the charge of the metal ion is indicated by a suffix. When a metal commonly occurs as either of two ions, *-ous* is used for the ion of lower charge and *-ic* for the ion of higher charge:

$FeCl_2$	Ferrous chloride
$FeCl_3$	Ferric chloride

Naming binary covalent compounds

The name of the compound is constructed from the names of the two elements.

- The *less electronegative element* is named first (just as with the metal in a binary ionic compound).
- The *more electronegative element* is named second and takes the suffix *-ide* (just as with the nonmetal in a binary ionic compound).
- *For each element,* a prefix indicates the number of atoms of that element in the formula. If only one atom of the first element is present, the prefix is generally omitted:

SF_6	Sulfur hexafluoride
N_2O_4	Dinitrogen tetroxide

Naming polyatomic ionic compounds

The compound is named the way a binary ionic compound is, except that the anion name is the name of the polyatomic ion.

- The metal ion is named first, using the rules for binary ionic compounds.
- The polyatomic ion is named second. The names of the common polyatomic ions are given in Table 5.5.

$NaNO_2$	Sodium nitrite
$NaNO_3$	Sodium nitrate
$Cu(NO_3)_2$	Copper(II) nitrate or cupric nitrate

Naming acids

- For now, an acid is any substance that dissociates in water to yield an H^+ ion. (You will learn more in Chapter 15.)
- The naming rules given here are used when the acid is dissolved in water.
- When not dissolved in water, acids of this type are either binary compounds or polyatomic ionic compounds and are named by the appropriate rules.

When the formula does not contain oxygen,

add the prefix *hydro-* and the suffix *-ic acid* to the name of the nonhydrogen element or polyatomic ion:

HF	Hydrofluoric acid
HCN	Hydrocyanic acid
H_2S	Hydrosulfuric acid

When the formula contains oxygen (these acids are all polyatomic ionic compounds),

and the name of the anion ends in *-ite,*

replace the suffix *-ite* with *-ous acid*:

H_2SO_3 (sulfite ion)	Sulfurous acid
HNO_2 (nitrite ion)	Nitrous acid

and the name of the anion ends in *-ate,*

replace the suffix *-ate* with *-ic acid*:

H_2SO_4 (sulfate ion)	Sulfuric acid
HNO_3 (nitrate ion)	Nitric acid

Concept Questions

1. Regarding an ionic bond:
   (a) It forms due to transfer of one or more electrons from one atom to another.
   (b) It is the name given to the attractive force between oppositely charged ions that are in contact.
   (c) It is what holds an ionic lattice together.
   (d) All of the above.

2. NaCl is not considered to be a molecular substance because:
   (a) It is an ionic compound with no discrete molecules in its solid lattice.
   (b) It is a molecular substance. That's why we can write its formula.
   (c) It is made from more than one element.
   (d) Only one of the elements is an ion in the compound.

3. A covalent bond:
   (a) Is the name given to the attractive force between atoms in a molecule.
   (b) Is strong because the electrons in the bond can be attracted to two nuclei instead of just one.
   (c) Is formed only by the sharing of valence electrons.
   (d) All of the above.

4. When two electrons are shared between atoms in a covalent bond, we say:
   (a) Each atom owns one of the shared electrons.
   (b) Each atom owns both of the shared electrons.
   (c) Neither atom owns the shared electrons.
   (d) One atom owns both electrons and the other owns none of the electrons.

5. According to the octet rule:
   (a) Bonds will form between elements separated by eight spaces on the periodic table.
   (b) Only atoms with eight valence electrons can form covalent bonds to other atoms.
   (c) A sufficient number of bonds will form so as to give all atoms a valence octet of electrons.
   (d) None of the atoms in group VIII of the periodic table can form covalent bonds.

6. A good rule of thumb for forming covalent bonds between atoms is:
   (a) Unpaired electrons are for sharing (bonding); electrons pairs are not.
   (b) Electron pairs are used for sharing (bonding); unpaired electrons are not.
   (c) Representative elements cannot form covalent bonds.
   (d) Electron pairs on an atom are the ones most commonly shared between atoms.

7. A multiple bond:
   (a) Results from the sharing of more than two electrons between two atoms.
   (b) Is always stronger than a single bond.
   (c) Should be drawn if, after sharing all unpaired electrons, an atom still has less than an octet.
   (d) All of the above.

8. Electronegativity is:
   (a) A measure of how negative an electron is.
   (b) An indication of how effective an atom is at attracting shared electrons to itself.
   (c) Highest on the left side of the periodic table and lowest on the right side.
   (d) Lowest for N, O, Cl, and F.

9. If two atoms with unpaired electrons and vastly different electronegativities approach one another:
   (a) It will be impossible for a bond to form between them.
   (b) The unpaired electrons will be transferred from the atom of low electronegativity to the one of higher electronegativity.
   (c) The unpaired electrons will be shared between the atoms.
   (d) A covalent bond will be formed.

10. A useful rule of thumb is that if ΔEN for the atoms involved in a bond is ≤ 0.4 (as it is for a C-H bond), this is close enough to zero and the bond can be considered to be nonpolar. This rule makes sense because:
   (a) the electrons in the bond are shared very unequally between the atoms.
   (b) the electrons in the bond are shared almost equally between the atoms.
   (c) such a low ΔEN means that there is no electron sharing.
   (d) only for a low ΔEN value will the $\delta+$ charge cancel the $\delta-$ charge.

Answers: 1a; 2a; 3d; 4a; 5c; 6a; 7d; 8b; 9b; 10b

Ionic Bonding

5.37 Draw dot diagrams for the following simple ions:
(a) I^- (b) O^{2-} (c) Cl^- (d) H^+

5.38 What do you call the attractive force that results when a negative ion comes into contact with a positive ion?

5.39 In terms of valence electrons, what must happen for an ionic bond to form between initially neutral atoms? What kind of atoms are likely to do what you describe? Give an example.

5.40 If potassium and oxygen were to bind, describe with words and dot diagrams what would happen to the valence electrons.

5.41 Predict the formulas of the ionic compounds that result from combining the following:
(a) Mg (solid metal) with Br_2 (liquid)
(b) Be (solid metal) with O_2 (gas)
(c) Na (solid metal) with I_2 (solid)

5.42 Predict the formula of the ionic compound formed by each of the following pairs of elements:
(a) Ca, I (b) Ca, O (c) Al, S (d) Ca, Br

5.43 Predict the formula of the ionic compound that would result from the reaction between lithium and nitrogen.

5.44 Suppose you come across a solid substance and experiments show no long-range order among the particles that make up that solid. What type of solid do you have?

5.45 What insights does Coulomb's Law give into ionic lattice energy?

5.46 Should the lattice energy for Al_2O_3 be greater or less than the lattice energy for Na_2O? By how much? Justify your answers.

5.47 Explain the term *lattice energy* in terms of melting an ionic lattice.

5.48 A simple drop on the floor of an ionic substance may cause it to fracture along a perfectly straight line. Explain why.

Molecules—What Are They? Why Are They?

5.49 Give a general definition of a molecule.

5.50 Are all molecules also compounds? Explain.

5.51 Suppose you could see into solid objects at the atomic level. Explain, based on what you see, why solid NaCl would not be called a molecular substance whereas solid H_2O would.

5.52 What is it about molecules that makes them worth preparing? Give some examples.

5.53 What common diatomic molecules exist in our atmosphere? What triatomic molecules exist in our atmosphere? Give the formulas for all your answers.

5.54 Ethanol (C_2H_6O) is consumed in alcoholic beverages, but another alcohol, methanol (CH_4O), is toxic. What explanation can you give for this?

5.55 Can a molecule also be an elemental substance? If so, give some examples.

Holding Molecules Together—The Covalent Bond

5.56 The attractive force between atoms that results from the sharing of two valence electrons between those atoms has a special name. What is it?

5.57 Chemists will often say that a covalent bond is two electrons shared between two atoms. Technically, this is not correct. They should really say "A covalent bond is ______________ that results from ______________." Fill in the blanks.

5.58 Why does the sharing of two electrons between two atoms bond the atoms to each other?

5.59 In terms of *energy*, why is an H_2 molecule more stable than two isolated H atoms?

5.60 In terms of *interaction* between the atoms, why is an H_2 molecule more stable than two isolated H atoms?

5.61 The bond distance in an H_2 molecule is 0.74 Å. Why isn't it shorter than this? Why isn't it longer than this?

5.62 Is energy released or absorbed when a covalent bond forms between two atoms?

5.63 Why are an atom's valence electrons the only electrons involved in bonding?

5.64 Suppose one of the electrons from the covalent bond in H_2 suddenly vanished. Why would the bond between the atoms weaken?

5.65 A student decides to boil water to produce hydrogen gas and oxygen gas. Will this work? Explain your answer.

5.66 What do we mean by "double-counting" when it comes to counting the electrons around an atom in a molecule?

5.67 What is the accepted shortcut for drawing a shared pair of electrons (a covalent bond) in a molecular drawing?

5.68 How many electrons does an atom gain for each covalent bond that it forms in a molecule?

Molecules, Dot Structures, and the Octet Rule

5.69 A student draws the following dot diagrams for N, O, and F. What, if anything, is wrong with each?

5.70 For any representative element, what is the relationship between the element's Roman-numeral group number and the number of dots in the Lewis dot diagram?

5.71 Draw dot diagrams for the following atoms:
(a) S (b) I (c) He (d) B (e) Ne

5.72 Chlorine (Cl), neon (Ne), and helium (He) all exist as gases, but only one of them is diatomic. Which is it, and why is it diatomic while the others are monatomic?

5.73 Valence electrons in an isolated atom can exist in either of two ways. What are they?

5.74 Why isn't the formula for water HO?

5.75 Hydrogen (H) and sulfur (S) form the toxic compound hydrogen sulfide, a gas that smells like rotten eggs and is spewed from volcanoes. Predict the formula of hydrogen sulfide starting with dot diagrams for the atoms.

5.76 Phosphorus (P) and bromine (Br) form a compound. Predict the formula of this compound starting with dot diagrams for the atoms.

5.77 Nitrogen (N) forms a compound with bromine (Br) similar to that in Problem 5.76. How is it similar? Explain why this is so.

5.78 Ethers are compounds of C, H, and O that are often used as solvents. One particular ether molecule has the formula C_2H_6O. The structure is such that both carbons are attached to the oxygen atom, and there are no O—H bonds. Starting with dot diagrams for the atoms, draw a dot diagram for this ether molecule. How many bonding pairs does the molecule have? How many lone pairs?

5.79 Ethanol, the alcohol found in alcoholic beverages, has the same formula as the ether of Problem 5.78. However, it has a different structure, in which only one carbon atom is bound to the oxygen atom. Starting with dot diagrams for the atoms, draw a dot diagram for the ethanol molecule. Indicate the total number of bonding pairs and lone pairs of electrons. Also comment on the fact that one form of C_2H_6O is a solvent (poisonous) but another form is drinkable and what this says about the role of structure in determining the chemical properties of a molecule.

5.80 The molecule HCl is known, but the molecule HeCl is not. Explain why this is so.

5.81 The valence electrons in a helium atom are paired, but suppose they were not. Postulate what might form if oxygen atoms reacted with helium atoms that had their valence electrons unpaired.

5.82 Oxygen, in almost all of its compounds, forms two bonds. Explain why this is so.

5.83 How many bonds will an atom from group VA generally form? Explain why this is so.

Multiple Bonds

For all the problems in this section, you must draw all the resonance forms if they exist.

5.84 Draw a dot diagram for ethylene, C_2H_4.

5.85 Is the bond between the carbon atoms in the ethylene molecule of Problem 5.84 stronger or weaker than the bond in acetylene, C_2H_2 (WorkPatch 5.7)? Explain fully.

5.86 Explain what is meant by resonance forms.

5.87 Draw a dot diagram for the molecule SO_2, sulfur dioxide, a gas that comes from burning coal and is responsible for acid rain. (*Hint:* Sulfur is in the middle.)

5.88 Draw a dot diagram for the molecule O_3, ozone, the molecule in our upper atmosphere that protects us from the Sun's harmful ultraviolet electromagnetic radiation. (*Hint:* Only the middle oxygen atom forms bonds with the two other oxygen atoms.)

5.89 Should there be any similarities between the dot diagrams of Problems 5.87 and 5.88? Explain your answer.

5.90 A student claims that the bonds in ozone are really not double bonds or single bonds but somewhere between (roughly 1.5 bonds between each oxygen). Justify this statement.

5.91 Experiments show that it takes more energy to break the bond between oxygen atoms in the O_2 molecule than in the O_3 molecule. How can you explain this?

5.92 Is anything wrong with the following dot diagram for carbon monoxide (CO)? If so, explain what it is and fix it.

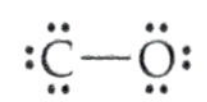

5.93 Is anything wrong with the following dot diagram for acetone (C_3H_6O)? If so, then fix it.

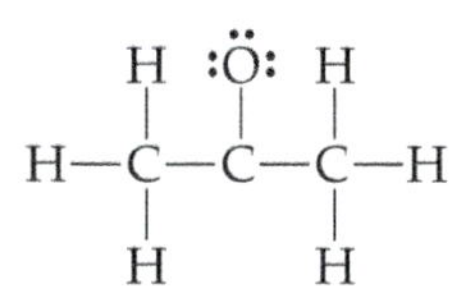

5.94 How many covalent bonds do you think would form between the phosphorus atoms in the molecule P_2? Explain your answer and draw a dot diagram.

5.95 Draw a dot diagram for the nitrate (NO_3^-) ion.

5.96 Draw a dot diagram for the hypothetical O_2^{2+} ion.

5.97 Acetaldehyde has the formula C_2H_4O. Draw a dot diagram for it given the following atomic connections.

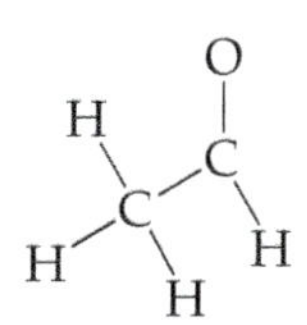

5.98 Acetic acid, a major component of vinegar, has the formula $C_2H_4O_2$. Draw a dot diagram for it given the following atomic connections.

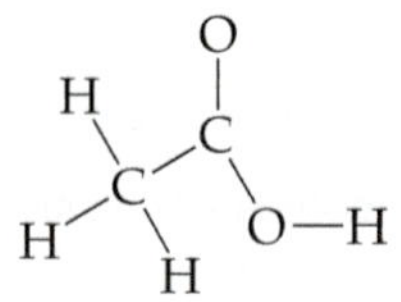

5.99 The acetate ion has the formula $C_2H_3O_2^-$. Draw a dot diagram for it given the following atomic connections. (*Hint:* Watch for resonance forms.)

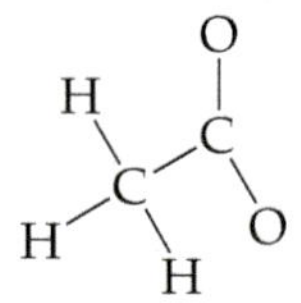

5.100 Draw a dot diagram for perchloric acid, $HClO_4$. The chlorine is the central atom to which all the oxygens are attached, and the hydrogen is attached to one of the oxygens.

5.101 Draw a dot diagram for the NO_2^+ cation.

5.102 Consider benzene, a pleasant-smelling but carcinogenic (cancer-causing) liquid. Benzene molecules have the formula C_6H_6, with the atoms connected as shown. Complete the dot diagram for this molecule and include any resonance forms.

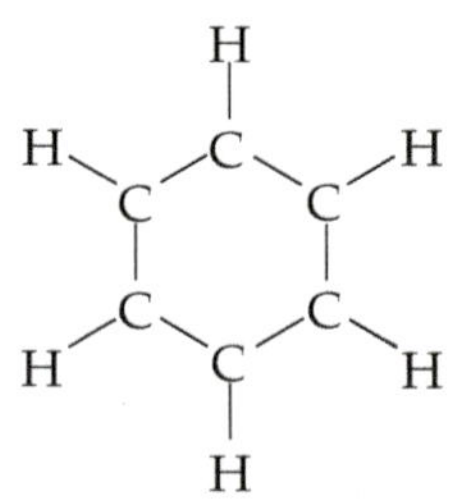

5.103 How many valence electrons are present on each carbon atom and each hydrogen atom in the benzene molecule of Problem 5.102?

5.104 Hydrogen isothiocyanate has formula HNCS, and its atoms are connected in the order written. Draw dot diagrams showing all the valid resonance forms.

Equal versus Unequal Sharing of Electrons—Electronegativity and the Polar Covalent Bond

5.105 There is one very important word missing from the following definition of electronegativity. What is it? "Electronegativity is an indication of an atom's ability to attract electrons to itself."

5.106 Why did we call the elements N, O, Cl, and F electron "hogs" in Section 5.6?

5.107 What is the most electronegative element in the periodic table? What is the least electronegative element? What kind of compound would form if these two elements were brought together?

5.108 How do the categories metal and nonmetal relate to electronegativity?

5.109 What are the trends for electronegativity in the periodic table:
(a) Down a group?
(b) Across a period from left to right?
(c) Going from the bottom left corner to the upper right corner?

5.110 What are the guidelines for using electronegativity to predict which type of bond forms between two atoms?

5.111 How does a polar covalent bond differ from a covalent bond? Give examples of diatomic molecules that contain a polar covalent bond and of diatomic molecules that contain a covalent bond.

5.112 Would you classify the C—H bonds in ethane (C_2H_6) as covalent, polar covalent, or ionic? Explain.

5.113 Lithium is a metallic element. Consider the hypothetical species dilithium, Li_2. Predict whether the bonding in such a species would be covalent, polar covalent, or ionic. Explain fully.

5.114 Which molecule has bonds that are the most polar covalent?
(a) H_2 (b) CO (c) H_2S (d) H_2O

5.115 Chemists sometimes think of molecules with polar covalent bonds as being part covalent and part ionic. How can a bond be both covalent and ionic?

5.116 In the following molecule, atoms X and Y are from the same period, meaning they are approximately the same size. The difference in the size of the spheres indicates a difference in the amount of time the shared electrons spend near each atom. Which atom is more electronegative, and how do you know?

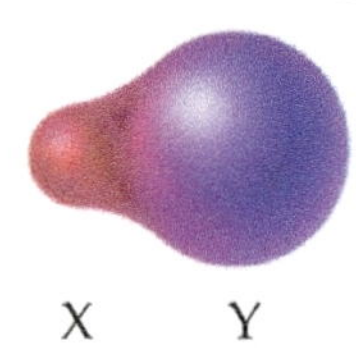

5.117 In the following molecule AB, one of the atoms is more electronegative than the other. Which is more electronegative, and how do you know?

$\delta-$ $\delta+$
A—B

5.118 How would you change the electronegativities of the atoms at either end of a polar covalent bond to make the bond ionic? To make the bond covalent?

5.119 Consider the following:
Cl_2 CF_4 MgO NaCl PH_3 SCl_2
(a) Which of these substances is ionic?
(b) Which of these substances has nonpolar bonds?
(c) Which of these substances has polar covalent bonds?

5.120 Without knowing the electronegativity values, in what situation can you be absolutely sure that the bonding between two atoms will be purely covalent?

5.121 Classify the bonds in each of the following as ionic, covalent, or polar covalent. Explain each choice.
(a) Br_2 (b) PCl_3 (c) LiCl (d) ClF (e) $MgCl_2$

Nomenclature—Naming Chemical Compounds

5.122 What is a binary compound? Is atmospheric oxygen an example of a binary compound?

5.123 What does the suffix *-ide* mean, and which element in the name of a binary ionic compound gets it?

5.124 Which element in a binary covalent compound gets the suffix *-ide*? Why?

5.125 Any ionic compound has an overall charge of zero. How does this fact help to determine its formula?

5.126 Consider the two binary compounds Al_2O_3 and N_2O_3.
(a) Which is a binary ionic compound and which is a binary covalent compound? Explain.
(b) Al_2O_3 is properly named aluminum oxide, but N_2O_3 is named dinitrogen trioxide. Explain fully why one name uses Greek prefixes and the other does not.

5.127 Give the proper name for the following binary ionic compounds:
(a) Ca_3N_2 (b) AlF_3 (c) Na_2O (d) CaS

5.128 Give the formulas of the following binary ionic compounds:
(a) Calcium bromide
(b) Sodium sulfide
(c) Potassium nitride
(d) Lithium oxide

5.129 What "monkey wrench" do transition metals throw into the picture when naming binary compounds? What do we include in the nomenclature system to accommodate them?

5.130 Give the name of the following binary compounds of transition metals using both the modern and older systems:
(a) CuCl and $CuCl_2$
(b) $Fe(OH)_2$ and $Fe(OH)_3$

5.131 Give the proper name for the following compounds:
(a) Na_2SO_4 (b) $(NH_4)_3PO_4$
(c) KClO (d) $CaCO_3$
(e) $Al(NO_3)_3$

5.132 Give the formulas for the following compounds:
(a) Ammonium acetate
(b) Ammonium carbonate
(c) Iron(II) nitrate
(d) Ferric hydroxide
(e) Calcium hypochlorite

5.133 Give the proper name for the following binary covalent compounds:
(a) PCl_3 (b) SO_2 (c) N_2O_4 (d) P_5O_{10}

5.134 Consider the sulfate, sulfite, nitrate, nitrite, chlorate, and chlorite ions. What information do the *-ate* and *-ite* suffixes communicate?

5.135 The oxide ion is O^{2-}. How does this ion differ from the peroxide ion, O_2^{2-}? Draw dot diagrams for both.

5.136 A polyatomic ion not listed in Table 5.5 is iodate, IO_3^-. What is the formula for the periodate ion, and what is the formula for magnesium periodate? (*Hint*: Look at the oxyanions of chlorine in Table 5.5 on p. 199.)

5.137 Students commonly confuse ammonia with ammonium.
(a) Write the formula and draw dot diagrams for both species, and describe the difference between them.
(b) Based on your answer to (a), write the formulas for phosphine and phosphonium, complete with charges if there are any.

5.138 A student writes the formula for magnesium hydroxide as $MgOH_2$. What is wrong with that formula? How should the student fix it?

5.139 A compound consisting of one oxygen atom and two nitrogen atoms is written N_2O and is called dinitrogen monoxide. It is *not* written as ON_2, and it is *not* called oxygen dinitride. Explain why.

5.140 N_2O is called dinitrogen monoxide, whereas, Na_2O is called sodium oxide, not disodium oxide. Explain why, and also explain how the name sodium oxide still lets us know that there are two sodium atoms in the formula.

5.141 Many acids can be thought of as anions that have an H^+ ion attached to them. Fill in the following table:

Anion	Anion name	Acid formula	Acid name
F^-			
NO_3^-			
Cl^-			
$C_2H_3O_2^-$			
NO_2^-			

5.142 When naming acids, when do you use the *-ic* suffix and when do you use the *-ous* suffix?

5.143 What is the name of the polyatomic ion NO_3^-? What are the names of the acids HNO_3 and HNO_2?

5.144 Acetic acid, sulfurous acid, and phosphoric acid can each be thought of as an anion with one or more H^+ ions attached. For each acid, how many H^+ ions have been attached to the anion?

5.145 Give the molecular formula for hypochlorous acid and perchloric acid.

5.146 What is wrong with the name *hypochloric acid*?

Additional Problems

5.147 What is an ionic bond? How does it differ from a covalent bond, and how is it similar to a covalent bond?

5.148 Draw dot diagrams for Cl_2 and for $MgCl_2$ (put Mg in the middle) that agree with the type of bonding occurring in each. (That is, show electrons as being shared or transferred, and if they are transferred, show the charges of the resulting ions.)

5.149 What is a shortcut rule for determining the number of covalent bonds a representative element from group IVA, VA, VIA, VIIA, or VIIIA can form?

5.150 What is incorrect about this statement? "Electron–electron electrostatic forces hold molecules together."

5.151 Why are the noble gases monatomic?

5.152 How many lone pairs of electrons are on the P atom of PCl_3?

5.153 Which one of these four molecules contains a triple bond: F_2, O_3, HCN, or H_2CO?

5.154 Which is the correct Lewis dot diagram for carbonyl fluoride, COF_2? What is wrong with the other two?

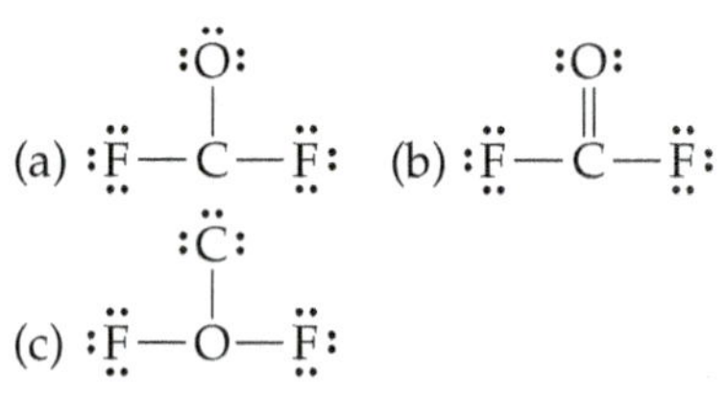

5.155 Which is the correct Lewis dot diagram for N_2H_2? What is wrong with the other two?

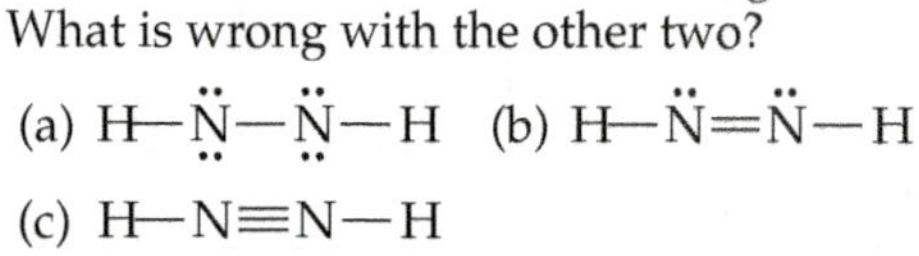

5.156 Draw dot diagrams for all resonance forms of the NCO^- ion.

5.157 Complete the dot diagram for this molecule to determine (a) how many double bonds it has and (b) how many lone pairs of electrons it has.

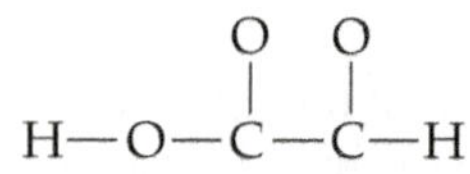

5.158 Which one of the following is expected to have resonance forms? Explain your choice.
(a) NH_4^+ (b) HCN (c) CO (d) NO_2^-

5.159 How many lone pairs of electrons are there on Br in BrF_2^+?

5.160 Which of the following represents the correct resonance forms for SO_2? Explain what is wrong with the sets you did not choose.

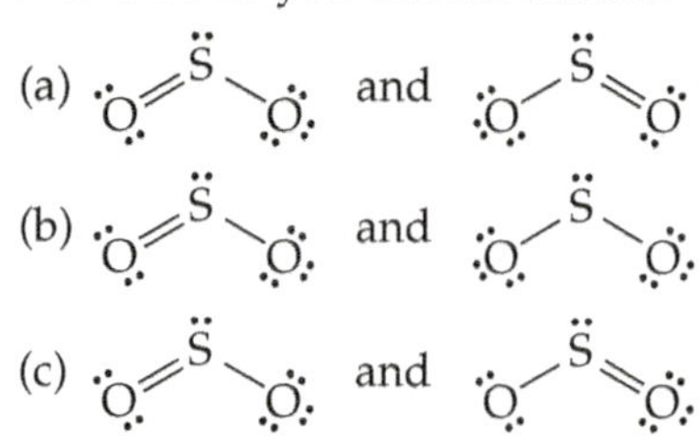

5.161 Which element has the lowest electronegativity, and how can you answer this question without looking up electronegativity values?
(a) Mg
(b) Cl
(c) Ca
(d) Br

5.162 Which one of the following statements is true about BeH_2?
(a) It is ionic with H as the anion.
(b) It is ionic with H as the cation.
(c) It has polar covalent bonds with a partial negative charge on H.
(d) It has polar covalent bonds with a partial positive charge on H.

5.163 The greater the electronegativity difference between two bonded atoms, the greater the percent ________ character of the bond.

5.164 The phosphorus atom in PCl_3 should have a
(a) δ^+ charge
(b) δ^- charge
(c) 3+ charge
(d) 3− charge

Explain.

5.165 Iodine atoms in I_2 should have a
(a) I^- charge
(b) δ^- charge
(c) δ^+ charge
(d) No charge

5.166 Arrange in order of increasing ionic character: CsBr, KBr, PBr_3, $MgBr_2$.

5.167 Complete this table of ionic compounds. Give both old and new names where you can.

Name	Formula
	$AgNO_3$
Aluminum selenide	
Lithium oxide	
Ammonium iodide	
	$CuSO_4$
	$KMnO_4$
	$Ca(ClO_3)_2$

5.168 Complete this table of ionic compounds. Give both old and new names where you can.

Name	Formula
Sodium bicarbonate	
Magnesium acetate	
Barium hypochlorite	
	$Fe(NO_3)_3$
	$(NH_4)_2SO_4$
Calcium phosphate	
	$Co_2(CrO_4)_3$

5.169 Complete this table of molecular compounds.

Name	Formula
	CdTe
Nitrogen triiodide	
	SiI_4
Bromine trifluoride	
Hydroiodic acid	
	HI (as a pure gas)
	S_4N_4

5.170 Complete this table of molecular compounds.

Name	Formula
	XeF_4
	XeF_2
Iodine monochloride	
Bromine trichloride	
Diboron hexahydride	
	N_2O
	S_4O_2

5.171 Which of the following substances are ionic and which are molecular? Name each substance. Draw Lewis dot diagrams for each molecular substance and for the anion of each ionic substance.
(a) Cl_2
(b) HCl(g)
(c) NaCl
(d) $Mg(ClO_2)_2$
(e) CH_3OH (methanol)
(f) $Fe(NO_3)_3$
(g) $Pb(C_2H_3O_2)_2$

5.172 How many electrons does the nickel atom lose when forming nickel(II) nitrate? When forming nickel(II) sulfate?

5.173 Why is the oxygen atom listed first in OF_2 but last in Br_2O?

5.174 The use of chlorofluorocarbons (CFCs) in air-conditioning and refrigeration systems has been linked to depletion of the ozone layer of the atmosphere. One of the most widely used CFCs, freon-12, has the formula CF_2Cl_2. It is now being replaced with a compound that has the formula CH_2FCH_3. Draw the dot diagrams for CF_2Cl_2 and CH_2FCH_3. (*Hint*: In CH_2FCH_3, the carbons bond to each other, and one of them bonds to fluorine.)

5.175 Why is HF not considered an ionic compound even though its percent ionic character is greater than 50%?

5.176 Give the formula for (a) calcium phosphate, (b) potassium hydrogen phosphate, (c) magnesium cyanide, (d) barium chlorate.

5.177 Give the proper name of these compounds:
(a) $Fe_2(SO_3)_3$ (b) $Au(NO_3)_3$
(c) NaH_2PO_4 (d) $Pb(CH_3CO_2)_2$

5.178 Hydroxylamine is a compound that contains one nitrogen atom and one oxygen atom connected with a single bond. The remaining atoms are hydrogens. Predict the formula for hydroxylamine and draw its dot diagram.

5.179 Using only the periodic table, arrange these sets of atoms in order of increasing electronegativity:
(a) Li, Be, K (b) O, Si, S (c) Br, I, Te

5.180 Give the proper name for each compound and indicate whether it is ionic or polar covalent:
(a) NH_4I (b) Cl_2O_7 (c) $SrCl_2$ (d) $Li_2Cr_2O_7$

5.181 Give the formula for each compound and indicate whether it is ionic or polar covalent:
(a) Copper(I) sulfide (electronegativity of Cu is 1.9)
(b) Aluminum carbide
(c) Diiodine pentoxide
(d) Chlorine trifluoride

5.182 Arrange the following sets of bonds in order of increasing ionic character. Use the symbols $\delta+$ and $\delta-$ to indicate partial charges, if any, in the bonds: O—Cl, C—F, N—Cl, O—H, S—O.

5.183 Hydrogen sulfide and hydrosulfuric acid have the same molecular formula. How are they different?

5.184 Using only the periodic table, determine which bond in each pair is more ionic:
(a) H—F or H—Cl
(b) O—F or C—F

5.185 The polyatomic ion MnO_4^{2-} is listed in your book as permanganate. What would be the name of MnO_3^{2-}?

5.186 According to the following diagram of an H_2 molecule, how many more positive–negative attractions are there relative to the case of two isolated hydrogen atoms? What do we call this extra positive–negative attraction?

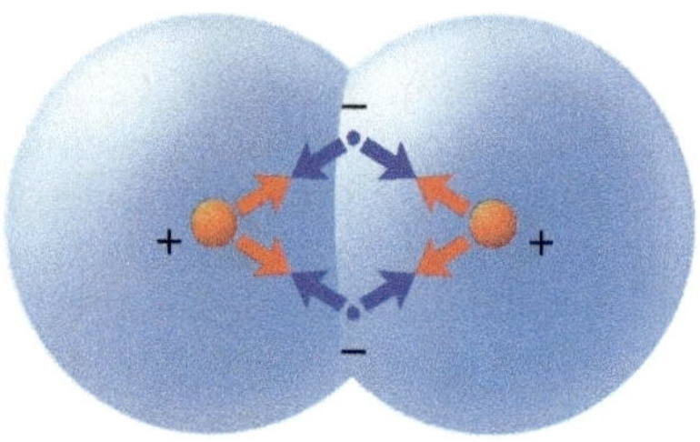

5.187 Nitrogen is a group VA representative element and thus has five valence electrons. A student draws a dot diagram for the nitrogen atom, shown below. What did the student do wrong?

5.188 A student draws the following dot diagram for the oxygen molecule and claims, "Each oxygen atom has an octet, so the diagram is correct." Part of her statement is true, and part of it is false. Which part is true, which part is false, and why is it false?

5.189 Atoms of different elements, each with at least one unpaired electron, come together to form a compound. According to the diagram below:
(a) Which atom is the more electronegative? Explain how you knew.
(b) Which atom has a partial positive ($\delta+$) charge and why?
(c) Suppose only one of the elements is a metal. Which element is it? Explain how you knew.
(d) What value would you have to calculate to decide if this bond is polar, covalent, or ionic?

5.190 Consider the following diagram:
(a) What do these four elements have in common?
(b) Where on the periodic table will you find elements that have the opposite property of these elements?
(c) What periodic trend or trends are identified with respect to these four elements?

5.191 In all three of the diagrams below, valence electrons have undergone redistribution compared to two isolated atoms. For each, describe the type of redistribution that has occurred and what it leads to.

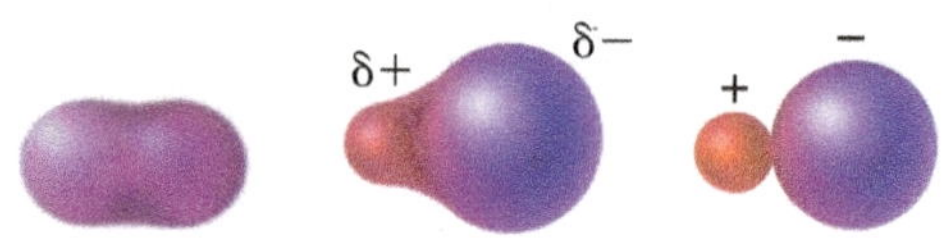

(a) (b) (c)

One Step Beyond

5.192 It is convenient to classify substances as ionic or covalent. How would you classify sodium sulfate? Explain fully.

5.193 P is immediately under N in the periodic table. Both react with three atoms of chlorine to form NCl_3 and PCl_3. However, phosphorus also forms PCl_5, whereas nitrogen does not. Draw dot diagrams for all three molecules. In doing so, you will have to violate the octet rule for PCl_5. That is OK. Do it. Then explain why phosphorus can do this but nitrogen can't. (*Hint:* Nitrogen has only 2*s* and 2*p* orbitals in its valence shell, which together can hold a maximum of eight electrons. Read section 5.8 about expanded octet atoms.) Then postulate what kind of atoms might be expected to exceed the octet rule.

5.194 Give an example of a covalent compound that, upon dissolving in water, yields ions. Write a reaction to show this transformation and classify the starting compound.

5.195 Draw a diagram that shows both the covalent and ionic nature of the compound sodium azide, formula NaN_3 (the N_3 portion is linear in shape). If there are resonance forms, then include them.

5.196 Benzene, formula C_6H_6, is a perfect closed hexagon of carbon atoms, each with one hydrogen atom attached. Draw a dot diagram for this molecule, and include any resonanace forms if they exist. Then explain why this compound is called delocalized (delocalized means that at least some of the electrons in the compound are free to move about many atoms).

5.197 Draw a dot diagram for selenium hexafluoride and explain how it violates the octet rule.

5.198 Consider the PO_4^{3-} phosphate anion.
(a) Draw a dot diagram that obeys the octet rule for all of the atoms.
(b) Suppose the actual bond order for the P—O bonds was 1.25. Draw a dot diagram that explains this, using the concept of expanded octet and resonance. Also state whether expanding the octet as you have shown would be allowed, and why.

5.200 Consider sulfur trioxide.
(a) Draw a dot diagram that obeys the octet rule for all of the atoms.
(b) Suppose the actual bond order for the S—O bonds was 2. Draw a dot diagram that explains this, using the concept of expanded octet. Also state whether expanding the octet as you have shown would be allowed, and why.

5.201 Does BeH_2 have all of its atoms obeying the octet rule? If not, indicate which atom or atoms violate the octet rule and why it is OK.

5.202 Consider two molecules, BH_3 and NH_3. Draw dot diagrams for both and indicate which one possesses and electron deficient atom.

5.203 Draw a dot diagram for PF_5. Which atom is an expanded octet atom?

5.204 Draw a dot diagram for BrF_5. (*Hint*: Br is an expanded octet atom in this molecule. Don't be afraid to stick lone pairs of electrons on it if you have to.)

5.205 Draw a dot diagram for XeF_4. (*Hint*: Xe is an expanded octet atom in this molecule. Don't be afraid to stick lone pairs of electrons on it if you have to.)

5.206 Draw a dot diagram for I_3^-. (*Hint*: Then central I atom is an expanded octet atom in this molecule. Don't be afraid to stick lone pairs of electrons on it if you have to.)

Group Exercises (Work on these exercises together with someone else in the class.)

1. Work together to finish the following statement: "You should expect a multiple bond when drawing a dot diagram for a molecule if..."

2. Work together to finish the following statement: "You don't need Greek numerical prefixes when naming simple ionic compounds because..."

Workpatch Solutions

5.1. By Coloumb's Law, the force of attraction between Mg^{2+} and O^{2-} ions should be four times greater than the force between Na^+ and Cl^- ions because the Mg^{2+} and O^{2-} ions both have an absolute charge q of 2 (and $q^+ \times q^- = 4$ when both q's are 2, compared to $q^+ \times q^- = 1$. when both q's are 1 as they are for Na^+ and Cl^-).

5.2. The molecules H_2O (water) and H_2O_2 (hydrogen peroxide) have very different chemical properties. For example, you drink water, but H_2O_2 is a harsh antiseptic and bleach. Different formulas mean different compounds with different properties, even when the elements in the formula are the same.

5.3. The carbon atom has eight valence electrons (two from each bond). Each hydrogen atom has two valence electrons (from the bond that attaches it to carbon).

5.4. Carbon normally has four valence electrons (it is a group IVA atom). In methane, it has eight valence electrons, so it has gained four electrons. That it forms four bonds to hydrogen tells you that it gains one valence electron with each bond it forms.

5.5. Helium atoms already have a filled valence shell (a duet) and so have no "desire" to gain more electrons by sharing with atoms of other elements. In addition, helium atoms have no unpaired electrons available for sharing and therefore cannot form covalent bonds.

5.6. Every atom in the NF_3 molecule has an octet of valence electrons (the bonding pairs must be double-counted).

5.7.

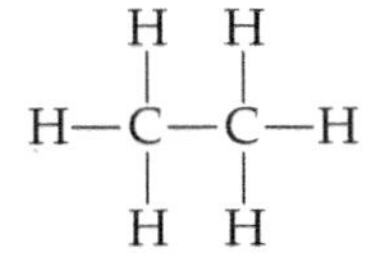

5.8. H—C≡C—H

5.9. By moving lone pairs into bonding positions from different atoms.

5.10. Gallium is a metal in group IIIA, and so it loses three electrons to become Ga^{3+}. Oxygen is a nonmetal in group VIA, and so it gains two electrons to become the oxide ion, O^{2-}. To obtain a neutral compound, we must combine two Ga^{3+} ions (for a +6 charge) with three O^{2-} ions (for a −6 charge). The formula is therefore Ga_2O_3.

5.11. The more electronegative element in a binary covalent compound has the partial negative charge, and the suffix *-ide* is associated with the anionic, or negative, part of a compound.

EQUAL
SWEETENER WITH NUTRASWEET
sweet as 2 tsp sugar
QUADE
peptide bond
CH2
HOOCCH2
Asp
Phe

CHAPTER 6

The Shape of Molecules

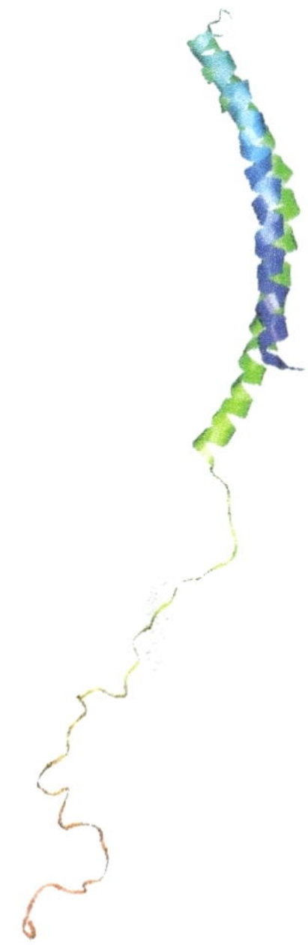

Meet alpha-synuclein, a protein molecule found in the human brain whose function is, as yet, unknown. Notice its extended and partially coiled (helical) shape.

Interestingly, when alpha-synuclein molecules change shape by folding in just the right way, they can start packing together to form aggregates. Some people have a gene encoded in their DNA that will activate this process as they age, allowing this shape change to occur. As the resulting aggregates accumulate in the brain, these people will either slowly lose control of their muscles (Parkinson's disease) or sink deeper and deeper into dementia (Alzheimer's disease).

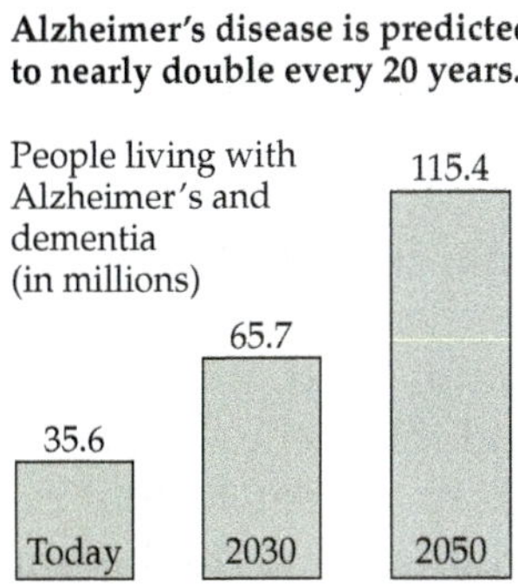

Is there a way to stop this protein from shape-shifting? Perhaps, but it will take more money and research to find out either how to stop the molecular shape change from occurring or how to keep aggregates from forming once the shape change has occurred.

6.1 The Importance of Molecular Shape

In Chapter 5, we discussed the forces that hold a molecule like methane together and why it has the formula CH_4. Now we'll take a closer look at the way we've been drawing molecules. So far, we've been using dot diagrams to represent molecules and drawing them as flat as a pancake. For example, we represented methane, CH_4, and ammonia, NH_3, as follows:

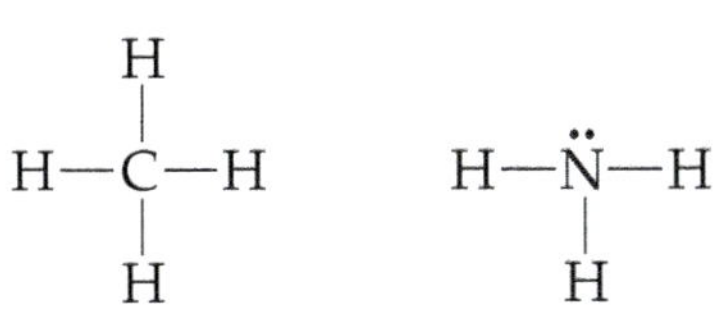

This method was fine because dot diagrams are meant to show only how atoms are connected in a molecule and how the valence electrons are arranged (into bonding pairs and lone pairs). However, neither methane nor ammonia is actually a flat, two-dimensional molecule. Both have much more interesting, three-dimensional shapes.

What do methane and ammonia molecules actually look like? In this chapter, we shall answer this question and learn how to predict the three-dimensional shape of a molecule.

At this point, you might be wondering why we should care about the shape of a molecule. After all, we can't see individual molecules. What possible relevance could their shape have? Quite simply, the shape of a molecule could save your life, and most likely it already has. Chemists figured this out shortly before World War II began. In all wars up to that point, the majority of combat deaths were caused not by the wounds inflicted in battle but by the infections that set in shortly thereafter. A soldier with a stomach full of shrapnel would lie in a makeshift field hospital recovering from abdominal surgery. Even if the operation went well, there was just no way to keep a few microscopic bacteria from entering his bloodstream and reproducing wildly. It was an all-too-common scenario. The soldier's body would try to fight the infection, but eventually he would die. During World War I, medics could do almost nothing to prevent this, and hundreds of thousands died. But World War II was different. For the first time, a medic could intervene with an effective defense—a drug called sulfanilamide, a trap meant to fool bacteria. The molecular structure of this sulfur-containing compound is similar to that of another molecule found in the bloodstream, *para*-aminobenzoic acid (PABA):

PABA (*para*-aminobenzoic acid), a metabolite in our bloodstream

Sulfanilamide, a molecule with a shape similar to PABA

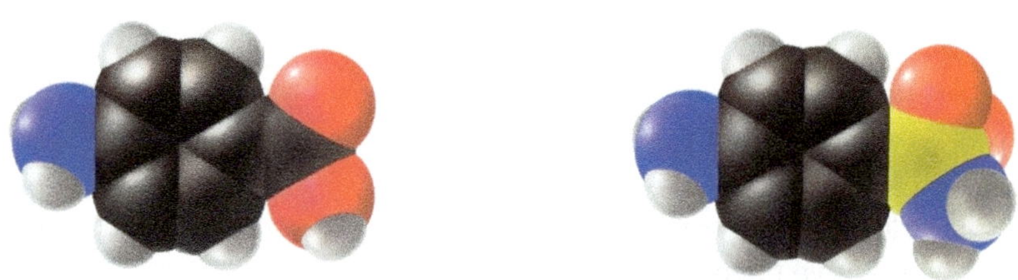

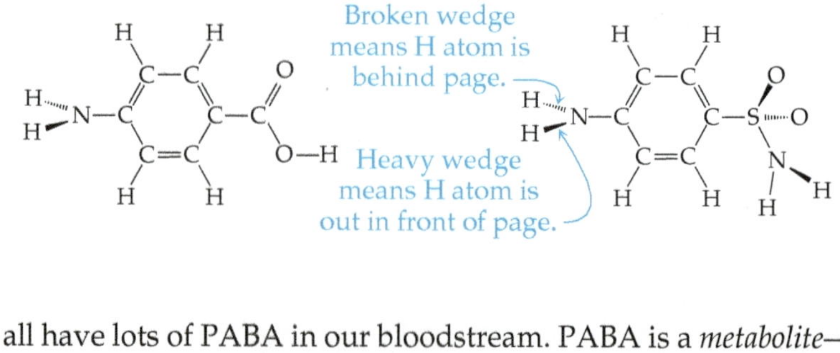

We all have lots of PABA in our bloodstream. PABA is a *metabolite*—that is, a natural product of our body's metabolism. Bacteria can use PABA to their own advantage by combining it with two other molecules also present in our

bloodstream. Once connected together, these three molecules form one of the B vitamins—folic acid:

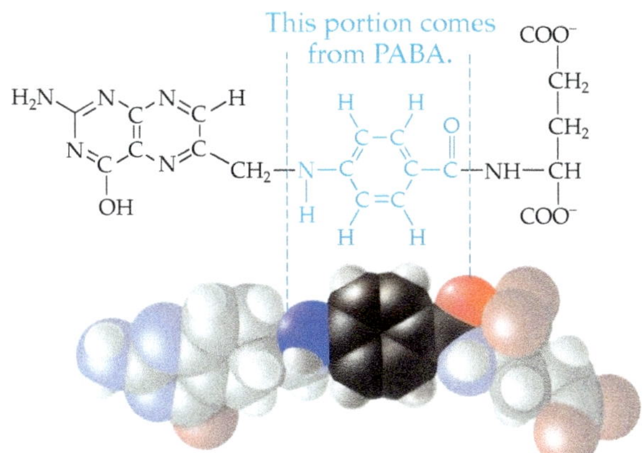

Folic acid, a B vitamin

This vitamin is essential for the health and continued functioning of the bacteria. Humans also need this vitamin in order to live, but we have lost the ability to make it ourselves and so must ingest it in the food we eat. It is this difference in the source of folic acid between bacteria and their human hosts that makes sulfanilamide and many other antibiotics the wonder drugs they are (read on).

The B vitamin folic acid is created inside the bacterial cell by a specific molecule called an *enzyme*. Enzymes are large protein molecules that control chemical reactions in living systems. They increase the speed of specific reactions that take place in the cells of the body. Enzymes have precisely shaped pockets (called *active sites*) on their surface, pockets that they use to bind with smaller molecules that have complementary shapes. The bacterial enzyme that constructs folic acid from its three precursor molecules has a pocket that is complementary to the PABA molecule.

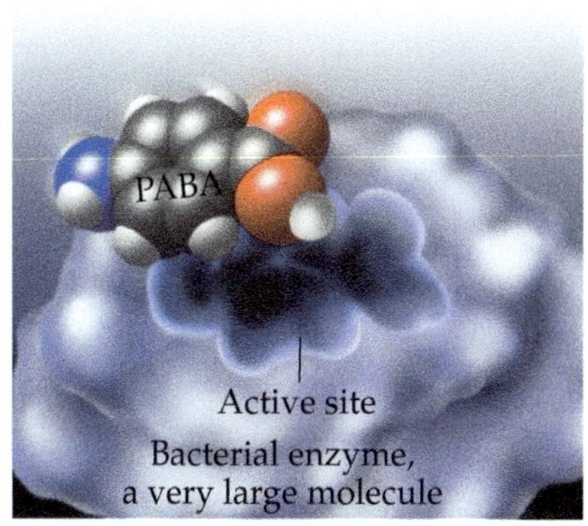

Efficiently manufacturing their own folic acid, the bacteria can thrive and reproduce in the human bloodstream. But something else happens when sulfanilamide molecules are waiting. Because sulfanilamide has almost the same shape as PABA, the sulfanilamide fits nicely into the pocket of the bacterial enzyme, which is thus tricked into making a defective form of folic acid, one that incorporates sulfanilamide instead of PABA. Starved for the real B vitamin, the bacteria stop replicating, and the infection subsides. Sulfanilamide saved many thousands of lives in World War II and was the precursor to a number of other antibiotics called sulfa drugs.

So, is the shape of a molecule important? Unquestionably! You can be sure that new antibacterial and antiviral agents that have yet to be discovered—including drugs that will one day free humanity of herpes, AIDS, hepatitis, meningitis, and cancer—will have molecular shape at the core of their ability to cure.

6.2 Valence Shell Electron Pair Repulsion (VSEPR) Theory

Of course, no one can build a molecule with a specific shape, or even predict what the shape a molecule will be without first understanding what it is that determines a molecule's shape. **Valence shell electron pair repulsion (VSEPR) theory** is the model most often used to predict molecular shape. It is based on the simple concept that the valence electrons in a molecule repel one another. Let's go back to methane to investigate this. We'll start by drawing a dot diagram for methane:

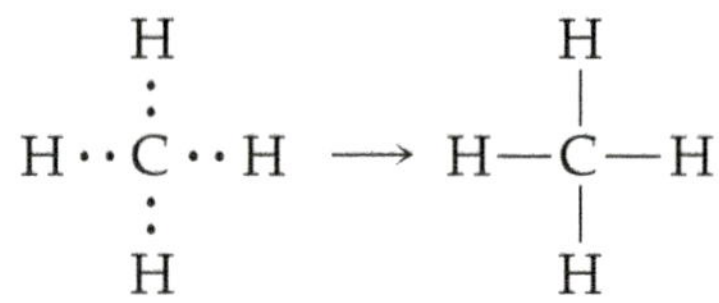

All electrons are negatively charged, and so each *electron group* (in this case, each bonding pair of electrons) represents a region of concentrated negative charge. An **electron group** is just some number of electrons that share the same region of space in a molecule. Because like charges repel, the electrons in the four electron groups repel one another. To minimize this repulsion, the electron groups try to get as far apart from one another as they can:

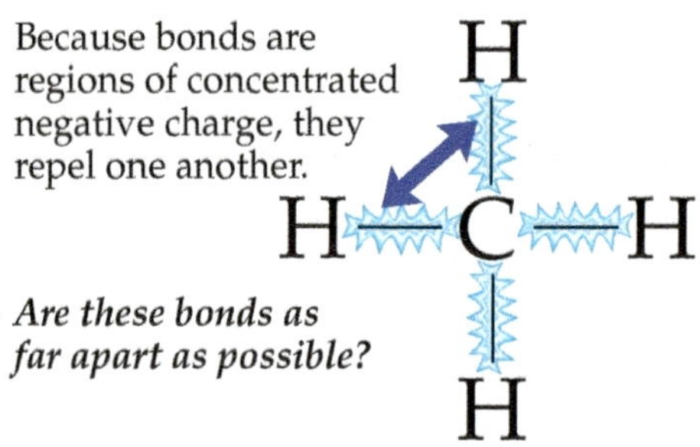

Our flat, square version of methane, with its 90° angles, would seem to accomplish that pretty well. Certainly, we can envision other ways of attaching four hydrogens to a central carbon atom that would make the repulsions between the electron groups worse. For example, we could clump the hydrogen atoms to make the angles between the bonds 60°:

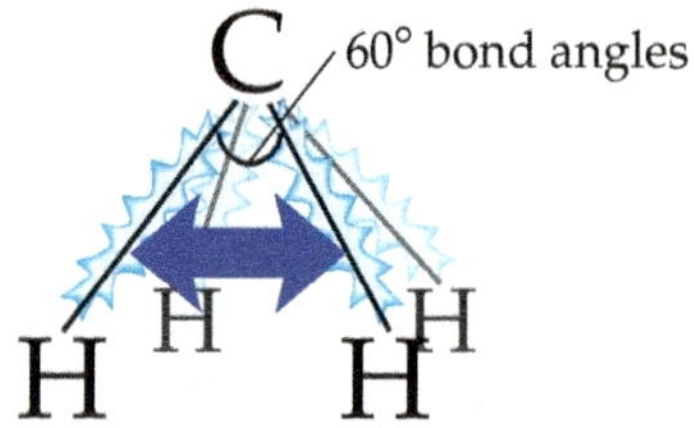

When you cram electron groups together, they REALLY repel one another!

A methane molecule would never assume this shape. With such small bond angles, the electron groups would be practically on top of one another, causing strong repulsion. The further away the electron groups can get from one another, the better!

While we can easily generate structures that are *worse* than our original square planar (flat) shape, the real question is, "Can we do any *better*?" The answer is "Yes, provided we don't insist that our methane molecule be flat." There is a better arrangement, one that opens up the H—C—H bond angles beyond 90° and maximizes the distance between the four electron groups surrounding the carbon atom. Here it is:

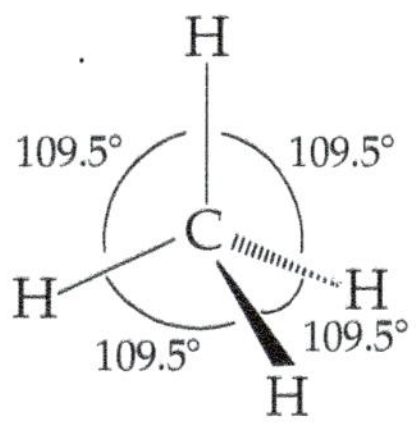

Structure diagram
- Bond projects out from page.
- Bond goes back behind page.
- Bond is in plane of page.

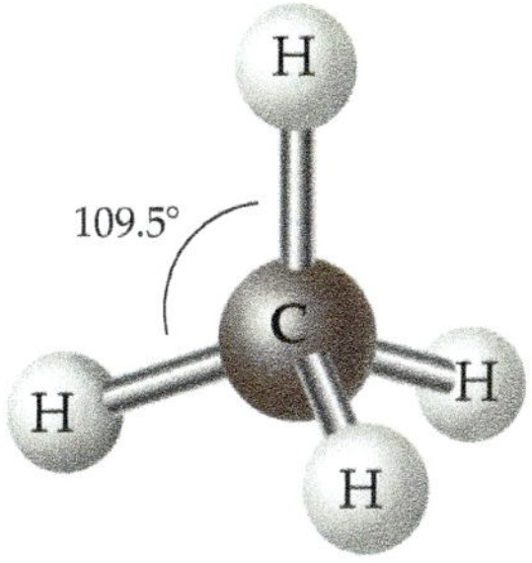

Ball-and-stick model emphasizes bonding pattern.

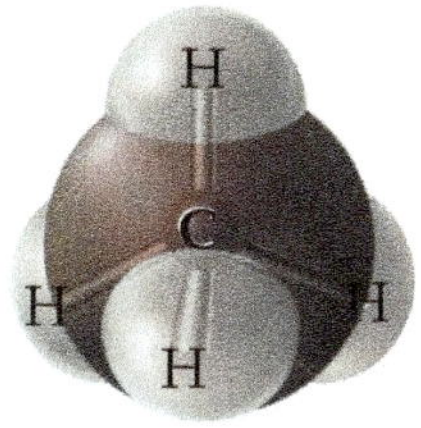

Space-filling model approximates actual molecular shape.

The methane molecule is said to have a *tetrahedral* shape. A **tetrahedron** is a regular four-sided polygon in which the four sides are identical equilateral triangles. The following figure shows a methane molecule inscribed within a tetrahedron:

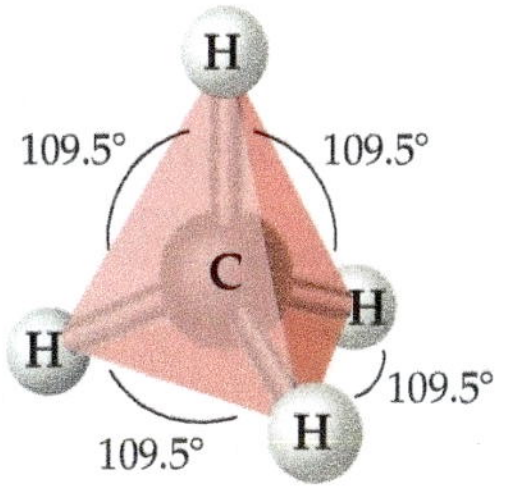

The H—C—H angles are all 109.5°. If we insist the bond angles be equal, then this is as large as they can be. When methane adopts this nonplanar (not flat) shape, the four bonding pairs are as far apart from one another as possible. This minimizes the repulsion between the electron groups and, thus, maximizes the stability of the molecule.

What we have said so far about methane is not specific to that molecule. Any time there are four separate electron groups around an atom in a molecule, a tetrahedron is the preferred arrangement. This is simply the best way to keep the four groups of electrons as far apart from one another as possible. Once you understand this, you understand one of the most critical factors that determine molecular shape.

Of course, we shall not always be dealing with four electron groups. Let's look at two other organic molecules, formaldehyde (used to preserve biological specimens) and acetylene (used in welding torches):

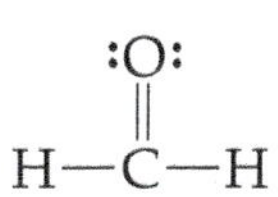

Formaldehyde, CH_2O

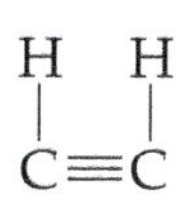

Acetylene, C_2H_2

The dot structures for these molecules have been drawn flat, with arbitrary right angles because, as noted earlier, dot structures are only meant to keep track of the bonds and lone pairs in a molecule, not its actual shape. Once the dot structure of a molecule has been determined, the question of how the bonds and atoms are oriented in space can be answered. We must arrange the electron groups in such a way as to minimize the repulsions between them. We start just as we did for methane, by counting the number of electron groups around a particular carbon atom, but there is a new twist here. Both of these molecules have multiple bonds (a double bond in formaldehyde, a triple bond in acetylene). The rule for counting multiple bonds for the purpose of VSEPR is to *treat each multiple bond as a single electron group,* just the way we treat a single bond. We do this because all the electrons in a multiple bond occupy roughly the same region of space. In formaldehyde, this means that there are just three electron groups surrounding the carbon atom—two single bonds and a double bond. Around each carbon atom in acetylene, there are two electron groups—one single bond and one triple bond. Chemists often refer to the number of electron groups about an atom in a molecule as the atom's **steric number (SN)**. The steric number for an atom in a molecule can be calculated by summing the number of atoms attached to that atom plus the number of lone pairs of electrons on that atom.

$$\begin{array}{c}\text{Steric number for an}\\ \text{atom in a molecule}\end{array} = \begin{array}{c}\text{Number of electron}\\ \text{groups around that atom}\end{array}$$

$$= \left[\begin{array}{c}\text{\# of other atoms}\\ \text{attached to that atom}\end{array} + \begin{array}{c}\text{\# of lone pairs}\\ \text{on that atom}\end{array}\right]$$

Once the counting of electron groups is done, the task of determining the shape of the formaldehyde molecule comes down to answering the question, "How can three electron groups (steric number 3) best arrange themselves around a central (carbon) atom?" In the case of acetylene, the question becomes, "How can two electron groups (steric number 2) best arrange themselves around a central (carbon) atom?" The answer in both situations involves a planar arrangement of the bonds. Can you determine the best angles between these bonds? Try WorkPatch 6.1. If you succeed, you've caught on to VSEPR theory.

WORKPATCH 6.1 Redraw each structure to get the electron groups as far apart as possible. What angles did you use?

Before going on, be sure to compare your answers with the solution given at the end of the chapter.

So far, we have looked at three specific molecules and used VSEPR theory to predict their shape. In methane, there are four electron groups to accommodate around the carbon atom, in formaldehyde there are three, and in acetylene there are two (counting each multiple bond as a single electron group). But there is nothing special about these molecules. The same results can be expected for any other molecules that contain an atom with these same numbers of electron groups around it, as shown in Table 6.1.

Table 6.1 VSEPR Arrangements of Electron Groups Around an Atom

Steric Number (Number of electron groups)	Geometry of electron groups	Corresponding molecular shape	Example
Two	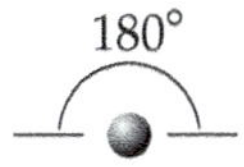 Bond angle 180°	 Linear	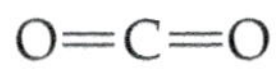  Carbon dioxide, CO_2
Three	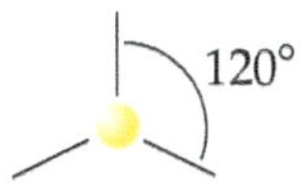 Bond angles all 120°	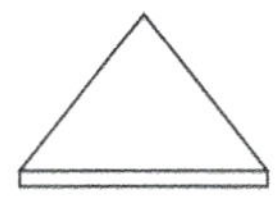 Trigonal planar	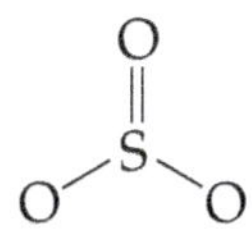 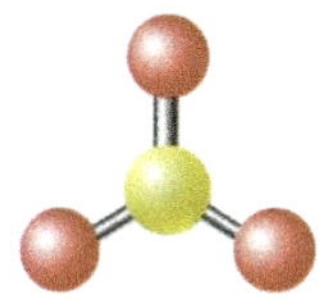 Sulfur trioxide, SO_3
Four	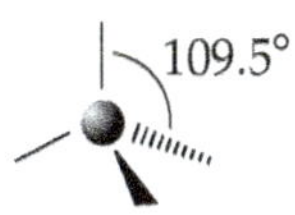 Bond angles all 109.5°	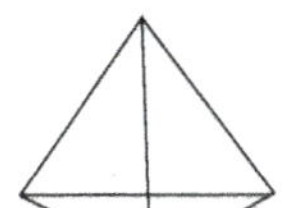 Tetrahedral	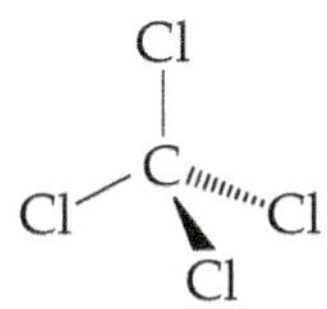  Carbon tetrachloride, CCl_4

Because determining the shape of a molecule requires knowing the number of electron groups that surround each atom, you really can't get started until you have a good dot diagram in front of you. This means that when you are asked to determine the shape of a molecule, you must first draw a dot diagram if one has not been supplied.

Practice Problems

For all these problems, describe the geometry of the electron groups and name the molecular shape resulting from that geometry. Also, draw the molecule, and label the size of all bond angles in your drawing.

6.1 CO_2.

Answer: First draw a dot diagram, then examine it via VSEPR:

Draw a dot diagram: Apply VSEPR: Determine the molecule's shape:

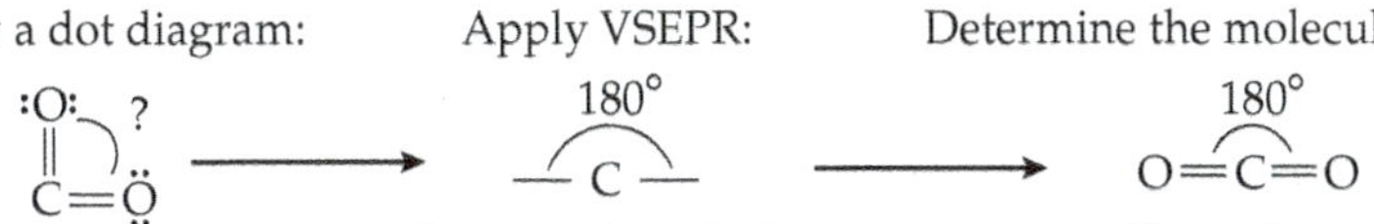

There are two electron groups 180° apart around the central C atom. (SN = 2)

The molecule is linear.

Note: Don't worry about the arrangement of the electron pairs about the peripheral (outer) atoms of a molecule. Only the electrons on the inner atoms affect the shape.

6.2 NH_4^+ (cation).

6.3 C_2Cl_2, which is connected Cl—C—C—Cl.

6.4 $COCl_2$, in which all atoms are connected to C, and there is a carbon–oxygen double bond.

So far, none of the molecules we have considered had lone pairs of electrons on the central atom. It's now time to consider these molecules. A good example is the ammonia molecule, NH_3. Ammonia is an interesting substance. It exists as a gas at room temperature, but you are probably more familiar with it dissolved in water as a household cleaner. It is the fifth most commonly produced chemical in the United States, used in greater quantities as an agricultural fertilizer than as a household grease cutter. What is the shape of the ammonia molecule? To answer this question, we have to decide how the electron groups are arranged in space around the central nitrogen atom, so we need a dot diagram to begin:

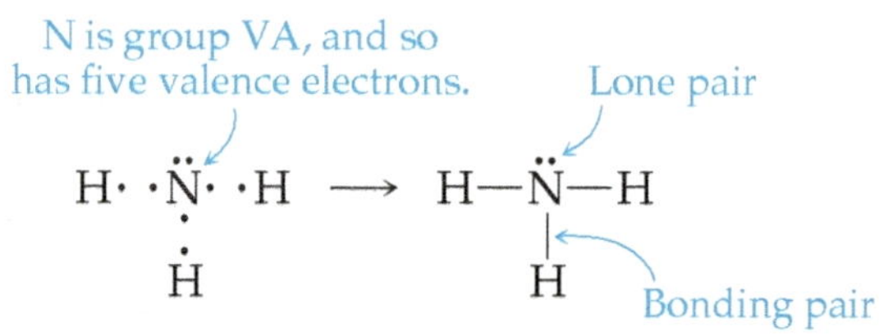

As usual, the dot diagram is drawn with a planar shape. Now focus on the central nitrogen atom and count the total number of electron groups that surround it: three uniquely located bonding pairs plus one lone pair. The lone pair must be included in the total count because it is negatively charged, just like a bonding pair, and thus it repels the other electron groups. The nitrogen atom in ammonia has four electron groups surrounding it, or a steric number of 4 (3 attached H atoms plus 1 lone pair). Therefore, much like the carbon in methane or any other atom with four electron groups around it, the electron groups on the nitrogen adopt a tetrahedral arrangement in order to minimize repulsion:

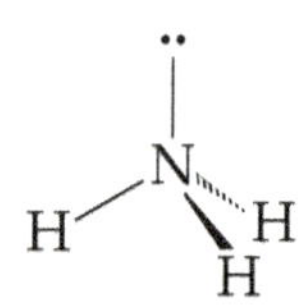

Count all electrons when determining the electron-group geometry.

So, the shape of the ammonia molecule is tetrahedral, right? Well, not quite. Ammonia is not considered to be a tetrahedral molecule. The above drawing is perfectly correct, but we describe the shape of a molecule differently when lone pairs are present. The convention is to include lone pairs in our initial count of electron groups in determining the spatial arrangement of these electron groups (in other words, how they are arranged in space relative to one another), *but when it is time to describe the molecule's shape, we ignore the lone pairs*. This is done because ultimately we want the name of the molecular shape to indicate the arrangement of the *atoms* in a molecule, not the electrons.

How then do we describe the shape of the ammonia molecule? Ignoring the lone pair of electrons, we see that the molecule is a tetrahedron with the top part removed. What's left looks likes a pyramid, and so the ammonia molecule is said to have a **pyramidal** shape:

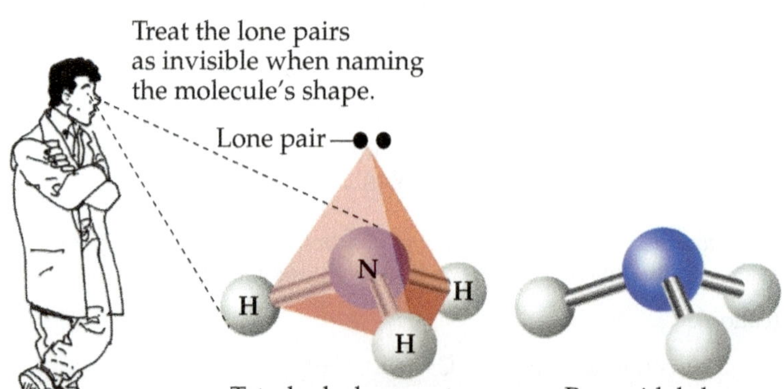

Tetrahedral geometry of electron groups

Pyramidal shape of molecule

(If you're having trouble seeing the difference between a tetrahedron and a pyramid, remember from our definition of a tetrahedron that all four sides are *identical* triangles. When we ignore the lone pair in NH_3, however, we essentially lop off the top of three triangles. Because we've not touched the base triangle, however, the four faces of our polyhedron are no longer identical. In our tetrahedral shape, all four faces are identical, but in our pyramidal shape, only three of the four faces are identical.)

We can therefore say that, even though the geometry of the electron groups about N is tetrahedral, *the shape of the molecule is not tetrahedral but rather pyramidal.* Don't forget: The shape of a molecule describes how the *atoms* are arranged, not how the electron groups are arranged. Therefore, we use VSEPR theory and *all* of the electron groups (bonding groups plus lone pairs, the steric number) to predict how the electrons should be arranged relative to one another, but then treat the lone pairs as if they were invisible when describing the shape of the molecule.

Let's take this one step further by asking what the H—N—H bond angles in ammonia are. To answer this question, think about where we started, before we ignored the lone pair. The four electron pairs surround the N atom in a tetrahedral geometry. Therefore, the H—N—H angles in ammonia should be close to the standard 109.5° tetrahedral angle:

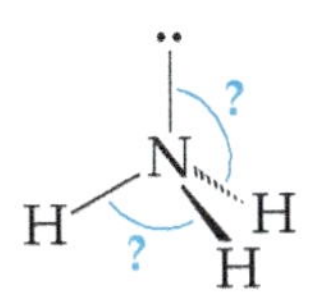

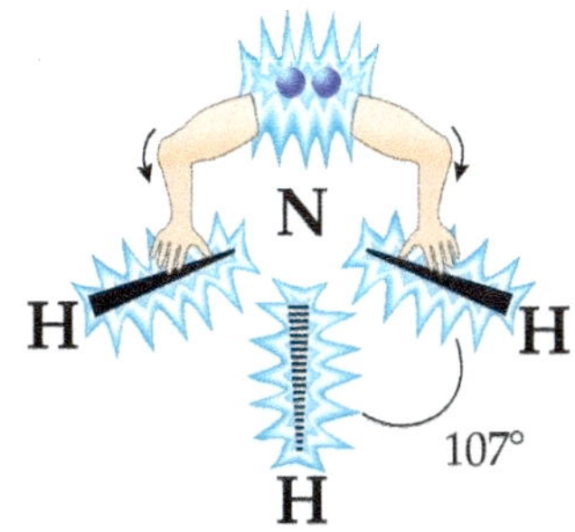

"Squeeze play"

The actual H—N—H bond angles are 107°, slightly smaller than the value for perfect tetrahedral geometry.

The experimentally determined H—N—H bond angles in ammonia are actually a little bit smaller than the ideal 109.5° tetrahedral angle because the electrons in a lone pair take up more space than the electrons in a bonding pair. This tends to squeeze down on the bonding pairs, compressing the angles. A useful rule of thumb is that each lone pair *on a period 2 atom* compresses the remaining bond angles around that atom by approximately 2°. Thus, we predict the H—N—H bond angles in ammonia to be approximately 107°, which is in good agreement with experimental results.

In summary, follow steps 1–4 to use VSEPR theory to determine the shape of a molecule:

Using VSEPR theory to determine molecular shape

Step 1	Draw a dot diagram for the molecule.
Step 2	Count the number of electron groups around the central atom. Include lone pairs, and count each multiple bond as a single electron group. This is called the steric number for the central atom.
Step 3	Using the steric number from step 2, decide on the best arrangement for the electron groups, including lone pairs, around the central atom.
Step 4	Pretend the lone pairs are invisible, and describe the resulting shape of the molecule.

Let's see if you've got it. We'll work part way through the determination of the shapes of water and ozone, and then we'll ask some questions. We begin as usual with electron dot diagrams:

Water Ozone

Focus on the central atom in each molecule, and count the total number of electron groups around it (remember that multiple bonds count only once). The oxygen atom in water is surrounded by four electron groups (two bonding pairs and two lone pairs), so the steric number is 4 (2 attached atoms plus 2 lone pairs). The central oxygen atom in ozone has three electron groups around it (a double bond, a single bond, and a lone pair), so the steric number is 3 (2 attached atoms plus 1 lone pair). How do these groups arrange themselves in space? VSEPR theory tells us what the arrangement must be:

Water

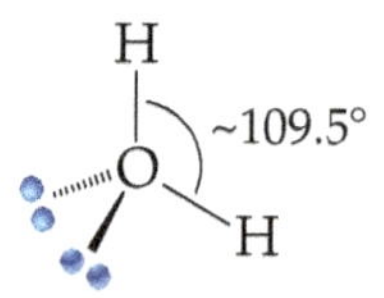

Four electron groups are arranged tetrahedrally around the oxygen atom.

Ozone

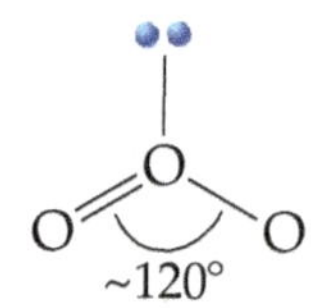

Three electron groups are arranged around the oxygen atom in a planar triangle.

WORKPATCH 6.2 How would you describe the shape of the water and ozone molecules? Follow the convention! Ignore the lone pairs on each central atom. Cover them up and use a word or phrase that best describes the shape created by the atoms. ●

What did you call the shapes of these molecules? Certainly, they are not linear because that would mean all three atoms must lie on a single straight line. They should not be called trigonal planar, our second choice from Table 6.1, because we reserve that term for cases when there are three atoms surrounding a central atom. Here the molecules are again without the lone pairs:

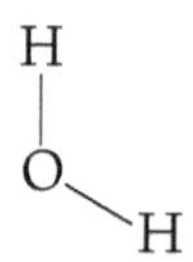

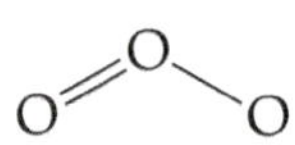

Check the WorkPatch answer at the end of the chapter to see how the shape of these molecules is most commonly described.

Finally, you should have noticed that the water and ozone molecules have essentially the same shape, but there is an important difference between them. The H—O—H angle in water is quite different from the O—O—O angle in ozone. If you can predict these two angles, you are ready to go on.

WORKPATCH 6.3 The drawings preceding WorkPatch 6.2 give the ideal values for the bond angles in the molecules. Recalling that there is roughly a 2° compression from the ideal VSEPR angle for every lone pair on the central atom, give a more accurate bond angle for each molecule. ●

Applying steps 1–4 for determining the shape of a molecule to all possible cases of four, three, and two electron groups, we obtain Table 6.2.

To make sure you understand how Table 6.2 was generated, try the next WorkPatch and the following practice problems.

WORKPATCH 6.4 We left three possible situations out of Table 6.2:

1. **Four electron groups (SN = 4): one bonding group and three lone pairs.**
2. **Three electron groups (SN = 3): one bonding group and two lone pairs.**
3. **Two electron groups (SN = 2): one bonding group and one lone pair.**

These are considered trivial cases. Why? What shape(s) do these bonding situations give rise to? ●

Table 6.2 VSPER Molecular Shape Table

Four electron groups (SN = 4)		**Three electron groups (SN = 3)**		**Two electron groups (SN = 2)**	
Electron-group arrangement	**Molecular shape**	**Electron-group arrangement**	**Molecular shape**	**Electron-group arrangement**	**Molecular shape**
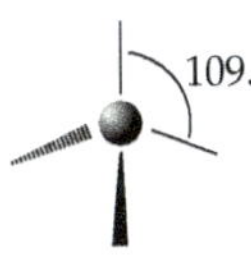	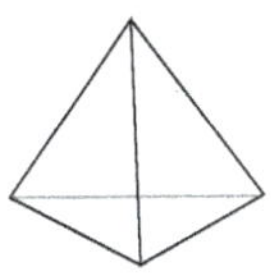	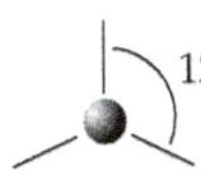	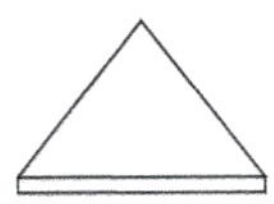	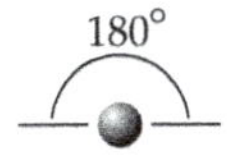	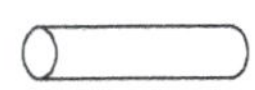
Four bonding groups No lone pairs Tetrahedral	Tetrahedral	Three bonding groups No lone pairs Trigonal planar	Trigonal planar	Two bonding groups No lone pairs Linear	Linear
Bond angles all 109.5°		Bond angles all 120°		Bond angle 180°	
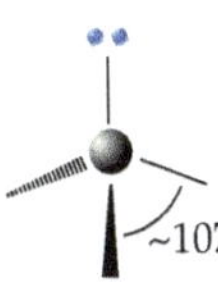	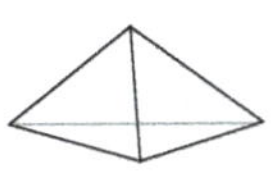	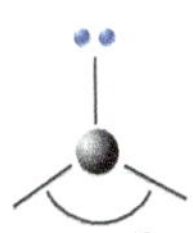			
Three bonding groups One lone pair Tetrahedral	Pyramidal	Two bonding groups One lone pair Trigonal planar	Bent		
Bond angles compressed to ~107° by lone pair*		Bond angle compressed to ~118° by lone pair*			
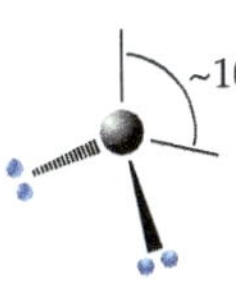					
Two bonding groups Two lone pairs Tetrahedral	Bent				
Bond angle compressed to ~105° by lone pair*					

Key: **SN = Steric Number = # of electron groups = # attached atoms around central atom and # of ion pairs**

● **Central atom**

Bonding electron group (single, double, or triple covalent bond)

•• **Lone pair of electrons**

* For elements in period 2, the bond angles around a central atom are compressed by about 2° for every lone pair.

Practice Problems

For each molecule or polyatomic ion in the following five problems, (a) draw the electron-group geometry, (b) name the shape of the molecule, and (c) estimate the size of all the bond angles in the molecule or ion. (Some of the molecules and ions have lone pairs on the central atom.)

6.5 PBr_3.

Answer: First draw a dot diagram (which must show 26 electrons):

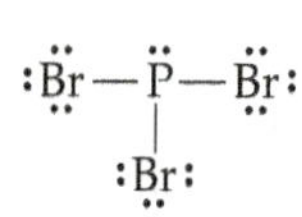

Next, count electron groups about the central atom. There are four (three single bonds and one lone pair), indicating a tetrahedral geometry of these groups about the P atom. However, because you must ignore the lone pair on P, the molecule is pyramidal in shape, with a predicted Br—P—Br angle of less than 109.5°. (You might predict 107°, but P is not a period 2 element. The actual bond angle is 102°.)

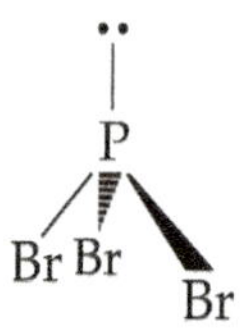

Pyramidal molecule

6.6 SO_4^{2-} (anion).

6.7 HCN, in which hydrogen and nitrogen are attached to carbon.

6.8 Sulfur dioxide, SO_2, a pollutant that comes from burning coal contaminated with sulfur.

6.9 HNO_3, in which hydrogen is attached to one of the oxygens and all oxygens are attached to nitrogen. To simplify, ignore the hydrogen when you name the molecule's shape.

6.3 Polarity of Molecules, or When Does 2 + 2 Not Equal 4?

We have already seen how the shape of a molecule can be used to trick bacteria. Molecular shape is important in many other ways as well. The chemical and physical properties of molecules are often dependent on their shape.

Molecular "stickiness" is one property that is very dependent on molecular shape. Molecules can "stick" to each other so tightly that they cannot easily be pulled apart from each other. Try it! The next time you have a drink with some ice in it, take out an ice cube, grab it with both hands, and try ripping it in half. You can't. The trillions of individual water molecules that make up your ice cube are all stuck to one another as if someone had used superglue. This simple experiment is proof of the existence of a large net attractive force between all the individual water molecules in the ice cube. Such forces between molecules are called **intermolecular attractive forces**, or *intermolecular forces* for short. If the intermolecular forces between water molecules could somehow be turned off, the ice cube would come apart and instantly turn into a gas (water vapor). It is because of these intermolecular forces that molecular substances can exist in condensed phases (liquid and solid). So where do these intermolecular forces come from? What factors determine

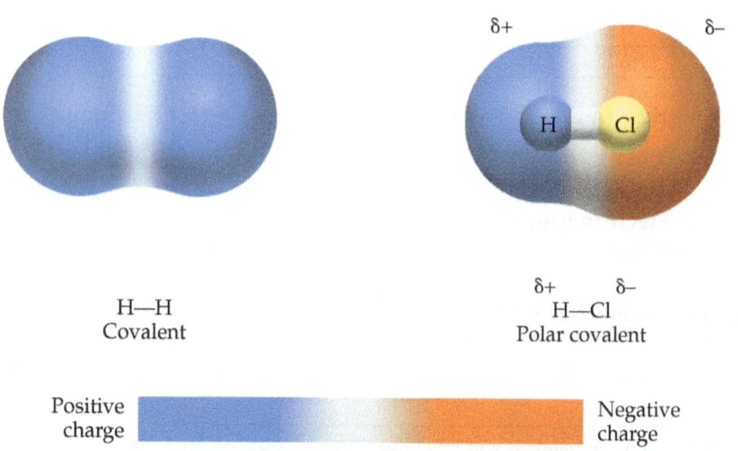

how strong or weak they are? Molecular shape plays a key role here, so let's look at a few examples.

We'll start with two simple molecules, H_2 and HCl. Both have a linear shape (all diatomic molecules are linear). The biggest difference between these two molecules, besides their elemental composition, is in their bonding. The H_2 molecule is held together by a covalent bond (equal sharing of electrons). In HCl, the very different electronegativities of hydrogen and chlorine give rise to a polar covalent bond (unequal sharing of electrons). This unequal sharing means that the bonding electrons spend more time around the Cl atom and less time around the H atom. That inequality produces a molecule with a partial negative charge ($\delta-$) on the more electronegative Cl atom, and a partial positive charge ($\delta+$) on the less electronegative H atom. If you are not sure about this, stop and review electronegativity and polar covalent bonding in Chapter 5.

Because of the presence of partial charges, the behavior of HCl molecules is significantly different from the behavior of H_2 molecules. One of these differences can be seen when samples of these molecules are placed between two oppositely charged metal plates. Here is what happens:

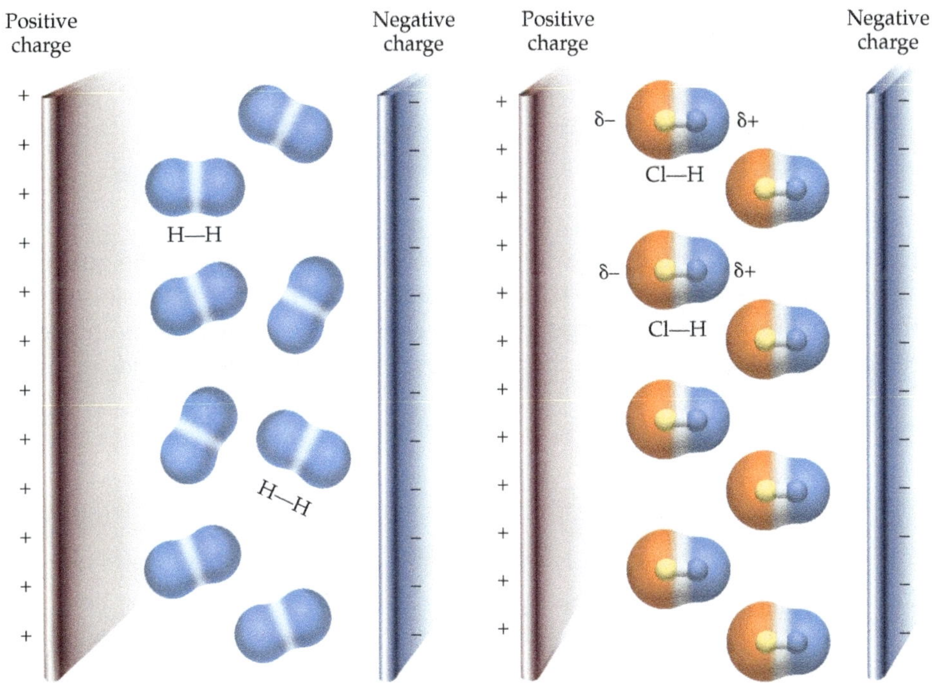

The H_2 molecules are not noticeably affected by the charged plates, and they orient themselves randomly between the plates. The situation is different for the HCl molecules. Here, the partially negative chlorines are attracted to the positive plate and the partially positive hydrogens are attracted to the negative plate, and the molecules all assume the same orientation. Based on this behavior, HCl is called a **polar molecule** and H_2 is called a **nonpolar molecule**. (This is an *operational definition*, which is a definition based on how something behaves in a particular situation.)

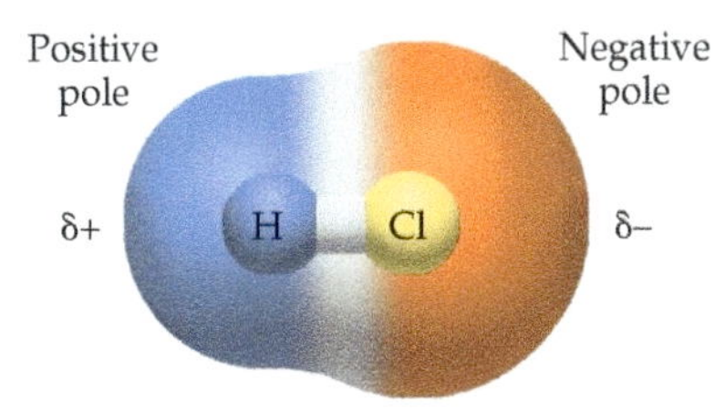

H—Cl
A polar molecule has two charged poles

Diatomic molecules like N_2 and O_2, the two most abundant molecular substances in our atmosphere, are nonpolar because the bonding electrons are shared equally between the atoms in each molecule. Diatomic molecules like HF and HCl are polar owing to the polar covalent nature of their bonds. Polar molecules are not all equally polar, however. To understand why this is so, look at the electronegativities of the atoms in HF and HCl:

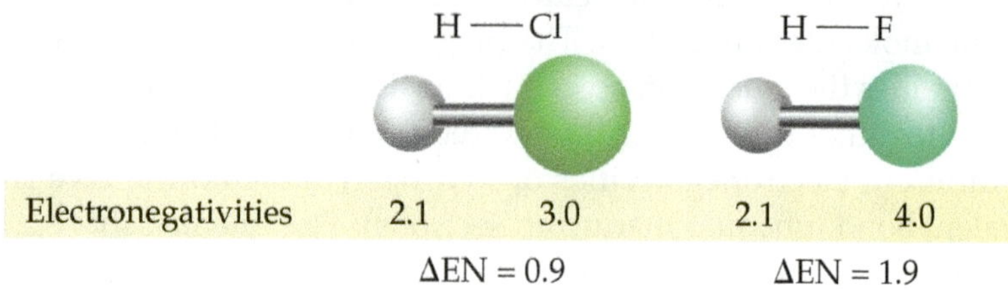

Both these molecules are polar, but HF is more polar than HCl. In both molecules, the bonding electrons are shared unequally, but fluorine is more electronegative than chlorine (recall that fluorine is the most electronegative element there is). The greater value of ΔEN for HF tells us that fluorine really hogs those shared electrons. Thus, the bonding electrons are shared more unequally in HF than in HCl, meaning that HF molecules have larger partial charges than do HCl molecules. HF molecules therefore are more polar than HCl molecules.

One way we might show this difference would be to use a larger type for the partial charges in HF:

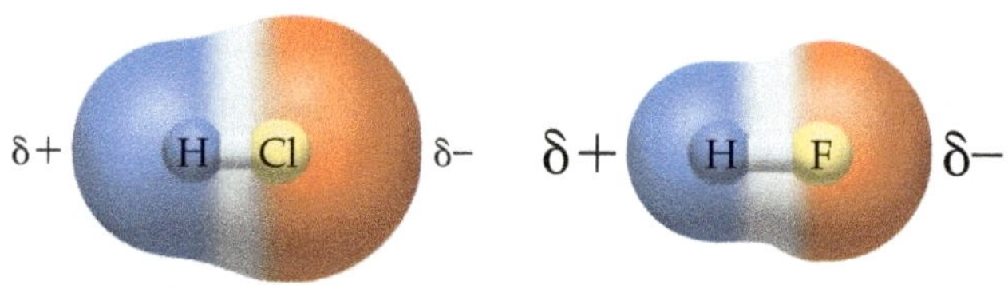

However, we don't do it that way. The convention for indicating the polarity of a bond (and even, as we shall see, of the molecule as a whole) is to use an arrow that points from the $\delta+$ charge to the $\delta-$ charge. The arrowhead points to the negative end of the bond, and the arrow is crossed at the positive end:

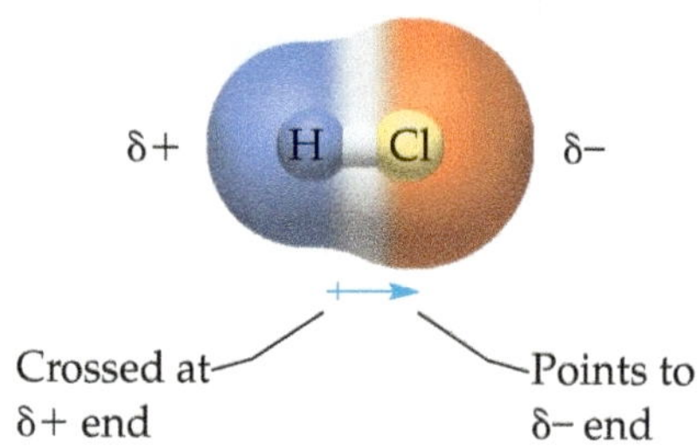

This arrow represents a property of the bond called the **dipole moment**, which is simply a measure of how polar the bond is. The longer the arrow (the larger the dipole moment), the more polar the bond (the larger the $\delta+$ and $\delta-$ charges). Physicists refer to these arrows as *vectors*. (A **vector**, usually represented by an arrow, can best be thought of as a number that has a direction associated with it.)

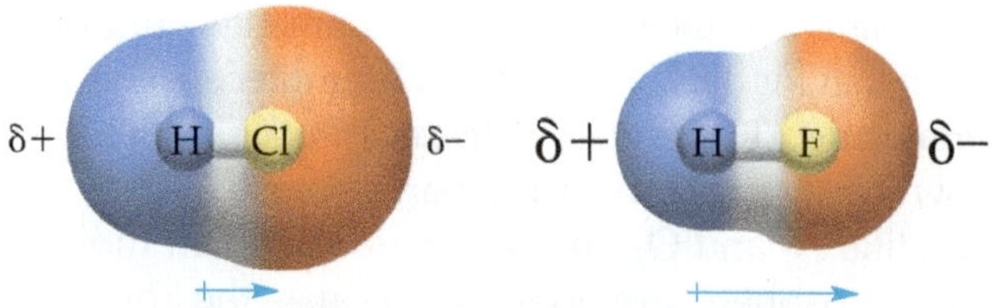

A shorter dipole moment vector means a less polar bond.

A longer dipole moment vector means a more polar bond.

WORKPATCH 6.5 When does a bond have a dipole moment of zero? Here is a hint for the WorkPatch. The dipole moment will be zero only when the two atoms have no partial charges. When does this happen? What does a dipole moment of zero mean for the bond? Check your answer against ours at the end of the chapter. ●

Notice that the WorkPatch asked you about the dipole moment (polarity) associated with a bond. There can also be a dipole moment associated with an *entire molecule*. It is called a **molecular dipole moment**, and it points from the $\delta+$ end of the entire molecule to the $\delta-$ end. If a molecule is polar (if it lines up between charged plates), then it has a molecular dipole moment. For simple molecules such as H_2, HCl, and HF, considering the polarity of the bond is the same as considering the polarity of the entire molecule because there is only one bond in the molecule. But what about molecules with more than one bond? In these cases, determining whether the entire molecule is polar and therefore has a molecular dipole moment becomes a bit more involved, and molecular shape plays a role!

Let's consider two molecules that have more than one polar bond—water, H_2O, and dichloroacetylene, C_2Cl_2:

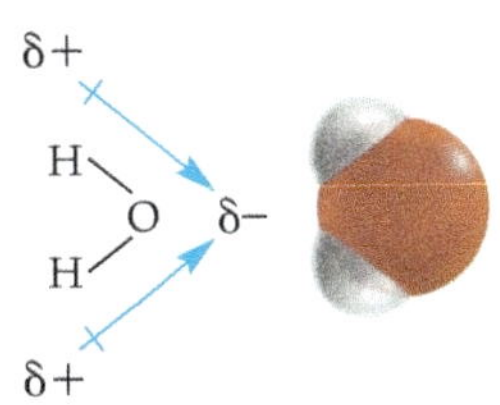

Water

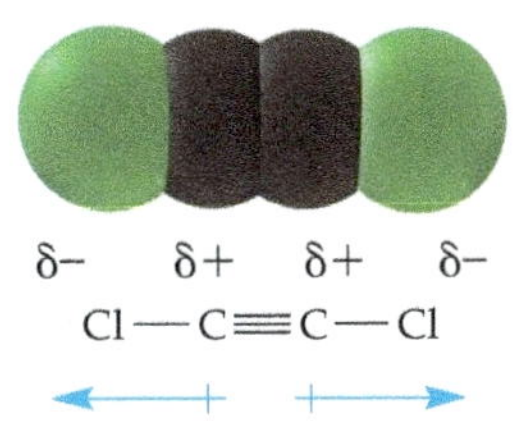

Dichloroacetylene

Notice a few things about these examples. As always, the individual bond dipole moment vectors point from the $\delta+$ to the $\delta-$ end of the bond, and the end of the arrow over the $\delta+$ atom is crossed, like a plus sign. There is no bond dipole moment vector over the carbon–carbon triple bond in dichloroacetylene because this bond is nonpolar (electrons in this bond are shared equally). Both molecules contain two polar covalent bonds: the O—H bonds in water and the C—Cl bonds in dichloroacetylene. So, are both molecules polar?

The only way to truly know is to put both molecules between oppositely charged plates and see what happens. Here are the results of just such an experiment:

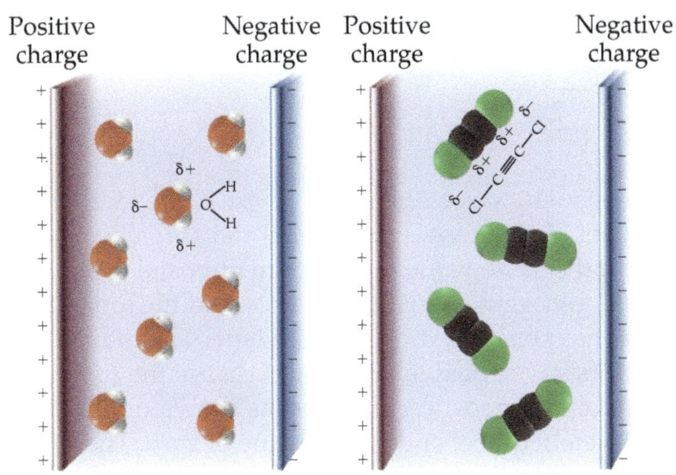

The molecules all assume the same orientation between the charged plates, meaning H_2O must be polar.

The molecules are oriented randomly, meaning C_2Cl_2 must be nonpolar.

Notice how the water molecules all assume the same orientation. The dichloroacetylene molecules, on the other hand, remain randomly oriented. By our operational definition of polar molecules, water is considered a polar molecule but dichloroacetylene is nonpolar. Nonpolar? Can dichloroacetylene be nonpolar even though it contains polar covalent bonds with $\delta+$ and $\delta-$ ends? The answer is yes. A molecule can be nonpolar even though it contains polar bonds, and you must be able to predict when this will occur. Making this prediction requires that you add up all the individual bond dipole moments.

The trick in doing this is to recall that each individual bond dipole moment is a vector quantity, which is a quantity that has both a numeric value and a direction associated with it (the direction of the vector). When you add the numbers 2 + 2, you always get 4. But when you add two vectors, each of length 2, the answer you get depends on which way the two vectors are pointing relative to each other. In vector arithmetic, 2 + 2 can equal anything from 0 to 4:

Vectors pointing in the same direction add:

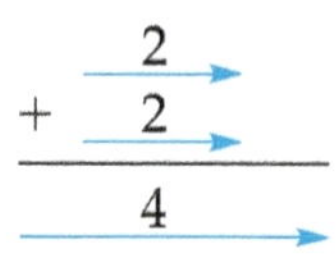

Vectors pointing in opposite directions cancel:

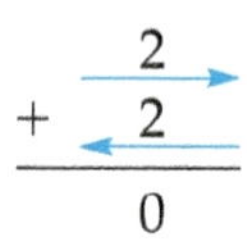

Vectors at an angle to each other add to give a *resultant vector* having a length that can have any value between zero and the sum of the lengths of the two original vectors. This resultant-vector length depends on the angle between the two original vectors:

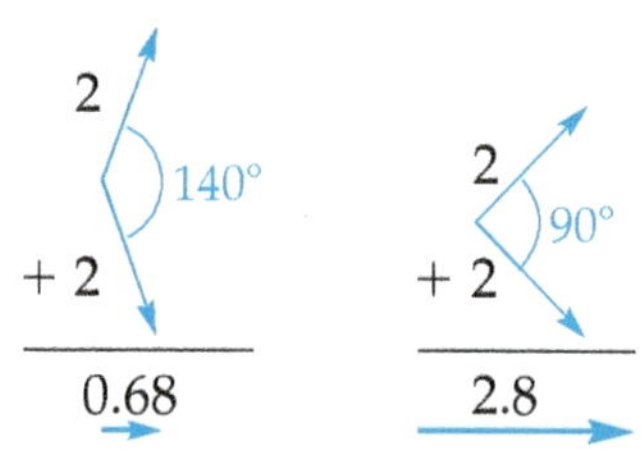

You need not worry about how to calculate the length of the resultant vectors. Just note that the length in both cases is between 0 and 4.

If the vectors have the same value but point in opposite directions, they cancel each other and give zero as a result, as shown at the top of this page. This is exactly what happened in dichloroacetylene. The individual bond dipole moment vectors have equal lengths, but because the molecule has a linear shape, they point in opposite directions, as shown on the next page. They therefore completely cancel and give a value of zero. Therefore, even though two of the bonds in the dichloroacetylene molecule are polar, we predict that the molecule as a whole will be nonpolar (which it is).

A case where individual bond dipole moment vectors cancel:

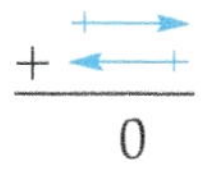

Nonpolar
The dipole moment vectors add up to zero (they cancel each other).

In the case of the water molecule, because of its bent shape, the two O—H vectors do not cancel each other. Instead, they add to give a resultant vector that points in the direction shown below—between the two original vectors. This vector represents the molecular dipole moment for the entire molecule. Because this vector is not zero, we predict that the molecule will be polar (which it is), and we call water a polar molecule.

A case where individual bond dipole moment vectors do not cancel:

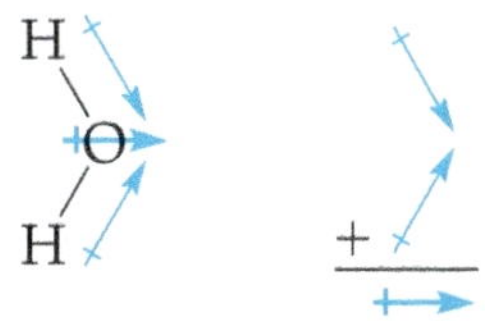

The dipole moment vectors for the O—H bonds add to give the dipole moment vector for the entire molecule.

The molecular dipole moment vector points from the $\delta+$ side of the entire molecule to the $\delta-$ side of the entire molecule as shown at right.

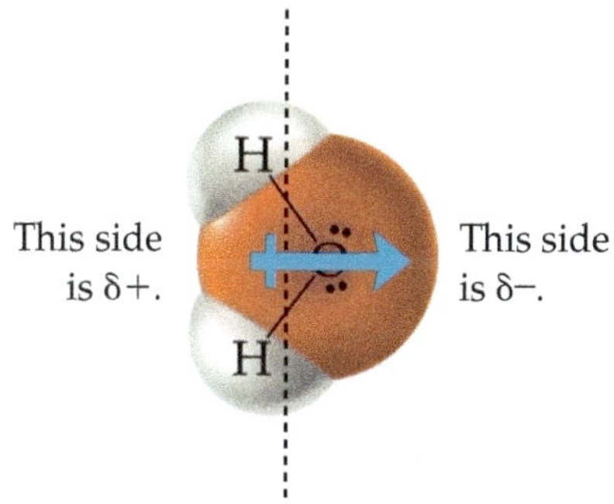

So molecular shape is important once again. For molecules with more than two atoms, just having polar bonds is not enough to make the entire molecule polar. For these molecules, molecular polarity also depends on molecular shape. Try the following WorkPatch to see if you are catching on to these concepts.

WORKPATCH 6.6 Which molecule is more polar? [Be careful!]

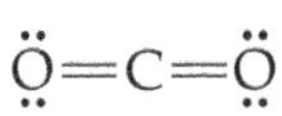

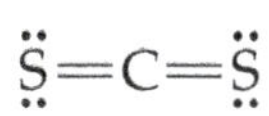

Which molecule is more polar?

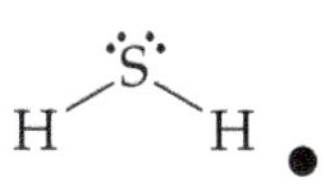

The top two molecules in the WorkPatch are a bit tricky. Did you catch it? The bottom two molecules should have you thinking about the length of the vectors drawn over the bonds.

Adding up all the individual bond dipole moment vectors can be tricky for molecules with more than three atoms. However, there is a quick way to determine whether or not they cancel by doing a "tug-of-war test." To perform this test, imagine you are standing at the central atom. Let each bond dipole moment vector be a rope that tugs on you in the direction of the vector. The longer a vector is, the harder that rope pulls on you. Now consider two molecules, SO_3 and $COCl_2$. Both have the same shape (trigonal planar), and both have three polar covalent bonds:

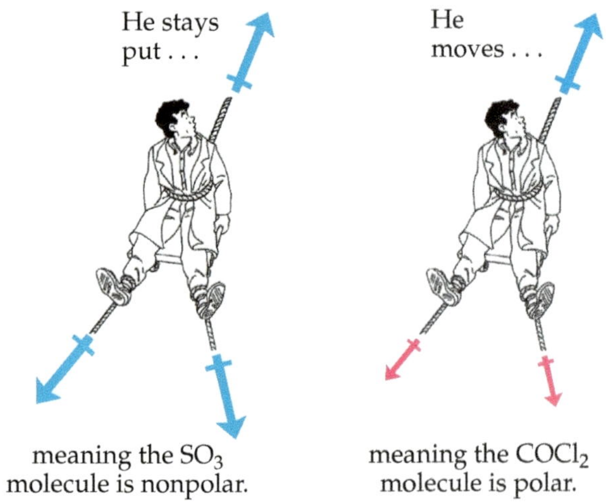

Don't worry about the bonds being single or double. Focus only on the vectors (the "ropes"). As all the ropes pull on you simultaneously, ask the question "Will I move?" If the answer is no, the bond dipoles cancel and the entire molecule is nonpolar. If the answer is yes, the bond dipoles do not cancel and the molecule is polar. For SO_3, the three equal-strength tugs to the corners of an equilateral triangle leave you sore but unmoved. That means the individual bond dipole moments cancel. Therefore, SO_3 is a nonpolar molecule even though all of its bonds are polar! However, the tugs for $COCl_2$ are not of equal strength. The stronger tug along the C—O bond is not canceled by the two weaker tugs along the C—Cl bonds. You move because of this difference. Because the individual bond dipole moments do not cancel, $COCl_2$ is a polar molecule.

Because the length of a bond dipole moment vector determines the strength of the tug for our tug-of-war test, it is critical that you don't forget what it is that determines the vector length.

WORKPATCH 6.7 Why is the upward tug in $COCl_2$ slightly stronger than either of the downward tugs? ●

This WorkPatch is extremely important. Don't go on until you understand the answer.

The above examples teach you that three equal-length dipole moment vectors in a trigonal planar molecule cancel to leave the entire molecule nonpolar. What about tetrahedral molecules? To answer this question, let's look at carbon tetrachloride, CCl_4, a tetrahedral molecule:

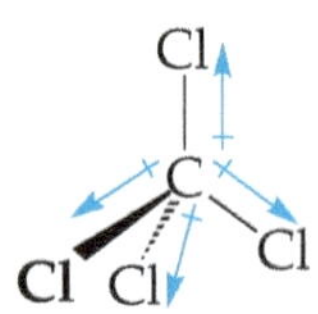

Carbon tetrachloride

The electronegativity difference between C and Cl (remember that Cl is one of our four electron hogs from Chapter 5—see page 188) ensures that all four bonds are polar covalent. In addition, all four bond dipole moment vectors have equal length. Do the dipole moments cancel? To find out, let's look at our molecule in a different orientation:

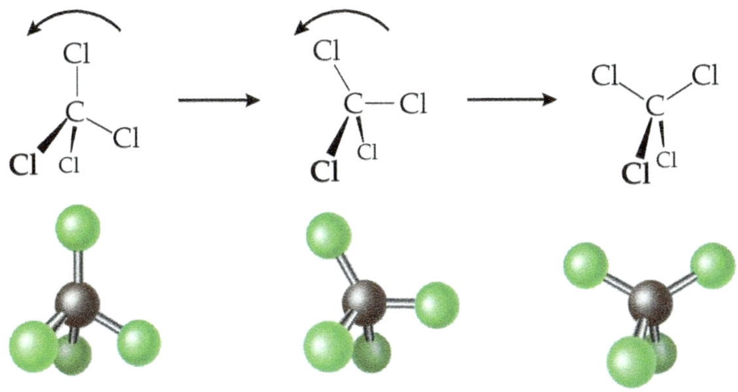

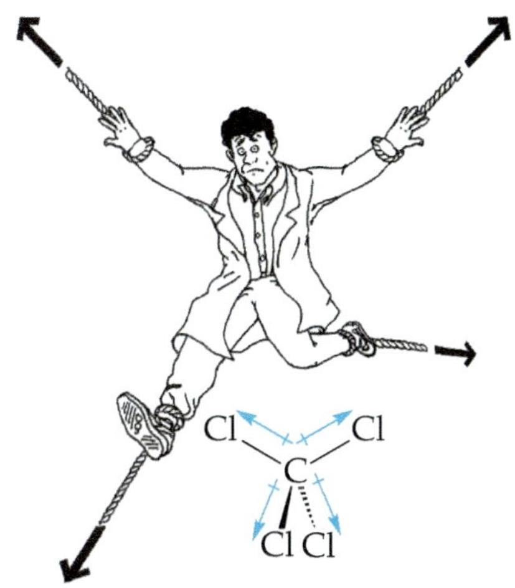

Drawn this way, two tugs are directed upward (to the left and right) and two are directed downward (toward the front and back). Our assistant at the right kindly demonstrates this for you. Once again, such combinations of tugs cancel (our assistant does not move). Carbon tetrachloride therefore is not polar even though its individual bonds are. Remove one of those tugs, however, and the story changes, as you'll see in Practice Problem 6.10.

The tug-of-war test can be very useful in determining the polarity of larger molecules. However, the test is useless unless you can first determine the shape of the molecule and estimate the relative strengths of the different tugs. Try the following practice problems, keeping in mind that for purely covalent bonds, the dipole moment is zero. In addition, you may approximate and treat C—H bonds as nonpolar because C and H have very similar electronegativities.

Practice Problems

6.10 Is phosphorus trichloride, PCl_3, a polar molecule? If it is, draw the dipole moment vector for the entire molecule and show where the δ+ and δ− regions of the molecule are.

Answer: Yes, it is a polar molecule. PCl_3 should have 26 electrons.

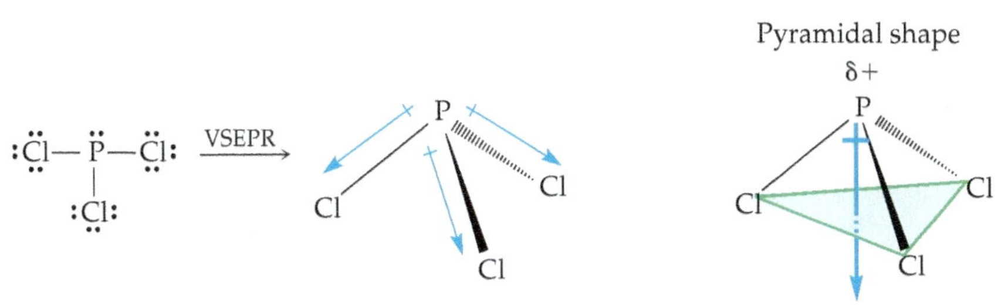

The P atom feels three equal downward tugs, and so by the tug-of-war test the molecule is polar.

The three individual bond dipole moment vectors add up to give the dipole moment vector for the entire molecule.

6.11 Is chloroform, $CHCl_3$, a polar molecule? If it is, draw the dipole moment vector for the entire molecule and show where the $\delta+$ and $\delta-$ regions of the molecule are.

6.12 Show how the molecules of Problems 6.10 and 6.11 would arrange themselves if they were placed between two plates, one carrying a positive electrical charge and the other carrying a negative electrical charge.

When we started this section, we asked, "What makes molecules stick to one another?" or, to put it another way, "Why can't you rip an ice cube in half?" As we promised, molecular shape plays a key role in answering this question. This is because the way molecules are attracted to one another depends on whether the molecules are polar, and as we have just seen, molecular polarity depends on molecular shape. We are now prepared to take a closer look at intermolecular forces.

6.4 Intermolecular Forces—Dipolar Interactions

How do water molecules stick to one another to form liquid water or solid ice? We now know enough about the water molecule to begin answering this question.

First of all, we know that the water molecule is V-shaped (bent), and because of this shape and its polar covalent bonds, the molecule itself is polar. The dipole moment vector for the molecule shows a partially negative area located at the oxygen atom and a partially positive portion located at the hydrogen atoms. What happens if we bring a group of water molecules together? Because opposite charges attract, the $\delta+$ portion of one molecule is attracted to the $\delta-$ portion of another, causing them to come close to each other:

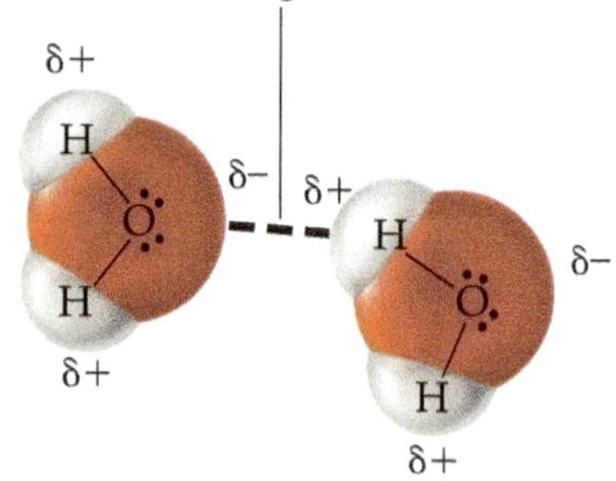

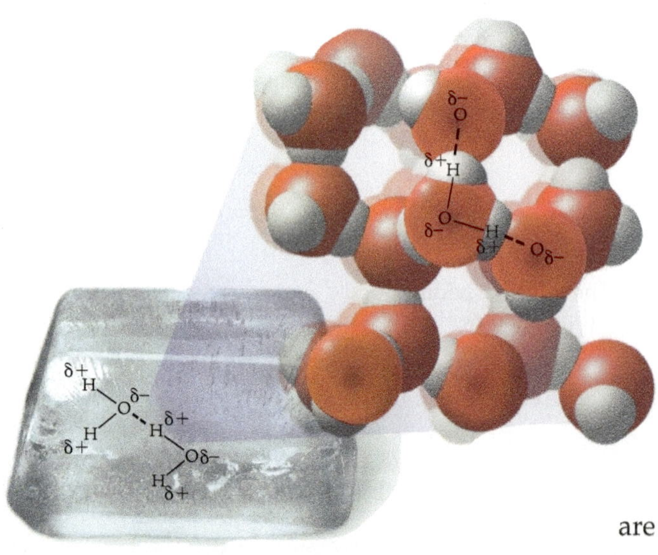

In a piece of ice, billions of water molecules are attracted to one another by these *intermolecular forces*. These forces are the origin of the attractive force between water molecules—the reason these molecules stick together so strongly in ice.

Intermolecular forces between polar molecules are called either **dipolar forces** or **dipole–dipole forces**, indicating an attraction between dipole moments of different molecules. On a per-atom basis, dipole–dipole forces are among the strongest that exist between molecules.

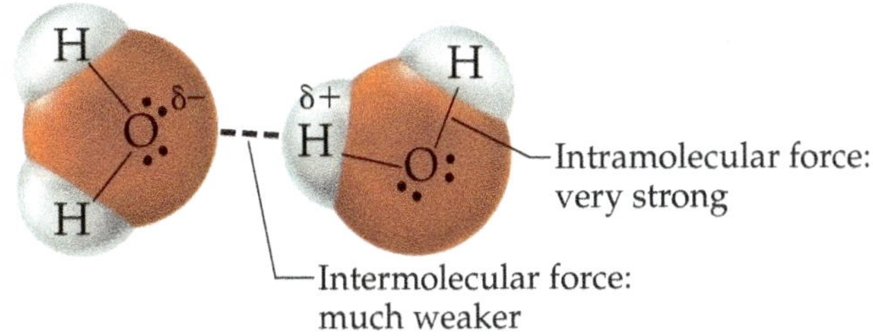

Even so, these forces are nowhere near as strong as the ionic and covalent bonds, sometimes referred to as *intramolecular forces* (*intra-* meaning "within"), that we discussed earlier.

There are other types of intermolecular forces besides dipole–dipole attractions. For example, the hydrogen molecule, H_2, is nonpolar. It has no partial charges, and yet NASA fills fuel tanks on the space shuttle with *liquid* hydrogen. The H_2 molecules must stick to each other to form a liquid, but why? We shall take a more in-depth look at intermolecular forces in Chapter 7.

For now, you shouldn't lose sight of the importance of molecular shape in all of this. If water were a linear molecule instead of bent, it would have no dipole moment (the individual bond dipole moments would cancel). Water molecules would no longer have strong attractive forces between them, and all H_2O on this planet would exist as vapor. Life, at least as we now know it, could not exist if this were so.

You should get comfortable with being able to predict the shape of a molecule from its molecular formula, such as H_2O or PCl_3. You can't answer questions about molecular polarity and intermolecular forces unless you know the shape of the molecule. And remember, you can't determine a molecule's shape without a dot diagram.

The process for going from molecular formula to molecular polarity

Formula

1. Determine the total number of valence electrons (use roman-numeral group numbers to help you).
2. Connect the atoms using single bonds.
3. Put in remaining dots as lone pairs to complete octets.
4. If all octets are not complete, create multiple bonds.

Dot diagram

Determine the molecule's shape by applying the concepts of VSEPR to the interior atom(s).

Shape

1. Draw vectors representing all bond dipole moments (consider the electronegativities).
2. Determine the molecular dipole moment, if any.

Molecular polarity
Intermolecular forces

The dot diagram is the key starting point in this process. Start with it, stick to the outline above, and keep practicing.

+1 One More Thing

6.5 VSEPR Theory for Molecules Possessing Expanded Octet Atoms

The octet rule of chemical bonding was, and still is, an important cornerstone around which our understanding of molecule formation was created. First proposed in the 1920s by G. N. Lewis, it helped to make sense of the patterns that were observed in hundreds of known molecules and allowed for predictions of the formulas and structures of many molecules yet to be created. Yet, as important as the octet rule is to our understanding of compound stability, we now know that it is an incomplete description of what is happening at the molecular level. In Chapter 5, we discovered that there are many stable molecules that do not obey the octet rule. Some of these molecules are electron deficient and have fewer than 8 electrons around one or more of the atoms in the molecule. Many more molecules are known to have more than 8, most typically 10 or even 12, electrons around one or more atoms in the molecule. These molecules are often referred to as hypervalent and are said to possess an "expanded octet." In some cases, the theoretical justification for the formation and stability of these molecules is still the subject of vigorous debate among chemists. But this much we do know: These molecules really *do* exist. We even know what they look like. And after a small modification to the VSEPR theory we encountered earlier in this chapter, so will you.

In applying the VSEPR theory to "well-behaved" molecules, we saw that there were only a few geometric arrangements possible around any atom in a molecule. These arrangements depended on the total number of electron pairs that had to be accommodated around the atom and the desire to keep these as far apart as possible.

We came up with these possible atomic arrangements for the well-behaved, octet-obeying atoms:

	Three electron pairs		Four electron pairs		
Molecular model	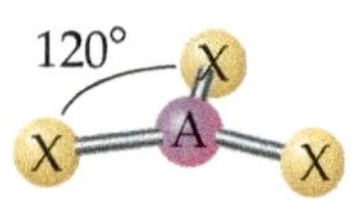 No lone pairs	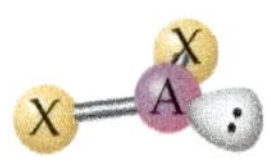 One lone pair	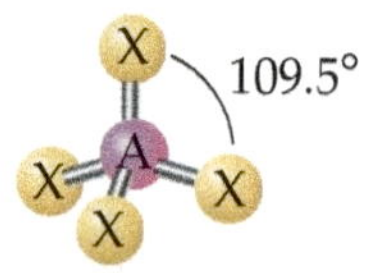 No lone pairs	 One lone pair	 Two lone pairs
Electron-pair geometry	Trigonal planar	Trigonal planar	Tetrahedral	Tetrahedral	Tetrahedral
Molecular geometry	Trigonal planar	Angular	Tetrahedral	Trigonal pyramidal	Angular

Now let's examine some compounds that possess atoms with expanded octets. Consider, for example, the molecules PCl_5 and SF_6. If we use the same considerations we used earlier in determining the possible electron arrangements around atoms that possess an expanded octet with 10 or 12 electrons (five or six electron groups), we get the following arrangements around the central atom:

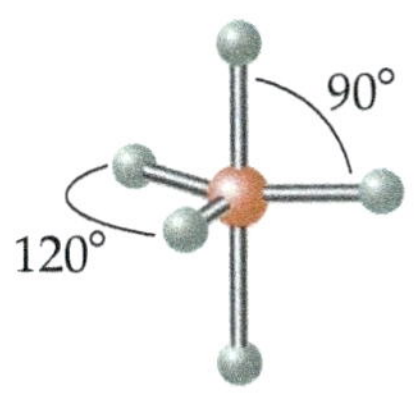

No lone pairs

PCl_5
A trigonal bipyramid

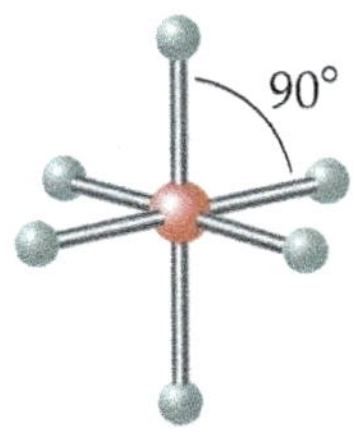

No lone pairs

SF_6
An octahedron

These are the shapes taken on by the five and six electron groups around the central P or S atom in each molecule. So we have to add these two additional basic shapes to the tetrahedron and planar triangle arrangements from the octet-obeying VSEPR rules.

And, just as with their better-behaved cousins, these expanded octet molecules can have one or more of the surrounding atoms replaced with lone pairs and, as before, this changes the description we give of the molecular shape but not of the overall electron arrangement. For five electron pairs, we can replace one, two, or three attached atoms with lone pairs. We still have a trigonal bipyramid overall, but the description of the molecule's shape is different.

	Five electron pairs			
Molecular model	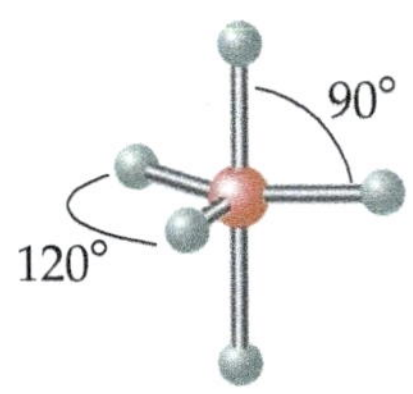 No lone pairs	 One lone pair	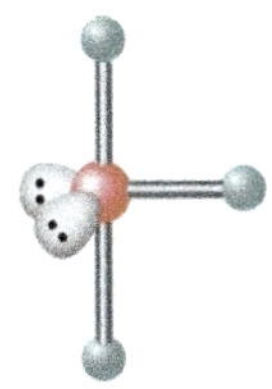 Two lone pairs	 Three lone pairs
Example	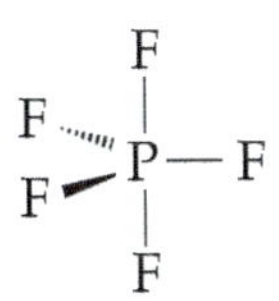	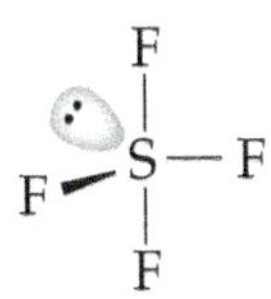	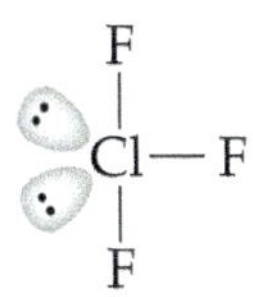	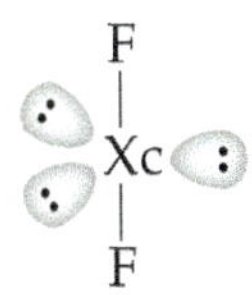
Electron-pair geometry	Trigonal bipyramidal	Trigonal bipyramidal	Trigonal bipyramidal	Trigonal bipyramidal
Molecular geometry	Trigonal bipyramidal	Seesaw	T-shaped	Linear

You might have observed from the structures above that, in the case of five electron groups, lone pairs have a preferred position in the final structure. They tend to go into the triangular part of the trigonal bipyramid since there is more room, and therefore, there is less repulsion between them when they occupy these positions.

With six electron groups around an atom, we can replace anywhere from 1 to 4 attached atoms with lone pairs. This gives rise to the following shapes and descriptions:

	Six electron pairs		
Molecular model	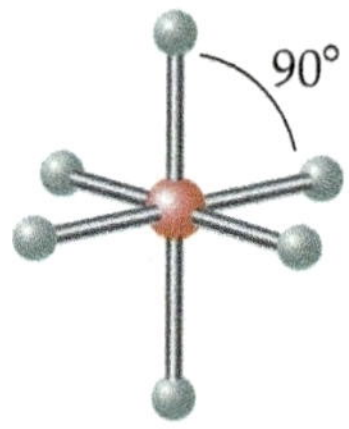 No lone pairs	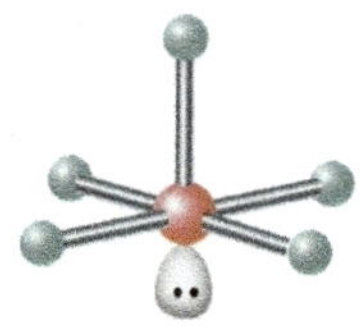 One lone pair	 Two lone pairs
Example	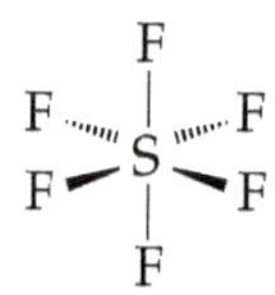	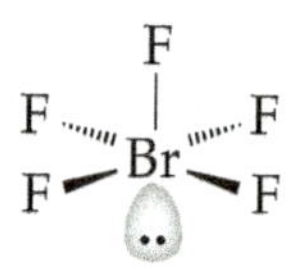	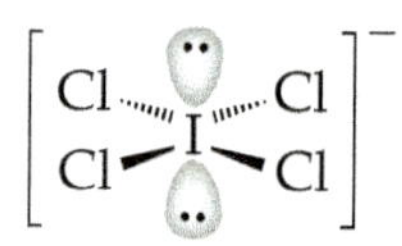
Electron-pair geometry	Octahedral	Octahedral	Octahedral
Molecular geometry	Octahedral	Square pyramidal	Square planar

So, as you can see, with just a small tweaking of the octet rule, one that allows five or six electron pairs around a central atom, we have greatly expanded the list of possible molecular geometries and with it the variety and versatility of the science of chemistry itself. And we did all of this without even mentioning the examples of super-expanded octets, with up to 20 electrons (10 pairs) around an atom.

Have You Learned This?

Valence shell electron pair repulsion (VSEPR) theory (p. 219)
Electron group (p. 220)
Tetrahedron (p. 221)
Steric number (p. 222)
Pyramidal (p. 224)
Intermolecular attractive forces (p. 228)
Polar molecule (p. 229)
Nonpolar molecule (p. 229)
Dipole moment (p. 230)
Vector (p. 230)
Molecular dipole moment (p. 231)
Dipolar forces (p. 236)
Dipole–dipole forces (p. 236)

Learning Outcomes

6.1	The Importance of Molecular Shape	Explain how molecular shape is important with regard to disease and drugs that might cure it. [13, 14]
6.2	Valence Shell Electron Pair Repulsion (VSEPR) Theory	Explain the principles behind VSEPR theory. [15, 16, 17, 20, 21, 22, 23, 26] Draw and name a molecule's shape starting from a dot diagram. [18, 19] Draw the molecular shape for molecules possessing multiple bonds. [24, 25, 27, 28]
6.3	Polarity of Molecules, or When Does 2 + 2 Not Equal 4?	Explain polarity in terms of individual bonds and an entire molecule. [29, 30, 31, 33, 34, 35] Predict the relative polarity of a molecule. [32, 37, 38, 39, 40] Explain how molecular polarity leads to intermolecular attraction. [42, 43, 44, 45, 46] Deduce and draw a dipole moment for an entire molecule. [41, 48, 49, 52, 54, 55]
6.4	Intermolecular Forces—Dipolar Interactions	Describe how intermolecular attractions arise between dipolar molecules. [58, 59, 87, 98]
6.5	VSEPR Theory for Molecules Possessing Expanded Octet Atoms	Draw dot diagrams and describe the shapes of molecules possessing central atoms with an expanded octet of electrons. [107–111]

Skills to Know

VSEPR

1. For the purposes of VSEPR, a single bond, a multiple bond, and a lone pair each count as a single electron group.

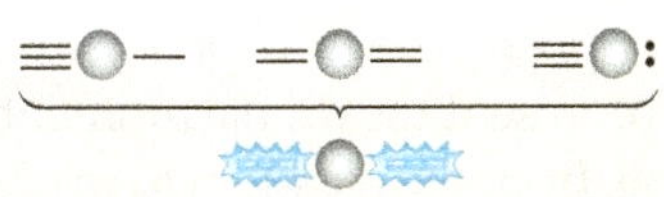

Three ways in which an atom can have two VSEPR electron groups

2. The VSEPR method of determining molecular shape is based on the geometrical arrangement of electron groups around the central atom(s) of a molecule. This geometry depends on the number of electron groups (the steric number, SN), as shown at right.

   Know your bond angles. As shown at right, the linear bond angle is 180°, the trigonal planar bond angle is 120°, and the tetrahedral bond angle is 109.5°.

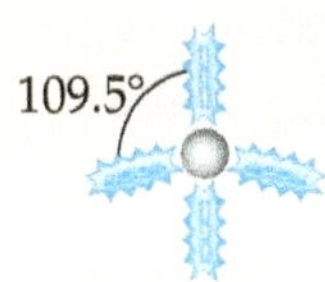

Four electron groups (SN = 4) Tetrahedral electron-group geometry, bond angles all 109.5°

Three electron groups (SN = 3) Trigonal planar electron-group geometry, bond angles all 120°

Two electron groups (SN = 2) Linear electron-group geometry, bond angle 180°

Using VSEPR to find the shape of a molecule

Example: What is the molecular shape of PCl_3?

1. Draw a dot diagram for the molecule.

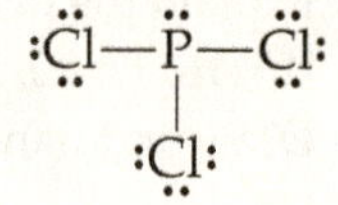

2. Count the number of electron groups around the central atom. Include lone pairs, and count each multiple bond as a single electron group. This is the steric number (SN) for the central atom.

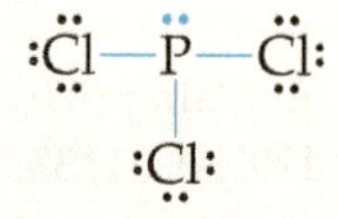

Four electron groups around P atom **(SN = 4)**

3. Using the steric number (SN) from step 2, use Table 6.2 to determine how the electron groups are arranged around the central atom. Include lone pairs.

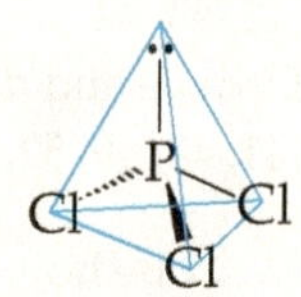

Tetrahedral electron-group geometry

4. Pretend the lone pairs are invisible and describe the resulting shape of the molecule.

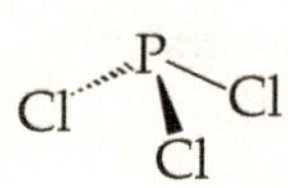

Pyramidal molecular shape

Bond polarity and molecular polarity

1. Every polar covalent bond has a dipole moment. We represent the dipole moment by an arrow that points from the $\delta+$ atom to the $\delta-$ atom of the bond. The longer this arrow, the more polar the bond. The strength of the partial charges, and thus the polarity of the bond, are determined by the size of ΔEN, the difference between the electronegativities of the bonded atoms.

Dipole moment:

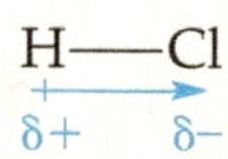

2. A molecule that contains polar bonds may have an overall polarity. To decide whether the molecule is polar, do the tug-of-war test. If the individual bond polarities do not cancel each other out, the molecule is polar overall.

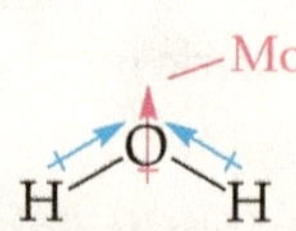

Bond dipoles do not cancel

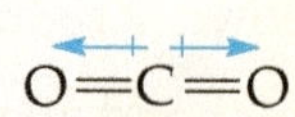

Bond dipoles cancel

The process for going from molecular formula to polarity

Example: Is PCl_3 polar?

Formula

1. Determine the total number of valence electrons (use roman-numeral group numbers to help you).
2. Connect the atoms using single bonds.
3. Add remaining dots as lone pairs to satisfy octets.
4. If all octets are not satisfied, send adjacent lone pairs to create multiple bonds.

From molecular formula to dot diagram:

PCl_3

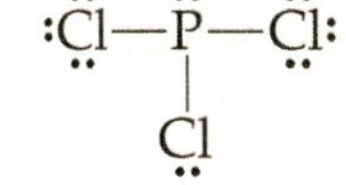

Dot diagram

Determine the molecule's shape by applying the concepts of VSEPR to the interior atom(s).

From dot diagram to molecular shape:

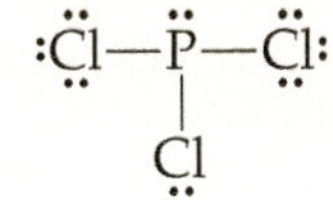

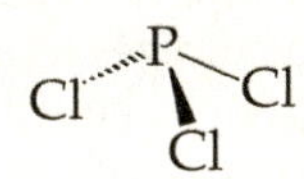

Pyramidal shape

Shape

1. Draw vectors representing all bond dipole moments (consider the electronegativities).
2. Determine the molecular dipole moment, if any (do tug-of-war test).

Determine molecular dipole moment (if any):

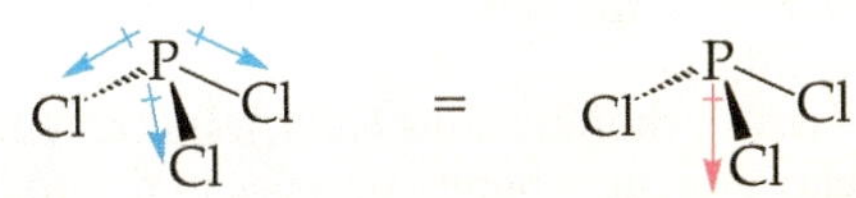

Three bond dipole moments that do not cancel = Molecular dipole moment

Molecular polarity
Intermolecular forces

Concept Questions

1. To indicate the three-dimensional shape of a molecule
   (a) bonds that come out of a page are drawn with a solid wedge.
   (b) bonds that go back behind the page are drawn with a broken wedge.
   (c) bonds that are in the plane of the page are drawn with a solid line.
   (d) all of the above.

2. VSEPR theory
   (a) stands for valence shell electron proton repulsion theory.
   (b) states that bonds form to minimize a molecule's energy.
   (c) states that the shape of a molecule derives from repulsion of valence electrons for each other.
   (d) states that the shape of a molecule is that which occupies the smallest volume.

3. The steric number of an atom
   (a) is equal to the number of electrons on that atom.
   (b) is equal to the number of electron pairs on that atom.
   (c) is equal to the number of lone pairs on the atom.
   (d) is equal to the number of lone pairs + attached atoms.

4. If a molecule has a steric number of 4, then its shape
   (a) must be trigonal.
   (b) must be tetrahedral.
   (c) must be pyramidal.
   (d) none of the above.

5. To determine the shape of a molecule, the first thing you should do is
   (a) determine the steric number of the central atom.
   (b) count the number of lone pairs on the central atom.
   (c) locate the central atom on the periodic table.
   (d) draw a dot diagram of the molecule.

6. A molecule's shape is one that
   (a) allows it to be the most reactive.
   (b) minimizes repulsions between electron groups about atoms in the molecule.
   (c) maximizes repulsions between electron groups about atoms in the molecule.
   (d) comes as close as possible to a tetrahedral geometry.

7. The VSEPR ideal shapes
   (a) are idealized and never actually occur.
   (b) are those that minimize repulsions between electron groups.
   (c) are linear, trigonal planar, and pyramidal.
   (d) all of the above.

8. A molecule is polar if
   (a) one or more bonds has a $\Delta EN > 0.4$ and all of the bond dipoles perfectly cancel.
   (b) all of the bonds are polar.
   (c) it possesses highly electronegative atoms.
   (d) one or more bonds has a $\Delta EN > 0.4$ and the bond dipoles do not cancel.

9. To be polar means that a molecule
   (a) has a partially positive and a partially negative region separated from one another.
   (b) will assume a unique orientation between oppositely charged electrical plates.
   (c) has a dipole moment, which is a vector quantity, associated with it.
   (d) all of the above.

10. A dipolar interaction is
   (a) a repulsive force between dipolar molecules.
   (b) an attractive force between nonpolar molecules.
   (c) stronger than a chemical bond.
   (d) an attractive force between polar molecules.

Answers: 1d; 2c; 3d; 4d; 5d; 6b; 7b; 8d; 9d; 10d

Shapes of Molecules and VSEPR Theory

6.13 Cisplatin is a molecule with a platinum (Pt) atom in the middle that is bound to two chlorine atoms and two ammonia atoms (NH_3). It is an effective anticancer drug (chemo) because in the body the two chlorines fall off and the Pt atom binds to the two strands of the DNA double helix of malignant tumor cells. Therefore, the DNA cannot split apart and replicate, which keeps the tumor cells from growing.

Notice that the bonding geometry about the Pt atom is flat (a square), which projects the ammonia atoms *away* from the DNA strands and allows the Pt to get close enough to bind the DNA strands. What do you think would happen if the geometry about the Pt atom were not flat such that the ammonia atoms were more pointed *toward* the DNA strands?

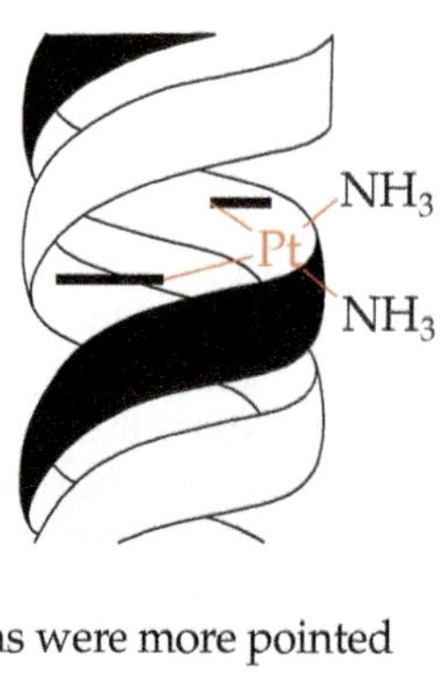

6.14 How do you think a bacterium species might evolve to develop resistance to a drug like sulfanilamide that is designed to kill it?

6.15 Why is the theory that governs the shape of molecules called VSEPR and not just EPR?

6.16 Consider the methane molecule.
   (a) Draw the methane molecule showing its tetrahedral shape, using lines, solid wedges, and dashed wedges to show three-dimensionality.
   (b) Draw the methane molecule inscribed inside a tetrahedron such that the H atoms touch the vertices of the tetrahedron.
   (c) Why is it better for methane to have 109.5° bond angles rather than 90° bond angles?

6.17 Draw and name the three-dimensional shape for $SOCl_2$. Indicate the values of the bond angles assuming they are ideal.

6.18 Shown below are dot diagrams for some simple molecules:

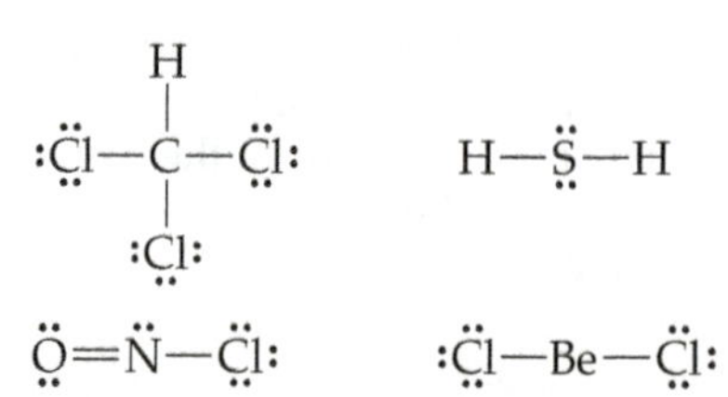

(a) Draw the three-dimensional shape for each molecule. Use lines, solid wedges, and dashed wedges as necessary. Indicate the numeric value of all bond angles.
(b) For each molecule, name both the electron-group geometry around the central atom and the molecular shape.

6.19 Shown below are dot diagrams for some simple molecules and polyatomic ions:

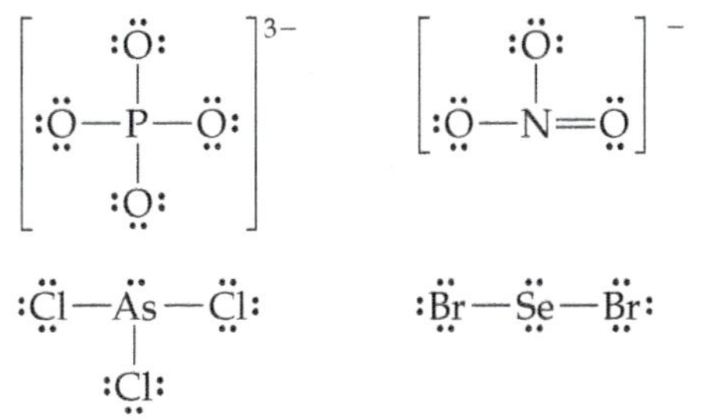

(a) Draw the three-dimensional shape for each molecule. Use lines, solid wedges, and dashed wedges as necessary. Indicate the numeric value of all bond angles.
(b) For each species, name both the electron-group geometry around the central atom and the molecular shape.

6.20 Why is it reasonable to treat a multiple bond as a single electron group in VSEPR shape determination?

6.21 Two molecules may both be correctly described as bent, even though one has a bond angle of approximately 118° and the other has a bond angle of approximately 105°. How is this possible?

6.22 Consider the CH_3^- anion.
(a) Draw and name its three-dimensional shape and indicate the predicted values of its bond angles.
(b) Explain how the structure of this anion would compare with that of NH_3 and why both would have bond angles that are less than ideal.

6.23 Ammonia has four pairs of electrons around the central nitrogen atom, and yet we don't call it a tetrahedral molecule. Why not? What is the shape of this molecule?

6.24 Consider the following molecule. The diagram shows how the atoms are connected, but it is not a complete dot diagram.

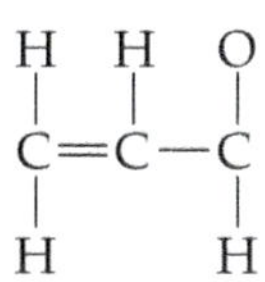

(a) Complete the dot diagram. (*Hint*: Work on the C atoms one at a time. If the C you are working on is bonded to an atom that itself is bonded to other atoms, treat all those atoms as a single unit. Each C here, for instance, has three bonding groups, no lone pairs.)
(b) Redraw the molecule showing its three-dimensional shape. Use lines, solid wedges, and dashed wedges if necessary.
(c) Indicate the numeric value of all bond angles.

6.25 Consider the following molecule. The diagram shows how the atoms are connected, but it is not a complete dot diagram.

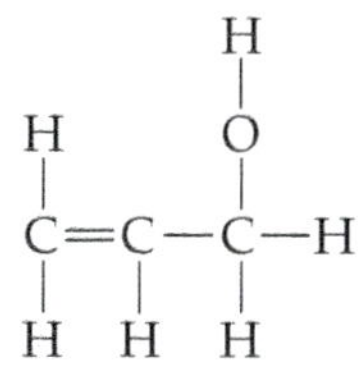

(a) Complete the dot diagram. (*Hint*: Work on the three C atoms and the O atom one at a time. If the atom you are working on is bonded to an atom that is itself bonded to other atoms, treat all those atoms as a single unit. The O, for instance, has two bonding groups, two lone pairs.)
(b) Redraw the molecule showing its three-dimensional shape. Use lines, solid wedges, and dashed wedges if necessary.
(c) Indicate the numeric value of all bond angles.

6.26 Consider CH_3^+ and CH_3^-. Using these as examples, explain why it is necessary to draw a correct Lewis dot diagram before trying to predict the shape of a molecule. Use the concept of steric number in your explanation.

6.27 A hexagonal molecule has angles of 120° inside the ring and is flat. The following two molecules are often drawn as flat hexagons, but only one is truly flat. Which one is it and why?

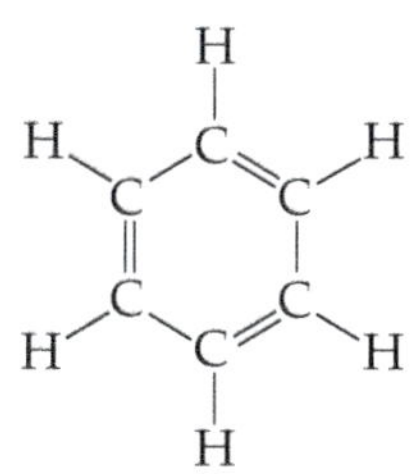

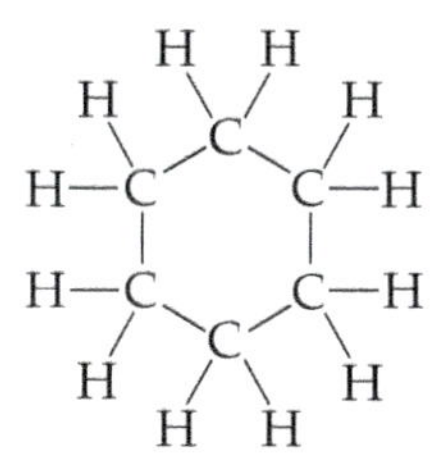

6.28 *Proteins* are long-chain molecules that make up our skin, hair, muscles, and enzymes, among other things. Shown is a dot diagram of the part of a protein known as a *peptide bond*. Redraw the diagram to show the three-dimensional shape around each carbon atom and around the nitrogen atom, using lines, solid wedges, and dashed wedges as needed. Name the shape around each of these four atoms, and give the numeric value of all bond angles.

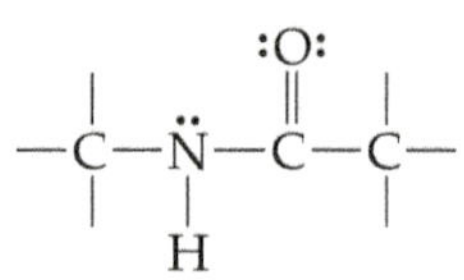

Polarity of Molecules

6.29 In terms of an operational definition, when is a molecule considered to be polar?

6.30 Is it possible to predict whether a molecule will be polar? Explain completely what you would need to know about the molecule.

6.31 True or false? All molecules that contain polar bonds must be polar. Explain your answer.

6.32 Consider the molecules HCl and HBr.
(a) Which molecule has the larger bond dipole moment? Explain why.
(b) Which molecule is more polar? Explain why.

6.33 What does the magnitude of a bond dipole moment (the length of the arrow) tell you about the bond?

6.34 Why do we use an arrow to represent a bond dipole moment? Why not just use a number?

6.35 Consider an individual bond dipole moment.
(a) What does electronegativity have to do with an individual bond dipole moment?
(b) Explain how an individual bond dipole moment is drawn.

6.36 Which of these molecules is the least polar?
(a) CH_2O (b) F_2 (c) CH_3F (d) CH_3Cl

6.37 Consider SO_2 and CO_2. Both have polar covalent bonds. One of these molecules is polar and the other is nonpolar. Which is which and why?

6.38 Which of each pair is most polar? Explain.
(a) Chlorine and phosphorus trichloride
(b) Carbon disulfide and sulfur difluoride
(c) Nitrogen trifluoride and phosphorus trifluoride
(d) Silicon tetrabromide and hydrogen cyanide (HCN)
(e) Nitrogen trichloride and oxygen dichloride

6.39 Consider all the hydrogen halide molecules HX, where X is a group VIIA atom.
(a) Which is the most polar? Why?
(b) Which is the least polar? Why?
(c) Draw all these molecules, showing their relative bond dipole moments.

6.40 Consider the two molecules CO and CO_2. They are both made of the same elements, and yet only one is polar. Why is this so?

6.41 Consider the following molecules. For those that are polar, draw the molecular dipole moment.
(a) $CHBr_3$ (b) CH_3Br (c) H_2S (d) NOCl
(e) C_2Cl_2 (connected Cl—C—C—Cl)

6.42 What is a dipole–dipole force? Give an example.

6.43 Are dipole–dipole forces between molecules as strong as the forces between oppositely charged ions? Explain.

6.44 What do we mean by intermolecular forces? What evidence is there that they exist?

6.45 Draw two HCl molecules and show how they would be attracted to each other. Show the partial charges and dipole moment vectors for both molecules, and orient the molecules properly with respect to each other.

6.46 Draw two ammonia molecules in their three-dimensional shape and show how they would be attracted to each other. Show the partial charges and individual bond dipole moment vectors for both molecules, and orient the molecules properly with respect to each other.

6.47 Why is it important to always show the lone pairs in a Lewis dot diagram?

6.48 The dot diagram

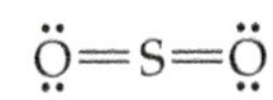

for SO_2 is incorrect.
(a) Draw the correct dot diagram.
(b) How do the correct and incorrect dot diagrams differ in their prediction of molecular shape and polarity?

6.49 Consider the molecule $SiCl_4$.
(a) Draw the dot diagram.
(b) Draw the molecule's three-dimensional shape, and label the numeric value of all bond angles.
(c) What is the shape of this molecule?
(d) Draw in the individual bond dipole moments.
(e) Is the molecule polar? If yes, draw the molecular dipole moment vector.

6.50 Consider the molecule AsF_3.
(a) Draw the dot diagram.
(b) Draw the molecule's three-dimensional shape, and label the numeric value of all bond angles.
(c) What is the shape of this molecule?
(d) Draw in the individual bond dipole moments.
(e) Is the molecule polar? If yes, draw the molecular dipole moment vector.

6.51 Consider the molecule SO_3.
(a) Draw the dot diagram.
(b) Draw the molecule's three-dimensional shape, and label the numeric value of all bond angles.
(c) What is the shape of this molecule?
(d) Draw in the individual bond dipole moments.
(e) Is the molecule polar? If yes, draw the molecular dipole moment vector.

6.52 The atoms in the molecule HSCN are connected in the order given in the formula.
(a) Draw the dot diagram.
(b) Draw the molecule's three-dimensional shape, and label the numeric value of all bond angles.
(c) Draw in the individual bond dipole moments.
(d) Is the molecule polar? If yes, draw the molecular dipole moment vector.

6.53 Consider the molecule hydrazine, N_2H_4, used in rocket fuel. Each nitrogen is bonded to two hydrogens, and the nitrogens are bonded to each other.
(a) Draw the dot diagram.
(b) Draw the molecule's three-dimensional shape, and label the numeric value of all bond angles.
(c) What is the shape of this molecule?
(d) Draw in the individual bond dipole moments.
(e) Is the molecule polar? If yes, draw the molecular dipole moment vector.

6.54 Consider the molecule HNF_2 (N is the central atom in the molecule).
(a) Draw the dot diagram.
(b) Draw the molecule's three-dimensional shape, and label the numeric value of all bond angles.
(c) What is the shape of this molecule?
(d) Draw in the individual bond dipole moments.
(e) Is the molecule polar? If yes, draw the molecular dipole moment vector.

6.55 Consider the molecule N_2O (connected N—N—O).
(a) Draw the dot diagram.
(b) Draw the molecule's three-dimensional shape, and label the numeric value of all bond angles.
(c) What is the shape of this molecule?
(d) Draw in the individual bond dipole moments.
(e) Is the molecule polar? If yes, draw the molecular dipole moment vector.

6.56 Some molecules pose special challenges to the rules for obtaining the correct Lewis dot diagram. Consider the molecule NO_2 (N is the central atom).
(a) What challenge does it present?
(b) Suppose you were allowed to violate the octet rule for the N atom. What would the dot diagram look like?
(c) Based on your answer to part (b), what is the shape of such a molecule? Is the molecule polar?

6.57 Consider the phosphonium ion, PH_4^+.
(a) Draw the dot diagram.
(b) Draw the ion's three-dimensional shape, and label the numeric value of all bond angles.
(c) What is the shape of this polyatomic ion?
(d) Draw in the individual bond dipole moments.

6.58 Imagine that you could vary the magnitude of the molecular dipole moment of all molecules in a substance. What would happen to the strength of the intermolecular forces between a group of molecules as the dipole moment of each molecule is increased? Explain.

6.59 Consider the PX_3 molecule, where X is either H or F.
(a) For X = H, the entire molecule is nonpolar. Why is this so? (*Hint*: Consider electronegativities.)
(b) For X = F, the entire molecule is polar. Draw two molecules next to one another in proper orientation so as to yield a dipolar intermolecular attraction, and explain why the attraction occurs.

Additional Problems

6.60 An atom has no lone pairs of electrons on it and four other atoms bound to it. Why is 109.5° the bond angle adopted by this molecule?

6.61 Using lines, solid wedges, and dashed wedges, draw the three-dimensional shape of ethane, C_2H_6. Indicate the numeric value of all bond angles.

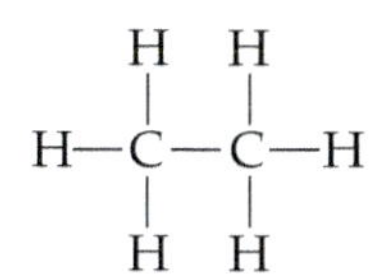

6.62 Using lines, solid wedges, and dashed wedges, draw the three-dimensional shape of acetaldehyde, C_2H_4O. Indicate the numeric value of all bond angles.

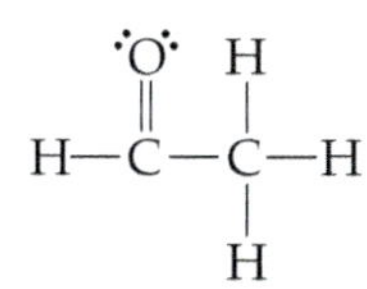

6.63 Using lines, solid wedges, and dashed wedges, draw the three-dimensional shape of diethyl ether, C_2H_6O. Indicate the numeric value of all bond angles.

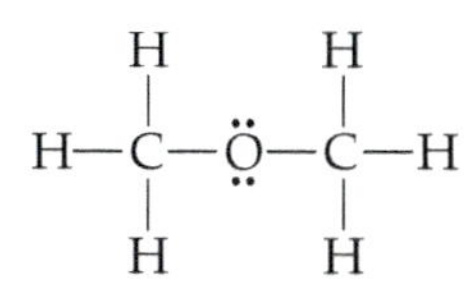

6.64 Using lines, solid wedges, and dashed wedges, draw the three-dimensional shape of the hydronium ion, H_3O^+. Indicate the numeric value of all bond angles.

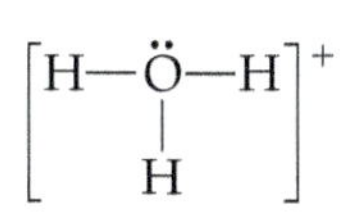

6.65 Consider the molecules CH_4, CH_3Cl, CH_2Cl_2, $CHCl_3$, and CCl_4. Which are polar and which are nonpolar?

6.66 Using lines, solid wedges, and dashed wedges, draw the three-dimensional shape of chloromethylacetylene, C_3H_3Cl. Indicate the numeric value of all bond angles.

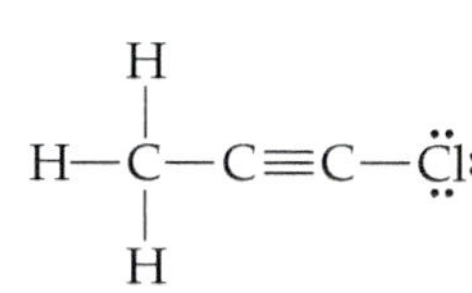

6.67 The connections in $C_2N_2H_2O$ are:

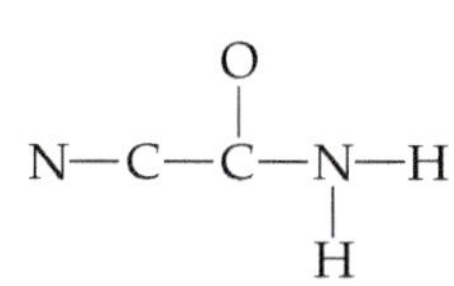

(a) Put in the remaining valence electrons to complete the Lewis dot diagram.
(b) Using lines, solid wedges, and dashed wedges, draw the three-dimensional shape of this molecule. Indicate the numeric value of all bond angles.

6.68 The connections in CH_3NO_2 are:

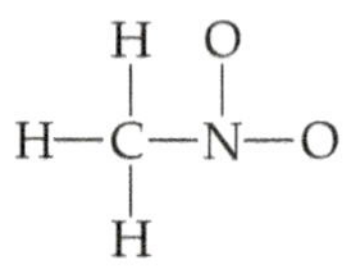

(a) Put in the remaining valence electrons to complete the Lewis dot diagram.
(b) Using lines, solid wedges, and dashed wedges, draw the three-dimensional shape of this molecule. Indicate the numeric value of all bond angles.

6.69 Up until now, you've been drawing first a Lewis dot diagram (two-dimensional, shows lone electron pairs) and then a separate diagram showing molecule shape (three-dimensional, no lone pairs shown). Now begin combining the two, drawing one structure showing both three-dimensional shape and all lone pairs. Draw such a structure for each of the following polyatomic ions. Name each shape, and indicate whether the ion has an overall dipole moment. If so, draw the dipole moment vector. (*Hint*: It's a good idea to continue to draw a regular Lewis diagram first, even though you do not show it in your final answer.)
(a) CN^- (b) ClO_4^- (c) PCl_4^+ (d) NO_2^-

6.70 Draw a combined Lewis dot, molecular-shape diagram for each of the following species. Name each shape, and indicate whether the molecule or ion has an overall dipole moment. If so, draw the dipole moment vector.
(a) Cl_2O (b) NOF (c) PF_3 (d) ICl_2^+

(*Hint*: See problem statement and hint for Problem 6.69. *Hint for (b)*: N=O bonds are common.)

6.71 Draw a combined Lewis dot, molecular-shape diagram for each of the following species. Name each shape, and indicate whether the molecule or ion has a dipole moment. If so, draw the dipole moment vector. (*Hint*: See problem statement and hint for Problem 6.69.)
(a) $CHBr_3$ (b) NF_3 (c) ClO_3^- (d) CS_2

6.72 Draw dot diagrams for the carbonate anion, CO_3^{2-}, and for the sulfite anion, SO_3^{2-}. What is the electron-group configuration around the central atom in each anion? What is the molecular shape of each anion? Is either ion polar? If so, which one? Explain your choice.

6.73 Which, if any, of these molecules do you expect to be polar: CO_2, CS_2, or CSO (carbon is the central atom in all three molecules)? Explain your answer.

6.74 Which, if any, of these molecules are polar? For any molecule you classify as polar, show both the individual bond dipole moment vectors and the overall molecular dipole moment vector. Explain your answers. (*Hint*: For the tug-of-war test, place yourself midway between the carbons.)

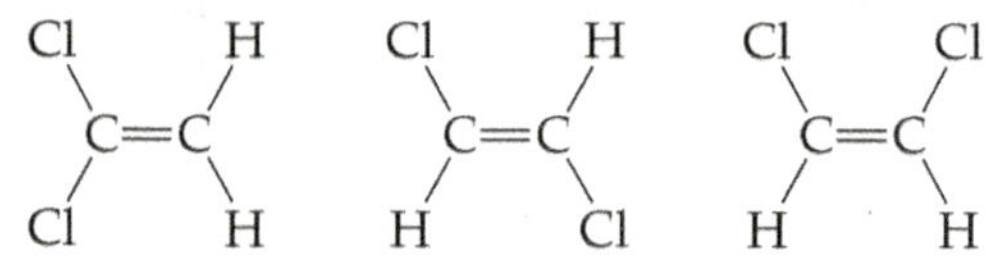

6.75 Draw the three-dimensional shape of each molecule shown below. Label the numeric value of all bond angles, and indicate whether the molecule is polar or nonpolar. If polar, draw the molecular dipole moment vector.

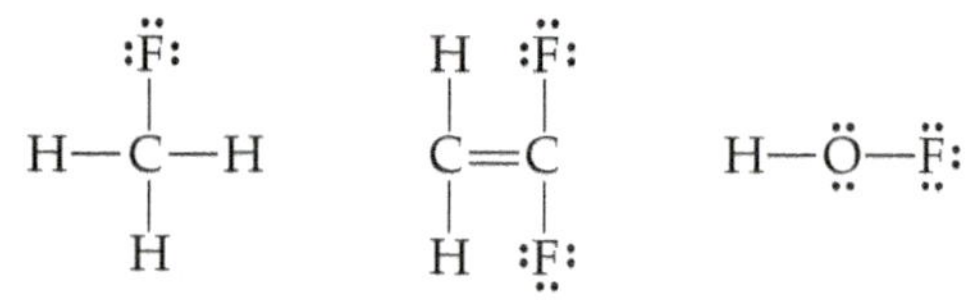

6.76 List the steps you take to decide (a) whether or not a covalent bond is polar and (b) whether or not a molecule is polar.

6.77 Which should have the largest molecular dipole moment: H_2, CO_2, CH_3F, or CH_3I?

6.78 What is the molecular shape of NCl_3? What is the electron-group geometry around N? Is this a polar molecule?

6.79 When asked the shape of NH_2^-, a student begins by drawing two dot diagrams:

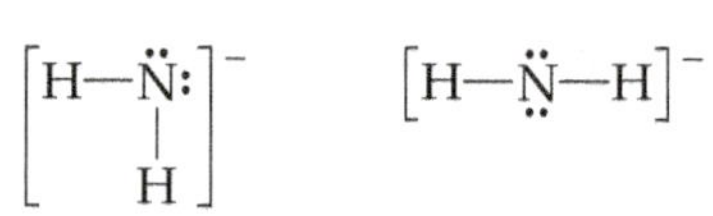

She then answers that the shape can be either bent or linear, depending on the arrangement of the hydrogen atoms. What is wrong with this answer? What is the correct shape?

6.80 What is the difference between intermolecular forces and intramolecular forces? Which are stronger?

6.81 Draw the three-dimensional shape of S_2F_2 (connected F—S—S—F) and N_2F_2 (connected F—N—N—F). Indicate the numeric value of all bond angles.

6.82 Although Lewis dot diagrams are useful in understanding how the valence electrons in a molecule are arranged, what important aspect of the molecule do they neglect to show?

6.83 Draw the three-dimensional shape of methanol, CH_3OH. Indicate the numeric value of all bond angles. Is this molecule polar? If so, draw the molecular dipole moment vector.

6.84 Hydrogen peroxonitrite has the formula HNO_3, with the atoms connected O—N—O—O—H.
(a) Draw the dot diagram for this molecule.
(b) Draw the three-dimensional shape of the molecule, showing the numeric value of all bond angles.
(c) Describe the electron-group geometry and molecular shape around the N and around each O to the right of the N.

6.85 Explain the difference between electron-group geometry and molecular shape. How do you use electron-group geometry when deciding what shape a molecule has?

6.86 Draw each ion or molecule showing its three-dimensional shape and valence electrons. For each species, name the electron-group geometry around the central atom and the molecular shape. Indicate whether each species has a molecular dipole moment. If so, draw the dipole moment vector.
(a) NSF (b) CNO^- (c) SF_3^+

6.87 A student forgets that the N in ammonia, NH_3, has a lone pair as well as its three single bonds. After checking Table 6.2, he mistakenly draws the molecule—three bonding groups, no lone pairs—as having a trigonal planar shape. If the ammonia molecule really were trigonal planar, how would the intermolecular forces differ from what they actually are?

6.88 Under what conditions is the electron-group geometry for a molecule the same as the molecular shape?

6.89 What is the shape of the N_3^- anion? What is the numeric value of each bond angle in the anion?

6.90 Which should have the largest dipole moment: O_3, H_2O, or OF_2?

6.91 Covalent molecules that contain an O—H, N—H, or F—H bond have very strong intermolecular forces. Explain why.

6.92 Draw the three-dimensional shape of H_2O_3. Label the numeric value of all bond angles. In this molecule, the three oxygen atoms form a chain with a hydrogen atom at each end.

6.93 Draw each ion or molecule showing its three-dimensional shape and valence electrons. For each species, name the electron-group geometry around the central atom (or atoms) and the molecular shape. Indicate whether each species has a molecular dipole moment. If so, draw the dipole moment vector.
(a) NCCN (b) BF_4^- (c) $NClO_2$ (d) SeH_2

6.94 White phosphorus, P_4, consists of four phosphorus atoms, with each atom bonded to the other three. Draw the three-dimensional shape of this molecule, showing all valence electrons and bond angles. (*Hint*: Begin by drawing one atom bonded to the three others using VSEPR theory, and then make the remaining connections.)

6.95 BH_3 and PH_3 each contain four atoms, with the three hydrogens surrounding the central atom. Do the two molecules have the same electron-group geometry? Do they have the same molecular shape? Name the geometry and the shape for both molecules. You know from Problem 6.59 that PH_3 is nonpolar. What about BH_3? Is it polar or nonpolar? Explain.

6.96 Nature supplies examples that violate VSEPR. Consider the two molecules shown below. Their correct shapes are indicated.

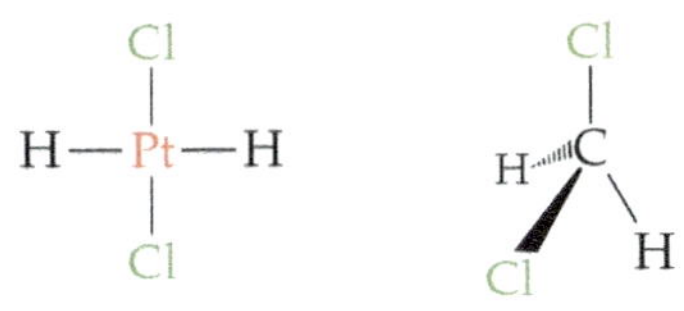

(a) (b)

(a) What is the steric number for the central atom for both molecules?
(b) Which molecule violates VSEPR?
(c) Which molecule is polar and which is nonpolar? Explain why.

6.97 A student draws the following for ozone and claims it is a polar molecule. Is the student correct? Explain why.

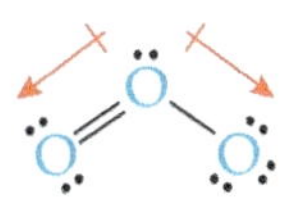

6.98 A student postulates that when two NI_3 molecules approach one another, they do so with the preferential orientation shown below. Would you argue for or against his postulate? Explain why.

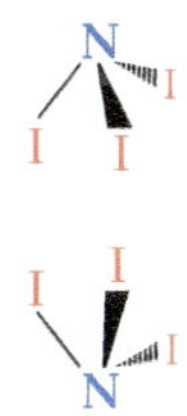

6.99 One of the molecules below is C_2Cl_2 (linear, connectivity Cl—C—C—Cl) and the other is F_2C_2 (linear, connectivity F—C—C—F). The bond dipole moment vectors are shown:

(a) (b)

(a) Which molecule contains Cl and which contains F? Explain how you knew from the diagram.
(b) Could you distinguish between samples of the two molecules by putting them between charged plates to see which one lines up? Explain why.

6.100 Some molecules have central atoms with steric numbers greater than 4. For example, the central sulfur atom in SF_6 has a steric number of 6. Its actual geometry (called octahedral), is shown below (all the bond angles are 90° or 180°).

(a) Are the bonds in SF_6 polar, polar covalent, or ionic? Explain.
(b) Would you predict the SF_6 molecule to be polar or nonpolar? Justify your answer.

One Step Beyond

6.101 There are exceptions to the predictions of VSEPR. Consider CH_3, known as a methyl radical.
(a) Create a dot diagram for the methyl radical. How is it fundamentally different from other dot diagrams you have done?
(b) Use VSEPR to predict the shape of the methyl radical and draw it with that shape (treat the odd electron as a single electron group).
(c) The methyl radical is known to be planar with 120° H—C—H angles. What steric number is being employed here, and what is the carbon atom doing with respect to the odd electron in determining molecular shape?
(d) The CF_3 radical does obey VSEPR. Draw it according to its VSEPR-predicted shape. What steric number is being employed here?
(e) The C—H bond is shorter than the C—F bond. When bonds are short, the atoms at the ends of the bonds can bang into each other (this is called steric congestion) unless a geometry is adopted to get around this. Use this knowledge to explain why CH_3 violates VSEPR, but CF_3 does not.

6.102 Given the electronegativities of F and I, what would have to be true for IF_2^+ to be nonpolar? The fact that it is polar means what?

6.103 Cyclohexane, formula C_6H_{12}, is often drawn as a puckered (not flat) ring as shown below (the purple and blue atoms are the hydrogens). Explain how drawing it as a puckered ring makes it consistent with VSEPR, and explain why the purple atoms are called axial hydrogens and the blue atoms are called equatorial hydrogens.

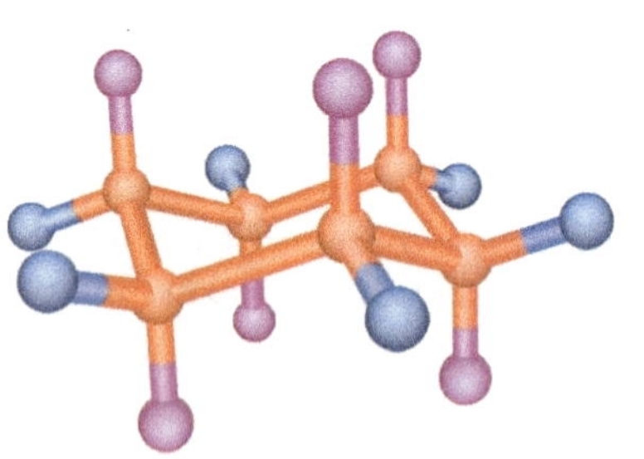

6.104 Consider 1,2,3-trichlorobenzene and 1,3,5-trichlorobenzene, both shown below (the carbon atoms are numbered as shown). Both are flat molecules with 120° bond angles, yet one of them is polar and the other nonpolar. Which is which, and why? For the polar one, indicate which part of the molecule is $\delta+$ and which is $\delta-$.

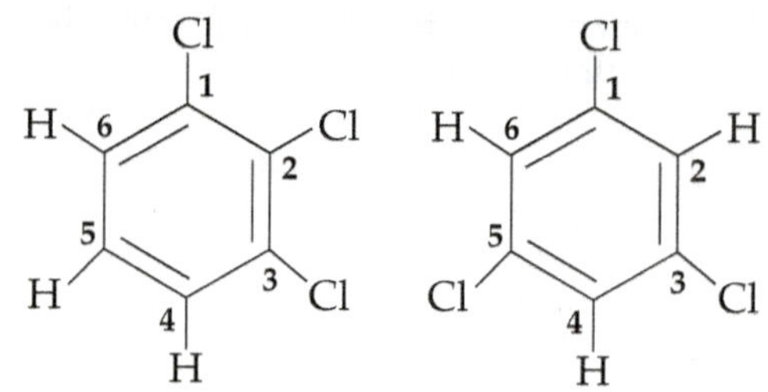

6.105 Cubane, C_8H_8, is an interesting molecule shown below. Comment on it relative to VSEPR. Then imagine you had dichloro cubane, $C_8H_6C_{12}$. Draw all the unique molecules with this formula (you are drawing isomers, molecules with identical formulas but different structures). Indicate which of them are polar.

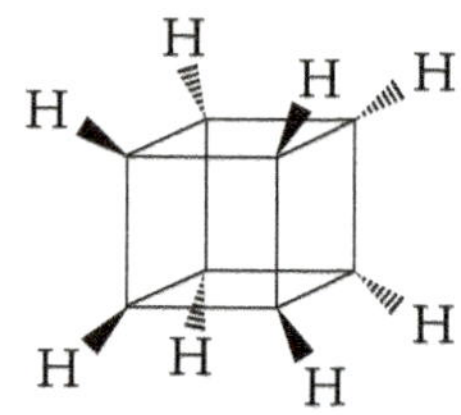

6.106 Using your knowledge of VSEPR theory and its preferred bond angles, suggest a reason for the fact that the cyclopropane molecule (pictured below) is not a very stable molecule and is highly reactive, while the cyclopentane molecule is not nearly as reactive or unstable.

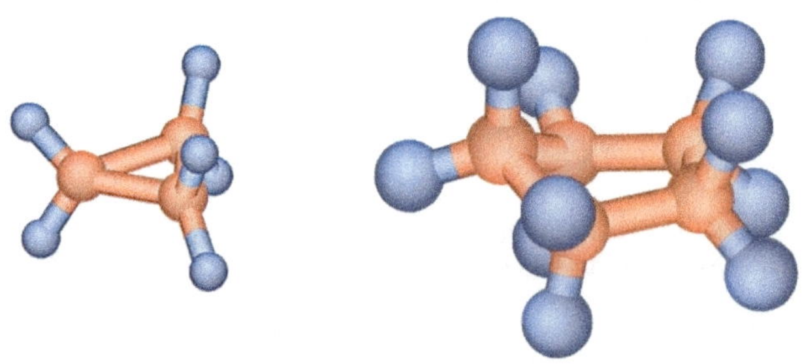

6.107 What is it about the trigonal bipyramidal shape that distinguishes it from all the other molecular shapes that we have covered? (*Hint*: Think about lone pairs of electrons.)

6.108 Consider the molecule $BrCl_3$.
(a) Draw a dot diagram for the molecule.
(b) Predict the molecule's electron group geometry.
(c) Predict the molecule's shape (name the shape and draw it using wedges, lines, and broken lines, and indicate the ideal value of the bond angles).
(d) Predict if the molecule is polar or nonpolar. If polar, draw the dipole moment for the molecule.

6.109 SeF_4 and SeF_4^{2-} have very different shapes. Explain why, and also draw and name the shapes for both.

6.110 One student claims XeF_4 will be polar. Another claims it will be nonpolar. Which is right, and why?

6.111 Is ClF_5 polar or nonpolar? Fully explain your answer.

Group Exercises (Work on these exercises with someone else in the class.)

1. The diagram below shows how bent water molecules in solid ice orient themselves with respect to each other. Notice the large openings in the lattice of water molecules.

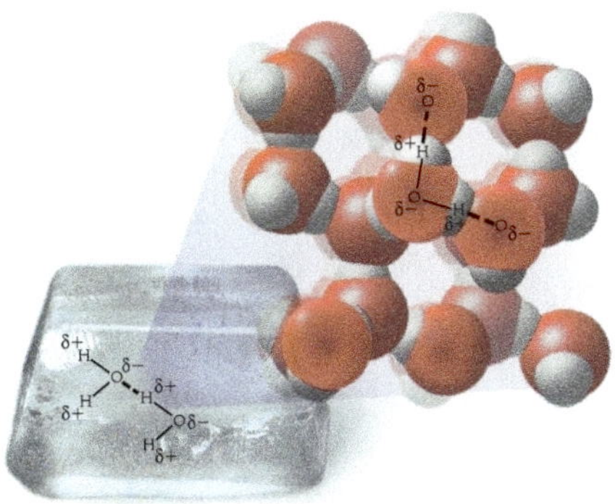

   (a) Discuss why it takes energy to melt ice, and then write a statement that tells what the energy is being used for.
   (b) Solid ice floats on top of liquid water. In other words, solid ice is less dense than liquid water. Based on the diagram above, come up with an explanation for why this is so and what must be true about the liquid to make it more dense than the solid.

2. Explain to your partner how a molecule can have polar bonds but still be nonpolar. Make sure your explanation includes the drawing of a hypothetical molecule that goes along with your explanation. Then your partner must make a change to the molecule that results in it now being polar.

WorkPatch Solutions

6.1

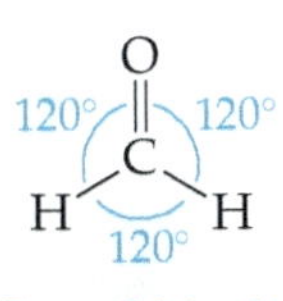

Formaldehyde

Acetylene

6.2 Both molecules would be called either bent or V-shaped.

6.3 In H_2O, expect a roughly 4° compression from 109.5° because there are two lone pairs on the oxygen:

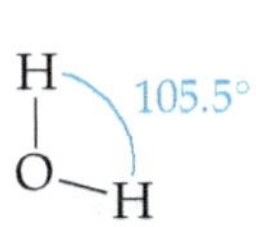

In O_3, expect a roughly 2° compression from 120° because of the one lone pair on the central oxygen:

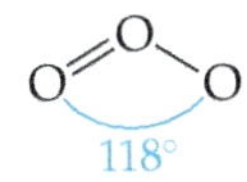

6.4 These are trivial because they all represent a central atom with only one other atom attached (we know this from the information that there is only one bonding group in each case). When a molecule consists simply of two atoms, the only shape it can have is linear. No other shape is possible.

6.5 A bond is guaranteed to have a zero dipole moment when the electrons forming the bond are shared absolutely equally by the two atoms. This occurs when the two atoms have identical electronegativities (this is assured when the atoms are of the same element).

6.6 The electronegativity values are C 2.5, O 3.5, and S 2.5. Thus, the individual bonds in CO_2 are more polar because the electronegativity difference between C and O ($\Delta EN = 1.0$) is greater than the difference between C and S ($\Delta EN = 0.0$). However, because both molecules are linear, the bond dipole moments in both of them cancel. Therefore, neither molecule is polar.

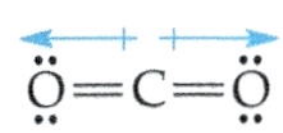

Nonpolar

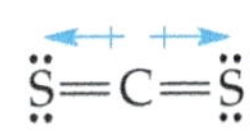

Nonpolar

The electronegativity of H is 2.1. Therefore, the differences are now $\Delta EN = 1.4$ for the O—H bonds and $\Delta EN = 0.4$ for S—H bonds. Because the bonds in H_2O are more polar, H_2O is the more polar molecule.

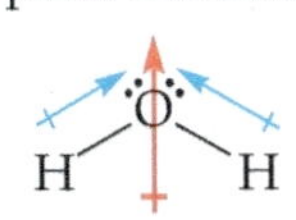

More polar

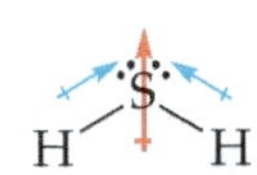

Less polar

6.7 The upward tug in the $COCl_2$ molecule is slightly stronger because the C—O bond is more polar than the C—Cl bonds. This is because the electronegativity difference between O and C ($\Delta EN = 3.5 - 2.5 = 1.0$) is greater than the electronegativity difference between Cl and C ($\Delta EN = 3.0 - 2.5 = 0.5$). If the three tugs were equal, and assuming the bond angles were 120° (trigonal planar), the tugs would cancel and the molecule would be nonpolar.

CHAPTER 7

Intermolecular Forces and the Phases of Matter

In Canada's Northwest Territories during the winter months, "ice-road truckers" drive their big rigs over the surface of frozen lakes on what are known as ice roads. These roads service mines are the only way to get goods in and out during the winter. Trucks hauling more than 40 tons ply these roads day and night.

The average thickness of the ice is 125 cm (a little over 4 feet thick), but the ice road can be opened for travel at 22 cm (8.7 inches) thickness, and still it supports crushing loads. Drivers have to be constantly ready for a break in the ice. A monotony can set in, driving hundreds of miles on ice against the white landscape, but it never quite numbs the driver to the reality of the ever-present danger. Indeed, many seasoned veterans forgo seatbelts because if the surface gives way, a trucker will have just seconds to jump clear.

The only thing between the driver and certain death are the tenacious attractive forces that exist between the water molecules in the ice. Why do water molecules attract one another? Read on.

7.1 Why Does Matter Exist in Different Phases?

It's the middle of July, 34 °C, and you feel as if you're covered with a wet blanket. Sound familiar?

These are the sweltering, high-humidity days of summer when there is just too much water in the air. You can't see this water because it exists in the gas, or vapor, phase, but you can certainly feel it. It makes you feel miserable. Today's solution to this problem is, of course, the air-conditioner, which cools the air by drawing it in with a fan and passing it over cold refrigeration coils. If you were to look at the coils inside an air-conditioner, you would notice they are dripping wet. Cooling humid air causes the water vapor in the air to *condense* to a liquid on the cold coils, and in this way the air is dried as it is "conditioned." This is why air-conditioned air feels so good on a humid summer day—it's not just cooler, it's also drier. By setting an air-conditioner on its highest setting, you can sometimes cause the refrigeration coils to get so cold that the liquid water on them freezes. Thus, simply by cooling the air, an air-conditioner can drive atmospheric water through three phases—from the *gas phase* through the *liquid phase* and into the *solid phase*.

We touched briefly on the different states or phases of matter in Section 1.3. In this chapter, we'll run through this subject matter again but from a different point of view. This time, we'll look at what happens on the molecular level. Why should cooling air cause the water in it to change phase? Exactly what are the water molecules doing in each phase?

And let's not forget about the other gases in the atmosphere, nitrogen (N_2, 78.1% by volume), oxygen (O_2, 20.1% by volume), argon (Ar, 0.934% by volume), and carbon dioxide (CO_2, 0.033% by volume). Unlike water, these gases don't change phase in an air-conditioner even though they can exist in liquid and solid forms (see below). Why do these substances remain in the gas phase while water vapor condenses and eventually freezes? What is so different about water molecules that could account for this difference in behavior?

Gases

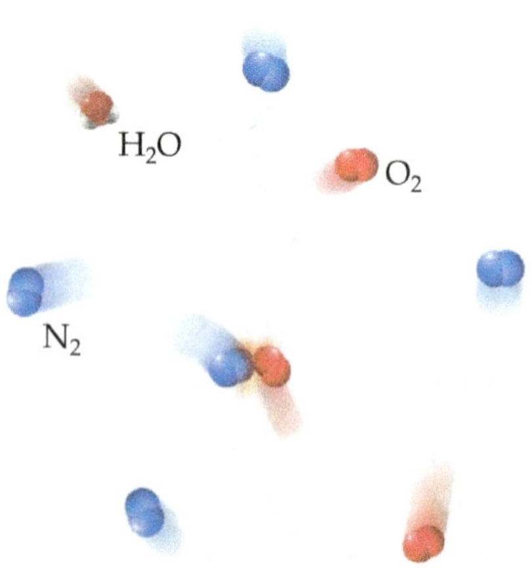

Air consists of fast-moving, independent molecules and atoms.

If you could see the individual molecules that make up the atmosphere (mostly N_2 and O_2), you would see that they are in constant motion. They travel in all directions, at speeds around 500 miles per hour. The molecules are either colliding with one another or zipping past one another.

Something very interesting happens when we cool air by passing it over the refrigeration coils of an air-conditioner—the molecules slow down. *This is a very important piece of information*—the average speed of the molecules in a gas depends on temperature. When you heat a collection of gas molecules, the molecules move faster, on average. When you cool a gas, the molecules in the gas move more slowly, on average. (The gas molecules in our sample are not all moving with the same speed. There is actually a distribution of speeds in the sample. That's why we focus on the average speed in our discussion.)

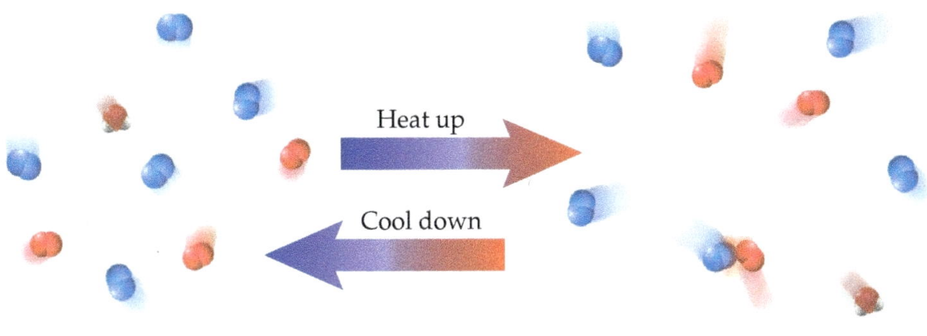

Cold air: molecules move slowly on average.

Hot air: molecules move fast on average.

Condensed Phases—Solids and Liquids

So maybe the reason water molecules condense out of the gas phase when air is cooled is that they have slowed down. But this cannot be the whole story. If it were, the O_2, N_2, and other gas molecules in the air would also condense because they also slow down as heat is removed.

There must be something else involved. Think back to Chapter 6, where we suggested you try ripping an ice cube in half. It can't be done because the water molecules in the ice cube are "stuck" tightly to one another by the attractive, intermolecular forces between them. The $\delta+$ end of one polar water molecule is attracted to the $\delta-$ end of another. These forces are responsible for the ability of water to exist in the solid phase (ice).

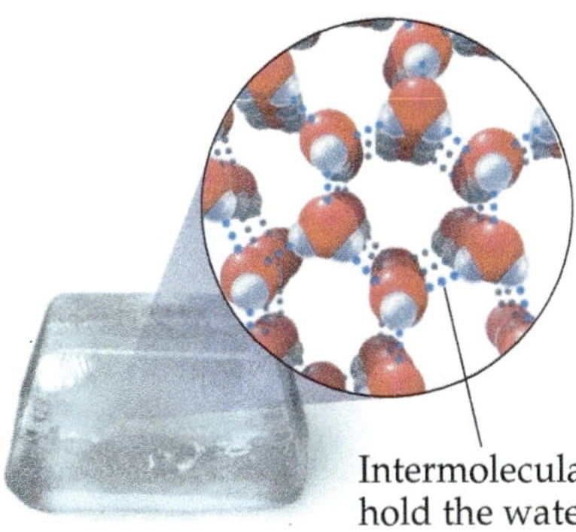

Intermolecular forces hold the water molecules together.

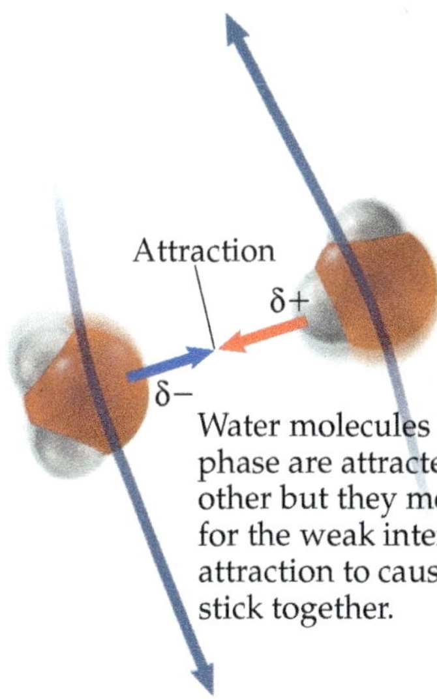

Water molecules in the gas phase are attracted to each other but they move too fast for the weak intermolecular attraction to cause them to stick together.

Water molecules are polar no matter what phase they are in. Even in the gas phase, two H_2O molecules are attracted to each other when they are close. In a gas, however, these attractive forces are too weak to cause the fast-moving molecules to stick together. Their kinetic energy (the energy associated with their motion) is great enough to overcome their mutual attraction, and they shoot on past each other.

However, as any gas is cooled (energy is removed), the molecules slow down. At some temperature, the molecules are finally moving slowly enough for the attractive intermolecular forces to take hold and overcome the molecules' kinetic energy. At that temperature, many of the gas-phase molecules start to stick to one another and form the liquid phase.

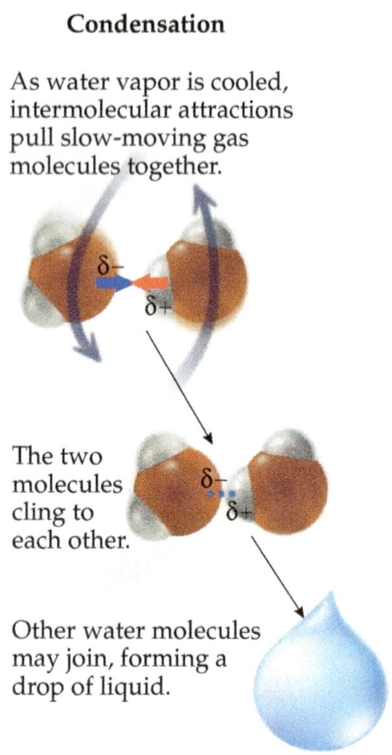

When a substance changes from the gas phase to the liquid phase, we say **condensation** has occurred. If you reverse this process—start with a liquid and add heat—the molecules in the liquid move faster and faster until they finally overcome the intermolecular attractions and leap into the gas phase. This reverse process, going from liquid to gas, is called **vaporization**:

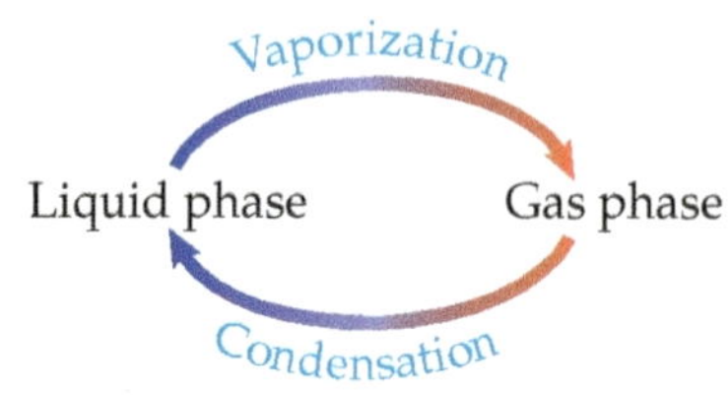

The highest temperature to which a liquid can be heated, and the temperature where it vaporizes most rapidly, is called the **boiling point**. Because condensation is just the reverse of vaporization, the boiling point is also the temperature below which a gas condenses.

Continued cooling of a liquid (that is, continued removal of energy) eventually causes the liquid to freeze—in other words, to enter the solid phase. To see why, let's take a close look at liquid water. A molecular view shows that the molecules in the liquid phase are still moving, even though they are all stuck together.

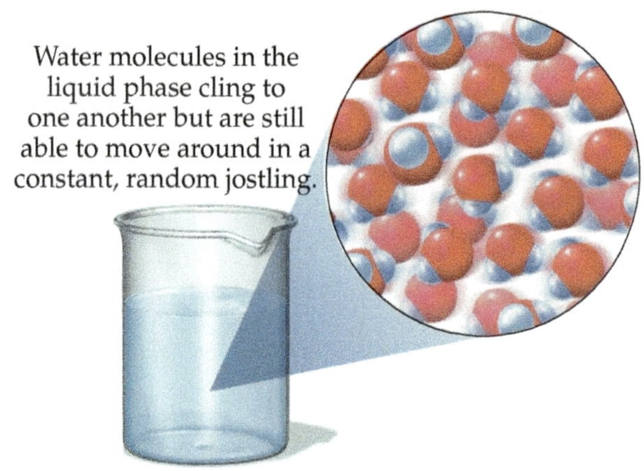

This is as close as you'll probably ever come to feeling like a molecule in a liquid.

Even though the molecules in a liquid are in a much more crowded environment than they would be in a gas, they still have enough kinetic energy to jostle past one another, each molecule continually breaking the attractive force to one molecule and re-establishing it with another. You can come close to experiencing what it's like to be a molecule in the liquid phase by visiting a shopping mall on the day after Thanksgiving. You can continue to move throughout the mall and get your shopping done but all the while you keep getting pushed, shoved, and bumped as you travel.

As the temperature of a liquid drops, the molecules continue to slow down. Soon they are no longer moving fast enough to overcome intermolecular attractive forces and jostle past one another. At that temperature (called the **freezing point**), the molecules are effectively frozen in place, with only small vibrations around their fixed positions allowed. They have arrived at the solid phase.

Structure of Solid Water (Ice)

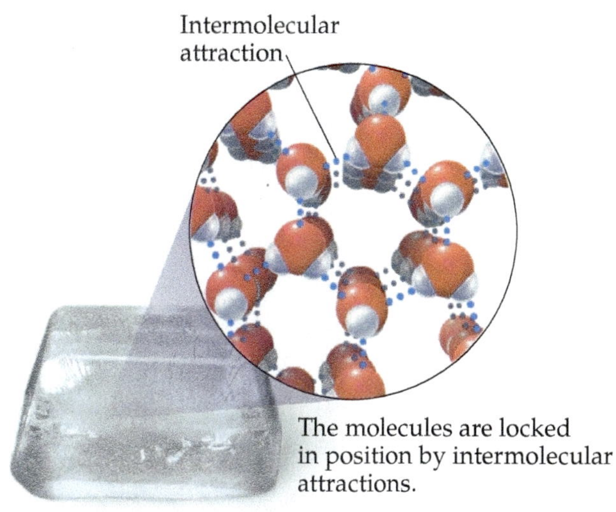

Overview of the Three Phases

A molecular view helps us understand why water exists in three phases and how it changes phases. The key points:

1. Molecules move because they have kinetic energy. Adding heat (energy) increases this kinetic energy. The more heat added, the faster the molecules move and the higher the temperature.
2. Intermolecular attractive forces tend to attract molecules to each other. When molecules are moving fast enough, they have sufficient kinetic energy to overcome these forces and remain in the gas phase. Cooling eventually slows them down enough so that they stick together and form the liquid phase. Further cooling slows them down even more, allowing them to form the solid phase.

In the gas phase, the molecules are essentially independent of one another and are, on average, relatively far apart. In the liquid and solid phases, the molecules are quite close to one another (they essentially "touch"). It is for this reason that the liquid and solid phases are often referred to as **condensed phases**.

Now that we understand what happens with water, we can consider the other gases in the atmosphere, the ones that do not change phase when the air is cooled with our air-conditioners. Why don't blocks of, say, solid carbon dioxide clog the coils of our air-conditioners? Why don't pools of liquid nitrogen and liquid oxygen form at our feet? One possibility is that these molecules experience no attractive intermolecular forces. This is wrong, however, because all molecules exert some attraction on one another. They just don't do so with equal strength. The attractive force between two CO_2 molecules (or two N_2 molecules or two O_2 molecules) is much weaker than the attractive force between two H_2O molecules. Thus, in order to get CO_2 or N_2 or O_2 molecules to stick to one another, you have to cool them (slow them down) much more than you have to cool water for the attractions to "take hold." It takes a temperature of −78 °C (at normal atmospheric pressure) to slow CO_2 molecules to the point where their mutual attraction can overcome their kinetic energy, allowing them to form a solid. (Normal atmospheric pressure is discussed in Section 11.1.) It takes a temperature of −183 °C (at normal atmospheric pressure) to slow oxygen molecules to the point where they liquefy, and only when the temperature gets down to −196 °C (at normal atmospheric pressure) does nitrogen begin to form a liquid. This is extremely cold, so cold that materials like rubber, plastic, and even bananas lose their flexibility when immersed in liquid nitrogen and become as rigid and fragile as glass.

A banana that was chilled in liquid nitrogen (−196 °C) is hit with a hammer.

WORKPATCH 7.1 The normal boiling point of ethanol, the alcohol in beer and wine, is 78 °C. The normal boiling point of water is 100 °C. (*Normal* means that the phase change occurs under the atmospheric pressure usually found at sea level.) What can you say about the strength of the intermolecular forces in ethanol relative to those in water? ●

The WorkPatch is a reminder that physical properties such as boiling point and freezing point are related to the strength of the attractive forces between molecules.

Practice Problems

7.1 Draw a picture showing how H—Cl molecules are attracted to one another, using dotted lines to represent intermolecular forces.

Answer:

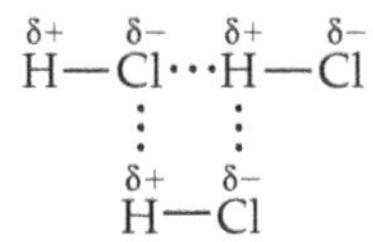

Other arrangements of intermolecular forces are possible. Just make sure the δ+ end of one molecule approaches the δ− end of another molecule.

7.2 Based on the strength of their molecular dipole moments, which compound should have the higher boiling point, HF or HBr? Explain.

7.3 Draw a picture showing how NH_3 molecules attract one another.

7.4 Based on the strength of their molecular dipole moments, which compound should have the higher boiling point, H_2S or H_2O? Explain.

We stated earlier that all molecules are attracted to one another under the right conditions. Even though extremely low temperatures are required, the fact that oxygen and nitrogen can exist as liquids is proof of this statement. For any gas to condense to the liquid phase, there must be some attractive force causing the molecules to stick together. The nature of this force in water is obvious—water molecules are polar, each molecule having permanent $\delta+$ and $\delta-$ ends. The partial positively charged portion of one molecule is attracted to the partial negatively charged portion of another molecule. But what about nonpolar molecules? Molecules like CO_2, N_2, and O_2 are completely nonpolar. Since they do not have permanent $\delta+$ and $\delta-$ ends, how can they attract one another?

7.2 Intermolecular Forces

In Chapter 6, we introduced the dipole–dipole forces that exist between polar molecules. However, these are obviously not the only types of intermolecular forces. To begin our study of some other types, let's consider carbon dioxide. In Chapter 6, we determined that although this molecule contains polar bonds, it is linear and therefore nonpolar.

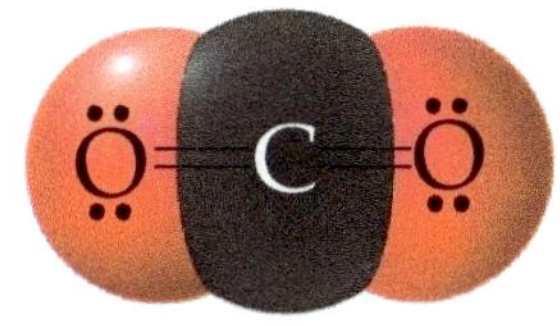

Because the carbon dioxide molecule is nonpolar, it contains no δ+ or δ− ends.

At −78 °C and normal atmospheric pressure, CO_2 molecules go directly from the gas phase to the solid phase. The fact that CO_2 solidifies at a much lower temperature than water (which freezes at 0 °C) indicates that the attractive forces between CO_2 molecules are considerably weaker than those between H_2O molecules. But the real question remains: "Why are CO_2 molecules attracted to one another if they are truly nonpolar and therefore have no permanent $\delta+$ and $\delta-$ ends?" The answer comes from a model proposed more than 100 years ago by the chemists Fritz London and J. D. van der Waals. Today, we refer to the relatively weak attractive forces these men studied as either **London forces** or **dispersion forces**. Although each London force is weak, there can be many such interactions in a large molecule, with the result that the total attraction can be very strong. Nonetheless, on a per-atom basis, it is the weakest intermolecular force there is.

Basically, the original model that is used to explain London forces treats any atom or molecule as a spherical container of electrons. The electrons inside the sphere are assumed to be in constant, random motion. According to this simplified model, the hydrogen molecule, H_2, is represented by a sphere containing two electrons and the methane molecule, CH_4, by a larger sphere containing ten electrons (six from the carbon and one from each hydrogen):

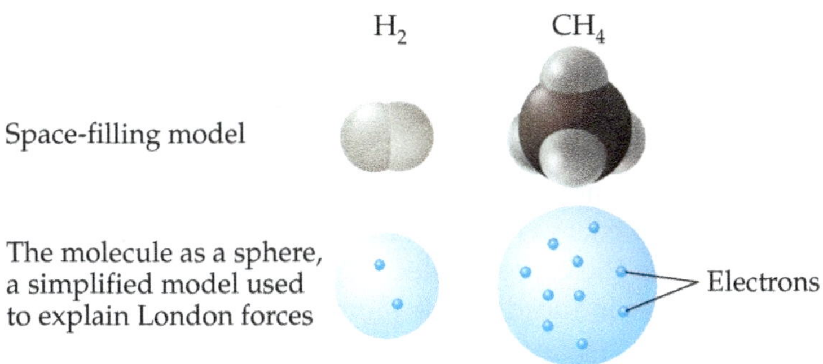

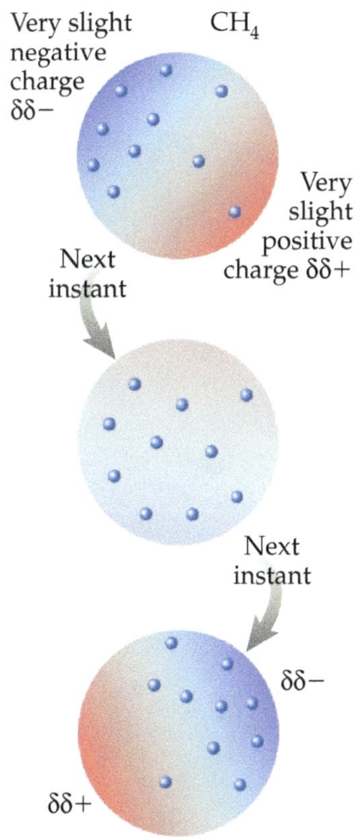

Let's focus on our methane molecule. Remember that its electrons are in constant motion. Imagine that, for a brief instant, this motion happens to put most of the electrons on one side of the molecule, as shown in the top diagram at left. For that instant, the side of the molecule with the excess of electrons has a slight negative charge, and the side momentarily drained of electrons is slightly positive. In the next instant, this momentary imbalance disappears, but soon a new imbalance appears randomly in some other part of the molecule.

According to this model, all atoms and molecules exhibit these brief, constantly shifting electron imbalances. Therefore, even though a nonpolar molecule has no *permanent* dipole moment, it does have weak, randomly oriented dipole moments that appear and disappear from one instant to the next. Because the electron imbalances that produce these dipole moments are typically small (our diagrams exaggerate them), the temporary charges are much weaker than the partial $\delta+$ and $\delta-$ charges in a polar molecule. To signify this extreme weakness, we write these charges as $\delta\delta+$ (partial partial positive) and $\delta\delta-$ (partial partial negative).

This constantly shifting distribution of electrons—and therefore of charges—is what causes nonpolar molecules to attract one another. To really understand this, we must look at more than one molecule at a time. Therefore, imagine a row of three methane molecules. Let's assume they happen to have no charge imbalances at the instant we first see them:

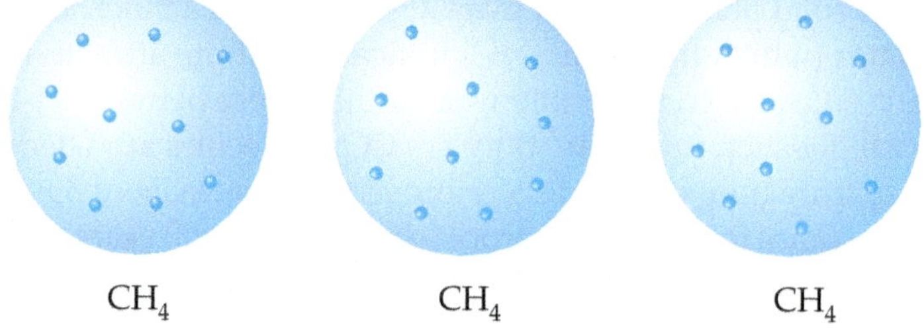

Now suppose the electrons in the center molecule shift left. How do the other molecules respond? A buildup of electrons (and negative charge) on the left side of the center molecule repels the electrons in the molecule on the left. Likewise, the partial positive charge on the right side of the center molecule attracts the electrons in the molecule on the right.

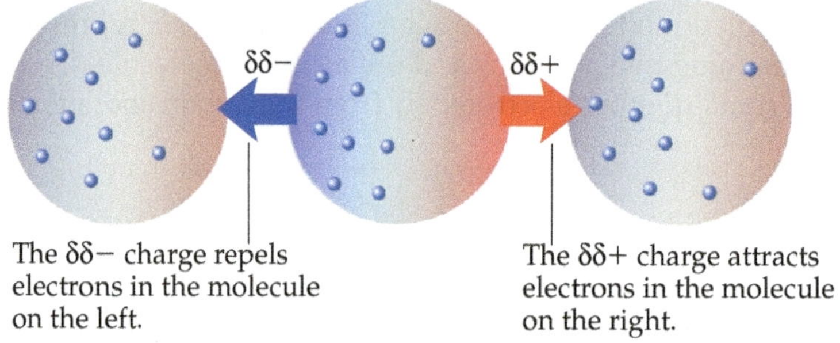

The electrons in the left and right molecules shift in response to these forces. This causes $\delta\delta+$ and $\delta\delta-$ charges to form in both the left and right molecules (see below). We say that the center molecule *induces* dipole moments in the adjacent molecules. These induced dipole moments are precisely oriented in a way that creates an attractive force between adjacent molecules—the $\delta\delta+$ end of each molecule faces the $\delta\delta-$ end of the next:

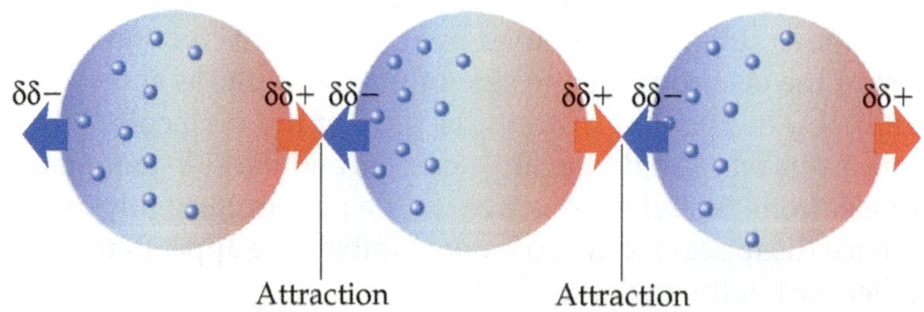

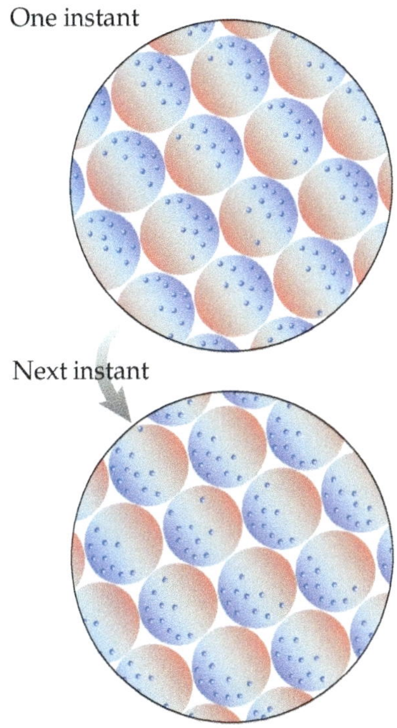

The original temporary dipole moment in the center molecule created, or induced, dipole moments in the adjacent molecules, resulting in attraction. The outcome is that the methane molecules stick together even though they are not polar—that is, even though they don't have permanent $\delta+$ and $\delta-$ ends. Of course, with billions of molecules instead of just the three shown, we get billions of instantaneous induced dipole moments, aligned so as to "stick" all the methane molecules to one another. This pattern lasts for only an instant. In the next instant, it is replaced by another aligned pattern that is equally effective at attracting molecules to one another.

London forces exist between any two molecules. Polar molecules experience both dipole intermolecular forces and London intermolecular forces at the same time, whereas nonpolar molecules experience only London intermolecular forces.

How weak (or strong) are London forces? Remember, we have been exaggerating the electron imbalances in our diagrams. In actuality, these imbalances are very slight, and the partial dipole moments that develop are very small. Thus, for molecules containing *only a few electrons* (small molecules made from relatively light atoms, such as N_2, O_2, H_2O, or CH_4), London forces are extremely weak. The dipole–dipole forces that exist between small polar molecules are quite a bit stronger than the London forces that exist between the molecules.

WORKPATCH 7.2 Why are the attractive intermolecular forces between water molecules much stronger than the attractive intermolecular forces between oxygen molecules, O_2? Which do you think would experience stronger London forces, CH_4 or H_2? Explain. ●

The answer to the WorkPatch gets right to the heart of intermolecular forces: They depend on the existence of partial charges. Even though water molecules have both dipole–dipole forces and London forces going for them, the London forces can be ignored because they are dwarfed by the dipole forces. This is not the case for nonpolar methane, CH_4. All it has attracting its molecules to one another are weak London forces. For this reason, methane gas must be cooled to −164 °C (its normal boiling point) before it liquefies. You really have to slow methane molecules down a lot before the weak London forces between them have any effect. Liquefying hydrogen, H_2, requires even lower temperatures. Because there are fewer electrons in each molecule than in methane molecules, the attractive London forces between these molecules are so weak that almost any motion can overcome them. As a result, you must cool hydrogen gas to −253 °C at normal atmospheric pressure before it becomes a liquid.

On the other hand, London forces can be quite substantial for large molecules containing many atoms and therefore many electrons. Thus, although nonpolar methane is a gas at room temperature, nonpolar carbon tetrachloride, CCl_4, with its large chlorine atoms each containing 17 electrons, is a liquid with a boiling point of 77 °C. This is more than 200 °C higher than the boiling point of methane, attesting to the fact that the London forces between CCl_4 molecules are much stronger than those between CH_4 molecules.

Practice Problems

7.5 Carbon tetrabromide, CBr_4, has a melting point of 88–90 °C. Explain why this makes sense when compared with the boiling points of methane and carbon tetrachloride.

Answer: Because CBr_4 has a melting point well above room temperature (which is usually about 25 °C), CBr_4 is a solid at room temperature. At room temperature, CCl_4 is a liquid and CH_4 is a gas. This makes perfect sense. All three molecules are tetrahedral and nonpolar, which means their only intermolecular forces are London forces. Because CBr_4 has the most electrons (146), it has the strongest London forces of the three molecules.

7.6 The compound ethylene, $H_2C{=}CH_2$, is a gas at room temperature, but the compound polyethylene, made from a large number of ethylene units bonded together, is a solid at room temperature. Explain why.

7.7 (a) Draw a dot diagram for NH_3 and one for PH_3.
(b) Is either molecule polar? (The electronegativities of N, P, and H are 3.0, 2.1, and 2.1, respectively.)
(c) In which substance are the London forces stronger? Explain.
(d) Basing your answer solely on the London forces in the two substances, which substance would you expect to have the higher boiling point? Explain.

7.3 A Closer Look at Dipole Forces—Hydrogen-Bonding

We cannot complete a discussion of intermolecular forces without discussing a special type of attractive force that exists between certain polar molecules. This attractive force is special because it is considerably stronger than the average dipole–dipole attraction. To illustrate this, we shall consider two polar molecules, acetone and water.

Acetone, the main ingredient in many brands of nail polish remover, has the molecular formula C_3H_6O. Because of the structure of the molecule and the large electronegativity difference between the carbon and oxygen atoms, the acetone molecule has a large permanent dipole moment:

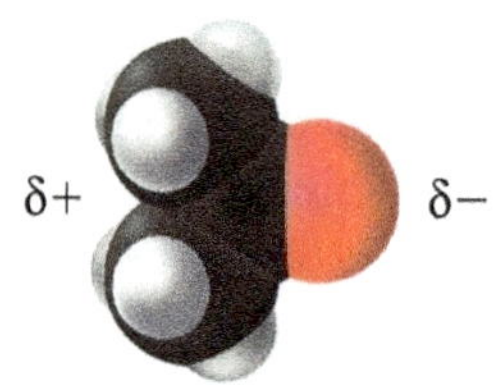

Acetone

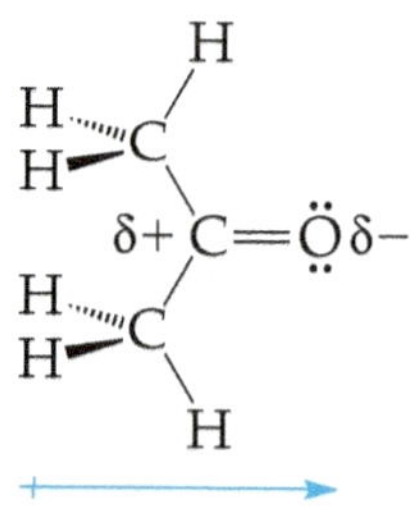

Dipole moment vector

The resulting attractive polar forces between acetone molecules make acetone a liquid at room temperature. At normal atmospheric pressure, it boils at 56 °C.

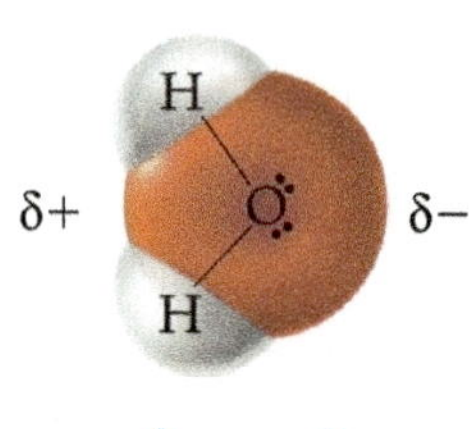

A water molecule is also polar, having a permanent dipole moment that is somewhat smaller than that of acetone. Yet water boils at 100 °C at normal atmospheric pressure, an unusually high boiling point for such a small molecule. If water boiled at the temperature one would expect for a molecule of its size and composition (see Problem 7.38), Earth would be a very different place indeed because virtually all the water on this planet would be in the gas phase, making life as we know it impossible.

The fact that water boils at 100 °C and acetone boils at 56 °C tells us that the intermolecular attractive forces between water molecules must be stronger than those between acetone molecules. Since the dipole moment of water is smaller than the dipole moment of acetone, there must be another reason for this difference. The reason has to do with the presence of hydrogen atoms in these molecules. The dipole moment in a water molecule is a result of the polar O—H bonds in the molecule. The dipole moment in an acetone molecule is primarily the result of the polar C=O bond. Even though acetone does contain hydrogen atoms, they are bonded to an almost equally electronegative carbon atom. These hydrogens therefore have no significant partial charges and contribute very little to the overall dipole moment of the acetone molecule:

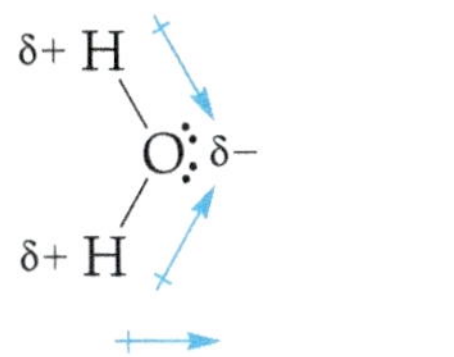

The dipole moment comes from polar O — H bonds.

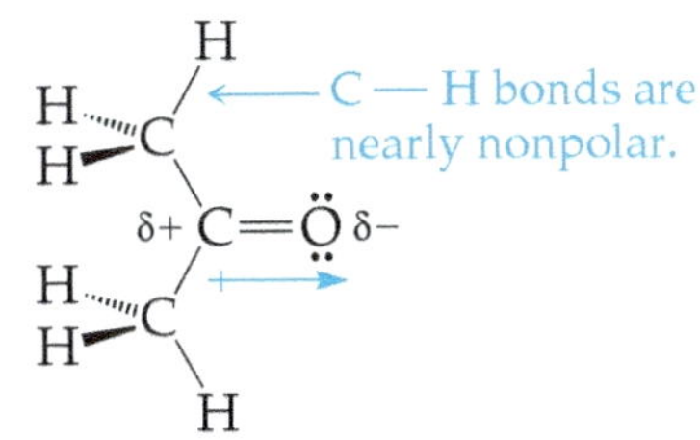

The dipole moment comes from the C = O bond.

This structural variation makes all the difference in the world to the strength of the polar forces. To understand why, we need to examine the hydrogen and oxygen atoms more closely. When a hydrogen atom is covalently bonded to oxygen, the oxygen atom attracts the lion's share of the two shared electrons, and, as a result, the hydrogen atom develops a sizable δ+ charge. This is not unexpected. Remember, oxygen is one of the four highest electronegative electron "hogs." Just about any element bound to oxygen develops a δ+ charge because all atoms except fluorine have a lower electronegativity than oxygen. However, there is something unique about the hydrogen atom—it has no other electrons. Thus, when the two electrons of the O—H bond in water are pulled toward oxygen, the hydrogen atom ends up looking almost like a naked proton sticking out in space. The picture at left shows what a water molecule looks like to another molecule. The oxygen nucleus is buried deep inside the (blue) cloud of negative charge. Each hydrogen nucleus (a proton) is covered by only a thin veil of negative charge because the shared electrons spend little time on the hydrogen atom.

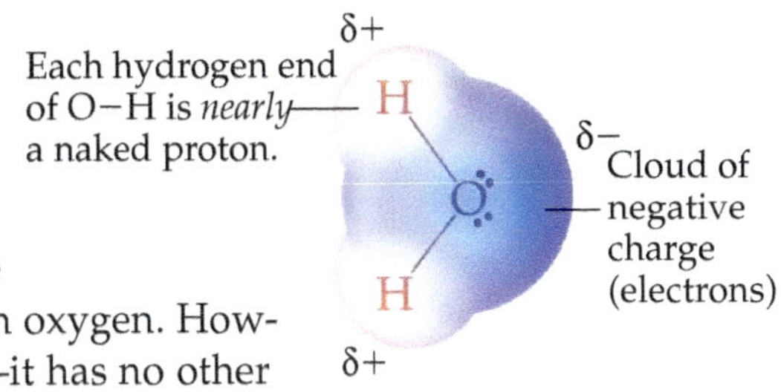

The nearly naked protons have a strong tendency to seek out electrons wherever they can find them. One obvious source is a lone pair of electrons on the oxygen atom in another water molecule. The nearly naked proton can get very close to the lone pair because the proton has no other electrons to be repelled by the lone-pair electrons. This combination of the strong δ+ charge on the hydrogen of one water molecule and the short distance between it and an oxygen lone pair in another water molecule leads to an unusually strong attractive force between the two. These polar interactions are given a special name to indicate their uniqueness. They are called **hydrogen bonds** and are usually represented by a dotted line between the atoms involved.

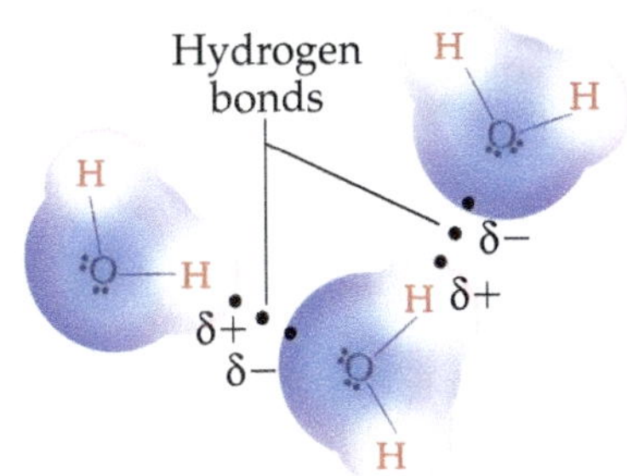

Calling this polar interaction a bond is somewhat misleading because even though this attractive force between water molecules is stronger than a normal dipole–dipole interaction, it is still nowhere near as strong as a covalent bond. The strength of the O—H covalent bond in a water molecule is greater than 400 kJ/mole (that is, it takes more than 400 kJ of energy to break 1 mole of O—H covalent bonds; a mole is a very large number of molecules, and you will learn about a mole in the next chapter). The hydrogen "bond" between two water molecules is only a fraction as strong—around 16 kJ/mole. This means that it would take roughly 16 kJ to separate 1 mole of water molecules stuck to one another by hydrogen bonds. Despite the dubious choice of the word *bond* in describing this interaction, it is still different enough from normal dipole–dipole attractions (whose strength typically ranges from 4 to 8 kJ/mol) to warrant being treated as a special case.

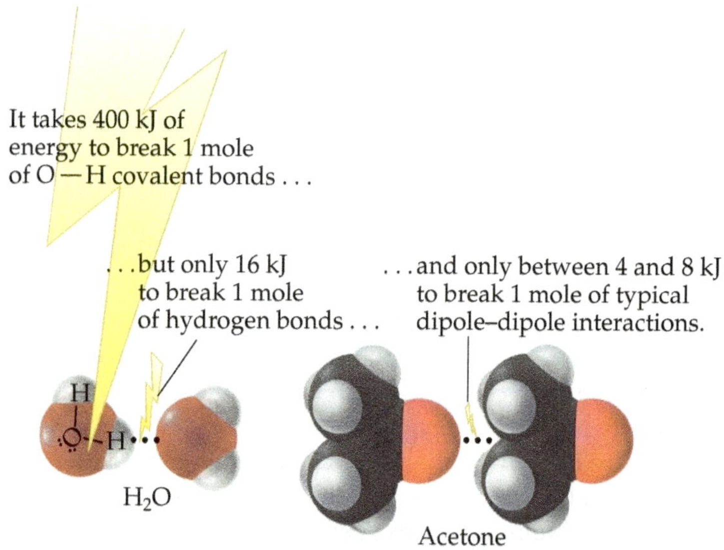

*Intra*molecular attractions (bonds) are much stronger than *inter*molecular attractions!

Water is an excellent example of hydrogen bonding, but it doesn't tell the complete story. The fundamental requirement for the existence of a hydrogen bond between molecules is a nearly naked proton sticking out in space. This requirement is met whenever a hydrogen atom is directly bonded to a small atom having an electronegativity much higher than that of hydrogen. Only oxygen, nitrogen, and fluorine fill the bill. Thus, if there is an O—H, N—H, or F—H covalent bond in a molecule, there can be hydrogen bonding between the molecules.

To form hydrogen bonds, a molecule must contain at least one of these covalent bonds:

···H—N< (or H—N=)

···H—O—

···H—F

Recall that O, N, and F are three of the four electron "hogs" we discussed in Chapter 5 (chlorine is the fourth). These are the four most electronegative elements in the periodic table. So it's not so strange that when a hydrogen atom is bonded to an atom of one of these elements, the hydrogen is practically stripped of its electron and conditions are set for hydrogen bonding. (Chlorine is usually not included in the set of elements that lead to hydrogen bonding.)

WORKPATCH 7.3 The boiling point of one of these compounds is −24 °C, and the boiling point of the other is 78 °C:

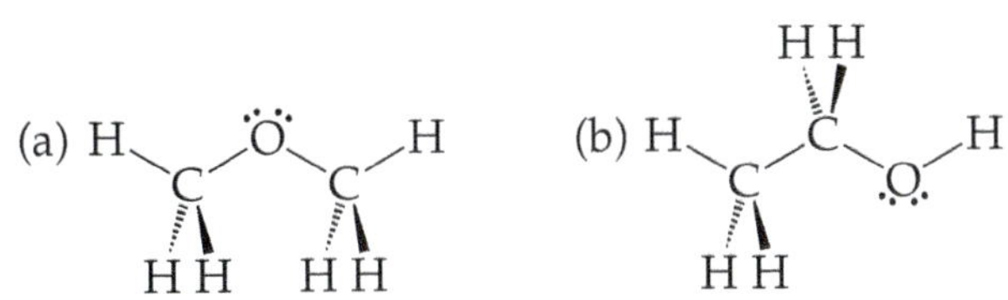

Which boils at which temperature? Why? If hydrogen bonds are involved in either case, make a drawing using dotted lines to show the hydrogen bonding. ●

Both molecules in the WorkPatch have the molecular formula C_2H_6O, which means the difference between their boiling points must be a consequence of their different structures. What is it about their structures that leads to the difference? Make sure you know before going on.

In a very real sense, hydrogen bonding is responsible for life. DNA—the substance that carries the "blueprint" for your body—depends on hydrogen bonding to do its work. A DNA molecule is like a very long, twisted zipper made of two strands held together by a ladder of teeth. There are four kinds of teeth, and the information carried by the DNA is encoded in the sequence of teeth. When a cell needs information from a portion of its DNA, it unzips the two DNA strands in that region and reads the sequence of teeth. Then the strands zip back up.

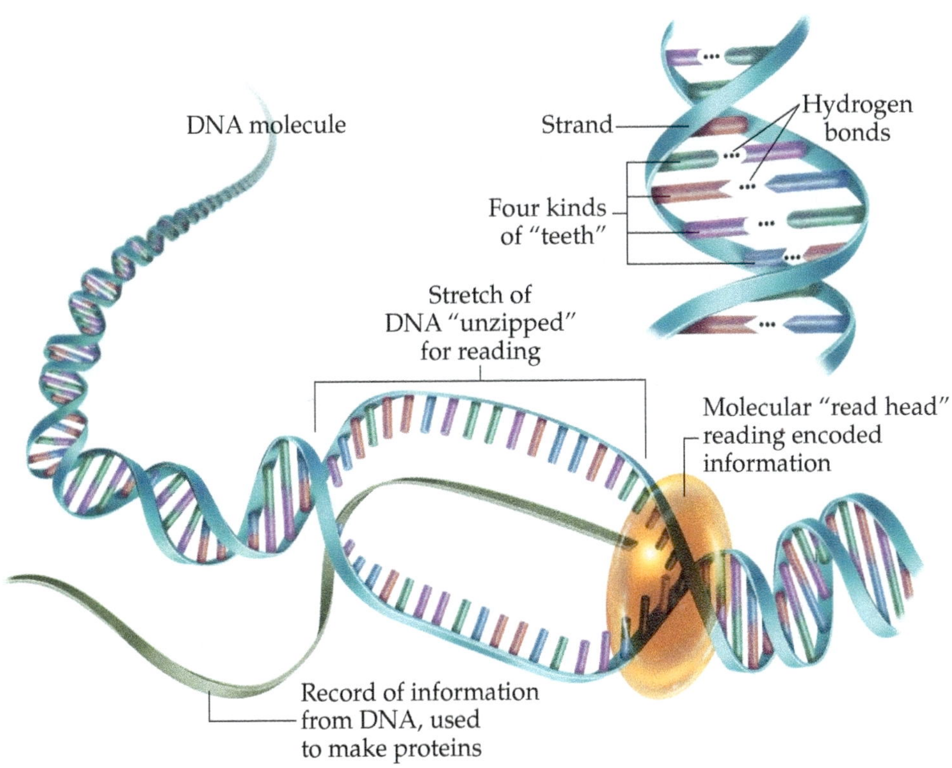

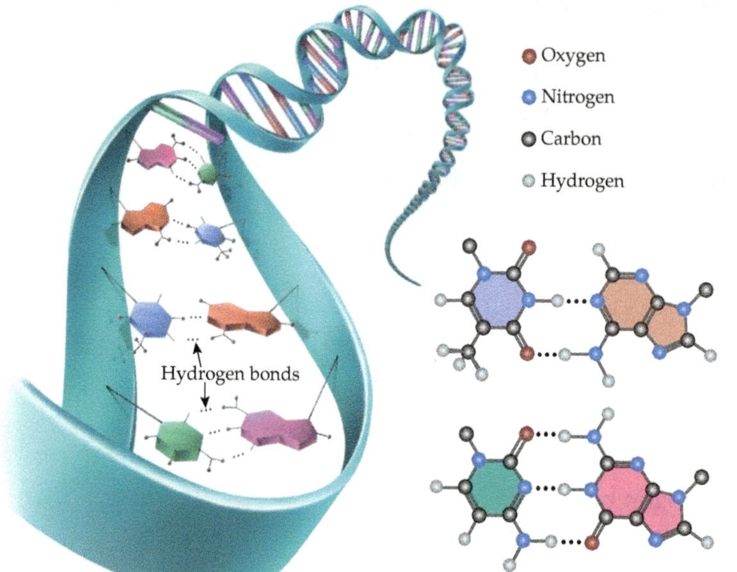

Like a real zipper, a DNA molecule has to be easy to zip and unzip, but when it is zipped, it should stay zipped. Unlike a real zipper, however, the teeth of DNA are held together by chemical interactions. Hydrogen bonds are perfect for this job because covalent bonds would be too strong and London forces would be too weak. The hydrogen bonds are broken when the DNA is unzipped, and they form again when the DNA zips back up.

Practice Problems

7.8 In Practice Problem 7.7, you should have predicted that, based solely on London forces, PH_3 (18 electrons) has a higher boiling point than NH_3 (10 electrons). In fact, this is not the case. Explain why.

Answer: Ammonia molecules can hydrogen-bond (H to N). The hydrogen bonds between ammonia molecules, being much stronger than the London forces acting between PH_3 molecules, give ammonia the higher boiling point.

7.9 Draw some molecules of ammonia, NH_3, and use dotted lines to show the hydrogen bonding between them.

7.10 For the hydrogen halides, the order of boiling points is HF > HI > HBr > HCl.

(a) Why does HF have the highest boiling point?
(b) Why is the boiling point of HI greater than that of HBr and HCl?

7.4 Nonmolecular Substances

So far in this chapter, the solids we have looked at consist of individual molecules held together by intermolecular forces (London forces, dipole–dipole forces, or hydrogen bonds). As you learned in Section 5.5, a solid of this type is called a *molecular solid*. Ice, for instance, is a molecular solid consisting of water molecules attracted to one another by hydrogen bonds. If you could shrink yourself down to the size of a molecule and enter an ice cube, you would see an ordered arrangement of water molecules "locked" in place by hydrogen bonds. Candle wax is another molecular solid, made of individual hydrocarbon molecules, such as $C_{20}H_{42}$, attracted to each other by London forces.

Portion of a NaCl Lattice

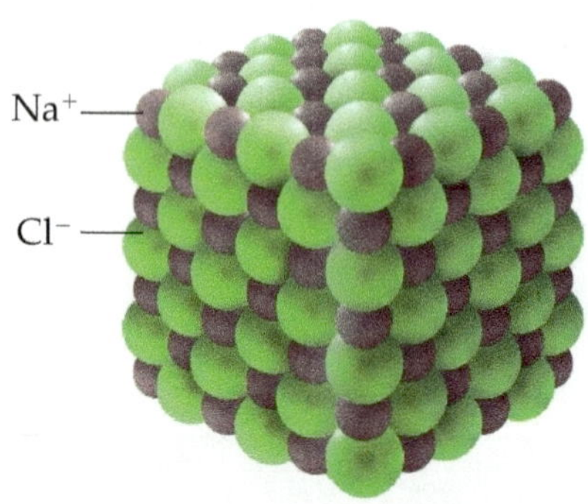

However, not all solids are molecular solids formed because of relatively weak intermolecular forces. As you also learned in Chapter 5, some solids contain no molecules at all. Instead, they owe their ability to exist in a condensed phase to far stronger covalent or ionic bonding forces. Sodium chloride is one example of such a *nonmolecular ionic solid*. As we saw in Chapter 5, NaCl is made up of a vast array (lattice) of positive sodium ions, Na^+, and negative chloride ions, Cl^-. In such a lattice, there is nothing you can point to and call a molecule. There is only a repeating pattern of positive and negative ions. Any solid that doesn't consist of discrete molecules is called a **nonmolecular solid**.

All simple ionic compounds are like this. Calcium chloride, $CaCl_2$, sodium bromide, NaBr, and magnesium oxide, MgO, are all lattices of ions. Because there is nothing in these solids that we can identify as being a molecule, it would be incorrect to call the forces holding them together *intermolecular* because this word means "between molecules," and there are no molecules. Instead, these compounds are held together by the much stronger ionic bonds we talked about in Chapter 5. Melting such compounds means breaking these ionic bonds. Because overcoming these strong attractions requires tremendous amounts of energy (heat), many ionic compounds have melting points in excess of 1000 °C. Sodium chloride, for instance, has a melting point of 801 °C, much higher than the 0 °C melting point of water.

WORKPATCH 7.4 What is wrong with the reasoning behind this incorrect statement? The fact that water melts at 0 °C and NaCl melts at 801 °C indicates that ionic bonds are much stronger than covalent bonds. ●

The incorrect statement in the WorkPatch is a common misconception among chemistry students. After all, water molecules do have covalent bonds in them. If you see why the statement is incorrect, you understand the fundamental difference between melting a molecular solid and melting a nonmolecular solid. Check your answer against the one given at the end of the chapter.

There are nonmolecular solids that are not ionic. Some examples are silicon dioxide (quartz), SiO_2, from which glass is made, and diamond, a form of pure carbon. These compounds are often called either **network solids** or **network covalent substances** because they consist of a large network of atoms held together by covalent bonds.

Two Network Solids

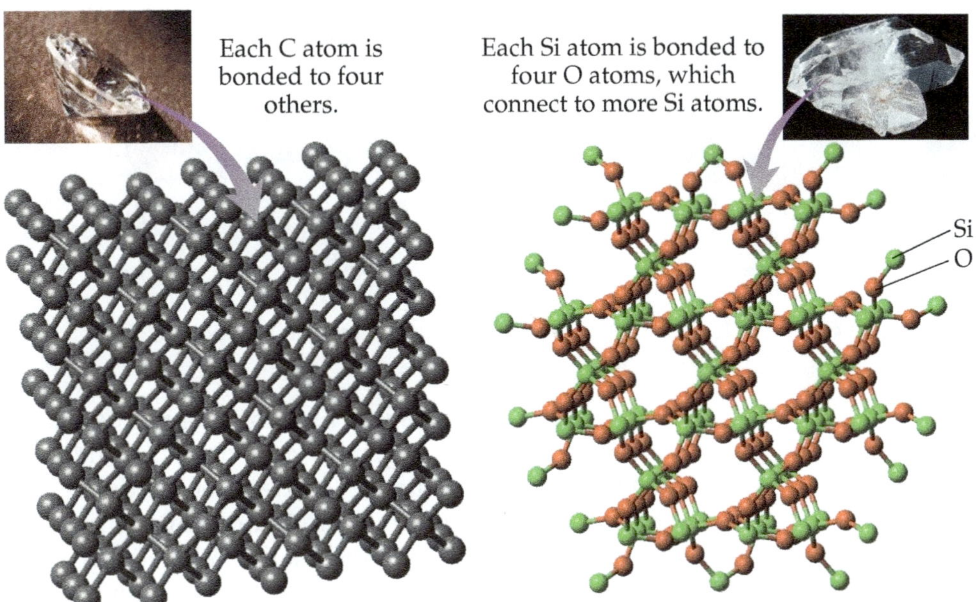

Diamond, C

Quartz, SiO_2

Diamond is a vast network of carbon atoms, with each atom covalently bonded to four other carbon atoms. Quartz is a vast network of silicon atoms, each covalently bonded to four oxygen atoms, which are in turn covalently bonded to more silicon atoms. Just as with an ionic lattice, there is nothing in these network solids that stands out as an individual molecule. What we have now is a lattice of atoms instead of ions, all interconnected by very strong covalent bonds. Melting these compounds involves breaking covalent bonds and consequently requires a lot of heat. Diamond melts above 3550 °C, and quartz requires temperatures in excess of 1700 °C before it begins to liquefy.

It is not always easy to determine whether a particular solid substance is molecular (like solid H_2O) or nonmolecular (like SiO_2 or NaCl) just by looking at the chemical formula. If the formula includes a group IA (1) or IIA (2) metal plus a nonmetal (like NaCl or $MgBr_2$), it is a good bet the solid is ionic and therefore nonmolecular. For other compounds, such as SiO_2, the formula does not contain enough information to tell you whether the solid is molecular or nonmolecular. If we hadn't told you that quartz is a network solid, for instance, you would have to determine its internal structure by experiment. The melting point of a solid can often provide a good hint, however, because melting point is directly related to the strength of the forces holding the solid together. If a melting point is tremendously high—1000 °C or above, say—the solid is most likely nonmolecular. If a melting point is much lower, weaker forces are present, and the solid is more likely molecular.

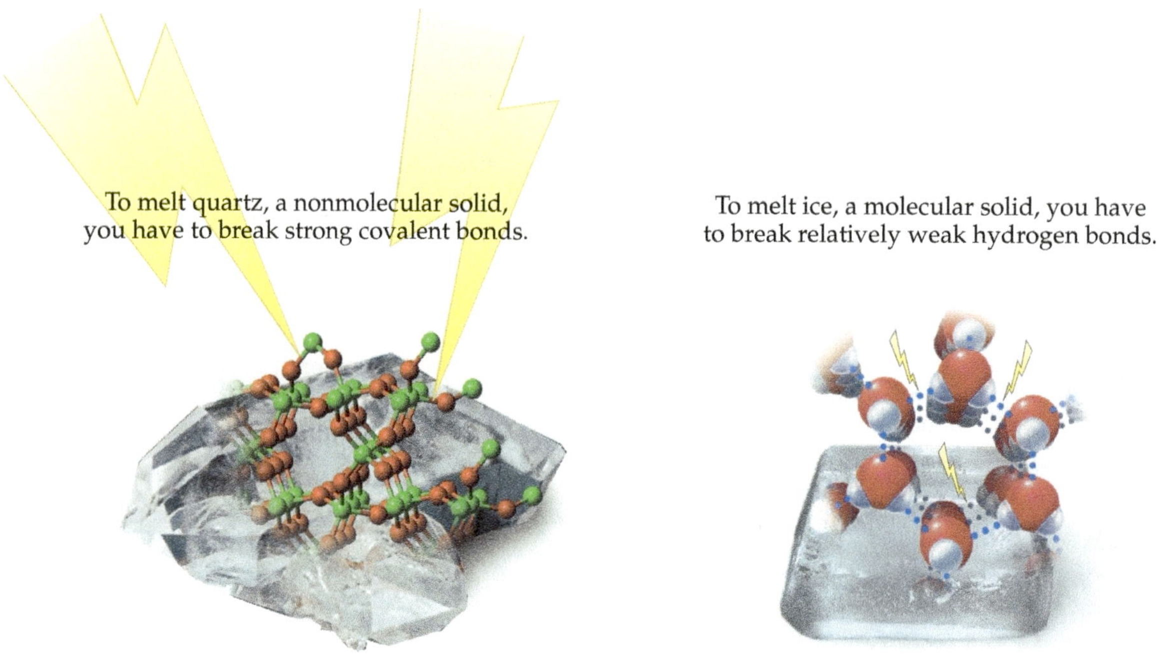

Quartz, melting point 1710 °C

Ice, melting point 0 °C

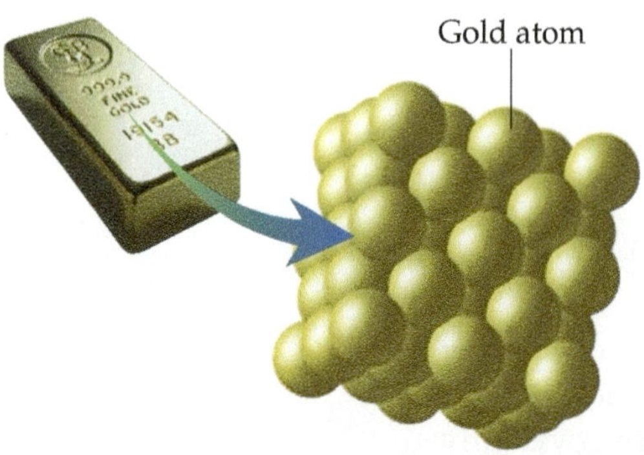

Another class of solids having high melting points are the pure metals, such as silver, Ag, iron, Fe, copper, Cu, and gold, Au. These nonmolecular solids are made up of a lattice of neutral metal atoms held together by **metallic bonds**. Like covalent bonds, metallic bonds are the result of shared electrons, but there is one important difference. In a piece of metal, the valence electrons on the metal atoms are free to move about and be shared by and be attracted to *all* the nuclei in the piece, giving rise to extra attractive forces between atoms. (This mobility of valence electrons explains why metals conduct electricity and heat so well.)

Like their covalent and ionic counterparts, metallic bonds are true chemical bonds, which means that, in general, they are very strong. As you now know, strong interparticle bonds in a solid usually mean high melting points. Gold, Au, has a melting point of 1064 °C, for example, and iron, Fe, has a melting point of 1535 °C. These are very high temperatures, reflecting the high strength of metallic bonds.

As usual, there are exceptions to this generalization. Sodium metal, Na, has a melting point of only 98 °C, and mercury, Hg, is a liquid at room temperature, with a melting point of −39 °C. For reasons that go beyond the scope of this book, these are two cases where metallic bonding is quite weak even though these substances are nonmolecular solids.

The following figure summarizes the types of solids we have looked at.

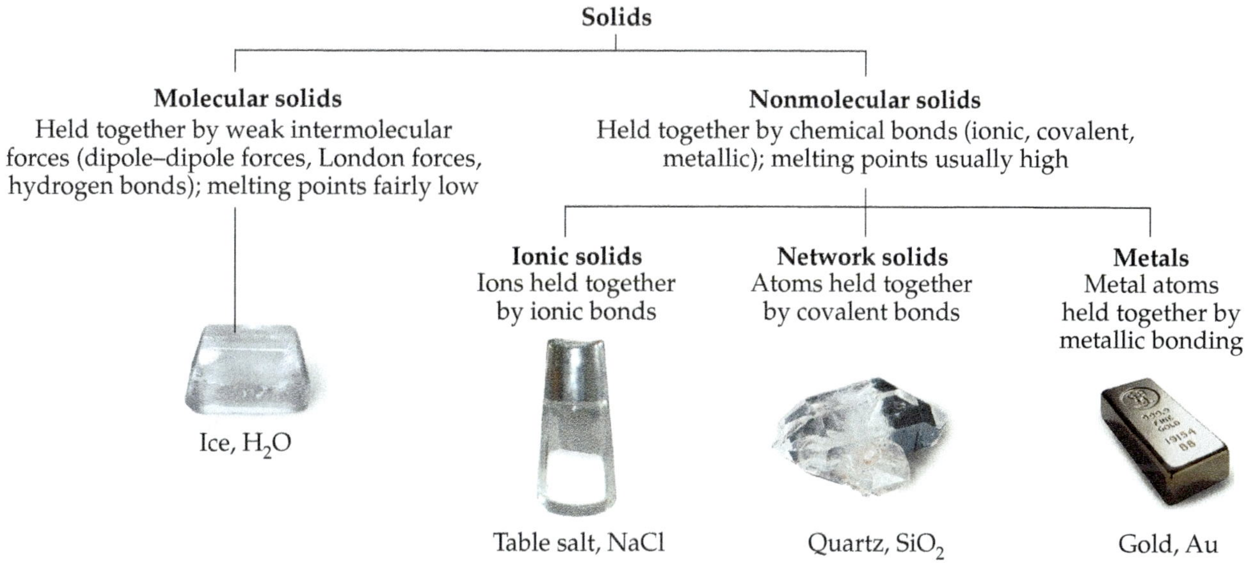

Practice Problems

7.11 Sodium oxide, Na_2O, is a white-gray powder that sublimes (goes directly from solid to gas) at 1275 °C. Is sodium oxide a molecular or nonmolecular solid?

Answer: First, sodium oxide is made from a group IA (1) metal and a nonmetal. This practically ensures that the compound is a nonmolecular ionic solid. Second, this compound sublimes at a tremendously high temperature. This also indicates that the compound is nonmolecular (the high temperature is required to break ionic bonds in the lattice of Na^+ and O^{2-} ions).

7.12 How is a piece of iron similar to a piece of sodium chloride in terms of the forces that must be overcome to melt it?

7.13 Elemental sulfur is a yellow solid that melts at 113 °C. Studies show sulfur to be a molecular solid consisting of S_8 molecules held together in a lattice. What forces must be overcome to melt the solid?

7.14 The following are solids at room temperature (about 25 °C). Classify them as molecular, ionic, network, or metallic:

(a) Zirconium, Zr, mp = 1852 °C
(b) Lead, Pb, mp = 328 °C
(c) Calcium nitride, Ca_3N_2, mp = 1195 °C
(d) Graphite form of carbon, C, sublimes at 3652 °C
(e) Yellow phosphorus, P_4, mp = 44 °C

+1 One More Thing

7.5 Vancomycin—The Antibiotic of Last Resort and Its Five Life-Saving Hydrogen Bonds!

Modern medicine has increased average life expectancy by more than 30 years since the turn of the last century. This dramatic and unprecedented extension is due in large part to the introduction and widespread use of chemical compounds called antibiotics in the treatment and prevention of many diseases. The availability and widespread use of sulfa drugs (discussed in Chapter 6, p. 219) in the bombed-out cities of Western Europe during World War II are credited with saving more lives than have been lost in all the wars that have ever been fought in the history of the planet. The introduction of even more powerful antibiotics, like penicillin, streptomycin, and the cephalosporins, in the mid-twentieth century, began what is considered by many to be the advent of the "Golden Age of Medicine."

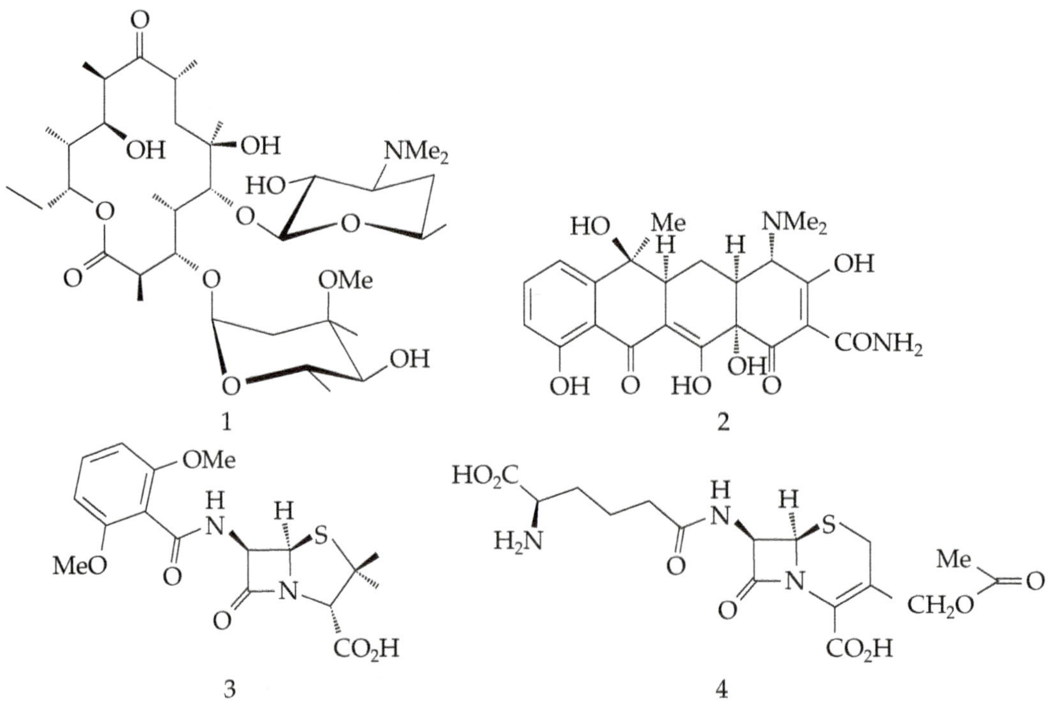

Classic antibiotics: erythromycin B (1), tetracycline (2), methicillin (3), and cephalosporin C (4). [Me = CH_3]

The phenomenal success these and many similar antibiotics have had in treating disease has led to their widespread use well outside of the doctor's office. Present today in everything from antibacterial hand soap to cattle feed, the extensive use of these wonder drugs has reduced their effectiveness in many life-and-death situations. The overuse of these drugs has accelerated the creation of large populations of disease-carrying bacteria that have developed an immunity to their effects, and these strains of antibiotic-resistant bacteria are increasing at an alarming rate.

To its credit, the medical profession has seen this coming for some time and responded by reserving a small handful of very potent antibiotics for use only in extreme cases of bacterial infection in humans. One of the most valuable of these is the antibiotic vancomycin.

Vancomycin:

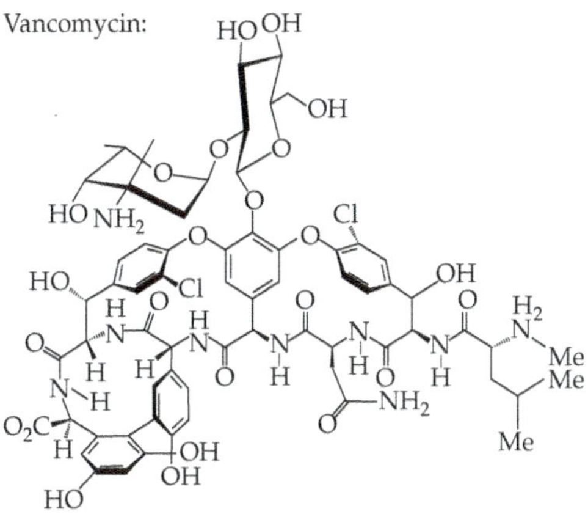

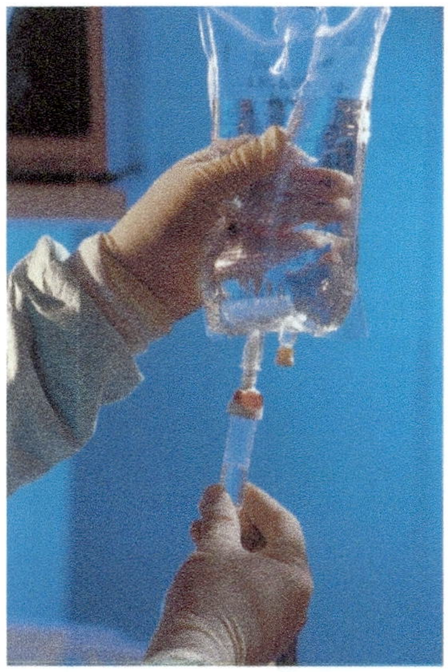

Vancomycin, like penicillin, works by interfering with the construction of the cell walls of bacteria, weakening these structures so effectively that the cells often burst open from their own internal pressure. Vancomycin does this through the formation of five separate hydrogen bonds between itself and a key component (called a glycopeptide) of the newly forming bacterial cell wall.

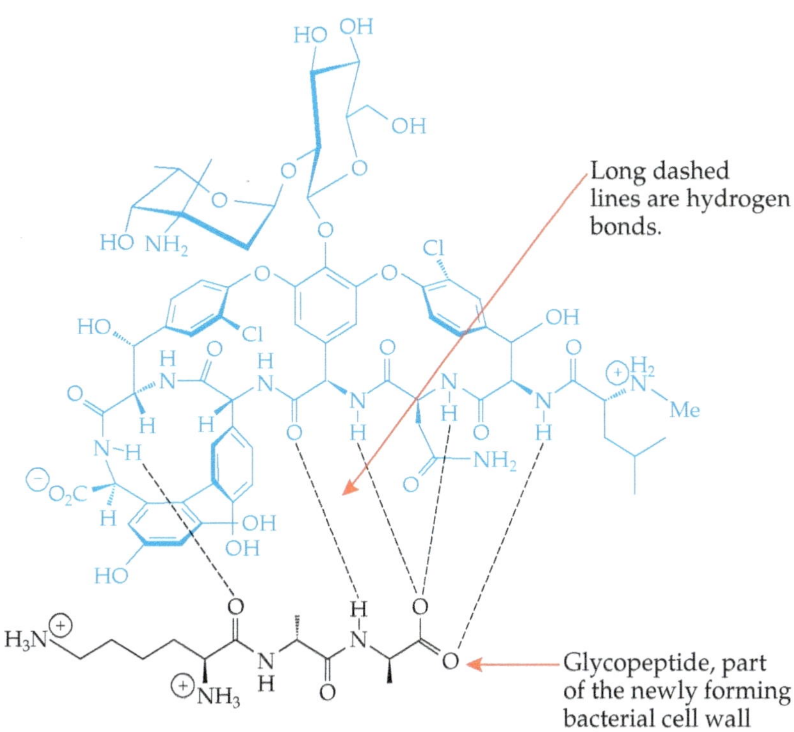

This strategy of reserving vancomycin for only the most serious cases of antibiotic-resistant infections has worked well for nearly 20 years, but time is running out. Since the 1990s, the number of reported cases of hospital staph infections caused by bacteria that are vancomycin resistant has increased dramatically. These bacteria have become immune to the effects of vancomycin by making one simple change in their cell wall machinery. The replacement of a single NH with an O atom is all it takes. This destroys one of the key hydrogen-bonding interactions and decreases the effectiveness of vancomycin by nearly 1000-fold. For want of this single hydrogen bond, many lives have been lost.

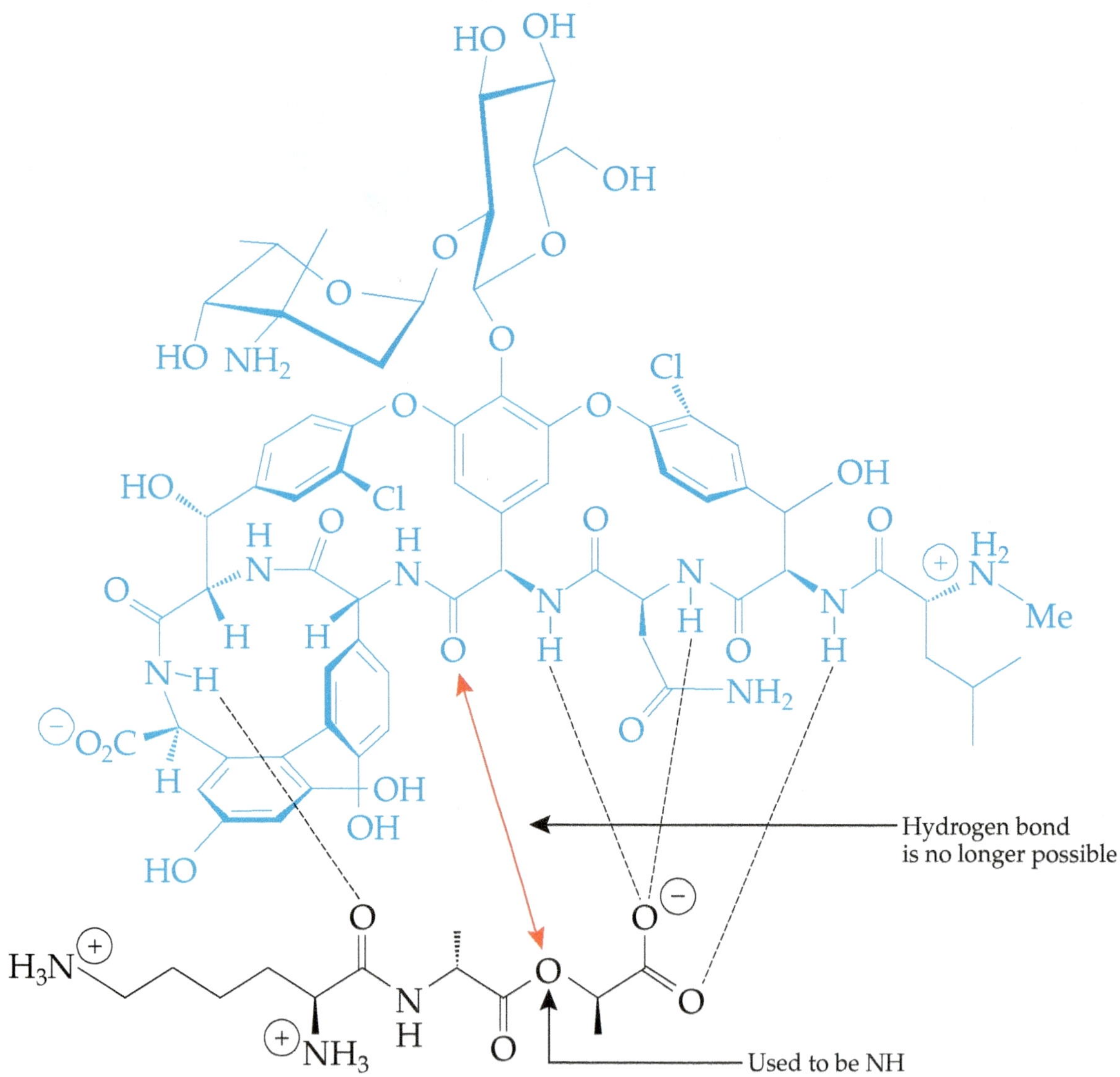

Drug companies are working furiously, around the clock, to develop new and unique antibiotics to replace vancomycin, but the drug development and approval process can take many years. In the meantime, chemists are working to modify vancomycin in different ways to foil these resistant bacteria. One approach has been to recognize that the cell wall of a bacterium is primarily a nonpolar or hydrophobic environment, and a variation of vancomycin has been produced that has a small nonpolar tail attached to it (see next page).

When this vancomycin derivative attaches to the cell wall of a resistant bacterium with its now weakened quartet of hydrogen bonds, this tail can interact and "dissolve" (like dissolves like) in the cell wall itself and help to anchor the vancomycin molecule more firmly. This, and other minor modifications of vancomycin, will hopefully increase the effectiveness of the drug long enough for the drug industry to get some of the newer antibiotics out of the development pipelines and on to the front lines of the battle.

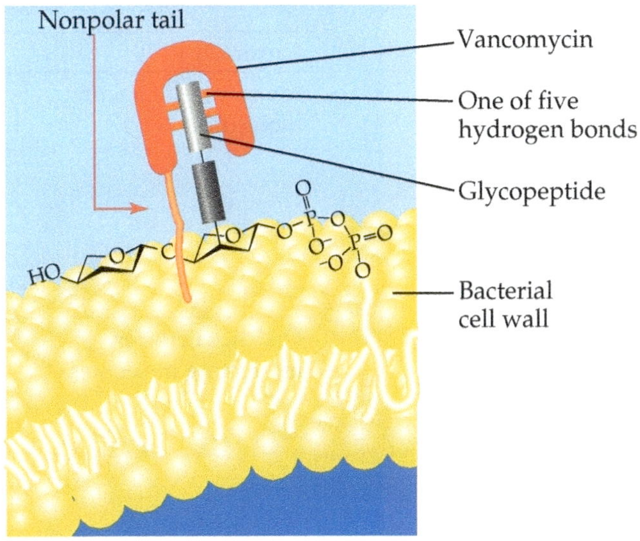

We end this discussion of intermolecular forces with a cautionary note. It's easy to get in the habit of thinking that hydrogen bonds are always the strongest intermolecular forces, dipole–dipole forces always the next strongest, and London forces always the weakest. Although this is *generally* true, it is not *always* true. For example, antimony trihydride, SbH_3, is an essentially nonpolar molecule because Sb and H have comparable electronegativities. Nevertheless, SbH_3 has a higher boiling point than does ammonia, NH_3, which forms hydrogen bonds. This seeming discrepancy occurs because antimony is a big atom, containing 51 electrons. This large size results in London forces between SbH_3 molecules that are stronger than the hydrogen bonds between NH_3 molecules. So, as always, respect the general trend but keep an open mind.

This chapter began by considering the random motion of molecules in the gas phase. Intermolecular forces have little effect on gas behavior because the molecules are too far apart and moving much too fast to feel more than brief, faint forces of attraction and repulsion for one another. We shall have considerably more to say about this state of affairs in Chapter 11.

Once we slowed down the gas particles and allowed them to feel various intermolecular forces, we saw a distinct change in both properties and appearance of the collection of molecules. First, the gas phase becomes the liquid phase with its less chaotic, more ordered depiction of molecules collected together, touching one another, still moving but much more slowly, sliding over and around one another. And finally the liquid phase gives way to the solid phase, the ultimate in order, where each particle is assigned a distinct location in a lattice and the only allowed motions are slight vibrations about this position. The ordering of the chaotic state of affairs that existed in the gas phase was due to intermolecular forces. To understand the behavior of matter in a condensed phase (solid or liquid), you must be able to accurately describe the interactions between the molecules in that phase. These interactions are difficult to describe mathematically because they depend on many variables. For one thing, they depend on the identities of the molecules involved. Are they polar or nonpolar, large or small? These interactions can also depend on how far apart the molecules are and on how they are oriented relative to one another. In a condensed phase, the interactions even depend on a molecule's location—that is, is the molecule on the surface or in the interior of the condensed phase? Table 7.1 summarizes our discussion.

Table 7.1 Factors Influencing Strength of Intermolecular Forces

Factor	Stronger force	Weaker force
Identity of interacting molecules	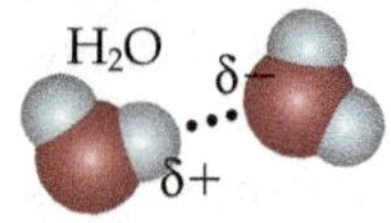 Hydrogen bonding between water molecules	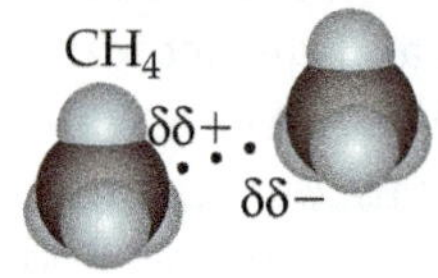 London forces between methane molecules
Distance between interacting molecules	 Force between close molecules	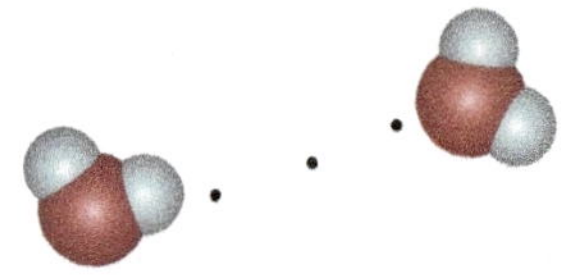 Force between distant molecules
Relative orientation of interacting molecules	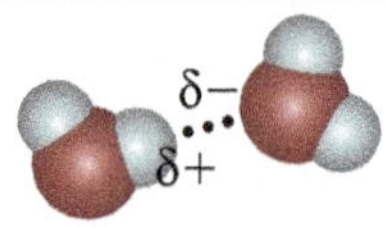 Attraction	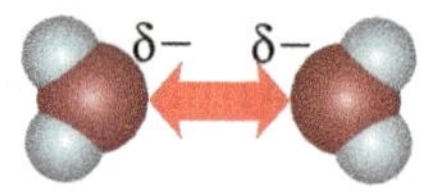 Repulsion
Location of molecule in liquid or solid	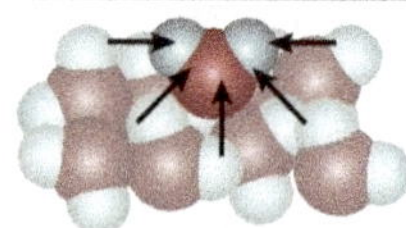 Surface molecule feels forces only from below.	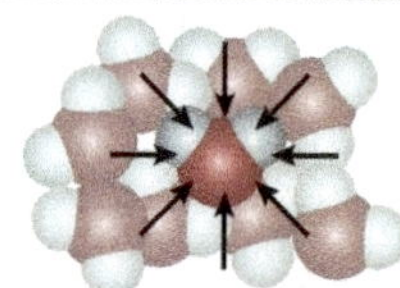 Interior molecule feels forces from all around.

All these factors that affect intermolecular forces make understanding and predicting the behavior of condensed phases very complicated. This makes it nearly impossible to develop a single mathematical equation that can describe the behavior of all liquids and solids. Just imagine, then, how easy it might be to explain the behavior of matter if we could ignore intermolecular forces. For gases, we can! Stay tuned for the next chapter.

Have You Learned This?

Learning Outcomes

7.1	Why Does Matter Exist in Different Phases?	Describe the three phases of matter from a molecular viewpoint. [15, 18, 21, 22, 23] Explain what is necessary for one phase to transform into another. [16, 17, 19, 24, 25]
7.2	Intermolecular Forces and Hydrogen Bonding	Describe London forces and their magnitude. [26, 27, 29, 39]
7.3	A Closer Look at Dipole Forces—Hydrogen-Bonding	Describe hydrogen bonding and predict when it can occur. [31, 32, 35 36, 40] Predict the most important intermolecular force present and its effect on physical properties. [33, 37, 38]
7.4	Nonmolecular Substances	Describe what a nonmolecular substance is on an atomic level. [44, 45, 46, 48] Explain why nonmolecular compounds typically have very high melting points. [49, 50, 51]
7.5	Vancomycin—The Antibiotic of Last Resort and Its Five Life-Saving Hydrogen Bonds	Describe some interesting and important effects due to hydrogen bonding. [89, 91]

Concept Questions

1. A condensed phase is
(a) a shortened (condensed) version of a substance.
(b) the gas form of a substance.
(c) the liquid and solid forms of a substance.
(d) formed by evaporation.

2. The molecules in a gas phase will
(a) speed up upon heating the gas.
(b) slow down upon heating the gas.
(c) change phase upon sufficient heating of the gas.
(d) form a condensed phase upon sufficient heating.

3. Condensation occurs when
(a) the temperature of a substance is adjusted to its melting point.
(b) the kinetic energy of the molecules is not sufficient to overcome intermolecular attractions.
(c) the molecules are moving in random directions.
(d) the kinetic energy of the molecules is sufficient to overcome intermolecular attractions.

4. A difference between the solid phase and the other two phases of a pure substance is
(a) only the solid phase has molecules in fixed positions.
(b) only the solid phase can experience melting.
(c) only the solid phase is at or below the freezing point of the substance.
(d) all of the above are correct.

5. Which of the following statements is true regarding London forces?
(a) They are present only in the gas phase.
(b) They are present only in the liquid phase.
(c) They are present only in the solid phase.
(d) They are present in all phases of matter.

6. Which of the following statements is true regarding London forces?
(a) They have nothing to do with the electrons in a molecule.
(b) They get smaller as the number of electrons in a molecule increases.
(c) They get larger as the number of electrons in a molecule increases.
(d) They get larger as the electrons get closer to the nucleus.

7. Which of the following statements is true regarding London forces?
(a) They are due primarily to valance electrons.
(b) They can never be stronger than dipolar attractions.
(c) They can be quite strong for molecules that possess many electrons.
(d) All of the above.

8. Hydrogen bonding is
(a) the name given to the chemical bond in the H—H or H_2 molecule.
(b) a particularly strong type of dipolar interaction.
(c) as strong as a chemical bond.
(d) when hydrogen bonds to any other element.

9. A hydrogen bond between molecules can occur
   (a) only if the molecules involved are nonpolar.
   (b) only if the molecules possess a hydrogen atom bound to O, N, or F.
   (c) only if full charges develop on the atoms instead of partial charges.
   (d) only in the absence of dispersion forces.

10. A network solid is
   (a) the easiest of solids to melt.
   (b) made up of ions bound to one another via ionic bonds.
   (c) made up of molecules attracted to one another.
   (d) made up of atoms covalently bound to one another.

Answers: 1c; 2a; 3b; 4d; 5d; 6c; 7c; 8b; 9b; 10d

Why Does Matter Exist in Different Phases?

7.15 Describe as completely as you can how small molecules like N_2 and O_2 behave at the molecular level in a warm room.

7.16 Consider cooling a gas so that it gets colder and colder.
   (a) Explain why this would eventually cause the gas to condense into a liquid.
   (b) Explain what would happen to this gas if you could somehow turn off all intermolecular forces.

7.17 What does the kinetic energy of molecules have to do with changing phases?

7.18 Is it incorrect to say that molecules are motionless in the liquid phase? Explain.

7.19 Explain in molecular terms how heating causes a liquid to change to the gas phase.

7.20 Water vapor liquefies when cooled below 100 °C. Gaseous nitrogen liquefies when cooled below −196 °C. What does this information tell you about the relative strengths of the intermolecular forces for these molecules?

7.21 Why are liquids and solids referred to as condensed phases?

7.22 Why does a gas expand to fill the container it is in, but a liquid and a solid do not?

7.23 On the molecular level, describe each phase of matter with respect to the amount of order present.

7.24 For this question, you should consider only polarity. Molecule A—B has a smaller dipole moment than molecule C—D.
   (a) Which molecule has the higher freezing point? Explain why.
   (b) Which molecule has the higher boiling point? Explain why.

7.25 Draw a picture that shows how three polar HBr molecules in the gas phase would attract one another. What kind of intermolecular force is involved?

Intermolecular Forces and Hydrogen Bonding

7.26 Explain what gives rise to London forces and when they occur.

7.27 Would you expect CCl_4 or CBr_4 to have the higher boiling point? Explain your answer.

7.28 Chloromethane (CH_3Cl) has a much higher boiling point than methane (CH_4). Give two reasons for this.

7.29 Propane (C_3H_8) is a gas at room temperature, whereas octane (C_8H_{18}) is a liquid. Explain why this is so.

7.30 What is wrong with the statement "London forces are always weaker than dipole–dipole forces"?

7.31 Consider the molecules HF and HCl. The electronegativities of the atoms involved are H, 2.1; Cl, 3.0; F, 4.0.
   (a) Which of the two molecules is more polar? Explain your answer.
   (b) For which substance are the dipole–dipole attractions between the molecules stronger? Explain your answer.
   (c) Which gas would, upon cooling, liquefy first? Explain your answer. Also, indicate which compound would have the higher boiling point based on your answer.
   (d) Do both molecules also experience London forces of attraction? If yes, which would have the greater London forces?

7.32 When discussing the intermolecular forces between methanol molecules, chemists usually ignore any London forces between them. Why are they justified in doing this?

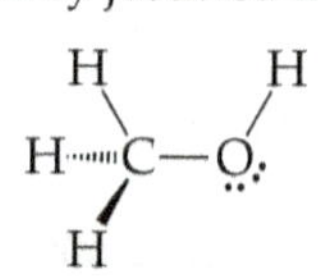

Methanol

7.33 Predict which compound in each pair will have the higher melting point, and explain why for each pair.

(a) CS_2 or CCl_4
(b) Cl_2 or F_2
(c) CH_4 or NH_3
(d) CH_3Cl or Cl_4

7.34 Long-chain hydrocarbon molecules of the type $CH_3(CH_2)_{20}CH_3$ are solids and are used for things like waxes. The CH bonds are essentially nonpolar. Why are waxes solid at room temperature?

7.35 Which of the molecules below would *not* form hydrogen bonds?
(a) CH_3OH
(b) CH_3OCH_3
(c) CH_3COOH
(d) NH_3
(e) None of these would form hydrogen bonds.

7.36 What is so special about hydrogen atoms that makes hydrogen bonding possible?

7.37 Two different compounds have the same elemental composition, C_3H_8O. One has a low boiling point and the other a much higher boiling point. What attractive force must be present in one of these compounds that is not present in the other?

7.38 Consider molecules of the type H_2A, where A is a group VIA atom (O, S, Se, Te). Their relative boiling points are shown below.

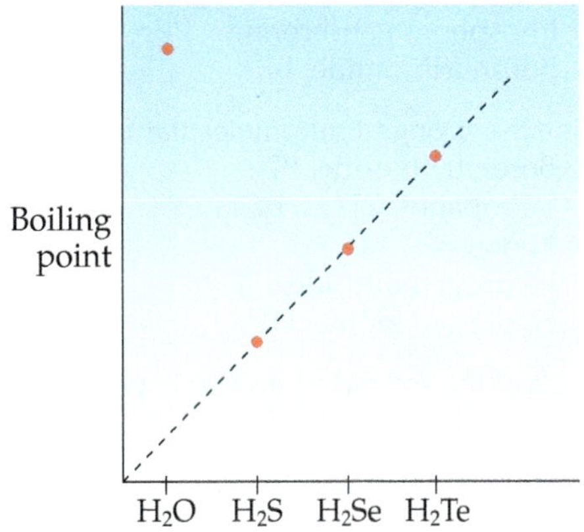

(a) Why is the boiling point trend $H_2S < H_2Se < H_2Te$?
(b) The boiling point of H_2O is well above the line that the others are on. Explain why this is so.

7.39 What do we mean by *induced dipole* when discussing London forces?

7.40 Why do we use dotted lines rather than solid lines to represent hydrogen bonds?

7.41 Examine the diagram of DNA on page 265 to answer this question: Why can hydrogen bonds form in DNA?

7.42 Why would either covalent bonds or London forces be inappropriate for attaching the two strands of DNA to each other?

7.43 How is it possible for a nonpolar molecule to have a higher boiling point than a polar one?

Nonmolecular Substances

7.44 What is the fundamental difference between a molecular substance and a nonmolecular substance?

7.45 Is it sometimes, always, or never possible to tell from a compound's formula whether it is a molecular or nonmolecular substance?

7.46 What do we mean by a "lattice of ions"?

7.47 In general, nonmolecular solids have much higher melting points than molecular solids. Why is this so?

7.48 What is a network covalent substance? Give an example.

7.49 What evidence is there that metallic bonds can be as strong as ionic or covalent bonds?

7.50 Predict which compound in each pair will have the higher melting point, and explain why for each pair.
(a) HI or KI
(b) Na_2O or H_2O
(c) CaF_2 or HF
(d) SiO_2 or CO_2

7.51 All of the following are solids at room temperature. Classify them as molecular, ionic, network, or metallic.
(a) Potassium (K), mp = 64 °C
(b) Potassium chloride (KCl), mp = 770 °C
(c) Red phosphorus (P), mp = 590 °C
(d) Boron triiodide (BI_3), mp = 50 °C

7.52 How does the attraction between two molecules depend on the distance between them?

7.53 Why should the attraction between two water molecules depend on how they are oriented with respect to each other? Use drawings to illustrate your answer.

Additional Problems

7.54 Describe how the molecules in a liquid behave as the temperature of the liquid increases.

7.55 Solid ice is less dense than liquid water. What does this say about how many water molecules there are per unit volume of ice relative to how many water molecules there are per unit volume of liquid water?

7.56 Which of the following best describes a liquid:
(a) The phase of matter in which particles are typically separated by the least distance

(b) The phase of matter in which particles are in a fixed, rigid arrangement
(c) The phase of matter in which particles completely fill the volume of their container
(d) The phase of matter in which particles are in a loose, changeable arrangement but do not completely fill the volume of their container

7.57 At 25 °C, fluorine, F_2, and chlorine, Cl_2, are gases but bromine, Br_2, is a liquid. What does this say about the intermolecular forces in bromine relative to those in fluorine and chlorine?

7.58 In which of the following are there no dipole–dipole forces between molecules? Justify your answer.
(a) AsH_3
(b) CO_2
(c) H_2O
(d) $SeCl_2$

7.59 Which is more likely to be a gas at room temperature, CH_4 or CH_2Cl_2? Justify your answer.

7.60 Which of the following does not form hydrogen bonds? Justify your choice.
(a) Methyl alcohol, CH_3OH
(b) Hydrofluoric acid, HF
(c) Ammonia, NH_3
(d) Methane, CH_4

7.61 Which of the following would you expect to have the highest boiling point? Justify your choice.
(a) Propane, C_3H_8
(b) Carbon dioxide, CO_2
(c) Ethyl alcohol, CH_3CH_2OH
(d) Methyl fluoride, CH_3F

7.62 Which of the following would most likely be a gas at room temperature? Justify your choice.
(a) NaCl
(b) C_2H_2
(c) Na metal
(d) CH_3F

7.63 (a) Name the strongest intermolecular force in CH_3OH, CH_3Cl, CH_3CH_3, and $CH_3CH_2CH_3$.
(b) Rank these molecules from lowest to highest boiling point.

7.64 Acetic acid,

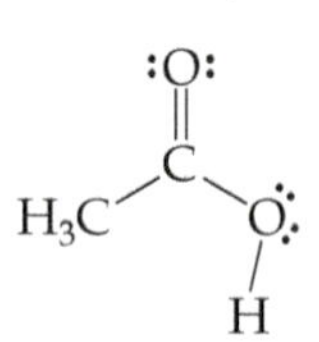

is very soluble in water because of the many hydrogen bonds that can form. Add water molecules to this drawing and show the hydrogen bonds possible.

7.65 Show how two CH_2Cl_2 molecules in the liquid phase are oriented with respect to each other and explain why the molecules align this way.

7.66 (a) Rank these molecules in order of increasing boiling point:

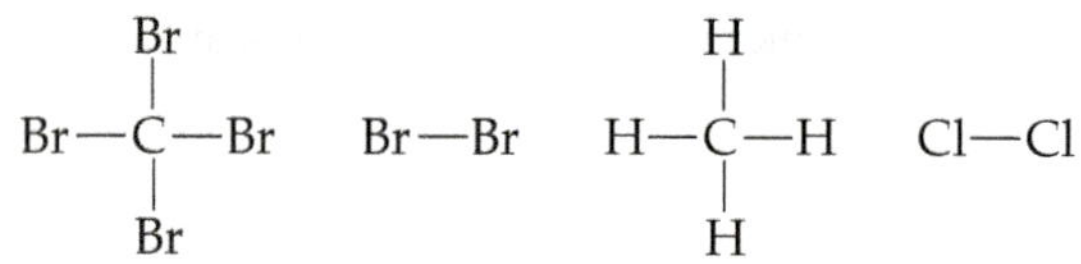

(b) State your reason for the order you chose in part (a).

7.67 (a) Rank these molecules in order of increasing boiling point:

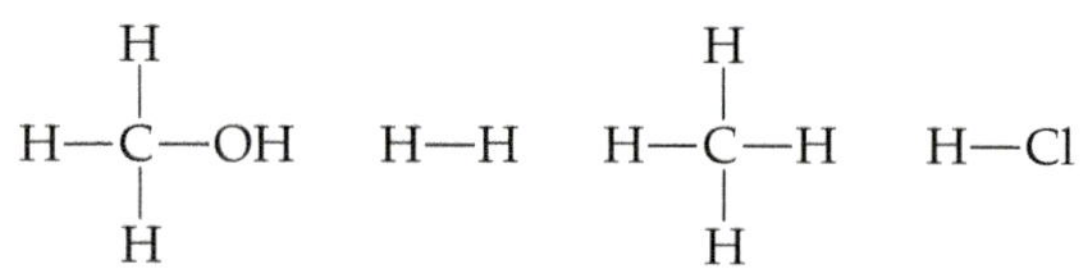

(b) State your reason for the order you chose in part (a).

7.68 Which compounds would you expect to experience only London forces? Explain your choices.
(a) Carbon tetrabromide, CBr_4
(b) Methyl bromide, CH_3Br
(c) Phosphorous tribromide, PBr_3
(d) Boron tribromide, BBr_3

7.69 Name the strongest intermolecular force expected in
(a) Boron trifluoride, BF_3
(b) 1-Propanol, $CH_3CH_2CH_2OH$
(c) Xenon, Xe
(d) Hydrogen fluoride, HF
(e) Hydrogen iodide, HI

7.70 Which of the following do you expect to be nonmolecular solids?
(a) Sodium hydroxide, NaOH
(b) Solid ethanol, CH_3CH_2OH
(c) Iron, Fe
(d) Solid silane, SiH_4

7.71 SiO_2, the main component of glass, is a solid at room temperature, and CO_2 is a gas at room temperature. How do the structures of these compounds explain this fact?

7.72 Arrange these substances in order of increasing melting point: CH_3OH, SiO_2, C_2H_6, NaCl.

7.73 Ethylene glycol, CH_2OHCH_2OH, and pentane, C_5H_{12}, have approximately the same molar mass. Nevertheless, one of these compounds boils at 198 °C, and the other boils at 36 °C.
Which compound boils at which temperature?

Use an argument based on intermolecular forces to justify your choice.

7.74 Sketch the hydrogen bonds present in the liquid 1-propanol, $CH_3CH_2CH_2OH$.

7.75 Both diamond and graphite are network solids consist solely of carbon atoms:

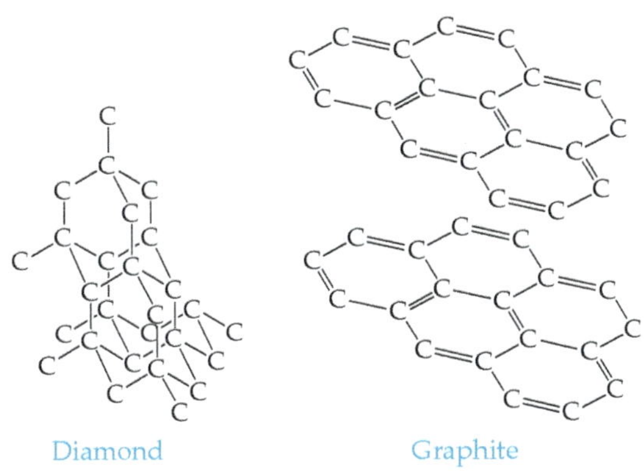

In diamond, every carbon atom is bonded to four other carbon atoms. In graphite, every carbon atom is bonded to three other carbon atoms, creating sheets of atoms that lie on top of one another.

(a) Which type of intermolecular forces exist between the sheets of carbon atoms in graphite?

(b) Diamond is an extremely hard substance, whereas graphite is a soft, slippery substance often used as a lubricant. Use intramolecular and intermolecular forces to explain this difference in physical properties.

7.76 When considering which is stronger, a covalent bond or an ionic bond, why is it fairer to contrast the melting points of diamond and NaCl than to contrast the melting points of ice and NaCl?

7.77 Metallic bonding is often described as a lattice of metal cations in a sea of valence electrons. What is it about metallic bonding that allows it to be described in this fashion?

7.78 As you go down the column of noble gases in the periodic table, the boiling point increases. Explain this trend.

7.79 Eicosane, $C_{20}H_{42}$, has a higher melting point and a higher boiling point than water, even though water has hydrogen bonds between molecules and eicosane does not. Explain how this can be.

7.80 Most covalent molecular substances have much lower melting points than ionic substances. What does this say about the strength of a covalent bond relative to the strength of an ionic bond? Explain your answer.

7.81 Arrange in order of increasing boiling point: CO_2, SO_2, CH_3CH_2OH, Al.

7.82 Consider the compounds titanium tetrachloride ($TiCl_4$) and zirconium tetrachloride ($ZrCl_4$). Zirconium is in the same group as titanium on the periodic table, and both Ti–Cl and Zr–Cl polar covalent bonds are quite strong. $TiCl_4$ has a melting point of −25 °C (i.e., it is a liquid at room temperature). $ZrCl_4$ has a melting point of 437 °C (i.e., it is a solid at room temperature). Based on this information, which diagram goes with which compound? Justify your explanation.

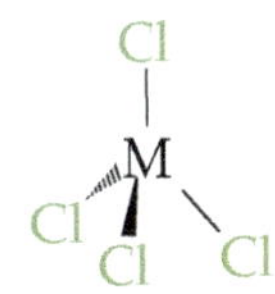

(a)

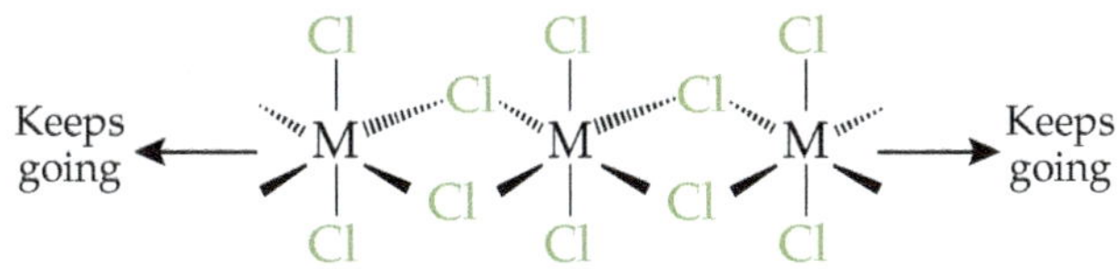

(b)

7.83 The three molecules below increase steadily in molar mass from left to right. However, their boiling points do not. Explain why the boiling points do not also steadily increase from left to right.

	NH_3	NF_3	NCl_3
Molar mass	17.0 g/mol	71.0 g/mol	120.4 g/mol
Boiling point	−33.3 °C	−85.1 °C	71.0 °C

7.84 What is wrong with the following diagram when it comes to explaining London forces?

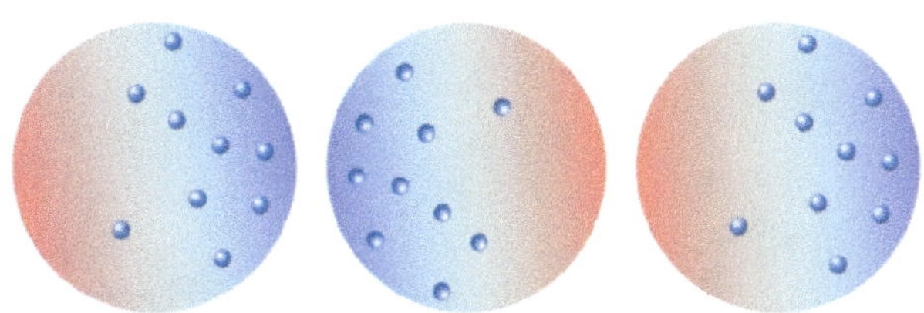

7.85 Consider the two square-planar molecules shown below. Which would you expect to have a higher boiling point, and why?

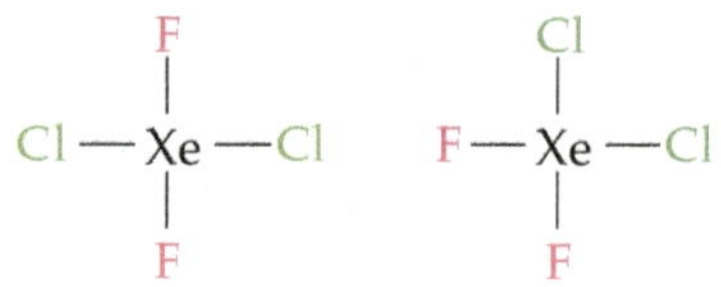

(a) (b)

7.86 CO_2 is a linear molecule, but just suppose it was possible to do the following (see below). If you could, what would you predict would happen to its melting point? Explain.

One Step Beyond

7.87 For each part of this question, choose two different species from the list below, and illustrate the indicated intermolecular interaction with a dotted line. Be sure you show all charges involved in each interaction. (Use $+, -, \delta+, \delta-, \delta\delta+$, and $\delta\delta-$ where appropriate.)

If it is not possible to illustrate the specified interaction using these species, write NA (not applicable) for the answer. (Note: You may use a species from the list more than once as long as you use different species in each part).

HBr, Br_2, CH_4, Na^+, Cl^-, S^{2-}, 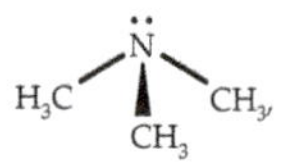 and $CH_3—O—CH_3$

(a) an ion—dipole interaction
(b) a dipole—induced dipole interaction
(c) a dipole—dipole interaction
(d) London forces
(e) hydrogen bonding

7.88 For each part of this problem, redraw both of the molecules, orienting them in a way that allows you to clearly show (using a dotted line) a hydrogen bond between them. If it is not possible to form a hydrogen bond between the two molecules, then write NA (not applicable).

(a)

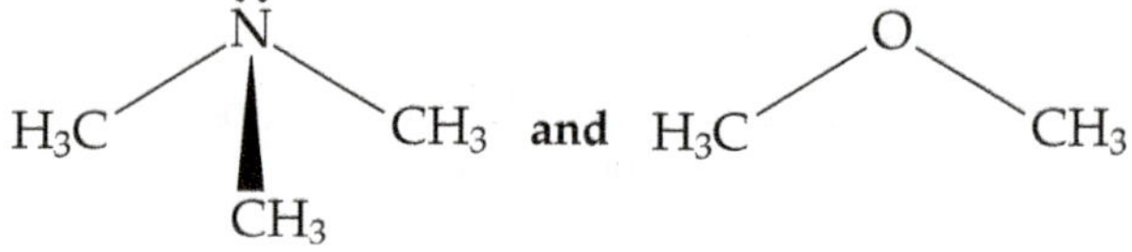

(b)

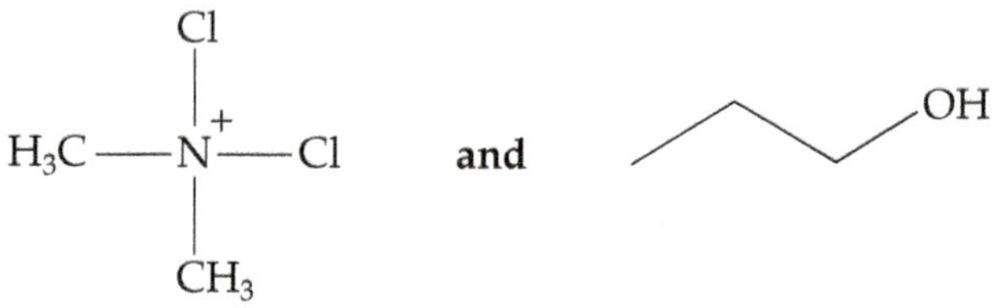

(c)

7.89 Shown below are the structures for two isomers of hydroxy-benzaldehyde. Structure (a) is 2-hydroxy-benzaldehyde and structure (b) is 4-hydroxy-benzaldehyde. The melting point for isomer (a), 2 °C, is significantly lower than the melting point of isomer (b), 118 °C.

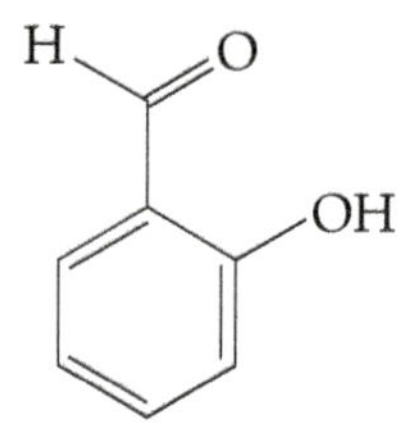

(a)

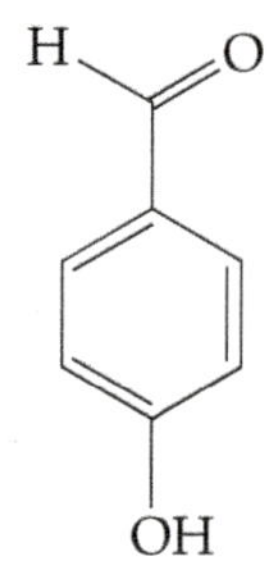

(b)

Based on your knowledge of intermolecular forces, suggest a substantive reason for this difference in melting points. (*Hint*: Consider the intermolecular forces that must be overcome to melt each compound and postulate a reason for the difference in the strengths of those forces).

7.90 Rank the molecules below in order of increasing boiling point.

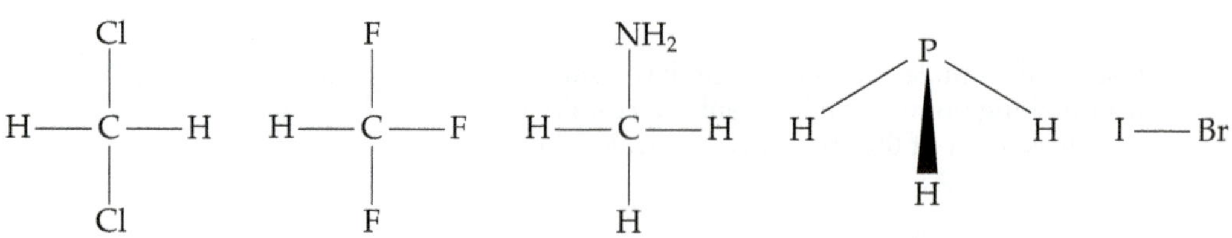

1 2 3 4 5

7.91 Some bacteria have evolved in such a way to render the antibiotic Vancomycin ineffective. How did they accomplish this (make sure to explain the role of intermolecular forces)?

Group Exercises (Work on these exercises together with someone else in the class.)

1. Consider the two well-dressed businessmen below. As they walk past each other, they clasp hands. This prevents them from walking on past each other.

What if they tried to do the same thing as they were running at full speed, and how does this businessman exercise relate to heating a liquid?

2. Consider the two nonpolar hydrocarbon molecules, both of formula C_5H_{12}.

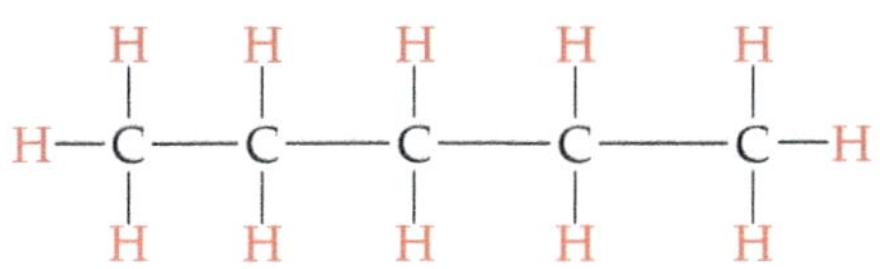

Pentane
boiling point 35–36 °C

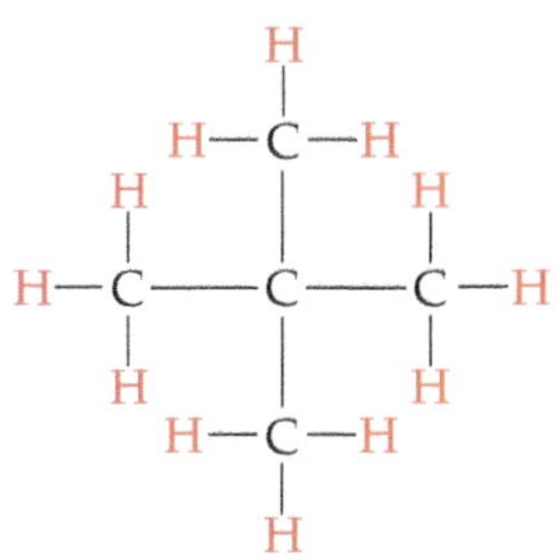

Tertiary-pentane
boiling point 9.6 °C

Having exactly the same formula, they have exactly the same number of electrons, yet pentane has a higher boiling point than tertiary-pentane. Discuss why and come up with a theory that explains why.

WORKPATCH SOLUTIONS

7.1 The fact that ethanol has a lower boiling point means it takes less heat (energy) to boil ethanol. Therefore, the intermolecular forces in ethanol must be weaker than those in water.

7.2 Being a nonpolar molecule, O_2 has no permanent $\delta+$ and $\delta-$ charges that can create strong attractive forces between molecules. All it has are the weak, instantaneous $\delta\delta+$ and $\delta\delta-$ partial charges that result in weak London forces between molecules. Being a polar molecule, H_2O has relatively large, permanent $\delta+$ and $\delta-$ charges that can produce relatively strong attractive forces between molecules. The London forces are stronger in CH_4 because there are more electrons in CH_4 (ten electrons in the molecule) than there are in H_2 (two electrons in the molecule).

7.3 Compound (b) boils at 78 °C, and compound (a) boils at −24 °C. The reason can't be London forces because both compounds have the same molecular formula, C_2H_6O, and therefore the same number of electrons to be unbalanced. Both molecules are polar, but only (b) has a hydrogen atom directly attached to an oxygen atom. This H—O combination allows for the formation of hydrogen bonds between molecules, giving compound (b) the higher boiling point:

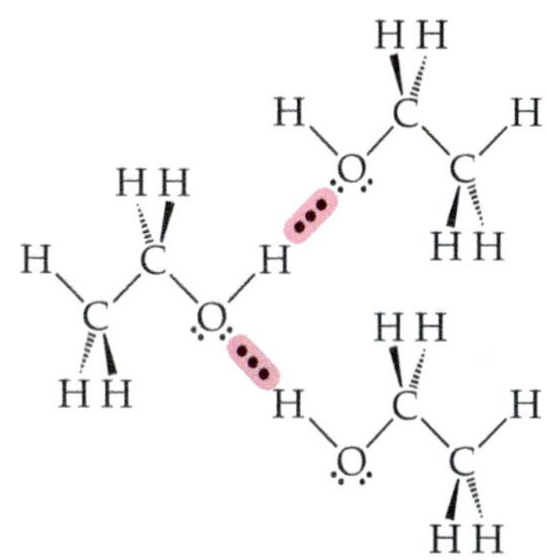

7.4 The reasoning is specious because covalent bonds have nothing to do with ice melting. Melting ice involves overcoming the hydrogen bonds between water molecules. The covalent bonds in the water molecules remain intact. Thus, comparing the melting points of NaCl and water involves comparing the relative strengths of ionic bonds and hydrogen bonds, not of ionic bonds and covalent bonds. Hydrogen bonds are always going to be weaker than actual chemical bonds (ionic bonds and covalent bonds, which are of roughly equal strength).

Chemical Reactions

Time lapse photography—Coast Guard surveillance video of the coast off of Jacksonville, Florida, December 17, 2007, capturing the explosion of the T2 Chemical Plant due to a runaway chemical reaction, killing four people and injuring many others.

Chemical reactions can be dangerous unless you know what you are doing. Here, a reaction performed inside a 2500 gallon steel vessel ran away, violently releasing energy equivalent to 1400 pounds of TNT. In this chapter, you will learn what chemical reactions are and how they can be classified into various types. Later, in Chapter 13, you will learn where all that energy came from.

8.1 What Is a Chemical Reaction?

We touched briefly on the subject of chemical reactions in Section 1.4, but now we want to look at them in more detail.

In the mid-1980s, chemists combined just the right amounts of two metal oxides—yttrium oxide, Y_2O_3, and copper(II) oxide, CuO—with barium carbonate, $BaCO_3$, and heated the combination in a stream of oxygen gas, O_2. The result was a new ceramic compound, $YBa_2Cu_3O_7$, that possessed extraordinary properties. For example, when this compound is cooled to 90 K (−183 °C, −298 °F), a magnet placed above it floats as if defying gravity!

This new ceramic compound is called a *superconductor*, a class of materials that conduct electricity with zero resistance. In addition, magnetic field lines cannot penetrate a superconductor (as they do ordinary materials), and this is the reason a magnet floats above one. Magnetically levitated (maglev) trains are just one of the uses envisioned for superconductors. Just put some strong magnets on the bottom of a train and replace the steel railroad tracks with tracks made from $YBa_2Cu_3O_7$. Then cool the tracks to 90 K (−183 °C) with liquefied nitrogen, a relatively cheap industrial commodity, and the train floats virtually frictionless above the tracks. This is the near

Japan's demonstration Maglev train has reached 581 km/h (361 mph), faster than any wheeled trains. At the heart of the maglev system are onboard niobium titanium superconducting magnets that must be cooled to below −263 °C to be superconducting. This requires the use of expensive liquid helium, which is at −269 °C. Newer trains are being tested that use superconducting magnets made from the above ceramic compounds. These become superconducting at higher temperatures and can be cooled using much less expensive liquid nitrogen (liquid nitrogen is at −196 °C). Whoever invents a "room-temperature" superconductor will win a Nobel Prize.

future, brought to you by a chemical reaction that turned a few piles of otherwise not very useful metal oxides into a brand-new compound.

In creating this ceramic superconductor, a chemical reaction has taken place because we created an entirely new substance that has vastly different properties than the starting materials. In a **chemical reaction**, one or more substances are converted into new substances with compositions and properties that are different from those of the starting materials. We call the starting substances **reactants** and the newly created substances **products**. We represent a chemical reaction by writing a **chemical equation**, which is something like a mathematical equation in that it shows us two things (or two groups of things) that are related to each other. Instead of numbers, however, a chemical equation uses chemical formulas, and instead of an equal sign, it uses an arrow. We write a chemical equation with the reactants to the left of the arrow and the products to the right:

$$\text{Reactants} \rightarrow \text{Products}$$

Remember (Section 1.4), the arrow means "react to give." That is, reactants *react* with each other *to give* products.

It is important that you understand what constitutes a chemical reaction and what doesn't. Try the WorkPatch to see whether or not you do.

WORKPATCH 8.1 Which chemical equations represent chemical reactions?

(a) $H_2O(l) \rightarrow H_2O(g)$ **[Remember, (*l*) means "liquid" and (*g*) means "gas."]**
(b) $CO_2 + H_2O \rightarrow H_2CO_3$
(c) $CO_2 + O_2 \rightarrow O_2 + CO_2$ ●

Did you get the correct answer? There has been a change in (a), true, but a phase change is not a chemical reaction. To be a chemical reaction, at least one new substance must be formed, but how does this happen? How do reactants become products? It is this *how* that we are now prepared to consider.

8.2 How Are Reactants Transformed into Products?

To see how chemical reactions take place, we'll look at a reaction that is much simpler than the one that makes a superconductor. Imagine you have a flask filled with two different diatomic molecules, hydrogen, H_2, and iodine, I_2. Hydrogen is a colorless gas. Iodine is a dark solid at room temperature, but if we heat the flask, we can drive the iodine into the gas phase and produce a purple vapor of iodine molecules. Now, with both reactants in the gas phase, the hydrogen and iodine molecules will react with each other to form another colorless gas, hydrogen iodide, HI. The chemical and physical properties of this product are completely different from those of the reactants, as this figure shows:

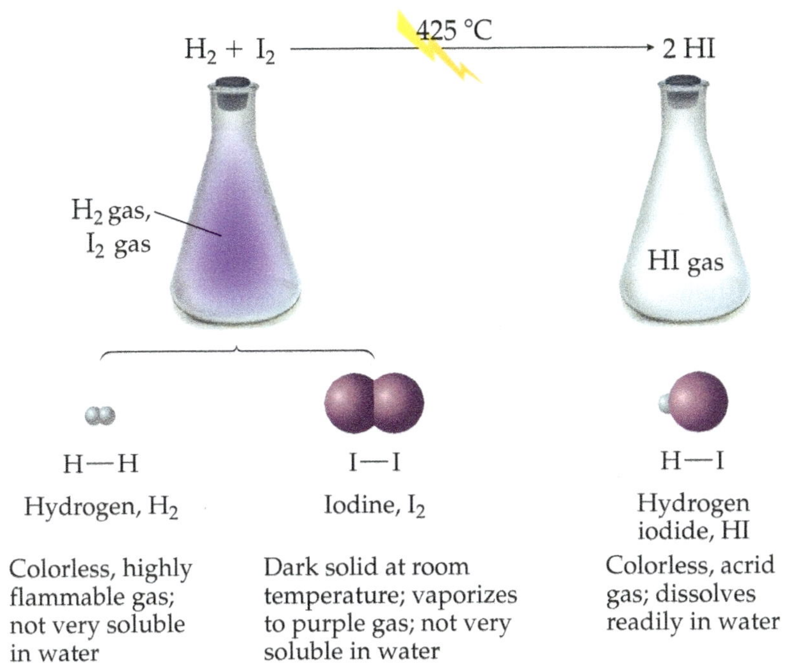

Because both reactants are gases, we say that the reaction between them is a **gas-phase reaction**. But exactly how does a mixture of H_2 and I_2 molecules become a collection of HI molecules? To get some insight, let's change our point of view. Imagine you are so small you can see individual atoms and molecules. At the very beginning of the reaction, you can clearly see the H_2 and I_2 molecules, each with a strong single covalent bond between the atoms. An instant later, the reaction is over. In the blink of an eye, all the H—H and I—I covalent bonds have broken and new H—I covalent bonds have formed.

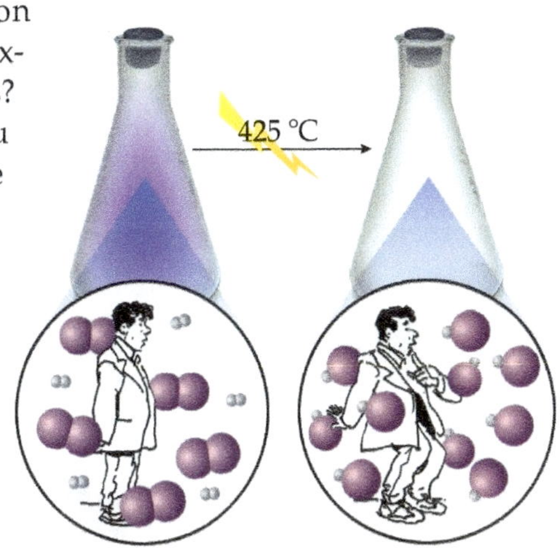

At first you might be puzzled. Why did the H—H and I—I bonds break? It couldn't be that they were weak and ready to break anyway because, as we've seen, covalent bonds are among the strongest chemical forces. And strong bonds certainly would not break for no reason at all. That would be like the legs on a solid oak table suddenly breaking off for no reason. (When was the last time you saw that happen?) You should be just as surprised to see all those H—H and I—I covalent bonds break. To find out why they did, let's change our point of view once again. This time, picture yourself *on* an H_2 or I_2 molecule in the gas-filled flask. You'd better hold on tight because the molecules are not sitting still. At room temperature, you would be zipping about at hundreds of miles per hour!

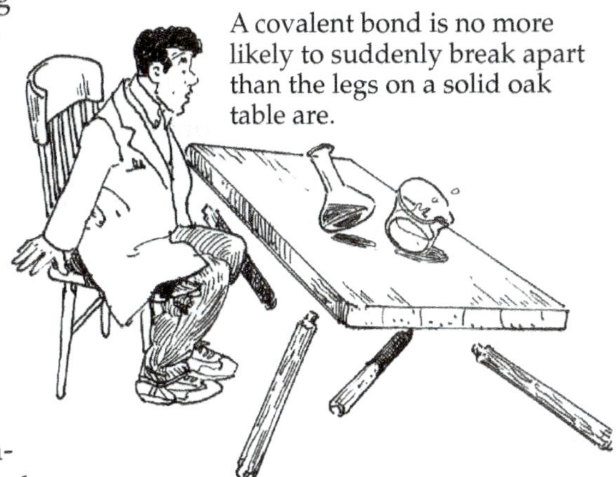
A covalent bond is no more likely to suddenly break apart than the legs on a solid oak table are.

This is true for all small molecules in the gas phase. For example, at room temperature, the oxygen, O_2, and nitrogen, N_2, molecules in the air you are breathing are traveling at speeds close to 500 mph. (Even faster if the temperature is increased.) So why don't you feel like you are in a hurricane at this very moment? The answer is that although the molecules in the air are moving rapidly, they are moving in random directions. They are

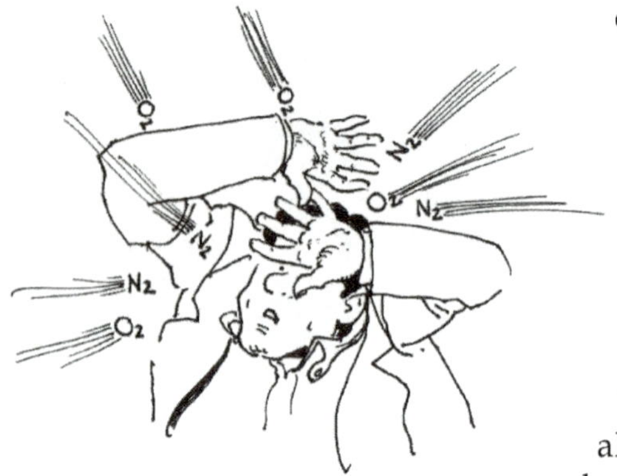

crashing into you at high speeds, each pushing you a little bit, but in different directions. The net result is that all the pushes cancel one another and the air feels still.

Now, let's return to the flask of fast-moving H—H and I—I reactant molecules. Sooner or later, an H_2 and an I_2 molecule will collide with each other. As with two speeding automobiles involved in a collision, the energy associated with their motion (their kinetic energy) gets absorbed by the colliding molecules. We are all familiar with the result when cars absorb all that energy. Windows break, tires fly off, and steel crumples. In the case of colliding H_2 and I_2 molecules, the energy absorbed on impact can begin to break their covalent bonds. The fundamental concept here is that:

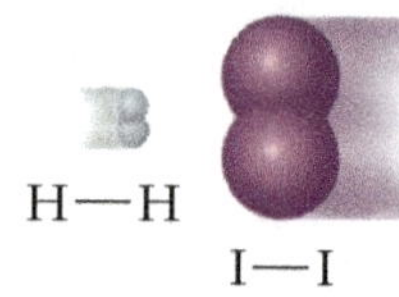

Collision Course!

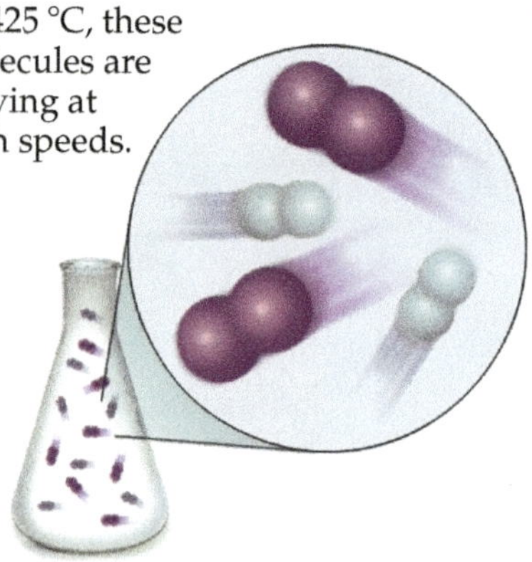

At 425 °C, these molecules are moving at high speeds.

It always takes energy to break a chemical bond.

There are no exceptions to this rule—ever. And the opposite is also true: Energy is always released when a chemical bond forms.

So, for our gas-phase $H_2 + I_2$ reaction, the energy needed to break bonds in the reactant molecules comes primarily from collisions between the fast-moving molecules. Immediately after these collisions, however, the situation differs radically from a crash between automobiles. When automobiles crash and break apart, they stay that way. When H_2 and I_2 molecules collide and break apart, new H—I covalent bonds can form and a new substance is formed. A chemical reaction has occurred.

A Chemical Reaction Taking Place

This description of a chemical reaction is a simplified and idealized one. For example, it is possible to reform the original bonds and restore the H_2 and I_2 reactant molecules. In Chapter 14, we shall deal with this possibility. For now, we shall focus on cases in which collisions between reactant molecules result in the direct formation of product molecules.

WORKPATCH 8.2 Imagine a collision between an oxygen, O_2, molecule and a hydrogen, H_2, molecule to give the product hydrogen peroxide, H_2O_2. Draw Lewis dot diagrams for all the molecules, and then indicate which bonds must be broken and which bonds must form. (*Hint*: The atoms in hydrogen peroxide are connected H—O—O—H.) ●

Did you notice in the WorkPatch that two reactant molecules collide to give just one molecule of product? This was not the case in the $H_2 + I_2$ reaction.

This observation sets us up for what comes next. So far, we have been talking about what a chemical reaction is in terms of bond breaking and bond making and what causes the bonds to break. When we represent chemical reactions in writing, however, we must always go one step further and *balance* the chemical equation.

8.3 Balancing Chemical Equations

If we combine a certain amount of H_2 with a certain amount of I_2, how much HI will we make? It is easy to answer this question on a molecular level because we can "see" it in the illustration above. If we take one H_2 molecule and bang it into one I_2 molecule with a force sufficient to break the bonds, we can form two HI molecules. We can represent this reaction with a *balanced chemical equation*:

$$H_2 + I_2 \rightarrow 2\,HI$$

This equation is shorthand for "one molecule of H_2 and one molecule of I_2 react to give two molecules of HI." Note that the number of atoms to the left of the arrow is equal to the number of atoms to the right of the arrow, meaning the equation is *balanced*. A *balanced chemical equation* is one in which the law of conservation of matter is obeyed. If we start with an unbalanced equation, say $H_2 + I_2 \rightarrow HI$, we have to insert numerical *coefficients* in front of the reactants and/or products to balance the equation. (If no numerical coefficient is written in front of a reactant or product, the coefficient is assumed to be 1.) A chemical equation is balanced when the numbers of atoms of each element are equal on both sides of the arrow:

H_2	+	I_2	$\longrightarrow$	2 HI
H—H		I—I		H—I H—I
Two H atoms		Two I atoms		Two H atoms, two I atoms

On each side of the arrow there are two H atoms and two I atoms. This means we start with two H atoms and two I atoms, and we end up with two H atoms and two I atoms. Nothing is lost or gained. Because any chemical reaction must obey the law of conservation of matter, all chemical equations must be balanced.

Here is another version of the HI reaction. Is this equation balanced?

$$2\,H_2 + 2\,I_2 \rightarrow 4\,HI$$

It is certainly not the same as the equation above, but it is still balanced. There are now four of each type of atom on each side of the arrow. There is nothing wrong with writing the balanced equation this way, but it is traditional to use the smallest possible whole numbers as coefficients.

What if you come across an equation that is unbalanced, like this one?

$H_2 + I_2 \xrightarrow{?} HI$ Not balanced

You can (you must) balance it! To do this, you adjust the coefficients (which are also called *balancing coefficients*). In this case, you simply insert a balancing coefficient 2 in front of the HI:

$H_2 + I_2 \longrightarrow 2\,HI$ Balanced

To balance a chemical equation, you can always insert balancing coefficients. What you should never, *never* do is adjust the subscripts in a molecular formula. For example, you should never balance the HI equation this way:

$H_2 + I_2 \not\longrightarrow H_2I_2$ Impossible!

Changing the subscripts from HI to H_2I_2 does balance the equation (which now obeys the law of conservation of matter), but what we did is *absolutely wrong* because the reaction of H_2 and I_2 makes HI, not H_2I_2. (The latter, if it existed, which it doesn't, would be a compound completely different from HI, with a whole different set of chemical and physical properties.) One last time, then: When balancing a chemical equation, *never* change the subscripts in a molecular formula. If you do, you will have changed the identity of the substance and the nature of the chemical reaction.

Many of the chemical equations we deal with in this book can be balanced by the method of inspection, which is just a formal way of saying a trial-and-error approach. The best way to proceed is to consider one element at a time, adjusting the coefficient in front of every molecular formula that contains that element so that there are equal numbers of atoms of that element on each side of the arrow. Then move on to another element in the equation and repeat the procedure.

To illustrate this method, let's balance this equation for the production of ammonia from nitrogen and hydrogen:

$N_2 + H_2 \rightarrow NH_3$ Not balanced
Ammonia

In scanning this equation from left to right, the first element we come to is nitrogen. There are two nitrogen atoms on the left, which means we must have two nitrogen atoms on the right. We can achieve this by putting a coefficient 2 in front of the NH_3:

$N_2 + H_2 \rightarrow 2\,NH_3$

Now the nitrogen atoms are balanced. The next element to balance is hydrogen. There are two H atoms on the reactant (left) side of the equation. How many hydrogen atoms are there on the product (right) side? If you said three, you need to look again. There are three H atoms in each NH_3 molecule, but there are now two NH_3 molecules, meaning a total of six H atoms on the right side. We need to fix this by increasing the total number of hydrogen atoms on the left side to six. To do this, we place a coefficient 3 in front of the H_2:

$N_2 + 3\,H_2 \rightarrow 2\,NH_3$ Balanced

Now there are six hydrogen atoms on each side. Because both the nitrogen and the hydrogen are balanced, the overall equation is also balanced.

Try balancing the following unbalanced chemical equations. Here's a helpful hint. If the equation has a pure elemental substance in it, such as H_2, N_2, O_2, or Fe, save this element for last when balancing by inspection. Doing things this way sometimes makes the job of balancing a bit easier.

Practice Problems

Balance the following chemical equations.

8.1 $CH_4 + O_2 \rightarrow CO_2 + H_2O$

Answer: Because oxygen appears as a pure elemental substance, we save it for last. Starting with carbon (an arbitrary choice), we note that there is one carbon on each side of the equation, meaning C is balanced. There are four hydrogen atoms on the left and only two on the right. To fix this, we put a balancing coefficient 2 in front of the H_2O*:*
$CH_4 + O_2 \rightarrow CO_2 + 2H_2O$

This balances the hydrogen. Now it's time to balance the oxygen. There are two oxygen atoms on the left and four on the right (two from the CO_2 *molecule and two from the two* H_2O *molecules). To balance the equation, we put a 2 in front of the* O_2*:*
$CH_4 + 2O_2 \rightarrow CO_2 + 2H_2O$

8.2 $CaH_2 + H_2O \rightarrow Ca(OH)_2 + H_2$

8.3 $C_6H_{12}O_6 + O_2 \rightarrow CO_2 + H_2O$

8.4 $HCl + Na_2CO_3 \rightarrow NaCl + CO_2 + H_2O$

Of course, things don't always go as smoothly as in the preceding practice problems. For example, consider the combustion of liquid octane, C_8H_{18}, a component of gasoline. The unbalanced equation is:

$$C_8H_{18}(l) + O_2(g) \rightarrow CO_2(g) + H_2O(g) \qquad \text{Not balanced}$$

We'll use the trick of saving the elemental substance (O_2) for last. To balance carbon, we need an 8 in front of the CO_2:

$$C_8H_{18}(l) + O_2(g) \rightarrow 8\,CO_2(g) + H_2O(g) \qquad \text{Not balanced}$$

To balance hydrogen, we need a 9 in front of the H_2O:

$$C_8H_{18}(l) + O_2(g) \rightarrow 8\,CO_2(g) + 9\,H_2O(g) \qquad \text{Not balanced}$$

That leaves oxygen. The right side has 25 oxygen atoms (16 from the 8 CO_2 molecules plus 9 from the 9 H_2O molecules). How are we going to get 25 oxygen atoms on the left side? Here's the easiest way:

$$C_8H_{18}(l) + \tfrac{25}{2}\,O_2(g) \rightarrow 8\,CO_2(g) + 9\,H_2O(g) \qquad \text{Balanced}$$

We now have 25 oxygen atoms on the left because $\left(\frac{25}{2}\right) \times 2 = 25$ (the 2 comes from the molecular formula O_2). Can we do this? Can we use fractional coefficients to balance equations? Absolutely! However, even though our equation is now balanced, many chemists often take one more step to clear out the fractions. To do this, we multiply all of the coefficients by a number that removes any fractions present. In this case, all we need to do is multiply through by 2:

$$2\,C_8H_{18}(l) + 25\,O_2(g) \rightarrow 16\,CO_2(g) + 18\,H_2O(g) \qquad \text{Balanced}$$

We must multiply all the coefficients by the same number so as not to change the relationship between them. Try the following practice problems to see if you're getting the hang of this.

Practice Problems

8.5 Balance the chemical equation for the combustion of hexane, $C_6H_{14}(l)$. The unbalanced equation is:

$$C_6H_{14}(l) + O_2(g) \rightarrow CO_2(g) + H_2O(g)$$

Answer:

Balance C: $C_6H_{14}(l) + O_2(g) \rightarrow 6\,CO_2(g) + H_2O(g)$

Balance H: $C_6H_{14}(l) + O_2(g) \rightarrow 6\,CO_2(g) + 7\,H_2O(g)$

Balance O: $C_6H_{14}(l) + \frac{19}{2}O_2(g) \rightarrow 6\,CO_2(g) + 7\,H_2O(g)$

Multiply through by 2 to get rid of the fractional coefficient:

$$2\,C_6H_{14}(l) + 19\,O_2(g) \rightarrow 12\,CO_2(g) + 14\,H_2O(g)$$

8.6 Balance the chemical equation:

$$C_6H_{12}(l) + O_2(g) \rightarrow H_2C_6H_8O_4(l) + H_2O(l)$$

8.7 Balance the chemical equation:

$$CH_3NH_2(g) + O_2(g) \rightarrow CO_2(g) + H_2O(g) + N_2(g)$$

Once a chemical equation is balanced, it is a valuable tool for answering all sorts of "how much?" questions. How much of each reactant must I mix together to make a certain amount of product? How much product will I make if I mix a certain amount of reactants together? Will any of the reactants be left over (unused), and if so, which ones and how much?

We should point out that a number of chemical equations are extremely difficult, if not impossible, to balance by the method of inspection. A variety of methods have been developed to balance such reactions, but these methods are beyond the scope of this book. As we come across these reactions in later chapters, we shall present their equations in balanced form.

8.4 Types of Reactions

Our planet teems with life, from amoebae to zebras. Consider then the plight of the early biologists who set out to understand all living organisms. How were these scientists ever going to make sense of it all? Their answer was to create a biological classification scheme that grouped every living thing into one or another kingdom, followed by one or another phylum, then class, order, family, genus, and finally species. Using this scheme, a modern biologist can tell you that the only relationship between a koala "bear" (order Marsupialia) and a grizzly bear (order Carnivora) is that both are mammals (they are in the same class, Mammalia). In other words, the classification scheme became a tool for organizing and understanding life.

Chemists faced a similar problem understanding chemical reactions because hundreds of thousands of seemingly unrelated reactions have been studied over the years. By categorizing these chemical reactions into various types, chemists could begin to see patterns and to uncover the underlying common factors behind them. Questions such as "What is this reaction doing?" and "Are these two reactions related?" could be answered. The goal in this and Sections 8.5 and 8.6 is to introduce you to some of the reaction types chemists have identified so that you can recognize some common features of various chemical reactions as well. For example, suppose you come across a reaction that you have never seen before, such as the one between calcium ions and carbonate ions:

$$Ca^{2+}(aq) + CO_3^{2-}(aq) \rightarrow CaCO_3(s)$$

[The (*aq*) means "aqueous solution" (the ion is dissolved in water) and serves the same purpose as the (*s*), (*l*), (*g*) designations used to indicate physical state.] This reaction is used to soften "hard" water. How does it soften water? What is the reaction doing? Can other reactions also be used to soften water? You can get some insight into the answers to these questions by assigning this reaction a reaction type. But don't be surprised if a reaction can be assigned more than one reaction type. Reaction types are not meant to be exclusive or exact. Rather, they are meant to help us understand what a reaction is doing, and some reactions do a few things at once. Let's look at some reaction types.

Have you ever tossed a ball to a friend? Some reactions work like this and are called *single-replacement reactions*. Consider the reaction between iron oxide, Fe_2O_3 (rust), and aluminum powder, Al, called the thermite reaction:

$$2\,Al(s) + Fe_2O_3(s) \rightarrow Al_2O_3(s) + 2\,Fe(l)$$

The iron and aluminum are "playing ball" with the oxygen. As a result, aluminum has replaced iron in the oxygen compound. In a **single-replacement reaction**, one element replaces another element in a compound.

Now consider the reaction:

$$Mg(s) + 2\,HCl(aq) \rightarrow H_2(g) + MgCl_2(aq)$$

At first glance, this reaction seems unrelated to the thermite reaction. Look closely, though, and you'll see that the chlorine compound has replaced its H atom with a Mg atom (hydrogen and magnesium are "playing ball" with the chlorine). Therefore, this is also a single-replacement reaction. By recognizing this, you begin to understand the reaction.

Now let's get fancier. Suppose you have a red ball and a friend has a blue ball. You each toss the other your ball. This is what happens in a *double-replacement reaction*. In a **double-replacement reaction**, two reactants AB and CD react to form products AD and CB. For example, consider this reaction between phosphorus trichloride and silver fluoride:

$$PCl_3(l) + 3\,AgF(s) \rightarrow PF_3(g) + 3\,AgCl(s)$$

Phosphorus, P, "tossed" its chlorine to silver, Ag, and received fluorine in return. Another example of a double-replacement reaction is:

$$Ca(NO_3)_2(aq) + 2\,NaF(aq) \rightarrow CaF_2(s) + 2\,NaNO_3(aq)$$

This is also an example of a *precipitation reaction*, which we will see shortly in Section 8.5.

Next we come to *decomposition reactions*. A **decomposition reaction** is one in which a compound breaks down (decomposes) into two or more simpler substances. The molecules in living organisms decompose after death to give simpler compounds that serve as nutrients for new life. For example, bacteria cause the decomposition of protein molecules that make up skin and muscles into simple amino acids. Another way to decompose compounds is to heat them. Heat the red compound mercury(II) oxide, for instance, and it decomposes into elemental mercury and oxygen:

$$2\,HgO(s) \xrightarrow{\text{Heat}} 2\,Hg(l) + O_2(g)$$

Combination reactions (also called *synthesis reactions*) are those in which two or more substances combine to give another substance. (In other words, combination reactions are the reverse of decomposition reactions.) Some examples are:

$$H_2(g) + I_2(g) \rightarrow 2\,HI(g)$$
$$2\,Na(s) + Cl_2(g) \rightarrow 2\,NaCl(s)$$
$$4\,Ni(s) + 4\,Al(s) + S_8(s) \rightarrow 2\,Ni_2Al_2S_4(s)$$

Combination reactions are commonly used by pharmaceutical companies to prepare many of the drugs that have revolutionized modern medicine.

If at this point you are feeling a little lost, take a look at the following visual summary of the reaction types we have covered so far, and then try the practice problems.

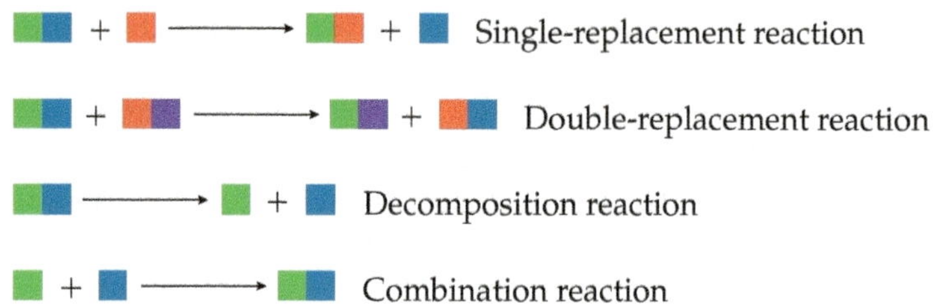

You can use this visual summary to help you think about reactions. For example, consider a *combustion reaction*. A **combustion reaction** is the reaction of any substance with oxygen, O_2 (another term for combustion is burn—that is, when something combusts, it burns). Consider the combustion of methane (CH_4, natural gas) to give carbon dioxide and water.

$$CH_4(g) + 2\,O_2(g) \rightarrow CO_2(g) + 2H_2O(g)$$

Can it also be classified as one of the above reaction types? The best fit is double-replacement (the carbon atom in methane gives its hydrogens to oxygen, and in turn, O_2 gives some O atoms to carbon).

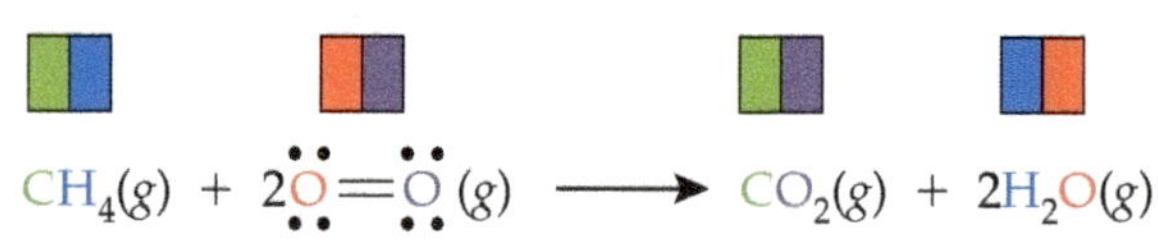

Metals also burn to give metal oxides M_xO_y (they are called ox**ides** since electrons have been transferred from the metal atom(s) to oxygen, forming a metal cation and the oxide anion—remember, **ide** means negative). Is the combustion of a metal also a double replacement reaction like the combustion of methane? To answer, you need to examine the balanced combustion reaction for a metal. Consider the combustion of iron.

$$4\,Fe(s) + 3\,O_2(g) \rightarrow 2\,Fe_2O_3(s)$$

The combustion of a metal yields only one product. This is a combination reaction.

Practice Problems

Balance each equation and classify the reaction as combination, decomposition, single-replacement, or double-replacement.

8.8 $KBr(aq) + Cl_2(aq) \rightarrow KCl(aq) + Br_2(aq)$

Answer: The balanced equation is

$2\,KBr(aq) + Cl_2(aq) \rightarrow 2\,KCl(aq) + Br_2(aq)$

This is a single-replacement reaction because the potassium compound replaces its bromine with chlorine.

8.9 $Pb(NO_3)_2(s) \xrightarrow{Heat} PbO(s) + NO(g) + O_2(g)$

8.10 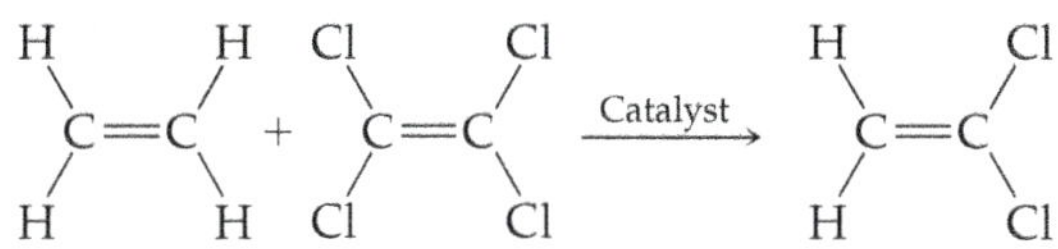

8.5 Solubility and Precipitation Reactions

The ambulance had just pulled up to the hospital's emergency room. The paramedic was shouting out her status report to the doctor: "26-year-old lab technician! Chemical burns to the right arm and both legs. Doused with hydrofluoric acid."

Hydrofluoric acid, an aqueous solution of HF gas, is a good source of fluoride ion, F^-. In less than a minute, the medical team had wheeled the patient into the ICU and begun intravenous fluids. The doctor called out instructions. "He needs calcium chloride, 10% solution. Set up an IV drip, 1 g every 10 min. Stat!"

The patient—awake, conscious, and panicked—pummeled the doctor with questions.

Patient: My whole arm is tingling, doctor. Is that bad?
Doctor: Right now we're more worried about the fluoride from the acid. Fluoride ions can remove calcium ions from your cellular fluids, and without calcium ions your heart will stop beating.

The doctor was referring to the Ca^{2+} ions dissolved in the aqueous fluids of our cells. Fluoride ions combine with calcium ions to form solid calcium fluoride, CaF_2, an insoluble salt.

Patient: Is that why you're giving me calcium?
Doctor: Yes, but everything depends on how much fluoride has entered your bloodstream. We'll just have to wait and see.

Within minutes, the patient's electrocardiogram stabilized.

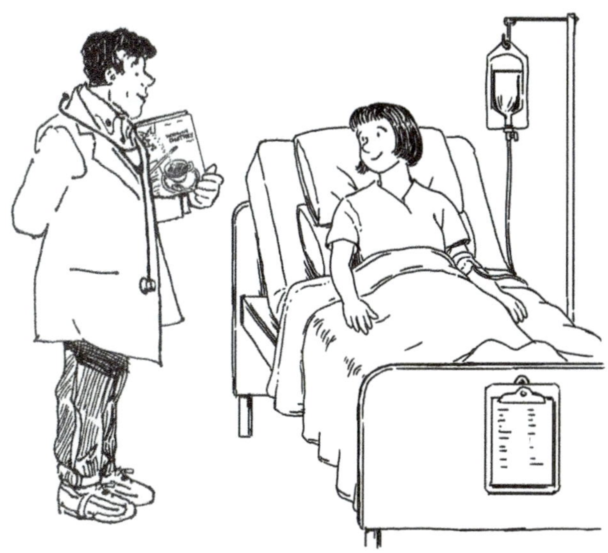

Doctor: Looks like you are going to be all right.
Patient: That's great! Thank you, doctor. You really know your chemistry!
Doctor: Don't thank me. Thank my college chemistry instructor.

Well, what did you expect? This *is* a chemistry text, after all! The reaction occurring in the patient's cellular fluid is called a **precipitation reaction**, defined as one in which an insoluble solid (the **precipitate**) forms when two or more aqueous solutions are mixed. The particular precipitation reaction here is:

$$Ca^{2+}(aq) + 2\,F^{-}(aq) \rightarrow CaF_2(s)$$

The insoluble calcium fluoride, $CaF_2(s)$, is the precipitate. It is an ionic compound, and ionic compounds are often called *salts* (that's why sodium chloride, NaCl, is called table salt). Of course, a big difference between CaF_2 and NaCl is that NaCl is very soluble in water but CaF_2 is not.

Solubility of Ionic Compounds

What happens when an ionic compound dissolves in water? Remember from Section 5.1 that ionic solids consist of a lattice of ions held together by ionic bonds. When an ionic solid dissolves, its lattice comes apart to yield cations and anions in solution, as shown in the figure at the bottom of the page. This breaking up of the lattice is called **dissociation** (to dissociate means to separate). When an ionic compound dissolves in water, it dissociates into ions. This is in sharp contrast to most covalent compounds. For example, consider sucrose (table sugar, $C_{12}H_{22}O_{11}$), a covalent compound. When it dissolves in water, you get intact, dissolved sucrose molecules, $C_{12}H_{22}O_{11}(aq)$, with no ions in solution. Keep this in mind because you will often see the molecular formulas for ionic compounds written with an (*aq*) label, as in NaCl(*aq*). The notation NaCl(*aq*) *does not mean there are intact NaCl units dissolved in water*. Rather, it means that $Na^{+}(aq)$ ions and $Cl^{-}(aq)$ ions are present. It is up to you to recognize this.

Of course, not all ionic compounds are soluble in water. Some are incredibly insoluble. We'll discuss why in Chapter 12, but for now it is useful for you to know which ionic compounds are soluble and which are not. Thus, what we need is a set of solubility rules that gives us this information. Just such a set of rules—all based on observations—has been developed over the years and is shown in Table 8.1. You can see from the table that all sodium salts and all nitrates are soluble in water (this is something worth remembering). Even though this table is not exhaustive (e.g., fluorides are not included), you are going to find it very useful for what comes next.

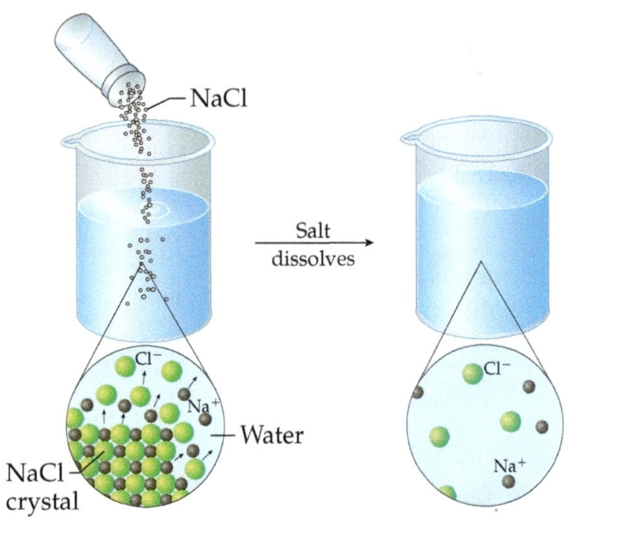

Ions break away from the dissolving crystal.

The solution consists of aqueous Na^{+} and Cl^{-} ions.

Table 8.1 Solubility Rules

Soluble in water		
All sodium, potassium, and ammonium salts		
All acetates and nitrates		
Most halides (chlorides, bromides, iodides)	*except*	Halides of lead(II), silver(I), and mercury(I)
Most sulfates	*except*	Sulfates of calcium, barium, lead(II), and strontium

Insoluble in water		
Most phosphates, carbonates, and sulfides	*except*	Sodium, potassium, and ammonium salts; calcium sulfide
Most hydroxides	*except*	Sodium, potassium, calcium, and barium hydroxides

Precipitation Reactions—Writing Net Ionic Equations

The hydrofluoric acid accident described earlier put the unfortunate laboratory technician in such grave danger because of a precipitation reaction in which CaF_2 was the precipitate. Precipitation reactions can be unwanted, or they can be quite desirable. Until recently, for example, lead salts were used as the coloring pigments in many paints. (Lead compounds are used much less frequently today because of their toxicity.) For example, lead(II) iodide, PbI_2, is a bright yellow powder. Suppose you wanted to make some yellow paint using $PbI_2(s)$ as the pigment. You go to your laboratory to get some PbI_2 but find you have run out. No problem. You'll just make some via a precipitation reaction. On the shelf are bottles of lead(II) nitrate, $Pb(NO_3)_2$, and sodium iodide, NaI, both white powders. After checking the solubility rules and confirming that both these salts are soluble, you prepare two colorless solutions:

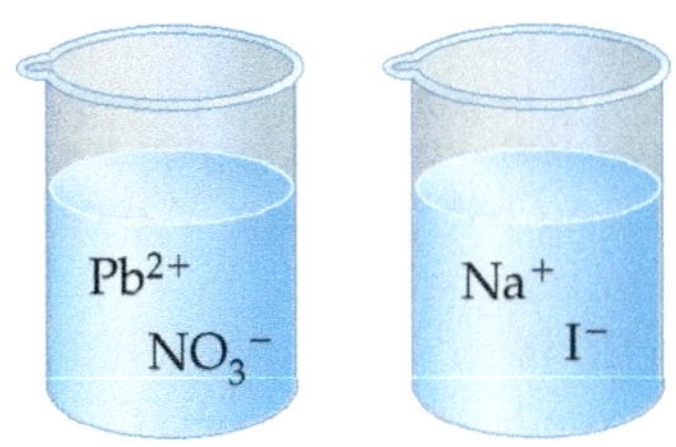

$Pb(NO_3)_2$ solution   NaI solution

If you combine these two solutions, all of the cations and anions suddenly find themselves in the same solution. This makes possible some new combinations of cations and anions, as indicated here by the crisscross lines:

Possible new combinations of cations and anions

This crisscross method of considering new cation–anion combinations tells you which precipitates *might* form. According to the lines, the possible precipitates are lead(II) iodide, PbI_2, and sodium nitrate, $NaNO_3$. *You must be careful with molecular formulas here!* Ionic compounds are always electrically neutral. This means the only precipitate possible when Pb^{2+} is combined with I^- is PbI_2 and *not* PbI. The crisscross method can tell you which ions might form precipitates, but it's up to you to account for charge neutrality to get the correct molecular formula.

So, will you get a precipitate when you combine the two solutions? To predict whether or not you will, simply consult the solubility rules regarding PbI_2 and $NaNO_3$. If both salts are soluble in water, no reaction takes place, but that is not the case here. The solubility rules state that the halides of lead(II) are insoluble, which means yellow PbI_2 precipitates. All you need to do now is isolate the PbI_2 by filtering the solution, and you're ready to mix up a batch of yellow paint. The story is not complete quite yet, however. We know what happened to the $Pb^{2+}(aq)$ ions and $I^-(aq)$ ions, but what happened to the $Na^+(aq)$ ions and $NO_3^-(aq)$ ions? The answer is that absolutely nothing happened to them. These ions started out in solution, and they end up in solution. No $NaNO_3(s)$ precipitates because it is a water-soluble salt.

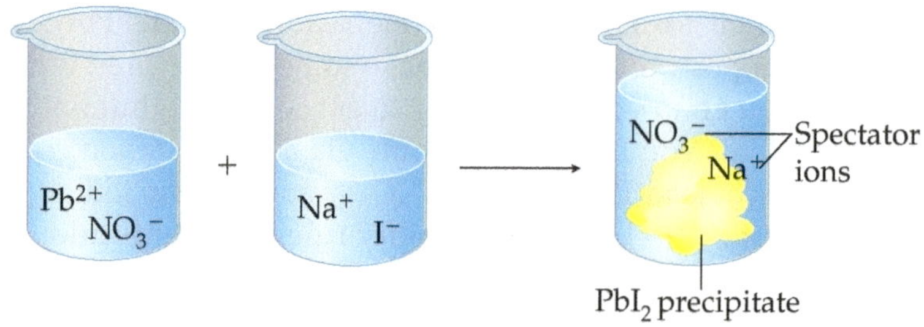

In other words, the $Na^+(aq)$ ions and $NO_3^-(aq)$ ions did not participate in the reaction; they just went along for the ride. We call such ions **spectator ions** because they take no part in the reaction.

All that is left for us to do is to represent what happened with a balanced chemical equation. There are a few ways to do this. We can write it with every species intact rather than dissociated:

$$Pb(NO_3)_2(aq) + 2\,NaI(aq) \rightarrow PbI_2(s) + 2\,NaNO_3(aq)$$

However, to show the actual species present during the reaction, we can write a **complete ionic equation**, which is one that shows all the ions, both those that take part in the reaction and the spectator ions:

$$Pb^{2+}(aq) + 2\,NO_3^-(aq) + 2\,Na^+(aq) + 2\,I^-(aq) \rightarrow$$
$$PbI_2(s) + 2\,Na^+(aq) + 2\,NO_3^-(aq)$$

Finally, we can cancel the spectator ions from the two sides of the complete ionic equation. After all, a reaction is defined as a process in which reactants *change* into products, and spectator ions do not change. The result is a chemical equation called a **net ionic equation**:

$$Pb^{2+}(aq) + 2\,I^-(aq) \rightarrow PbI_2(s)$$

Whichever of the three ways we choose to represent this reaction, we can see that it is a precipitation reaction because an insoluble solid precipitates from solution. However, if we write our equation in the form

$$Pb(NO_3)_2(aq) + 2\,NaI(aq) \rightarrow PbI_2(s) + 2\,NaNO_3(aq)$$

we can see that it is also a double-replacement reaction, and the net ionic equation form,

$$Pb^{2+}(aq) + 2\,I^{-}(aq) \rightarrow PbI_2(s)$$

shows us it is also a combination reaction. As we said earlier, don't be surprised if one reaction can be assigned more than one reaction type.

The following chart summarizes the steps you should use to predict whether a precipitate will form and what it will be.

Identifying the precipitate(s) in a potential precipitation reaction

Step 1: Identify all ions present when the starting solutions are mixed. (These are the ions produced when the starting materials dissolve.)

Step 2: Identify all possible cation–anion combinations that are different from the ones in the starting materials. These combinations represent possible precipitates.

Step 3: Use the solubility rules to decide if any of these new cation–anion combinations represent insoluble salts. Any combination that does will form a solid and produce a precipitate. If no combinations represent insoluble salts, no precipitation reaction will occur.

Step 4: If a precipitation reaction occurs, write the net ionic equation for it. Make sure the molecular formula you write for the precipitate is electrically neutral.

Now try the following problems. As usual, the first one is worked out for you.

Practice Problems

8.11 A solution of sodium sulfate is combined with a solution of barium chloride. Does a precipitation reaction occur? If so, which salt precipitates?

Answer:

Step 1: Identify the ions present:

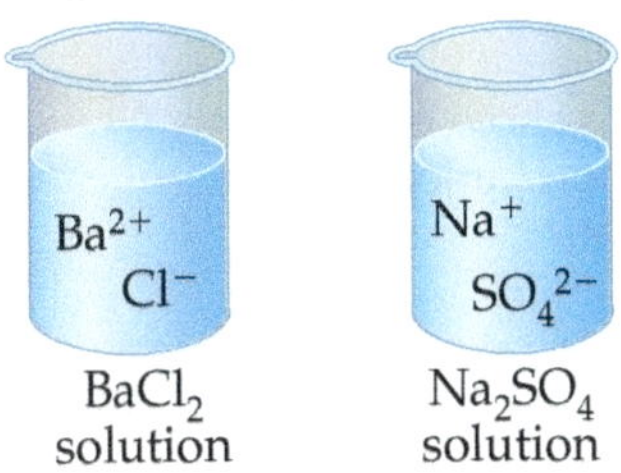

Step 2: Identify the possible new combinations of ions (candidates for precipitation):

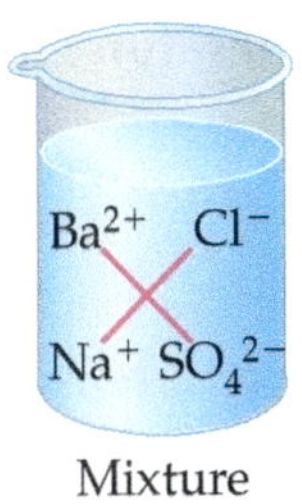

Step 3: Check the solubility rules to find out whether either candidate for precipitation is insoluble in water. The rules tell you NaCl is soluble (all sodium salts are soluble) and $BaSO_4$ is insoluble. Therefore $BaSO_4$ precipitates when the solutions are mixed.

Step 4: Write the net ionic equation for the precipitation reaction:

$$Ba^{2+}(aq) + SO_4^{2-}(aq) \longrightarrow BaSO_4(s)$$

8.12 Aqueous solutions of lead nitrate and sodium sulfate combine to produce a precipitate.

(a) What are the spectator ions?
(b) Write the net ionic equation for this precipitation reaction.

8.13 Aqueous solutions of nickel(II) nitrate and ammonium phosphate combine to produce a precipitate.

(a) What are the spectator ions?
(b) Write the net ionic equation for this precipitation reaction.

Now that you are getting the hang of writing net ionic equations, let's pull out all the stops! Let's try combining *three* solutions of ionic compounds.

WORKPATCH 8.3 Consider three separate aqueous solutions of sodium nitrate, aluminum sulfate, and sodium hydroxide. Draw a beaker for each, and show which ions are present in each solution. ●

Will there be a precipitate when the three solutions of this WorkPatch are combined? To find out, draw a beaker containing all the ions shown in the three beakers and apply the crisscross method.

WORKPATCH 8.4 Write the molecular formulas for all potential precipitates in the beaker below.

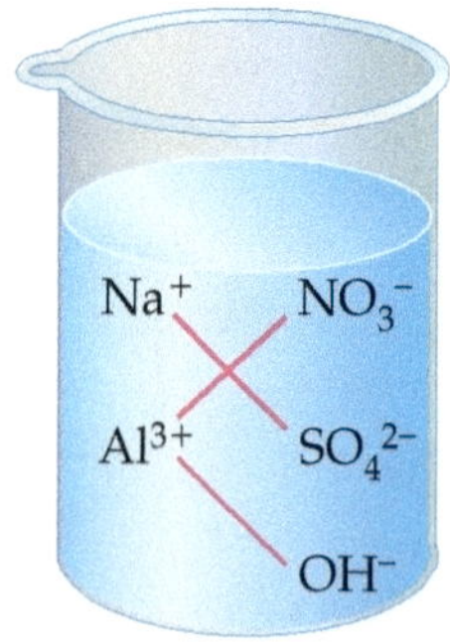

Red lines show only new cation/anion combinations.

Check your formulas against those at the end of the chapter. Of the three precipitates possible, only $Al(OH)_3$ is insoluble in water. Therefore, the net ionic equation for the precipitation reaction is:

$$Al^{3+}(aq) + 3\,OH^-(aq) \longrightarrow Al(OH)_3(s)$$

If you feel comfortable with this three-solution problem, you are ready to tackle any other precipitation problem. ●

8.6 Introduction to Acid–Base Reactions

Do you like the taste of aqueous protons—$H^+(aq)$? Chances are you do. Sour candies, lemons, and other citrus fruits owe their sourness to the $H^+(aq)$ ions they deposit on your tongue. The principal compound responsible for this eye-squinting, mouth-puckering sensation is citric acid:

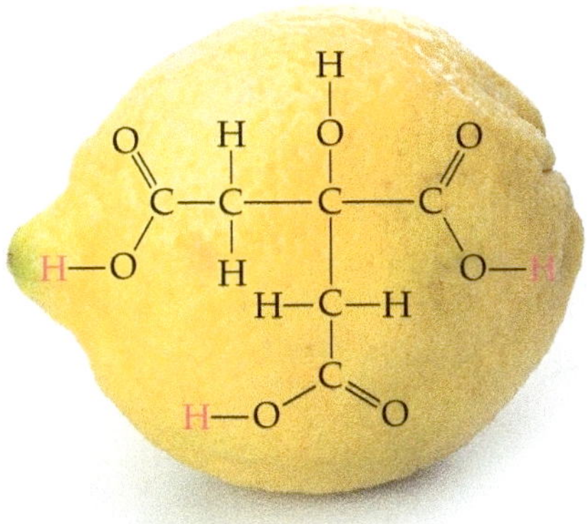

Lemons are loaded with citric acid, $C_3H_4OH(COOH)_3$.

When added to water, a molecule of citric acid dissolves and dissociates to produce a maximum of three $H^+(aq)$ ions:

$$C_3H_4OH(COOH)_3 \xrightarrow{H_2O(l)} C_3H_4OH(COO)_3^{3-}(aq) + 3\,H^+(aq)$$

Any compound that produces $H^+(aq)$ ions when added to water can be called an **acid**. Some other acids present in the food you eat are acetic acid (in vinegar), ascorbic acid (vitamin C), and phosphoric acid (in many soft drinks). There are also many acids you would never eat:

HCl	Hydrochloric acid (also called muriatic acid, used to clean steel)
HNO_3	Nitric acid (used to dissolve many metals)
H_2SO_4	Sulfuric acid (used in car batteries)
HF	Hydrofluoric acid (highly toxic, used to etch glass)

All these compounds are acids because they dissociate to produce $H^+(aq)$ ions when dissolved in water. For instance, if we add HCl gas to water, the HCl dissociates into H^+ and Cl^- ions:

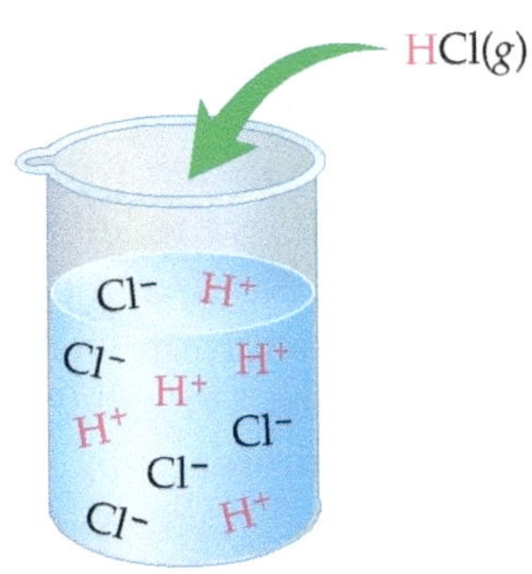

HCl is an acid because it dissociates in water to produce $H^+(aq)$ ions.

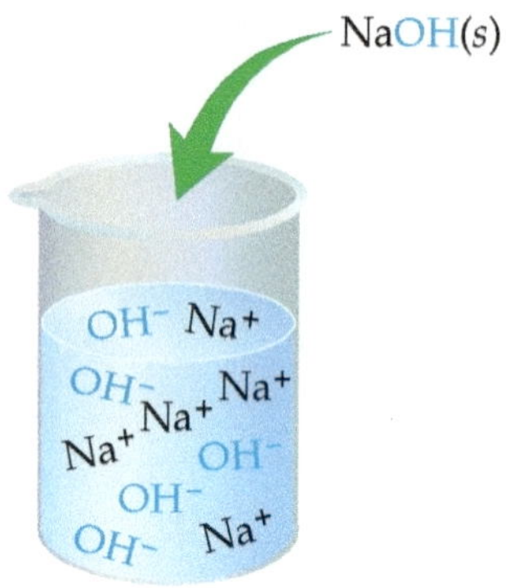

NaOH is a base because it produces $OH^-(aq)$ ions when added to water.

You can glance back at the very end of Section 5.7 to see a few more examples of acids. The story doesn't end here, though, because acids have "opposites" called *bases*.

A **base** can be thought of as any substance that produces aqueous hydroxide ions, $OH^-(aq)$, when added to water. A well-known base is sodium hydroxide, NaOH, also called lye.

These ionic compounds are some typical bases:

NaOH	Sodium hydroxide
KOH	Potassium hydroxide
$Ca(OH)_2$	Calcium hydroxide

An interesting thing happens when an aqueous solution of an acid meets an aqueous solution of a base, as shown here:

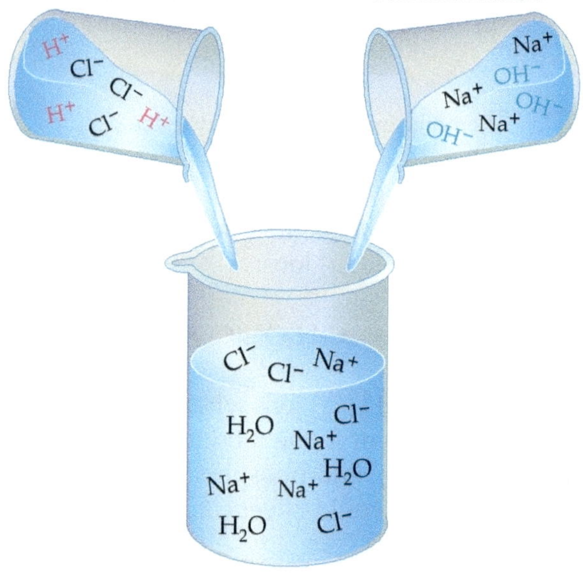

We could write the equation for the reaction that occurs with all compounds intact,

$$HCl(aq) + NaOH(aq) \rightarrow NaCl(aq) + H_2O(l)$$

or we could write the net ionic equation:

$$H^+(aq) + OH^-(aq) \rightarrow H_2O(l)$$

Either way, the bottom line is that the $H^+(aq)$ ions and $OH^-(aq)$ ions react with each other to form water. We are left with a solution of the two spectator ions Na^+ and Cl^-—in other words, we're left with salty water, just as though we had dissolved some table salt, NaCl, in the beaker.

Any time you react an acidic solution with a basic solution, you end up with a solution of the spectator ions. We call the reaction between an acid and a base an **acid–base neutralization reaction**. The acid *neutralizes* the base by removing the OH^- ions of the base. At the same time, the base *neutralizes* the acid by removing the H^+ ions of the acid. If you were feeling dramatic, you might even say that the $H^+(aq)$ ions and $OH^-(aq)$ ions killed each other. If you combine equal quantities of H^+ ions and OH^- ions, you will end up with a solution that is *neutral* (neither acidic nor basic). This neutralization is why we think of acids and bases as opposites. We shall have much more to say about acids and bases in Chapter 15.

Practice Problems

8.14 (a) Write the intact-compound equation and the net ionic equation for the neutralization reaction that occurs between aqueous solutions of sulfuric acid and potassium hydroxide.
(b) What salt would be isolated if you evaporated the water from this solution after neutralization?

Answer:

(a) Intact-compound equation:

$H_2SO_4(aq) + 2\,KOH(aq) \rightarrow K_2SO_4(aq) + 2\,H_2O(l)$

Net ionic equation:

$H^+(aq) + OH^-(aq) \rightarrow H_2O(l)$

(b) K_2SO_4, *potassium sulfate.*

8.15 It takes one molecule of NaOH to neutralize one molecule of HCl, but it takes two molecules of NaOH to neutralize one molecule of H_2SO_4. Explain why this is true.

8.16 Write the intact-compound equation and the net ionic equation for the neutralization of aqueous hydrofluoric acid by aqueous calcium hydroxide. What is the name of the salt that forms?

This chapter has given you a feel for what chemical reactions are, how they occur, some of the types that exist, and how to write and balance equations representing them. In the next chapter, we shall look at chemical reactions again but from a quantitative point of view, dealing with such questions as "How many grams of reactants must be combined to get a desired mass of product?" You will learn how to use a balanced chemical equation to answer questions like this. And, as you might have guessed, you will need your calculator.

+1 One More Thing

8.7 Chemical Synthesis

Making new substances with improved and useful properties is one of the major pursuits of modern chemistry. From plastics to nanowires to semiconductors to new pharmaceuticals, when a chemist sets out to make something new, some sort of plan or recipe is going to be needed. After all, you don't just mix carbon, hydrogen, and oxygen together and hope you can pull out a useful drug like aspirin.

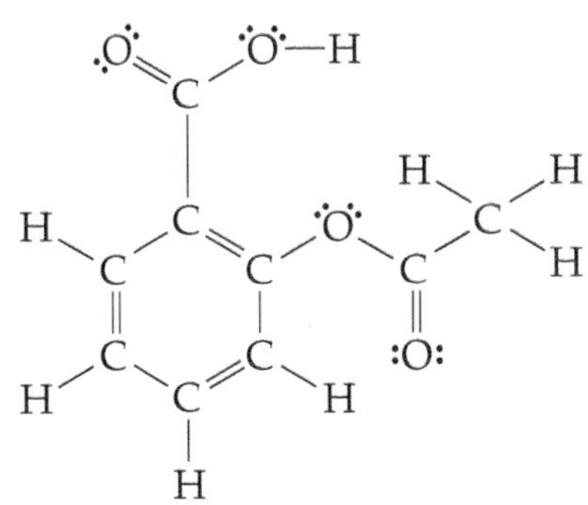

Aspirin, $C_9H_8O_4$

The work of planning, performing, and perfecting the preparation of a new compound is called **chemical synthesis**. A chemical synthesis starts with one or more starting materials and converts them to products in one step if possible, but more often in a multistep procedure that requires the isolation and purification of intermediates along the way. Quite often, solvents such as water or organic liquids are required to dissolve the reactants so that they may come together and react. Additional requirements are often sufficient stirring, heating, cooling, isolation from solvents and by-products (purification), identification (characterization), and recycling or disposal of hazardous wastes. All of this costs money and time.

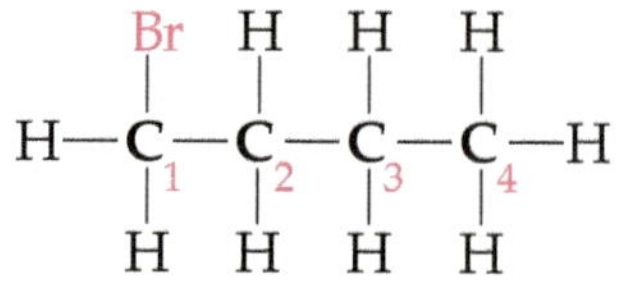

1-Bromobutane

Coming up with a plan for a chemical synthesis of a target molecule requires an intimate knowledge of how different molecules are likely to react with one another. Students of chemistry generally begin learning this knowledge in a course called Organic Chemistry, but we thought we would give you a taste of this now. Imagine you were given the task to make the target molecule shown on the left, called 1-bromobutane (C_4H_9Br).

A synthetic chemist would know a few things. For example, it has been well established over the years that the readily available compound hydrogen bromide (HBr), a gas, will react with a carbon–carbon double bond (C=C) in a combination-type reaction.

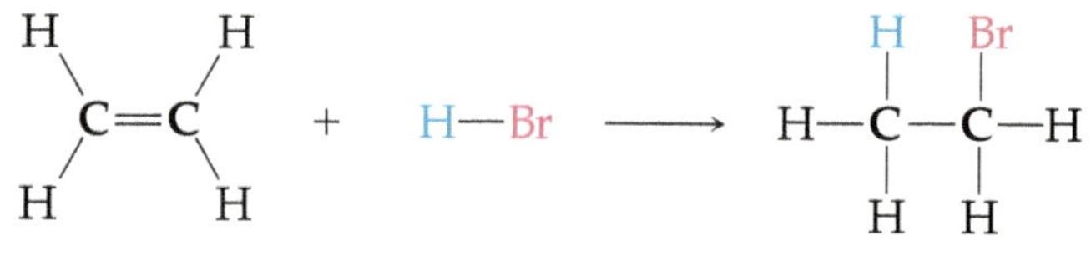

A synthetic chemist could use this knowledge to postulate some possible reactant molecules to start with.

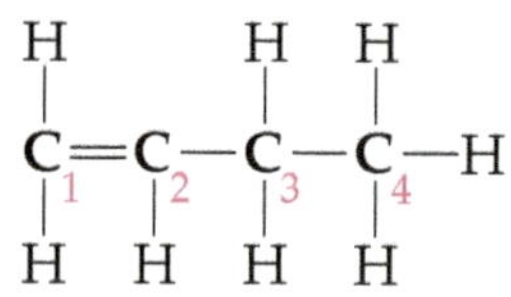

1-Butene

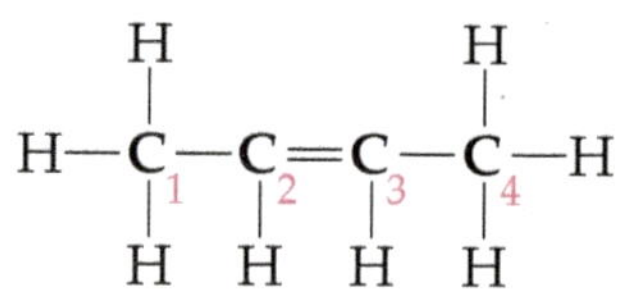

2-Butene

Why these? Well, they are both quite similar to the target compound (they are both four carbon atoms long with hydrogen atoms attached to the carbon atoms), and they both have one C=C double bond that we can add HBr across. In fact, both are commercially available, so which should we choose? Can you see that one of them would not do? Our desired product has the bromine atom on the number one carbon, and only 1-butene gives us this possibility.

Next, a synthetic chemist would have to worry about which way the HBr will add across the double bond of 1-butene. There are two possibilities.

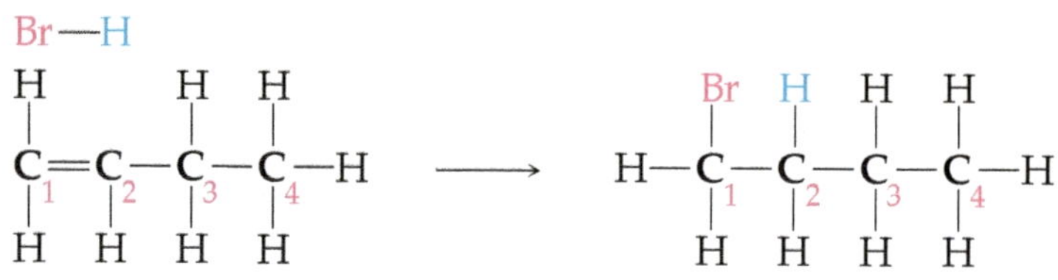

1-Butene 1-Bromobutane

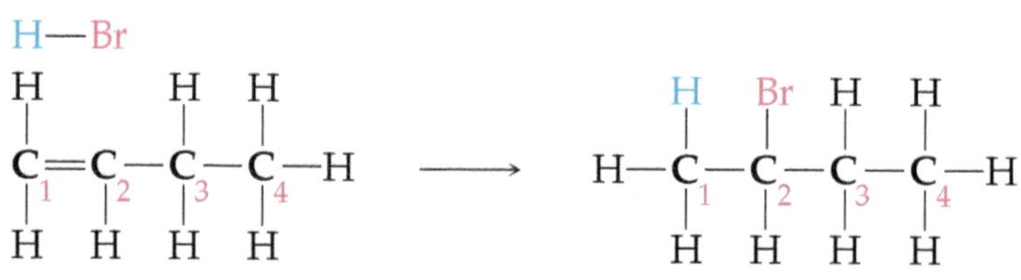

1-Butene 2-Bromobutane

Which way will it add? Will it add both ways? If it does, then we will have to try and separate the different products from one another. Once again, a synthetic chemist would make use of previous knowledge. In this case, it is known that HBr can be made to add almost exclusively either way depending on conditions.

For example, if HBr is used by itself, we will get the undesired 2-bromobutane, but if we use HBr in combination with another reagent called a peroxide, we will get the desired 1-bromobutane. Why? The answer has to do with the mechanisms of these reactions (how they actually occur at the atomic level), which a synthetic chemist would have learned. In Chapter 13, we introduce you to reaction mechanisms and also to the energy changes that accompany chemical reactions. Our goal in this section is to give you a taste of the utility of chemical reactions and some of the planning that occurs before implementing them.

Have You Learned This?

Chemical reaction (p. 284)
Reactants (p. 284)
Products (p. 284)
Chemical equation (p. 284)
Gas-phase reaction (p. 285)
Single-replacement reaction (p. 291)
Double-replacement reaction (p. 291)
Decomposition reaction (p. 291)
Combination reaction (p. 291)
Combustion reaction (p. 292)
Precipitation reaction (p. 293)
Precipitate (p. 293)
Dissociation (p. 294)
Spectator ions (p. 296)
Complete ionic equation (p. 296)
Net ionic equation (p. 296)
Acid (p. 299)
Base (p. 300)
Acid–base neutralization reaction (p. 300)
Chemical synthesis (p. 301)

Learning Outcomes

8.1	What Is a Chemical Reactions?	Describe what a chemical reaction is in general terms and in terms of chemical bonds. [17, 18, 21, 24, 25]
8.2	How Are Reactants Transformed into Products	Explain how collisions play a role in chemical reactions. [20, 22, 23]
8.3	Balancing Chemical Equations	Write and balance chemical equations. [28, 29, 30, 31, 32, 33]
8.4	Types of Reactions	Describe and identify the different reaction types. [35, 36, 37, 38, 39, 40, 41]
8.5	Solubility and Precipitation Reactions	Predict the precipitate and write a chemical equation for precipitation reactions. [43, 44, 45, 46, 47, 48, 49, 50]
8.6	Introduction to Acid–Base Reactions	Write chemical reactions that involve acids and bases. [55, 56, 57, 58, 59, 60, 62]
8.7	Chemical Synthesis	Plan a simple chemical synthesis. [144, 145]

Skills to Know

Balancing a chemical equation

1. An equation is balanced when the same numbers of atoms of each element are present on the reactant side and on the product side.

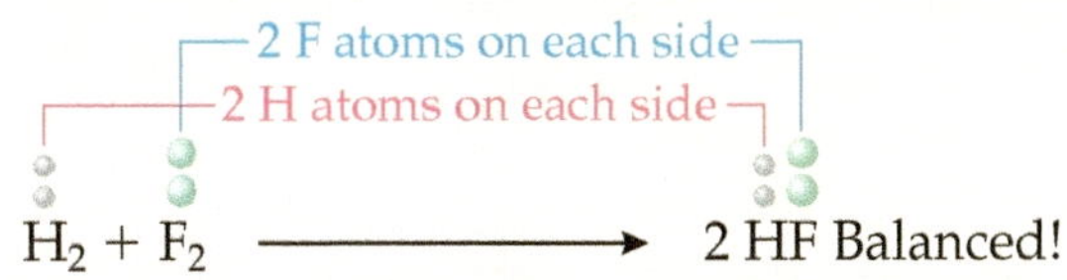

2. Balance an equation by adjusting the balancing coefficients in front of the formulas. Never alter the subscripts in the formulas!

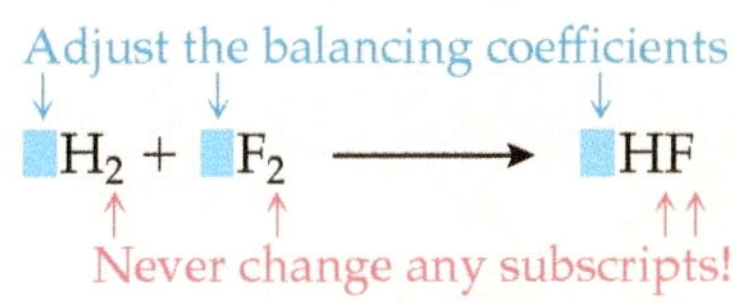

3. If the equation is complex, balance it one element at a time. First, choose an element. Adjust the coefficients until the equation is balanced for that element. Then, choose another element and repeat the process.

See example below.

4. If the equation contains a pure elemental substance, save that element for last. This sometimes makes the job of balancing easier.

Pure elemental substance, so balance O last in this equation

$$C_6H_{12}O_6 + 6\ O_2 \longrightarrow 6\ H_2O + 6\ CO_2$$

5. If your balanced equation contains a fractional coefficient, it is customary to convert the fraction to a whole number. To do that, multiply all the coefficients on both sides of the equation by the smallest number that will turn the fraction into a whole number.

Balanced equation with fractional coefficient:

$$C_8H_{18} + {}^{25}\!/_2\ O_2 \longrightarrow 9\ H_2O + 8\ CO_2$$

Multiply all coefficients by 2:

$$2\ C_8H_{18} + 2({}^{25}\!/_2)\ O_2 \longrightarrow 2(9)\ H_2O + 2(8)\ CO_2$$

$$2\ C_8H_{18} + 25\ O_2 \longrightarrow 18\ H_2O + 16\ CO_2$$

Example: Balance the equation for the combustion of ethanol, C_2H_5OH:

We shall balance this equation one element at a time, saving O for last because the elemental substance O_2 is present (guideline 4 above).

$$C_2H_5OH + O_2 \longrightarrow H_2O + CO_2$$

1. We choose to start with C. We can balance it by adding a coefficient 2 for CO_2.

2 C atoms on each side

$$C_2H_5OH + O_2 \longrightarrow H_2O + 2\ CO_2$$

2. Now H. Notice that ethanol has 6 H atoms! We balance H by adding a coefficient 3 for H_2O.

6 H atoms on each side

$$C_2H_5OH + O_2 \longrightarrow 3\ H_2O + 2\ CO_2$$

3. Now O. We add a coefficient 3 to O_2. (We don't alter the coefficient of C_2H_5OH because that would unbalance C and H.) A check verifies that the equation is balanced.

7 O atoms on each side

$$C_2H_5OH + 3\ O_2 \longrightarrow 3\ H_2O + 2\ CO_2$$

2 C + 6 H + 7 O atoms on each side

Balanced!

Solubility rules

Soluble in water	All sodium, potassium, and ammonium salts		
	All acetates and nitrates		
	Most halides (chlorides, bromides, iodides)	*except*	Halides of lead(II), silver(I), and mercury(I)
	Most sulfates	*except*	Sulfates of calcium, barium, lead(II), and strontium
Insoluble in water	Most phosphates, carbonates, and sulfides	*except*	Sodium, potassium, and ammonium salts; calcium sulfide
	Most hydroxides	*except*	Sodium, potassium, calcium, and barium hydroxides

Identifying the precipitate(s) in a potential precipitation reaction

Example: A solution of barium chloride is mixed with a solution of sodium sulfate. Will a precipitation reaction occur? If so, what will precipitate?

1. Identify all the ions that will be present when the starting solutions are mixed. (These are the ions produced when the starting materials dissolve.)

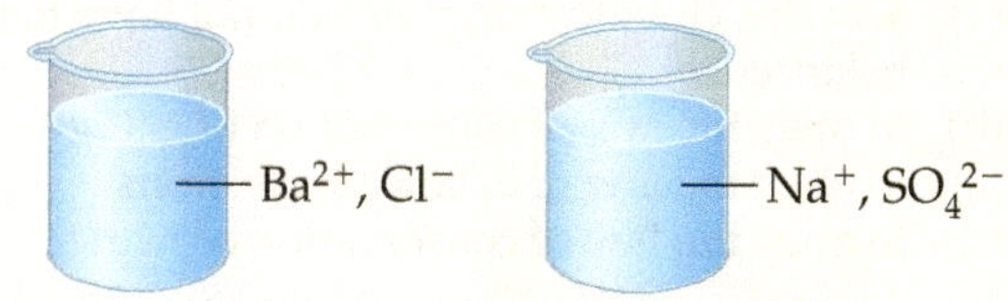

$BaCl_2$ solution Na_2SO_4 solution

2. Identify all the possible cation–anion combinations that are different from the ones in the starting materials. These combinations represent possible precipitates.

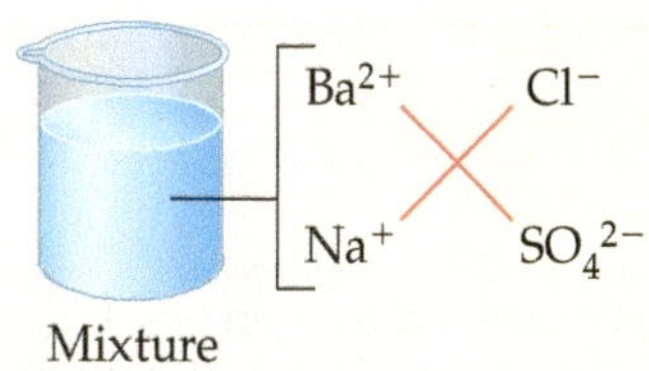

Mixture

Possible new combinations:
$BaSO_4$
$NaCl$

3. Use the solubility rules to decide if any of the ion combinations represent an insoluble salt. Any combination that does will precipitate out of the solution. If none of the combinations represents an insoluble salt, no precipitation reaction will occur.

$BaSO_4$: Insoluble. Will precipitate.

$NaCl$: Soluble (like all sodium salts). Will not precipitate.

4. If a precipitation reaction occurs, identify it by writing the net ionic equation. Make sure the formula for the precipitate is correct (electrically neutral).

Net ionic equation:

$$Ba^{2+}(aq) + SO_4^{2-}\ (aq) \longrightarrow BaSO_4(s)$$

Concept Questions

1. Which one of the following is true about chemical reactions?
   (a) They accomplish only physical changes.
   (b) They are represented with a chemical equation.
   (c) Products are converted into reactants.
   (d) The arrow means "heat and stir."

2. Which one of the following is true about chemical bonds?
   (a) It takes energy to break a bond only for polar compounds.
   (b) Energy is released when breaking weak bonds.
   (c) It always takes energy to break chemical bonds.
   (d) They can survive molecular collisions of any amount of energy.

3. The energy required to yield a chemical change in a gas-phase reaction
   (a) comes from high-energy electrons.
   (b) comes from collisions between molecules.
   (c) is supplied by the products.
   (d) has nothing to do with the reactant molecules.

4. The reason we balance chemical equations is
   (a) actually, chemical reactions do not have to be balanced.
   (b) to obey the law of conservation of energy.
   (c) to obey the law of balancing electrons.
   (d) to obey the law of conservation of matter.

5. A good hint for balancing chemical equations is
   (a) save the elemental substances for last.
   (b) balance the elemental substances first.
   (c) divide all the balancing coefficients by a number that makes them all whole.
   (d) take random guesses until you are right.

6. A combination reaction is a reaction in which
   (a) the chemical properties of the reactants are combined.
   (b) decomposition occurs.
   (c) atoms are exchanged between reactants.
   (d) different reactant molecules are combined to yield products.

7. A precipitation reaction is
   (a) a single-replacement reaction.
   (b) a double-displacement reaction.
   (c) a combination reaction.
   (d) (b) and (c).

8. A precipitation reaction will occur when
   (a) the reactants and products are incompatible.
   (b) all of the reactants are soluble.
   (c) at least one of the possible products is insoluble.
   (d) all of the reactants are insoluble.

9. To write a net ionic equation, you must
   (a) write every species in solution.
   (b) include the spectator ions.
   (c) omit the spectator ions.
   (d) write the precipitate with (*aq*) after it.

10. An acid is any substance that
   (a) produces $H^+(aq)$ ions when added to water.
   (b) produces hydroxide $OH^-(aq)$ ions when added to water.
   (c) contains hydrogen, H, atoms in its formula.
   (d) forms a precipitate when added to a soluble compound.

Answers: 1b; 2c; 3b; 4d; 5a; 6d; 7c; 8c; 9c; 10a

Chemical Reactions

8.17 Tap water contains dissolved oxygen gas, $O_2(aq)$. Adding heat to tap water causes the dissolved oxygen to leave. Which of the two equations below represents a chemical reaction, and which does not? Defend your answer.
(a) $2\,H_2O_2(l) \rightarrow 2\,H_2O(l) + O_2(g)$
(b) $H_2O(l) + O_2(aq) \xrightarrow{\text{Heat}} H_2O(l) + O_2(g)$

8.18 Someone claims that a substance has undergone a chemical reaction. What would they have to demonstrate in order to prove this?

8.19 Recall the reaction to make the superconductor described at the beginning of this chapter. What could be done to show that the product of the chemical reaction is truly a new compound and not just a heterogeneous mixture of the reactants?

8.20 Under what circumstances (if any) does it *not* take energy to break a chemical bond?

8.21 Consider the following reaction. Describe it in terms of which bonds must be broken and which bonds must be formed.

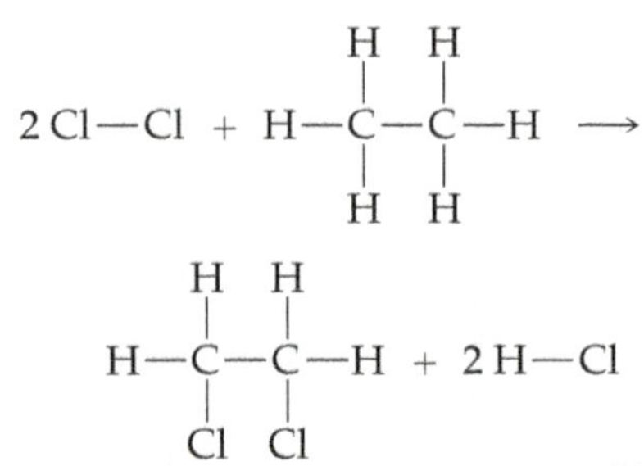

8.22 For most gas-phase chemical reactions, where does the energy come from to break covalent bonds in reactant molecules?

8.23 Consider a gas-phase reaction.
(a) Cooling the mixture of reactants can slow and even stop the chemical reaction. Explain why this is so.
(b) Increasing the number of reactant molecules in the flask makes the reaction go faster. Explain why this is so.

8.24 Describe how you would explain to someone how a chemical reaction changes one set of compounds into another.

8.25 A student writes the following chemical equation. Is it a valid chemical equation? Explain.
$2\,Fe_2O_3(s) + 3C(s) \rightarrow 3\,CO_2(g) + 2\,Au(s)$

Balancing Chemical Equations

8.26 Why must a chemical equation be balanced?

8.27 What is wrong with changing the subscripts in a chemical formula to balance a chemical equation?

8.28 Balance this chemical equation by inspection:
$C_2H_4 + O_2 \rightarrow CO_2 + H_2O$

8.29 Translate this balanced chemical equation into words:
$CH_4 + 2\,O_2 \rightarrow CO_2 + 2\,H_2O$
Methane

8.30 Balance this chemical equation by inspection:
$Fe_2O_3 + C \rightarrow Fe + CO_2$

8.31 Balance this chemical equation by inspection:
$Al_2Cl_6 + H_2O \rightarrow Al(OH)_3 + HCl$

8.32 Balance this chemical equation by inspection:
$KClO_3 \rightarrow KCl + O_2$

8.33 Consider the following chemical equation:
$C_2H_2 + O_2 \rightarrow H_2O + CO_2$
(a) Balance this chemical equation by inspection.
(b) Translate the balanced equation into words. The compound C_2H_2 is called acetylene and is used as a fuel for torches that cut through steel.

8.34 John writes the chemical equation $O_2 + O_2 \rightarrow 1\frac{1}{3}O_3$. Sue claims that is not a chemical reaction because there is no chemical change, saying "Oxygen is on both sides of the arrow!" (*Hint*: O_2 is oxygen that you breathe; O_3 is ozone, which is quite toxic.) Sue also says it's not balanced. Which of Sue's complaints is correct? Explain.

Types of Reactions

8.35 Balance this chemical equation and assign it a reaction type:
$N_2O_5(g) \rightarrow NO_2(g) + O_2(g)$

8.36 Balance this chemical equation and assign it a reaction type:
$CO(g) + NO(g) \rightarrow CO_2(g) + N_2(g)$

8.37 Balance this chemical equation and assign it a reaction type:
$Fe(NO_3)_3(aq) + Na_2S(aq) \rightarrow Fe_2S_3(s) + NaNO_3(aq)$

8.38 Balance this chemical equation and assign it a reaction type:
$SO_2(g) + O_2(g) \rightarrow SO_3(g)$

8.39 Balance this chemical equation and assign it a reaction type:
$Li(s) + N_2(g) \rightarrow Li_3N(s)$

8.40 Consider the following unbalanced chemical equation:
$C_5H_{10}O_2(l) + O_2(g) \rightarrow CO_2(g) + H_2O(g)$
(a) Balance this equation.
(b) Classify the reaction type and explain.
(c) Because one of the reactants is oxygen, what can you call this type of reaction?

8.41 Under the proper conditions, hydrogen peroxide, $H_2O_2(l)$, can be converted to water and oxygen gas. Write a balanced equation for this reaction and assign it a reaction type.

8.42 Compare and contrast combination reactions and decomposition reactions. Give an example of each.

Solubility and Precipitation Reactions

8.43 "Hard" water is hard because it contains $Ca^{2+}(aq)$ and $Mg^{2+}(aq)$ ions. It can be softened by adding borax powder, which is sodium tetraborate, $Na_2B_4O_7$. Adding borax removes the $Ca^{2+}(aq)$ and $Mg^{2+}(aq)$ from solution. Write a net ionic equation for each of these two reactions. What ion replaces the $Ca^{2+}(aq)$ and $Mg^{2+}(aq)$ ions in the water?

8.44 Draw a beaker that shows the result of dissolving methanol, CH_3OH, in water and a second beaker that shows the result of dissolving sodium phosphate in water. Indicate all species present in solution in each beaker. Describe the difference between how these two substances behave when added to water.

8.45 Fill in this table:

Name	Formula	Soluble or insoluble in water?
Sodium phosphate		
	$Ba(CH_3CO_2)_2$	
	$(NH_4)_2S$	
Iron(II) carbonate		
Mercury(II) chloride		
	$Co(OH)_2$	
	$HgCl$	

8.46 Write a net ionic equation for the precipitation reaction, if any, that occurs when aqueous solutions of the following ionic compounds are mixed:
(a) $Pb(NO_3)_2$ and NaCl
(b) $Ba(NO_3)_2$ and $NiSO_4$
(c) NaCl and KNO_3

8.47 Write a net ionic equation for the precipitation reaction, if any, that occurs when aqueous solutions of the following ionic compounds are mixed:
(a) $Bi(NO_3)_3$ and NaOH
(b) K_2SO_4 and SrI_2
(c) $Cu(CH_3COO)_2$ and Na_3PO_4

8.48 Draw a beaker that shows the result of combining aqueous solutions of Na_2S and nickel(II) nitrate. Indicate which ions are in solution, and write the formula for the precipitate that forms. Write a net ionic equation for the precipitation reaction.

8.49 Aqueous solutions of iron(III) sulfate and barium hydroxide are combined. Does a precipitate form? If yes, write a net ionic equation for the precipitation reaction.

8.50 Aqueous solutions of calcium chloride and potassium carbonate are combined. Does a precipitate form? If yes, write a net ionic equation for the precipitation reaction.

8.51 You need calcium sulfate, but there is none in the lab. However, there are lots of other ionic compounds on the shelves.
(a) Propose a synthesis for calcium sulfate.
(b) Write a complete ionic equation for the synthesis.
(c) Write a net ionic equation for the synthesis.
(d) Write the equation you wrote for (b) as an English sentence.

8.52 You need nickel(II) hydroxide, but there is none in the lab. However, there are lots of other ionic compounds on the shelves. Propose a synthesis for nickel(II) hydroxide.

8.53 You dissolve some silver nitrate in your tap water from home, and the water turns cloudy. What chemical species might be in your tap water?

8.54 When an aqueous solution of ammonium sulfate is added to an aqueous solution of calcium nitrate, a precipitate forms.
(a) Write a net ionic equation for the precipitation.
(b) Write the precipitation reaction in a way that emphasizes that a double-replacement reaction has taken place.

Introduction to Acid–Base Reactions

8.55 An aqueous solution of hydrosulfuric acid, $H_2S(aq)$, reacts with an aqueous solution of $NaOH(aq)$.
(a) Write the complete ionic equation for the acid–base neutralization that occurs. (*Hint*: $S^{2-}(aq)$ ion is present in the neutralized solution.)
(b) What salt would remain behind if you evaporated the water from the neutralized solution?

8.56 When an aqueous solution of $H_2S(aq)$ is mixed with an aqueous solution of copper(II) sulfate, a black precipitate of CuS forms.
(a) Write and balance the full reaction equation for the reaction of $H_2S(aq)$ with $CuSO_4(aq)$.
(b) Explain how your answer to (a) demonstrates that $H_2S(aq)$ is an acid.
(c) Write the net ionic equation for the formation of $CuS(s)$.

8.57 Suppose you had oxalic acid, $H_2C_2O_4$, and you wanted to make sodium oxalate, $Na_2C_2O_4$. How might you do this?

8.58 Write (a) a reaction with all species intact and using (*aq*), (*s*), or (*l*) for the reaction between aqueous solutions of sulfuric acid and calcium hydroxide. (b) What salt is formed as a result of neutralization?

8.59 When aqueous H_2SO_4 is added to an aqueous solution of $Pb(NO_3)_2(aq)$, a precipitate forms.
(a) What is the precipitate?
(b) Write a net ionic equation for the precipitation reaction.
(c) Did adding $Pb(NO_3)_2(aq)$ to the sulfuric acid neutralize the acid? Explain.

8.60 Ammonium ions (NH_4^+) behave as an acid.
(a) Write an equation that shows $NH_4^+(aq)$ behaving as an acid when it reacts with aqueous hydroxide ion.
(b) What are the products of neutralizing an aqueous solution of NH_4Cl with NaOH?

8.61 You need to prepare some calcium nitrate.
(a) Write an intact-molecule balanced equation for an acid–base neutralization reaction that could be used to prepare calcium nitrate.
(b) Write the net ionic equation for the neutralization.

8.62 Write the complete ionic equation and the net ionic equation for the reaction that occurs when the following solutions are mixed, and name the salt formed in each case:
(a) $HBr(aq) + NaOH(aq)$
(b) $Ca(OH)_2(aq) + HNO_3(aq)$
(c) $Ca(OH)_2(aq) + HBr(aq)$
(d) $Mg(OH)_2(aq) + HCl(aq)$

8.63 The compound CH_3COOCH_3 (methyl acetate) is not an acid, but the compound CH_3COOH (acetic acid) is.
(a) Which of the hydrogen atoms in acetic acid is the acidic hydrogen (the one that becomes $H^+(aq)$ ion)?

(b) Speculate about why none of the hydrogen atoms in methyl acetate is acidic. (*Hint*: Look carefully at what the acidic hydrogen atom is bonded to in acetic acid.)

(c) The diagram below shows how electrons rearrange to make acetic acid behave as an acid. Notice the two different ways the resulting anion is drawn. This allows the negative charge to be spread among two oxygen atoms, which in turn makes the whole anion stable and helps acetic acid to lose an H^+ ion and be an acid. What do you call the two different forms of the anion? (*Hint*: see Section 5.5.)

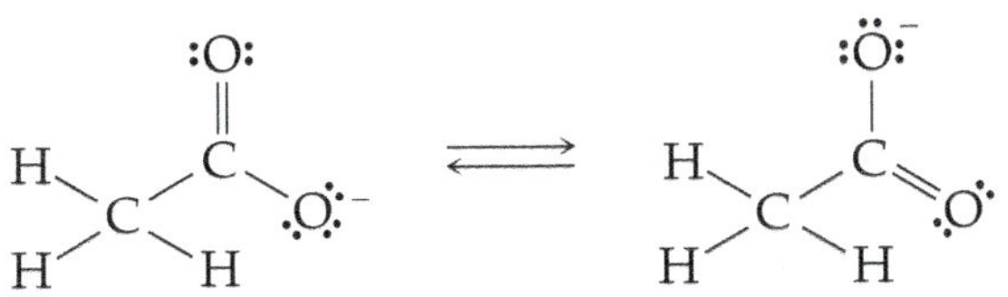

Additional Problems

8.64 When a precipitate forms in water, the water often becomes warmer. What is the source of the heat energy that warms the water?

8.65 Consider the covalent bond in H_2. Why must energy be added to break this bond (in other words, what is the added energy used for)?

8.66 Consider the ionic bonds between Na^+ ions and Cl^- ions in NaCl(s). Why must energy be added to break these bonds (in other words, what is the added energy used for)?

8.67 Balance each equation and assign each a reaction type:

(a) $SiO_2(s) + C(s) \rightarrow Si(s) + CO(g)$

(b) 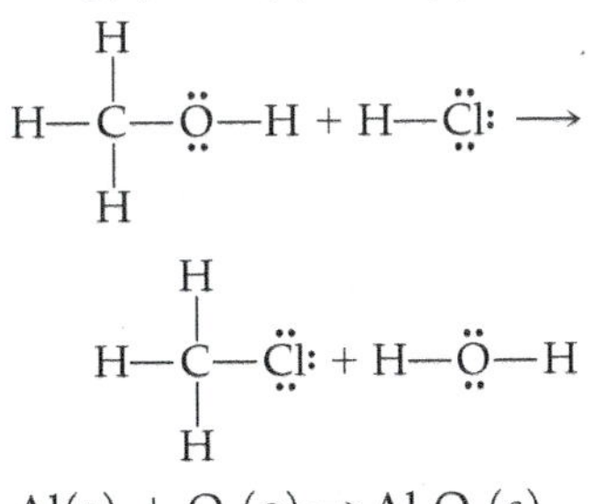

(c) $Al(s) + O_2(g) \rightarrow Al_2O_3(s)$

(d) $(NH_4)_2Cr_2O_7(s) \rightarrow N_2(g) + Cr_2O_3(s) + H_2O(g)$

8.68 Aqueous solutions of sodium sulfide and iron(III) nitrate are combined.

(a) Does a precipitation reaction occur? Explain.

(b) If it does, write a net ionic equation for the reaction.

8.69 Aqueous solutions of calcium chloride and potassium carbonate are combined.

(a) Write the formulas for both reactants.

(b) Does a precipitation reaction occur? Explain.

(c) If it does, write a net ionic equation for the reaction.

8.70 Describe what happens when $CaCl_2(s)$ and $NaNO_3(s)$ are added to the same beaker of water. What, if anything, dissolves, and what, if anything, precipitates?

8.71 Suppose you have an aqueous solution containing Cu^{2+} ions and Na^+ ions. What could you add that would remove the Cu^{2+} ions from solution without adding any new type of cations to the solution?

8.72 Suppose you have an aqueous solution of Pb^{2+} and Ba^{2+} ions.

(a) How could you remove just the Pb^{2+} ions from the solution?

(b) How could you remove both the Pb^{2+} and the Ba^{2+} ions from the solution at the same time?

8.73 Nitric acid consists of NO_3^- and a proton, H^+.

(a) Draw a Lewis dot diagram for the nitrate ion.

(b) Draw a Lewis dot diagram for nitric acid. (*Hint*: Nitric acid contains an O—H bond.)

(c) Which species, nitrate ion or nitric acid, is more likely to be found in aqueous solution? Explain.

8.74 Write the intact-molecule equation and the net ionic equation for the neutralization reaction between perchloric acid, $HClO_4(aq)$, and $NaOH(aq)$. What salt is in solution once the reaction is run?

8.75 $Na^+(aq)$, $Pb^{2+}(aq)$, and $Ba^{2+}(aq)$ are all present in the same solution. Fully describe a procedure whereby you could separate these ions from one another.

8.76 $Fe^{3+}(aq)$, $Ca^{2+}(aq)$, and $Ba^{2+}(aq)$ are all present in the same solution. Fully describe a procedure whereby you could separate these ions from one another.

8.77 Heptane, C_7H_{16}, is one of the components of gasoline. Burning heptane in air yields carbon dioxide and water. Balance the equation for this reaction:

$$C_7H_{16}(l) + O_2(g) \rightarrow CO_2(g) + H_2O(g)$$

8.78 The burning of ethanol, C_2H_5OH, in air yields carbon dioxide and water. Balance the equation for this reaction:

$$C_2H_5OH(l) + O_2(g) \rightarrow CO_2(g) + H_2O(g)$$

8.79 Sodium chloride reacts with bromine gas, Br_2, to produce sodium bromide plus chlorine gas. Write a balanced equation for this reaction.

8.80 Indicate which bonds are broken and which are formed when each reaction occurs:

(a) $N_2 + 3\,Br_2 \rightarrow 2\,NBr_3$

(b) $P_4 + 6\,H_2 \rightarrow 4\,PH_3$

(c) $KF + NaI \rightarrow KI + NaF$

(d) $CH_4 + Cl_2 \rightarrow CH_3Cl + HCl$

8.81 Balance the following equation and classify the reaction it represents as combination, decomposition, single-replacement, or double-replacement:

$Fe + CdCl_2 \rightarrow FeCl_3 + Cd$

8.82 Indicate whether each compound is soluble or insoluble in water:
(a) $BaCO_3$
(b) K_2S
(c) $CaSO_4$
(d) $Mg(NO_3)_2$
(e) $(NH_4)_3PO_4$

8.83 Would you expect a reaction in which all reactants are in the gas phase to go faster, slower, or remain at the same rate if the temperature is increased while everything else is held constant? Explain your answer.

8.84 Draw dot diagrams for the reactants and products in each equation, and indicate which bonds are broken and which bonds are formed:
(a) $CH_4 + I_2 \rightarrow CH_3I + HI$
(b) $C_2H_4 + F_2 \rightarrow FCH_2CH_2F$

8.85 Which equations represent chemical reactions:
(a) $CO_2(s) \rightarrow CO_2(g)$
(b) $3\,O_2(g) \rightarrow 2\,O_3(g)$
(c) $NaCl(s) + H_2O(l) \rightarrow NaOH(aq) + HCl(aq)$
(d) $H_2O(g) + CH_4(g) \rightarrow CH_4(g) + H_2O(g)$

8.86 Hydrogen gas is produced when elemental tin reacts with HF to produce hydrogen and tin(II) fluoride. Write the balanced equation for this reaction.

8.87 Classify each compound as acid, base, or salt:
(a) KCl
(b) CH_3COOH
(c) $Al(OH)_3$
(d) H_3BO_3
(e) $LiC_2H_3O_2$

8.88 Write the balanced chemical equation for the reaction between pentane, C_5H_{12}, and oxygen gas that produces carbon dioxide and water.

8.89 Write the balanced equation, the complete ionic equation, and the net ionic equation for the neutralization reaction that occurs when nitric acid, HNO_3, is mixed with calcium hydroxide, $Ca(OH)_2$.

8.90 Write the balanced chemical equation for the production of P_2O_5 from P and O_2.

8.91 Write the formula of the precipitate formed when these solutions are mixed:
(a) Aluminum nitrate and sodium hydroxide
(b) Potassium phosphate and calcium chloride
(c) Magnesium sulfate and sodium carbonate

8.92 Write the balanced equation for the preparation of iron(III) oxide from iron metal and oxygen gas. What type of reaction is this?

8.93 Balance this chemical equation, and classify it as combination, decomposition, single-replacement, or double-replacement:

$H_3PO_4 + Ca(OH)_2 \rightarrow Ca_3(PO_4)_2 + H_2O$

8.94 Ammonia, NH_3, reacts with chlorine gas, Cl_2, to form ammonium chloride, NH_4Cl, and nitrogen trichloride, NCl_3. Write a balanced equation for this reaction.

8.95 Classify each reaction as combination, decomposition, single-replacement, or double-replacement:
(a) $2\,NaCl \rightarrow 2\,Na + Cl_2$
(b) $Ba_3N_2 \rightarrow 3\,Ba + N_2$
(c) $Cl_2 + 2\,NaI \rightarrow 2\,NaCl + I_2$
(d) $KNO_3 + AgBr \rightarrow AgNO_3 + KBr$

8.96 What is the number of molecules of NaOH required to neutralize one molecule of:
(a) H_2SO_4
(b) HI
(c) H_3PO_4
(d) HNO_3
(e) CH_3COOH

8.97 Oxygen can be produced by the decomposition of potassium chlorate, $KClO_3$. The products of the reaction are KCl and O_2. Write a balanced equation for the reaction.

8.98 Balance each equation:
(a) $Zn + HCl \rightarrow ZnCl_2 + H_2$
(b) $Na_2O + H_2O \rightarrow NaOH$
(c) $CH_4 + H_2S \rightarrow CS_2 + H_2$
(d) $CO + H_2 \rightarrow CH_3OH$

8.99 Calcium hydroxide can be used as an antacid to neutralize HCl, the acid found in the stomach. Water and calcium chloride, $CaCl_2$, are the products of the reaction. Write the balanced equation for the reaction.

8.100 Carbon dioxide gas and calcium oxide are produced when calcium carbonate, $CaCO_3$, is heated strongly. Write the balanced equation for this process.

8.101 Balance this chemical equation and classify it as combination, decomposition, single-replacement, or double-replacement:

$K_2SO_3 + S_8 \rightarrow K_2S_2O_3$

8.102 Name the acid and base that combine to form:
(a) KNO_3
(b) $Ca_3(PO_4)_2$
(c) Li_2SO_4
(d) NaI

8.103 Mineral compounds containing sulfur are converted to oxygen-containing compounds by a process known as *roasting*, which involves heating the mineral in the presence of oxygen. In addition to the mineral oxide, sulfur dioxide is also produced. Write the balanced chemical equation for the roasting of zinc sulfide, ZnS, to produce zinc oxide, ZnO, plus sulfur dioxide.

8.104 Chemical reactions occur when new substances are produced as a result of bonds being ________ in the reactants and bonds being ________ in the products.

8.105 When propane gas, C_3H_8, reacts with oxygen gas, carbon dioxide gas and water are produced. Write the balanced equation for the reaction.

8.106 Balance this equation and classify it as combination, decomposition, single-replacement, or double-replacement:

$K_2CrO_4 + Al(NO_3)_3 \rightarrow Al_2(CrO_4)_3 + KNO_3$

8.107 The reaction of carbon monoxide gas with water vapor produces carbon dioxide gas and hydrogen gas. Write the balanced equation for the reaction.

8.108 When an aqueous acid is combined with an aqueous base, the products are a ________ and ________.

8.109 Balance this equation and classify it as combination, decomposition, single-replacement, or double-replacement:

$Mn + S \rightarrow MnS_2$

8.110 Which anions form compounds that are generally soluble in water:

$CO_3^{2-}, NO_3^-, Br^-, SO_4^{2-}, OH^-, PO_4^{3-}$?

8.111 Sulfur dioxide and water are formed when hydrogen sulfide, H_2S, reacts with oxygen gas. Write the balanced equation for the reaction.

8.112 Write the balanced equation for the decomposition of magnesium oxide, MgO, into Mg and O_2.

8.113 A solution containing which anion can be added to separate:
(a) $Na^+(aq)$ from $Ag^+(aq)$
(b) $Na^+(aq)$ from $Ca^{2+}(aq)$

8.114 Balance this equation and classify it as combination, decomposition, single-replacement, or double-replacement:

$H_2O_2 \rightarrow H_2O + O_2$

8.115 Write the net ionic equation for the reaction between solutions of:
(a) Aluminum nitrate and sodium hydroxide
(b) Potassium phosphate and calcium chloride
(c) Magnesium sulfate and sodium carbonate

8.116 Heating magnesium carbonate results in the production of carbon dioxide and magnesium oxide. Write the balanced equation for this process, and indicate the type of reaction.

8.117 Is the reaction $LiOH + HCN \rightarrow H_2O + LiCN$ a combination, decomposition, single-replacement, or double-replacement?

8.118 Indicate whether a precipitate forms when these solutions are mixed, and write the formula of the precipitate if one forms:
(a) Silver nitrate and potassium iodide
(b) Lithium sulfate and silver acetate
(c) Sodium chloride and ammonium sulfate

8.119 What does it mean when the label (*aq*) is attached to the formula for an ionic compound?

8.120 The ____________ ionic equation for a reaction is obtained by canceling the spectator ions in the ____________ ionic equation.

8.121 Write the balanced equation, the complete ionic equation, and the net ionic equation for the reaction between aqueous potassium chloride and lead(II) nitrate that produces lead chloride and potassium nitrate.

8.122 Acids produce _____ ions when dissolved in water, and bases produce ______ ions when dissolved in water.

8.123 The reaction between an acid and a base is called an acid–base ______________ reaction.

8.124 Write the balanced equation for the reaction that occurs when an aqueous solution of carbonic acid, H_2CO_3, is mixed with an aqueous solution of lithium hydroxide, LiOH.

8.125 Of the types of reactions discussed in the chapter, list three ways the reaction $Ca(OH)_2 + H_2SO_4 \rightarrow CaSO_4 + 2\,H_2O$ can be classified.

8.126 (a) True or false? It is possible for a reaction to be simultaneously a single-replacement reaction and a combination reaction.
(b) True or false? It is possible for a reaction to be simultaneously a double-replacement reaction and a neutralization reaction.

If either statement is false, explain why.

8.127 Would you expect a reaction in which all reactants are in the gas phase to go faster, slower, or remain at the same rate if the size of the container is increased while everything else is held constant? Explain.

8.128 Nitrous oxide, N_2O, is produced when ammonium nitrate, NH_4NO_3, is heated gently. Water is also produced in the reaction. Write the balanced equation for the reaction.

8.129 Write the complete ionic equation and the net ionic equation for the reaction that occurs between aqueous solutions of:
(a) Silver nitrate and potassium iodide
(b) Lithium sulfate and silver acetate

8.130 Write a balanced chemical equation for each reaction:
(a) Calcium metal reacts with solid phosphorus, P_4, to form calcium phosphide.

(b) Sodium metal reacts with water to produce aqueous sodium hydroxide and hydrogen gas.
(c) Ammonium nitrate powder explodes to form nitrogen gas, oxygen gas, and water vapor.

8.131 Ammonia gas reacts with fluorine gas to form ammonium fluoride and nitrogen gas.
(a) Write a balanced chemical equation for this reaction.
(b) Draw a picture representing the molecular view of the reactant mixture.
(c) Draw a picture representing the molecular view of the product mixture.

8.132 Write the complete ionic equation and net ionic equation for the neutralization reaction between aqueous hydrobromic acid and barium hydroxide.

8.133 One of the following diagrams shows a situation that is incorrect. Which is it and why is it incorrect?

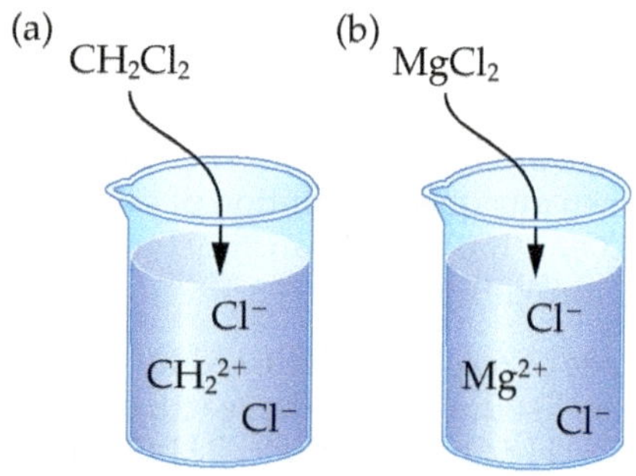

8.134 A student draws the diagram below and includes some reactions to represent the diagram.

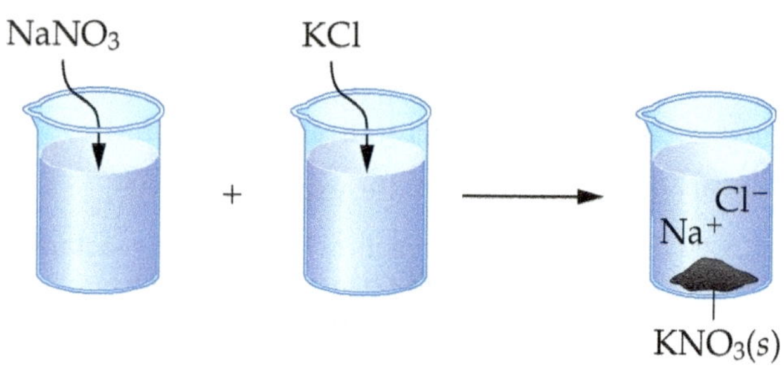

$NaNO_3(aq) + KCl(aq) \rightarrow KNO_3(s) + NaCl(aq)$
$K^+(aq) + NO_3^-(aq) \rightarrow KNO_3(s)$
(a) Which reaction is the net ionic reaction?
(b) Something is wrong with what the student wrote or drew. What is it?

8.135 Using molecular formulas, write a balanced equation for the following gas-phase reaction. Then translate the balanced reaction into an English sentence.

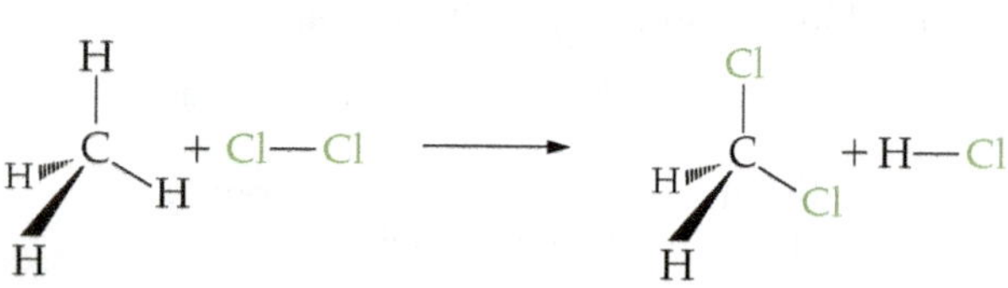

8.136 Complete the following reaction (the other product is a diatomic gas), and characterize the reaction type(s).

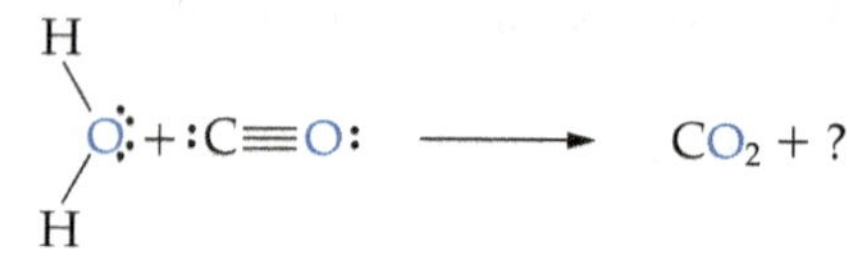

8.137 How would you classify the molecule in the left beaker?

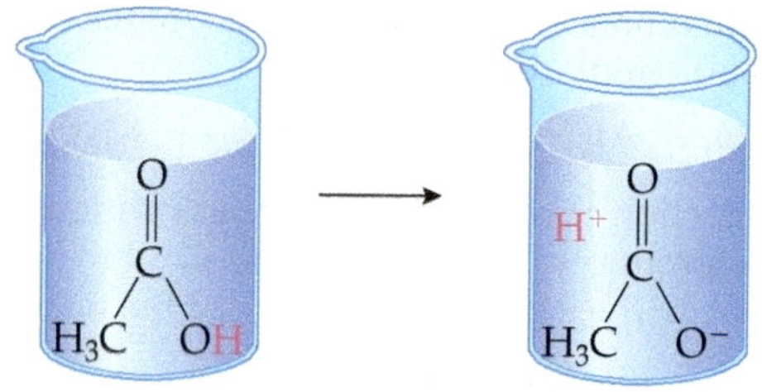

One Step Beyond

8.138 In the chapter, we saw the combustion reaction for methane. This was actually a *complete* combustion reaction because carbon ended up as carbon dioxide. If the same reaction is carried out where the oxygen supply is limited, incomplete combustion takes place, yielding toxic carbon monoxide, CO, instead of CO_2. Write the balanced reaction for the incomplete combustion of methane. Make sure it is balanced.

8.139 $Pb(NO_3)_2$ (lead (II) nitrate), upon sufficient heating, turns into PbO (lead (II) oxide), oxygen gas, and nitrogen dioxide gas. Write a balanced equation for this reaction, and classify the reaction type.

8.140 Simple soaps, like sodium laurate, $NaC_{12}H_{22}O_2$, are ionic compounds with small cations and large covalent anions. These soaps are somewhat soluble in water but form a soap scum if the water is "hard," which means it contains calcium ions Ca^{2+} or magnesium ions, Mg^{2+}. Write a net ionic equation for the formation of soap scum. Also classify the reaction type or types.

8.141 Ammonia, NH_3, is a gas, but it is very soluble in water, so when you bubble ammonia gas into water you produce an aqueous solution of ammonia. Interestingly, this solution is basic (hydroxide ions are formed). Postulate a reaction that occurs between ammonia and water to make the solution basic, and classify the reaction type.

8.142 When you strike a match, trace amounts of white phosphorus (P_4) are converted into an oxide of phosphorus, P_4O_{10}. This conversion releases sufficient heat to ignite the match head. Write a balanced reaction for this conversion, and classify the reaction type(s).

8.143 Barium ions are extremely toxic to most living things, especially humans. And yet, despite its toxicity, barium sulfate cocktails are often given to patients undergoing a CAT scan because barium readily absorbs X-rays and can therefore serve as a contrast medium which helps in the imaging of internal organs. Based on what you learned in this chapter, suggest a reason why barium sulfate may be safe to ingest despite the extreme toxicity of the Ba^{2+} ion. What do you think will be the fate of the person who is given a barium carbonate cocktail instead?

8.144 Starting from 1,3-butadiene, how would you prepare 2,3-dibromobutane?

8.145 Bromine (Br_2) can be made to add across a C = C double bond. Knowing this, postulate an alternative synthesis to the one you proposed in problem 8.144 for the preparation of 2,3-dibromobutane.

Group Exercises (Work on these exercises together with someone else in the class.)

1. Come up with five common objects that, when you change the subscript in their "formula," you can then show that what you have is no longer the same as what you started with. For example, the formula for a bicycle is $frame_1tires_2$. If you were to change the subscript to $frame_2tires_2$, would it still be a bicycle, or something else? How does this relate to changing the subscripts in a chemical formula? Then develop a rule that says when you can change the subscripts in a chemical formula while balancing a chemical equation.

2. Lead is a very toxic metal; if drinking water is contaminated with Pb^{2+} ions, you should not drink it. Design an experiment to confirm the absence or presence of Pb^{2+} ions in your drinking water using the information from Table 8.1 (the solubility rules). Also discuss whether a positive test would be absolute confirmation of Pb^{2+}.

Workpatch Solutions

8.1 (a) Not a chemical reaction because the same compound exists on both sides of the arrow. This is simply a phase change.
(b) A chemical reaction because a new compound, different from the reactants, is produced.
(c) Not a chemical reaction because the same substances appear on both sides of the arrow.

8.2 There are two correct answers to this question:

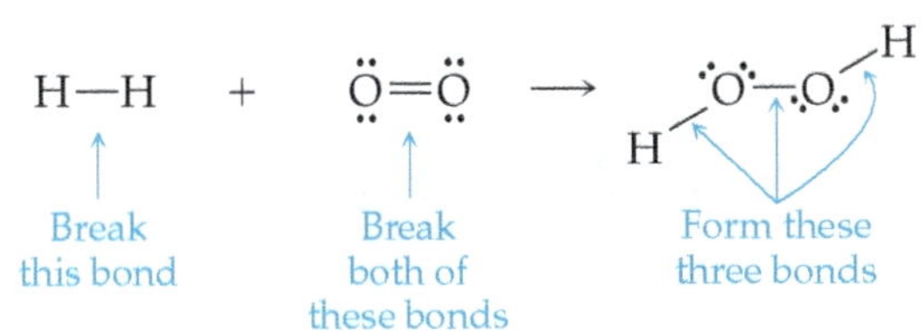

or

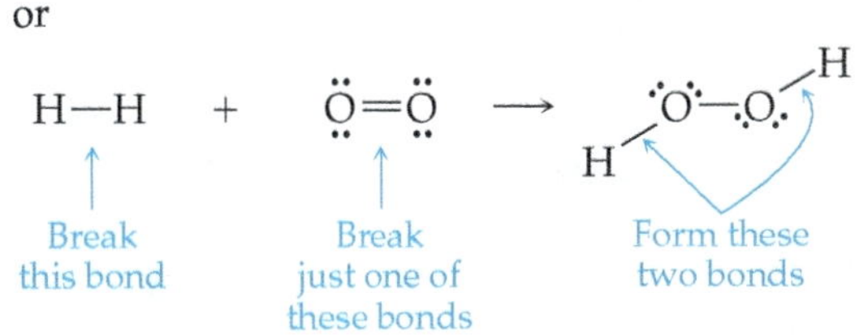

8.3

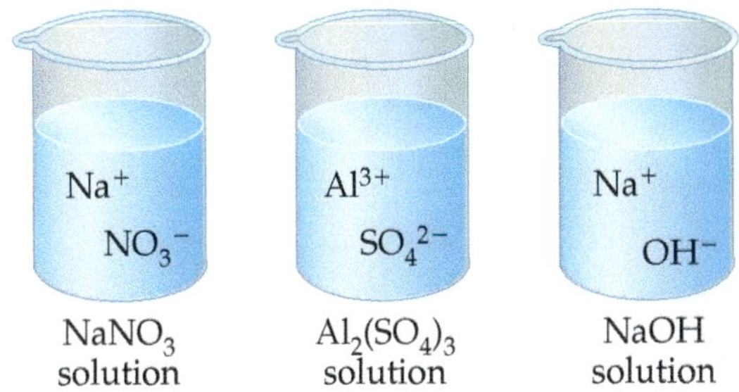

8.4 $Al(OH)_3$, $Al(NO_3)_3$, Na_2SO_4

$CH_4+2O_2 \longrightarrow CO_2+2H_2$
STOICHIOMETR

CHAPTER 9

Stoichiometry and the Mole

Chemistry has many branches—analytical (quantifying things), theoretical (understanding things), and physical (probing things), to name just a few. The one we want to deal with in this chapter is synthetic (making things). Of course, since this is chemistry, the things we will be making (synthesizing) are new compounds via chemical reactions. We will mix a starting set of reagents together, called reactants, to get one or more new products.

$$\text{Reactants} \rightarrow \text{Products}$$

In many ways, this is just like baking a cake. After all, you combine ingredients according to a recipe, and presto, a tasty cake.

Bake at 350°F for 45 minutes →

Recipe
(The Reactants)
2 eggs
1 cup milk
2 cups flour
1 cup sugar
1 pound butter
4 oz chocolate

However, there is a critical and fundamental difference between baking a cake and synthesizing a new compound. To understand this difference, imagine throwing caution to the wind and using more sugar than the recipe recommends. Why? Perhaps this is what your grandmother told you, and who knows better than Nana? What would the result be? You know the answer. It would still be a cake, just a different tasting one (sweeter, which is the way Nana likes 'em). However, if baking a cake behaved the way chemical reactions do, then the result would be as follows.

That extra cup of sugar would be rejected because it is in excess according to the original recipe. For chemical reactions, the recipe is king, and as you are about to learn, the recipe for a chemical synthesis comes in the form of a balanced chemical equation.

9.1 Stoichiometry—What Is It?

Lots of people love to cook—to mix ingredients and create a *pièce de résistance*. Indeed, some of the most popular books on the market these days are cookbooks. To use these cookbooks, you must follow their recipes. In this chapter, you are going to learn how to follow some unusual recipes. Now, you might think this is a strange topic for a chemistry book, but in fact it is not all that strange because chemists really do like to "cook." Many chemists spend their time mixing "ingredients" (reactants) to create a *pièce de résistance* (products). The recipes they follow come in the form of *balanced chemical equations*, and now you are going to learn how to use these equations—how to read them and how to do calculations based on them.

For example, we could whip up a batch of hydrogen iodide! Here is the recipe:

$$H_2 + I_2 \rightarrow 2\,HI$$

Balanced chemical equation for formation of hydrogen iodide

Suppose you wanted to go into the laboratory and make 10 g of HI. How much H_2 should you use? How much I_2? How could you do the reaction in a "balanced" way, using just the right amounts of H_2 and I_2 so that when the reaction was over, both reactants were completely gone and the only thing in the reaction vessel was 10 g of HI? These "how much" questions are the subject of **stoichiometry** (stoy-key-OM-e-tree), the study of the quantitative aspects of chemical reactions. The word is derived from two Greek roots, *stoicheon* (for "element") and *metron* ("to measure"). We use stoichiometry to calculate such things as the amount of reactant needed in order to make a certain amount of product, or how much product we can expect from a certain amount of reactant.

This is a good place to stop examining chemical recipes and return to cooking something edible because doing so will show you that you already know how to do stoichiometry calculations. Every time you follow a recipe, any recipe, you are using stoichiometry. For example, consider the recipe for cheesecake at right. It's an absolute diet-buster. According to this recipe, five eggs are required to make one cheesecake. So how many eggs would you need to make two cheesecakes? If you said ten, you have just done a successful stoichiometry calculation in your head. Now try doing a few more.

Cheesecake

3 blocks cream cheese
1 cup sugar
5 eggs

Practice Problems

Do the following stoichiometry problems based on the cheesecake recipe given above.

9.1 You have plenty of eggs and cream cheese but only 3 cups of sugar. How many cheesecakes can you make?

Answer: Because 1 cup of sugar makes one cake, 3 cups can make three cakes.

9.2 If you use 3 cups of sugar in making your cakes, how many eggs will you need?

9.3 If you use 21 blocks of cream cheese, how many eggs will you need?

You probably did all these problems in your head, but suppose you had to teach someone how to do the calculations on paper. How might you do it? As a budding chemist, you would first write the recipe in a form that makes it look more like a balanced chemical equation:

$$3 \text{ blocks cream cheese} + 5 \text{ eggs} + 1 \text{ cup sugar} \rightarrow 1 \text{ cheesecake}$$

It may look less appetizing this way, but it still says the same thing. Now, on paper, how do we solve the following problem: How many eggs are required to bake two cakes? We saw this type of problem in Chapter 2—it is a unit conversion problem, much like converting feet to meters. In this case, we want to convert from the given unit *cakes* to the unit we need in the answer, *eggs*. We begin just as we did in Chapter 2, by writing down the given information (number of cakes in this case) and then multiplying by the appropriate conversion factor:

$$\underbrace{2 \text{ }\cancel{\text{cakes}}}_{\text{Number}} \times \underbrace{\frac{5 \text{ eggs}}{1 \text{ }\cancel{\text{cake}}}}_{\text{Conversion factor}} = 10 \text{ eggs}$$

The conversion factor 5 eggs/1 cake came directly from the balanced cheesecake equation (the recipe). Of course, we could have written the conversion factor as 1 cake/5 eggs, but we want the unit *cakes* to cancel. (Remember that identical units cancel only when they are on both top and bottom in an equation.) What is important to realize here is that the conversion factor came from the balancing coefficients of the cheesecake equation.

To get practice in doing this type of conversion calculation, do the stoichiometry practice problems again. This time, though, don't do them in your head.

Practice Problems

Rework Practice Problems 9.1–9.3 by the method of unit conversion. Write out each solution in full, making sure to show your conversion factor and which units cancel.

9.1 You have plenty of eggs and cream cheese, but only 3 cups of sugar. How many cheesecakes can you make?

Answer: Because sugar is the ingredient in short supply, it is the amount of sugar available that determines how many cakes you can make. The recipe tells you that one cake requires 1 cup of sugar. Therefore, your conversion factors are:

$$\frac{1 \text{ cup sugar}}{1 \text{ cake}} \quad \text{and} \quad \frac{1 \text{ cake}}{1 \text{ cup sugar}}$$

Because you want the unit of your answer to be cakes, use the latter conversion factor:

$$\underbrace{3 \text{ cups sugar}}_{\text{Number}} \times \underbrace{\frac{1 \text{ cake}}{1 \text{ cup sugar}}}_{\text{Conversion factor}} = 3 \text{ cakes}$$

9.2 If you use 3 cups of sugar in making your cakes, how many eggs will you need?

9.3 If you use 21 blocks of cream cheese, how many eggs will you need?

As the answer to Practice Problem 9.1 notes, the balanced cheesecake equation is a source of conversion factors for doing stoichiometry problems. All you have to do to get your answer is write down the number given in the problem and multiply it by the appropriate conversion factor from the balanced equation. That's basically all there is to solving all stoichiometry problems.

Before we apply these techniques to a chemical reaction, let's try one more cheesecake stoichiometry problem. Suppose we are given 24 ounces of cream cheese. How many cakes can we make? Asking this question is throwing a curve ball because there is no way to answer it with the information given. The recipe (equation) deals with cream cheese in units of *blocks* and the problem is stated in terms of *ounces*. The only way we can solve this problem is to know (or be given) the relationship between ounces of cream cheese and blocks of cream cheese. For example, if we are told there are 8 ounces of cream cheese in each block, we can convert from ounces to blocks and then solve the problem. Using the conversion factor

$$\frac{1 \text{ block cream cheese}}{8 \text{ ounces cream cheese}}$$

we can translate the information we've been given into the language of the recipe:

$$24 \text{ ounces cream cheese} \times \frac{1 \text{ block cream cheese}}{8 \text{ ounces cream cheese}} = 3 \text{ blocks cream cheese}$$

Now we can finish the problem. We take the translated amount of cream cheese and multiply it by the appropriate conversion factor from the balanced equation:

$$3 \text{ blocks cream cheese} \times \frac{1 \text{ cake}}{3 \text{ blocks cream cheese}} = 1 \text{ cake}$$

We have our answer. We can make one cheesecake with 24 ounces of cream cheese.

This problem was complicated because we were given the amount of cream cheese in units that were different from the units in the balanced equation. This will often happen when we stop baking cheesecakes and start mixing chemicals. That's because balanced chemical equations *always* speak in terms of numbers of atoms or molecules, but our given information will usually be in terms of grams. Therefore, before we can use a balanced chemical equation as a recipe or source of conversion factors for stoichiometry problems, we have to learn how to translate back and forth between *grams* of atoms or molecules and *numbers* of atoms or molecules. This is where we get some help from the unit known as the *mole*.

The mole is used to convert numbers of atoms/molecules to grams of atoms/molecules. No, not this mole.

9.2 The Mole

Let's whip up another batch of hydrogen iodide. Suppose we need to prepare 10 g, following the "recipe":

$$H_2 + I_2 \rightarrow 2\,HI$$

A very common beginner's mistake is to assume that because the equation shows that H_2 and I_2 react in a one-to-one ratio, all we need to do is combine 5 g of H_2 and 5 g of I_2 to produce 10 g of HI:

$$5\text{ g }H_2 + 5\text{ g }I_2 \longrightarrow 10\text{ g HI}$$

This is not correct, however, because the balancing coefficients in a chemical equation *do not refer to masses*. Instead, they refer to the *number* of atoms or molecules that react and are produced. In other words, the balanced equation does *not* say that 1 g of H_2 reacts with 1 g of I_2 to produce 2 g of HI. What it *does* say is that one *molecule* of H_2 reacts with one *molecule* of I_2 to produce two *molecules* of HI.

Because the balanced equation "speaks" in terms of numbers of molecules, it would be easy to whip up a batch of ten HI molecules. This is just a matter of scaling up the recipe. Our task, however, is to prepare 10 g of HI, and for that task this recipe isn't much help. Plus, a recipe that required us to count out individual molecules would be pretty worthless in any case. Not only can we not see individual H_2 and I_2 molecules, but the numbers we would need to count in order to prepare 10 g of HI would be astronomical.

How can we use a balanced chemical equation written in terms of numbers of molecules when we are working with grams? This is where we get assistance from the mole. In concept, a mole is similar to a dozen, which always stands for 12 of anything. A mole is just a whole lot larger: one **mole** (mol) is 6.022×10^{23} of anything. This very large number is called **Avogadro's number**, in honor of Amadeo Avogadro (1776–1856), who first explained how equal volumes of different gases contain the same number of gas molecules. Just as you can have a dozen of something (donuts, dollars, molecules), you can have a mole of something (or Avogadro's number of something). A mole of anything (donuts, dollars, molecules)

. . . one billion and one, one billion and two, . . .

is always 6.022×10^{23} of that thing. Because this is so, we can always write:

$$\frac{1 \text{ mole of something}}{6.022 \times 10^{23} \text{ somethings}} \quad \text{or} \quad \frac{6.022 \times 10^{23} \text{ somethings}}{1 \text{ mole of something}}$$

These are useful conversion factors for doing calculations, as you will soon see.

Whereas a dozen is a conveniently sized number for things like donuts, a mole is a conveniently sized number when dealing with atoms and molecules. Because atoms and molecules are so tiny, you need a tremendous number of them to have a useful and measurable amount. You can get a feel for just how large a mole is by writing Avogadro's number in normal notation:

602,200,000,000,000,000,000,000

Let's put this number—almost a trillion trillion—in perspective. If you counted every grain of sand on every beach and ocean floor on this planet, you would only come close to 1 mole. If you were given 1 mole of dollars the day you were born and then spent a billion dollars a second for your entire life, by age 70, you would have spent only about 0.001% of your money!

Avogadro's number is huge, but atoms and molecules are the opposite—so tiny that it takes a lot of them to make a dent. One mole of water molecules is barely a swallowful, for instance, and 1 mole of sugar molecules will fit in your cupped hands. It might interest you to know that Avogadro had been in his grave for many years by the time the value of his number was experimentally determined with reasonable accuracy. He never knew the numeric value of the quantity named after him, but you should never forget it.

To see how the mole is going to help us with our problem of translating between grams or molecules and number of molecules, let's look once again at the balanced equation for making HI:

$$H_2 + I_2 \rightarrow 2\,HI$$

The balancing coefficients in this equation tell us the proportions in which hydrogen and iodine combine, and the key to solving our problem is that these coefficients can be interpreted in more than one way:

H_2	+	I_2	→	2 HI
1 molecule		1 molecule		2 molecules
12 molecules (1 dozen molecules)		12 molecules (1 dozen molecules)		24 molecules (2 dozen molecules)
6.022×10^{23} molecules (1 mole)		6.022×10^{23} molecules (1 mole)		$2 \times 6.022 \times 10^{23}$ molecules (2 moles)

So instead of reading the reaction exclusively as "1 H_2 molecule and 1 I_2 molecule react to give 2 HI molecules," we can also read it as "1 mole of H_2 molecules and 1 mole of I_2 molecules react to give 2 moles of HI molecules":

$H_2 + I_2 \longrightarrow 2\,HI$

says → 1 H_2 molecule and 1 I_2 molecule react to give 2 HI molecules.

OR

says → 1 mole of H_2 molecules and 1 mole of I_2 molecules react to give 2 moles of HI molecules.

This is a fundamental concept in chemistry. Any balanced chemical equation can always be read in terms of moles of molecules instead of individual molecules. This should make sense to you. After all, if the balanced equation says

> 1 H_2 molecule reacts with 1 I_2 molecule to give 2 HI molecules,

it also says

> 1 dozen H_2 molecules react with 1 dozen I_2 molecules to give 2 dozen HI molecules

and

> 1 mole of H_2 molecules reacts with 1 mole of I_2 molecules to give 2 moles of HI molecules.

This information by itself doesn't help us because we're still speaking in terms of numbers of molecules, even though in the laboratory we don't count out molecules directly. Rather, we determine masses. We go to the storage room, find a bottle of iodine, and measure out 10 g of it. If we use 10 g of I_2, how many grams of H_2 do we need? How many grams of HI will we make? To answer these questions, we need to learn one more thing about the mole. Just as we were able to translate from ounces of cream cheese to blocks of cream cheese before using the cheesecake recipe, knowing one more thing about the mole will allow us to translate from grams to numbers of molecules or from numbers of molecules to grams, so that we can use a balanced chemical equation as our recipe.

The mole can translate between numbers of atoms and grams of atoms because it is defined as the number of carbon atoms in exactly 12 g of $^{12}_{6}C$ Experiments over the years have determined that there are 6.022×10^{23} atoms of $^{12}_{6}C$ in 12 g of this carbon isotope. Knowing this fact allows us to "count" the $^{12}_{6}C$ atoms in a sample by determining the mass of the sample. To count out 6.022×10^{23} (1 mole) of $^{12}_{6}C$ atoms, all we have to do is measure out exactly 12 g.

Knowing this allows us to do a few things. For example, recall from Chapter 3 that atomic masses of elements are given in atomic mass units and that 1 amu = 1.66056×10^{-24} g. Where did this equivalent value for 1 amu come from? We can use the mole concept to answer this question. By definition, 1 mole of $^{12}_{6}C$ atoms has a mass of exactly 12 g, which means we can write:

$$\frac{12\text{ g}}{1\text{ mol } ^{12}_{6}\text{C}}$$

We can use this conversion factor, along with the knowledge that a mole of anything contains Avogadro's number of that thing, to calculate the mass of one $^{12}_{6}C$ atom:

$$\frac{12\text{ g}}{1\text{ mol } ^{12}_{6}\text{C}} \times \frac{1\text{ mol } ^{12}_{6}\text{C}}{6.022\times10^{23}\text{ atoms } ^{12}_{6}\text{C}} = 1.993 \times 10^{-23}\text{ g/atom } ^{12}_{6}\text{C}$$

The mole lets us count atoms by determining the mass of a sample.

$$12\text{ g of } ^{12}_{6}\text{C} = 1\text{ mole of } ^{12}_{6}\text{C} = 6.022 \times 10^{23}\text{ atoms of } ^{12}_{6}\text{C}$$

Finally, because 1 amu is defined as 1/12 the mass of a $^{12}_{6}C$ atom, we can say:

$$1 \text{ amu} = \frac{1}{12} \times 1.993 \times 10^{-23} \text{ g} = 1.661 \times 10^{-24} \text{ g}$$

Now let's get back to how the mole concept allows us to count atoms of any element simply by measuring masses. All we need to do is use the atomic masses discussed in Chapter 3. For example, what if we want 1 mole of magnesium, Mg, atoms? The average atomic mass of naturally occurring magnesium is 24.305 amu, as shown in the periodic table. This atomic mass value tells us that a magnesium atom is 24.305 amu/12 amu = 2.025 times more massive than a $^{12}_{6}C$ atom. So, if 1 mole of $^{12}_{6}C$ has a mass of 12 g, then 1 mole of magnesium has a mass 2.025 times that, which is 24.305 g. In other words, to get 1 mole of atoms of any element, the mole concept tells us to take its atomic mass, which is reported in atomic mass units, and change it to units of grams! Consider these elements and their atomic masses:

12
Mg
24.305

	1	6	92
	H	**C**	**U**
Atomic mass → (amu)	1.0079	12.011	238.029
	1 mole of naturally occurring H atoms has a mass of 1.0079 g.	1 mole of naturally occurring C atoms has a mass of 12.011 g.	1 mole of naturally occurring U atoms has a mass of 238.029 g.

We can generalize the information given for these three elements and say that 1 mole of any element has a mass equal to the element's atomic mass expressed in grams. We call the mass of 1 mole of an element or compound its **molar mass**. Therefore, the molar mass of hydrogen is 1.0079 g/mol, the molar mass of carbon is 12.011 g/mol, and the molar mass of uranium is 238.029 g/mol. Once you understand this concept, you can count the number of atoms present in any sample of an element simply by determining the mass of the sample. Said another way, you can convert back and forth between numbers of atoms or molecules and grams of atoms or molecules. Try the following WorkPatch to get some practice.

WORKPATCH 9.1 Consider the element gold, Au, a yellow metal. Suppose you obtained 393.934 g of it.

(a) What is the molar mass of gold?
(b) How many moles of gold do you have?
(c) How many atoms of gold do you have?

79
Au
196.967

Check your answers against ours at the end of the chapter, and do not go on until you agree with them. Understanding what follows depends on understanding this fundamental mole concept. ●

Practice Problems

Try to do the following problems in your head.

9.4 Suppose you have 0.5 mole of gold, Au, atoms.

(a) How many gold atoms do you have?
(b) What is the mass in grams of this much gold?

Answer:

(a) *Because 1 mole of gold atoms means* 6.022×10^{23} *atoms, there are* 3.011×10^{23} *gold atoms in 0.5 mole.*
(b) *Because 1 mole of gold has a mass of 196.967 g, 0.5 mole has a mass of 98.484 g.*

92
U
238.029

9.5 Suppose you have 0.10 mole of uranium, U, atoms.

(a) How many uranium atoms do you have?
(b) What is the mass in grams of this much uranium?

6
C
12.011

9.6 Suppose you have 120.11 g of carbon atoms.

(a) How many moles of carbon atoms do you have?
(b) How many carbon atoms do you have?

The understanding that 1 mole of atoms of an element has a mass equal to the element's atomic mass expressed in grams is critical to your being able to do chemical stoichiometry problems.

We can calculate the mass of 1 mole of molecules in the same way. For instance, what is the mass in grams of 1 mole of CO_2 molecules? The molecular formula tells us that one carbon dioxide molecule is made from one carbon atom and two oxygen atoms. Following the line of reasoning we developed earlier, it must also be true that 1 mole of CO_2 molecules is made from 1 mole of carbon atoms and 2 moles of oxygen atoms. This is very important! The subscripts in a molecular formula not only say how many atoms there are in one molecule; they also say how many moles of atoms there are in 1 mole of the molecule.

What the Subscripts Say

CO_2

Molecule message	One C atom per CO_2 molecule	Two O atoms per CO_2 molecule
Mole message	One mole of C atoms per mole of CO_2 molecules	Two moles of O atoms per mole of CO_2 molecules

The information contained in formula subscripts is similar to the information contained in the balancing coefficients in a chemical equation, which, as we've already seen, can always be interpreted as moles. To see if you are catching on, try the following WorkPatch. You should be able to do parts (a) through (c) in your head, but a calculator will help for (d).

WORKPATCH 9.2 Suppose you are given 1 mole of methane, CH_4.

(a) How many moles of carbon atoms do you have?
(b) How many moles of hydrogen atoms do you have?
(c) How many atoms of carbon do you have?
(d) How many atoms of hydrogen do you have?

(*Hint for (c) and (d)*: Remember that a mole of anything is 6.022×10^{23} units of it.) ●

Once you understand the principle behind the answers to the WorkPatch, it is a simple matter to figure out the molar mass of a molecule—that is, the mass in grams of 1 mole. Let's do this for CO_2. We know that 1 mole of CO_2 contains 1 mole of carbon atoms and 2 moles of oxygen atoms, and we also know the mass of 1 mole of each of these atoms. Thus, all we have to do is add up the molar masses:

	Molar mass
1 mol C atoms	12.011 g/mol
1 mol O atoms	15.999 g/mol
1 mol O atoms	15.999 g/mol
1 mol CO_2 molecules	44.009 g/mol

Or, because there are 2 moles of O atoms, we could have written:

	Molar mass
1 mol C atoms	12.011 g/mol
2 × 1 mol O atoms	2 × 15.999 g/mol
1 mol CO_2 molecules	44.009 g/mol

Either way, the molar mass of CO_2 is 44.009 g/mol. If you measure out 44.009 g of CO_2, you have 1 mole of it, which means 6.022×10^{23} molecules of it. Once again, you have counted by measuring masses.

Using the molar masses of the elements from the periodic table, we can calculate the molar mass for any chemical compound so long as we know its molecular formula. Here are some more examples:

Molecular formula	**Molar mass**
HI	H (1.0079 g/mol)+I (126.905 g/mol) = 127.913 g/mol
H_2	2 × H (1.0079 g/mol) = 2.0158 g/mol
I_2	2 × I (126.905 g/mol) = 253.810 g/mol
H_2O	2 × H (1.0079 g/mol)+O (15.999 g/mol) = 18.015 g/mol

Knowing about moles allows us to go back and forth between numbers of atoms or molecules and their masses in grams.

Practice Problems

9.7 If you have 1 mole of propane, C_3H_8, a gas used in outdoor grills and industrial torches:

(a) How many propane molecules do you have?
(b) How many hydrogen atoms do you have?
(c) What is the mass in grams of 1 mole of propane?

Answer:

(a) Having 1 mole of propane molecules means you have 6.022×10^{23} propane molecules. (That is the meaning of 1 mole.)

(b) You know that 1 mole of propane contains 8 moles of hydrogen atoms. (That is what the subscript 8 in the formula C_3H_8 tells you.) You also know there are 6.022×10^{23} H atoms per mole of them, which means you have:

$$1 \text{ mol propane} \times \frac{8 \text{ mol H atoms}}{1 \text{ mol propane}} \times \frac{6.022\times10^{23} \text{ H atoms}}{1 \text{ mol H atoms}} = 4.818 \times 10^{24} \text{ H atoms}$$

(c) $(3 \times 12.011\ g) + (8 \times 1.0079\ g) = 44.096\ g$

3 mol C — 8 mol H — Molar mass of propane (mass of 1 mol C_3H_8)

9.8 What is the mass in grams of 2 moles of propane?

9.9 How many moles of carbon atoms are there in 2 moles of propane?

9.10 Propane reacts with oxygen to produce CO_2 and H_2O, as shown in the balanced equation:

$$C_3H_8 + 5\,O_2 \rightarrow 3\,CO_2 + 4\,H_2O$$

Write this reaction in words using the word *mole(s)* four times.

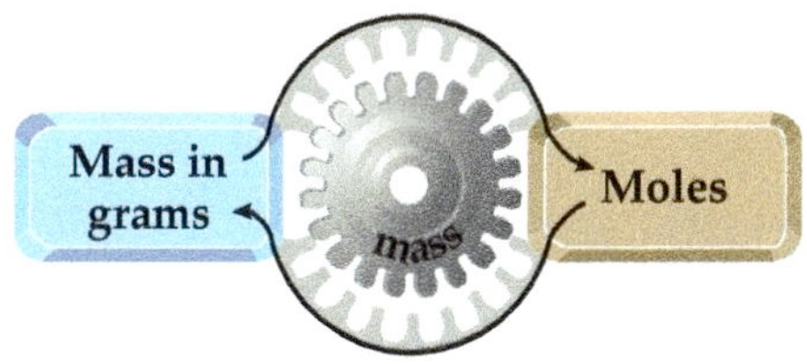

Molar mass is very important because it allows you to convert from grams to moles or from moles to grams.

For example, the molar mass of H_2O is 18.015 g/mol; in other words, 1 mole of H_2O has a mass of 18.015 g. Thus, you can write the molar mass of H_2O as if it were a conversion factor:

$$\text{Molar mass } H_2O = \frac{18.015\ g\ H_2O}{1\ mol\ H_2O} \qquad \text{and its inverse} \qquad \frac{1\ mol\ H_2O}{18.015\ g\ H_2O}$$

You can write the conversion factor either way. Which fraction you use depends on whether you want to convert from moles to grams or from grams to moles. Suppose you had 250.0 g of H_2O. To convert this mass to moles, you simply do the calculation:

$$250.0\ \cancel{g\ H_2O} \times \underbrace{\frac{1\ mol\ H_2O}{18.015\ \cancel{g\ H_2O}}}_{\text{Conversion factor, inverse of molar mass}} = 13.88\ mol\ H_2O$$

Doing the conversion this way is true to the method of unit analysis (note the cancellation of identical units on top and bottom). However, you will often see grams-to-moles conversions done by dividing the mass by the molar mass rather than multiplying by the inverse of the molar mass:

Mass → Divided by → Molar mass →

$$\frac{250.0\ g\ H_2O}{18.015\ g\ H_2O/mol} = 13.88\ mol\ H_2O$$

You get the same result either way. In other words, dividing something by a conversion factor is identical to multiplying it by the inverse of the conversion factor. We mention this only because many people prefer converting from grams to moles via the latter method.

Now, let's go the other way and convert from moles to grams. Suppose you have 13.88 moles of H_2O. How many grams do you have? Once again, the molar mass takes you to the answer:

$$13.88\ \cancel{mol}\ H_2O \times \underbrace{\frac{18.015\ g\ H_2O}{1\ \cancel{mol}\ H_2O}}_{\text{Conversion factor, molar mass}} = 250.0\ g\ H_2O$$

Armed with your knowledge of molar mass and Avogadro's number, you are now ready to convert from mass to moles to numbers of atoms or molecules at will. The following chart summarizes how. Now try some practice problems to see if you are catching on.

Conversions in Stoichiometry Calculations

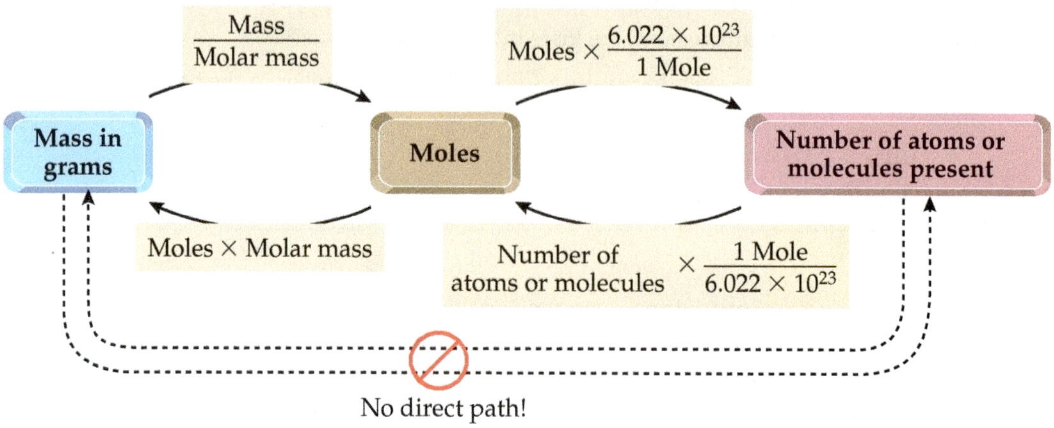

Practice Problems

9.11 Suppose you have 100.0 g of H_2O.

(a) How many moles of H_2O molecules do you have?
(b) How many H_2O molecules do you have?

Answer:

(a) $100.0\,\text{g H}_2\text{O} \times \frac{1\text{ mol H}_2\text{O}}{18.015\text{ g H}_2\text{O}} = 5.551\text{ mol H}_2\text{O}$

or

$$\frac{100.0\text{ g H}_2\text{O}}{18.015\text{ g H}_2\text{O/mol}} = 5.551\text{ mol H}_2\text{O}$$

(b) $5.551\text{ mol H}_2\text{O} \times \frac{6.022 \times 10^{23}\text{ molecules H}_2\text{O}}{1\text{ mol H}_2\text{O}} = 3.343 \times 10^{24}\text{ molecules H}_2\text{O}$

9.12 Suppose you have 0.565 moles of HI.

(a) How many grams of HI do you have?
(b) How many HI molecules do you have?

9.13 Suppose you have 5.000×10^{24} molecules of methane, CH_4.

(a) How many moles of methane do you have?
(b) How many grams of methane do you have?

9.3 Reaction Stoichiometry

We are now ready to use the mole in reaction stoichiometry problems. Let's go back to our HI reaction:

$$H_2 + I_2 \rightarrow 2\text{ HI}$$

Viewed as a recipe, this says, "Combine 1 mole of H_2 molecules with 1 mole of I_2 molecules, and they will react to give 2 moles of HI molecules."

We can finally go into the laboratory and carry out this reaction because we can now count molecules by measuring masses using a balance. One mole of H_2 molecules has a mass of 2.0158 g (which we got by doubling 1.0079 g/mol, the molar mass of the H atom), and 1 mole of I_2 molecules has a mass of 253.810 g (twice 126.905 g/mol, the molar mass of the I atom). If we combine these amounts of H_2 and I_2, the balanced reaction can tell us how many grams of product to expect. Because the molar mass of HI is 127.913 g/mol, we should produce 2 mol × 127.913 g HI/mol = 255.826 g of HI. This is called the **theoretical yield** of the reaction, the maximum amount of product you can hope to make from a given amount of reactants.

	H_2	+	I_2	$\longrightarrow$	2 HI
Reaction says:	1 mole of H_2	and	1 mole of I_2	React to give	2 moles of HI
But:	1 mole of H_2 has a mass of 2.0152 g (MM_{H_2})	and	1 mole of I_2 has a mass of 253.810 g (MM_{I_2})	and	1 mole of HI has a mass of 127.913 g (MM_{HI})
So:	2.0152 g H_2	and	253.810 g I_2	React to give	2 × 127.913 g HI = 255.826 g HI (Theoretical yield)

Interestingly, reactions do not always produce the theoretical yield, for a variety of reasons. Many reactions have competing *side reactions* that consume some of the reactants to produce unwanted products, called *side products*. Or, when it is time to recover the desired product, difficulties arise in collecting it or purifying it, causing some of it to be lost. Whatever the reason, it is very common to get something less than the theoretical yield of the product. The amount of product you actually end up with is called the **actual yield** of the reaction. Suppose, for example, we were able to isolate only 225.10 g of HI instead of the theoretical yield of 255.826 g. If this were so, the actual yield of the reaction would be 225.10 g of HI.

Actual yield allows us to define yet another quantity, the *percent yield* of the reaction, which is just what it sounds like. The **percent yield** (% yield) of a reaction is the percentage of the theoretical yield we are able to isolate:

$$\%\text{ yield} = \frac{\text{Actual yield}}{\text{Theoretical yield}} \times 100$$

For example, the calculation for our HI reaction is:

$$\%\text{ yield} = \frac{225.10\text{ g HI}}{255.826\text{ g HI}} \times 100 = 87.989\%$$

Our percent yield is almost 88%, which is not bad. Some reactions done in the chemical and pharmaceutical industries have smaller percent yields and are thus quite wasteful of their starting reactants. You end up absorbing this loss in the price you pay for pharmaceuticals and other useful chemicals.

Practice Problems

9.14 (a) Express the balanced chemical equation

$$CH_4 + 2\,O_2 \rightarrow 2\,H_2O + CO_2$$

in words, using the word *mole(s)* wherever appropriate.

(b) To produce 1 mole of CO_2 from this reaction, how many grams of CH_4 and O_2 must you combine?

(c) What is the theoretical yield in grams of H_2O for this reaction?

(d) If you recover 30.0 g of H_2O, what is the percent yield?

Answer:

(a) One mole of methane molecules and 2 moles of oxygen molecules react to give 2 moles of water molecules and 1 mole of carbon dioxide molecules.

(b) The molar mass of methane is (12.011 g/mol C)+(4 × 1.0079 g/mol H) = = 16.043 g/mol CH_4. This is the mass of 1 mole of methane, which is what the reaction calls for. The molar mass of oxygen, O_2, is (2 × 15.999 g/mol O) = 31.998 g/mol. This is the mass of 1 mole of O_2. Because the reaction calls for 2 moles, we multiply by 2 to get 63.996 g of O_2 needed.

(c) The most we can hope to form is 2 moles of water. The molar mass of water is (2 × 1.0079 g/mol H)+(15.999 g/mol O) = 18.015 g/mol H_2O. This is the mass of 1 mole of water. So the theoretical yield is just twice this, which is 36.030 g of H_2O.

(d) % yield = (30.0 g/36.030 g) × 100 = 83.3 %

9.15 (a) Balance this unbalanced equation by inspection:

$$\underset{\text{Benzene}}{C_6H_6} + \underset{\text{Hydrogen}}{H_2} \longrightarrow \underset{\text{Cyclohexane}}{C_6H_{12}}$$

(b) Express this reaction in words, using the word *mole(s)* wherever appropriate.

(c) To produce 1 mole of C_6H_{12} from this reaction, how many grams of C_6H_6 and H_2 must you combine?

(d) What is the theoretical yield in grams of C_6H_{12}?

(e) Suppose you recover 24.0 g of C_6H_{12}. What is the percent yield?

9.16 (a) Balance this unbalanced equation by inspection:

$$\underset{\text{Glucose}}{C_6H_{12}O_6} + O_2 \longrightarrow CO_2 + H_2O$$

(b) Express this reaction in words, using the word *mole(s)* wherever appropriate.

(c) To produce 6 moles of H_2O from this reaction, how many grams of glucose and O_2 must you combine?

(d) What is the theoretical yield in grams of CO_2?

(e) Suppose you recover 196.0 g of CO_2. What is the percent yield?

Up to now, we have been using a balanced equation like a recipe, following its instructions exactly. But recall that when we were baking cheesecakes, we sometimes wanted to depart from the recipe, doubling it to make two cakes and so forth. The same holds true for chemical reactions. Our balanced equation for hydrogen iodide is $H_2 + I_2 \rightarrow 2\,HI$, and it tells us how to make 2 moles of HI. But we rarely want to produce exactly 2 moles of HI (255.826 g, twice its molar mass). Suppose we want to make only 10.0 g of HI. How do we cut back on the recipe? The answer comes from a concept we covered in Chapter 2—conversion factors. Both a balanced chemical equation and a molecular formula are sources of conversion factors that are useful for doing stoichiometry problems. We simply read the conversion factors directly from the balanced equation or the formula, like this:

Conversion Factors from Balanced Equation $H_2 + I_2 \rightarrow 2\,HI$

$\frac{2\text{ mol HI}}{1\text{ mol H}_2}$	2 moles of HI are produced for every 1 mole of H_2 consumed.	Same, just inverted
$\frac{1\text{ mol H}_2}{2\text{ mol HI}}$	1 mole of H_2 is consumed for every 2 moles of HI produced.	
$\frac{2\text{ mol HI}}{1\text{ mol I}_2}$	2 moles of HI are produced for every 1 mole of I_2 consumed.	Same, just inverted
$\frac{1\text{ mol I}_2}{2\text{ mol HI}}$	1 mole of I_2 is consumed for every 2 moles of HI produced.	
$\frac{1\text{ mol I}_2}{1\text{ mol H}_2}$	1 mole of I_2 is consumed for every 1 mole of H_2 consumed.	Same, just inverted
$\frac{1\text{ mol H}_2}{1\text{ mol I}_2}$	1 mole of H_2 is consumed for every 1 mole of I_2 consumed.	

Conversion Factors from Molecular Formulas

$\frac{2\text{ mol H atoms}}{1\text{ mol H}_2\text{ molecules}}$	From formula H_2
$\frac{2\text{ mol I atoms}}{1\text{ mol I}_2\text{ molecules}}$	From formula I_2
$\frac{1\text{ mol H atoms}}{1\text{ mol HI molecules}}$	From formula HI
$\frac{1\text{ mol I atoms}}{1\text{ mol HI molecules}}$	From formula HI

WORKPATCH 9.3 This concept is so important that you need to practice it now.

(a) Express the balanced equation

$$\underset{\text{Aluminum}}{4\,Al} + \underset{\text{Oxygen}}{3\,O_2} \longrightarrow \underset{\text{Aluminum oxide}}{2\,Al_2O_3}$$

in words, using the word *moles* for every reactant and product.

(b) Write every conversion factor you can think of from the balanced chemical equation.

(c) Write all the conversion factors from the formula Al_2O_3. ●

Did you get six conversion factors in (b)? Check your answers before going on because being able to get these conversion factors is essential to doing stoichiometry problems.

A stoichiometry problem can often appear as a blizzard of numbers and words, so before beginning any stoichiometry problem, it is best to determine exactly what you are being asked to solve for. The easiest way to do this is to write the chemical equation (the recipe) and then, under each reactant and product for which information is given, write this information below the formula (these are your known quantities). Put a question mark under the ones for which no information is given (these are the unknowns that the problem can ask you to solve for). In our case, we want to make 10.0 g of HI. Write this information under the HI formula. No information is given about masses of H_2 and I_2. These are the unknown quantities. Write question marks under the H_2 and I_2 formulas.

$$\underset{?\,\text{g}}{H_2} + \underset{?\,\text{g}}{I_2} \rightarrow \underset{10.0\,\text{g}}{2\,HI}$$

Once you have done this preliminary work, the first step is to be sure the equation you wrote is balanced. If it isn't, then balance it.

Now, how do we use the balanced HI equation to determine how many grams of H_2 and I_2 we need to make 10.0 g of HI? Because the reaction speaks in moles, the next step is to translate the 10.0 g of HI into moles of HI. This is the job of the HI molar mass, 127.913 g/mol, used as a conversion factor:

$$10.0\ \cancel{\text{g HI}} \times \underbrace{\frac{1\ \text{mol HI}}{127.913\ \cancel{\text{g HI}}}}_{\text{Inverse of HI molar mass}}$$

By using this conversion factor, we have changed to units of moles of HI (because this is the only unit not crossed out).

The third step is to use the balanced equation to find out how many moles of H_2 and I_2 we need. Let's tackle H_2 first. From the balanced equation, we find the conversion factor that takes us from moles of HI to moles of H_2 and string it into our calculation:

$$10.0\ \cancel{\text{g HI}} \times \frac{1\ \cancel{\text{mol HI}}}{127.913\ \cancel{\text{g HI}}} \times \underbrace{\frac{1\ \text{mol}\ H_2}{2\ \cancel{\text{mol HI}}}}_{\text{From balanced equation}}$$

This step cancels moles of HI and leaves us with moles of H_2. Because we want to know how many grams of H_2 we need, we want our final answer in grams. So, the final step is to use the molar mass of H_2 to convert moles of H_2 to grams of H_2. Notice that the molar mass conversion factor for H_2 is written so that it makes the proper units cancel:

$$10.0\ \cancel{\text{g HI}} \times \frac{1\ \cancel{\text{mol HI}}}{127.913\ \cancel{\text{g HI}}} \times \frac{1\ \cancel{\text{mol}\ H_2}}{2\ \cancel{\text{mol HI}}} \times \underbrace{\frac{2.0158\ \text{g}\ H_2}{1\ \cancel{\text{mol}\ H_2}}}_{H_2\ \text{molar mass}}$$

(Surviving unit: g H_2)

We are done. The only unit not crossed out is the unit of the question we were asked, grams of H_2. All we have to do is perform the math and attach the surviving unit to it. Here is what you would enter into your calculator:

$$10.0 \div 127.913 \div 2 \times 2.0158 = 0.0788 \text{ g } H_2$$

This is the answer we are looking for. To make 10.0 g of HI, we need 0.0788 g of H_2. We leave the calculation of the amount of I_2 for you to do in WorkPatch 9.4. The following chart and diagram summarize the steps we have just done.

How to solve stoichiometry problems for reactions run with no reactant in short supply

Step 1: Write the balanced chemical equation.

Step 2: For each known reactant or product, convert the given information to moles.

Step 3: Use the balancing coefficients from the equation to find the relationship between the number of moles of the known reactant/product and the number of moles of the unknown reactant/product. Express this relationship as a conversion factor and use it to determine the number of moles of the unknown reactant/product.

Step 4: Convert the molar amount calculated in step 3 to the units asked for in the problem.

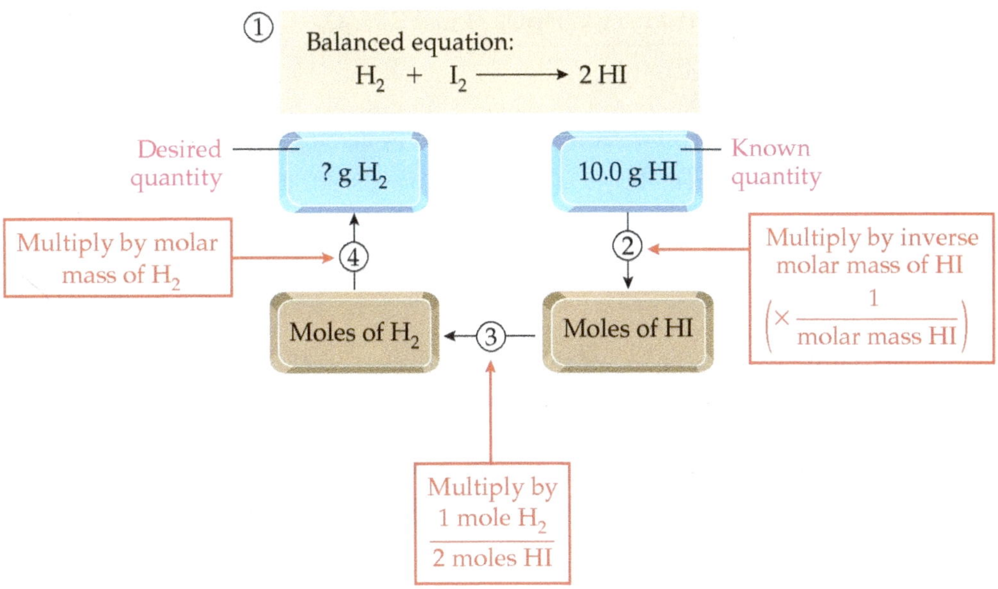

WORKPATCH 9.4 Using the four-step procedure, determine the mass in grams of I_2 required to make 10.0 g of HI. Write out all your steps and cross out all units that cancel. ●

Steps 1 and 2 in your solution should have been identical to steps 1 and 2 that we used when solving for the amount of H_2. Check your work against ours to see how you did.

Practice Problems

In the presence of O_2, glucose "burns" to give CO_2 and H_2O. The balanced equation and molar masses are:

	$C_6H_{12}O_6$	+	$6\ O_2$	$\xrightarrow{\text{React to give}}$	$6\ CO_2$	+	$6\ H_2O$
In words →	1 mol of glucose	+	6 mols of oxygen		6 mols of carbon dioxide	+	6 mols of water
Molar mass →	180.155 g/mol		31.998 g/mol		44.009 g/mol		18.015 g/mol

9.17 How many grams of oxygen does it take to burn 10.0 g of glucose?

Answer:

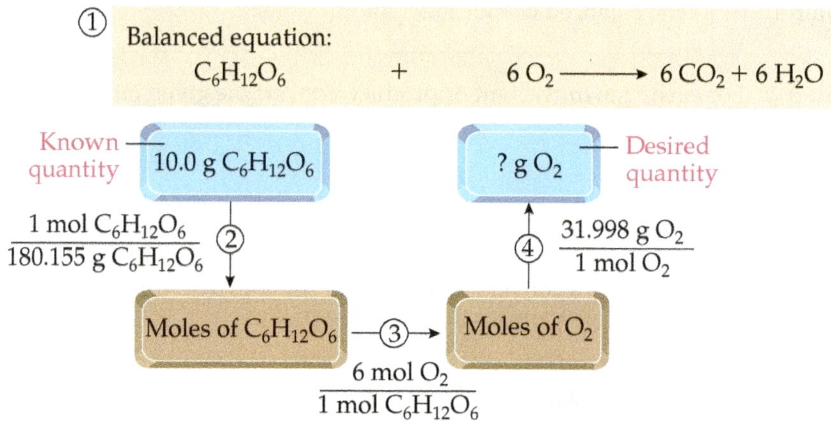

$$\underbrace{10.0\ \text{g}\ C_6H_{12}O_6 \times \frac{1\ \text{mol}\ C_6H_{12}O_6}{180.155\ \text{g}\ C_6H_{12}O_6}}_{②} \times \underbrace{\frac{6\ \text{mol}\ O_2}{1\ \text{mol}\ C_6H_{12}O_6}}_{③} \times \underbrace{\frac{31.998\ \text{g}\ O_2}{1\ \text{mol}\ O_2}}_{④} = 10.7\ \text{g}\ O_2$$

9.18 How many grams of water will be produced by burning 10.0 g of glucose?

9.19 What is the theoretical yield in grams of carbon dioxide if 10.0 g of glucose is burned?

9.20 If 10.0 g of carbon dioxide was produced, how many grams of glucose must have been burned?

Though we have stressed that the four-step method should be followed, it can be modified to suit the occasional oddball question. For example, the answer to Practice Problem 9.17 is that it takes 10.7 g of O_2 to burn 10.0 g of glucose. We could have been asked to go further and calculate how many O atoms are in 10.7 g of O_2. Our knowledge of the definition of mole (6.022×10^{23} of anything) and of what the molecular formula O_2 means (each oxygen molecule consists of two oxygen atoms) gives us more conversion factors, which we will use to find the answer in Practice Problem 9.21.

Practice Problems

9.21 How many O atoms are there in 10.7 g of oxygen, O_2, molecules?

Answer: Don't be afraid to string out conversion factors. They may come from a balanced equation, a molecular formula, or your knowledge of the mole, depending on the question you are trying to answer.

$$10.7\ \cancel{g\ O_2} \times \underbrace{\frac{1\ \cancel{mol\ O_2}}{31.998\ \cancel{g\ O_2}}}_{\text{Inverse of molar mass}} \times \underbrace{\frac{2\ \cancel{mol\ O\ atoms}}{1\ \cancel{mol\ O_2}}}_{\text{Conversion factor from formula } O_2} \times \underbrace{\frac{6.022 \times 10^{23}\ O\ atoms}{1\ \cancel{mol\ O\ atoms}}}_{\text{Conversion factor from knowledge of mole}} = 4.03 \times 10^{23}\ O\ atoms$$

9.22 How many aluminum atoms are there in 10.0 g of aluminum oxide, Al_2O_3?

9.23 How many molecules of water are there in 10.0 g of water?

With the proper conversion factors, you can use balanced equations to calculate "how much" when running a reaction at any scale. Because reactions speak in the language of moles, you can bet that for almost every stoichiometry problem, you will be converting to moles. Thus, our final advice on the subject is this—if in doubt, convert to moles.

9.4 Dealing With a Limiting Reactant

Let's try one more cheesecake problem. Here is the recipe again:

$$3 \text{ blocks cream cheese} + 5 \text{ eggs} + 1 \text{ cup sugar} \rightarrow 1 \text{ cheesecake}$$

Suppose you have 100 cups of sugar, 100 blocks of cream cheese, and 10 eggs. How many cheesecakes can you make? This stoichiometry question is a bit more complicated than our earlier ones because now one of the ingredients—eggs—is in short supply. However, once you realize this, it is easy to determine in your head that the correct answer is two cakes.

This is an example of a *limiting-reactant problem*. For this particular problem, chemists would call the eggs the *limiting reactant* because it is the eggs you'll run out of first, limiting the number of cakes you can make. It does not matter that you have a lot of extra sugar and cream cheese. Once the eggs run out, the cake-making (the reaction) stops. A **limiting reactant** can therefore be defined as any reactant that is present in short supply relative to the other reactants according to the demands of the balanced equation. In our example, the sugar and the cream cheese are referred to as **excess reactants**—reactants that are present in excess relative to the limiting reactant. When a reaction with a limiting reactant is carried out, the limiting reactant will be completely gone (used up), replaced by product and unused excess reactant(s). In our example, you would be left with no eggs, two delicious cheesecakes, 94 blocks of cream cheese, and 98 cups of sugar.

Up to now, we have done stoichiometry problems for reactions in which there are no limiting reactants. That is, we have combined exactly the right amounts of reactants so that, when the reaction is done, all of the reactants are used up and only product remains. Reactions performed in this manner are said to be done in a *stoichiometric*, or *balanced, fashion*. That is, the reactants are present in exactly the molar ratio dictated by the balanced equation. When a reaction is performed in a balanced fashion, it doesn't matter which reactant you use to calculate the theoretical yield of product. In reality, however, chemists often perform reactions so that some reactants are present in excess. In these cases, one reactant is a limiting reactant and the reaction is said to be run in a *limiting fashion*.

When a reaction is performed in a limiting fashion, you must use the limiting reactant to calculate the theoretical yield—for the same reason that only the number of eggs could tell us the cheesecake yield in our example. Therefore, if you are asked to determine the theoretical yield for a reaction performed in a limiting fashion, you must first identify which reactant is limiting.

Unfortunately, to identify the limiting reactant, you must usually do some calculations. In the case of our cheesecake-making, it was pretty obvious that eggs were a limiting reactant. However, when you do real chemical reactions—measuring amounts in grams—it is not obvious usually which reactant, if any, is limiting. The limiting reactant *would* be easier to identify if we always worked in moles instead of grams. To show you why that is so, we'll work with moles first and then switch to grams. Let's work with the simplest kind of reaction, one where all the reactants in the balanced equation are used in the same molar amount. Our HI reaction is a good example.

$$H_2 + I_2 \longrightarrow 2\,HI$$

Reactants used in same molar amount (H_2, I_2)

If we ran this reaction by combining 10 moles of H_2 with 10 moles of I_2, we would be running it in a balanced fashion because both reactants are present in the same molar amount, just what the balanced equation calls for. If we combined 10 moles of H_2 with 11 moles of I_2, we would be running the reaction in a limiting fashion. The H_2 would be the limiting reactant and the I_2 the excess reactant. There would be 1 mole of I_2 left at the end of the reaction. More important, if we wanted to know the theoretical yield of HI, we would have to use the H_2 to find out.

Using the limiting reactant to determine theoretical yield

$$\underbrace{10\ \text{mol}\ H_2}_{\text{Limiting reactant}} \times \underbrace{\frac{2\ \text{mol HI}}{1\ \text{mol}\ H_2}}_{\text{From balanced equation}} = \underbrace{20\ \text{mol HI}}_{\text{Theoretical yield}}$$

To determine whether this reaction was run in a stoichiometric fashion or in a limiting fashion and, if the latter, which reactant was limiting, all we had to do was compare moles of H_2 with moles of I_2 and see if one number was smaller than the other. The one present in the smaller molar amount was the limiting reactant. In actuality, you are much more likely to be told reactant amounts in grams rather than moles. If you are told you have 10.0 g of H_2 and 10.0 g of I_2, it is no longer obvious if the reaction is being run in a balanced or limiting fashion because the amounts are in grams but the equation speaks in moles. It's that old language problem again, but all we need to do is to translate grams to moles by using the appropriate conversion factors:

$$10.0\ \text{g}\ H_2 \times \underbrace{\frac{1\ \text{mol}\ H_2}{2.0158\ \text{g}\ H_2}}_{\text{Inverse of } H_2 \text{ molar mass}} = 4.96\ \text{mol}\ H_2$$

$$10.0\ \text{g}\ I_2 \times \underbrace{\frac{1\ \text{mol}\ I_2}{253.81\ \text{g}\ I_2}}_{\text{Inverse of } I_2 \text{ molar mass}} = 0.0394\ \text{mol}\ I_2$$

Now it's easy to see that the I_2 is present in a far smaller molar amount, and is therefore the limiting reactant. The theoretical yield of HI is thus twice the number of moles of I_2.

For reactions where all the reactants appear in the same molar amount in the balanced equation, finding the limiting reactant is just a matter of converting everything to moles and then looking for the reactant present in the smallest molar amount. However, for reactions where the reactants are required in different molar amounts, such as the combustion of propane,

$$C_3H_8 + 5\,O_2 \longrightarrow 3\,CO_2 + 4\,H_2O$$

Reactants used in different molar amounts

this method doesn't work. For example, suppose we try to run this reaction by combining 1 mole of propane, C_3H_8, with 4 moles of oxygen, O_2. If we blindly did things as we did for the HI reaction, we would notice that propane was present in the smaller molar amount and call it the limiting reactant. That would be wrong, though, because the balanced equation says that each mole of propane requires 5 moles of oxygen. We have 1 mole of propane but only 4 moles of oxygen. Therefore, the oxygen is in short supply and will run out first, making oxygen the limiting reactant.

For reactions that have unequal balancing coefficients, you can't just look for the reactant present in the smaller molar amount to determine which, if any, is limiting. You must also take the balanced equation into account. There is a simple way to do this. After you convert all the amounts of reactants to moles, determine the mole-to-coefficient ratios by dividing each molar amount by its corresponding coefficient from the balanced equation. The reactant that generates the smallest mole-to-coefficient ratio is the limiting reactant. If all the reactants generate the same number, the reaction is being run in a balanced fashion.

Why should this procedure work? We'll show you why with bicycles being assembled by our assistant:

The balanced bicycle equation is:

$$1 \text{ frame} + 2 \text{ wheels} \rightarrow 1 \text{ bicycle}$$

The latest workshop inventory reveals 101 frames but only 200 wheels. Now let's divide the number of each reactant by its balancing coefficient from the balanced bicycle equation to get number-to-coefficient ratios:

Number-to-coefficient ratio

$$\frac{101 \text{ frames}}{1} = 101$$

$$\frac{200 \text{ wheels}}{2} = 100$$ Smaller number-to-coefficient ratio

Unlike the number of wheels and frames, the quotients of these ratios *can* be compared to identify which reactant, if any, is limiting because dividing by the coefficients takes into account the needs of the balanced equation.

The smaller number-to-coefficient ratio is associated with the wheels, which makes them the limiting reactant.

This procedure also works when we deal with balanced chemical equations and moles of reactants because a mole of something is also a number of something. For example, let's return to our propane combustion reaction. Suppose we combine 6 moles of propane with 29 moles of oxygen. Are we running the reaction in a balanced or limiting fashion? If limiting, which reactant is the limiting reactant? To find out, we divide each molar amount by its balancing coefficient:

$$C_3H_8 \;+\; 5\,O_2 \longrightarrow 3\,CO_2 + 4\,H_2O$$

$$\frac{6 \text{ mol } C_3H_8}{1} = 6 \qquad \frac{29 \text{ mol } O_2}{5} = 5.8 \longleftarrow$$

Smaller mole-to-coefficient ratio, meaning O_2 is the limiting reactant

The fact that the ratios are not equal tells you the reaction is being run in a limiting fashion. The smaller mole-to-coefficient ratio is associated with O_2, making it the limiting reactant. This method always works, but be careful! *You must always do this with moles and never with grams.* If you are given the amounts of reactants in grams (which is almost always the case), you must first convert everything to moles before using this method.

The following chart summarizes the method for solving a stoichiometry problem when a limiting reactant is involved. It is essentially the earlier chart with an additional step (between steps 2 and 3) to calculate mole-to-coefficient ratios. Modifications to steps 2 and 3 are emphasized in boldface.

How to solve stoichiometry problems for reactions run in a limiting fashion

Step 1: Write the balanced chemical equation.

Step 2: For **all reactants**, convert the given information to moles.

Step 2a: Determine the mole-to-coefficient ratio for each reactant, using the appropriate coefficients from the balanced equation. The reactant with the smallest mole-to-coefficient ratio is the limiting reactant.

Step 3: Use the balancing coefficients and **the molar amount of the limiting reactant** to calculate the theoretical yield of the product in moles.

Step 4: Convert the molar amount calculated in step 3 to the units asked for in the problem.

Now try your hand at some limiting-reactant problems.

Practice Problems

In Practice Problems 9.24–9.26, use the method of dividing moles by coefficients to determine whether the reaction is being run in a balanced or limiting fashion. If it is being run in a limiting fashion and you are asked for theoretical yield, remember to use only the limiting reactant in your calculation.

9.24 In the burning of propane,

$$C_3H_8 + 5\,O_2 \rightarrow 3\,CO_2 + 4\,H_2O$$

suppose 10.0 g of propane is combined with 10.0 g of oxygen. What is the theoretical yield of CO_2, in grams?

Answer: Before we begin, we'll calculate the molar masses of everything because we'll probably need them:

$C_3H_8 = 44.096\ \text{g/mol}$ $\quad$ $O_2 = 31.998\ \text{g/mol}$

$CO_2 = 44.009\ \text{g/mol}$ $\quad$ $H_2O = 18.015\ \text{g/mol}$

Step 1: *The equation is already balanced.*

Step 2: *We convert the amounts of reactants to moles:*

$$10.0\ \text{g}\ C_3H_8 \times \frac{1\ \text{mol}\ C_3H_8}{44.096\ \text{g}\ C_3H_8} = 0.227\ \text{mol}\ C_3H_8$$

$$10.0\ \text{g}\ O_2 \times \frac{1\ \text{mol}\ O_2}{31.998\ \text{g}\ O_2} = 0.313\ \text{mol}\ O_2$$

Step 2a: To find which reactant is limiting we find the mole-to-coefficient ratios by dividing each molar amount by a balancing coefficient:

$$\frac{0.227\ \text{mol}\ C_3H_8}{1} = 0.227 \qquad \frac{0.312\ \text{mol}\ O_2}{5} = 0.0625$$

Smaller mole-to-coefficient ratio indicates O_2 as limiting reactant.

Steps 3 and 4: We use the balancing coefficients and the limiting-reactant molar amount to calculate the theoretical yield of CO_2 first in moles and then in grams:

$$\underbrace{0.313\ \text{mol}\ O_2 \times \frac{3\ \text{mol}\ CO_2}{5\ \text{mol}\ O_2}}_{\text{Step 3}} \times \underbrace{\frac{44.009\ \text{g}\ CO_2}{1\ \text{mol}\ CO_2}}_{\text{Step 4}} = 8.26\ \text{g}\ CO_2$$

The following diagrams show the strategy we used:

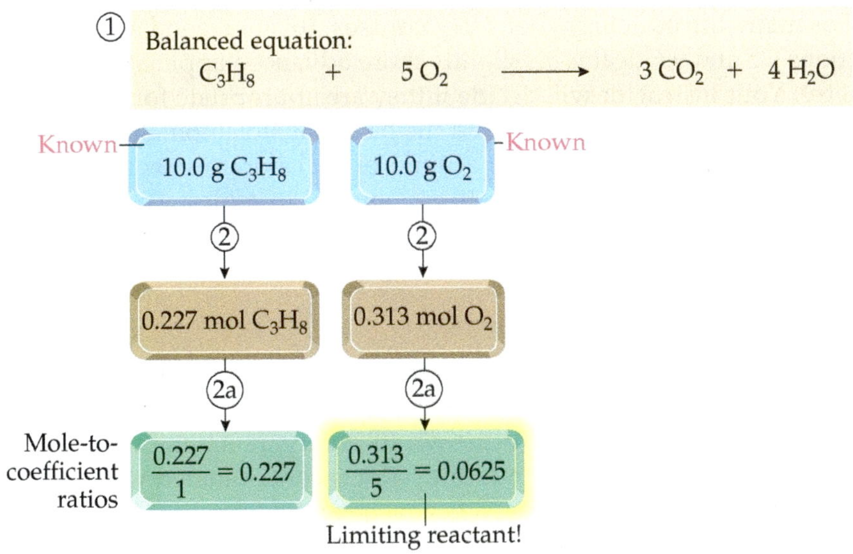

Once the limiting reagent is identified, the problem now becomes identical to the nonlimiting problems you did earlier. This means you can "block out" what you just did and continue on, using the moles of limiting reactant to calculate the theoretical yield.

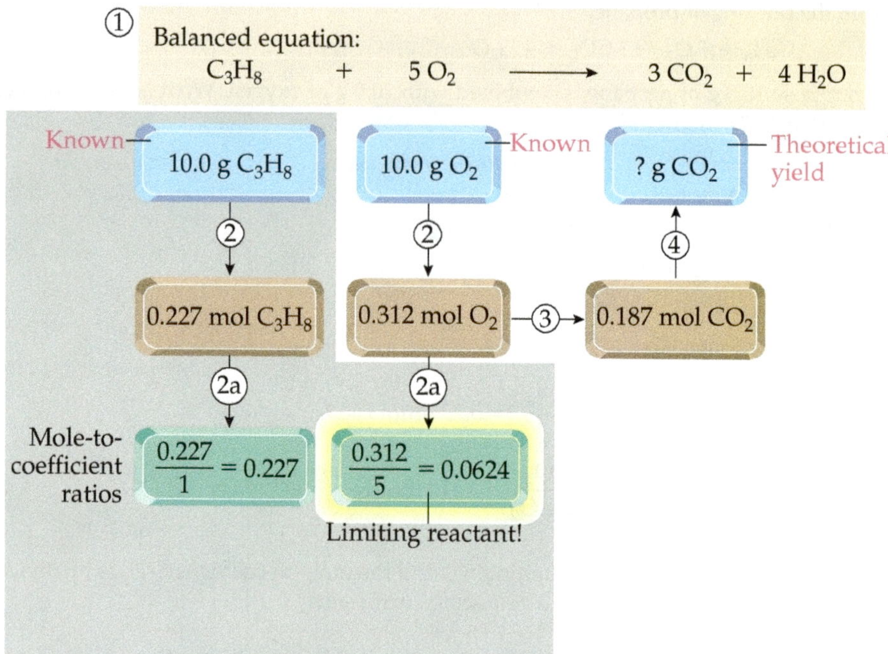

9.25 (a) What mass of oxygen is necessary to burn 100.0 g of propane in a balanced fashion?
(b) What is the theoretical yield of water, in grams?

9.26 Chemical treatment of zinc sulfide, ZnS, with oxygen, O_2, gives zinc oxide, ZnO, and sulfur dioxide gas, SO_2.

(a) Write a balanced equation for the reaction.
(b) If 10.0 g of ZnS is combined with 10.0 g of O_2, what is the theoretical yield of each product, in grams?
(c) How much of the excess reactant is left over, in grams?
(d) Suppose only 7.50 g of ZnO is recovered. What is the percent yield of ZnO?

For many introductory chemistry courses, this is a good stopping point. Sections 9.5 and 9.6 deal with slightly more advanced applications of stoichiometry. Your instructor will decide if they are appropriate for your course. If not, have no fear. We have covered the basics of stoichiometry, and you are well prepared for a next chemistry course or for laboratory work. Just remember that balanced chemical equations "speak" in moles. To use them, you must speak in moles also. If in doubt, convert to moles!

9.5 Combustion Analysis

When you see a written chemical equation and work stoichiometry problems based on it, you know what the products are because their formulas are neatly written on the right side of the arrow. In reality, reactions sometimes give totally unexpected products and/or products contaminated with impurities. For example, suppose you have a headache and a friend offers you a tablet, saying, "This is aspirin, I just synthesized it in a chemical reaction. Have some."

If enough impurities or unexpected products are in the aspirin, it could cure your headache forever! Before it is considered fit for consumption, a chemist would have to confirm both the aspirin's identity and its purity. One good test would be to determine the molecular formula of the product formed in the reaction run by your friend. Hopefully, it's $C_9H_8O_4$, which is aspirin. One of the best ways to determine the molecular formula of an unknown compound is to take a known mass of the compound and—believe it or not—burn it. This procedure is called **combustion analysis**. *Combustion* means "burning," and burning something means reacting it with oxygen. Therefore, when you burn a compound in excess oxygen, all the elements in it combine with oxygen. For example, the combustion reaction for aspirin is:

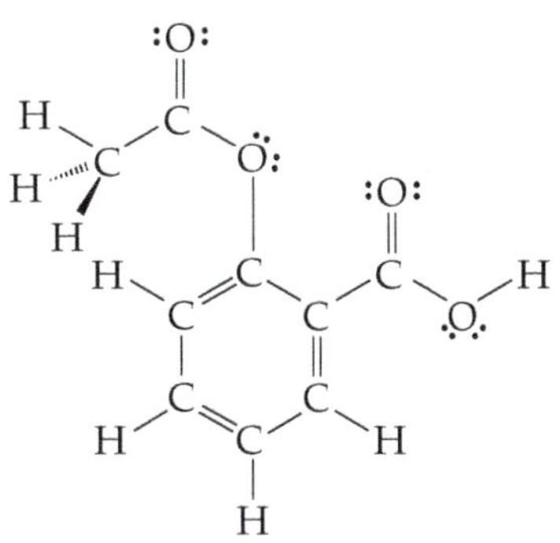

Aspirin, $C_9H_8O_4$

$$C_9H_8O_4 + 9\,O_2 \rightarrow 9\,CO_2 + 4\,H_2O$$

All the carbon ends up in CO_2, and all the hydrogen ends up in H_2O. You can think of the carbon dioxide and the water as "garbage cans" where all the carbon and hydrogen end up.

What Combustion Analysis Does

All C ends up in CO_2. All H ends up in H_2O.

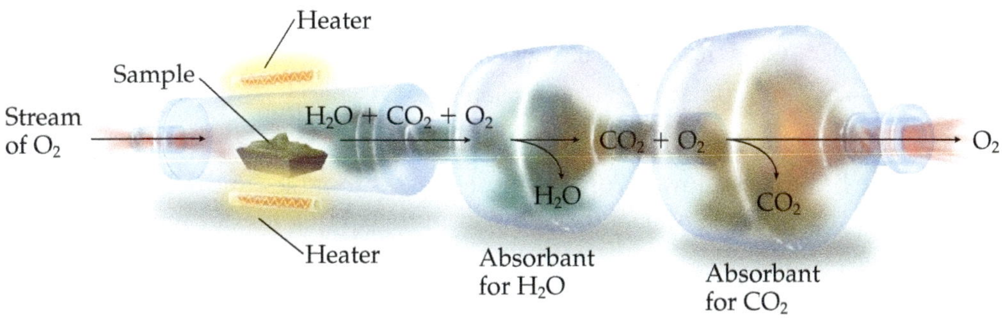

In the laboratory, a combustion analyzer like the one diagrammed above is commonly used to burn unknown substances and determine their molecular formulas. Different materials in the analyzer absorb the H_2O and CO_2 produced during the burning. The masses of the absorbants are determined before and after the combustion. The mass gain of each absorbant is equal to the mass of water or carbon dioxide produced. From these measured masses, the percent composition (by mass) of each element in the compound that was burned can be determined.

Here is the experimental data from the combustion analysis of a sample of your friend's "aspirin":

BURN-EM-UP

"we burn - you earn"

Report form 101

Dear Sir/Madam:

We burned 0.100 g of your white powder and obtained the following results:

60.0% C
4.48% H
35.5% O

Thank you for your business.

Sincerely,
L.M. Ent, Ph.D.

This means that of the 0.100 g sample that was burned, 60.0% of its mass was due to carbon, 4.48% of its mass was due to hydrogen, and 35.5% of its mass was due to oxygen. This is similar to saying that of your total mass, 15% is due to the mass of your head, 10% to the mass of your arms, 20% to the mass of your legs, and 55% to the mass of your torso. In both cases, if we accounted for all the elements in the compound (all the parts of your body), then the mass percents should add up to 100%, which they do. Our job now is to turn the data into a molecular formula. Right now, all we know is that the formula of this white powder is $C_xH_yO_z$; we don't know the values of any of the subscripts. We do know, however, that the subscripts can be thought of as numbers of moles, so we must somehow convert the mass percent of each element to moles of it. It turns out this is quite simple to do. Since the percents all add up to 100, then if we assume we have a 100 g sample of the compound on hand, then percent mass of each element is just the number of grams of that element present in the 100 g sample. This means if we have 100 g of our white powder, then in that 100 g are 60.0 g of carbon (C), 4.48 g of hydrogen atoms (H, not H_2), and 35.5 g of oxygen atoms (O, not O_2).

To summarize, in a 100 g sample:

60.0% C	means	60.0 g C are present
4.48% H	means	4.48 g H are present
35.5% O	means	35.5 g O are present

All we have to do now is convert these masses to moles and we have the values for the variables x, y, and z in the formula $C_xH_yO_z$ for our white powder.

convert to moles

$$60.0\,\cancel{\text{g}}\,\text{C} \times \frac{1\ \text{mol C}}{12.011\,\cancel{\text{g}}\,\text{C}} = \underbrace{5.00\ \text{mol C}}_{\text{Value of } x \text{ in } C_xH_yO_z}$$

$$4.48\,\cancel{\text{g}}\,\text{H} \times \frac{1\ \text{mol H}}{1.0079\,\cancel{\text{g}}\,\text{H}} = \underbrace{4.44\ \text{mol H}}_{\text{Value of } y \text{ in } C_xH_yO_z}$$

$$35.5\,\cancel{\text{g}}\,\text{O} \times \frac{1\ \text{mol O}}{15.999\,\cancel{\text{g}}\,\text{O}} = \underbrace{2.22\ \text{mol O}}_{\text{Value of } z \text{ in } C_xH_yO_z}$$

So the formula for our white powder is $C_{5.00}H_{4.44}O_{2.22}$. What's that? You don't like it? Neither do we, although it is technically correct because it expresses the correct mole ratios between the elements. However, being a molecular substance, an individual molecule is made from whole atoms so we want whole numbers in our formula. **Sometimes, when you do the above type calculation, you will have whole numbers (or numbers so close to whole that you can round them to whole), and you will be done.** This is not one of those cases. What do you do when the numbers are not integers. The answer is anything you want, so long as you don't change the mole ratios between the elements. We recommend in such cases that the first thing you do is divide all the molar amounts by the smallest mole amount (in this case, 2.221 moles of oxygen). By dividing all of them by the same number you will not be changing the mole ratios between them, but you might just turn the moles into whole numbers.

$$C_{\frac{5.00}{2.22}} H_{\frac{4.44}{2.22}} O_{\frac{2.22}{2.22}}$$ Divide through by the smallest molar amount

$$\downarrow$$

$$C_{2.25}H_2O_1$$

We almost did it! At this point, if you arrive at whole numbers, you are done. Unfortunately, 2.25 is not a whole number, and it is way too far from the closest integer value to simply round it to 2. When this sort of result occurs, we have a second recommendation. Find a factor that you can multiply the nonintegral number by to make it whole. Of course, you must once again do the same to all of the subscripts so as not to change the mole ratio between the elements. In this case, the obvious factor is 4.

$$C_{2.25 \times 4}H_{2 \times 4}O_{1 \times 4} \xrightarrow[\text{by factor of 4}]{\text{multiply through}} C_9H_8O_4$$

This indeed is the molecular formula of aspirin (check back a few pages).

Determining a formula from combustion analysis data

Step 1: Check that the percent by mass elemental compositions add up to 100% (they should).

Step 2: Assume you have a 100 g sample of the compound. This allows you to change the % mass of each element into grams.

Step 3: Convert the grams of each element to moles of that element. This gives you the subscript in the formula for that element.

Step 4: If the subscripts are integral or very close to integral (so you can round them), you are done. If not, divide them all by the smallest molar amount. If still non-integral, multiply them all by a factor that turns the nonintegral ones into integrals.

A "trick" question that might come your way is when the percents add up to substantially less than 100, and then the problem tells you that one more element might be present besides the ones reported (see Practice Problems 9.27 and 9.28). If you encounter one of these, just remember that the percent elemental compositions must always add up to 100 if all of the elements are accounted for. Therefore, if the reported percents do not add up to 100, the unreported percent for the additional element is just 100 minus the sum of all the other percents.

Practice Problems

9.27 A compound is known to contain carbon and hydrogen. It might also contain oxygen. A sample of the compound is burned. The results of the combustion analysis are 74.9% C and 25.1% H.

(a) What is the chemical formula for this compound?

(b) Write the balanced combustion reaction (reaction with O_2) for this compound.

Answer (not worked out on purpose—you do it):

(a) CH_4
(b) $CH_4 + 2\,O_2 \rightarrow 2\,H_2O + CO_2$

9.28 A compound is found to contain 52.4% carbon and 13.1% hydrogen. It might also contain oxygen.

(a) What is the chemical formula for this compound?
(b) Write the balanced combustion reaction (reaction with O_2) for this compound.

The combustion-analysis problems we have been doing so far have been chosen to avoid a complication we want to deal with now. Let's look at one more compound, but this time we'll tell you the molecular formula right at the start. It's the simple sugar glucose, formula $C_6H_{12}O_6$. When we burn a sample of glucose, we get the results shown below. We expect to get the formula $C_6H_{12}O_6$. Let's see if we can. The problem is worked out on the next page. Follow it through carefully.

BURN-EM-UP

"we burn - you earn"

Report form 101

Dear Sir/Madam:

We burned a sample of your white powder and determined the following results:

C 39.94%
H 6.72%
O 53.34%

Thank you for your business.

Sincerely,
L.M. Ent, Ph.D.

Step 1: Check that the reported percents add up to 100. They do, so there are no missing elements.

Step 2: Assume you have 100 g of the compound. This allows you to change the percent by mass of each element into grams of each element.

39.94 g C
6.72 g H
53.34 g O

Step 3: Convert grams of each element into moles of each element.

$$39.94\ \cancel{\text{g C}} \times \frac{1\ \text{mol C}}{12.011\ \cancel{\text{g C}}} = 3.33\ \text{mol C}$$

$$6.72\ \cancel{\text{g H}} \times \frac{1\ \text{mol H}}{1.0079\ \cancel{\text{g H}}} = 6.67\ \text{mol H}$$

$$53.34\ \cancel{\text{g O}} \times \frac{1\ \text{mol O}}{15.999\ \cancel{\text{g O}}} = 3.33\ \text{mol O}$$

$$\longrightarrow C_{3.33}H_{6.67}O_{3.33}$$

Step 4: If subscripts are not whole numbers, divide all of them through by the smallest one.

$$C_{\frac{3.33}{3.33}}H_{\frac{6.67}{3.33}}O_{\frac{3.33}{3.33}} \longrightarrow CH_2O$$

We have arrived at the formula of glucose, but wait a minute! We told you at the outset that the formula of glucose was $C_6H_{12}O_6$, and yet we just came up with CH_2O from the combustion analysis data for glucose. What went wrong? The answer is that nothing went wrong. In general, a combustion analysis does not give the *molecular* formula of a compound; it gives the *empirical formula*. An **empirical formula** (also called a *simplest formula*) is one that gives only the ratio of the elements present in a compound, with all subscripts being the smallest numbers possible. So how does an empirical formula differ from a *molecular formula*, a term you've been familiar with since Chapter 5? Simple—a molecular formula gives the *actual composition* of the compound. Thus, CH_2O is the empirical formula of glucose. It tells us that there are twice as many H atoms in the compound as C or O atoms, and that the C and O atoms are present in equal numbers. If you divide all the subscripts in the glucose molecular formula, $C_6H_{12}O_6$, by the smallest subscript, you get $C_1H_2O_1 = CH_2O$, the glucose empirical formula.

There are cases where the empirical formula of a compound is also the molecular formula (this was the case in all the examples and problems we did up to this point), but you can't count on this. More often than not, the empirical formula and the molecular formula are not the same.

That combustion analysis always gives the empirical formula presents a problem. How do you know if what you get is also the molecular formula? And if it is not, how can you figure out what the molecular formula is? To answer these questions, you need to know the molar mass of the substance. But wait a minute—if you don't know the molecular formula in the first place, you can't calculate the molar mass! As we will see in later chapters, there are other experiments that can be done to determine the molar mass of a compound (titrations, freezing point depression, etc.). For now, when you need a molar mass to go from an empirical formula to a molecular formula, we shall give it to you (based on values from experimental results).

For glucose, the molar mass is 180.155 g/mol. Now our job is easy. We calculate the molar mass of the empirical formula and compare it with the molar mass of the compound:

Empirical formula CH_2O

$$\text{Molar mass of } CH_2O = \underbrace{12.011\ \text{g/mol}}_{\text{C}} + \underbrace{(2 \times 1.0079\ \text{g/mol})}_{\text{H}} + \underbrace{15.999\ \text{g/mol}}_{\text{O}} = 30.026\ \text{g/mol}$$

Molecular formula $C_6H_{12}O_6$

$$\text{Molar mass of } C_6H_{12}O_6 = 180.155\ \text{g/mol}$$

Because these numbers are not the same, the empirical formula cannot be the molecular formula.

If the (known) molar mass of the compound and the calculated molar mass of the empirical formula are the same, the empirical formula is the molecular formula. When they are different, divide the molar mass of the compound by the molar mass of the empirical formula. Then multiply the subscripts in the empirical formula by this number to get the molecular formula. Here is how it works for glucose:

Converting an empirical formula to a molecular formula

Step 1: Divide the molar mass of the compound by the molar mass of the empirical formula:

$$\frac{\text{Molar mass of glucose}}{\text{Molar mass of empirical formula}} = \frac{180.155\ \text{g/mol}}{30.026\ \text{g/mol}} = 6\ \text{(approximately)}$$

Step 2: Multiply all subscripts in the empirical formula by the number calculated in step 1:

$$C_{1\times6}H_{2\times6}O_{1\times6} = C_6H_{12}O_6 \qquad \text{Molecular formula}$$

Now try the following practice problems.

Practice Problems

9.29 A compound is known to contain carbon and hydrogen and might also contain oxygen. A sample is burned yielding 54.6% C and 9.16% H.

(a) What is the empirical formula of the compound?
(b) The molar mass of the compound is 132.159 g/mol. What is the molecular formula?
(c) Write the balanced combustion reaction (reaction with O_2) for the compound.

Answer (not worked out on purpose—you do it):

(a) C_2H_4O
(b) $C_6H_{12}O_3$
(c) *To balance the equation, look at C and H first. Then balance the elemental substance (O_2) last:*

$$\underbrace{C_6H_{12}O_3}_{3\,O} + \underbrace{O_2}_{\text{Need 15 O here to match 18 on right}} \longrightarrow \underbrace{6\,CO_2 + 6\,H_2O}_{18\,O}$$

We make O_2 provide 15 O atoms by multiplying it by 7.5 (that is, $\frac{15}{2}$):

$$C_6H_{12}O_2 + \underbrace{7.5\,O_2}_{7.5 \times 2 = 15\text{ O atoms}} \longrightarrow 6\,CO_2 + 6\,H_2O$$

This is a perfectly correct balanced equation, but if you prefer the balancing coefficients to be whole numbers, you can multiply all of them by some number that makes them whole (in this case, 2):

$$2\,C_6H_{12}O_3 + 15\,O_2 \rightarrow 12\,CO_2 + 12\,H_2O$$

9.30 A compound is known to contain carbon and hydrogen and might also contain oxygen. A sample is burned yielding 85.62% C and 14.37% H. (Be careful here. Remember, there is always a little inaccuracy associated with measured numbers).

(a) What is the empirical formula for the compound?
(b) The molar mass is 28.054 g/mol. What is the molecular formula?
(c) Write the balanced combustion reaction.

9.6 Going from Molecular Formula to Percent Composition

In this section, you are going to exercise your developing stoichiometry skills. You just saw how to take percent by mass elemental compositions obtained from a combustion analysis and convert them into a formula. Now we want to show you how to go the other way, from the formula for a compound to its percent by mass elemental composition. Let's continue to use glucose, $C_6H_{12}O_6$, as our example. To go quickly from this (or any other) molecular formula to the mass percent of each element in the compound, all we have to do is assume we have 1 mole of the compound. The mass of 1 mole of any compound is its molar mass in grams.

If we have 1 mole of glucose, we have 6 moles of C atoms, 12 moles of H atoms, and 6 moles of O atoms (this information comes straight from the molecular formula $C_6H_{12}O_6$). To get the mass percents, we have to figure out the masses of these different moles of atoms, divide each by the total mass of 1 mole of glucose (180.155 g/mol, calculated earlier), and multiply by 100. Remember, the key to getting started is to assume you have 1 mole of the compound.

Going from molecular formula to elemental mass percents

Step 1: Assume you have 1 mole of the compound (glucose in this case). The molecular formula, in this example $C_6H_{12}O_6$, tells you how many moles of each element you have. Convert the moles of each element to mass in grams:

$$6 \text{ mol C} \times \frac{12.011 \text{ g C}}{1 \text{ mol C}} = 72.066 \text{ g C}$$

$$12 \text{ mol H} \times \frac{1.0079 \text{ g H}}{1 \text{ mol H}} = 12.0948 \text{ g H}$$

$$6 \text{ mol O} \times \frac{15.999 \text{ g O}}{1 \text{ mol O}} = 95.994 \text{ g O}$$

Step 2: Divide the mass of each element by the mass of the 1 mole of compound (in other words, divide by its molar mass) and multiply by 100:

$$\%\text{C} = \frac{72.066 \text{ g C}}{180.155 \text{ g}} \times 100 = 40.00\% \text{ C}$$

$$\%\text{H} = \frac{12.0948 \text{ g H}}{180.155 \text{ g}} \times 100 = 6.71\% \text{ H}$$

$$\%\text{O} = \frac{95.994 \text{ g O}}{180.155 \text{ g}} \times 100 = 53.28\% \text{ O}$$

$$99.99\%$$

Arbitrarily rounded to hundredths place

Mass percents add up to 100 after rounding.

Congratulations! You just went from molecular formula to elemental mass percents. You should now be able to quickly go from molecular formula to elemental mass percents, and from elemental mass percents to empirical formula. Just remember what initial assumption to make for each procedure.

Practice Problems

9.31 What is the mass percent of each element in hydrogen peroxide, H_2O_2?
Answer: 5.93% H, 94.07% O

9.32 What is the mass percent of each element in trinitrotoluene (TNT), $C_7H_5N_3O_6$?

9.33 A compound is found to have the following elemental mass percents: Cl = 89.09%, C = 10.06%, H = 0.84%. The molar mass of the compound is 119.378 g/mol. What are the empirical and molecular formulas?

9.34 A compound is known to contain C and H, and might also contain O. It is analyzed for C and H only, yielding the mass percents C = 54.53% and H = 9.15%. The molar mass of the compound is 88.106 g/mol. What are the empirical and molecular formulas?

+1 One More Thing

9.7 Nonstoichiometric Compounds

Shown below is an orange-colored mineral known as wüstite, which has formed on some green calcite. The mineral wüstite is an iron oxide, a compound with a chemical formula something like $Fe_{0.95}O$.

"Wait a minute," you may very well be asking yourself right now. Is that a typo? Don't you mean FeO or Fe_2O_3? After all that we covered in this chapter on stoichiometry, that is certainly not an unreasonable question to ask. Actually, FeO is the *ideal* molecular formula for wüstite. By ideal we mean stoichiometric. That is, there are only integral or whole numbers of atoms of each element present in the formula—in this case, one atom of iron and one atom of oxygen. Of course, this is what we expect. After all, you would think that you can't have fractions of an atom in a formula since a single atom is the smallest portion of an element you can have. So what does $Fe_{0.95}O$ mean then? Does it mean there are fractions of Fe atoms present where a "piece" has been chipped off of each iron atom? Of course not! Dalton was right. A single, whole atom is the smallest piece of an element possible. So what is going on? The answer is interesting and actually quite simple. In FeO, every iron atom has a +2 charge (think of them as being Fe^{2+} ions), and every oxygen atom has a −2 charge (think of them as O^{2-} ions). This would result in the expected, electrically neutral compound, FeO. However, nature can be tricky, and some of the Fe^{2+} ions present in the wüstite lattice have a +3 charge (Fe^{3+}). To maintain charge neutrality, for every two Fe^{3+} ions present, three Fe^{2+} ions must be replaced, leading to a deficiency of iron atoms. Therefore, the formula for wüstite is often written as $Fe_{1-x}O$, where x is a small number representing a deviation from the ideal formula. Such a compound is called **nonstoichiometric** because one or more of its subscripts is a non integral number, but it is important to realize that while there are compounds with a nonstoichiometric formula, it never means that fractions of atoms are present. Rather, some other explanation is at work.

You might think this is a rather esoteric thing to tell you about, but nonstoichiometric compounds are actually quite important, with their unusual and valuable properties dependent on the degree of nonstoichiometry. One example is the yttrium barium copper oxides of formula $YBa_2Cu_3O_{7-x}$, where x can be between 0 and 1. These materials have been shown to be high-temperature superconductors (go back and look at the beginning of Section 8.1 and you will see the ideal formula for this compound). The temperature at which this material superconducts is very dependent on the value of x.

Another place nonstoichiometry plays an important role is in semiconductors, which are at the heart of all of our modern electronic gadgetry. A semiconductor is a material that has electrical conductivity somewhere between that of a metal (conductor) and a nonmetal (insulator). These materials have either slight excesses or slight deficiencies of valence electrons. For example, let's return to the ideal FeO, iron oxide, discussed above. The nonstoichiometric version $Fe_{0.9}O$ is actually $(Fe^{2+})_{0.7}(Fe^{3+})_{0.2}O$. Since Fe^{3+} has one less electron than Fe^{2+}, the replacement of some Fe^{2+} ions with Fe^{3+} has the effect of draining valence electrons from the lattice, making it an electron-deficient semiconductor, known as a p-type semiconductor (p stands for positive; semiconductors with an excess of valence electrons are called n-type for negative).

Now think back to Chapter 3 when we introduced the law of constant composition, a law that became part of chemistry in the early 1800s and that states that samples of a compound taken from various locations will always contain the same percent by mass of each element making up the compound. Obviously, this does not hold true for samples of $Fe_{1-x}O$, as x will vary depending on where it is mined.

Stoichiometry problems of the type we have been doing in this chapter are the heart of chemical calculations. "How much?" and "What did I make?" are always important questions, whether you are baking cheesecakes or producing aspirin. Just keep in mind that our recipes—the balanced equations of chemical reactions and the molecular formulas in them—speak in moles. To use these recipes correctly, you must speak in moles also. If in doubt, convert to moles. It's almost always a good first step.

Have You Learned This?

Stoichiometry (p. 316)
Mole (p. 319)
Avogadro's number (p. 319)
Molar mass (p. 322)
Theoretical yield (p. 327)
Actual yield (p. 327)
Percent yield (p. 327)
Limiting reactant (p. 333)
Excess reactants (p. 333)
Combustion analysis (p. 339)
Empirical formula (p. 342)

Learning Outcomes

9.1	Stoichiometry – What is it?	Calculate amounts of ingredients using a recipe. [36, 37]
9.2	The Mole	Describe what a mole is and how it related to formulas of substances, molar mass, and chemical equations. [38–54, 66–72]
9.3	Reaction Stoichiometry	Balance chemical reactions and calculate the theoretical yield and percent yield for a reaction. [59–62, 73, 74]
9.4	Dealing with a Limiting Reactant	Identify a limiting reactant if present and calculate the theoretical and percent yield when a limiting reactant is present. [75–86]
9.5	Combustion Analysis	Calculate a compound's empirical and actual formula from percent by mass elemental compositions. [89–98]
9.6	Going from Molecular Formula to Percent Composition	Calculate percent by mass elemental compositions from a molecular formula. [99–102]
9.7	Nonstoichiometric Compounds	Describe how it is possible for a formula to have fractional subscripts. [186]

Skills to Know

The mole

A mole of anything is 6.022×10^{23} of them. This number is also called Avogadro's number. The definition of a mole gives rise to the useful conversion factors at right.	$\frac{1 \text{ mole of something}}{6.022 \times 10^{23} \text{ somethings}}$ or $\frac{6.022 \times 10^{23} \text{ somethings}}{1 \text{ mole of something}}$

Molar mass

The mass in grams of 1 mole of atoms or molecules is called the molar mass. Molar mass can be viewed as a conversion factor that converts mass to moles or moles to mass, as shown at right.	Moles × Molar mass = Mass → $\text{Moles} \times \frac{\text{Grams}}{\text{Mole}} = \text{Grams}$ $\frac{\text{Mass}}{\text{Molar mass}} = \text{Moles}$ → $\text{Grams} \times \frac{\text{Moles}}{\text{Gram}} = \text{Moles}$

Important conversions in stoichiometry calculations

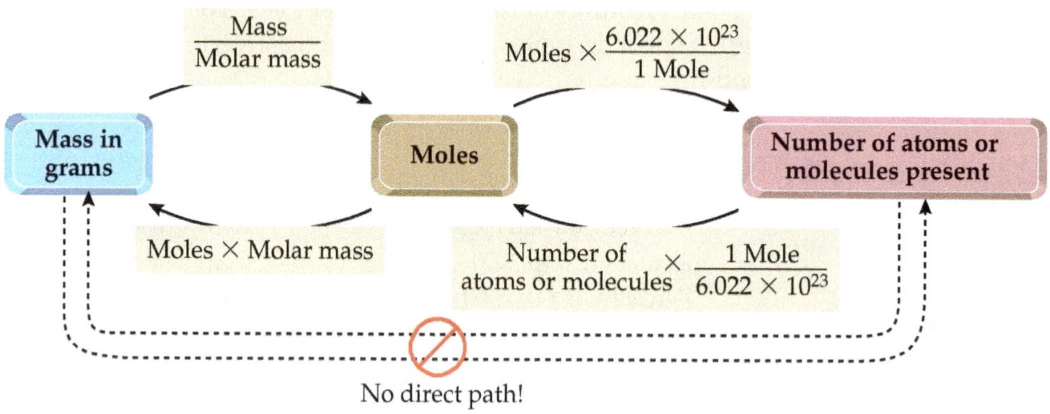

Mass → Moles

36.0 g of H_2O is how many moles?

$$36.0 \text{ g } H_2O \times \frac{1 \text{ mol } H_2O}{18.015 \text{ g } H_2O} = 2.00 \text{ mol } H_2O$$

Inverse of molar mass

Moles → Number

2.00 moles of H_2O is how many molecules?

$$2.00 \text{ mol } H_2O \times \frac{6.022 \times 10^{23} \; H_2O \text{ molecules}}{1 \text{ mol } H_2O}$$

$$= 1.20 \times 10^{24} \; H_2O \text{ molecules}$$

Moles → Mass

2.00 moles of H_2O is how many grams?

$$2.00 \text{ mol } H_2O \times \frac{18.015 \text{ g } H_2O}{1 \text{ mol } H_2O} = 36.0 \text{ g } H_2O$$

Molar mass

Number → Moles

1.20×10^{24} molecules of H_2O is how many moles?

$$1.20 \times 10^{24} \; H_2O \text{ molecules} \times \frac{1 \text{ mol } H_2O}{6.022 \times 10^{23} \; H_2O \text{ molecules}}$$

$$= 2.00 \text{ mol } H_2O$$

Tips for working stoichiometry problems

1. The coefficients from a balanced chemical equation and the subscripts from a formula can always be interpreted as moles and used to write conversion factors that are needed for doing stoichiometry problems.

   $H_2 + I_2 \longrightarrow 2\,HI$

   Can always be interpreted as moles

2. If a reaction is run in a balanced fashion, any reactant can be used to calculate the theoretical yield. If the reaction is performed in a limiting fashion, only the limiting reactant can be used to calculate theoretical yield.

3. For any stoichiometry problem, you will have to work in units of moles. That is because all balanced chemical equations and all formulas are written in terms of moles. If you are not sure how to start a stoichiometry problem, start by converting what you are given to moles.

How to solve a stoichiometry problem

Step 1: Begin with a balanced chemical equation.

Step 2: Identify a reactant or product whose mass you know. Convert its mass to moles by using its molar mass.

Step 3: Use the coefficients from the balanced equation to find the amount in moles of the reactant or product whose amount you need to know.

Step 4: Convert the amount from step 3 from moles to grams, using the molar mass of the reactant or product in question.

Example: You need to produce 10.0 g of HI by reacting H_2 with I_2. How many grams of H_2 will you need? (Assume you get the theoretical yield.)

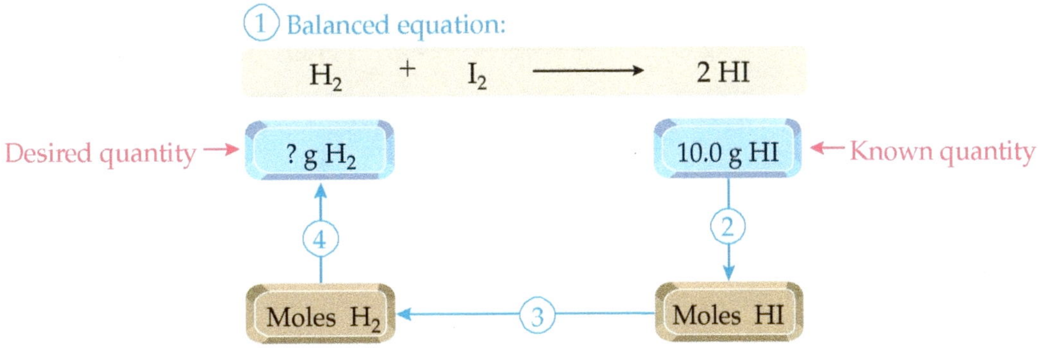

$$10.0\ \text{g HI} \times \frac{1\ \text{mol HI}}{127.913\ \text{g HI}} \times \frac{1\ \text{mol } H_2}{2\ \text{mol HI}} \times \frac{2.0158\ \text{g } H_2}{1\ \text{mol } H_2} = 0.0788\ \text{g } H_2$$

(2) (3) (4)

Solving a limiting-reactant stoichiometry problem

Example: If you react 5.00 g of methane, CH_4, with 12.00 g of oxygen, O_2, what is the theoretical yield of H_2O?	
Step 1: Begin with a balanced chemical equation.	$CH_4 + 2\,O_2 \longrightarrow CO_2 + 2\,H_2O$
Step 2: Convert the given masses of all the reactants to moles, using the molar masses of the reactants.	**CH_4:** $5.00\ \text{g}\ CH_4 \times \frac{1\ \text{mol}\ CH_4}{16.043\ \text{g}\ CH_4} = 0.312\ \text{mol}\ CH_4$ **O_2:** $12.00\ \text{g}\ O_2 \times \frac{1\ \text{mol}\ O_2}{31.998\ \text{g}\ O_2} = 0.375\ \text{mol}\ O_2$
Step 2a: Determine the mole-to-coefficient ratio for each of the reactants, using the appropriate coefficients from the balanced equation. The reactant with the smallest mole-to-coefficient ratio is the limiting reactant.	**CH_4:** $\frac{0.312}{1} = 0.312$ **O_2:** $\frac{0.375}{2} = 0.188$ ← This is the limiting reactant.
Step 3: Use the molar amount of the limiting reactant to calculate the theoretical yield of the desired product in moles. Do this by using the appropriate coefficients from the balanced equation.	**O_2:** $0.375\ \text{mol}\ O_2 \times \frac{2\ \text{mol}\ H_2O}{2\ \text{mol}\ O_2} = 0.375\ \text{mol}\ H_2O$
Step 4: Convert the theoretical yield from moles to grams, using the molar mass of the desired product.	**H_2O:** $0.375\ \text{mol}\ H_2O \times \frac{18.015\ \text{g}\ H_2O}{1\ \text{mol}\ H_2O} = 6.76\ \text{g}\ H_2O$

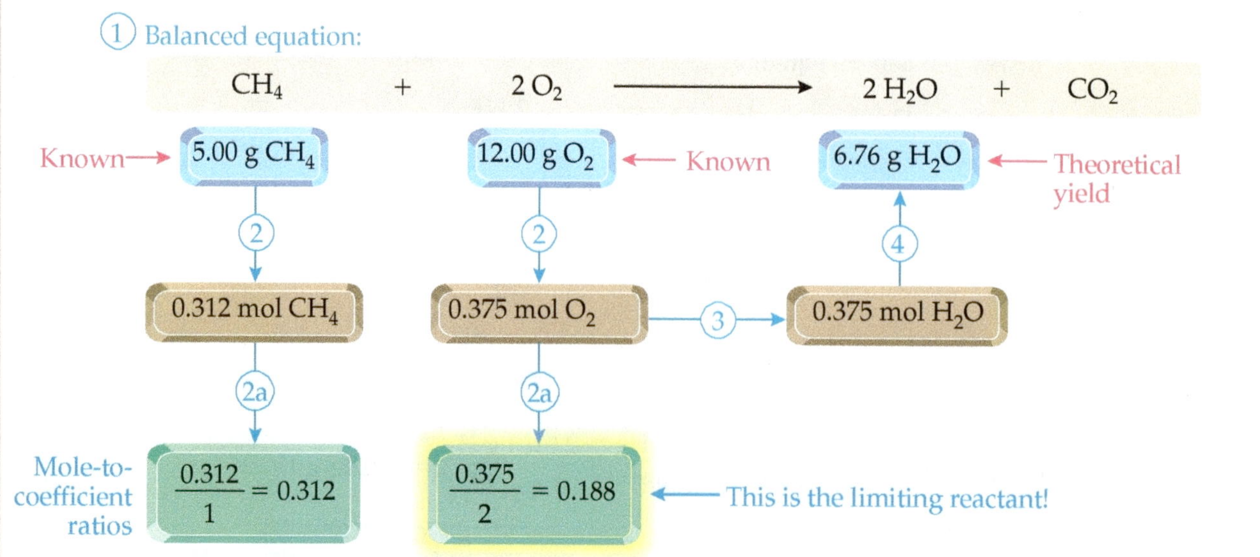

Concept Questions

1. Stoichiometry is the study of
   (a) the energetics of a reaction.
   (b) the quantitative aspects of a reaction.
   (c) how reactants turn into products.
   (d) *stoicheons* (Greek for elements) involved in a chemical reaction.

2. Conversion factors are
   (a) never written in fractional form (something over something else).
   (b) used to eliminate all units from a calculation.
   (c) used to improve accuracy and precision of a calculated result.
   (d) used to translate or convert from one unit to another unit.

3. The mole is
   (a) the number of ${}^{12}_{6}C$ atoms in exactly 12 g of ${}^{12}_{6}C$.
   (b) 6.022×10^{23} of anything.
   (c) also called Avogadro's number.
   (d) all of the above are correct statements.

4. For the reaction $CH_4 + O_2 \rightarrow CO_2 + H_2O$,
   (a) a2 is needed in front of H_2O.
   (b) a 2 is needed in front of O_2.
   (c) a 2 in needed in front of CH_4 and CO_2.
   (d) a 2 is needed in front of O_2 and H_2O.

5. One mole of an element weighs
   (a) its atomic mass divided by 6.022×10^{23}/mol.
   (b) its atomic mass times 6.022×10^{23}/mol.
   (c) its atomic mass listed on the periodic table in grams.
   (d) one-twelfth its atomic mass listed on the periodic table in grams.

6. In grams, one mole of a molecule weighs
   (a) the sum of its constituent element atomic masses, each multiplied by its subscript in the formula.
   (b) its molar mass in grams.
   (c) 6.022×10^{23} times the mass of one single molecule in grams.
   (d) all of the above are correct.

7. A balanced chemical equation can always be read
   (a) in terms of individual molecules.
   (b) in terms of moles of molecules.
   (c) in terms (as a source) of conversion factors.
   (d) all of the above are correct.

8. A limiting reactant
   (a) is always present in a reaction.
   (b) makes it impossible to write a balanced chemical equation.
   (c) ends up becoming an excess reactant.
   (d) limits the yield of a reaction.

9. To find if there is a limiting reactant requires
   (a) removing the balancing coefficients from the chemical equation.
   (b) dividing the moles of each reactant present by its balancing coefficient from the balanced chemical equation.
   (c) multiplying the moles of each reactant present by its balancing coefficient from the balanced chemical equation.
   (d) adding up the moles of products and reactants.

10. It is possible to go from a molecular formula to percent by mass composition for each element because
    (a) the formula subscripts can be interpreted as grams of each element present.
    (b) the formula subscripts can be interpreted as moles of each element present.
    (c) the empirical formula is a multiple of the actual formula.
    (d) it is not possible to go from formula to percent composition; it is only possible to go the other way.

Answers: 1b; 2d; 3d; 4d; 5c; 6d; 7d; 8d; 9b; 10b

Stoichiometry—What Is It?

9.35 How is a balanced chemical equation like a recipe?

9.36 Consider this recipe for 1 dozen cookies:

1 egg
2 cups flour
3 cups sugar
$\frac{1}{4}$ pound butter
1 cup milk

(a) Write at least five conversion factors that might be useful for solving cookie stoichiometry problems.
(b) Using unit analysis, calculate how many cups of flour are needed to make 30 cookies. Write down your method and cross out units that cancel.
(c) Why can't you answer the following question using just the information in the recipe: How many eggs are required to completely use up 1 container of milk?
(d) Given the conversion factor that 1 container of milk = 4 cups of milk, answer the question in part (c). Write down your method and cross out units that cancel.
(e) Write the recipe in chemical equation form.

9.37 One recipe for cheesecake is the following, which makes one cake:

3 blocks of cream cheese
5 eggs
1 cup sugar

(a) To bake three cakes, how many eggs do you need? Write down your method and cross out units that cancel.
(b) If you have 63 blocks of cream cheese, how many eggs do you need to completely use up the cream cheese? Write down your method and cross out units that cancel.
(c) Suppose you want to make 5 cakes and have 2 eggs left over for breakfast. How many eggs do you need? Write down your method and cross out units that cancel.
(d) You have 3.011×10^{24} eggs (that's a lot of eggs!). How many cakes can you make? Write down you method and cross out units that cancel.

The Mole

9.38 How is the concept of "1 mole" similar to the concept of "1 dozen"?

9.39 When is it allowed to insert the word *mole* into a chemical equation when translating the equation into words?

9.40 The human population of our planet is about 6 billion people. What percentage of a mole is 6 billion?

9.41 Consider bicycles where each wheel has 24 spokes.
(a) How many bicycles are there in 1 mole of bicycles?
(b) How many tires are there in 1 mole of bicycles?
(c) How many spokes are there in 1 mole of bicycles?
(d) How many bicycles would you need to have a total of 1 mole of spokes.

9.42 How many O_2 molecules are there in 1 mole of O_2 molecules? How many O atoms are there in 1 mole of O_2 molecules?

9.43 How many pennies are there in 2.5 moles of pennies? How many dollars does this equal? Answer both questions by using conversion factors, and show which units cancel.

9.44 How many years are there in 1 mole of seconds? Use conversion factors, and show which units cancel.

9.45 Translate the following balanced equation into words, first without using the word *moles*, then again with the word *moles*:

$$2\,SO_2 + O_2 \rightarrow 2\,SO_3$$

9.46 Sometimes using the word *moles* when translating a chemical equation into words can be very helpful. For example, consider the following *correctly balanced* equation. What difficulty do you run into when you try to translate it into words without using the word *moles*? How does using the word *moles* solve the difficulty?

$$C_2H_2 + \tfrac{5}{2}O_2 \rightarrow 2\,CO_2 + H_2O$$

9.47 In words, what is the mass of 1 mole of atoms of any element?

9.48 In words, what is the mass of 1 mole of any molecule?

9.49 How do you calculate the molar mass of a compound?

9.50 How does molar mass solve the "language problem" when a chemical equation is being used as a recipe?

9.51 How many atoms of the $^{12}_{6}C$ isotope are there in exactly 12 g of the isotope?

9.52 Consider a sample containing 12.011 g of naturally occurring carbon.
(a) How many carbon atoms are in this sample?
(b) How many dozens of carbon atoms are in this sample?
(c) How many moles of carbon atoms are in this sample?

9.53 If you have 1 mole of glucose ($C_6H_{12}O_6$):
(a) How many moles of carbon atoms do you have?
(b) How many moles of hydrogen atoms do you have?
(c) How many oxygen atoms do you have? (Note that we are not asking for moles here.)

9.54 Consider the ammonia (NH_3) molecule.
(a) If you have 1 mole of ammonia, how many moles of H atoms do you have?
(b) If you have 2 moles of ammonia, how many moles of H atoms do you have?
(c) If you have 2 moles of ammonia, how many N atoms do you have? (We are not asking for moles here.)

Reaction Stoichiometry

9.55 What do we mean by the theoretical yield of a reaction?

9.56 What do we mean by the actual yield of a reaction?

9.57 Why is the actual yield of a reaction often not equal to the theoretical yield?

9.58 What do we mean by the percent yield of a reaction?

9.59 A student runs a reaction to prepare 40.0 g of aspirin and yet recovers only 15.5 g. What is the percent yield?

9.60 Consider the unbalanced chemical equation

$$NO + O_2 \rightarrow NO_2$$

(a) Balance the equation.
(b) Translate the equation into words using the word *mole(s)* wherever you can.
(c) To produce 2 moles of NO_2 by the reaction you just wrote, how many grams of NO and O_2 must you combine?
(d) What is the theoretical yield in grams of NO_2?
(e) You carry out the reaction and recover 22.5 g of NO_2. What is the percent yield?

9.61 Consider the unbalanced chemical equation

$HCl + Zn \rightarrow H_2 + ZnCl_2$

(a) Balance the equation.
(b) Translate the equation into words using the word *mole(s)* wherever you can.
(c) To produce 1 mole of H_2 from the reaction you just wrote, how many grams of HCl and Zn must you combine?
(d) What is the theoretical yield in grams of H_2?
(e) You recover 2.00 g of H_2 after carrying out the reaction. What is the percent yield?

9.62 Consider the unbalanced chemical equation

$Na + Cl_2 \rightarrow NaCl$

(a) Balance the equation.
(b) Translate the equation into words using the word *mole(s)* wherever you can.
(c) To produce 1 mole of NaCl from the reaction, how many grams of Na and Cl_2 must you combine?
(d) What is the theoretical yield in grams of NaCl from the reaction you ran in part (c)?
(e) You carry out the reaction and recover 45.50 g of NaCl. What is the percent yield?

9.63 The formula $C_6H_{12}O_6$ is a source of many conversion factors.
(a) Write at least three of them.
(b) Suppose you have a sample of $C_6H_{12}O_6$, and it contains a total of 72,000 hydrogen atoms. How many carbon atoms does this sample contain?
(c) Suppose you have a sample of $C_6H_{12}O_6$, and it contains a total of 72,000 hydrogen atoms. What is the mass in grams of the entire sample?

9.64 The balanced chemical equation $2\,AgBr \rightarrow 2\,Ag + Br_2$ and the formulas of the substances in it are a source of many conversion factors. Write all that are possible using the word *mole(s)*.

9.65 The following chemical equation is unbalanced:

$I_2 + Cl_2 \rightarrow ICl_3$

(a) Balance the equation.
(b) The balanced equation is a source of many conversion factors containing the two reactants and one product. Write all of them that use the word *mole(s)*.
(c) The balanced equation is also a source of conversion factors involving the individual atoms and molecules. Write all of them that use the word *mole(s)*.

9.66 How many H atoms are there in 2.0158 g of H atoms?

9.67 Consider a sample of 24.0 g of O_2 molecules.
(a) How many moles of O_2 molecules are present?
(b) How many moles of O atoms are present?

9.68 Consider a sample of 92.5 g of O_2 molecules.
(a) How many moles of O_2 molecules are present?
(b) How many moles of O atoms are present?
(c) How many O atoms are present, and how much larger should this be than the number of O_2 molecules present?

9.69 Consider sulfuric acid, H_2SO_4, used in car batteries.
(a) What is the molar mass of sulfuric acid?
(b) What is the mass in grams of 1 mole of H_2SO_4?
(c) What is the mass in grams of 2.50 moles of H_2SO_4?
(d) What is the mass in grams of 1000 molecules of H_2SO_4? (*Hint*: Start by writing "1000 molecules H_2SO_4" and then apply conversion factors. Remember our admonition: If in doubt, convert to moles.)

9.70 How many O_2 molecules are there in 1.00 g of O_2 molecules?

9.71 Suppose you wanted 1 billion (1.00×10^9) water molecules and you didn't have time to sit and count them out. How many grams of water would you need to get 1 billion water molecules?

9.72 How many grams of glucose ($C_6H_{12}O_6$) would you need to get 5.00×10^{30} carbon atoms?

9.73 Consider the balanced chemical equation

$2\,H_2O_2 \rightarrow 2\,H_2O + O_2$

(a) Given 20.0 g of H_2O_2 (hydrogen peroxide), how many grams of water will the reaction yield?
(b) How many grams of hydrogen peroxide would you need to get 20.0 g of water?
(c) How many grams of hydrogen peroxide would you need to get 20.0 g of O_2?

9.74 Consider the balanced chemical equation

$SCl_4 + 2\,H_2O \rightarrow SO_2 + 4\,HCl$

(a) How many grams of H_2O will react with 5.000 g of SCl_4?
(b) How many grams of SO_2 can you make from 10.00 g of H_2O?
(c) Suppose you react 5.000 g of SCl_4 with the amount of H_2O you calculated in part (a). How many grams of HCl will you form?

Dealing With a Limiting Reactant

9.75 Suppose you run the reaction $A + 2\,B \rightarrow C$ with B as the limiting reactant and all of the reactants and products are solids. You desire to sell pure product C. What problem are you going to encounter that you will have to spend money on to fix?

9.76 With 3 cups of flour, 3 cups of sugar, and unlimited amounts of everything else, how many cookies can you make from the recipe in Problem 9.36? Write down your method and cross out units that cancel.

9.77 If you have 25 eggs, 9 blocks of cream cheese, and 4 cups of sugar, how many cakes can you make using the recipe of Problem 9.37? Write down your method and cross out units that cancel.

9.78 Consider the following balanced chemical equation:

$2\,H_2 + O_2 \rightarrow 2\,H_2O$

(a) How many grams of water are formed from 5.00 g of H_2 and an excess amount of O_2?
(b) How many grams of O_2 do you need to produce 5.00 g of H_2O?
(c) Given 100.0 g of H_2, how many grams of O_2 are required to run the reaction in a stoichiometric fashion?
(d) What is the theoretical yield in grams of water upon combining 50.0 g of O_2 with an excess amount of H_2?
(e) Express the answer to part (d) in terms of the number of water molecules.

9.79 Consider the following unbalanced chemical equation:

$P + O_2 \rightarrow P_2O_5$

(a) How many grams of phosphorus (P) are required to react completely with 20.0 g of O_2?
(b) What is the theoretical yield in grams if you combine the amounts of reactants in part (a)?

9.80 Consider the following unbalanced chemical equation:

$H_2 + N_2 \rightarrow NH_3$

(a) To run this reaction in a balanced fashion, how much nitrogen is required if you start with 10.0 g of H_2?
(b) How many grams of ammonia (NH_3) will you produce if you run the reaction with the masses calculated in part (a)?
(c) How many molecules of ammonia will you produce?

9.81 A gaseous mixture containing 10.079 g of H_2 and 7.00 moles of Br_2 react to form HBr.
(a) Write a balanced chemical equation for this reaction.
(b) Which reactant is limiting?
(c) What is the theoretical yield for this reaction in moles?
(d) What is the theoretical yield for this reaction in grams?
(e) How many moles of excess reactant are left over at the end of the reaction?
(f) How many grams of excess reactant are left over at the end of the reaction?

9.82 Chlorine (Cl_2) and fluorine (F_2) react to form ClF_3. A reaction vessel contains 2.50 moles of Cl_2 and 6.15 moles of F_2.
(a) Write a balanced chemical equation for this reaction.
(b) Which reactant is limiting?
(c) What is the theoretical yield for this reaction in moles?
(d) What is the theoretical yield for this reaction in grams?
(e) How many moles of excess reactant are left over at the end of the reaction?
(f) How many grams of excess reactant are left over at the end of the reaction?

9.83 5.00 g of solid sodium (Na) and 30.0 g of liquid bromine (Br_2) react to form solid NaBr.
(a) Write a balanced chemical equation for this reaction.
(b) Which reactant is limiting?
(c) What is the theoretical yield for this reaction in grams?
(d) How many grams of excess reactant are left over at the end of the reaction?
(e) When this reaction is actually performed, 14.7 g of NaBr is recovered. What is the percent yield of the reaction?

9.84 Chlorine (Cl_2) and fluorine (F_2) react to form ClF_3. A reaction vessel contains 10.00 g Cl_2 and 10.00 g F_2. (*Hint*: Refer to Problem 9.82.)
(a) Write a balanced chemical equation for this reaction.
(b) Which reactant is limiting?
(c) What is the theoretical yield for this reaction in grams?
(d) How many grams of excess reactant are left over at the end of the reaction?
(e) When this reaction is actually performed, 12.50 g of ClF_3 is recovered. What is the percent yield of the reaction?

9.85 Sodium (Na) reacts with hydrogen (H_2) to form sodium hydride (NaH). A reaction mixture contains 10.00 g Na and 0.0235 g H_2.
(a) Write a balanced chemical equation for this reaction.
(b) Which reactant is limiting?
(c) What is the theoretical yield for this reaction in grams?
(d) How many grams of excess reactant are left over at the end of the reaction?
(e) When this reaction is actually performed, 0.428 g of NaH is recovered. What is the percent yield of the reaction?

9.86 Butane (C_4H_{10}), used as the fuel in disposable lighters, reacts with oxygen (O_2) to produce CO_2 and H_2O. Suppose 10.00 g butane is combined with 10.00 g O_2.
(a) Write a balanced chemical equation for the combustion of butane.
(b) Which reactant is limiting?
(c) What is the theoretical yield of each product for this reaction in grams?
(d) How many grams of excess reactant are left over at the end of the reaction?
(e) How many additional grams of the limiting reactant are required to run this reaction in a balanced fashion?

Combustion Analysis

9.87 A compound has the empirical formula C_2H_4O. Its molar mass is about 90 g/mol. What is its molecular formula?

9.88 An organic compound of carbon and hydrogen has the empirical formula CH. What is its molecular formula if its molar mass is:
(a) 26 g/mol
(b) 52 g/mol
(c) 78 g/mol

9.89 A 1.540-g sample of a liquid is subjected to combustion analysis, yielding 40.00% C and 6.71% H. It may also contain oxygen.
(a) What is the empirical formula for this compound?
(b) The molar mass of this compound is determined to be about 30 g/mol. What is the molecular formula for this compound?

9.90 A 2.230-g sample of a solid is subjected to combustion analysis, yielding 76.59% C and 6.39% H. It may also contain oxygen.
(a) What is the empirical formula for this compound?
(b) The molar mass of this compound is determined to be about 94 g/mol. What is the molecular formula for this compound?

9.91 A 1.000-g sample of a liquid is subjected to combustion analysis, yielding 92.3% C and 7.7% H. It may or may not also contain oxygen.
(a) What is the empirical formula for this compound?
(b) The molar mass of this compound is determined to be about 78 g/mol. What is the molecular formula for this compound?

9.92 A 2.000 g sample of a liquid compound that contains only carbon and hydrogen in its formula is subjected to combustion analysis. From the result, the lab determines that 0.2874 g of the original sample's mass is due to hydrogen.
(a) What is the percent by mass hydrogen for this sample?
(b) What is the empirical formula for this compound?
(c) The molar mass of this compound is determined to be about 71 g/mol. What is the molecular formula for this compound?

9.93 A compound used as an insecticide that contains only C, H, and Cl is subjected to combustion analysis, yielding 55.55% C and 3.15% H.
(a) What is the empirical formula of this compound?
(b) What is the molecular formula if its empirical formula is 1/3 the mass of its actual formula?

9.94 A compound used as an insecticide that contains only C, H, and Cl is subjected to combustion analysis, yielding 24.78% C and 2.08% H.
(a) What is the empirical formula for this compound?
(b) What is the molecular formula if its actual formula is four times the mass of its empirical formula?

9.95 Determine the empirical formula of the compound with the following mass percents of the elements present: 66.63% C; 11.18% H; 22.19% O.

9.96 Determine the empirical formula of the compound with the following mass percents of the elements present: 58.5% C; 4.91% H; 19.5% O; 17.1% N.

9.97 Determine the empirical formula of the compound that is 26.4% by mass Na, 36.8% by mass S, and also contains oxygen.

9.98 Determine the empirical formula of the compound that is 43.2% by mass K, 39.1% by mass Cl, and also contains oxygen.

Going from Molecular Formula to Percent Composition

9.99 Ethanol, the alcohol in beer and wine, has the molecular formula C_2H_6O. Calculate the mass percent of each element in ethanol.

9.100 Ethylene glycol, used for antifreeze, has the molecular formula $C_2H_6O_2$. Calculate the mass percent of each element in ethylene glycol.

9.101 The compound P_4O_{10} has an empirical formula of P_2O_5. By what factor will the percent by mass composition differ for each element between these two formulas? Explain your answer.

9.102 The thyroid hormone thyroxine has the molecular formula $C_{15}H_{11}NO_4I_4$. Calculate the mass percent of each element in thyroxine.

Additional Problems

9.103 (a) What is the molar mass of ribose ($C_5H_{10}O_5$)?
(b) What is the mass of 3.87 moles of ribose?
(c) How many ribose molecules are there in 3.87 moles?
(d) How many oxygen atoms are there in 3.87 moles of ribose?
(e) What is the mass in grams of the oxygen atoms in part (d)?

9.104 Which has more mass:
(a) One atom of sulfur or one molecule of oxygen gas (O_2)?
(b) 0.125 mole of sulfur molecules (S_8) or 0.670 mole of ozone (O_3)?

9.105 Can the actual yield ever be greater than the theoretical yield for a chemical reaction?

9.106 Sodium metal reacts with water to form aqueous sodium hydroxide and hydrogen gas.
(a) Write a balanced chemical equation for this reaction.
(b) Which reactant is limiting if 100.0 g of Na and 4.00 moles of water are used?

9.107 Which has the greatest mass: 1 mole of ethylene gas (C_2H_4), 1 mole of carbon monoxide gas, or 1 mole of nitrogen gas (N_2)?

9.108 What is the maximum possible value of the percent yield of a chemical reaction?

9.109 (a) What is the molar mass of sucrose ($C_{12}H_{22}O_{11}$)?
(b) What is the mass of 1.25 moles of sucrose?
(c) How many sucrose molecules are in 1.25 moles?
(d) How many hydrogen atoms are in 1.25 moles of sucrose?
(e) What is the mass in grams of the hydrogen atoms in part (d)?

9.110 Aluminum metal burns in chlorine gas to form aluminum chloride.
(a) Write a balanced chemical equation for this reaction.
(b) Which reactant is limiting if 100.0 g of Al and 5.00 moles of Cl_2 are used?

9.111 Copper(I) oxide reacts with solid carbon to form copper metal. Carbon dioxide gas is the other product of this reaction.
(a) Write the balanced chemical equation for this reaction.
(b) Coke is a cheap, impure form of solid carbon that is often used industrially. If a sample of coke is 95% C by mass, determine the mass in kilograms of coke needed to react completely with 1.000 ton of copper(I) oxide. [1 ton = 2000 lb; 1 kg = 2.205 lb]

9.112 What is the empirical formula of a compound that is 17.552% Na, 39.696% Cr, and 42.752% O?

9.113 Consider the balanced chemical equation

$$Fe(CO)_5(s) + 2\,PF_3(l) + H_2(g) \rightarrow Fe(CO)_2(PF_3)_2H_2(s) + 3\,CO(g)$$

(a) How many grams of CO could be produced from 10.0 g of PF_3, excess $Fe(CO)_5$, and excess H_2?
(b) How many grams of CO could be produced from 5.0 moles of $Fe(CO)_5$, 8.0 moles of PF_3, and 6.0 moles of H_2?
(c) How many moles of CO could be produced from 25.0 g of $Fe(CO)_5$, 10.0 g of PF_3, and excess H_2?
(d) The density of hydrogen gas at room temperature and atmospheric pressure is 0.0820 g/L. When 5.00 L of hydrogen gas at room temperature and atmospheric pressure is mixed with excess $Fe(CO)_5$ and excess PF_3, the mass of CO collected is 13.5 g. What is the theoretical yield (in grams) and the percent yield of CO?

9.114 (a) Write a balanced chemical equation for the combustion of $CH_4(g)$ with $O_2(g)$ to form $CO_2(g)$ and $H_2O(g)$.
(b) Consider the reaction of three molecules of $CH_4(g)$ with four molecules of $O_2(g)$. The following boxes represent before and after pictures of the tiny, sealed container in which this reaction takes place. Draw pictures to show the contents of this container *before* and *after* the reaction happens:

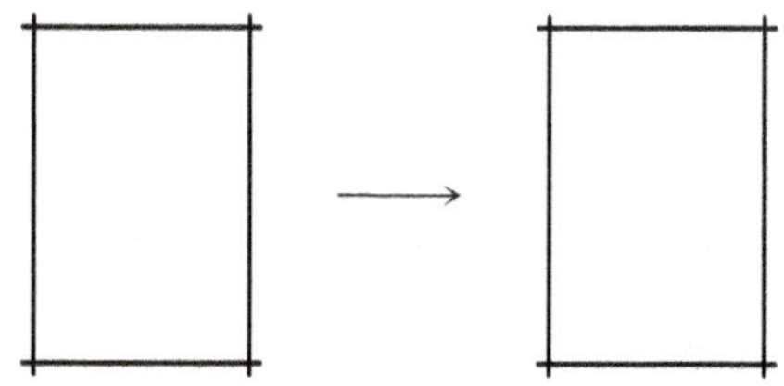

Mixture before reaction | Mixture after reaction

(c) Which is the limiting reactant, $CH_4(g)$ or $O_2(g)$?

9.115 Consider a 5.00-g sample of silver nitrate, $AgNO_3(s)$.
(a) How many moles of $AgNO_3$ are in this sample?
(b) How many moles of O are in this sample?
(c) How many grams of N are in this sample?
(d) How many Ag atoms are in this sample?

9.116 Nitrogen and fluorine react to form nitrogen trifluoride according to the balanced chemical equation

$$N_2(g) + 3\,F_2(g) \rightarrow 2\,NF_3(g)$$

For each of the following reaction mixtures, choose the limiting reactant:
(a) 0.50 mol $N_2(g)$ and 0.50 mol $F_2(g)$
(b) 12.0 mol $N_2(g)$ and 20.0 mol $F_2(g)$
(c) 2.5 mol $N_2(g)$ and 7.5 mol $F_2(g)$
(d) 100 molecules $N_2(g)$ and 500 molecules $F_2(g)$
(e) 5.00 g $N_2(g)$ and 15.0 g $F_2(g)$
(f) 20.0 mg $N_2(g)$ and 70.0 mg $F_2(g)$

9.117 For each part in Problem 9.116, calculate the amount of excess reagent left over at the end of each reaction.

9.118 When you bring your piggy bank of pennies into the bank, instead of counting them individually, the teller measures the mass of a single penny on a scale and then measures the mass of the entire contents of your piggy bank. If a single penny has a mass of 2.49 g and your entire penny collection has a mass of 5789.25 g, how much money in dollars do you receive?

9.119 For the reaction $Ca_3P_2 + H_2O \rightarrow Ca(OH)_2 + PH_3$,
(a) Balance the equation.
(b) How many grams of H_2O would you need to react with 60.0 g of Ca_3P_2?
(c) How many grams of PH_3 could theoretically be produced from the reactant amounts calculated in part (b)?

9.120 If you have 0.262 mole of iron(III) sulfate, how many oxygen atoms do you have?

9.121 Acetaminophen, used in many over-the-counter pain relievers, has a molecular formula of $C_8H_9NO_2$. Calculate the mass percentage of each element in acetaminophen.

9.122 Consider the unbalanced chemical equation

$KCl + O_2 \rightarrow KClO_3$

If you react 42.6 g of KCl and 36.5 g of O_2 and the reaction has a percent yield of 56.0%, how many grams of $KClO_3$ are produced?

9.123 Sucrose has the molecular formula $C_{12}H_{22}O_{11}$. If you were to completely burn 2.00 g of sucrose in a stream of oxygen, how many grams of CO_2 and H_2O would be produced?

9.124 If you have 44.6 g of carbon tetrachloride, how many atoms of chlorine do you have?

9.125 For the reaction $BF_3 + H_2O \rightarrow H_3BO_3 + HBF_4$,
(a) Balance the equation.
(b) If you react 24.2 g of BF_3 with an excess of water and generate 14.8 g of HBF_4, what is your percent yield?

9.126 The ingredients for making 1 dozen cupcakes are:

5 tablespoons butter
1 cup sugar
2 eggs
2 cups flour
1 cup milk

If you have only 5 eggs but plenty of all of the other ingredients:
(a) How many cupcakes can you make?
(b) How many cups of sugar do you need to make the number of cupcakes calculated in (a)?

9.127 Chrome yellow, a pigment used in paints, is 64.11% by mass Pb, 16.09% by mass Cr, and 19.80% by mass O. What is the empirical formula of this compound?

9.128 What is the mass of 1.62×10^{25} molecules of hydrogen sulfide?

9.129 Silicon nitride, Si_3N_4, is a ceramic material capable of withstanding high temperatures. It can be produced using the following unbalanced reaction:

$SiCl_4 + NH_3 \rightarrow Si_3N_4 + HCl$

If 64.2 g of $SiCl_4$ is reacted with 20.0 g of NH_3, how many grams of Si_3N_4 are produced if the reaction has a 96.0% yield?

9.130 Vitamin C, also known as ascorbic acid, contains carbon, hydrogen, and possibly oxygen. A 0.160 g sample of ascorbic acid is subjected to combustion analysis, yielding 40.93% C and 4.58% H. If the molar mass of ascorbic acid is approximately 176 g/mol, what is its molecular formula?

9.131 Consider the unbalanced chemical equation

$Cr_2O_3 + H_2S \rightarrow Cr_2S_3 + H_2O$

(a) How many grams of H_2S are needed to react completely with 13.6 g of Cr_2O_3?
(b) How many grams of Cr_2S_3 could theoretically be produced?

9.132 Determine the mass percent of each element in magnesium phosphate.

9.133 For the reaction $SiCl_4 + H_2O \rightarrow SiO_2 + HCl$,
(a) Balance the equation.
(b) If you want to make 120.0 g of HCl, how many grams of $SiCl_4$ and H_2O do you need?
(c) How many grams of SiO_2 are produced from the quantities calculated in (b)?

9.134 A compound is 91.77% by mass Si and 8.23% by mass H and has a molar mass of approximately 122 g/mol. What is its molecular formula?

9.135 Iron(III) oxide reacts with carbon monoxide to give iron metal and carbon dioxide. If you begin the reaction with 24.0 g of iron(III) oxide and 34.0 g of carbon monoxide, what is the theoretical yield in grams of carbon dioxide?

9.136 The compound ibuprofen, used in some pain relievers, has the molecular formula $C_{13}H_{18}O_2$. If 0.250 g of ibuprofen is burned in oxygen, how many grams of CO_2 and H_2O will be produced?

9.137 If you have 200.0 g of dinitrogen pentoxide, how many atoms of nitrogen do you have? How many atoms of oxygen do you have?

9.138 Potassium nitrate decomposes upon heating to form potassium oxide, nitrogen gas, and oxygen gas. If 19.6 g of potassium nitrate have decomposed, how many molecules of oxygen gas have been formed?

9.139 Tetraphosphorus decoxide, P_4O_{10}, reacts with water to form phosphoric acid. If 52.5 g of P_4O_{10} reacted with 25.0 g of water, how many grams of H_3PO_4 could theoretically be produced?

9.140 Determine the mass percent of each element in aluminum sulfate.

9.141 Consider the unbalanced chemical equation

$Cl_2O_7 + H_2O \rightarrow HClO_4$

The reaction is carried out at 82.0% yield and gives 52.8 g of $HClO_4$.
(a) What is the theoretical yield of $HClO_4$?
(b) How many grams of Cl_2O_7 and H_2O were consumed in the reaction?

9.142 The compound naphthalene, which is used in mothballs, contains carbon, hydrogen, and possibly oxygen. When 0.220 g of naphthalene is subjected to combustion analysis, it yields 93.66% C and 6.31% H. If the molar mass of naphthalene is approximately 128 g/mol, what is its molecular formula?

9.143 Ethylene gas (C_2H_4) reacts with fluorine gas (F_2) to form carbon tetrafluoride gas and hydrogen fluoride gas. If 2.78 g of ethylene reacted with an excess of fluorine, how many grams of each product could theoretically be produced?

9.144 How many grams of sulfur hexafluoride would you need to have 5.25×10^{24} fluorine atoms?

9.145 Consider the unbalanced chemical equation

$$HSbCl_4 + H_2S \rightarrow Sb_2S_3 + HCl$$

If 118.2 g of $HSbCl_4$ reacts with 47.9 g of H_2S and produces 41.6 g of HCl, what is the percent yield for the reaction?

9.146 Caffeine is made up of 49.48% by mass C, 5.19% by mass H, 28.85% by mass N, and 16.48% by mass O. If its molar mass is approximately 194 g/mol, what is the molecular formula of caffeine?

9.147 For the reaction $NH_3 + NO \rightarrow N_2 + H_2O$:
(a) Balance the equation.
(b) If you react 15.0 g NH_3 with 22.0 g NO and you produce 13.3 g N_2, what is your percent yield?

9.148 Nicotine has the formula $C_{10}H_{14}N_2$. Determine the mass percent of each element.

9.149 Consider the unbalanced chemical equation

$$CaC_2 + CO \rightarrow C + CaCO_3$$

When the reaction is complete, 135.4 g of $CaCO_3$ is produced and 38.5 g of CaC_2 is left over. Assuming the reaction had a 100% yield, what were the masses of the two reactants at the beginning of the reaction?

9.150 The beryllium mineral beryl is made up of 5.03% by mass Be, 10.04% by mass Al, 31.35% by mass Si, and 53.58% by mass O. If the molar mass is approximately 538 g/mol, what is the molecular formula for beryl?

9.151 Consider the unbalanced chemical equation

$$S + H_2SO_4 \rightarrow SO_2 + H_2O$$

(a) Balance the equation.
(b) If you react 4.80 g of sulfur with 16.20 g of H_2SO_4, how many grams of SO_2 can theoretically be produced?

9.152 The compound 3′-azido-3′-thymidine (AZT) is used in the treatment of AIDS. The molecular formula of AZT is $C_{10}H_{13}N_5O_4$. What is the mass percent of each element in AZT?

9.153 In a chemical reaction, the mass of the products is equal to the mass of the reactants consumed. Are the moles of product equal to the moles of reactant consumed? Explain your answer.

9.154 Consider the unbalanced chemical equation

$$N_2O_4 + N_2H_4 \rightarrow N_2 + H_2O$$

(a) Balance the equation.
(b) If 42.32 g of N_2 was produced and the reaction yield was 67.5%, how many grams of N_2O_4 and N_2H_4 were consumed in the reaction?

9.155 Ammonium fluoride reacts with calcium nitrate to produce calcium fluoride, dinitrogen monoxide, and water. If 22.8 g of ammonium fluoride reacts with 38.2 g of calcium nitrate, what is the theoretical yield in grams of each product?

9.156 Some calcium supplements contain calcium citrate, $Ca_3(C_6H_5O_7)_2$, and others contain calcium carbonate, $CaCO_3$. Which compound has the higher mass percent of calcium?

9.157 Lead(II) sulfide reacts with hydrogen peroxide to give lead(II) sulfate as shown in the unbalanced chemical equation

$$PbS + H_2O_2 \rightarrow PbSO_4 + H_2O$$

If 63.2 g of PbS is reacted with 48.0 g of H_2O_2,
(a) Which is the limiting reagent?
(b) How many grams of excess reactant remain after the reaction is complete?

9.158 The hormone estradiol contains carbon, hydrogen, and oxygen and has a molar mass of approximately 272 g/mol. When 1.15 g of estradiol is subjected to combustion analysis, the results yield 79.26% C and 8.88% H. What is its molecular formula?

9.159 Consider the unbalanced chemical equation

$$CaCN_2 + H_2O \rightarrow CaCO_3 + NH_3$$

If you began the reaction with 5.65 g of $CaCN_2$ and 12.2 g of H_2O, how many grams of NH_3 would be produced if the reaction had an 86.0% yield?

9.160 Saccharin, which is used as an artificial sweetener, is made up of 45.90% by mass C, 2.75% by mass H, 26.20% by mass O, 7.65% by mass N, and 17.50% by mass S. If the molar mass of saccharin is 183.19 g/mol, what is its molecular formula?

9.161 What is the mass in grams of 5.00×10^2 billion titanium atoms?

9.162 Copper reacts with nitric acid as shown in the balanced chemical equation

$$3\,Cu + 8\,HNO_3 \rightarrow 3\,Cu(NO_3)_2 + 2\,NO + 4\,H_2O$$

If you react 25.0 g of copper with an excess of nitric acid and produce 7.24 g of NO, what is the percent yield for the reaction?

9.163 The amino acid arginine is made up of 41.37% by mass C, 8.10% by mass H, 32.16% by mass N, and 18.37% by mass O. If the molar mass of arginine is approximately 174 g/mol, what is its molecular formula?

9.164 Consider the unbalanced chemical equation

$$As_4S_6 + O_2 \rightarrow As_4O_6 + SO_2$$

(a) How many grams of O_2 are needed to react completely with 58.9 g of As_4S_6?
(b) If 41.2 g of SO_2 is produced, what is the percent yield for the reaction?

9.165 Which contains a greater mass of copper, 11.0 g of copper(I) oxide or 12.6 g of copper(I) sulfide?

9.166 Consider the balanced chemical equation

$2\,A + B \rightarrow 2\,C + D$

When 8.0 g of A reacts completely with 6.0 g of B, 10.0 g of C and 4.0 g of D are produced. Assuming the yield is 100%,
(a) Which has a greater molar mass, A or C?
(b) Which has a greater molar mass, A or B?
(c) Which has a greater molar mass, A or D?
(d) If the molar mass of A is 24.0 g/mol, determine the molar mass of B, C, and D.

9.167 A sample of aluminum sulfide contains 4.25×10^{22} atoms of sulfur. What is the mass of the aluminum sulfide sample?

9.168 One of the most important industrial processes is the production of aluminum metal. The overall reaction is aluminum oxide reacting with carbon to give carbon dioxide and aluminum metal. If the reaction is performed at 94.7% yield and produces 284 kg of aluminum, what is the mass in kilograms of each reactant?

9.169 How many molecules of aspirin, $C_9H_8O_4$, would be in a tablet that contained 250 mg of aspirin? How many atoms of carbon would be in the aspirin in that tablet?

9.170 Calcium reacts with nitrogen gas to form calcium nitride. If 33.8 g of calcium react with 20.4 g of nitrogen gas,
(a) Which reactant is the limiting reagent?
(b) If the reaction has a 72.4% yield, how many grams of calcium nitride are formed?

9.171 The flavoring agent vanillin contains carbon, hydrogen, and possibly oxygen. When 0.450 g of vanillin is subjected to combustion analysis, the results are 63.08% C and 5.30% H. If the molar mass is approximately 152 g/mol, what is the molecular formula of vanillin?

9.172 Calcium hydroxide reacts with phosphoric acid to produce calcium phosphate and water.
(a) How many grams of phosphoric acid are needed to react completely with 34.6 g of calcium hydroxide?
(b) How many grams of calcium phosphate would theoretically be produced with the masses of part (a)?

9.173 Consider the unbalanced chemical equation

$Cr + S_8 \rightarrow Cr_2S_3$

If we need to produce 235.0 g of Cr_2S_3 and the reaction has a 63.80% yield, how many grams of each reactant do we need?

9.174 Consider the unbalanced chemical equation

$Al_2S_3 + H_2O \rightarrow Al(OH)_3 + H_2S$

If 56.0 g of aluminum sulfide reacts with 48.2 g of water,
(a) Which is the excess reactant?
(b) What mass in grams of the excess reactant remains after the reaction is complete?

9.175 Consider the following reaction. It is unbalanced.

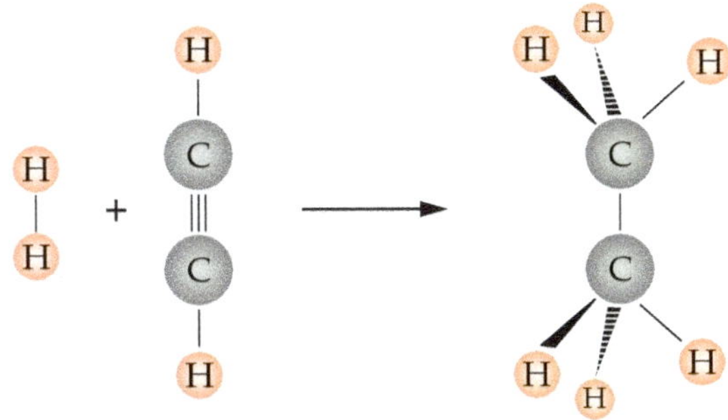

(a) Draw in another molecule(s) from among those already there that completes the reaction.
(b) Write a balanced equation using chemical formulas and balancing coefficients.
(c) Translate the reaction into an English statement using the word *moles* as many times as you can.
(d) Calculate how many grams of H_2 are required to make 10.0 g of product.

9.176 Consider the following decomposition reaction in which 47.20 g of some compound is decomposed into its elements.

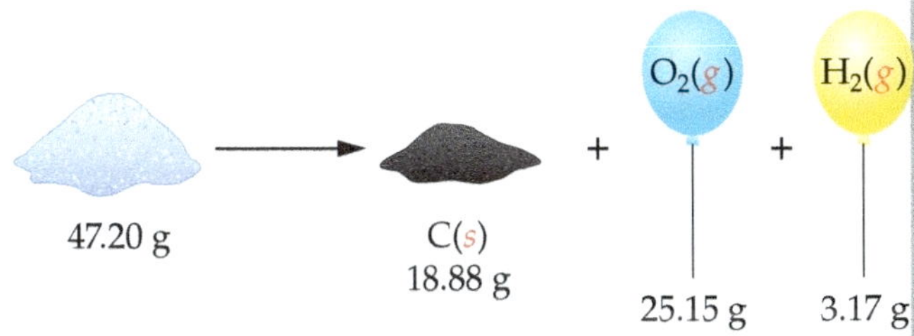

(a) What is the percent composition of each element in the original compound?
(b) What is the empirical formula of the compound?
(c) The molar mass of the compound is 60.05 g/mol. What is the actual formula of the compound?

9.177 If you take the ingredients below, in the amounts shown, you can make 12 dessert squares and you will have 2 eggs and 1 package of cream cheese left over.

(a) Write the recipe for dessert squares as a chemical reaction.
(b) How many dessert squares can you make if you double the number of eggs shown above?
(c) What is the limiting reagent?
(d) How many dessert squares can you make if you start with 637 cups of sugar, 2498 packages of cream cheese, and 4024 eggs? What will be left over?

9.178 The bicycle shown below has a formula of $frame_1tires_2$ and weighs 88.0 pounds (the frame includes the pedals, chain, and seat).

(a) Given that the mass percent of the frame is 75.0%, how much does each tire weigh and how much does the frame weigh?
(b) How many moles of bicycles would you need to obtain 1 mole of tires?
(c) If the frame were replaced with a carbon atom, and each tire was replaced by an oxygen atom, what would be the formula of the resulting compound?
(d) What is the molar mass of the above bicycle in pounds?

9.179 Consider the following unbalanced combustion reaction:

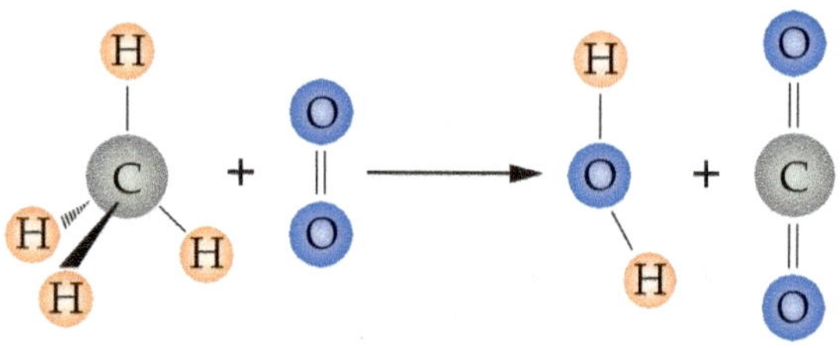

(a) Balance the equation.
(b) How many moles of CO_2 will be generated from using excess methane and 5.0 moles of O_2?
(c) How many molecules of CO_2 will be produced from using excess methane and 5.0 g of O_2?

One Step Beyond

9.180 When 490.0 mg of iron reacts with excess bromine, a mixture of $FeBr_2$ and $FeBr_3$ is produced. It is determined that 35.5% of the mixture is iron(II) (bromide). What is the total mass of $FeBr_2/FeBr_3$ mixture produced?

9.181 A mixture of methane (CH_4) and ethane (C_2H_6) is 25.6% methane by mass. When 1.50 g of this mixture is burned, what is the mass of carbon dioxide produced?

9.182 Succinic acid, an intermediate in the metabolism of certain foods, has a molecular mass of 118.1 g/mole. A 1.926 g sample of succinic acid (H_xSuc) reacts with exactly 1.25 g of NaOH according to the following balanced equation:

$H_xSuc + x\,NaOH \rightarrow Na_xSuc + x\,H_2O$

What is the value of x?

9.183 Ascorbic acid (otherwise known as vitamin C) prevents scurvy and may even help prevent certain cancers as well as the common cold. It is composed of 40.92% carbon (C), 4.58% hydrogen (H), and 54.50% oxygen (O) by mass. To the right is a ball-and-stick picture of ascorbic acid. What is the actual *molecular formula* of vitamin C?

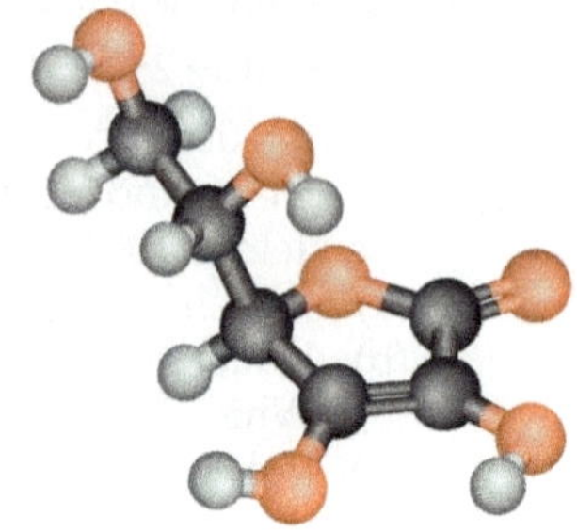

9.184 Due to its severity, bird flu, or avian flu, has caused great concern worldwide. The main drug for its treatment is Tamiflu. Tamiflu is an antiviral drug and acts to minimize the effects of bird flu by inhibiting the protein neuraminidase that lives on the surface of the flu virus. Analysis of Tamiflu shows it is composed of 61.5% carbon (C), 8.97% hydrogen (H), 8.93% nitrogen (N) and 20.6% oxygen (O) by mass. Below is a ball-and-stick picture of Tamiflu. What is the actual *formula* of Tamiflu?

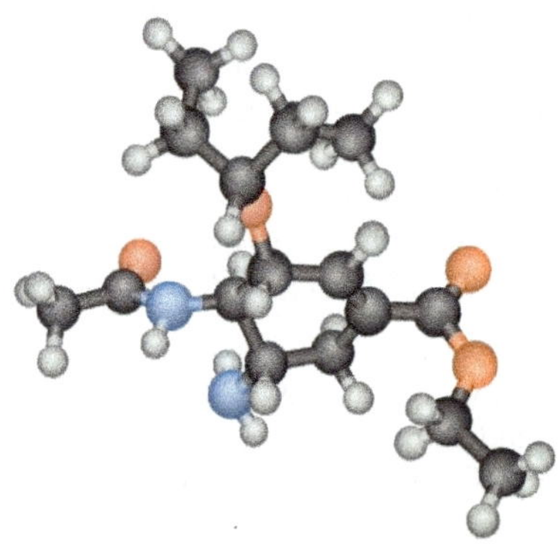

9.185 A 1.000-g sample of a liquid that contains only carbon and hydrogen burns in oxygen to produce 1.284 g of H_2O.
(a) What are the mass percents of the elements present in this sample?
(b) What is the empirical formula for this compound?
(c) The molar mass of this compound is determined to be about 71 g/mol. What is the molecular formula for this compound?

(*Hint*: When attempting this problem, understand that all of the carbon in the compound burned ends up as CO_2, and all of the hydrogen in the compound burned ends up as H_2O. Also, there is only 1 mole of C per mole of CO_2, but there are 2 moles of H per mole of H_2O.)

9.186 The semiconductor NiO, which consists of Ni^{2+} and O^{2-} ions, can be reacted with lithium to prepare "lithium-doped" $Ni_{1-x}Li_xO$. Suppose $x = 0.050$.
(a) How many lithium atoms would be present in one mole of the doped material? (Give your answer in moles and as an absolute number of atoms.)

(b) How is it possible to have fractional subscripts?
(c) In order to maintain charge neutrality, something happens to x moles of the Ni atoms. Describe what happens.
(d) How many lithium atoms are present in 10.0 g of the doped semiconductor?

Group Exercises (Work on these exercises together with someone else in the class.)

1. The word *mole* comes from the Latin word *moles*, which means "heap" or "pile." Like the word *dozen*, a mole has a numerical value (6.022×10^{23}) attached to it. Explain where this number comes from and how the ability to count individual atoms would help you to determine its value. Then, if you could count a million atoms per second, calculate how long in years it would take you to count a mole of atoms.

2. Balanced chemical reactions can be thought of as recipes. Write a short paragraph that explains why. Then, in the next paragraph, write down why such recipes can be thought of as being written in a different language from the one we speak. Finally, explain what does the translating between languages so that we can use the recipes. Finally, state how kitchen recipes are fundamentally different from balanced chemical reaction recipes.

Workpatch Solutions

9.1 (a) 196.967 g/mol (which is the atomic mass expressed in grams per mole).
(b) Because 1 mole of gold has a mass of 196.967 g and you have twice this (393.934 g), you have 2 moles of gold.
(c) You know that 1 mole of gold contains 6.022×10^{23} gold atoms, so 2 moles must contain $2 \times (6.022 \times 10^{23})$ atoms $= 1.204 \times 10^{24}$ atoms.

9.2 (a)

$$1 \text{ mol } CH_4 \text{ molecules} \times \underbrace{\frac{1 \text{ mol C atoms}}{1 \text{ mol } CH_4 \text{ molecules}}}_{\text{Conversion factor from formula } CH_4} = 1 \text{ mol C atoms}$$

(b)

$$1 \text{ mol } CH_4 \text{ molecules} \times \underbrace{\frac{4 \text{ mol H atoms}}{1 \text{ mol } CH_4 \text{ molecules}}}_{\text{Conversion factor from formula } CH_4} = 4 \text{ mol H atoms}$$

(c)

$$1 \text{ mol C atoms} \times \frac{6.022 \times 10^{23} \text{ C atoms}}{1 \text{ mol C atoms}} = 6.022 \times 10^{23} \text{ C atoms}$$

(d)

$$4 \text{ mol H atoms} \times \frac{6.022 \times 10^{23} \text{ H atoms}}{1 \text{ mol H atoms}} = 2.409 \times 10^{24} \text{ H atoms}$$

9.3 (a) Four moles of aluminum atoms and 3 moles of oxygen molecules react to give 2 moles of aluminum oxide molecules.

(b)

$$\frac{4 \text{ mol Al}}{3 \text{ mol } O_2} \quad \frac{3 \text{ mol } O_2}{4 \text{ mol Al}}$$

$$\frac{4 \text{ mol Al}}{2 \text{ mol } Al_2O_3} \quad \frac{2 \text{ mol } Al_2O_3}{4 \text{ mol Al}}$$

$$\frac{3 \text{ mol } O_2}{2 \text{ mol } Al_2O_3} \quad \frac{2 \text{ mol } Al_2O_3}{3 \text{ mol } O_2}$$

(c)

$$\frac{2 \text{ mol Al}}{1 \text{ mol } Al_2O_3} \quad \frac{1 \text{ mol } Al_2O_3}{2 \text{ mol Al}}$$

$$\frac{3 \text{ mol O}}{1 \text{ mol } Al_2O_3} \quad \frac{1 \text{ mol } Al_2O_3}{3 \text{ mol O}}$$

9.4 **Preliminary** $H_2 + I_2 \rightarrow 2\,HI$
? g (under I_2) 10.0 g (under 2 HI)

Step 1. $H_2 + I_2 \rightarrow 2\,HI$

Step 2. Mass of product HI known; convert to moles:

$$10.0 \text{ g HI} \times \frac{1 \text{ mol HI}}{127.913 \text{ g HI}}$$

Step 3. One mole of I_2 required to make 2 moles of HI:

$$10.0 \text{ g HI} \times \frac{1 \text{ mol HI}}{127.913 \text{ g HI}} \times \frac{1 \text{ mol } I_2}{2 \text{ mol HI}}$$

Step 4. Convert from moles of I_2 to mass of I_2:

$$10.0 \text{ g HI} \times \frac{1 \text{ mol HI}}{127.913 \text{ g HI}} \times \frac{1 \text{ mol } I_2}{2 \text{ mol HI}} \times \frac{253.810 \text{ g } I_2}{1 \text{ mol } I_2} = 9.92 \text{ g } I_2$$

solvent separation
ΔE solute separation
system energy

CHAPTER 12

Solutions

"Greetings, ugly bags of mostly water" (the greeting given to crew members of the starship Enterprise by an inorganic, crystalline life form in a first-season episode of *Star Trek: The Next Generation*).

A human body can be up to 78% water by mass, depending on body type. Your brain is about 70% water by mass, literally an aqueous computer; your lung tissue is 90% water by mass. Life as we know it cannot exist without water. The trillions upon trillions of chemical reactions that take place in your body and mind every minute take place in water. Approximately 70% of our planet is covered in water, amounting to an estimate of 326 million trillion gallons. So many things critical to the processes of life dissolve in water that it is often called the *universal solvent*. So you can imagine why astronomers recently became quite excited with the discovery of planet GJ 1214b in the constellation Ophiuchus, far outside our solar system. Indirect observations (for example, how its gravity affects the star) give strong support for the existence of liquid water on this planet. Does life exist there? Maybe. For life to exist, at least as we know it, it also would help for the planet's orbit to be in the star's so-called habitable (or "Goldilocks") zone, a range of distance from the star that allows the planet to be "just right," neither too hot nor too cold. Earth lies smack in the middle of our Sun's Goldilocks zone. Planet GJ 1214b is a little too close to its star, but who knows. Maybe, just maybe, someone there is measuring the effect of our planet's gravity on our star, using brains powered by chemical reactions taking place in water-based solutions in a mostly aqueous brain.

Artist rendition of planet GJ 1214b circling a dim red star called GJ 1214, just 40 light-years away in the constellation Ophiuchus (Permission granted by: David Aguilar).

12.1 What Is a Solution?

The next time you're at the supermarket, take a look at the detergent aisle. You'll find a huge array of products, all claiming that when added to water, they have the muscle to remove dirt and stains. Actually, water by itself does a pretty good job on most household dirt and stains. Consider a puddle of grape juice that has dried on a kitchen counter. Strenuous wiping with a dry cloth will generally not remove the dried juice, but a gentle wipe with a water-dampened cloth does the trick. The purple color on the cloth reveals that the juice has dissolved in the water. It appears that water's stain-removing muscle is due to its ability to dissolve things.

Of course, there are some stains that water alone will not remove. Recall, for instance, that salad-oil stain on your new T-shirt. In this chapter, we are going to answer two questions: (1) What exactly happens when a substance dissolves? and (2) Why is water so good at dissolving some things and so poor at dissolving others?

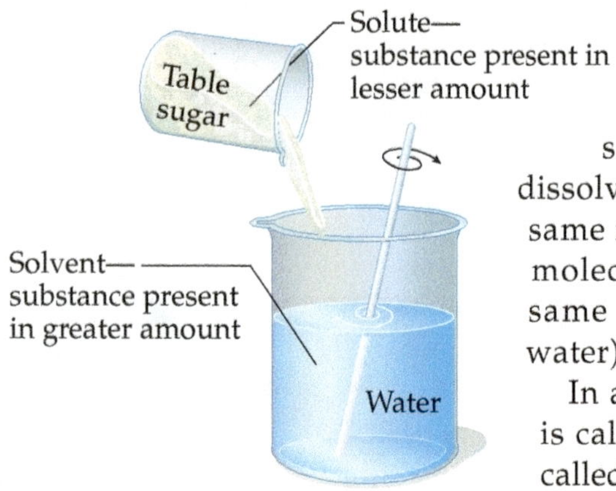

A **solution**, as we saw in Chapter 1, is a homogeneous mixture of two or more substances. By *homogeneous* we mean that the composition is the same throughout the solution. This means that when table sugar, for example, is dissolved in water, every tiny volume of the solution contains the same number of water molecules and the same number of sugar molecules. Another way to say this is that the solution has the same *concentration* (the same amount of sugar per amount of water) everywhere.

In any solution, the substance present in the greatest amount is called the **solvent** and all other substances in the mixture are called **solutes**. Thus, in our sugar–water example, water is the solvent and sugar the solute. Knowing this, we can now refine our definition to say that a solution is a homogeneous mixture of one or more solutes dissolved in a solvent. It is also important to note that once a solution forms, it will not of its own accord separate back into its components.

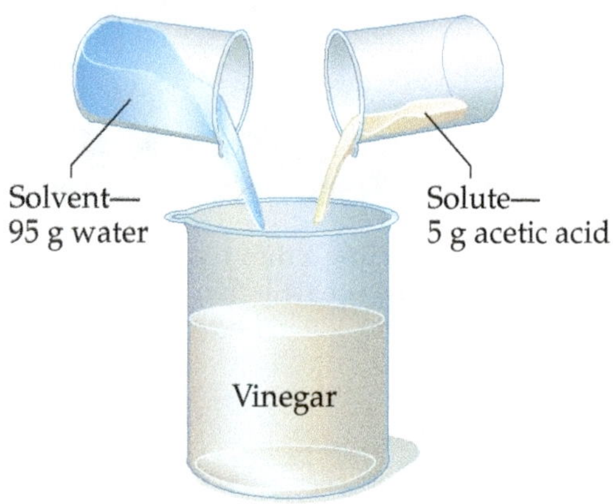

Solution of a liquid in a liquid

The word *solution* often brings to mind the image of a solid dissolved in a liquid, but not all solutions are like that. Solutions can also be combinations of liquids. Vinegar, for example, is a solution of a liquid called acetic acid mixed with water. Because vinegar is mostly water, water is the solvent and acetic acid the solute.

Solutions can be more exotic yet. Many metal alloys, for instance, are *solid solutions* consisting of one metal dissolved in another. The solutions are made when the metals are molten (liquid). For example, sterling silver is a solid solution consisting of a small amount of copper (the solute, Cu) dissolved in silver (the solvent, Ag). If the copper atoms are not evenly distributed among the silver atoms, it is not a solution and the result is not sterling silver.

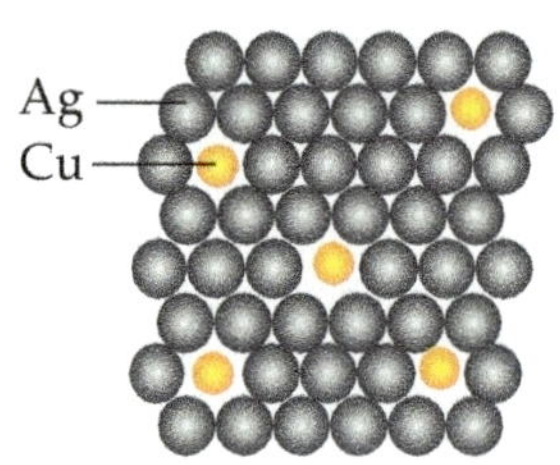

Sterling silver, a solid solution. The Cu atoms are evenly distributed among the Ag atoms.

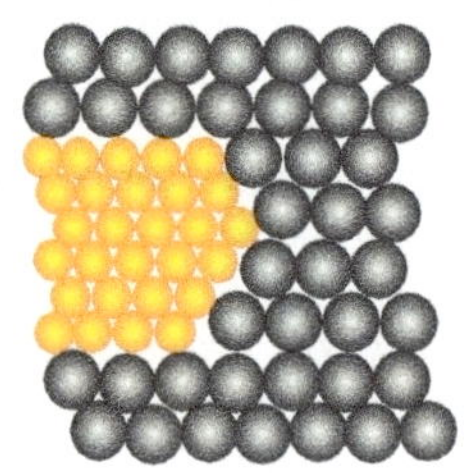

A heterogeneous mixture of copper and silver. This is not a solid solution because the Cu atoms are not evenly distributed among the Ag atoms.

As mentioned in Chapter 1, the air you are breathing is a solution of gases, primarily O_2 dissolved in N_2. Nitrogen, at 78%, is considered the solvent, and oxygen (21%) is the principal solute; other solute gases are also present in smaller amounts.

In most of the solutions we consider in this chapter, water is the solvent. Solutions in which water is the solvent are called **aqueous solutions** (from the Latin *aqua*, "water").

In ending this opening section, we want to look once more at the meaning of *homogeneous*. This topic was covered in Chapter 1, but it is worth another look.

You could put silicon dioxide (sand), SiO_2, sodium carbonate, Na_2CO_3, and calcium oxide, CaO, in a mortar and grind the mixture to a fine, pulverized dust. Even if you did this for a thousand years, the result would not be a true solution. It would be a heterogeneous mixture because there would still be tiny grains of pure silicon dioxide, pure sodium carbonate, and pure calcium oxide. Only by melting the components and stirring well would you achieve a molten solution that would become a solid solution (glass) when you allow it to cool and solidify. The same is true for a mixture of table salt and sugar. No amount of grinding these two solids together will yield a true solution, but dissolve them both in the same beaker of water and you get a homogeneous mixture—a solution—containing two solutes. Note that in both of these examples, we needed a liquid phase to achieve homogeneity. Simply grinding two solids together just doesn't do the job. In short, when dealing with solutions that involve solids, the presence of a liquid component makes homogeneity much easier to achieve. (You can also mix different gases together and get a homogeneous mixture—a solution—as we illustrated with air.)

WORKPATCH 12.1 From what you learned about the different phases of matter in Chapter 10, why do you think it is necessary to employ the liquid or gas phase to obtain the homogeneity needed to form a solution? ●

The models for liquids and gases used in Chapters 7 and 11 should help you understand what it takes to create a solution. The answer to WorkPatch 12.1 illustrates that chemistry works like a jigsaw puzzle, with the concepts and models you learn at different points fitting together to give you a better understanding of the whole.

Practice Problems

12.1 Classify the following as solutions or heterogeneous mixtures:

(a) A hot cup of instant coffee
(b) Chicken vegetable soup
(c) Unfiltered blood
(d) Filtered blood plasma
(e) A chromium-plated steel automobile bumper
(f) A stainless steel automobile bumper (stainless steel is prepared by combining iron, Fe, up to 30% chromium, Cr, and smaller amounts of nickel, Ni, and carbon, C, and heating the mixture until it becomes molten)

Answers: (a), (d), and (f) are solutions. (b), (c), and (e) are heterogeneous mixtures. (Unfiltered blood contains solids that separate out when the mixture is left undisturbed.)

12.2 Identify the solvent and solute or solutes in each solution:

(a) Nail-polish remover (30% acetone in water)
(b) Humid air
(c) Stainless steel (see Practice Problem 12.1)
(d) Aqueous solution of aspirin

12.3 The "proof" value of any liquor is equal to twice the percentage of alcohol; for example, a 50-proof liquor is 25% alcohol in water. Vodka is normally sold between 80 and 100 proof, but suppose you came across a bottle of 135-proof vodka. Would you be justified in calling the alcohol the solvent and the water the solute? Explain.

12.2 Energy and the Formation of Solutions

Liquid water is an excellent solvent, capable of dissolving many substances (solutes). The human body is approximately 80% water. Blood and cellular liquids are mostly water with small amounts of dissolved solutes (salts, gases, proteins, enzymes, and so forth). So many different substances are soluble in water that it is often referred to as a *universal solvent*.

What is it about water that makes it such a good solvent? To answer this question, let's reexamine the water molecule. We've seen that it is a polar molecule, a consequence of both its bent shape and of the large difference in electronegativity between its oxygen and hydrogen atoms:

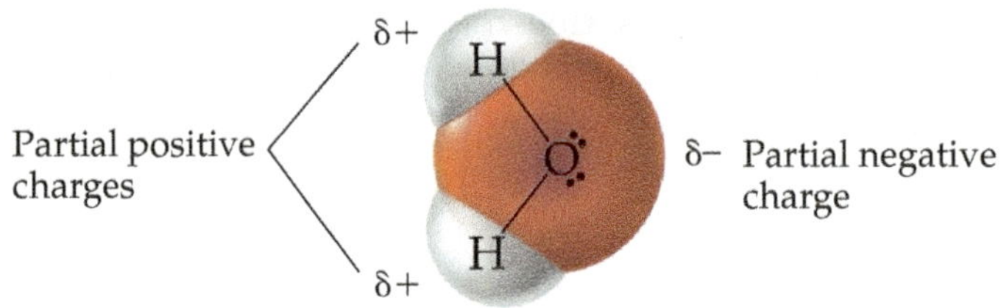

This polarity combined with the almost-naked-proton character of the H atoms in the molecule lead to formation of exceptionally strong dipolar attractions called hydrogen bonds (Chapter 7).

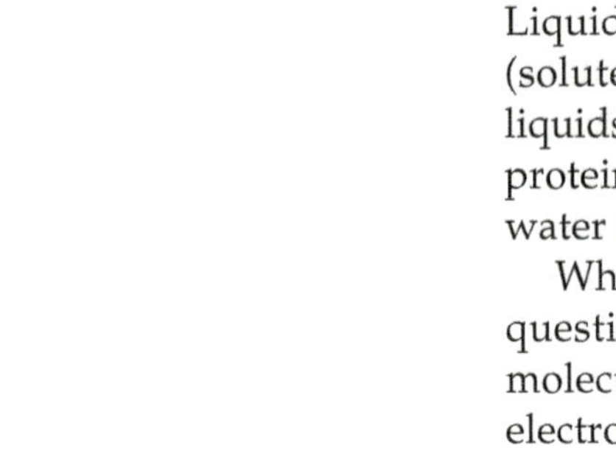

We saw in Chapter 7 that water molecules in the liquid state are in constant motion, moving in random directions but always strongly attracted to one another, always touching one another as they jostle past each other. Remember the analogy of a crowd during a sale in a shopping mall?

Now let's consider what happens when a solute is dissolved in the water. Because the ionic compound sodium chloride, NaCl, dissolves easily in water, we'll use it as our solute. We've seen that solid sodium chloride crystals are made up of relatively stationary Na^+ and Cl^- ions in a repeating three-dimensional pattern called a lattice. This lattice is held together by ionic bonds. This is why solid NaCl has to be heated to the extraordinarily high temperature of 801 °C before it will melt, becoming a liquid of mobile Na^+ and Cl^- ions. Yet, when we put this same compound in water at room temperature, the lattice comes apart easily to produce a solution of mobile Na^+ and Cl^- ions in water.

How can water break the ionic bonds of the crystal at room temperature (22 °C) when it takes 801 °C to do this without the water? To answer this question, we need to break the dissolving process down into its elementary steps.

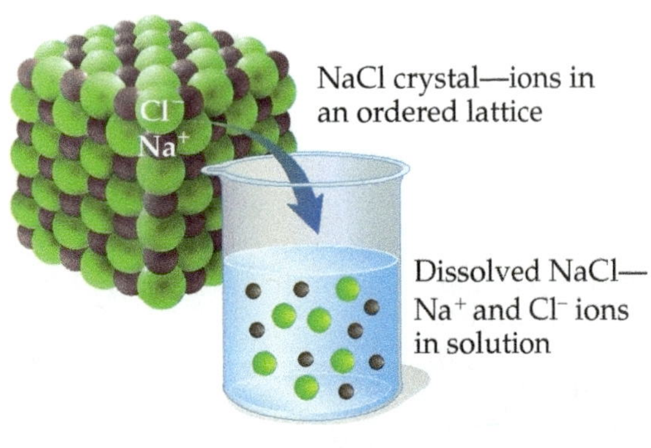

Three steps occur simultaneously as NaCl dissolves in water. (For the sake of clarity, we now talk about these steps one at a time, but always remember—they take place *simultaneously*.) First, the NaCl lattice has to be broken apart into individual Na^+ and Cl^- ions. We'll call this the *solute-separation step*. For NaCl, this breaking apart means overcoming the attractive forces (ionic bonds) between the ions in the lattice. Because ionic bonds are strong, this step requires an input of energy.

Second, spaces must be opened up in the solvent to make room for the solute particles. We'll call this the

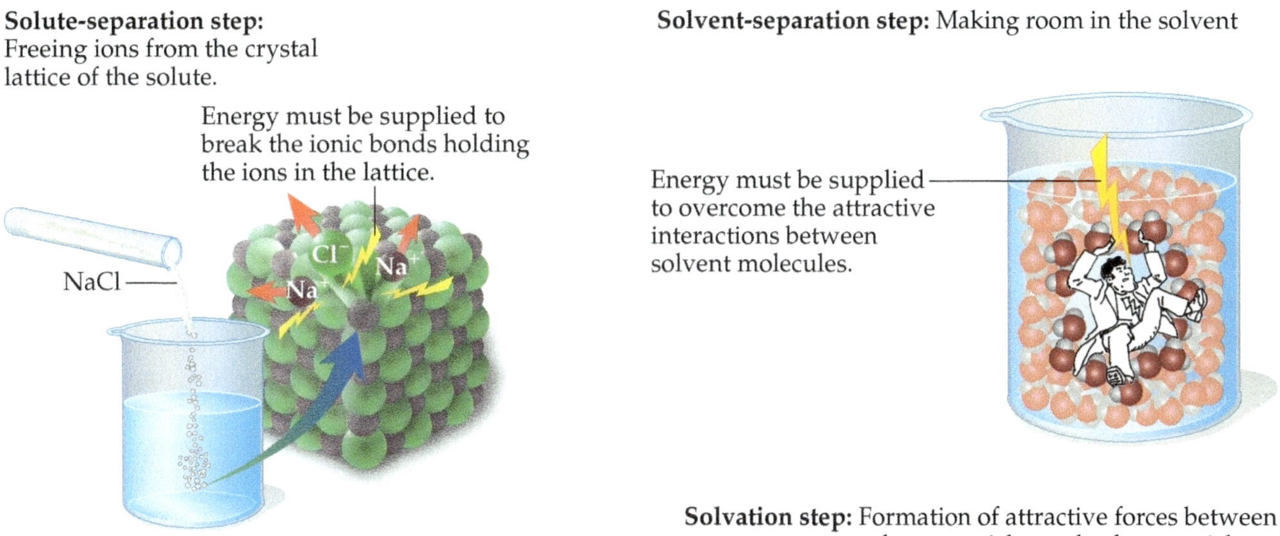

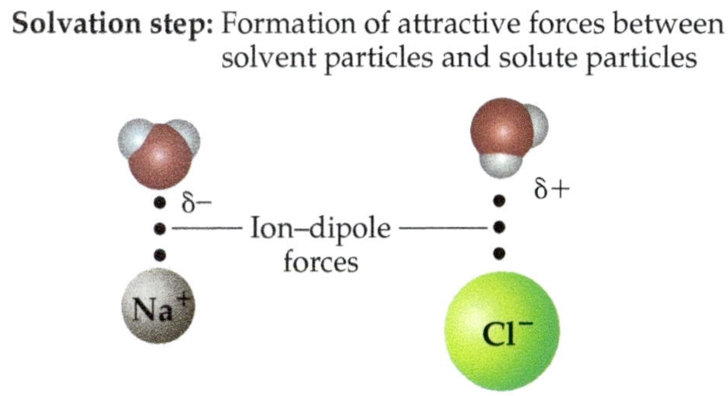

solvent-separation step. Recall that the solvent molecules are attracted to one another and touching one another. For a solute particle to dissolve, therefore, the solvent molecules have to move away from one another. Because this moving apart requires overcoming the attractions between solvent molecules, this step also requires an input of energy.

The third simultaneous step—called *solvation*—places the individual Na^+ and Cl^- ions into the "holes" or spaces created in the solvent, often called a **solvent cage**. Now the charged solute ions can interact with the polar solvent molecules that surround them, resulting in attractive forces called **ion–dipole forces**. Energy is *released* in this step because energy is released anytime an attractive force develops.

WORKPATCH 12.2 In the above figure, the water molecules are drawn so that an oxygen atom faces the Na^+ ion and a hydrogen atom faces the Cl^- ion. Explain why. ●

The WorkPatch answer explains why the Na^+ and Cl^- ions are attracted to the water molecules. In fact, there is room for more than one water molecule about each ion. Both Cl^- and Na^+ surround themselves with up to six water molecules, making for many ion–dipole attractions per ion:

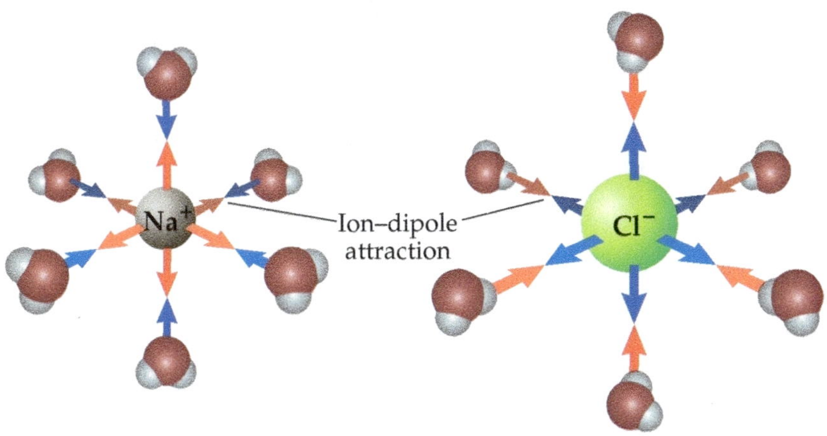

This means that each Cl^- ion and each Na^+ ion in the solution is surrounded by a shell of water molecules. The surrounding of solute ions by many solvent molecules along with the development of attractive solvent–solute forces is called **solvation** (hence the name of this step). When the solvent is water, this phenomenon can also be called **hydration**, so this step can be called the *hydration step* as well. When we write the equation

$$NaCl(s) \xrightarrow{H_2O} Na^+(aq) + Cl^-(aq)$$

the symbol (*aq*), standing for "aqueous," reminds us that each dissolved ion is surrounded by and attracted to many water molecules. The ions have been solvated (hydrated).

The hydration step is different from the other two. Solute separation and solvent separation require overcoming attractive forces and thus require an input of energy. However, energy is released anytime attractive interactions form, which means energy is released as the ions are hydrated. The energy released by this hydration is called—you guessed it!—**hydration energy**. The fact that this step in the dissolving process releases energy is very important because it supplies the bulk of the energy required by the other two steps. However, hydration can supply sufficient energy only if the solvent–solute attractive forces formed are strong enough—that is, only if their formation liberates enough energy to do the job.

We can now finally address the question "When does a solute dissolve in a solvent?" The answer comes from looking at the relative magnitudes of the energies involved in the three steps. The short answer is that the hydration energy released in the third step must be large enough to supply the energy needs of the first two steps. The following problems will give you some practice in determining the magnitude of the hydration energy. As you work them, keep Coulomb's law in mind:

> **When two objects of opposite charge attract each other, the attractive force between them increases as the magnitude of the charges increases and as the distance between the objects decreases.**

Coulomb's law is very useful for explaining many phenomena, including hydration energy.

Practice Problems

12.4 Consider the ionic compound magnesium chloride, $MgCl_2$. Do you think the hydration energy for this compound is greater than, less than, or about equal to that of NaCl?

Answer: The hydration energy released for $MgCl_2$ is greater than that for NaCl because $MgCl_2$ contains Mg^{2+} ions. The attractive force between oppositely charged particles increases as the magnitude of the charges increases. The +2 charge, which is greater than the +1 charge of the Na^+ ions in NaCl, leads to stronger ion–dipole attractions with water molecules, releasing more hydration energy than for NaCl.

12.5 Imagine you are trying to dissolve NaCl in liquid carbon tetrachloride, CCl_4. Would the energy released in the solvation step be greater than, less than, or about equal to that released when NaCl dissolves in water? Explain.

12.6 Based on the answer to Practice Problem 12.4, is the energy required for solute separation greater for NaCl or $MgCl_2$? Explain.

Very often the combination of a solute (or solutes) and a solvent is referred to as a *system*. In considering the dissolving process for any particular solvent/solute combination, what is important is how the energy of the system changes. By examining the change in energy at each step, we can determine whether a solute dissolves in a solvent. The symbol we use to represent the system's energy changes is ΔE. (Recall from Section 5.6 that the Greek letter delta, Δ, means "change in.")

When we supply energy to a system (as we do for the solute-separation and solvent-separation steps), the energy of the system increases, and we represent this increase with a positive ΔE value. When the system releases energy (as it does in the hydration step), the energy of the system decreases, and we represent this decrease with a negative ΔE value. In the following energy analysis for forming a solution, we use a schematic diagram to represent the process, with the overall process represented by two filled circles. The diagram to the right represents the entire dissolution process. The next three diagrams that follow illustrate each step in the process.

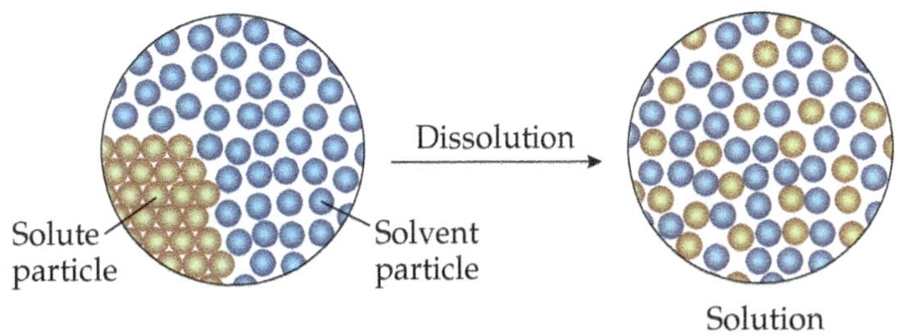

Solute-separation step: Here we must overcome the solute–solute attractive forces. Because this takes energy, we must feed energy into the system and therefore ΔE is positive. Note that if the solute is a liquid, the energy requirement is considerably less than that for an ionic solid (where the lattice must be separated into ions); for a gas, the energy needed in this step is essentially zero.

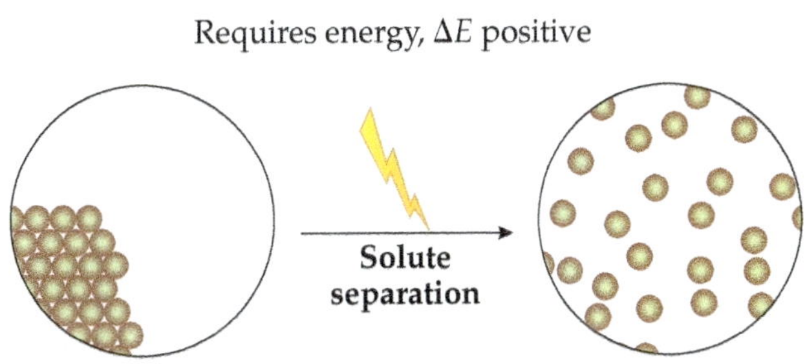

Solvent-separation step: We must overcome the solvent–solvent attractive forces and pry solvent particles apart to make room for solute particles. This takes energy, which again means ΔE is positive.

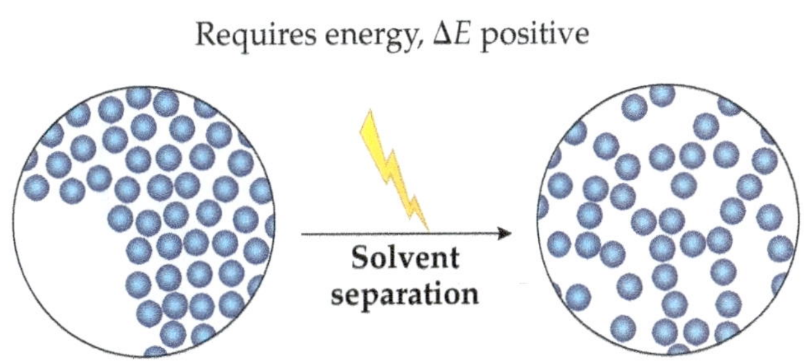

Solvation step: As solute–solvent attractive forces develop, energy is released, making ΔE negative.

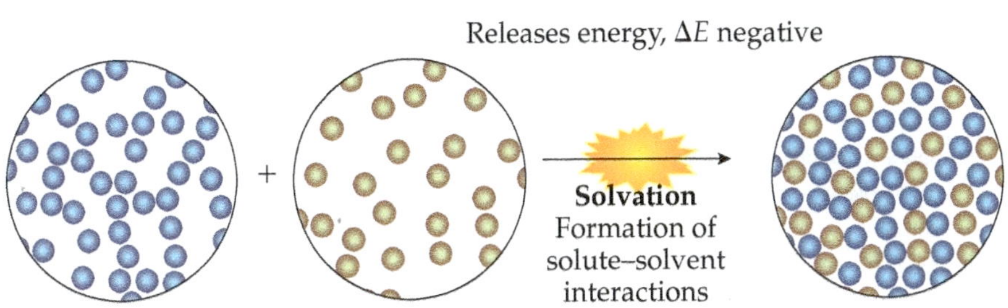

The combined energy changes are represented on an energy diagram:

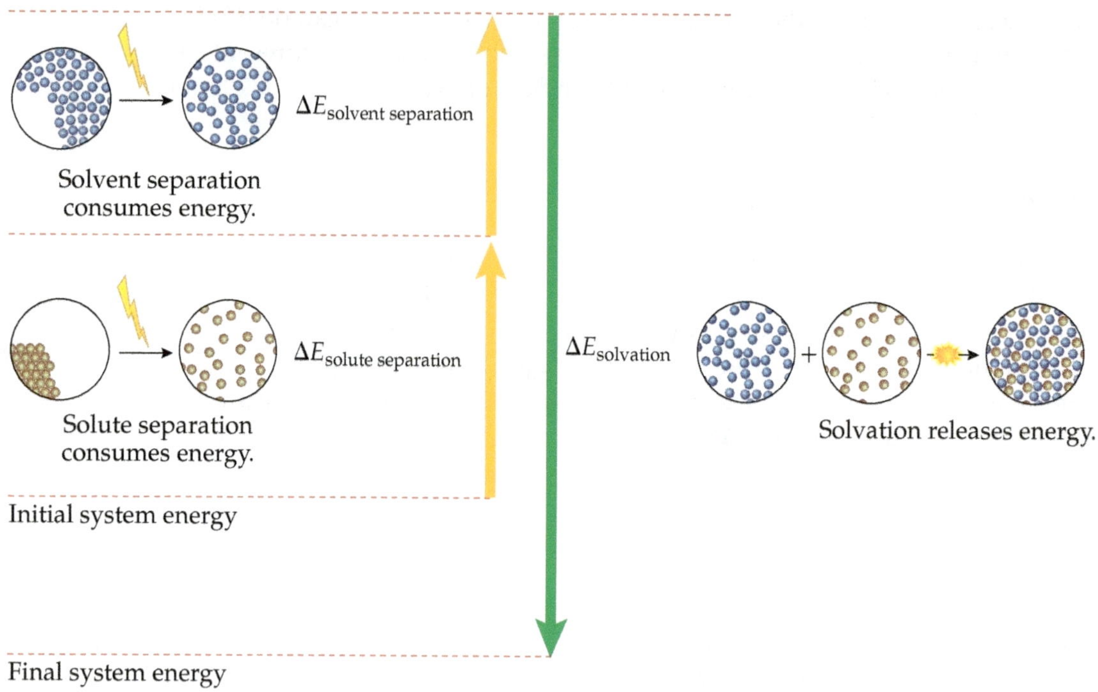

Whether or not a solute dissolves in a solvent depends on the total energy change for the process. This total energy change is the algebraic sum of the energy changes for the three steps:

$$\Delta E_{\text{total}} = \Delta E_{\text{solute separation}} + \Delta E_{\text{solvent separation}} + \Delta E_{\text{solvation}}$$

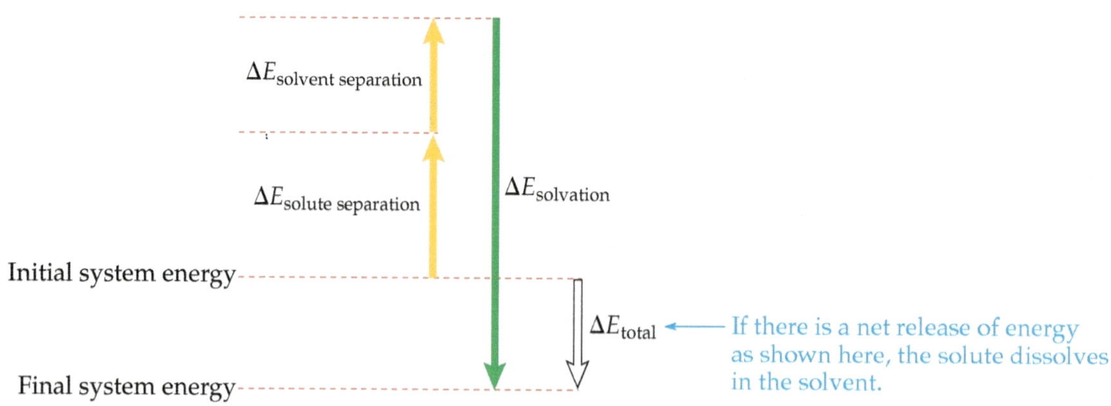

For example, when 1 mole of LiCl dissolves in water (shown at left),

$$\Delta E_{\text{solute separation}} = +853 \text{ kJ}$$
$$\Delta E_{\text{solvent separation}} = +96 \text{ kJ}$$
$$\Delta E_{\text{solvation}} = -986 \text{ kJ}$$
$$\Delta E_{\text{total}} = 853 \text{ kJ} + 96 \text{ kJ} + (-986 \text{ kJ}) = -37 \text{ kJ}$$

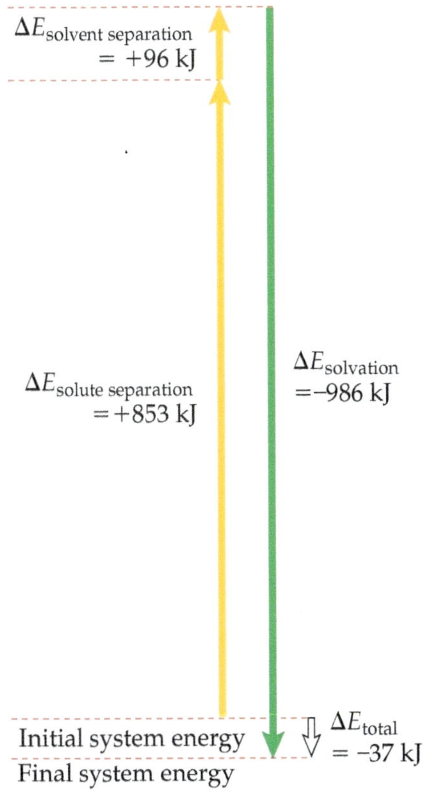

We get a negative value for ΔE_{total}, which means that, overall, energy is released. In other words, the energy produced in the solvation step is more than the energy required in the solute-separation and solvent-separation steps combined. In this case, the solvation step provides plenty of energy to drive the solute-separation and solvent-separation steps. In fact, solvation provides 37 kJ more energy than is needed. As a general rule, we can say

If $\Delta E_{\text{total}} < 0$, the substance will dissolve.

Dissolving LiCl in water results in an energy surplus, but this is not always the case. Sometimes the energy it takes for the solute-separation and solvent-separation steps is greater than the energy released during the solvation step. The result is an energy deficit, which means that solvation can't feed enough energy to the two separation steps to make them occur. The result is that ΔE_{total} is positive (overall, energy is absorbed), and the substance probably does not dissolve. As a general rule, we can say

If $\Delta E_{\text{total}} > 0$, the substance will probably not dissolve.

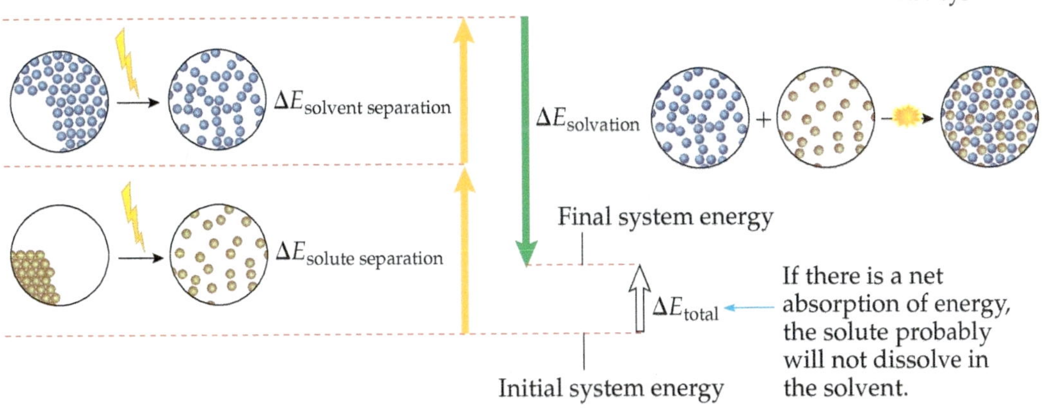

Why "probably"? We shall answer that question in Section 12.4.

Practice Problems

12.7 For a given solute in water, the energy changes are $\Delta E_{\text{solute separation}} = 835$ kJ, $\Delta E_{\text{solvent separation}} = 98$ kJ, and $\Delta E_{\text{solvation}} = -805$ kJ. Will this solute dissolve in water? Explain your answer, both numerically and in terms of the three-step model for a solute's dissolving in a solvent.

Answer: $\Delta E_{\text{total}} = +835\text{ kJ} + 98\text{ kJ} + (-805\text{ kJ}) = +128\text{ kJ}$ *Because* ΔE_{total} *is positive (there is an energy deficit), the solute probably will not dissolve. The first two steps take energy (both energies are positive). The third step releases energy, but not enough.*

12.8 Suppose we want to dissolve a gaseous solute in water. Would you expect $\Delta E_{\text{solute separation}}$ to be larger for the gaseous solute or for a solid ionic solute? Explain.

12.9 The more negative $\Delta E_{\text{solvation}}$ is, the more likely a solute will dissolve. Explain.

12.10 Suppose you want to dissolve some $MgCl_2$ in water.

(a) How do we know that magnesium cations have a +2 charge?
(b) $\Delta E_{\text{solute separation}}$ for $MgCl_2$ is much more positive than $\Delta E_{\text{solute separation}}$ for NaCl. Explain what this means and why it might be so.
(c) $MgCl_2$ is more soluble in water than NaCl. Explain how this is possible in light of the information given in (b). (*Hint:* Consider the answer to Practice Problem 12.4.)

A general rule for determining whether or not a solute will dissolve is often expressed as "like dissolves like," where *like* refers to a similarity in polarity. This rule means that polar solutes dissolve best in polar solvents, and nonpolar solutes dissolve best in nonpolar solvents. This is reasonable because it is most likely for the solvation step to release enough energy to feed the two separation steps when the solvent particles and the solute particles have similar polarities. For example, NaCl, a very polar substance, dissolves in polar solvents such as water because the ion–dipole interactions of the solvation step help compensate for the energy requirements of the two separation steps. What if we try to dissolve NaCl in a nonpolar solvent like carbon tetrachloride, CCl_4, though? What would happen?

WORKPATCH 12.3 For the (attempted) dissolution of NaCl in carbon tetrachloride, CCl_4, choose the best answer for the energy required in each step. (*Hint:* Consider your answer to Practice Problem 12.5.) •

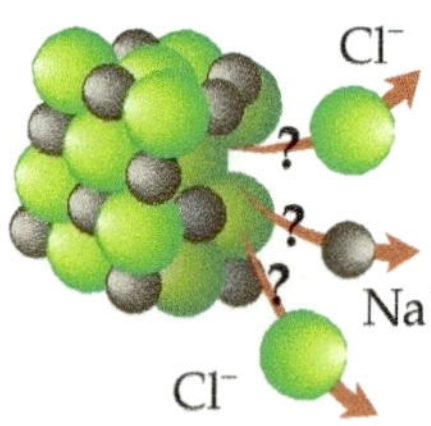

Solute-separation step, overcoming solute–solute interactions. Does it require a

small amount of energy?
moderate amount of energy?
large amount of energy?

Solvent-separation step, overcoming solvent–solvent interactions. Does it require a

small amount of energy?
moderate amount of energy?
large amount of energy?

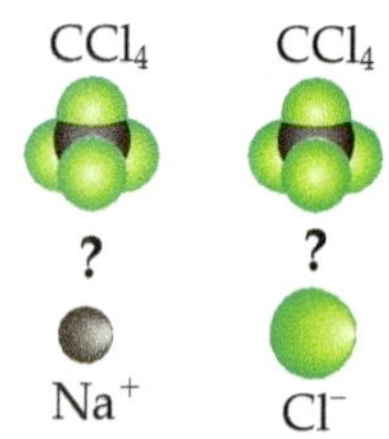

Solvation step, creating solvent–solute interactions. Does it release a

small amount of energy?
moderate amount of energy?
large amount of energy?

In the WorkPatch you should have concluded that the solvation step releases a small amount of energy (because there will be no strong ion–dipole forces between the CCl_4 molecules and the Na^+ and Cl^- ions) and that the solute-separation step requires a large amount of energy (breaking ionic bonds). Because the energy released during solvation (small amount) is less than the energy required for the two separation steps (large amount + moderate/small amount), this hypothetical solution process has a large energy deficit. Consequently, NaCl does not dissolve in CCl_4.

Now let's look at a different situation, dissolving a nonpolar solute in a nonpolar solvent. According to our general rule—like dissolves like—nonpolar solutes should dissolve well in nonpolar solvents. Consider the forces involved in the three steps in this case:

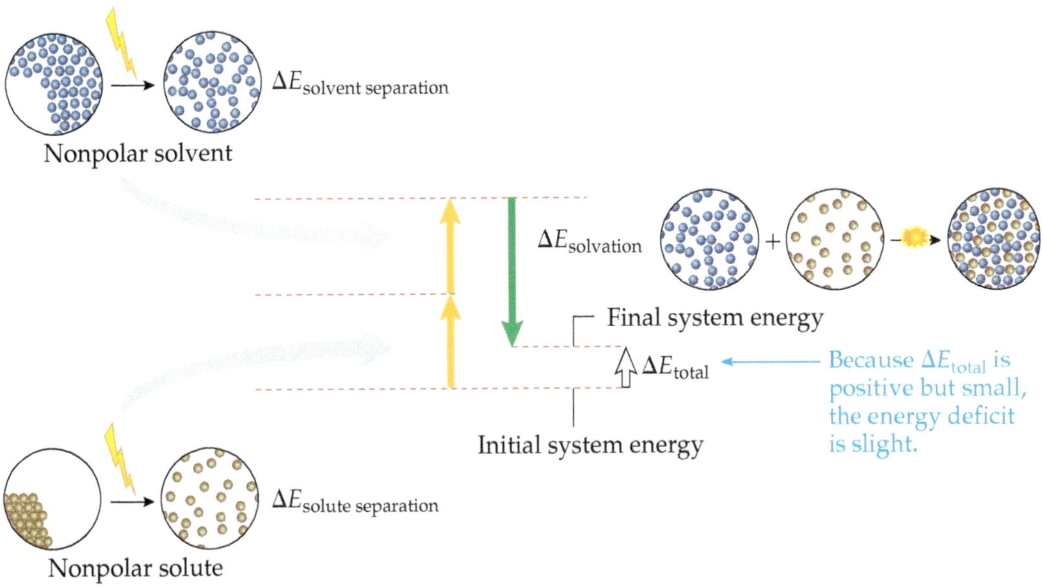

Here, only London forces that attract the nonpolar solute molecules to one another must be broken during solute separation. Also, only London forces attracting the **solvent** molecules to one another must be broken during solvent separation. The solvent–solute interactions that form during solvation are also London forces. Thus, all three steps involve comparable energies and ΔE_{total} will be small, regardless of whether it's positive or negative.

In fact, even if ΔE_{total} is positive for such a process—that is, even if there's a slight energy deficit—a nonpolar solute is likely to dissolve well in a nonpolar solvent. This is true because there is another factor besides ΔE_{total} that can help a solute dissolve. This factor, called *entropy*, can help tip the balance in favor of dissolution when the energy deficit is small.

12.3 Entropy and the Formation of Solutions

There are two natural tendencies that determine the outcome for any process that occurs in the universe. One is the tendency for things to go in a direction that lowers their energy. A ball released on the side of a hill *always* rolls downhill (which lowers its potential energy) and is never observed to roll spontaneously uphill (towards a higher potential energy). When this occurs, we say that the system under consideration has gone to a lower energy state. But there is another tendency that must be considered when predicting the outcome of an event. This other tendency is for the system to go to a more probable situation or arrangement rather than a less probable one. When this occurs, we say that the system under consideration has undergone an increase in **entropy**.

Let's take a minute and look at this tendency toward increased entropy a little more closely. Entropy is really just a measure of the relative probability of one situation occurring compared to another. For example, consider the simple act of putting away your socks (three socks, each a different color) and shirts (three shirts, each a different color) after doing the laundry. One possible way of doing this would be to keep your socks and shirts separate from one another—socks in the bottom drawer, shirts in the top drawer. There are many different ways to achieve this result depending on how we order the colors, as shown below.

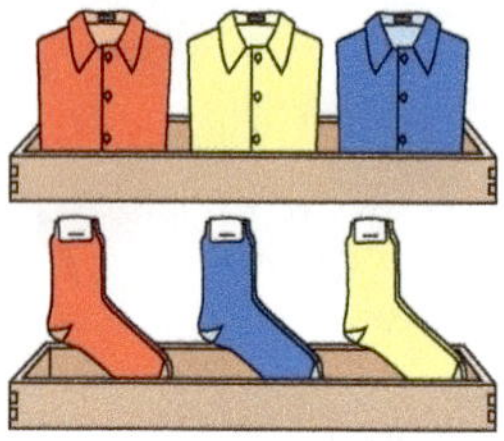

or

or

and many more different arrangements

Although there are many different ways to arrange your socks and shirts this way, the number of arrangements is finite and countable. Now consider another possible way of putting your clothes away, one that allows the socks and shirts to be mixed together in the two drawers. It should be obvious that there are many more ways to put your clothes away under these rules than when they were kept separate—all the previous possibilities (above) are still allowed in addition to a number of new, mixed-up possibilities, such as:

Consider what this means if you put your laundry away the way a typical busy college student would do it, by tossing all of the shirts and socks in the general direction of the open dresser drawers. The chances of ending up with all of the socks in one drawer and all of the shirts in another drawer are not very great.

Why? Because there are so many more ways to end up with the socks and shirts mixed between drawers that the mixed result is much more probable. In this example, the first sorted, neater, less random arrangement of clothing would have a lower entropy than the latter mixed arrangement, which would have a higher entropy. It is much more likely that you would end up with the second situation than the first. Thus, a situation with a higher entropy represents a more probable way of finding the particles in a system. This is the statistical interpretation of entropy (there is an energy-based interpretation of entropy that we will not cover). In any real-world collection of atoms and molecules, there are some "arrangements" that are "more probable" than others. The natural tendency is for the collection of atoms and molecules to assume the most probable arrangement, that is, to go to the state with a higher (increased) entropy.

Since nature favors outcomes that occur with a decrease in energy and an increase in entropy, any process that occurs with both a decrease in energy and an increase in entropy would be primed to happen (the

A Highly Unlikely Event

process is said to be *spontaneous*). Dissolving a solid solute into a solvent always results in an increase in entropy of the system (the system being defined as the solute plus the solvent). For example, consider dissolving solid LiCl in water. When solid LiCl dissolves, the ordered rows of Li^+ and Cl^- ions are replaced by a much more random situation in which the ions are scattered throughout the solution.

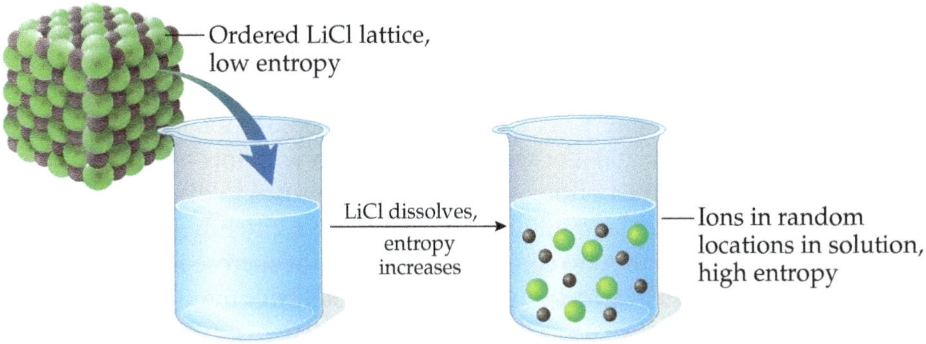

If the dissolving process also results in an overall decrease in the energy of the system (if the solvation step releases more energy than is required by the solvent-separation and solute-separation steps), then it will spontaneously dissolve in water. For the process of LiCl dissolving in water, the system decreases in energy and increases in entropy. Both favor spontaneity, so we can predict that this process will be spontaneous (LiCl will dissolve in water).

If the dissolving process results in a net gain of energy (the solvation step does not compensate for the separation steps), dissolving can still be spontaneous if the net gain in energy *is not too large*. This is because the increase in entropy, which favors spontaneity, can win out. This is sometimes the case for dissolving a nonpolar solute in a nonpolar solvent (where ΔE_{total} for the system can be positive but small). Thus, the entropy increase explains why nonpolar substances are almost always soluble in nonpolar solvents. This is also the case for dissolving NaCl in water, where ΔE_{total} ends up being slightly positive at room temperature and yet is soluble. However, when attempting to dissolve a polar substance like NaCl in a nonpolar solvent like CCl_4, ΔE_{total} is so largely positive that the increase in entropy cannot compensate, and the NaCl will not dissolve.

WORKPATCH 12.4 An H-shaped tank contains gas A in one arm and gas B in the other arm. When a valve separating the two arms is opened, the gases will spontaneously mix if given enough time. The reverse process—the gases spontaneously separating into all A in one arm and all B in the other—will never happen, no matter how long you wait. Why? (Assume ΔE_{total} for the mixing is zero.) ●

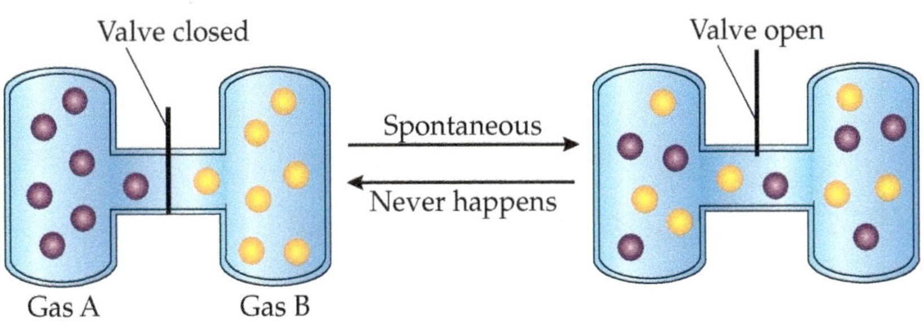

Did you consider which situation is more probable due to the number of ways it can be arrived at? If you did, then you are beginning to understand entropy.

12.4 Solubility, Temperature, and Pressure

We now know what must occur on the molecular level for a solute to dissolve in a solvent. However, just knowing that a particular substance dissolves in a solvent is only part of the story. When chemists are working with solutions, they also need to know *how much* solute can be dissolved in a given amount of solvent. The maximum amount of a solute that can be dissolved in a given amount of solvent is called the **solubility** of the solute. Solubility is often reported as the maximum number of grams of solute that will dissolve in 100 g of a solvent. An interesting aspect of solubility is that it can be changed. For example, try to dissolve 200 g of sucrose in 100 g of water at 0 °C. Even with lots of stirring, some of the sugar just sits at the bottom of the glass and refuses to dissolve. Heat the water to 100 °C, however, and all the solid dissolves. Evidently, temperature is one variable that can be used to control the solubility of sucrose in water. As the temperature of the water increases, so does the solubility of the sucrose:

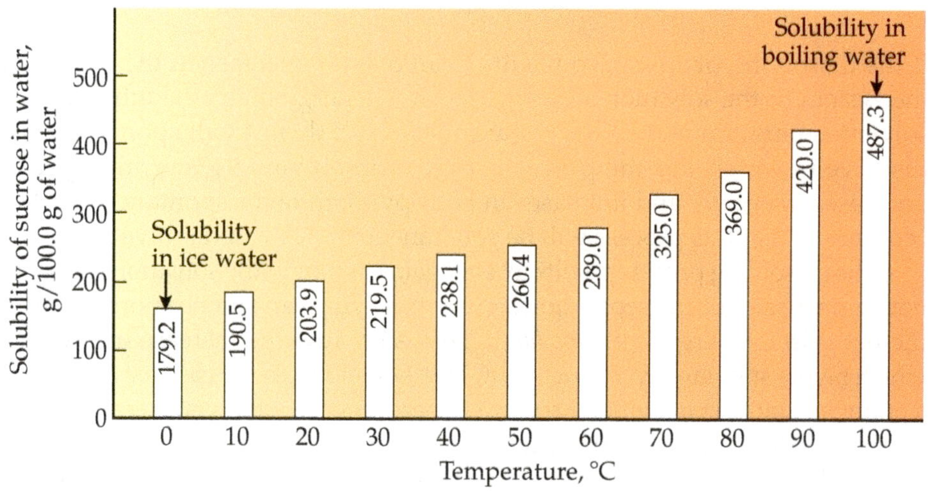

For most molecular solids (such as sucrose), an increase in temperature causes an increase in solubility. The same is true for most ionic solids, as you can see from this graph:

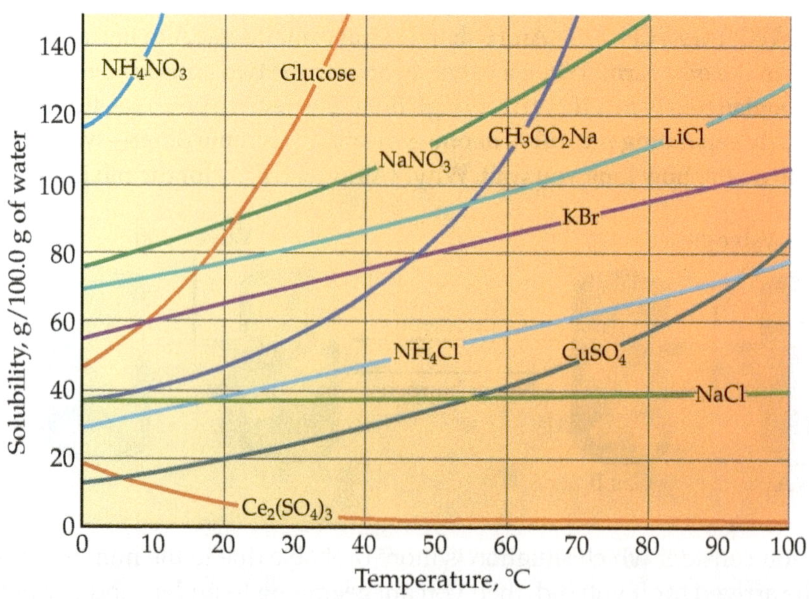

The increase in solubility with increasing temperature is small for some compounds (NaCl) and quite dramatic for others (CH_3CO_2Na, sodium acetate). Note that cerium(III) sulfate, $Ce_2(SO_4)_3$, is an exception and becomes *less* soluble with increasing temperature.

Although it might be tempting to relate the way solubilities change with temperature to the energy changes that take place during the dissolving process, you should not do this. For example, LiCl dissolves with an energy surplus and NH_4NO_3 dissolves with an energy deficit. Yet the solubility of both compounds increases with increasing temperature. The best way to determine how temperature affects solubility is by experiment.

Another exception occurs in systems in which a gas dissolves in a liquid. Here, solubility always *decreases* with increasing temperature. Many familiar aqueous solutions fit this profile. The familiar fizz in soft drinks is the result of dissolved CO_2 gas. Raising the temperature of the solvent decreases the solubility of the gas solute. Think of what it's like to drink warm soda pop. It tastes flat because most of the dissolved CO_2 has escaped from the solution.

Another variable that affects the solubility of a gas in a liquid is pressure. Increasing the pressure of the gas in the space above a liquid surface always forces more gas into the liquid—that is, the increase in gas pressure increases the solubility of the gas. This pressure–solubility relationship is known as *Henry's law*.

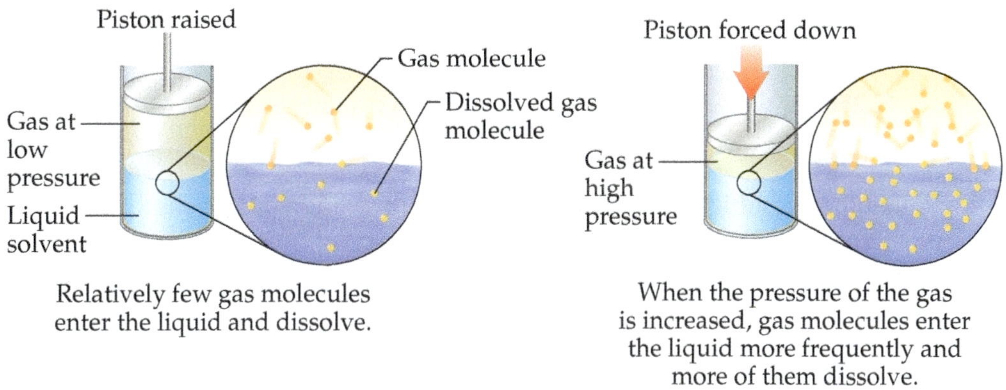

One way to increase gas pressure is by pushing on a piston above the gas, as shown in the accompanying drawing. Pushing down on the piston decreases the gas volume and thus increases its pressure (recall the ideal gas law from Chapter 11). The gas is essentially being forced into the solvent. This is how carbonated beverages are made. Beverage makers increase the solubility of CO_2 by forcing it into solution using high pressure (2–5 atm), well above the normal pressure of 1 atm. When you open a can of soda, the pressure above the liquid surface immediately drops back to 1 atm, drastically decreasing the solubility of the CO_2 and causing it to come bubbling out of solution. If you leave the can open long enough, most of the CO_2 will escape and the drink will taste flat.

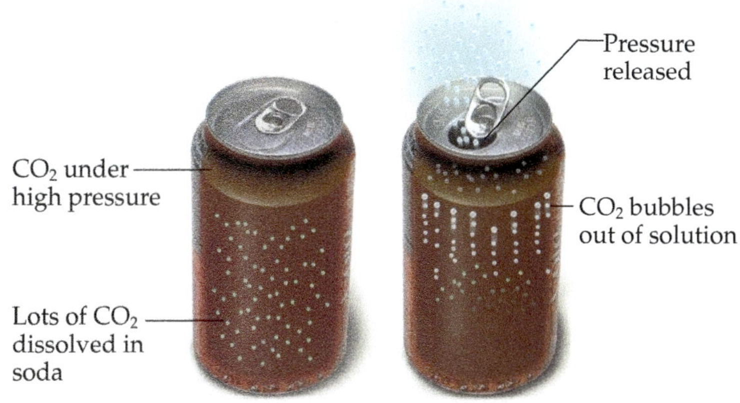

A similar thing happens when divers rise too quickly from a deep-sea dive. Deep under water, the increased pressure increases the solubility of the gas the divers are breathing in their blood. This gas is mostly N_2 if they are breathing compressed air. Coming up too rapidly is like popping the top on a can of soda. The rapid decrease in pressure lowers the solubility of the nitrogen in the blood, and the nitrogen comes out of solution and forms bubbles in the blood. This condition, known as the bends, can be fatal.

The most effective treatment for the bends is to place the afflicted diver as quickly as possible in a high-pressure chamber (called a *hyperbaric chamber*). The pressure in the chamber is raised to a level that causes the gas to redissolve in the blood, temporarily curing the bends. Then the pressure is reduced slowly enough to allow the extra dissolved gas to come out of solution slowly, so it can escape through the diver's lungs instead of forming bubbles in the blood.

Solubility values like those given in the two graphs shown on page 462 represent the maximum amount of solute that dissolves in a given amount of solvent. When the maximum amount of solute is dissolved, we say the solution is a **saturated solution**. Knowing the solubility enables us to calculate how much solute it takes to saturate a given amount of solvent. For example, suppose we have 250.0 g of water and want to use it to prepare a saturated sucrose solution at 20.0 °C. How much sucrose do we need? From the sucrose-solubility graph, we find that the solubility of sucrose at 20.0 °C is 203.9 g sucrose/100 g water. To answer the question, we use the solubility as a conversion factor and multiply it by the amount of solvent we have:

$$250.0\ \cancel{\text{g water}} \times \frac{203.9\ \text{g sucrose}}{100.0\ \cancel{\text{g water}}} = 509.8\ \text{g sucrose}$$

Remember, solubility values are *maximums*, meaning you can dissolve 509.8 g of sucrose in 250.0 g of water at 20.0 °C and no more.

Try the following practice problems to gain some experience using solubility data.

Practice Problems

12.11 How many grams of sucrose will it take to saturate 1 ton of water at 20.0 °C? (1 ton = 2000 lb; 1 lb = 453.6 g)

Answer:

$$1.00\ \cancel{\text{ton}}\ \cancel{\text{water}} \times \frac{2000\ \cancel{\text{lb}}}{1\ \cancel{\text{ton}}} \times \frac{453.6\ \cancel{\text{g}}}{1\ \cancel{\text{lb}}} \times \frac{203.9\ \text{g sucrose}}{100.0\ \cancel{\text{g water}}} = 1.85 \times 10^6\ \text{g sucrose}$$

12.12 How many grams of potassium chloride, KCl, will it take to prepare a saturated solution in 500.0 mL of boiling water? (The solubility of KCl at 100 °C is 56.7 g/100.0 g water; assume the density of water is 1.000 g/mL.)

12.13 On the basis of the three steps involved in the formation of a solution, what is the biggest difference between dissolving a gas and dissolving a solid?

12.5 Molarity

The solubility graphs we looked at earlier tell us the maximum amount of a particular solute we can dissolve in a fixed amount of solvent. For example, at 20 °C we can dissolve a maximum of 203.9 g of sucrose in 100.0 g of water.

This sucrose–water solution is saturated and would be quite sweet. We could also make a less sweet solution by dissolving less than the maximum, say 20 g of sucrose, in 100 g of water. The **concentrations** of these two solutions—which means the amount of solute either per amount of solvent or per amount of solution—are different.

There are times when knowing the concentration of a solution takes on a whole new level of importance.

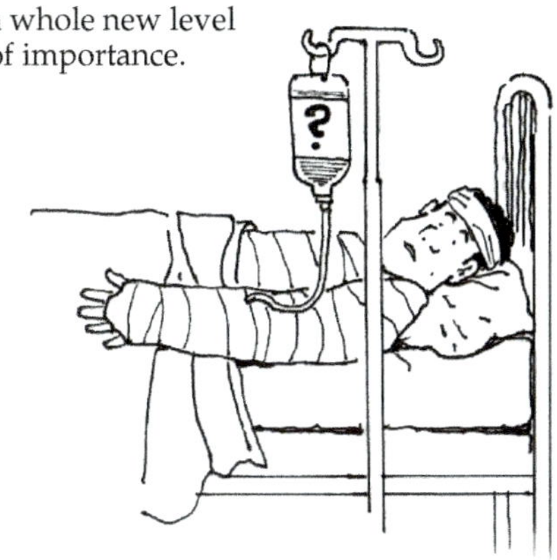

In this and the next section (Section 12.6), we look at the quantitative ways of describing the concentration of a solution, but first we'll take a brief look at ways of describing concentration qualitatively. For example, you take a swallow of some watery, light-brown liquid purchased at a snack-bar coffee machine and mutter, "This coffee's pretty weak." Or you sip some dark, expensive cappuccino at a classy restaurant and proclaim, "This is strong!" *Weak* and *strong* are qualitative terms describing how much coffee (solute) is dissolved in the water (solvent). Instead of *weak* and *strong* you could have said *dilute* and *concentrated,* other terms often used to describe concentration. These latter terms are also qualitative because they don't describe the composition of the solution in numeric terms.

There will be times when you need quantitative information about how much solute is dissolved in a given amount of solution. For example, suppose you have been in an automobile accident and find yourself in immediate need of an intravenous saline (salt) solution. Too concentrated a solution will kill you, and too dilute a solution won't help. It's at a time like this that you seriously hope the attending health professional knows something more about concentration than just "weak" or "strong." Considering that your life depends on it, an accurate quantitative value of the concentration would be nice.

Saturated is a quantitative term. As we mentioned earlier, a solution is said to be saturated when the maximum amount of solute is dissolved in it. If you add more solute to a saturated solution, it just settles to the bottom of the container and doesn't dissolve. In fact, a good way to prepare a saturated solution is to add an excess of solute so that some doesn't dissolve and then filter the solution.

The Easy Way to Make a Saturated Solution

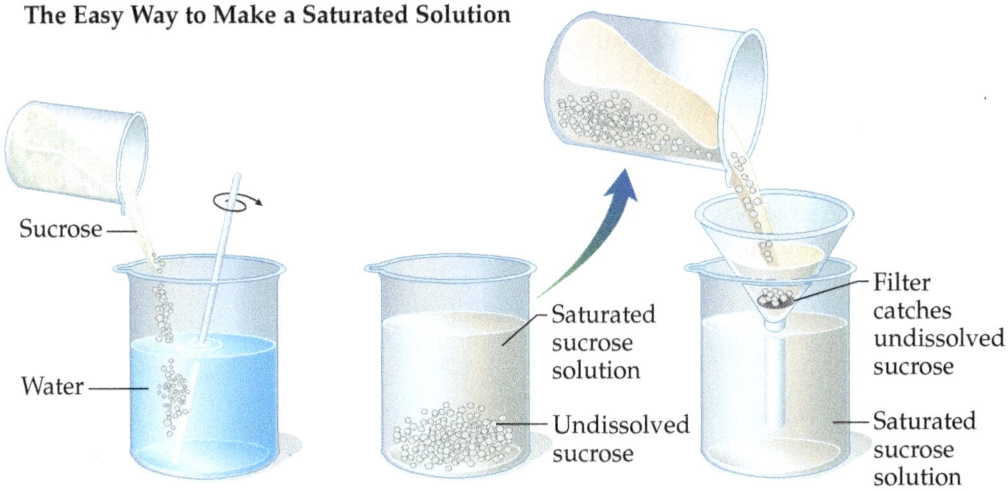

At 0 °C, you can dissolve a maximum of 37.5 g of NaCl in 100 g of water. No more will dissolve. The word *saturated* therefore has a quantitative meaning, telling you the solubility of the solute at the temperature of the solution.

Did you think you were rid of me?

Unfortunately, just saying the word *saturated* doesn't tell you how much salt is dissolved in a saturated NaCl solution. You would have to look up the solubility in a table or graph. It would be much more useful to have some quantitative description of the concentration that could instantly tell you the amount of solute present in the solvent. One such description commonly used by chemists is called *molarity*. From the name, you can probably guess what's coming.

When describing the amount of solute dissolved, chemists most often think in terms of moles and express the concentration in terms of molarity. (If you need to review the concept of the mole, look back at Section 9.2.) The **molarity** of a solution is defined as the number of moles of solute dissolved in 1 L of solution:

$$\text{Molarity} = \frac{\text{Moles of solute}}{1\text{ L of solution}}$$

For example, if you dissolve 1 mole (58.443 g) of NaCl in enough water to give exactly 1 L of solution, you will have a 1-molar (1 M) solution of NaCl:

How to Prepare a 1 M NaCl Solution

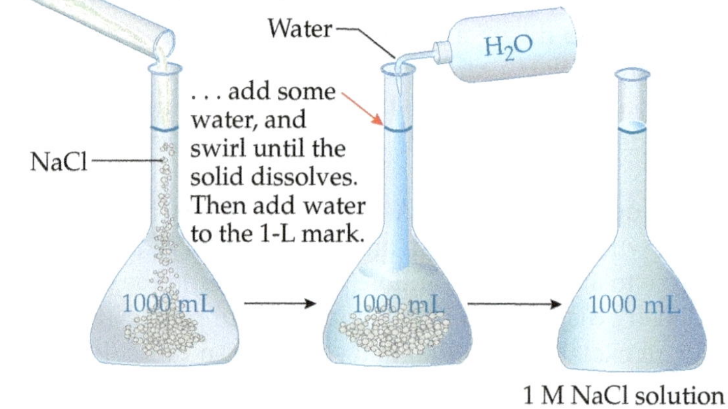

Molarity gives us what we wanted—a quantitative measure of concentration. If someone hands you a bottle of saline solution labeled 1 M, you quantitatively and absolutely know its concentration. No longer is it weak or strong, saturated or unsaturated. It's 1 M. Every liter of that solution contains exactly 1 mole (58.443 g) of NaCl. A 2 M NaCl solution is twice as concentrated as a 1 M solution, meaning every liter of the 2 M solution contains 2 moles (116.886 g) of NaCl. And a 0.5 M solution is half as concentrated as a 1 M solution—every liter of a 0.5 M solution contains 0.5 mole (29.222 g) of NaCl.

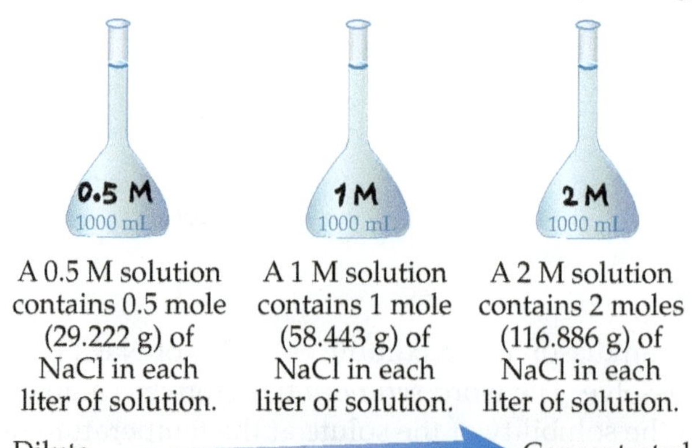

In fact, knowing the molarity, we can quickly determine how much solute we have for any volume of solution. We just use molarity as a conversion factor. For example, suppose you have a flask that contains 200.0 mL of a 1.00 M NaCl solution. If all the water evaporated, solid sodium chloride would remain, but how much? In other words, what mass of NaCl is contained in 200.0 mL of a 1.00 M solution?

How many moles of NaCl are there in the flask? How many grams?

We can answer these questions by using unit analysis (Section 2.8). We can treat the molarity as a conversion factor because it has two units (moles of solute over liters of solution). We treat the given volume, which has just one unit (mL), as the measurement. (This is not to say that the molarity is not a measured quantity. It is, but it can serve as a conversion factor.)

Step 1: Write down the measurement: 200.0 mL solution

Step 2: Since our conversion factor contains the unit *liter*, we'll first convert our measurement from milliliters to liters. Then we multiply the result by the molarity conversion factor to give final units of moles:

$$\underbrace{200.0\ \text{mL solution}}_{\text{Measurement}} \times \underbrace{\frac{1\ \text{L solution}}{1000\ \text{mL solution}}}_{\substack{\text{Conversion factor}\\ \text{to change mL to L}}} \times \underbrace{\frac{1.00\ \text{mol NaCl}}{1\ \text{L solution}}}_{\substack{\text{Molarity}\\ \text{conversion factor}}} = 0.200\ \text{mol NaCl}$$

So 200.0 mL of a 1.00 M NaCl solution contains 0.200 mole of NaCl. To convert the answer to grams, we multiply by the appropriate conversion factor, the molar mass of NaCl:

$$0.200\ \text{mol NaCl} \times \underbrace{\frac{58.443\ \text{g NaCl}}{1\ \text{mol NaCl}}}_{\substack{\text{Molar mass}\\ \text{conversion factor}}} = 11.7\ \text{g NaCl}$$

There are 11.7 g of NaCl in our 200.0 mL of 1.00 M solution. Of course, we did not have to do this last calculation separately. We could have just strung the molar mass conversion factor onto the end of the preceding calculation.

Armed with molarity as a conversion factor, you are now ready to practice some numeric problems. Remember, you can always use the inverse of a conversion factor to make units cancel.

Practice Problems

12.14 How many milliliters of a 1.500 M solution of NaCl do you need to obtain 100.0 g of NaCl?

Answer: As usual, we write the measurement and multiply it by appropriate conversion factors:

$$\underbrace{100.0\ \text{g NaCl}}_{\text{Measurement}} \times \underbrace{\frac{1\ \text{mol NaCl}}{58.443\ \text{g NaCl}}}_{\substack{\text{Conversion factor}\\ \text{to change to moles}\\ \text{(inverse of NaCl}\\ \text{molar mass)}}} \times \underbrace{\frac{1\ \text{L solution}}{1.500\ \text{mol NaCl}}}_{\substack{\text{Conversion factor}\\ \text{(inverse of}\\ \text{molarity)}}} \times \underbrace{\frac{1000\ \text{mL solution}}{1\ \text{L solution}}}_{\substack{\text{Conversion factor}\\ \text{to change L to mL}}} = 1141\ \text{mL solution}$$

Notice that the first thing we did was convert to moles. Remember our general advice for stoichiometry problems in chemistry. Convert to moles as quickly as you can. It's almost always the right thing to do.

12.15 How many milliliters of a 2.55 M solution of glucose, $C_6H_{12}O_6$, molar mass = 180.155 g/mol, do you need to obtain 25.0 g of glucose?

12.16 How many grams of ethanol, C_2H_6O, are there in 200.0 mL of a 2.00 M aqueous solution of ethanol?

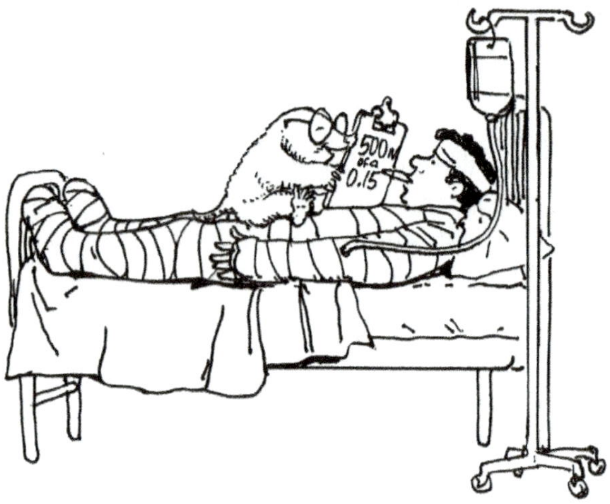

There is one other type of problem involving molarity you should know how to do—a practical problem that asks "How do you prepare a specified amount of a solution with a specific molarity?" For example, let's go back to you lying in the hospital hooked up to a bottle of intravenous saline solution. Noticing that the bottle is just about empty, your nurse says, "I need to prepare 500.0 mL of a 0.15 M NaCl solution to refill this bottle." Realizing that the nurse studied introductory chemistry some time ago, whereas you just recently completed the course, you remind him how to do this. Here's how it's done.

There are two ways to prepare this life-saving solution. One is to measure out the proper amount of NaCl solid and dissolve it in the proper amount of water. We'll call this method preparing a solution from scratch. A second way is to take a more concentrated solution of NaCl off the shelf and dilute it to the proper concentration. Because both ways are commonly used, we'll go over both.

Method 1 Preparing a Solution from Scratch

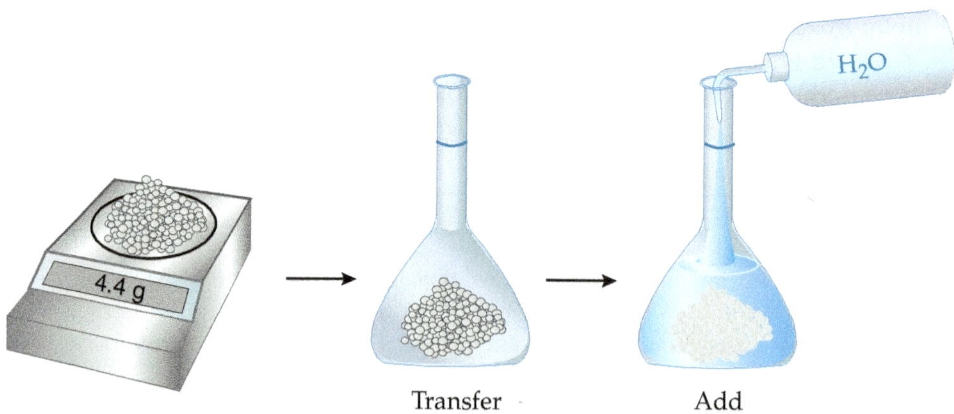

Measure out correct amount of solid NaCl. → Transfer to volumetric flask. → Add correct amount of water to bring it to the desired volume.

Method 1: **Preparing a solution from scratch.** Our task is to prepare 500.0 mL of a 0.15 M NaCl solution. Our first step is to calculate how much solid NaCl we need. Just use the usual method of unit analysis:

$$\underbrace{500.0\ \text{mL solution}}_{\text{Measurement}} \times \underbrace{\frac{1\ \text{L solution}}{1000\ \text{mL solution}}}_{\text{Change mL to L}} \times \underbrace{\frac{0.15\ \text{mol NaCl}}{1\ \text{L solution}}}_{\text{Molarity conversion factor}} = 0.075\ \text{mol NaCl}$$

The answer is in moles, but because we measure amounts of solids in grams, we need to convert to grams:

$$0.075\ \text{mol NaCl} \times \underbrace{\frac{58.443\ \text{g NaCl}}{1\ \text{mol NaCl}}}_{\text{Molar mass conversion factor}} = 4.4\ \text{g NaCl}$$

So to prepare 500.0 mL of a 0.15 M NaCl solution from scratch, place 4.4 g of NaCl in a 500-mL volumetric flask, add some water, swirl to dissolve the solid, and then add enough water to get exactly 500.0 mL of solution. (Remember, do *not* just add 500 mL of water! To get exactly 500 mL of solution you cannot just add 500 mL of water since the dissolved solid will add some volume as well). See figure below.

Preparation of a 0.15 M NaCl Solution from Scratch

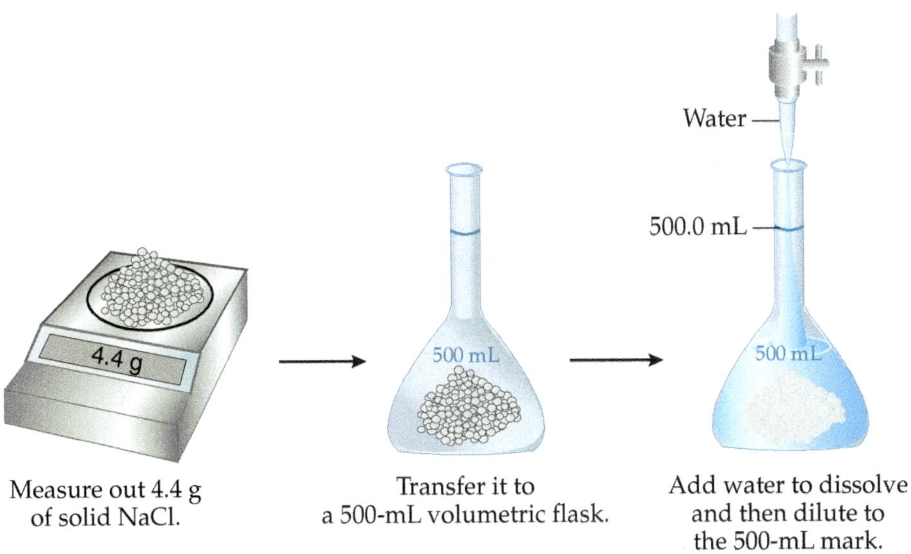

Method 2: Preparing a solution by diluting a more concentrated solution. On the shelf is a bottle of 0.50 M NaCl solution. This is our "reservoir" of solution, what is typically called a *stock solution*. We can use it to prepare the 500.0 mL of 0.15 M solution we need. Because the stock solution is more concentrated than the solution we need, we have to dilute some amount of it by adding additional water. How much stock solution should we use, and how much water do we add to dilute it to the desired concentration?

Method 2 Preparing a Solution by Diluting a More Concentrated Solution

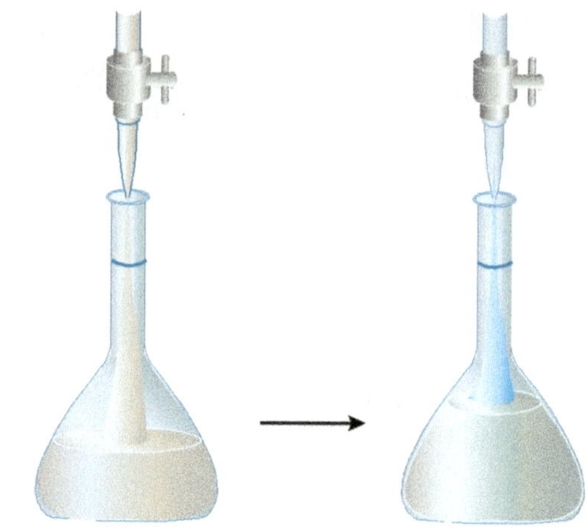

Add correct amount of concentrated solution. Add correct amount of water.

At first glance, this might seem more difficult than preparing the solution from scratch because now we have to consider two solutions, the stock solution and the one we want to prepare. To start, though, forget about the stock solution. Just worry about the solution we want to make (500.0 mL of a 0.15 M NaCl solution). When we prepared this solution from scratch, the first question we asked was "How much NaCl do we need?" We ask *exactly* the same question here. How much NaCl do we need to prepare 500.0 mL of a 0.15 M solution? Because this is the same question we asked in method 1, we know that the answer is 0.075 mole of NaCl. Here is where the stock solution comes into play because this time we will get our 0.075 mole of NaCl from the stock solution. The question now becomes "What volume of 0.50 M stock solution contains 0.075 mole of NaCl?" Look closely at this question. There is a measurement (0.075 mol NaCl) and a conversion factor (0.50 M). It's time to use unit analysis again:

$$0.075\ \cancel{\text{mol NaCl}} \times \underbrace{\frac{1\ \text{L stock solution}}{0.50\ \cancel{\text{mol NaCl}}}}_{\substack{\text{Conversion factor}\\ \text{(inverse of stock}\\ \text{solution molarity)}}} = 0.15\ \text{L stock solution}$$

This calculation tells us that 0.15 L of stock solution (150 mL) contains 0.075 mole of NaCl. To prepare our solution, we put 0.15 L of stock solution in a 500-mL volumetric flask. This puts 0.075 mole of NaCl in the flask. Then we add enough water to get 500.0 mL of solution. Once again, we have 0.075 mole of NaCl dissolved in a total volume of 500.0 mL of solution, which is a 0.15 M NaCl solution.

Method 2 can be considered a dilution problem because we are diluting a stock solution to a desired concentration. An equivalent way to solve such a problem is to use this **dilution equation**:

$$M_{\text{stock solution}} \times V_{\text{stock solution}} = M_{\text{diluted solution}} \times V_{\text{diluted solution}}$$

For example, let's repeat our preparation of 500.0 mL of 0.15 M NaCl from a 0.50 M NaCl stock solution. We are still looking for the same thing, the volume of stock solution we need to use. This is $V_{\text{stock solution}}$. All we need to do is solve the dilution equation for $V_{\text{stock solution}}$:

$$V_{\text{stock solution}} = \frac{M_{\text{diluted solution}} \times V_{\text{diluted solution}}}{M_{\text{stock solution}}}$$

Now we plug in the numeric values for the three factors on the right and we have a value for $V_{\text{stock solution}}$:

$$V_{\text{stock solution}} = \frac{\overbrace{0.15\text{ M}}^{M_{\text{diluted solution}}} \times \overbrace{500.0\text{ mL}}^{V_{\text{diluted solution}}}}{\underbrace{0.50\text{ M}}_{M_{\text{stock solution}}}} = 150\text{ mL stock solution}$$

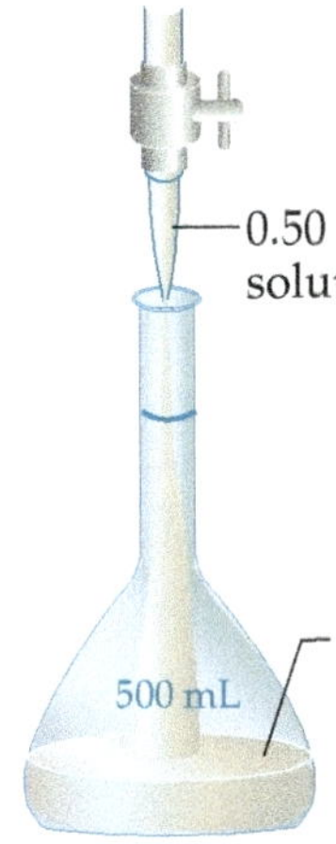

Add 150 mL of stock solution.

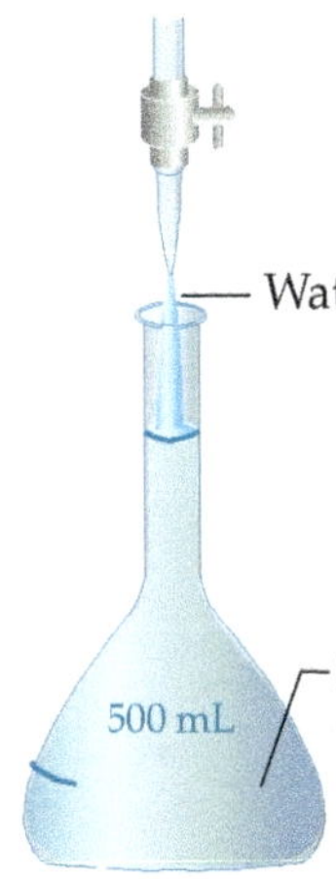

Add water until the flask contains 500.0 mL of solution. The final concentration is 0.075 mol/0.5000 L = 0.15 M.

This is the same answer we arrived at a moment ago but now done in one quick step.

The dilution equation does not work because of magic. Rather, it works because it is based on a simple fact—when you dilute some volume of stock solution, the number of moles of solute contained in that volume does not change!

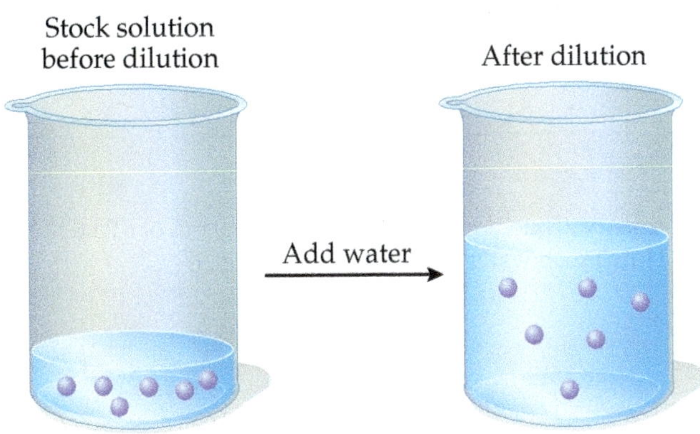

The two beakers contain the same number of moles of solute.

When you multiply the molarity of a solution by its volume, the product is number of moles of solute ($M \times V$ = moles):

$$\overbrace{\frac{\text{Moles of solute}}{\cancel{\text{Liters of solution}}}}^{M} \times \overbrace{\cancel{\text{Liters of solution}}}^{V} = \text{Moles of solute}$$

Therefore, the dilution equation is just the mathematical version of the above diagram. $M_{\text{stock solution}} \times V_{\text{stock solution}}$ equals moles of solute in the left beaker, and $M_{\text{diluted solution}} \times V_{\text{diluted solution}}$ equals moles of solute in the right beaker, and these are equal to each other. So feel free to use the dilution equation anytime a dilution problem comes up.

A summary of what we just did appears below.

Preparing various volumes of solutions having particular molarities, both from scratch and by diluting stock solutions, is routine practice in chemical laboratories, medical offices, and hospitals throughout the world. Now it's time to convince yourself you can do it.

How to prepare a solution of desired molarity

Step 1: Determine the number of moles of solute needed to prepare the solution. This quantity equals the desired volume in liters multiplied by the desired molarity:

moles of solute needed = Volume of solution × Molarity of solution

Step 2: **If preparing the solution from scratch:** Convert number of moles of solute to grams, using the molar mass as a conversion factor. Place this number of grams of solute in the appropriate volumetric flask.

If diluting a stock solution*: Convert number of moles of solute to volume of stock solution, using the stock solution molarity as a conversion factor. (Multiply the number of moles of solute by the inverse of the stock solution molarity.) Place this volume of stock solution in the appropriate volumetric flask.

Step 3: Add enough water to reach the desired volume.

*Alternatively, you can combine steps 1 and 2 and use the dilution equation, as described in the text.

Practice Problems

12.17 What mass of solid NaCl do you need to prepare 400.0 mL of a 2.00 M NaCl solution from scratch? How much water do you add to the NaCl?

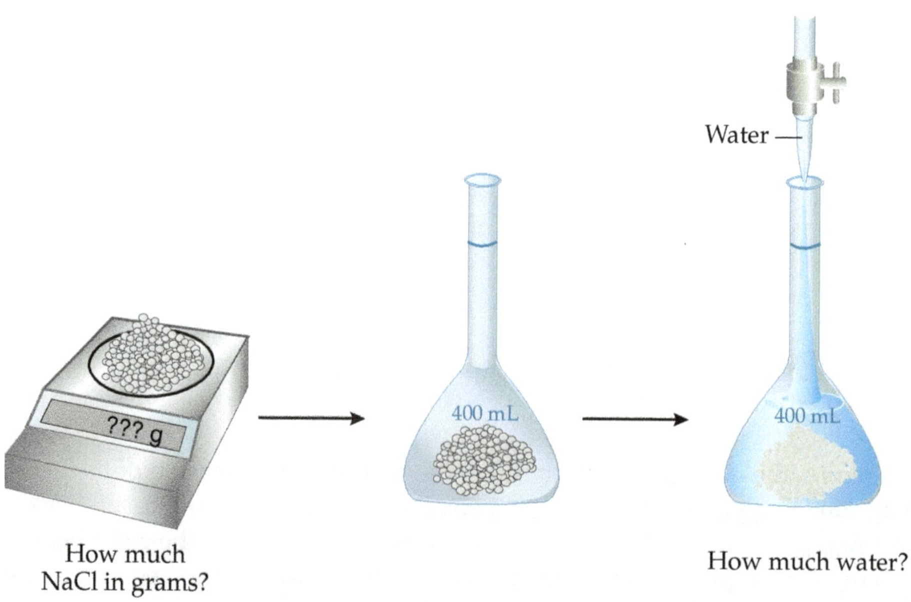

12.18 You have 1 L of a 3.00 M stock solution of NaCl and need to prepare 400.0 mL of a 2.00 M solution. Describe how you would do it.

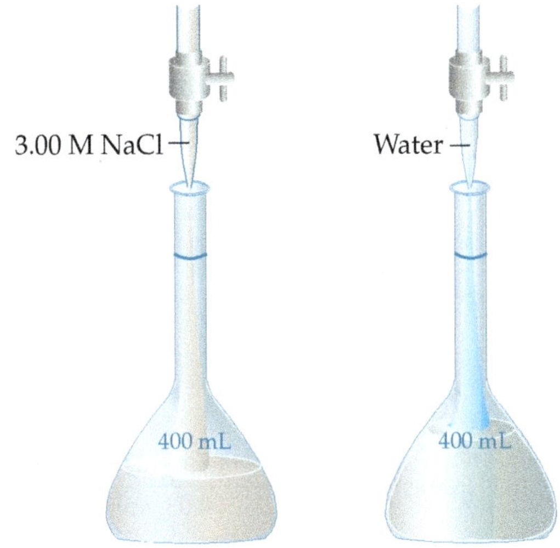

12.6 Percent Composition

Molarity is an excellent way to specify solution concentration quantitatively, and chemists use it routinely. However, solutions encountered in hospitals, doctors' offices, and grocery stores often have concentrations expressed as *percent composition*. In a fundamental sense, percent composition is similar to molarity. In fact, most of the ways of numerically reporting solution concentrations are based on the definition:

$$\text{Concentration} = \frac{\text{Amount of solute}}{\text{Amount of solution}}$$

For molarity, the numerator (amount of solute) is expressed in moles and the denominator (amount of solution) is expressed in liters. Percent composition uses the same definition of concentration but with different quantities in the numerator and the denominator.

There are three types of **percent composition,** depending on whether the solute and solution values are given by mass or by volume:

Percent Composition by Mass	Percent Composition by Volume	Percent Composition by Mass/Volume
$\frac{\text{Grams of solute}}{\text{Grams of solution}} \times 100$	$\frac{\text{Volume of solute}}{\text{Volume of solution}} \times 100$	$\frac{\text{Grams of solute}}{\text{Volume of solution}} \times 100$

Percent composition by mass is often used to describe the concentration of a solution prepared by dissolving a solid in a liquid solvent. Let's try one example. If you dissolve 10.0 g of sucrose in enough water to make a final solution with a mass of 100.0 g, the percent-by-mass concentration is 10.0 mass %:

Percent Composition by Mass

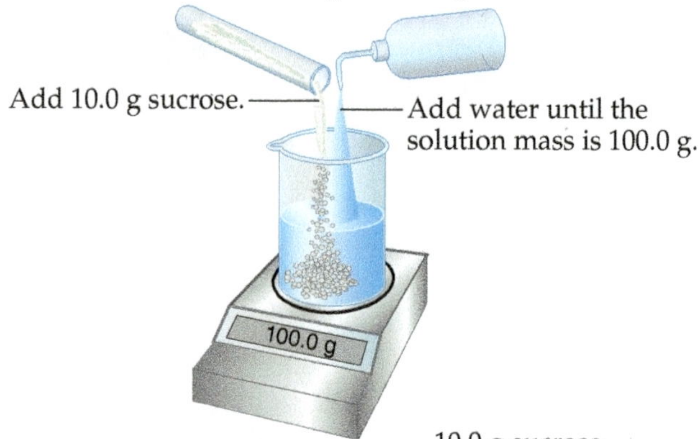

$$\text{Percent by mass} = \frac{10.0 \text{ g sucrose}}{100.0 \text{ g solution}} \times 100$$
$$= 10.0 \text{ mass } \%$$

When the solute is a liquid, percent composition by volume is often used. For example, dissolving 25.0 mL of the liquid ethanol in enough water to give a total solution volume of 100.0 mL produces a solution that has a concentration of 25.0 vol %:

Percent Composition by Volume

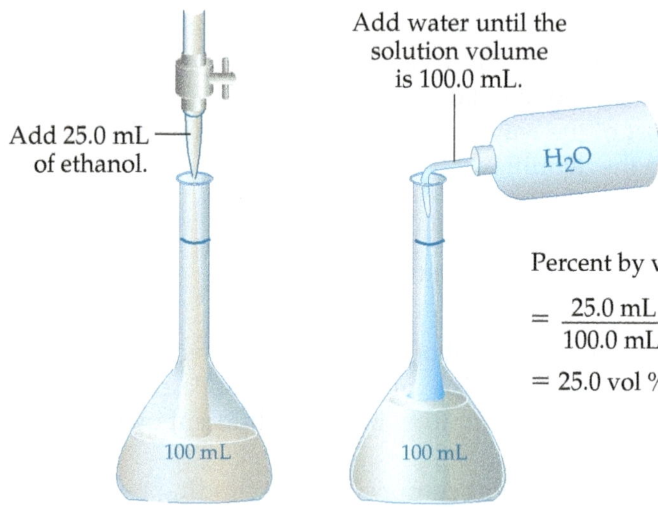

Percent by volume

$$= \frac{25.0 \text{ mL ethanol}}{100.0 \text{ mL solution}} \times 100$$
$$= 25.0 \text{ vol } \%$$

Interestingly, you cannot prepare this solution by adding 75 mL of water to the 25 mL of alcohol because when you mix different liquids together, the volumes are usually not exactly additive. In the case of ethanol and water, a solution made by mixing 25 mL and 75 mL, respectively, has a volume measurably less than 100 mL. That is why the diagram says "Add water until the solution volume is 100 mL" instead of "Add 75 mL of water."

The examples we used to calculate percent compositions are a bit unrealistic because we set the denominator equal to 100.0 to make the calculations easy. In real situations, it's unlikely you will have exactly 100.0 g or 100.0 mL of a solution. For instance, suppose you have 273.0 g of a solution that you know contains 35.0 g of NaCl. What is its percent composition by mass? To find out, you use the same equation we used for our simpler example—percent by mass = (grams of solute/grams of solution) × 100:

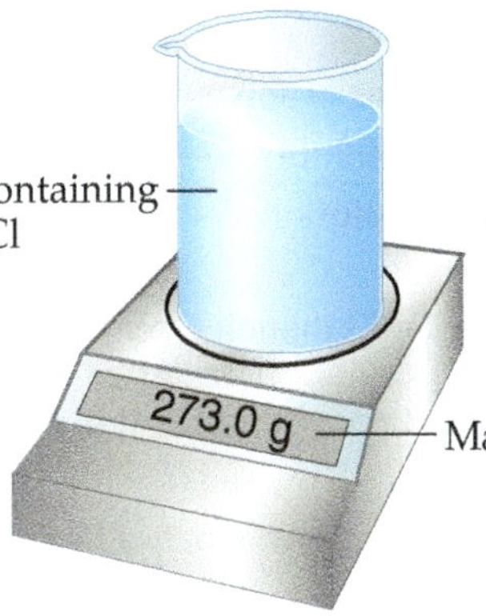

$$\text{Percent by mass} = \frac{35.0\text{ g NaCl}}{273.0\text{ g solution}} \times 100 = 12.8\text{ mass }\%$$

The percent composition for a solution can serve as a conversion factor, just as molarity can. For example, suppose you have 265.5 mL of an aqueous ethanol solution that is 30.0 vol % ethanol. The calculation for determining how much ethanol is in the bottle is:

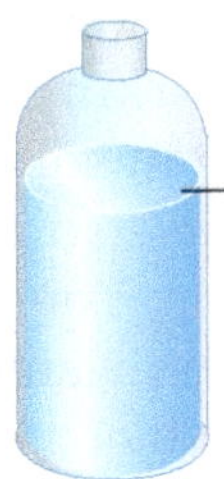

$$\underbrace{265.5\text{ mL solution}}_{\text{Measurement}} \times \underbrace{\frac{30.0\text{ mL ethanol}}{100\text{ mL solution}}}_{\text{Vol \% conversion factor}} = 79.7\text{ mL ethanol}$$

WORKPATCH 12.5 The density of ethanol is 0.7893 g/mL. Use this information to determine the number of grams of ethanol in the 79.6 mL of 30.0 vol % solution just discussed. ●

The WorkPatch is one more reminder that many quantities in chemistry can serve as conversion factors.

The third type of percent composition (percent mass/volume) is less frequently seen and uses grams for amount of solute and milliliters for amount of solution.

Because there are three types of percent composition, it is always important to state which definition you are using. For example, a bottle of an ethanol–water solution labeled 5% ethanol in water does not give you enough information about the solution's composition. If the 5% is by mass, 100 g of the solution contains 5 g of ethanol. If the 5% is by volume, 100 mL of the solution contains 5 mL of ethanol, which is only 3.9 g. This potential for ambiguity is one reason chemists prefer to use molarity.

Practice Problems

12.19 A solution is prepared by dissolving 25.0 g of sucrose in 175.0 g of water. Characterize its concentration by the appropriate percent composition.

Answer: The appropriate percent composition is by mass because both measurements are given in grams.

$$\text{Grams of solution} = \text{Grams of solute} + \text{Grams of solvent} = 25.0\text{ g} + 175.0\text{ g} = 200.0\text{ g}$$

$$\text{Percent composition by mass} = \frac{\text{Grams of solute}}{\text{Grams of solution}} \times 100 = \frac{25.0\text{ g}}{200.0\text{ g}} \times 100 = 12.5\text{ mass }\%$$

12.20 Gasohol is a solution of gasoline and ethanol. Every liter of gasohol contains 90.0 mL of ethanol dissolved in gasoline. Characterize the solute concentration by the appropriate percent composition. (*Hint:* Assume you have 1 L of gasohol solution.)

12.21 Suppose you have 65.0 g of the solution from Practice Problem 12.19. How many grams of sucrose do you have?

12.7 Reactions in Solution

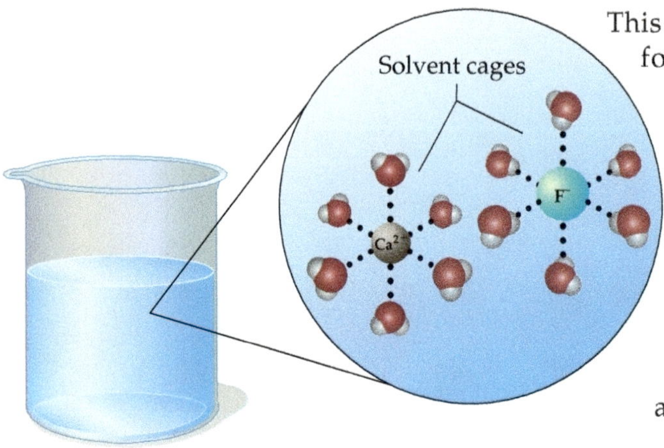

Calcium fluoride solution

This chapter has been all about solutions—why they form, how they form, their concentrations—and we've saved the best for last—working with chemical reactions that take place in solution. This is an important topic because so many chemical reactions occur in solution. Your own body is approximately 80% by mass water, for instance, meaning that most of the biochemical reactions that occur in your cells do so in aqueous solution.

Let's return to the precipitation reaction that almost killed the laboratory technician in Chapter 8:

$$Ca^{2+}(aq) + 2F^-(aq) \rightarrow CaF_2(s)$$

and the subject of stoichiometry, our old friend from Chapter 9, so we can answer questions like "How much precipitate forms for reactions done in solution?"

Stoichiometry

You will be pleased to know that the strategy for doing solution-based stoichiometry calculations is the same as you learned earlier. After all, we are still dealing with balanced chemical equations that "speak" in moles, which means we must convert from grams or liters to moles in order to make use of the equations (*if in doubt, convert to moles!*). When we worked with pure substances, molar mass was the key, serving as a conversion factor between mass and moles. When it comes to solutions, molarity is the key, giving us a way of moving back and forth between a solution volume and the number of moles of reactant/product in that volume. All we need to do is treat molarity like a conversion factor:

$$\overset{V}{\text{Liters of solution}} \times \overset{M}{\frac{\text{Moles of substance}}{\text{Liters of solution}}}$$

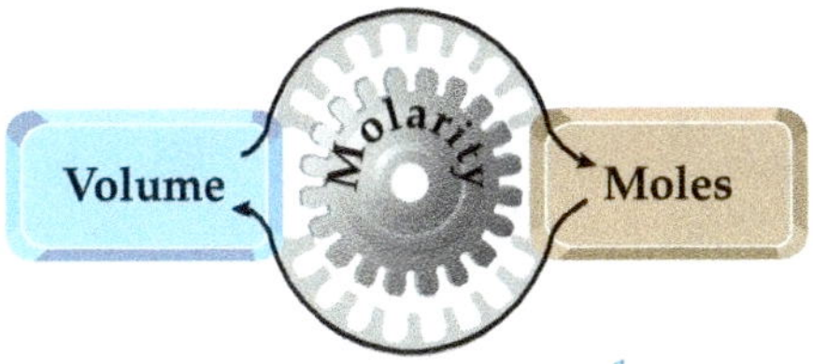

Going from
volume to moles
or
moles to volume

$$\overset{\text{Moles}}{\text{Moles of substance}} \times \overset{\frac{1}{M}}{\frac{\text{Liters of solution}}{\text{Moles of substance}}}$$

In other words, molarity times volume is moles ($M \times V =$ number of moles), and moles divided by molarity is volume (number of moles$/M = V$). This is very useful stuff when it comes to doing solution stoichiometry calculations, which means you should try exercising your skills now.

Practice Problems

12.22 How many moles of CaF_2 are there in 25.0 mL of 0.350 M $CaF_2(aq)$?

Answer:

$$\overset{V}{0.0250\ \text{L solution}} \times \overset{M}{\frac{0.350\ \text{mol CaF}_2}{1\ \text{L solution}}} = 0.00875\ \text{mol CaF}_2$$

Notice how the units cancel, leaving moles. Also notice that we converted the volume from milliliters to liters before multiplying by molarity. You must do this or else the units won't match! Never multiply molarity by milliliters if you are trying to calculate number of moles!

12.23 What volume of 0.350 M CaF_2 solution is required to obtain 0.00875 mole of CaF_2?

Answer: This is just the reverse of Practice Problem 12.22:

$$\overset{\text{Moles}}{0.00875\ \text{mol CaF}_2} \times \overset{\frac{1}{M}}{\frac{1\ \text{L solution}}{0.350\ \text{mol CaF}_2}} = 0.0250\ \text{L solution} = 25.0\ \text{mL}$$

12.24 How many moles of glucose are there in 255.0 mL of a 0.998 M solution of glucose?

12.25 What volume of 0.350 M $BaCl_2$ solution is required to obtain 0.500 mole of $BaCl_2$?

12.26 What volume of 0.350 M $BaCl_2$ solution is required to obtain 0.500 mole of $Cl^-(aq)$?

Did you see the warning in Practice Problem 12.22? Heed it well. Make sure you also do Practice Problems 12.25 and 12.26. The difference between them is subtle but important!

Armed with this new understanding of molarity as a conversion factor, you are ready to do some solution stoichiometry problems. The best way to show you how is to do a few for you. Follow them through.

Suppose we want to synthesize $CaF_2(s)$ by combining these solutions:

Combining them causes insoluble CaF_2 to precipitate (Chapter 8):

$$Ca^{2+}(aq) + 2F^-(aq) \rightarrow CaF_2(s)$$

How much of each solution do we need to produce 1.00 g of $CaF_2(s)$ in a balanced fashion? The first step is to write the balanced equation for the reaction, which we did. The second step is to convert the mass of product we want, 1.00 g of $CaF_2(s)$, to moles (no surprise here).

Step 2: Convert the mass of the desired product to moles, using the product's molar mass as the conversion factor:

$$1.00\,\cancel{\text{g CaF}_2} \times \frac{1\text{ mol CaF}_2}{78.074\,\cancel{\text{g CaF}_2}} = 0.0128\text{ mol CaF}_2$$

Now speaking the right language, we can "ask" the balanced equation how many moles of each reactant we need. To do this, we use the result from step 2 along with conversion factors derived from the balanced equation.

Step 3: Determine the number of moles of each reactant, obtaining conversion factors from the balanced equation:

$$0.0128\,\cancel{\text{mol CaF}_2} \times \frac{1\text{ mol Ca}^{2+}}{1\,\cancel{\text{mol CaF}_2}} = 0.0128\text{ mol Ca}^{2+}$$

$$0.0128\,\cancel{\text{mol CaF}_2} \times \frac{2\text{ mol F}^-}{1\,\cancel{\text{mol CaF}_2}} = 0.0256\text{ mol F}^-$$

The final step is to calculate the volume of each reactant solution required to obtain these molar amounts. This is the job of molarity.

Step 4: Calculate the volume of each solution needed, using the molecular formulas and molarities of the solutes as conversion factors.

For the $Ca(NO_3)_2$:

$$\underbrace{0.0128\,\cancel{\text{mol Ca}^{2+}}}_{\text{Moles of reactant needed}} \times \underbrace{\frac{1\,\cancel{\text{mol Ca(NO}_3)_2}}{1\,\cancel{\text{mol Ca}^{2+}}}}_{\text{From reactant molecular formula}} \times \underbrace{\frac{1\text{ L Ca(NO}_3)_2\text{ solution}}{0.100\,\cancel{\text{mol Ca(NO}_3)_2}}}_{1/M_{\text{reactant}}} = 0.128\text{ L Ca(NO}_3)_2\text{ solution}$$

For the NaF:

$$\underbrace{0.0256\ \cancel{\text{mol F}^-}}_{\substack{\text{Moles of}\\\text{reactant}\\\text{needed}}} \times \underbrace{\frac{1\ \text{mol NaF}}{1\ \cancel{\text{mol F}^-}}}_{\substack{\text{From reactant}\\\text{molecular formula}}} \times \underbrace{\frac{1\ \text{L NaF solution}}{0.100\ \cancel{\text{mol NaF}}}}_{1/M_{\text{reactant}}} = 0.256\ \text{L NaF solution}$$

Combining 128 mL of the calcium nitrate solution with 256 mL of the sodium fluoride solution will yield 1.00 g of $CaF_2(s)$, which we can isolate via filtration. There will be no unused $Ca^{2+}(aq)$ or $F^-(aq)$ ions in the solution since we performed the reaction in a stoichiometric fashion.

We can also turn things around a bit. Instead of specifying how much product we want to make, we'll give you a specified amount of reactant solution and ask you to calculate the theoretical yield. For example, what is the maximum amount of $Ba(OH)_2$ produced by combining 120.0 mL of 0.250 M NaOH with an excess amount of $BaCl_2$? The strategy is always the same—begin by converting to moles.

Step 1: Write the balanced chemical equation:

$$BaCl_2(aq) + 2\,NaOH(aq) \rightarrow Ba(OH)_2(s) + 2\,NaCl(aq)$$

Step 2: Convert known amount of NaOH solution to moles:

$$\overset{V}{0.1200\ \cancel{\text{L solution}}} \times \overset{M}{\frac{0.250\ \text{mol NaOH}}{1\ \cancel{\text{L solution}}}} = 0.0300\ \text{mol NaOH}$$

Step 3: Use molecular formulas and the balanced equation to obtain conversion factors for determining number of moles of unknown—$Ba(OH)_2$:

$$0.0300\ \cancel{\text{mol NaOH}} \times \frac{1\ \text{mol Ba(OH)}_2}{2\ \cancel{\text{mol NaOH}}} = 0.0150\ \text{mol Ba(OH)}_2$$

Step 4: Convert from moles to appropriate units:

$$0.0150\ \cancel{\text{mol Ba(OH)}_2} \times \frac{171.341\ \text{g Ba(OH)}_2}{1\ \cancel{\text{mol Ba(OH)}_2}} = 2.57\ \text{g Ba(OH)}_2$$

If you think about it, we are doing solution stoichiometry calculations exactly as we did our stoichiometry calculations in Chapter 8, the only difference being that molarity has joined the party as a conversion factor. Here's the summary chart.

How to solve solution stoichiometry problems

Step 1: Write the balanced chemical equation.

Step 2: For each known reactant or product, convert the given information to moles.

Step 3: Use molar amounts from step 2 and conversion factors derived from molecular formulas and the balanced equation to calculate number of moles of desired reactant or product.

Step 4: Convert molar amounts from step 3 to desired units, using molarities and/or molar masses as conversion factors.

Now try your hand at some practice problems.

Practice Problems

12.27 How would you prepare 9.70 g of $PbCl_2(s)$ from a 0.100 M solution of $Pb(NO_3)_2$ and a 0.200 M solution of $CaCl_2$?

Answer: **Step 1:** Write the balanced chemical equation:

$$Pb(NO_3)_2(aq) + CaCl_2(aq) \rightarrow PbCl_2(s) + Ca(NO_3)_2(aq)$$

Step 2: Convert given information to moles:

$$9.70\ \text{g PbCl}_2 \times \frac{1\ \text{mol PbCl}_2}{278.10\ \text{g PbCl}_2} = 0.0349\ \text{mol PbCl}_2$$

Step 3: Determine moles of each reactant required:

$$0.0349\ \text{mol PbCl}_2 \times \frac{1\ \text{mol Pb(NO}_3)_2}{1\ \text{mol PbCl}_2} = 0.0349\ \text{mol Pb(NO}_3)_2$$

$$0.0349\ \text{mol PbCl}_2 \times \frac{1\ \text{mol CaCl}_2}{1\ \text{mol PbCl}_2} = 0.0349\ \text{mol CaCl}_2$$

Step 4: Convert from moles to desired:

$$0.0349\ \text{mol Pb(NO}_3)_2 \times \frac{1\ \text{L}}{0.100\ \text{mol Pb(NO}_3)_2} \times \frac{1000\ \text{mL}}{1\ \text{L}} = 349\ \text{mL Pb(NO}_3)_2$$

$$0.0349\ \text{mol CaCl}_2 \times \frac{1\ \text{L}}{0.200\ \text{mol CaCl}_2} \times \frac{1000\ \text{mL}}{1\text{L}} = 175\ \text{mL CaCl}_2$$

Combine 349 mL of the lead(II) nitrate solution with 175 mL of the calcium chloride solution, and then use a funnel and filter paper to isolate the 9.70 g of $PbCl_2$ that forms.

12.28 Suppose you did as required in Practice Problem 12.27 but, because of poor filtering technique, isolated only 8.24 g of $PbCl_2$. What is the percent yield for your synthesis?

12.29 How would you prepare 20.0 g of iron(III) hydroxide from a 0.250 M solution of $Fe(NO_3)_3$ and a 0.150 M solution of $Ba(OH)_2$? How much product would you have isolated if your percent yield were 67.5%?

Titrations

Now we want to discuss a useful experimental procedure for determining solution concentrations, a procedure you will do many times if your chemistry course includes a laboratory section. The process is called **titration**, and it entails using a known amount of a solution of known concentration to determine the concentration of some other solution.

A common application of titrations is in determining acid concentrations. Recall that in Chapter 8 we said that acids and bases are compounds that produce $H^+(aq)$ and $OH^-(aq)$ ions in water, respectively. You also learned that an acid and base neutralize each other to form water, in what is called an acid–base neutralization reaction:

$$H^+(aq) + OH^-(aq) \rightarrow H_2O(l)$$

We can use this reaction to determine the concentration of an acid via titration. For example, suppose we have a hydrochloric acid solution of unknown concentration. To titrate it, we use a *buret*, which is a calibrated tube with a valve (called a *stopcock*), positioned above a collection flask. The buret is filled with 0.100 M NaOH. *To titrate an acid, it is important to know the exact concentration of the base in the buret!* Exactly 50.00 mL of the hydrochloric acid solution of unknown concentration is in the flask. *To titrate an acid, it is important to know the exact volume of acid in the flask!* We have also added a few drops of a colorless compound called an *indicator*, to the flask. We will describe an indicator in a moment.

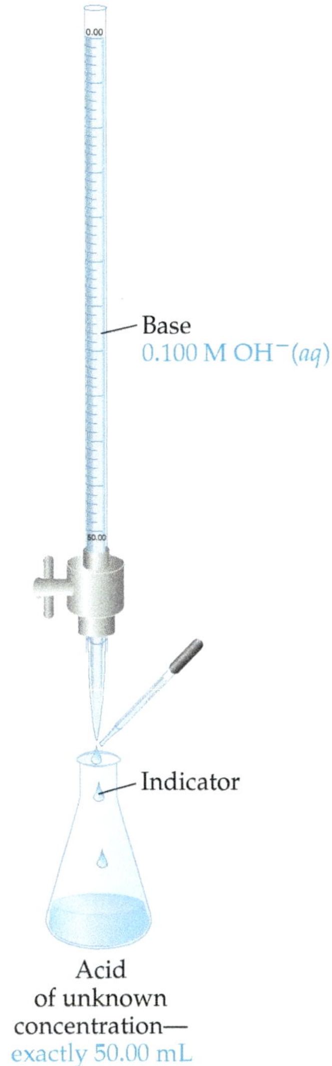

To do the titration, we open the stopcock and begin adding the base to the acid. Because the buret is calibrated and we filled it exactly to the 0.00-mL mark, we can measure the volume of base we drain from the buret. As it enters the flask, the base reacts with the acid, converting $H^+(aq)$ ions to water. If we add sufficient base, we can convert all of the acid's $H^+(aq)$ ions to water. When this occurs, we say the acid has been neutralized. *The key to determining the molarity of the acid is to determine how many moles of $OH^-(aq)$ base were required to accomplish this neutralization.*

This is where the indicator comes in. We added a common indicator called phenolphthalein, which is colorless in acidic solution but red in basic solution. Therefore, the solution in the flask starts out colorless because the flask contains only our acid. After we've added enough base to neutralize the acid, just one more drop will make the solution basic, and the phenolphthalein turns red, alerting us that neutralization has been accomplished. At this point, called the *equivalence point* (because at this point the number of moles of base we've added is equivalent to the number of moles of acid in the flask), we record the volume of base added. Once we know this volume, we can multiply it by the molarity of the base to calculate the moles of base required for neutralization:

$$V_{\text{base added}} \times M_{\text{base}} = \text{Moles of base required for neutralization}$$

For simple acids like HCl(aq) that can produce only 1 mole of $H^+(aq)$ per mole of acid, this is also the number of moles of acid that were in the flask. Remember this!

OK, let's do the titration. The buret is filled with our 0.100 M NaOH to the 0.00-mL mark, and the collection flask contains 50.00 mL of our acid of unknown concentration plus a few drops of phenolphthalein solution. We begin draining base from the buret to neutralize the acid. At a buret reading of 22.50 mL, the solution in the flask changes from colorless to pale red:

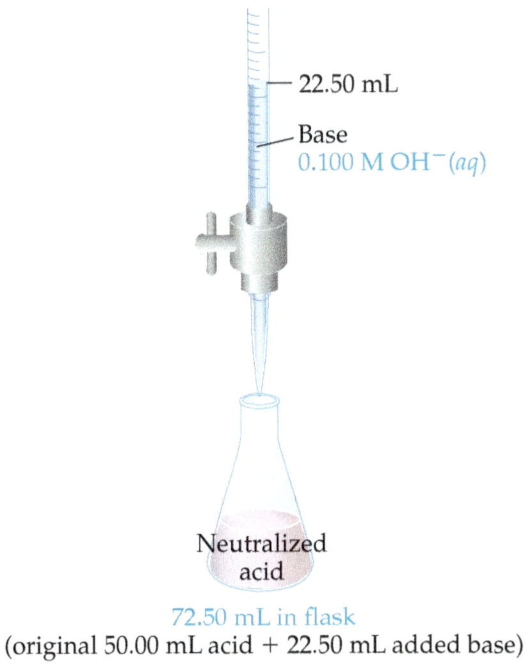

This change in color tells us it took 22.50 mL of base to reach the equivalence point ($V_{\text{base added}} = 22.50\text{ mL}$). The calculation we can now do to find the concentration of the acid is just like any other stoichiometry calculation.

Step 1: Write the balanced chemical equation:

$$HCl(aq) + NaOH(aq) \rightarrow H_2O + NaCl(aq)$$

Step 2: Convert given information to moles. In this case, convert the volume of NaOH solution required to reach the equivalence point to moles of OH^-:

$$\overset{V_{\text{base}}}{0.02250\ \cancel{\text{L NaOH}}} \times \overset{M_{\text{base}}}{\frac{0.100\text{ mol NaOH}}{1\ \cancel{\text{L NaOH}}}} = 0.00225\text{ mol NaOH}$$

Step 3: Use moles of OH^- from step 2 and conversion factors from molecular formulas and the balanced equation to find moles of HCl in the flask:

$$0.00225\ \cancel{\text{mol NaOH}} \times \underbrace{\frac{1\ \cancel{\text{mol HCl}}}{1\ \cancel{\text{mol NaOH}}}}_{\text{From the balanced equation}} = 0.00225\text{ mol HCl}$$

This is how many moles of acid were in the flask!

Step 4: Convert moles from step 3 to desired units. In this case, wanting the acid molarity, we divide moles of HCl by the volume of acid we originally put in the collection flask:

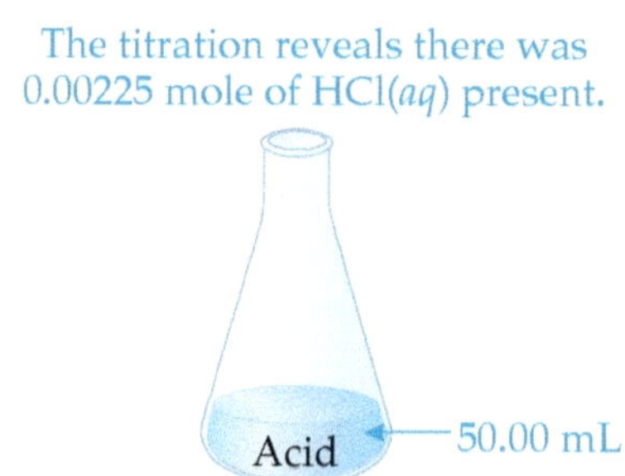

Original acid solution

$$M_{\text{acid}} \times \frac{0.00225\text{ mol HCl}}{0.05000\text{ L}} = 0.0450\text{ M}$$

So the concentration of our unknown acid was 0.0450 M.

Try the following practice problems involving titration. The first one takes you through a titration of sulfuric acid, H_2SO_4, which can produce two $H^+(aq)$ ions per molecule of acid. Then Practice Problem 12.31 drills you on this concept.

Practice Problems

12.30 A 30.00-mL volume of aqueous sulfuric acid, H_2SO_4, is titrated using 0.200 M NaOH and an indicator that turns color only after all the sulfuric acid protons have reacted with OH^- ions. The solution turns color when 48.32 mL of base has been added. What is the molar concentration of the sulfuric acid?

Answer: **Step 1:** Write the balanced neutralization equation:

$$H_2SO_4(aq) + 2NaOH(aq) \rightarrow 2H_2O + Na_2SO_4(aq)$$

Step 2: Convert the given amount of NaOH to moles of OH^-:

$$0.04832 \text{ L NaOH} \times \frac{0.200 \text{ mol NaOH}}{\text{L NaOH}} = 0.00966 \text{ mol NaOH}$$

Step 3: Use moles of NaOH from step 2 to calculate moles of H_2SO_4:

$$0.00966 \text{ mol NaOH} \times \frac{1 \text{ mol } H_2SO_4}{2 \text{ mol NaOH}} = 0.00483 \text{ mol } H_2SO_4$$

Step 4: Convert moles of H_2SO_4 to molarity:

$$\frac{0.00483 \text{ mol } H_2SO_4}{0.03000 \text{ L}} = 0.161 \text{ M } H_2SO_4$$

12.31 A 50.00-mL volume of aqueous phosphoric acid, H_3PO_4, is titrated using 0.100 M NaOH and an indicator that turns color only after all the phosphoric acid protons have reacted with OH^- ions. The solution turns color when 38.60 mL of base has been added. What is the molar concentration of the phosphoric acid?

12.32 A 1.65-g sample of an acid that has one acidic proton per molecule is dissolved in water to give 25.00 mL of solution. It takes 27.48 mL of 1.000 M NaOH to neutralize the acid.

(a) What is the molar concentration of the acid?
(b) What is the molar mass of the acid?
(c) The empirical formula of the acid is CH_2O. What are the molecular formula and name of the acid?

12.8 Colligative Properties of Solutions

We end this chapter with a discussion of some very interesting properties of solutions. To be sure you have the proper background to understand these properties—called *colligative* properties—let's take a look first at some things happening in pure solvents rather than in solutions.

Pure Solvents

You think you know something, and then all of a sudden you find out you don't. For example, suppose you are asked what happens to something when you heat it. This sounds like it could be the easy first question on some television quiz show.

What happens to something when you heat it?

A. It gets warmer.

B. It gets colder.

C. It turns into a chicken.

D. It depends.

"It gets warmer—A. That's my final answer!"

"Oh! I'm sorry! The answer is D, it depends."

It depends? Yes, it depends. To show you why, we need to heat a substance, and to keep things uncomplicated, let's work with one you are familiar with: water. The two physical properties of water we are interested in here are its boiling point and its freezing point. The boiling point of water is 100 °C, and the freezing point is 0 °C. These numerical values are more properly called the **normal boiling point** and **normal freezing point**, with the word *normal* indicating that these are the freezing and boiling points at standard pressure (760 mm Hg = 1 atm).

What you may be less familiar with is water's behavior at these temperatures. For example, consider a beaker of water initially at room temperature (around 22 °C) on a day when atmospheric pressure is exactly 760 mm Hg. If we place the beaker in a 200 °C oven, the temperature of the water increases. Certainly this is no surprise. After all, things get warmer when you add heat to them. Or do they? When the water reaches 100 °C, its temperature stops increasing even though it is still being heated in the 200 °C oven.

Water in a 200 °C Oven

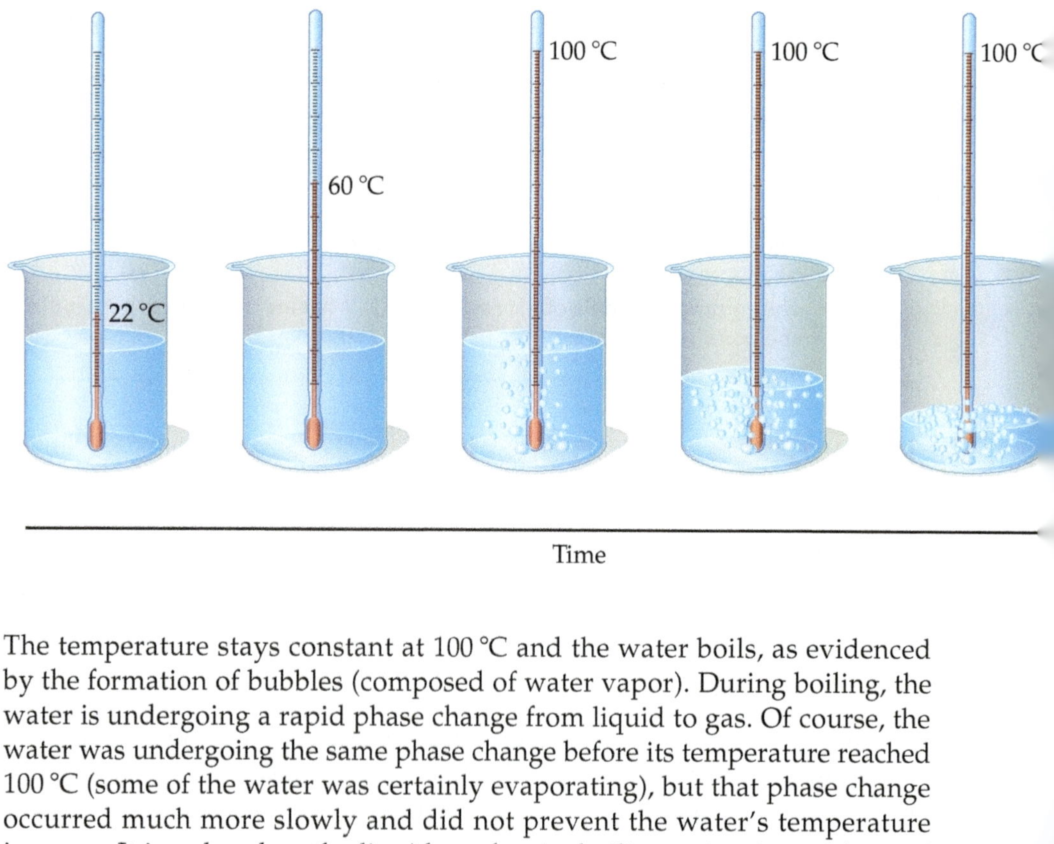

The temperature stays constant at 100 °C and the water boils, as evidenced by the formation of bubbles (composed of water vapor). During boiling, the water is undergoing a rapid phase change from liquid to gas. Of course, the water was undergoing the same phase change before its temperature reached 100 °C (some of the water was certainly evaporating), but that phase change occurred much more slowly and did not prevent the water's temperature increase. It is only when the liquid reaches its boiling point that additional heating does not increase its temperature as this phase change occurs.

Once the liquid water is all converted to vapor at 100 °C, the temperature of the vapor starts to increase, and the vapor is ultimately heated to the temperature of the oven. Why is this so? How is it possible to continuously add heat to liquid water and not have the water get hotter? The answer is that below the boiling point, the heat energy flowing into the water is mostly being used to increase the kinetic energy of the water molecules making them move faster.

(As we'll see in Section 13.3, temperature is a measure of the average kinetic energy of a collection of molecules.) At the boiling point, however, adding heat no longer increases the molecules' kinetic energy. Now all the heat added is used to overcome the attractive forces (hydrogen bonds) between water molecules and separate them from one another so that they can escape into the gas phase.

We get this same type of temperature behavior in the other direction. If you place liquid water initially at room temperature (22 °C) in a freezer set at −20 °C, the temperature of the water drops until it reaches 0 °C. At this point, the water temperature stops decreasing as the water changes phase from liquid to solid. Only when the phase change is complete does the temperature of the ice begin to decrease again.

This behavior is shown below on a plot of temperature versus heat added, called a *heating curve*.

So if anyone asks you whether or not the temperature of a particular substance increases when you heat the substance, your answer should be, "It depends! Even though a substance is being heated, its temperature remains constant anytime it is changing phase at its freezing (melting) point or at its boiling point. That's my final answer!"

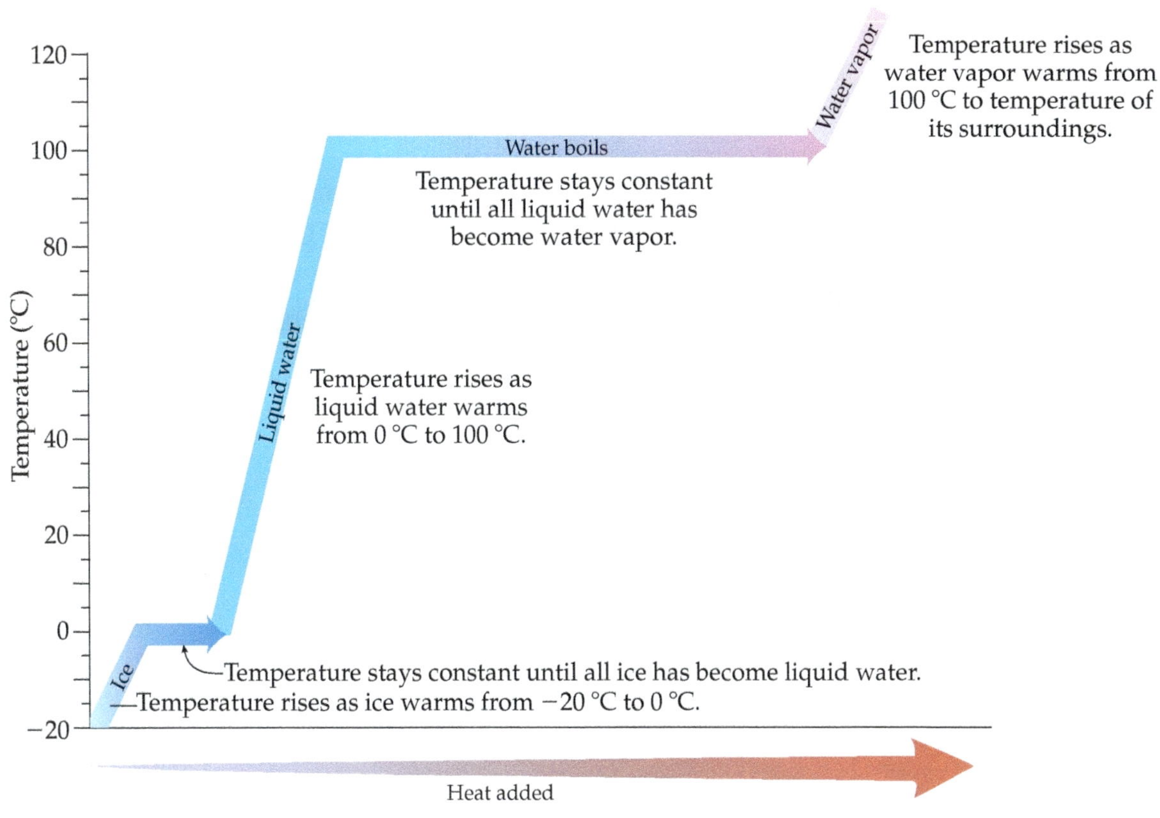

Vapor Pressure

An interesting property of liquids is their "urge" to enter the gas phase even at temperatures well below their normal boiling points. To demonstrate this property, let's again use water and run a simple experiment with it. First we remove all the air from a flask connected to a vacuum pump, as shown in the drawing, and then close the stopcock.

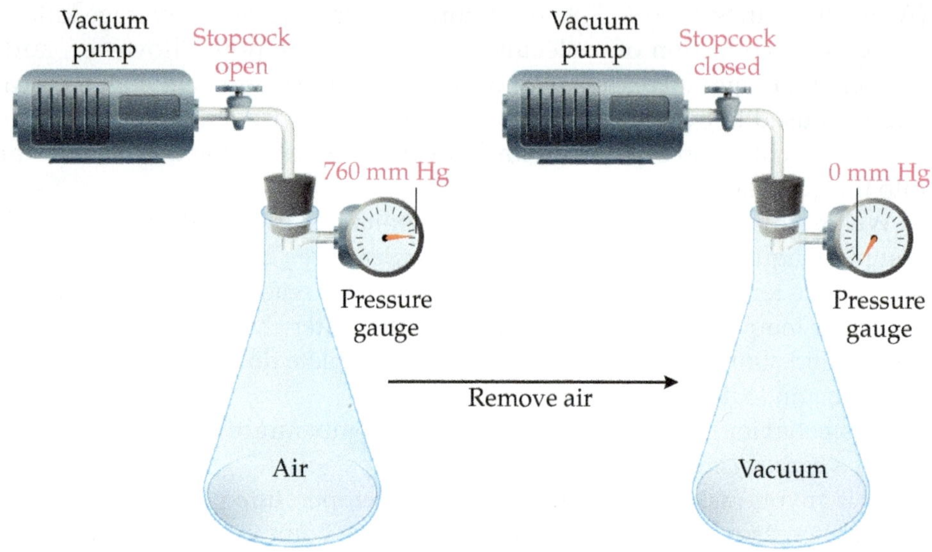

Because we removed all the air, the pressure inside the flask is 0 mm Hg. From this point on, we keep the stopcock closed so that no air can get back into the flask. Next, we use a syringe to place a small amount of water at 25 °C in the flask. (The syringe lets no air into the flask.)

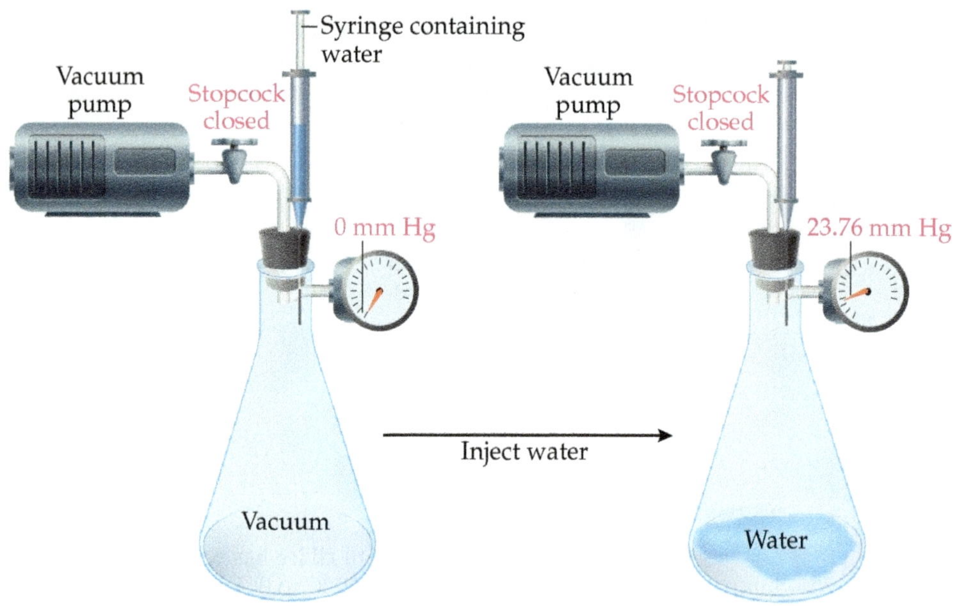

We keep injecting until a small puddle of water has formed at the bottom of the flask. At this point, the pressure inside the flask will be 23.76 mm Hg. If we continue to add water, the pressure remains constant at 23.76 mm Hg. The fact that the pressure in the flask is no longer zero tells us there must be some gas in the flask. That gas is water vapor. Evidently, some of the liquid water evaporated into the gas phase, causing the pressure to increase.

Why did the pressure in the flask stop increasing when it reached 23.76 mm Hg? The answer is that evaporation is not the only thing going on.

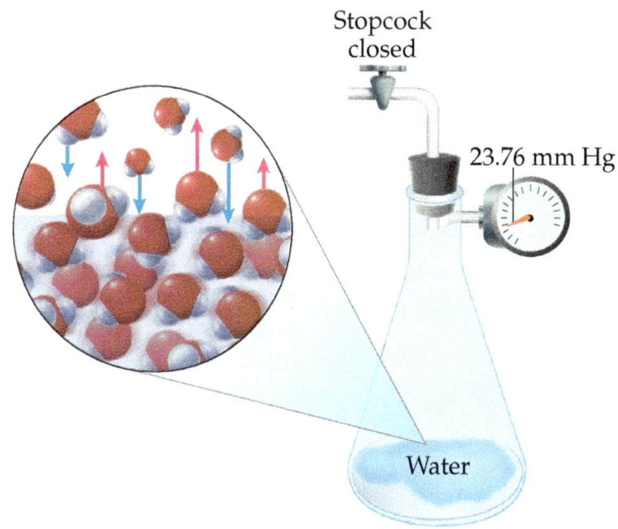

At the same time evaporation is occurring, some of the gas-phase water molecules collide with the liquid surface and re-enter the liquid phase (condensation). At 23.76 mm Hg, the number of water molecules in the gas phase is enough to make the condensation rate equal to the evaporation rate.

When these two rates are equal, the pressure above the liquid no longer changes. We call such a situation in which the rate of a forward process equals the rate of the reverse process a **dynamic equilibrium** and represent it by a set of two arrows:

$$H_2O(l) \rightleftharpoons H_2O(g)$$

The word *dynamic* reminds us that things are happening (in this case, evaporation and condensation) even though the situation appears static.

With this definition of dynamic equilibrium under our belts, we can see why the pressure in our flask stays constant at 23.76 mm Hg. Because there is no air in the flask, that pressure is exerted solely by water molecules in the gas state. Once the pressure reaches 23.76 mm Hg, which is the pressure at which the rate of the change from $g \rightarrow l$ is the same as the rate of the change from $l \rightarrow g$, the number of gas-phase H_2O molecules no longer increases and so the pressure no longer increases. When the liquid and gas states of a substance are in dynamic equilibrium in this way, we say that the pressure exerted by the gas molecules is the **vapor pressure** of the liquid. (Strange but true—the pressure is caused by the gas phase but is said to be a characteristic of the liquid phase.)

Vapor pressure is a function of temperature, as Table 12.1 shows. Had we run our experiment at a different temperature, we would have found a different vapor pressure for water.

Table 12.1 Vapor Pressure of Water at Various Temperatures

Temperature (°C)	Vapor pressure (mm Hg)	Temperature (°C)	Vapor pressure (mm Hg)
0	4.58	55	118.0
5	6.54	60	149.4
10	9.21	65	187.5
15	12.79	70	233.7
20	17.54	75	289.1
25	23.76	80	355.1
30	31.82	85	433.6
35	42.2	90	525.8
40	55.3	95	633.9
45	71.9	100	760.0
50	92.5	105	906.1

WORKPATCH 12.6 What happens to the vapor pressure of liquid water as its temperature is increased? ●

A good way to think about the vapor pressure of a liquid is as a measure of its "urge" to evaporate. The higher the vapor pressure, the greater the tendency for liquid molecules to escape into the gas phase. You can picture the vapor pressure as an arrow that "lives" in the liquid and points up. The length of this arrow is a measure of the vapor pressure of the liquid. Now imagine a second arrow, one above the liquid and pointing down. This arrow represents any external pressure pushing down on the liquid. (For an open container, this external pressure is simply atmospheric pressure.) For example, we can envision the two arrows for an open beaker of water at 25 °C as:

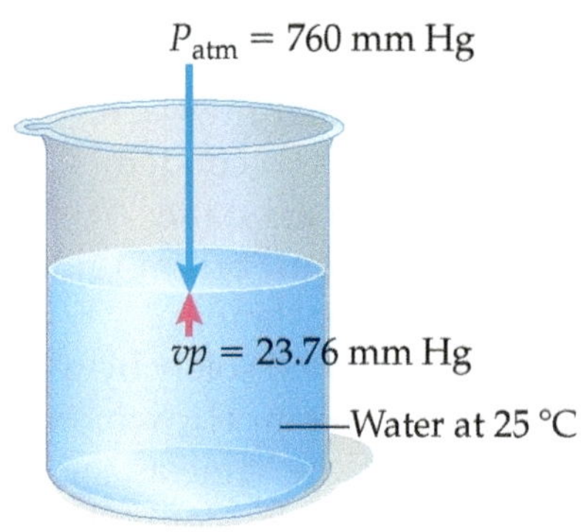

Getting the water to boil is a matter of dueling arrows. As we heat the water, the vapor pressure arrow gets longer until, at 100 °C, its length indicates a vapor pressure of 760 mm Hg (Table 12.1). This means that at 100 °C, the length of the vapor pressure arrow matches the length of the external pressure arrow. It is at this point of equal pressures that the water boils.

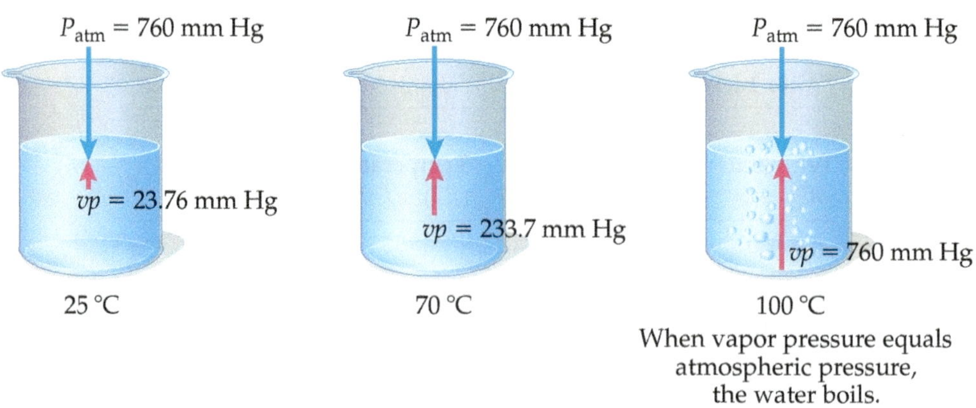

Any liquid boils when its vapor pressure becomes equal to or greater than the external pressure. Knowing this allows us to give another definition of normal boiling point. The normal boiling point of a liquid is the temperature at which its vapor pressure equals standard atmospheric pressure, 760 mm Hg.

It is important to realize that water is changing phase from liquid to gas at all temperatures shown in the above diagram (as well as at all the intermediate temperatures not shown). As it is heated, the water evaporates at a faster and faster rate because its vapor pressure increases. However, only when the water reaches its boiling point and the vapor pressure equals the external pressure does boiling occur.

Once you understand how to make a liquid boil in terms of dueling arrows, you should realize that a liquid can boil at temperatures other than its normal boiling point. All you have to do is vary the external pressure. For example, in Denver, a city 1609 m above sea level, the average atmospheric pressure is only 0.83 atm (631 mm Hg). From Table 12.1 you can see that water has a vapor pressure of 631 mm Hg at approximately 95 °C. Thus, you have to heat water to only 95 °C in Denver to get it to boil (to get the arrows to be of equal lengths). Thus, in boiling water that is 5 °C cooler than it would be at sea level, your noodles will take a bit longer to cook.

All liquids have a vapor pressure. For example, chloroform, CH_3Cl, is a useful industrial solvent. At 25 °C, it has a vapor pressure of 190 mm Hg, much greater than the vapor pressure of water at the same temperature (23.76 mm Hg). The result is that, at any given temperature, chloroform has a much greater "urge" to evaporate than water, and we say that chloroform is more volatile than water. (The word *volatile* means readily vaporizable.) This fact implies that the boiling point of chloroform should be lower than the boiling point of water, and that is indeed the case. The normal boiling point of chloroform is 61.7 °C.

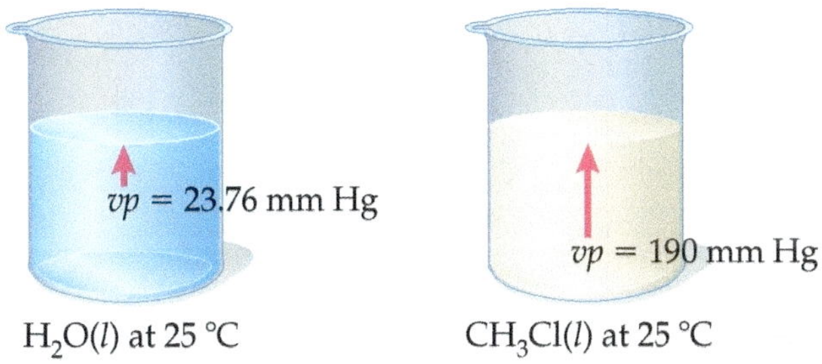

WORKPATCH 12.7 What is the vapor pressure of chloroform at 61.7 °C? ●

Make sure you check your answer against the one at the end of the chapter. If you answered the WorkPatch correctly, you understand what has to happen to the vapor pressure of a liquid in order to make the liquid boil.

Generally speaking, the greater a liquid's vapor pressure at a given temperature, the more volatile the liquid is (the faster it evaporates).

Finally, we can understand this difference between chloroform and water by looking back to the material in Chapters 7 and 11 on intermolecular forces. The attractive forces between chloroform molecules are of the dipole–dipole and London type, both types weaker than the hydrogen-bonding attractions between H_2O molecules in liquid water. Therefore, chloroform molecules have an easier time overcoming one another's grasp and escaping into the vapor phase.

Solutions

It had been slightly below freezing for the past week, and Jim couldn't wait to try out his new ice skates. The freshwater pond in the backyard had frozen solid. It was a very small pond, though, and Jim wished for a much larger expanse of ice. There was a nearby ocean inlet, but it was still liquid. Why, he wondered, hadn't it frozen yet? Luckily, Jim's grandfather was a chemist, and he answered Jim's question by explaining that the ocean is not pure water. Rather, it is a solution of various salts (mostly NaCl).

"Jim," he said, "the physical properties of a liquid change as soon as you dissolve a solute in it. The boiling point goes up (*boiling-point elevation*), the freezing point goes down (*freezing-point depression*), and the vapor pressure decreases (*vapor-pressure lowering*). *Stay awake, Jim!*"

Thanks, Grandpa; we'll take it from here. The amount that the boiling point goes up and the freezing point goes down are represented by the symbols ΔT_b and ΔT_f, respectively:

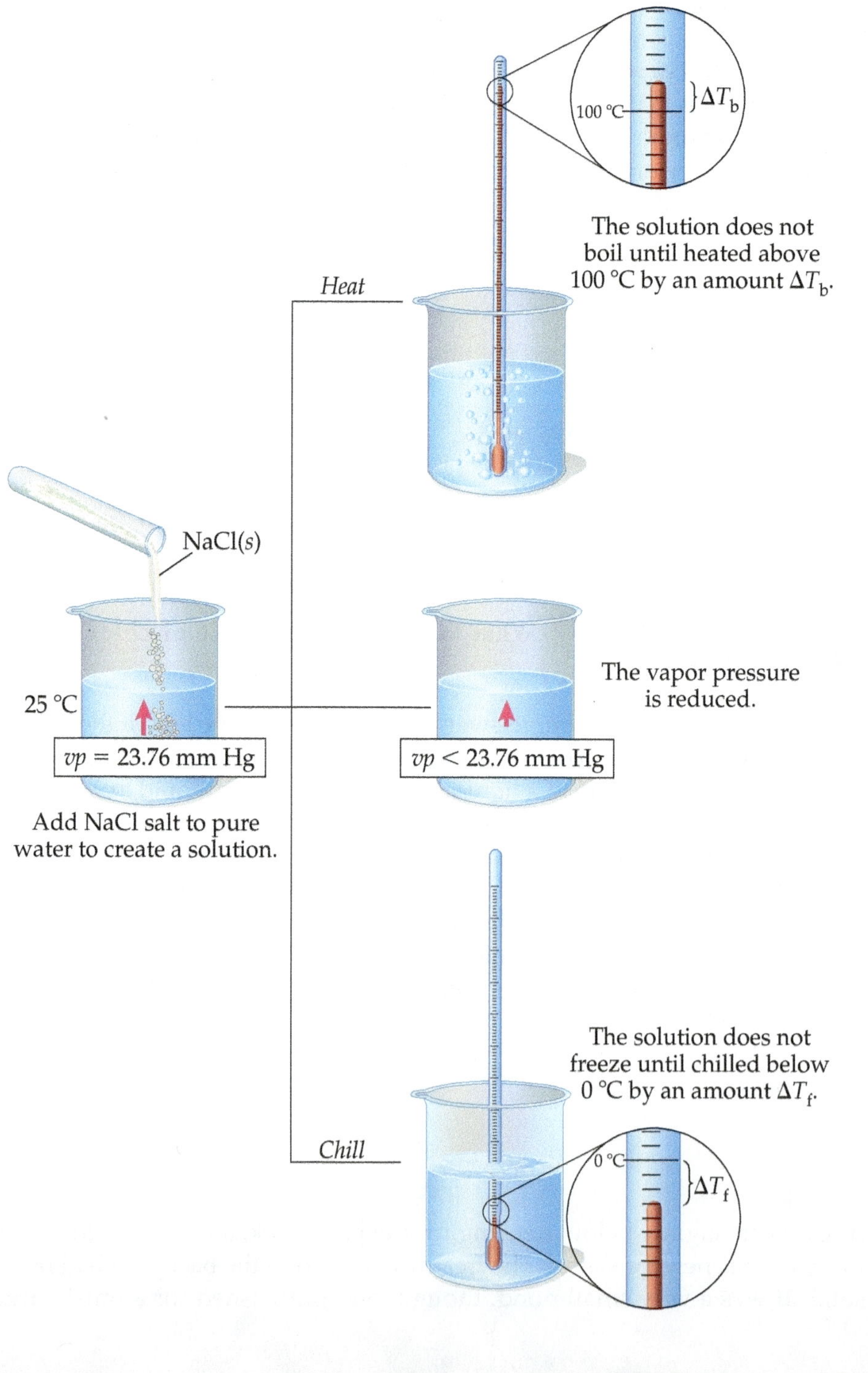

An aqueous solution is therefore more difficult to boil (you have to heat it to above 100 °C) and more difficult to freeze (you have to cool it to below 0 °C) than pure water. The elevation of the boiling point, the depression of the freezing point, and the lowering of the vapor pressure are called *colligative properties* of a solution. A **colligative property** is one that depends on the number of solute particles in a solution but not on their identity. *This is an important point!* How much the boiling point goes up, how much the freezing point goes down, and how much the vapor pressure goes down depend *only on the number of dissolved solute particles*, not on their identity. For example, consider two beakers containing equal volumes of water, and pay close attention to the amount of solute added to each:

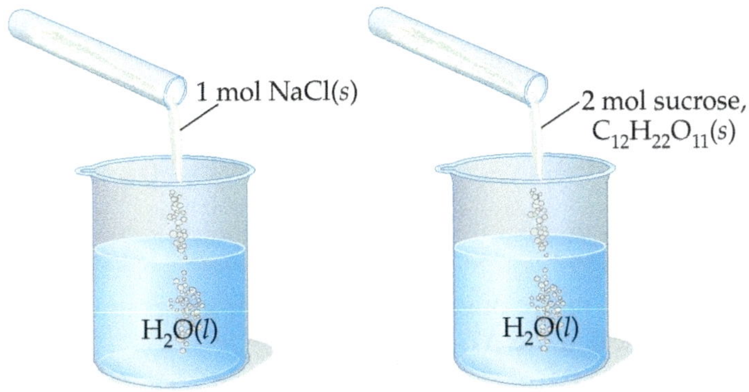

WORKPATCH 12.8 How many moles of dissolved solute particles (molecules or ions) are present in each beaker in the above diagram? (*Hint:* Don't forget that soluble salts dissociate into ions when they dissolve.) ●

The two solutions contain equal numbers of dissolved solute particles. Therefore, the two solutions experience the same increase in boiling-point ΔT_b and the same decrease in freezing-point ΔT_f. The fact that the solute particles are aqueous sugar molecules in one solution and aqueous $Na^+(aq)$ and $Cl^-(aq)$ ions in the other is not important. The only thing that is important is the *total number* of dissolved solute particles in a given volume of solution.

Finally, it is possible to calculate ΔT_b and ΔT_f for a solution by using the equations:

$$\Delta T_f = \frac{K_f \times \text{Moles of solute particles}}{\text{Kilograms of solvent}}$$

$$\Delta T_b = \frac{K_b \times \text{Moles of solute particles}}{\text{Kilograms of solvent}}$$

The K_f and K_b are called the *freezing-point constant* and *boiling-point constant*, respectively, and each solvent has its own values for these constants, as Table 12.2 shows. We can use these equations to calculate the freezing and boiling points of ocean water. On average, ocean water contains about 35.0 g (0.599 mole) of NaCl per kilogram of water. This means there are

Table 12.2 Freezing-Point and Boiling-Point Constants for Some Common Solvents

Solvent	K_f $\left(\frac{°C \cdot kg\ solvent}{Moles\ solute}\right)$	Normal freezing point (°C)	K_b $\left(\frac{°C \cdot kg\ solvent}{Moles\ solute}\right)$	Normal boiling point (°C)
Acetic acid, CH_3CO_2H	3.9	16.7	3.07	118
Benzene, C_6H_6	5.12	5.53	2.53	80.1
Camphor, $C_{10}H_{16}O$	40	179.5	—	—
Carbon tetrachloride, CCl_4	31.8	−22.96	5.03	76.5
Cyclohexane, C_6H_{12}	20.0	6.47	2.79	80.7
Ethanol, C_2H_5OH	1.99	−115	1.22	78.4
Methanol, CH_3OH	—	−97.8	0.80	64.7
Naphthalene, $C_{10}H_{18}$	6.9	80.2	5.65	218
Phenol, C_6H_6O	7.40	43	3.56	181.8
Tertiary butanol, C_4H_9OH	8.3	25.6	—	82.4
Water, H_2O	1.86	0.00	0.52	100.0

2 × 0.599 mole = 1.20 moles of solute particles per kilogram of water. Knowing this allows us to calculate ΔT_f and ΔT_b:

$$\Delta T_f = \frac{K_f \times \text{Moles of solute particles}}{\text{Kilograms of solvent}}$$

$$= \frac{\frac{1.86\ °C \cdot \text{kg solvent}}{\text{mol solute particles}} \times 1.20\ \text{mol solute particles}}{1\ \text{kg solvent}} = 2.23\ °C$$

$$\Delta T_b = \frac{K_b \times \text{Moles of solute particles}}{\text{Kilograms of solvent}}$$

$$= \frac{\frac{0.52\ °C \cdot \text{kg solvent}}{\text{mol solute particles}} \times 1.20\ \text{mol solute particles}}{1\ \text{kg solvent}} = 0.62\ °C$$

These are the changes to the normal freezing and boiling points. To find the depressed freezing point, we must subtract ΔT_f from the normal freezing point of pure water:

$$\text{Depressed freezing point} = \text{Normal freezing point} - \Delta T_f$$
$$= 0.00\ °C - 2.23\ °C = -2.23\ °C$$

To find the elevated boiling point, we must add ΔT_b to the normal boiling point of pure water:

$$\text{Elevated boiling point} = \text{Normal boiling point} + \Delta T_b$$
$$= 100.00\ °C + 0.62\ °C = 100.62\ °C$$

Try the following practice problems. The first two are worked out for you, and the second one is very interesting. It shows how we can calculate the molar mass of a solute without knowing its molecular formula (similar to what we did for gases using the ideal gas law in Chapter 11). This is an important use of colligative properties.

Practice Problems

12.33 Automotive antifreeze is essentially the compound ethylene glycol, $C_2H_6O_2$, molar mass 62.07 g/mol. Suppose you fill your car radiator with 6.00 L of water and then add 4.00 kg of ethylene glycol. At what temperature will this mixture freeze? (The density of water is 1.00 g/mL.)

Answer:

$$\Delta T_f = \frac{\overbrace{\frac{1.86\ ^\circ\text{C} \cdot \text{kg H}_2\text{O}}{\text{mol C}_2\text{H}_6\text{O}_2}}^{K_f} \times \overbrace{4.00\times10^3\ \text{g C}_2\text{H}_6\text{O}_2 \times \frac{1\ \text{mol C}_2\text{H}_6\text{O}_2}{62.07\ \text{g C}_2\text{H}_6\text{O}_2}}^{\text{Moles of solute}}}{\underbrace{6.00\ \text{L H}_2\text{O} \times \frac{1000\ \text{mL H}_2\text{O}}{1\ \text{L H}_2\text{O}} \times \frac{1.00\ \text{g H}_2\text{O}}{1\ \text{mL H}_2\text{O}} \times \frac{1\ \text{kg H}_2\text{O}}{1000\ \text{g H}_2\text{O}}}_{\text{Kilograms of solvent}}} = 20.0\ ^\circ\text{C}$$

Therefore the new freezing point = 0 °C − 20 °C = −20 °C

12.34 A white powder is analyzed via combustion analysis. Results indicate its empirical formula is CH_2O. To determine its molecular formula, a chemist dissolves 117.5 g into 1.175 kg of water, and the resulting solution freezes at −1.03 °C. What is the molecular formula?

Answer: The molar mass of the empirical formula CH_2O is 30.026 g/mol. If we can determine the molar mass of the compound, we can divide it by 30.026 g/mol to find out how many empirical units there are in one molecular formula (Section 9.5). To do this, all we need do is solve the freezing-point-depression equation for moles of solute (we know everything else):

$$\text{Moles of solute} = \frac{\Delta T_f \times \text{kg solvent}}{K_f} = \frac{1.03\ ^\circ\text{C} \times 1.175\ \text{kg H}_2\text{O}}{1.86\ ^\circ\text{C} \cdot \text{kg H}_2\text{O/mol solute}} = 0.651\ \text{mol solute}$$

Thus, we added 0.651 mole of solute to the water, but we also know that the mass of this number of moles of the compound is 117.5 g. To find the molar mass of the compound, we divide this mass by the number of moles:

$$\text{Molar mass} = \frac{117.5\ \text{g}}{0.651\ \text{mol}} = 180\ \text{g/mol}$$

Finally, 180 g/mol ÷ 30.026 g/mol = 6. Therefore, the molecular formula consists of six empirical formulas, which means the molecular formula must be $C_6H_{12}O_6$ (it's glucose!).

12.35 A solution is prepared at sea level (external pressure = 1 atm) by dissolving 100.0 g of calcium nitrate in 450.0 g of water. At what temperature will this solution have a vapor pressure of 760 mm Hg? (*Hint*: What does an aqueous solution at sea level do when its vapor pressure is 760 mm Hg?)

12.36 A student dissolves 45.0 g of an unknown solid in 225.0 g of cyclohexane. It is known to dissolve without dissociating. She cools the solution and finds that the temperature remains at 2.70 °C while the solution changes phase from liquid to solid. What is the molar mass of the solid?

We have spent most of this chapter examining solutions—what they are, why solutes dissolve, how to quantify the amount of solute dissolved, how the properties of solvents change when solutes are added, and how to do stoichiometry calculations for reactions that occur in solution. This emphasis on solutions is certainly understandable given that we are aqueous beings living on a planet covered with water, one of the best all-purpose solvents known. So the next time you dissolve some sweetener in your iced tea, add some antifreeze to a car radiator, or wipe a stain clean with a wet cloth, think about all that is happening as solute and solvent mingle, and you will have a newfound respect for solutions.

+1 One More Thing

12.9 Getting Unlikes to Dissolve—Soaps and Detergents

"Dry"-cleaning

We've been talking so far about like dissolving like. However, much of the dirt that accumulates on our bodies and our clothing—such as grease and oil—is nonpolar, which means washing with polar water won't accomplish much in the way of cleaning. In order to dissolve these nonpolar substances, we need a nonpolar solvent. We could use a nonpolar liquid like carbon tetrachloride, but its expense and toxicity make this approach to cleaning impractical at home. (This is effectively what happens when you send your clothes to be dry-cleaned, however. The clothes are immersed in a nonpolar liquid called perchloroethane in which nonpolar grease and oil dissolve. Dry-cleaning is "dry" only in the sense that no water is used.)

Water—abundant, cheap, nontoxic, and recyclable—is still the solvent of choice for household cleaning, but how do we make nonpolar dirt dissolve in polar water? To find out how, let's go back to the detergent aisle of the supermarket. Getting "unlikes" to dissolve is the job of soaps and detergents. Each soap/detergent molecule consists of a long nonpolar "tail" of carbon and hydrogen (called a *hydrocarbon tail*) and a highly polar "head":

A Typical Soap Molecule

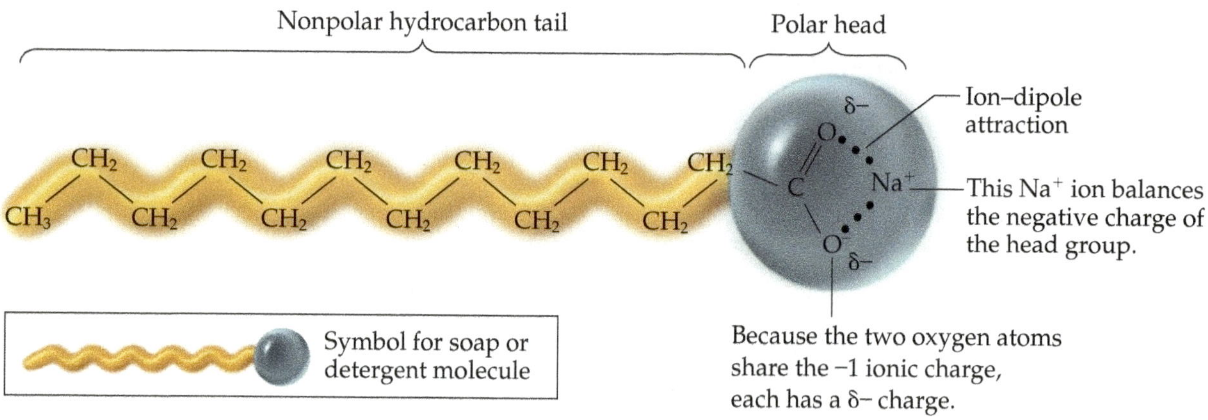

Being polar because of its ionic charges, the head of a soap/detergent molecule is soluble in polar water and is called **hydrophilic** ("water-loving"). The nonpolar hydrocarbon tail is insoluble in water and is referred to as **hydrophobic** ("water-fearing"). (The major difference between soaps and detergents is that soaps contain a carbon–oxygen group at the polar head and detergents contain a sulfur–oxygen group at the polar head. For our purposes in this discussion, you can think of the two as behaving identically in how they clean.)

When soap/detergent molecules are added to water, an interesting spherical structure called a *micelle* forms. The outside of the micelle is formed from the polar heads and is therefore hydrophilic. The inside is made up of the nonpolar tails, all attracted to one another via London forces, and is therefore hydrophobic. A micelle thus presents a polar surface to the surrounding water molecules, which, being polar themselves, are attracted to the micelle via intermolecular forces. The polar heads contain O atoms carrying a partial negative charge (δ^-), and because of this, these O atoms form hydrogen bonds with the water molecules. As a result of these attractions to water, the micelle is soluble in water.

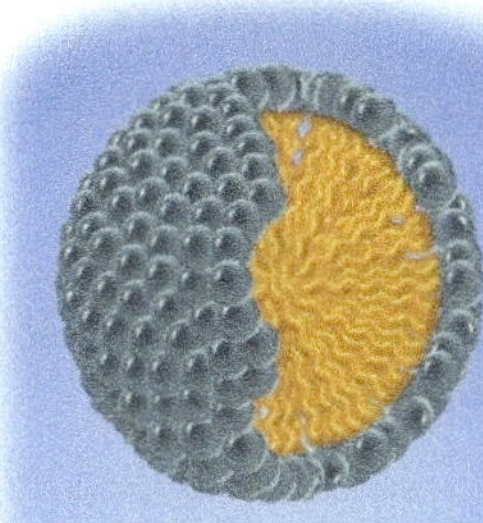

Partial cutaway view of micelle in water

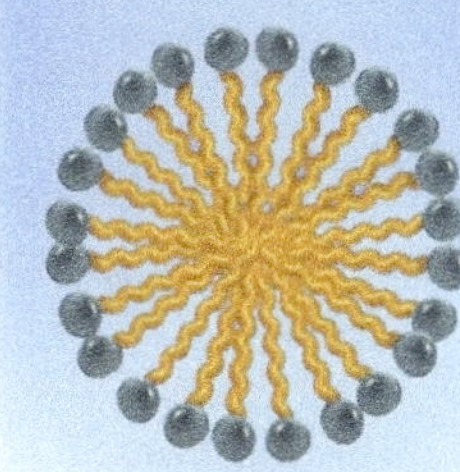

Cross-sectional view of micelle

It's a different story inside the micelle, where, among the conglomeration of nonpolar tails, the environment is strictly hydrophobic. This is just the right kind of environment for dissolving nonpolar grease and oil molecules (like dissolves like). Grease/oil molecules that encounter the micelle quickly dissolve in its interior. Because the grease/oil dissolves in the micelle and the micelle dissolves in water, the result is that the soap/detergent dissolves the grease/oil in the water—a clever way of getting unlikes to dissolve in one another.

The first soaps were prepared from animal fats cooked with potash (potassium hydroxide, KOH) from the ashes of wood fires (see top of following page). Animal fat and bacon grease provided the hydrocarbon tail, and the potash helped to create the polar head.

Today, the chemical industry spends a great deal of time trying to improve soaps and detergents, making grease dissolve even better in water, attempting to whiten and brighten your socks until your feet practically glow when you put them on.

WORKPATCH 12.9 Suppose you put detergent in a dry-cleaning machine that uses carbon tetrachloride. Assuming micelle-type structures form,

(a) Draw the micelle-type structures that would form and explain your diagram.
(b) Explain what kind of dirt these structures would be good at removing. ●

Indeed, there are many dry-cleaning detergents. The micelle you drew to answer the WorkPatch explains how they might work.

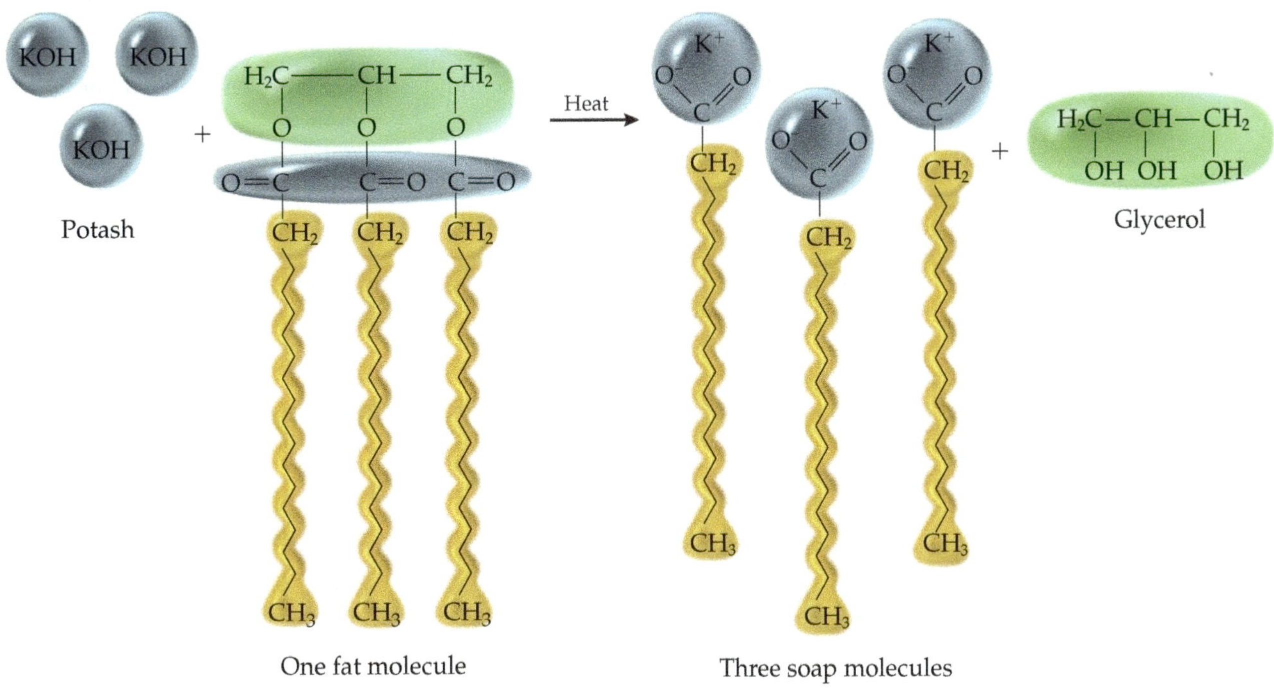

Have You Learned This?

Learning Outcomes

12.1 What Is a Solution?	Describe a solution in terms of solvent, solute, and homogeneity. [37–45]
12.2 Energy and the Formation of Solutions	Describe the energetics and intermolecular attractions associated with creating a solution. [50–77]
12.3 Entropy and the Formation of Solutions	Define entropy and its natural tendency. [78–80, 85] Describe the role of entropy in forming solutions. [71–84]
12.4 Solubility, Temperature, and Pressure	Describe how temperature and pressure affect the solubility of gaseous and solid solutes. [86–92] Calculate the solubility of a solute or the grams of solute required for saturation. [93–97]
12.5 Molarity	Calculate and then describe how to prepare a specific volume of a solution that has a specified molar concentration. [100–104, 109, 110] Perform calculations on solutions that make use of molarity. [105–108, 111]
12.6 Percent Composition	Perform calculations on solutions that involve its concentration in terms of percent composition. [113–119]
12.7 Reactions in Solution	Perform calculations regarding reactions that occur in solution that involve reaction stoichiometry and reactant concentrations. [122–128]
12.8 Colligative Properties of Solutions	Describe the vapor pressure of a liquid and how it depends on temperature. [132–137] Calculate vapor-pressure lowering, freezing-point depression, boiling-point elevation, and solute molar mass for solutions via colligative property laws. [139–145]
12.9 Getting Unlikes to Dissolve—Soaps and Detergents	Describe the two parts of a soap molecule and how it functions to make unlikes dissolve. [146–151]

Skills to Know

Energetics of dissolving

A substance will dissolve in a given solute if the energy released by solvation ($\Delta E_{solvation}$) is greater than the energy required for solute separation and solvent separation. If $\Delta E_{solvation}$ is slightly less than the amount required, the solution is likely still to dissolve, owing to the effect of entropy.

These results can be summed up as "like dissolves like."

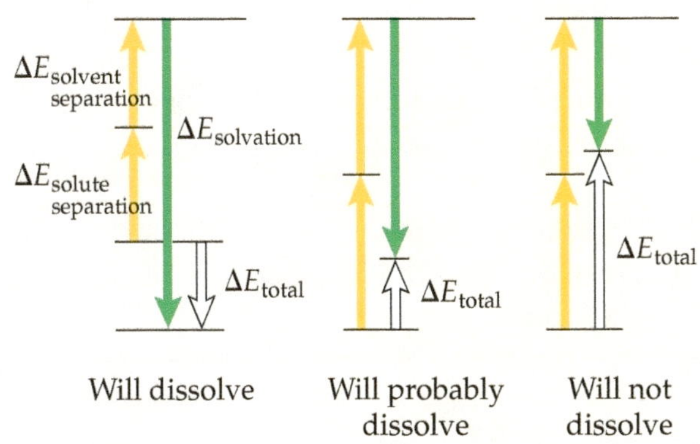

Molarity
Molarity (M) is one of the most common ways to characterize a solution's concentration.

$$\text{Molarity} = \frac{\text{Moles of solute}}{\text{Liters of solution}}$$

Molarity times volume always gives the moles of solute present.

$$\text{Molarity} \times \text{Volume} = \text{Moles of solute}$$

How to prepare a solution of desired molarity

Example: Prepare 250.0 mL of a 1.50 M solution of NaCl.

1. Determine the moles of solute that you need in order to prepare the solution. This quantity equals the desired volume in liters times the desired molarity:

$$\text{Moles of solute needed} = \text{Volume of solution} \times \text{Molarity of solution}$$

$$0.2500\ \text{L} \times \frac{1.50\ \text{mol NaCl}}{\text{L}} = 0.375\ \text{mol NaCl}$$

2. **From scratch**: Convert the moles of solute to grams by using the molar mass as a conversion factor.

   By diluting a stock solution*: Convert moles of solute to volume of stock solution by using the molarity of the stock solution as a conversion factor. (Multiply the moles of solute by the inverse of the stock solution's molarity.)

$$0.375\ \text{mol NaCl} \times \frac{58.443\ \text{g NaCl}}{1\ \text{mol NaCl}} = 21.9\ \text{g NaCl}$$

Using 2.00 M NaCl stock solution:

$$0.375\ \text{mol NaCl} \times \frac{1\ \text{L}}{2.00\ \text{mol NaCl}} = 0.188\ \text{L stock solution} = 188\ \text{mL}$$

*Alternatively, you can combine steps 1 and 2 and use the dilution equation, as described in the text.

3. Add enough water to reach the desired volume.	Add water to make 250.0 mL of solution.

Solving solution stoichiometry problems

Example: Given 0.100 M solutions of $Ca(NO_3)_2$ and NaF, how much of each solution would you need to prepare 1.00 g of $CaF_2(s)$?

1. Write the balanced equation for the reaction.	Net ionic reaction: $Ca^{2+}(aq) + 2\,F^-(aq) \rightarrow CaF_2(s)$
2. Convert all given amounts to moles, using molarities and/or molar masses as conversion factors.	$1.00\ \cancel{g\ CaF_2} \times \frac{1\ mol\ CaF_2}{78.074\ \cancel{g\ CaF_2}} = 0.0128\ mol\ CaF_2$
3. Use the known molar amounts from step 2 to calculate the molar amounts of the substances the question asks for. Use conversion factors derived from the formulas and the balanced equation.	$0.0128\ \cancel{mol\ CaF_2} \times \frac{1\ mol\ Ca^{2+}}{1\ \cancel{mol\ CaF_2}} = 0.0128\ mol\ Ca^{2+}$ $0.0128\ \cancel{mol\ CaF_2} \times \frac{2\ mol\ F^-}{1\ \cancel{mol\ CaF_2}} = 0.0256\ mol\ F^-$
4. Convert the molar amounts from step 3 to the desired units, using molarities and/or molar masses as conversion factors.	$\overbrace{0.0128\ \cancel{mol\ Ca^{2+}}}^{\text{Moles of reactant needed}} \times \overbrace{\frac{1\ \cancel{mol\ Ca(NO_3)_2}}{1\ \cancel{mol\ Ca^{2+}}}}^{\text{From formula of reactant}} \times \overbrace{\frac{1\ L\ solution}{0.100\ \cancel{mol\ Ca(NO_3)_2}}}^{1/M_{reactant}}$ $= 0.128\ L\ solution$ $\underbrace{0.0256\ \cancel{mol\ F^-}}_{\text{Moles of reactant needed}} \times \underbrace{\frac{1\ \cancel{mol\ NaF}}{1\ \cancel{mol\ F^-}}}_{\text{From formula of reactant}} \times \underbrace{\frac{1\ L\ solution}{0.100\ \cancel{mol\ NaF}}}_{1/M_{reactant}} = 0.256\ L\ solution$

Colligative effects

Adding a solute to a solvent has the following colligative effects on the solvent's physical properties: • The freezing point is depressed by an amount ΔT_f. • The boiling point is elevated by an amount ΔT_b. • The vapor pressure is decreased.	$\Delta T_f = \frac{K_f \times \text{Moles of solute particles}}{\text{kg solvent}}$ $\Delta T_b = \frac{K_b \times \text{Moles of solute particles}}{\text{kg solvent}}$ where ΔT_f and ΔT_b are in °C, and K_f and K_b are constants for the given solvent and have units of (°C · kg solvent)/(moles of solute particles).

Concept Questions

1. In an aqueous solution of sugar:
   (a) Sugar is the solvent and water is the solute.
   (b) Water is the solvent and sugar is the solute.
   (c) There is no solvent because *aqueous* means water.
   (d) The composition of the solution is nonhomogeneous.

2. A finely ground mixture of solid sugar with solid salt is:
   (a) A heterogeneous mixture.
   (b) A homogeneous mixture.
   (c) A solution.
   (d) A nonaqueous solution.

3. In order for a solute to dissolve in a solvent, there must be:
   (a) Ion-dipole forces between solvent molecules.
   (b) Repulsion between solvent and solute molecules.
   (c) Intermolecular attraction between solvent and solute molecules.
   (d) Weak intermolecular forces between solute molecules.

4. Solvation is a step in the dissolving process that:
   (a) Always takes some energy.
   (b) Always releases some energy.
   (c) Is independent of any energy changes.
   (d) Has little to do with whether or not a solution will form.

5. For a solute to dissolve in a solvent, it is best that:
   (a) Solvation requires more energy than solvent separation and solute separation.
   (b) Solvation releases as little energy as possible.
   (c) The sum of the energy required for solute and solvent separation be huge.
   (d) Solvation releases more energy than solvent separation and solute separation.

6. With respect to entropy, when a solution forms:
   (a) Entropy increases.
   (b) Solute particles become more randomized.
   (c) We obtain a more probable distribution of solute particles.
   (d) The increase in entropy can sometimes overcome a positive ΔE_{total} for dissolving.
   (e) (a), (b), (c), and (d) are true.
   (f) ΔE_{total} for dissolving will always be the only consideration for dissolving.

7. A saturated solution is a solution:
   (a) Where the concentration of solute is at a maximum.
   (b) Where the molarity is greater than the maximum possible.
   (c) Prepared by using excess solvent to dissolve the solute.
   (d) Where the concentration is given in terms of moles of solute per liter of solution.

8. To prepare 500.0 mL of a 2.0 M solution, you would:
   (a) Dissolve 2 moles of solute in enough solvent to give 500.0 mL of solution.
   (b) Dissolve 1.0 mole of solute in enough solvent to give 500.0 mL of solution.
   (c) Dissolve 1.0 mole of solute in 500.0 mL of solvent.
   (d) Dissolve 2.0 moles of solute in enough solvent to give 1.00 L of solution.

9. If you have 500. mL of a 2.0 M solution and you add enough solvent to dilute it to a total new volume of 2.0 L, its new concentration will be:
   (a) 4.0 M.
   (b) 1.0 M.
   (c) 0.50 M.
   (d) 0.25 M.

10. When a solute is added to a solvent:
   (a) The vapor pressure of the solvent increases.
   (b) The vapor pressure of the solvent decreases.
   (c) The solvent's boiling point and freezing point increase.
   (d) The solvent's boiling point and freezing point decrease.

Answers: 1b; 2a; 3c; 4b; 5d; 6e; 7a; 8b; 9c; 10b

What Is a Solution?

12.37 Define the terms *solute*, *solvent*, and *solution*.

12.38 Suppose you mixed a small amount of table sugar with a large amount of flour and then spent hours grinding the mixture to a very fine powder. Is this mixture a solution? Explain.

12.39 Both a soft drink and the atmosphere we breathe are properly called solutions. For each, state why this is true, and also state the solvent for each.

12.40 Suppose there is a dust storm due to the winds blowing dry soil into the air. Would a sample of this dusty atmosphere be a mixture? Would it be a solution? Explain.

12.41 Vinegar is a common household solution that we consume. What is the solvent and what is the solute in vinegar?

12.42 What is a solid solution? Give some examples.

12.43 Classify the following as solutions or heterogeneous mixtures:
   (a) 14 karat gold (prepared by mixing 10 parts molten copper with 14 parts molten gold and then allowing the substance to cool until it solidifies)

(b) Filtered ocean water
(c) A piece of wood
(d) Your exhaled breath
(e) A bottle of salad oil and vinegar shaken extremely well

12.44 Suppose you had a 50:50 homogeneous mixture of oxygen gas in helium gas. Which would you call the solvent and which would you call the solute?

12.45 A solution of table salt and table sugar in water is allowed to evaporate, leaving behind the two solid solutes. Is what remains a solution or a heterogeneous mixture? Explain your answer. (*Hint:* As solids, both solutes exist as lattices that would exclude each other.)

Energy and the Formation of Solutions

12.46 Is there anything wrong with the following diagram? Explain your answer.

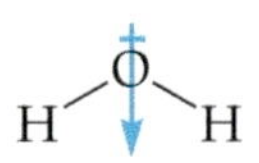

12.47 What evidence is there in your everyday experience to indicate that intermolecular attractive forces must exist between water molecules?

12.48 If there were no attractive forces between water molecules, what phase or phases of water would you expect to be prevalent at room temperature? Explain your answer.

12.49 In what way is melting a piece of NaCl similar to dissolving it in water?

12.50 Why does it take such a high temperature to melt NaCl but a much lower temperature to dissolve it in water?

12.51 When an ionic substance such as NaCl dissolves, the crystal lattice has to break apart to release the individual ions into the solution. Does this part of the dissolving process absorb energy or release energy? Explain your answer.

12.52 In dissolving any solute, room must be made in the solvent to accommodate solute particles. Does making this room absorb energy or release energy? Explain why.

12.53 Consider the process known as hydration during aqueous solution formation.
(a) What is hydration?
(b) Does it release or absorb energy? Explain.

12.54 Anhydrous calcium sulfate [$CaSO_4$]; *anhydrous* means without any water in its lattice, will absorb water (will hydrate) to give calcium sulfate hydrate. The water molecules end up in "holes" or spaces in the ionic lattice. This process releases a considerable amount of energy in the form of heat. Explain why.

12.55 List the three steps that occur as NaCl dissolves in water, and illustrate each with a diagram.

12.56 Solid sucrose consists of individual sucrose molecules fixed in a lattice. The molecules are hydrogen-bonded to one another via OH groups. Knowing this, comment on which takes more energy—breaking up an NaCl lattice or breaking up a sucrose lattice. Justify your answer.

12.57 When NaCl dissolves, what helps keep the dissolved Na^+ and Cl^- ions from coming back together and reforming the lattice, precipitating the solid?

12.58 What is the name of the attractive force between dissolved Na^+ ions and water molecules? Diagram this force, showing how a water molecule would approach an Na^+ ion. Do the same for a Cl^- ion.

12.59 Some alcohols are quite soluble in water. For example, isopropyl alcohol, shown below, is sold as an aqueous solution we call rubbing alcohol. Show how a water molecule would be attracted to isopropyl alcohol, and name the strongest intermolecular force involved. Would you call this interaction hydration?

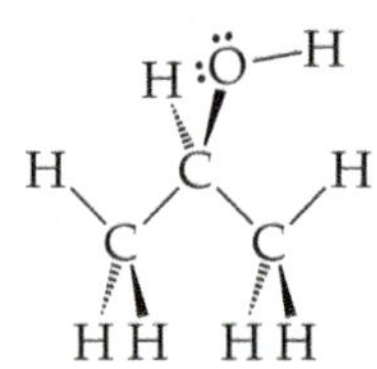

12.60 Why is the amount of energy associated with the solvation step of the dissolving process so critical to whether or not a solute will dissolve?

12.61 Consider an ionic solid dissolving in a liquid. Suppose the energy released as solute–solvent interactions take place is substantially less than the energy required to break up the lattice and to make room in the solvent. Would the solid be very soluble in this liquid? Explain your answer fully.

12.62 Consider dissolving an ionic salt in room-temperature water.
(a) Some salts dissolve in water to yield a much warmed solution. How is this possible?
(b) Some salts dissolve in water to yield a cold solution. How is this possible?

12.63 When solid sodium hydroxide NaOH(*s*) pellets are dissolved in water, the water can get extremely hot and almost boil to give NaOH(*aq*) solution. What is the source of this heat?

12.64 Suppose the hydration energy for an ionic compound is much less than the energy required to pull apart the lattice. Is such a compound likely to be soluble or insoluble in water? Explain your answer.

12.65 Sodium chloride is very soluble in water but insoluble in liquid hexane, C_6H_{14}. Why is this so?

12.66 Even though NaCl does not dissolve in hexane, C_6H_{14}, if we imagine the dissolving process for this system, there is one step that would take less energy than the corresponding step for NaCl dissolving in water. Which step is it? Why would it require less energy?

12.67 Which has the greater hydration energy, $AlCl_3$ or NaCl? Explain your answer.

12.68 In theory, it is possible for one ionic compound to have a greater hydration energy than another and still be the less soluble of the two. What would have to be true for this to be so?

12.69 What is meant by the rule of thumb "like dissolves like"?

12.70 What do we mean by the total energy change ΔE_{total} for the dissolving process, and why is it important to know about this energy change?

12.71 Consider the following diagram and answer the questions that follow:

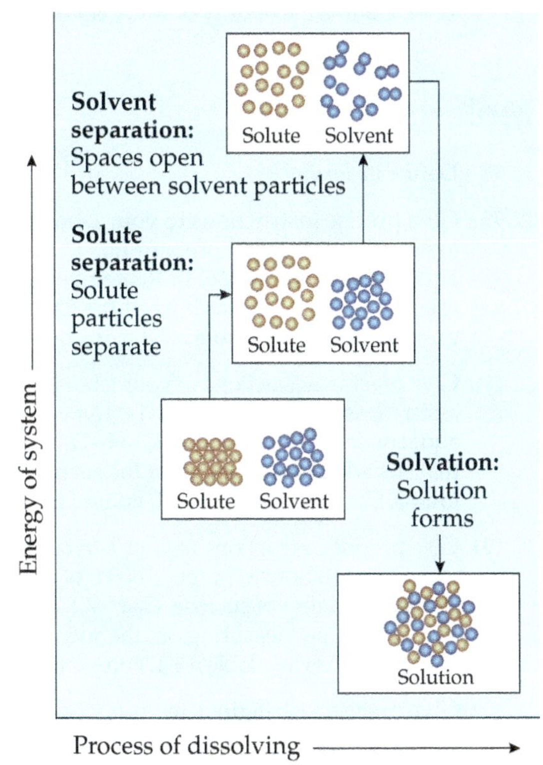

(a) Why does solute separation require energy?
(b) Why does solvent separation require energy?
(c) Why does solvation release energy?
(d) On the basis of this diagram, would you expect the solute to be soluble or insoluble in the solvent? Explain.
(e) Assuming the solute does dissolve, would you expect the solution to get warmer or colder? Explain.

12.72 When we write (*aq*) for a dissolved ion, such as $Na^+(aq)$, exactly what does it mean? Answer both in words and with a diagram.

12.73 Suppose you have two ionic compounds, A and B. The only significant difference between them is that the ionic bonds that hold the lattice together in compound A are stronger than those in compound B. Which compound would you expect to be more soluble in water? Explain your answer.

12.74 When a gaseous solute dissolves in water, which step in the dissolving process is essentially skipped? Explain why.

12.75 When a liquid solute dissolves in water, there is still a solute-separation step that absorbs energy, but the step doesn't require breaking up a crystal lattice as for a typical solid solute. What happens in this step with a liquid solute, and why does the step absorb energy?

12.76 These three substances are all liquids at room temperature:

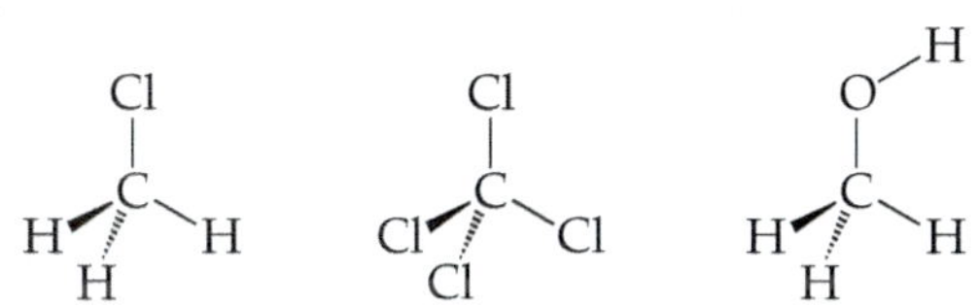

CH_3Cl CCl_4 CH_3OH

Which do you expect to be least soluble in water? Most soluble in water? Explain your answers fully.

12.77 Oil is insoluble in water. Which step in the three-step process for dissolving is responsible for this? Explain why this step is the problem.

Entropy and the Formation of Solutions

12.78 What is entropy?

12.79 The natural tendency for systems is to decrease in energy. What is the natural tendency for entropy?

12.80 What is the reason for your answer to the previous problem? Make sure your answer uses the word *probable*.

12.81 When you release a drop of blue food dye into a beaker of water, the drop eventually dissolves to give a homogeneous light-blue solution. No matter how long you wait, the dye molecules will never regroup to form the original drop. Why not?

12.82 In terms of total energy change, when is entropy an important factor in determining whether or not a solute dissolves in a solvent? When is it not an important factor?

12.83 True or false? Whenever something spontaneously dissolves in water, the entropy has increased. Justify your answer.

12.84 Certainly if NaCl dissolved in liquid hexane, C_6H_{14}, the entropy associated with the Na^+ and Cl^- ions would increase. Nevertheless, NaCl does not dissolve in hexane. Why not?

12.85 If the natural tendency of systems is more towards states of increased probability, then what do you predict for the ultimate fate of the universe?

Solubility, Temperature, and Pressure

12.86 How does increasing temperature affect the solubility in water (a) of most solids and (b) of gases?

12.87 What effect does increasing pressure have on the solubility of gaseous solutes? Explain your answer.

12.88 How is the medical condition known as the bends related to solubility?

12.89 Examine the bottom graph on page 462 showing solubility in water as a function of temperature. What is the trend for most of the ionic substances shown?

12.90 One ionic compound in the bottom graph on page 462 shows almost no temperature dependence, and one clearly violates the general trend. Identify these two ionic compounds.

12.91 Aquatic life is often damaged when hot water is discharged from power stations into rivers and lakes. What might this have to do with gas solubility in water?

12.92 When a can of soda pop is opened, the pressure of carbon dioxide gas that is above the pop in the can drops from high pressure to 1 atm. What does this have to do with the pop going flat when the can is left open?

12.93 What is the mass in grams of sucrose necessary to saturate 100.0 lb of water at 90 °C? (Use the top graph on page 462; 1 lb = 453.6 g.)

12.94 (a) Use these data to plot solubility as a function of temperature for KCl and Li_2SO_4:

	Solubility (g/100 g water)	
Temp (°C)	**KCl**	**Li_2SO_4**
0	27.6	35.3
10	31.0	35.0
20	34.0	34.2
30	37.0	33.5
40	40.0	32.7
50	42.6	32.5
100	56.7	29.9

(b) Using the plot, estimate the solubility of both compounds in water at 70 °C.

(c) How much of each compound can be dissolved in a beaker containing 75 g of water at 70 °C?

12.95 How many more grams of Li_2SO_4 can you dissolve in 250.0 g of water at 10.0 °C than at 50.0 °C? (See Problem 12.94.)

12.96 A saturated solution of KCl was prepared at 40.0 °C using 5.00 lb of water. How many grams of KCl were required to prepare this solution? (See Problem 12.94; 1 lb = 453.6 g.)

12.97 A saturated solution of KCl was prepared at 50.0 °C using 2.00 L of water. How many grams of KCl were required to prepare this solution? (See Problem 12.94; assume the density of water is 1.00 g/mL.)

Molarity

12.98 Define *molarity*.

12.99 Give precise instructions to your laboratory assistant as to how to prepare 1.00 L of a 1.00 M aqueous solution of $CaCl_2$. Remember that your assistant will be measuring out the $CaCl_2$ in grams. She has available a 1-L volumetric flask.

12.100 Give precise instructions to your laboratory assistant as to how to prepare 1.00 L of a 1.00 M aqueous solution of sucrose, $C_{12}H_{22}O_{11}$. Remember that she will be measuring out the sucrose in grams. She has available a 1-L volumetric flask.

12.101 Give precise instructions to your laboratory assistant as to how to prepare 1.00 L of a 0.250 M aqueous solution of sucrose, $C_{12}H_{22}O_{11}$. Remember that she will be measuring out the sucrose in grams. She has available a 1-L volumetric flask.

12.102 Give precise instructions to your laboratory assistant as to how to prepare 0.500 L of a 1.50 M aqueous solution of sucrose, $C_{12}H_{22}O_{11}$. Remember that she will be measuring out the sucrose in grams. She has available a 0.50-L volumetric flask.

12.103 Your assistant tells you she measured out 2.50 moles of NaCl and then added enough water to get 500.0 mL of solution to prepare a 5.00 M solution of NaCl.

(a) What was the mass of the NaCl in grams?
(b) Did she successfully prepare a 2.5 M solution? Prove your answer.

12.104 Your assistant tells you he measured out 116.886 g of NaCl and then added exactly 1.00 L of water to it to prepare a 2.00 M solution of NaCl. Do you fire him or give him a promotion? Explain.

12.105 You have 2500.0 mL of a 0.250 M solution of NaCl.
(a) How many moles of NaCl are present in this solution?
(b) How many moles of ions are present in this solution?
(c) How many grams of NaCl would you recover if you evaporated all of the water off of this solution?

12.106 You have 45.0 mL of a 0.250 M solution of sucrose, $C_{12}H_{22}O_{11}$.
(a) How many moles of $C_{12}H_{22}O_{11}$ are present in this solution?
(b) How many grams of sucrose would you recover if you evaporated all of the water off of this solution?
(c) A student says that if you did part (b) and recovered all of the evaporated water as a liquid, you would get 45.0 mL of liquid water. Is this student correct? Explain.

12.107 How many milliliters of a 1.00 M solution of NaCl are required to obtain 5.00 g of NaCl?

12.108 How many milliliters of a 0.250 M solution of glucose, $C_6H_{12}O_6$, are required to obtain 100.0 g of glucose?

12.109 There is a bottle of 0.500 M sucrose stock solution in the laboratory. Give precise instructions to your assistant on how to use the stock solution to prepare 250.0 mL of a 0.348 M sucrose solution.

12.110 There is a bottle of 4.50 M NaCl solution in the laboratory. Give precise instructions to your assistant on how to use the stock solution to prepare 100.0 mL of a 4.00 M NaCl solution.

12.111 A student plans to divide the molarity of his solution by its volume to determine the number of moles of solvent in it. He is making two mistakes here. Identify the mistakes and correct both of them.

Percent Composition

12.112 Consider the three types of percent composition of a solution.
(a) What are the names of these three types of percent compositions?
(b) Which term is missing from all three types of percent compositions: *solvent*, *solute*, or *solution*?
(c) Give the mathematical definition of each type of percent composition.

12.113 In Practice Problem 12.3, you learned that "proof" for an alcoholic drink equals twice the percentage of alcohol in the drink. The complete definition of proof is that it is twice the percentage *by volume* of alcohol. Knowing this, exactly what does it mean to have a 90-proof drink?

12.114 A solution of ethanol is prepared by combining 22.5 g of ethanol with 49.6 g of water. What is the percent composition by mass of alcohol in this solution?

12.115 A solution of a particular solid solute in water has a concentration of 25.0 mass %.
(a) Given 100.0 g of this solution, how many grams of solute do you have?
(b) Given 48.0 g of this solution, how many grams of solute do you have?
(c) How many grams of this solution do you need to obtain 56.5 g of solute?

12.116 How would you prepare 2.00 kg of an NaCl solution that is 30.0 mass % NaCl?

12.117 You want 1.00 L of an alcohol–water solution that is 5.00 vol % alcohol.
(a) How would you prepare this solution?
(b) Suppose the alcohol is ethanol, C_2H_6O, which has a density of 0.789 g/mL and an MM = 46.07 g/mol. What would be the molarity of the solution?

12.118 An alcohol–water solution is 35.00 vol % alcohol. How much solution is required to obtain 200.0 mL of alcohol?

12.119 An aqueous solution is 25.0 mass % NaCl. The density of the solution is 1.05 g/mL. What is the molarity of the solution? (*Hint:* Assume you have 100.0 g of solution.)

Reactions in Solution

12.120 What is a solvent cage?

12.121 What do we mean when we say that solute particles *diffuse* through a solution?

12.122 You have two solutions, one 1.50 M sodium sulfide and the other 1.00 M $Pb(NO_3)_2$.
(a) Write a net ionic equation for the precipitation reaction that occurs when these solutions are combined.
(b) How many milliliters of the two solutions must be combined to prepare 10.00 g of precipitate?
(c) Suppose you filter off the precipitate and find that your percent yield is 50.0%. What volumes of the solutions should you have combined to isolate 10.00 g of precipitate?

12.123 You have two solutions, one 0.755 M barium nitrate and the other 1.250 M calcium hydroxide.

(a) Write a net ionic equation for the precipitation reaction that occurs when these solutions are combined.
(b) How many milliliters of the two solutions must be combined to prepare 5.00 g of precipitate?
(c) Suppose you filter off the precipitate and find that your percent yield is 85.0%. What volumes of the solutions should you have combined to isolate 5.00 g of precipitate?

12.124 You have two solutions, one 0.650 M iron(III) nitrate and the other 1.500 M ammonium carbonate.
(a) Write a net ionic equation for the precipitation reaction that occurs when these solutions are combined.
(b) If you pour 200.0 mL of each solution into the same flask, what is the theoretical yield of the precipitate in grams?
(c) What is the molar concentration of the excess reactant ion?

12.125 You have two solutions, one 0.800 M sodium phosphate and the other 0.800 M lead(II) acetate.
(a) Write a net ionic equation for the precipitation reaction that occurs when these solutions are combined.
(b) If you pour 100.0 mL of the sodium phosphate solution and 50.0 mL of the lead(II) acetate solution into the same flask, what is the theoretical yield of the precipitate in grams?
(c) What is the molar concentration of the excess reactant ion?

12.126 A 25.00 mL sample of aqueous hydrobromic acid of unknown concentration is neutralized by 43.28 mL of 0.1001 M NaOH(*aq*).
(a) Write a net ionic equation for this acid–base neutralization reaction.
(b) How many moles of NaOH did it take for the neutralization?
(c) How many moles of hydrobromic acid were neutralized?
(d) What is the molar concentration of the hydrobromic acid?

12.127 A 25.00 mL sample of aqueous sulfuric acid of unknown concentration is neutralized by 27.55 mL of 1.0002 M NaOH(*aq*).
(a) Write a net ionic equation for this acid–base neutralization reaction.
(b) How many moles of NaOH did it take for the neutralization?
(c) How many moles of sulfuric acid were neutralized?
(d) What is the molar concentration of the sulfuric acid?

12.128 Propionic acid has one acidic proton per molecule. A solution is prepared by dissolving 0.273 g of propionic acid in enough water to yield 100.0 mL of solution. This solution is neutralized by 36.82 mL of 0.1001 M NaOH(*aq*).
(a) What is the molar concentration of the propionic acid?
(b) What is the molar mass of propionic acid?

Colligative Properties of Solutions

12.129 A substance has the following heating curve:

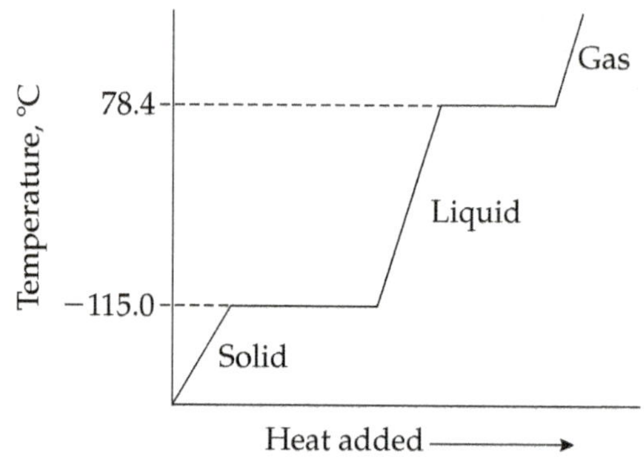

(a) What is the melting point of the substance?
(b) What is its freezing point?
(c) What is its boiling point?
(d) At what temperature does the vapor condense?
(e) Why isn't the temperature increasing as the substance is being heated at the horizontal portions of the curve?
(f) According to Table 12.2, what is this substance?

12.130 When it is warmed, a piece of plastic changes from a solid phase in which its molecules are positioned in an ordered arrangement to a solid phase in which its molecules are arranged randomly. How could you determine the temperature at which this phase change occurs?

12.131 When water boils, its temperature does not increase even though it is being heated. If the heat energy isn't being used to increase the temperature of the water, what is it being used for?

12.132 If liquid water is introduced into an evacuated (zero pressure) flask, the pressure will begin to rise.
(a) Eventually, the pressure will stop rising and reach a constant value. Why does it stop rising?
(b) What do we call this pressure?
(c) What will happen to this pressure if we increase the temperature of the flask? Why does it happen?

12.133 Define *vapor pressure* and include the word *equilibrium* in your definition.

12.134 Hexane, which is a liquid at room temperature, is more volatile than water.
(a) What does this mean?
(b) Which liquid has the higher vapor pressure at a given temperature?

(c) Which liquid most likely has the higher boiling point?
(d) Why is hexane more volatile than water?

12.135 How could you make water boil at 20 °C (approximately room temperature)? (*Hint*: Use information from Table 12.1.)

12.136 Use the dueling-arrows model to explain what heating does to a liquid that eventually causes the liquid to boil.

12.137 What is the vapor pressure of benzene at 80.1 °C? (*Hint*: Consult Table 12.2.)

12.138 What does the word *colligative* mean when referring to colligative properties of a substance? Also, give three examples of colligative properties.

12.139 A student dissolves 28.7 g of $NaCl(s)$ in 2.00 kg of water. How many grams of sucrose, $C_{12}H_{22}O_{11}$, would she have to dissolve in 2.00 kg of water to make the sucrose solution have the same freezing point as the NaCl solution?

12.140 A student dissolves 36.9 g of calcium nitrate in 500.0 g of water. How many grams of glucose, $C_6H_{12}O_6$, would he have to dissolve in 500.0 g of water to make the glucose solution have the same boiling point as the calcium nitrate solution?

12.141 A student dissolves 45.6 g of sodium sulfate in 550.0 g of water at a given temperature. How many grams of sodium chloride would she have to dissolve in 550.0 g of water at the same temperature to make the chloride solution have the same vapor pressure as the sulfate solution?

12.142 What are (a) the boiling point and (b) the freezing point of a solution made by dissolving 38.60 g of sodium phosphate in 200.0 mL of water? (The density of water is 1.000 g/mL.)

12.143 What are (a) the boiling point and (b) the freezing point of a solution made by dissolving 5.75 g of solid cetyl alcohol, $C_{16}H_{34}O$, in 100.0 mL of benzene? (The density of benzene is 0.874 g/mL.)

12.144 A solution is prepared by dissolving 5.00 g of caffeine in 100.0 g of carbon tetrachloride. The solution is cooled and the temperature plotted over time:

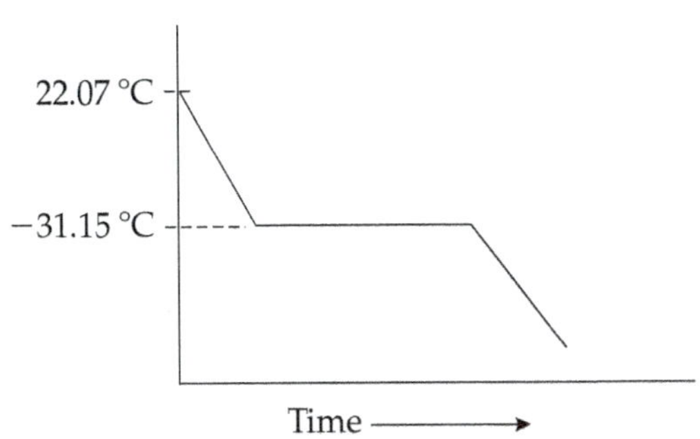

(a) What is the molar mass of caffeine?
(b) Combustion analysis reveals that the empirical formula of caffeine is $C_4H_5N_2O$. What is the molecular formula?

12.145 When 1.56 g of cholesterol is dissolved in 50.0 mL of cyclohexane, the resulting solution freezes at 4.40 °C. What is the molar mass of cholesterol? (The density of cyclohexane is 0.779 g/mL.)

Getting Unlikes to Dissolve—Soaps and Detergents

12.146 Soap molecules have two portions with very different properties, called hydrophobic and hydrophilic.
(a) Draw a typical soap molecule.
(b) State how a typical soap molecule is related to fat (animal fat) molecules.
(c) Discuss the properties of the two portions of a soap molecule.

12.147 When soap molecules are dissolved in water in a pool, the soap molecules migrate to the surface of the water and actually coat the surface, with the polar heads being dissolved in the water and the non-polar tails sticking out into the air above the surface of the water.
(a) Draw the scenario just described.
(b) Using the rule of thumb "like dissolves like," explain why the soap molecules at the water's surface align as described.
(c) Water bugs can "walk" on the surface of a pool's water, but if you put soap in the water, they fall through. Why do you think this is? (*Hint*: Water molecules at the surface are attracted to each other very strongly via hydrogen bonds.)

12.148 Micelles are spherical, although they are usually drawn as a flat cross section (as shown on page 495). Why wouldn't micelles exist in water as flat, two-dimensional structures?

12.149 Soap molecules have a hydrophobic portion and yet they dissolve in water. Explain how they accomplish this.

12.150 Explain how soaps allow water to wash oily, nonpolar dirt off clothes and skin.

12.151 A common molecule used in detergents is sodium lauryl sulfate:

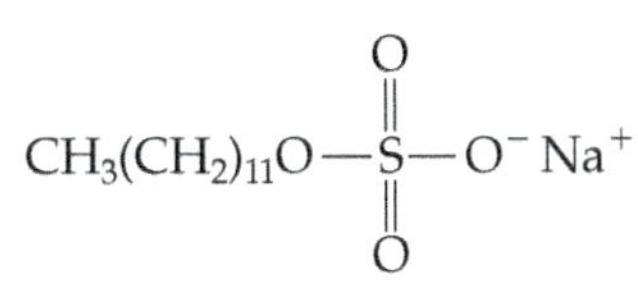

Identify the hydrophobic and hydrophilic portions of this molecule.

Additional Problems

12.152 Which would you expect to release the most hydration energy when dissolved in water: $KCl(s)$, $Mg(OH)_2(s)$, or $CO_2(g)$? Which would you expect to release the least hydration energy? Explain your answers.

12.153 Which would you expect to release the most hydration energy when dissolved in water: $CH_3CH_2OH(l)$, $CH_3Cl(l)$, or $C_8H_{18}(l)$? Which would you expect to release the least hydration energy? Explain your answers.

12.154 The ionic lattice in which compound requires the most energy to break: KCl, $Mg(OH)_2$, or $NaNO_3$? Explain your answer.

12.155 Which one of the following steps in the dissolving process must have a negative value for ΔE? Explain your answer.
(a) The physical separation of solute particles
(b) The formation of solvent–solute interactions
(c) The physical separation of solvent particles
(d) None of the above

12.156 Which one of the following statements is true?
(a) Gases are more soluble in liquids when the pressure is decreased.
(b) Pressure has no effect on how soluble a gas is in a liquid.
(c) Gases are more soluble in liquids when the pressure is increased.
(d) Gases are less soluble in liquids when the pressure is increased.

12.157 How many moles of potassium permanganate, $KMnO_4$, are there in 28.86 mL of a 5.20×10^{-3} M solution of $KMnO_4$?

12.158 (a) How many grams of NaOH are needed to prepare 500.0 mL of a 0.300 M NaOH solution?
(b) Describe how you would make this solution, including the equipment needed.

12.159 How would you prepare 250.0 mL of a 0.350 M NaOH solution from a 6.00 M NaOH stock solution?

12.160 Use the bottom graph on page 462 to determine what mass of water is required to dissolve 50.0 g of $NaNO_3$ at 40.0 °C.

12.161 Calculate the number of moles of each ion present in 2.00×10^2 cm^3 of (a) 0.200 M NaCl, (b) 0.350 M K_3PO_4, (c) 1.44 M $Al(NO_3)_3$.

12.162 Complete the table:

Solute	Solute mass (g)	Solution volume (L)	Molarity (mol/L)
NaBr	3.96	0.150	
$Ba(OH)_2$	2.58		0.0800
$(NH_4)_2SO_4$	8.65	2.40	
NH_4Cl		4.20	0.420

12.163 4.70 g of $CuSO_4$ is added to enough water to make 150.0 cm^3 of solution.
(a) What is the molarity of the solution?
(b) How many moles of $CuSO_4$ are there in 1.00 mL of this solution?
(c) What is the percent by mass of $CuSO_4$ of this solution? (The density of the solution is 1.01 g/mL).

12.164 Calculate the freezing point and boiling point of each solution:
(a) 18.4 g of glucose, $C_6H_{12}O_6$, in 95.5 g of water
(b) 15.00 g of urea, N_2H_4CO, in 75.0 g of water

12.165 Combustion analysis reveals vitamin C to be 40.9% by mass C and 4.58% by mass H. The only other element present is oxygen. A solution of 19.40 g of vitamin C in 100.0 g of water freezes at 22.05 °C. What is the molecular formula of vitamin C?

12.166 The freezing point of a solution prepared by dissolving 0.200 mole of HF(*g*) in 2.00 kg of water is −0.19 °C. Is HF primarily intact in solution, existing as HF(*aq*), or has it dissociated to $H^+(aq)$ and $F^-(aq)$ ions? Does this mean HF is a weak or a strong acid?

12.167 How many gallons of 24-proof wine would you have to drink to consume 0.100 gallon of alcohol?

12.168 Draw pictures illustrating the solute–solvent interactions and the solvent–solvent interactions for methanol, CH_3OH, in water. Is this a case of like dissolves like? Explain.

12.169 After 125.0 mL of water is added to 50.0 mL of a 0.250 M solution of ammonium phosphate, (a) what is the molar concentration of ammonium phosphate in the diluted solution?

Once in solution, the ammonium phosphate exists not as intact ammonium phosphate but rather as ammonium ions and phosphate ions. What are the molar concentrations (b) of ammonium ions in the diluted solution and (c) of phosphate ions in the diluted solution?

12.170 After 50.0 mL of a 0.250 M solution of calcium nitrate is combined with 100.0 mL of a 0.835 M solution of calcium nitrate, (a) what is the molar concentration of $Ca(NO_3)_2(aq)$ in the combined solution?

Once in solution, the calcium nitrate exists not as intact calcium nitrate but rather as calcium ions and nitrate ions. What are the molar concentrations (b) of $Ca^{2+}(aq)$ in the combined solution and (c) of $NO_3^-(aq)$ in the combined solution?

12.171 Suppose 250.0 mL of a 0.600 M solution of barium nitrate is combined with enough 0.500 M sodium sulfate solution to precipitate the maximum amount of barium sulfate. What volume of the sodium sulfate solution is required?

12.172 A bottle contains 1.00 L of a stock solution of $Fe(NO_3)_3$ of unknown concentration. A lab technician dilutes 5.00 mL of the stock solution to 100.0 mL with water. He then determines the $Fe(NO_3)_3$ concentration of this solution to be 0.0478 M. What is the $Fe(NO_3)_3$ concentration of the stock solution?

12.173 What volume of a 0.245 M solution of NaI would you need to add to 100.0 mL of a 0.300 M solution of lead acetate to precipitate out all the lead? Which ions are spectators?

12.174 Citric acid can produce 3 $H^+(aq)$ ions per molecule. A solution of citric acid is prepared by dissolving 0.177 g of solid citric acid in enough water to yield 100.0 mL of solution. When this solution is titrated with 0.1001 M $NaOH(aq)$, the indicator turns color after 27.55 mL of $NaOH(aq)$ has been added.
(a) What is the molar concentration of citric acid?
(b) What is the molar mass of citric acid?

12.175 A 25.00-mL sample of a hydrochloric acid solution of unknown concentration is titrated with 0.1004 M NaOH. However, before the acid is titrated, 27.65 mL of water is added to it. The phenolphthalein indicator turns red after 28.70 mL of NaOH has been added. What is the concentration of the 25.00-mL sample of hydrochloric acid?

12.176 Identify the solvent and the solute in each solution:
(a) Brass, which is 60 to 80% Cu, and 18 to 40% Zn
(b) Household ammonia cleaner, which is 1% by mass ammonia gas in water
(c) 2.59 g of sucrose in 1.00 g of water at 50 °C

12.177 Carbon dioxide, CO_2, is a compound of carbon and oxygen. Does this mean that a sample of pure carbon dioxide can be considered to be a solution of carbon dissolved in oxygen? Explain your answer.

12.178 Describe at the molecular level what is happening when solid iodine, $I_2(s)$, dissolves in carbon tetrachloride liquid, $CCl_4(l)$.

12.179 Would you expect solid iodine, $I_2(s)$, to be more or less soluble in $CCl_4(l)$ than in $H_2O(l)$? Explain your answer in terms of energy.

12.180 What is the difference between hydration and solvation? Does one process always release more energy than the other? Explain.

12.181 How would you explain the fact that Na_3PO_4 is quite soluble in water but $AlPO_4$ is essentially insoluble?

12.182 Consider dissolving these two molecules in water:

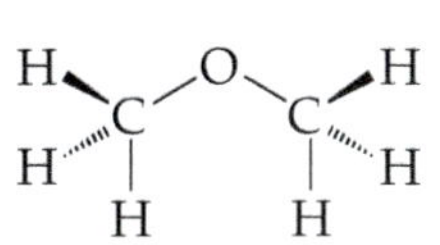

Diethyl ether, C_2H_6O

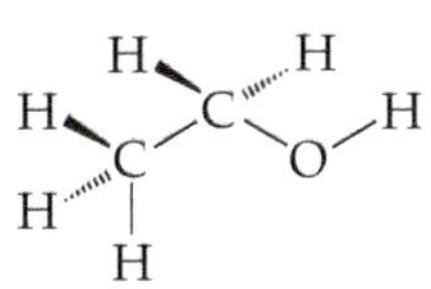

Ethanol, C_2H_6O

Even though the solute-separation step for ethanol requires more energy than does the solute-separation step for diethyl ether, ethanol is more soluble in water. Explain why.

12.183 Whether or not a solute dissolves in a solvent is decided not only by changes in energy but also by changes in ________________.

12.184 Correct the two errors in this statement: A solute can dissolve even when the energy absorbed by the solute–solvent interactions is less than the energy absorbed by the solute–solute and solvent–solvent interactions because the entropy of the system decreases.

12.185 If the dissolution of a particular solute in water is endothermic, what must be true about (a) ΔE_{total} and (b) the change in entropy for the system as a result of the dissolution?

12.186 Henry's law tells us that the solubility s of a gas in a liquid increases as the pressure P of the gas increases. Which of the following mathematical expressions of Henry's law is correct? Explain your choice and explain why the other two expressions are incorrect. (*Hint*: The k is just a constant of proportionality and can be ignored.)
(a) $s = kP$
(b) $s = k/P$
(c) $s = kP/P$

12.187 Could 32 g of NaCl be dissolved in 75 g of water at 20.0 °C? (*Hint*: Refer to the bottom graph on page 462.)

12.188 Could 75 g of sodium nitrate be dissolved in 90.0 g of water at 60.0 °C? (*Hint*: Refer to the bottom graph on page 462.)

12.189 What is the mass in grams of glucose dissolved in 60.0 g of water at 20.0 °C if the solution is saturated? (*Hint*: Refer to the bottom graph on page 462.)

12.190 Why do most solids become more soluble in water with increasing temperature? (*Hint:* Think about what happens to the water molecules as temperature increases.)

12.191 Phospholipids are naturally occurring soaplike molecules present in the membranes of living cells. A typical phospholipid structure is:

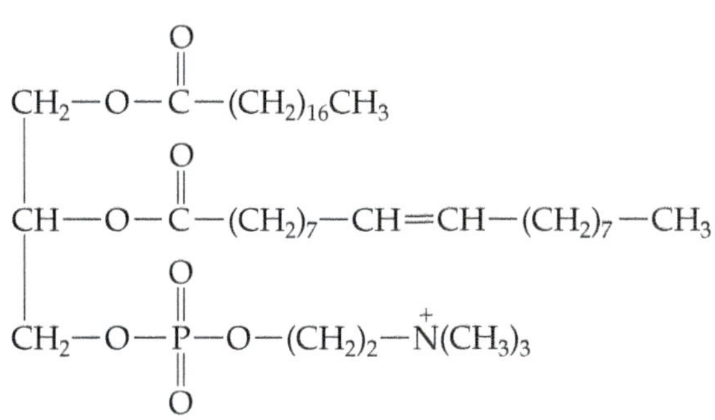

Identify the hydrophobic and hydrophilic portion(s) of this molecule.

12.192 When a small amount of soap is added to a beaker of water, the soap molecules end up positioning themselves at the surface (hence soaps are often called *surfactants*):

Explain why the soap molecules migrate to the surface, and why they orient themselves with their hydrocarbon tails sticking out of the water.

12.193 Draw the structure of the fat molecule that, when it reacts with KOH, forms glycerol and a soap molecule containing 14 carbon atoms.

12.194 Consider the soap molecules at the top of page 495 and the ionic compound potassium acetate.

(a) Draw a dot diagram for the acetate ion, and then place a K^+ ion next to the O bonded to C by a single bond.

(b) What is the similarity between potassium acetate and a soap molecule? How do they differ?

(c) Hard water often contains calcium ions, which react with soap molecules to form a solid soap scum. Draw a soap scum molecule, using a soap molecule containing ten C atoms. What is the molecular formula of this soap scum molecule?

12.195 Soap molecules not only form spherical micelles in water, they also form spherical vesicles, which you can picture as thick-walled hollow spheres. Here is a cross section of such a vesicle, with the blue regions representing water:

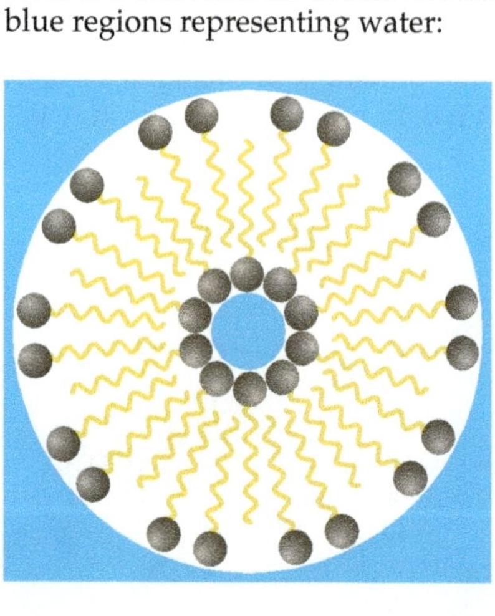

Unlike a micelle, a vesicle traps water in its interior. What gives a vesicle this ability? (If you are having trouble with the difference between micelles and vesicles, think of a baseball and a hollow rubber ball. The baseball, with no empty space inside, is analogous to the micelle, and the hollow ball is analogous to the vesicle.)

12.196 A student dilutes 75.0 mL of a 2.00 M solution of iron(III) nitrate with sufficient water to prepare 2.00 L of solution. (a) What is the molar concentration of iron(III) nitrate in the diluted solution?

Once in solution, the iron(III) nitrate exists not intact but rather as dissociated ions. What are the molar concentrations (b) of $Fe^{3+}(aq)$ in the diluted solution and (c) of $NO^{3-}(aq)$ in the diluted solution?

12.197 A student dilutes 45.0 mL of a 0.500 M solution of aluminum sulfate with sufficient water to prepare 1.50 L of solution. (a) What is the molar concentration of aluminum sulfate in the diluted solution?

Once in solution, the aluminum sulfate exists not intact but rather as dissociated ions. What are the molar concentrations (b) of $Al^{3+}(aq)$ in the diluted solution and (c) of $SO_4^{2-}(aq)$ in the diluted solution?

12.198 A student combines 60.0 mL of 0.250 M NaOH with 60.0 mL of 0.125 M NaOH. What is the NaOH molar concentration in the resulting solution?

12.199 A student combines 60.0 mL of 0.250 M NaOH with 60.0 mL of 0.125 M $Ba(OH)_2$. What is the hydroxide ion molar concentration in the resulting solution?

12.200 What is the mass in grams of the nitrogen atoms in 100.0 mL of 1.00 M $Ca(NO_3)_2$ solution?

12.201 How many grams of NaOH are needed to prepare 500.0 mL of 0.300 M NaOH solution?

12.202 What is the molarity of 3.69 mL of solution containing 0.0025 mole of calcium chloride?

12.203 How many moles of potassium permanganate, $KMnO_4$, are there in 28.68 mL of a 5.20×10^{-3} M solution of $KMnO_4$?

12.204 What is the volume in milliliters of 0.0150 M NaOH solution required to neutralize 50.0 mL of 0.0100 M $HNO_3(aq)$?

12.205 What is the volume in milliliters of 0.0150 M NaOH solution required to neutralize 50.0 mL of 0.0100 M $H_2SO_4(aq)$?

12.206 A 50.00-mL sample of hydrochloric acid of unknown concentration was neutralized by 47.35 mL of 0.01020 M sodium hydroxide solution. Calculate the concentration of the original HCl solution.

12.207 A solution is prepared by dissolving 5.00 g of sucrose, $C_{12}H_{22}O_{11}$, in 1.00 L of water. What is the percent by mass concentration of sucrose? (The density of water is 1.00 g/mL).

12.208 A solution is prepared by combining 4.00 mL of hexane with a sufficient volume of ethanol to obtain 250.0 mL of solution. What is the percent by volume concentration of hexane?

12.209 Air is approximately 21% by volume oxygen, O_2. How many grams of oxygen are present in 200.0 L of air at 1.00 atm and 25 °C? (*Hint*: Use the ideal gas equation from Chapter 11.)

12.210 When 5.00 g of a solute is dissolved in sufficient solvent to produce 100.0 mL of solution,
(a) What is the percent by mass/volume concentration of the solute?
(b) What additional information do you need to calculate the percent by mass concentration of the solute, and how would you calculate that concentration once you had the missing information?

12.211 A sample of stainless steel, which is an alloy of iron, chromium, and minor amounts of other components, is 11.5% by mass chromium. How many grams of chromium are there in 250.0 lb of stainless steel?

12.212 Your boss says, "Prepare 0.5000 kg of a 1.00% solution of hexane (C_6H_{14}, a liquid) in dichloromethane (CH_2Cl_2, also a liquid)."
(a) What is wrong with his request?
(b) How would you prepare this solution if he meant a 1.00% by mass solution?

12.213 When 26.5 mL of 0.100 M $Ca(NO_3)_2$ is combined with 49.8 mL of 0.100 M NaF,
(a) What is the theoretical yield of $CaF_2(s)$ in grams?
(b) What is the molar concentration of Ca^{2+} (excess reactant) in the combined solution?

12.214 Suppose 200.0 mL of a 2.50 M solution of sodium hydroxide is combined with 100.0 mL of a 1.50 M solution of iron(III) nitrate.
(a) Write a net ionic equation for the formation of the expected precipitate.
(b) What is the theoretical yield of precipitate in grams?
(c) Suppose only 10.95 g of precipitate is isolated. What is the percent yield for the reaction?
(d) What is the molar concentration of the excess reactant in the combined solution?

12.215 Suppose 20.0 g of NaBr(*s*) is added to 50.0 mL of a 2.00 M silver nitrate solution.
(a) Write a net ionic equation for the formation of the expected precipitate.
(b) What is the theoretical yield of precipitate in grams?
(c) Suppose only 15.0 g of precipitate is isolated. What is the percent yield for the reaction?
(d) What is the molar concentration of the excess reactant in the combined solution?

12.216 Consider the following diagram:

(a) Which step in the dissolving process does it represent?
(b) Does this step take energy or release energy?
(c) Would more energy be involved with this step if the solvent were CCl_4 instead of water?

12.217 The flask below has been evacuated, so its internal pressure might be expected to read zero, yet it does not. Why does the pressure not equal zero, and why does it stay stopped at the fixed values shown?

12.218 The following situation does not represent boiling. Name two things you could do to the situation to cause the water to boil.

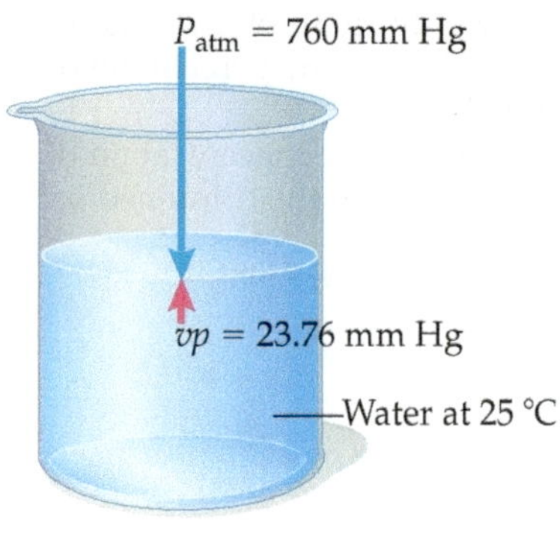

12.219 Consider the following diagram:
(a) Which situation represents a soluble solute?
(b) Which situation represents an insoluble solute?
(c) Which situation represents a case where an entropy increase probably results in a soluble solute?

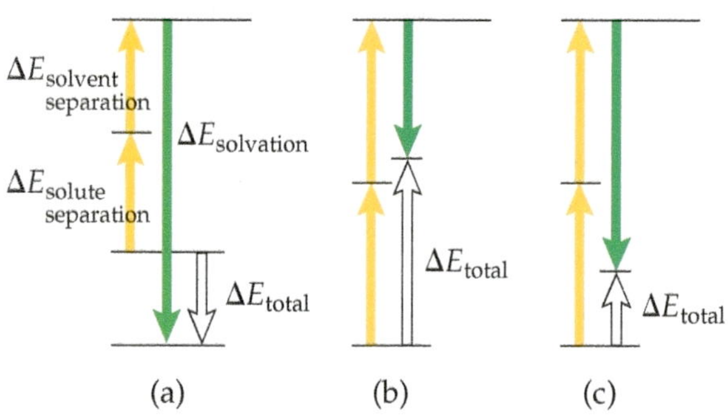

12.220 In a titration of HCl with NaOH, you read the buret to see how much base it took to neutralize the acid (turn the indicator pink). You do this titration to determine the molar concentration of the acid.

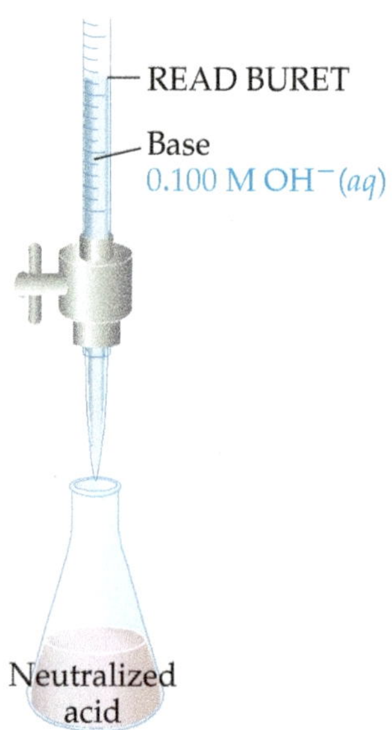

Why is it important to accurately know the amount (volume) of acid that you originally put into the flask?

One Step Beyond

12.221 Calculate the molarity of pure water at 4.0 °C. The density of water at 4.0 °C is 1.0000 g/mL.

12.222 When methanol (CH_3OH) dissolves in water, both initially at 22 °C, heat is released and the resulting solution is warmer than 22 °C. Explain why this is so. (What must be true?)

12.223 The solvent cyclohexane is in Table 12.2. The compound methyl lithium (CH_3Li) is a solute that dissolves readily in cyclohexane. A simple way to think of methyl lithium is as an ionic compound made up of methyl anions (CH_3^-) and lithium cations (Li^+). Dissolving 0.100 moles of it in 0.100 kg of cyclohexane yields a freezing point of 3.97 °C. Explain how this is consistent with the known structure of methyl lithium in solution, which is shown below.

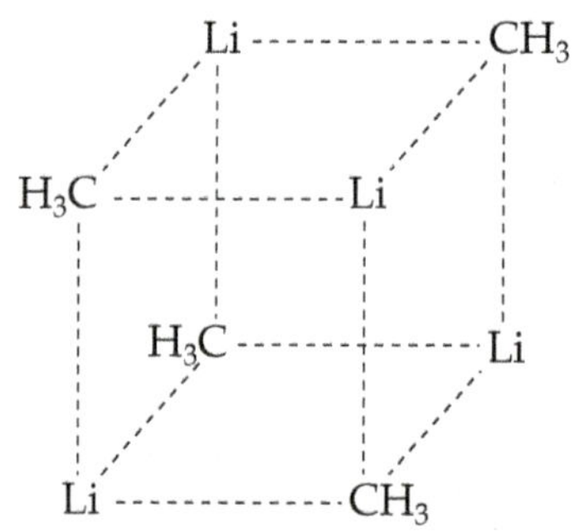

12.224 At 0.00 °C and sea level, dry air can be assumed to have a molar volume of 22.414 L. Given that air is 21.0% by volume oxygen, what is the molar concentration of oxygen in the atmosphere at sea level at 0.00 °C?

12.225 The amount to which a solvent's boiling point is elevated and vapor pressure is lowered depends on the number per volume (concentration) and not the chemical identity of solute particles dissolved in it. Explain why this is so in terms of entropy.

12.226 The dissolving of a solute (like NaCl) into a solvent (like water) can be thought of as a three-step process.
(a) Briefly name or describe these three steps.
(b) What are the energy implications of each step? (exothermic, endothermic, depends on the particular solvent/solute)
(c) Which of the three steps is primarily responsible for the "like dissolves like" rule for solubility?

12.227 The dissolving of a solute into a solvent can be thought of as a three-step process. The drinking alcohol, ethanol (CH_3CH_2-OH), is infinitely soluble in water. However, another alcohol called pentanol, with the molecular formula $CH_3CH_2CH_2CH_2CH_2$-OH, is nearly insoluble in water.

(a) Which of the three steps in the solubility model is **most** responsible for this difference in solubility? Describe the step you choose (specify its name).
(b) Give a **BRIEF** explanation for your choice.

12.228 Our task is to measure the volume of blood in a live hippopotamus. One way to do this would be to drain its blood into a suitable container and measure its volume. Aside from the obviously disastrous effect this method would have on the hippo, it would not be very accurate since some blood would still remain behind in the tissues. Instead, we will inject the hippo with 2.00 mL of a 2.00 M dye solution which the hippo will not appreciably metabolize or excrete in one hour, and then measure the dye concentration in the bloodstream after 30 minutes (which should be sufficient time for the dye to thoroughly mix in the bloodstream). After 30 minutes, the concentration of the dye in a 78.0 mL blood sample was found to be 0.0000125 M. What is the volume of blood in the hippo?

Group Exercises (Work on these exercises together with someone else in the class.)

1. Water bugs can walk on the surface of the water in a swimming pool, but if you add a little soap, they no longer can (they fall through). Why do you think this is so? (*Hint:* Consider Problem 12.192.)

2. Suppose you want to make a 1 M ethanol solution in water (ethanol is a clear, colorless liquid). What is wrong with adding 1 mole of ethanol to 1 liter of water to make this solution? What wrong assumption is involved in doing this?

WORKPATCH SOLUTIONS

12.1 Homogeneity can occur only when the molecules are free to move about randomly, and such motion is possible only in the liquid or gas phase.

12.2 The water molecule is polar, with its partial negative charge located on the oxygen and its partial positive charges located on the hydrogens. Because opposite charges attract, the δ^- oxygen atoms are attracted to Na^+ ions and the δ^+ hydrogen atoms are attracted to Cl^- ions.

12.3 The solute-separation step requires a large amount of energy because ionic bonds are very strong. The solvent-separation step requires a moderate or small amount of energy because these are nonpolar solvent molecules, meaning that the only forces to be overcome are weak London forces. The solvation step releases a small amount of energy. This is a case of very different particles trying to attract one another. CCl_4 is nonpolar, but Na^+ and Cl^- ions are charged. The only force that can exist between a CCl_4 molecule and either ion is a weak London force. Development of this weak attractive force releases only a small amount of energy.

12.4 Because ΔE_{total} is zero, any mixing is due solely to entropy. In the spontaneous direction (from left to right in the drawing), we go from a less random, less probable situation to a more random, more probable situation (like the socks and shirts, there are more ways to arrive at the mixed arrangement). The natural tendency is towards increased entropy (increased probability), so unmixing will not spontaneously occur.

12.5 Use the density as a conversion factor:

$$79.6 \text{ mL ethanol} \times \frac{0.7893 \text{ g ethanol}}{1 \text{ mL ethanol}} = 62.8 \text{ g ethanol}$$

12.6 The vapor pressure increases, as Table 12.1 shows.

12.7 The definition of normal boiling point is the temperature at which a liquid's vapor pressure is 760 mm Hg. Thus, chloroform's vapor pressure must be 760 mm Hg at 61.7 °C because, as you were told in the text, this temperature is the normal boiling point of chloroform.

12.8 There are 2 moles of solute particles in each beaker—1 mole of $Na^+(aq)$ ions plus 1 mole of $Cl^-(aq)$ ions in the left beaker, and 2 moles of $C_{12}H_{22}O_{11}(aq)$ molecules in the right beaker.

12.9 (a) The detergent molecules would arrange themselves with the polar heads pointing inward, exactly the opposite of the arrangement shown earlier for a micelle forming in water. This way, the nonpolar tails are presented to the nonpolar solvent, leading to a maximum attractive force. The polar heads buried inside avoid the nonpolar solvent that surrounds the structure.

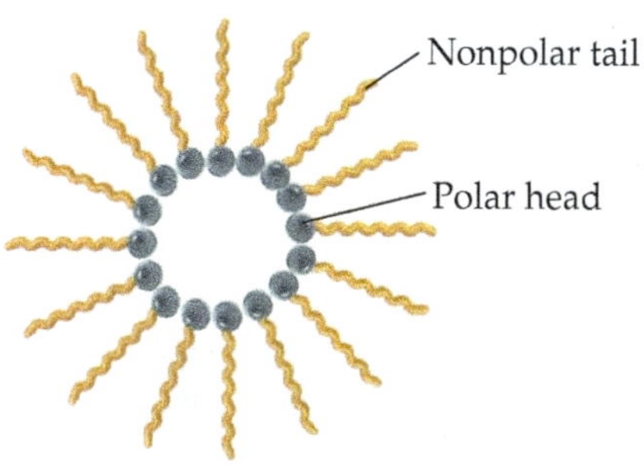

(b) This structure would be good at removing polar dirt from clothes being dry-cleaned.

CHAPTER 10

Electron Transfer in Chemical Reactions

Wouldn't it be great if you could continually run your iPod, laptop, or cell phone for months at a time without ever having to go near a wall socket? While we may not be there yet, things are progressing. All batteries make use of a special class of chemical reactions that produce electricity as reactants are converted into products. Think about the battery in your car, called a lead storage battery. It is the size of a large shoe box and weighs more than a bowling ball. Can you imagine having to haul that around on your back in order to power your cell phone? Instead, your cell phone uses a much smaller, much lighter lithium ion battery. Two sources of electricity in two very different packages. What's the difference? It comes down to energy density, the amount of energy in mega joules (MJ) that can be released per kilogram of reactant for the electricity-producing chemical reaction. A lead storage battery has an energy density of 0.14 MJ/kg, whereas a lithium ion battery has an energy density up to 0.72 MJ/kg. With an energy density more than five times higher, the lithium ion battery can supply sufficient power for your cell phone and still weigh very little. Are there other substances with even higher energy densities? You bet. A battery made from lithium and fluorine would have an energy density of 23.75 MJ/kg, which is almost 170 times greater than a car battery. So, why don't you have one of those batteries? Well, fluorine is an extremely corrosive and toxic gas, so you can just imagine the problems that might arise should such a battery spring a leak. Indeed, the substance with the highest energy density (this is for all you *Star Trek* fans) is antimatter, which clocks in with an energy density of 89.9×10^9 MJ/kg (that's almost 100 billion MJ/kg). Now that would make for one tiny, everlasting battery, but unfortunately, antimatter annihilates all normal matter (you, me, your iPod) that it comes into contact with. Talk about dangerous!

10.1 What Is Electricity?

Did you ever accidentally bite down on a piece of aluminum foil? Did it hurt?

The pain is due to a particular type of chemical reaction, one that produces an electric jolt that goes straight to the nerve endings in your teeth. Electricity from a chemical reaction? Yes, some reactions can produce electricity. However, before we study such reactions, let's quickly review what electricity is.

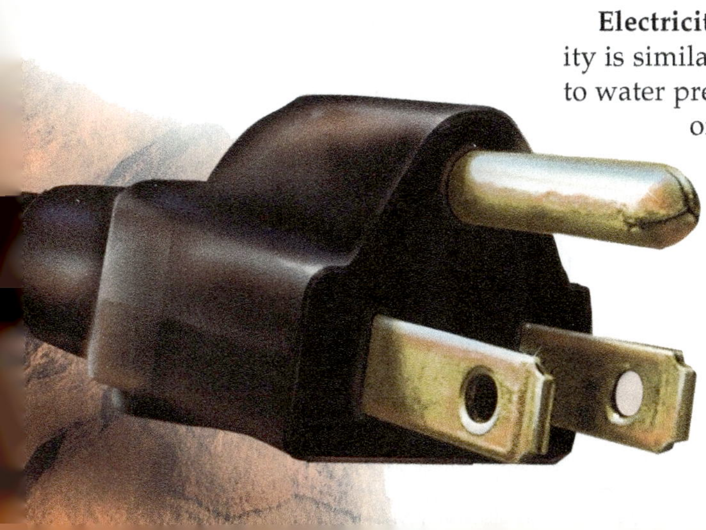

Electricity is simply the flow of electrons. In many ways, electricity is similar to water flowing through a hose. **Voltage** is analogous to water pressure. The higher the voltage, the greater the "pressure," or "push," behind the electrons that causes them to flow. (Remember, we represent an electron by the symbol e^-.)

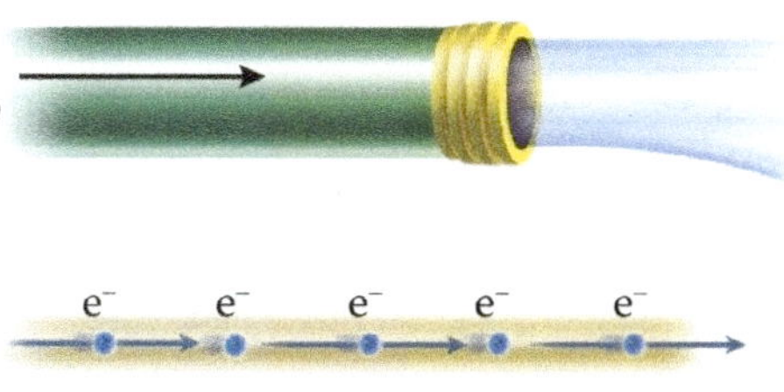

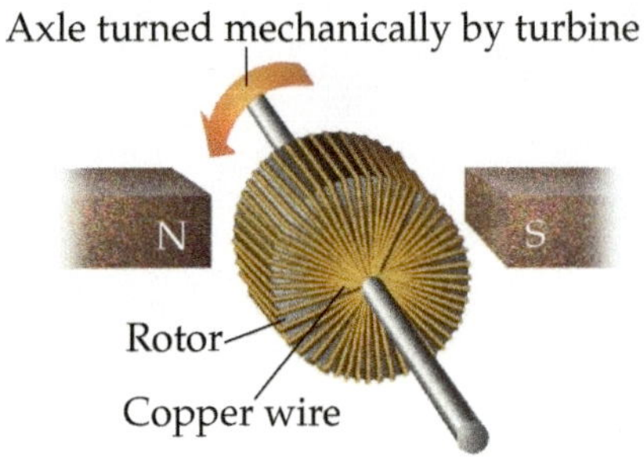

The push begins at a power plant, where electricity is produced mechanically by a generator consisting of copper wire wound on a rotor that is spun between the poles of a magnet. The rotor is spun by a turbine that is usually powered by steam that is generated as water is vaporized by burning oil, coal, or natural gas or by the flow of water in a hydroelectric plant. As the copper wire moves through the magnetic field that exists between the poles of the magnet, electrons in the wire are pushed and made to flow, producing electricity. What you buy from the power company is the right to use their flowing electrons. They flow through your electric devices and do work and then return to the power company to be pushed again.

Electrons can also be made to flow through a wire by chemical means. The chemical reactions that do this often come packaged in containers called *batteries*. In this chapter, you will learn how some chemical reactions can generate electricity and how to recognize these reactions.

10.2 Electron Bookkeeping—Oxidation States

All chemical reactions convert reactants into products, but some chemical reactions also transfer one or more electrons from one atom to another as part of the conversion process. For example, the reaction of calcium metal with acid (aqueous H^+ ions) occurs as follows:

$$\text{Ca:} + 2\,\text{H}^+ \longrightarrow \text{Ca}^{2+} + \text{H{:}H}$$

This pair of valence electrons starts out belonging to Ca . . . but ends up belonging to (and being shared between) the H atoms.

During this reaction, there must be a transfer of electrons from the calcium atom to the hydrogen ions. Another way to say this is that the calcium atom loses some electrons and the H^+ ions gain them. It is this type of reaction that can be used to push electrons through a wire and thereby produce electricity. These reactions are known as **electron-transfer reactions**. Note that only valence electrons are transferred in these reactions. Core electrons are too tightly held to their respective nuclei to be easily transferred.

Not all chemical reactions involve electron transfer. Some reactions proceed without any atom gaining or losing electrons. Because electron-transfer reactions are the only ones that can be used to produce electricity, it is important to be able to distinguish which reactions involve electron transfer and which do not. This is fairly easy to do if the reaction is written as shown above, with the valence electrons explicitly shown. With the reaction written this way, you can see that the calcium lost two electrons because its lone pair disappeared, and you can see that the two hydrogens gained two electrons because a covalent bond appeared between them.

The problem is that you will rarely see reactions written this way. Rather, you will be presented with just the balanced reaction, which for our example is:

$$\text{Ca} + 2\,\text{H}^+ \rightarrow \text{Ca}^{2+} + \text{H}_2$$

Now how can you tell that electrons are transferred as reactants become products? Perhaps you feel you can tell by the fact that charges are changing (for example, calcium goes from a charge of 0 to +2). For calcium to change from the neutral atom to the +2 ion, it must lose two electrons, so electrons must have been transferred. But consider the reaction

$$H_2 + Cl_2 \rightarrow 2\,HCl$$

No apparent changes in charge here. The reactants and product are all neutral molecules. Nevertheless, this reaction is also an electron-transfer reaction.

So, given just a balanced reaction, how are we to tell whether electron transfer is occurring? What we need is information about electron ownership. For example, suppose we wrote the $H_2 + Cl_2$ reaction this way:

$$H_2 + Cl_2 \longrightarrow 2\,HCl$$

Valence-electron ownership information	H_2: Each H atom owns one electron.	Cl_2: Each Cl atom owns seven electrons.	HCl: H atom owns no electrons.	HCl: Cl atom owns eight electrons.

With the ownership information given below the balanced equation, we can now tell that electron transfer has occurred. According to the ownership information, the H atoms in the H_2 molecule each own one valence electron, but the atoms end up with no electrons in the HCl molecules, telling us that each H atom has lost an electron. The ownership information also indicates that each Cl atom has gained one electron. Evidently, electrons moved from H atoms to Cl atoms during this reaction, so it can be classified as an electron-transfer reaction.

Unfortunately, balanced chemical reactions don't come with electron ownership information written below each atom. We need a way, therefore, to determine the number of electrons owned by every atom on both sides of a chemical equation. This can be done in much the same way that we keep track of our personal finances, by doing some bookkeeping. For this, chemists have created an electron-bookkeeping method known as the *oxidation-state method.* Its goal is to assign a number to every atom that tells us how many valence electrons each atom owns. These assigned numbers are called either an *oxidation number* or an *oxidation state.* After assigning oxidation states, we can look at a balanced reaction and tell whether electron transfer occurs.

The best way to understand how oxidation states are assigned to atoms in a molecule or in a polyatomic ion is to use electron dot diagrams. We'll demonstrate with a water molecule:

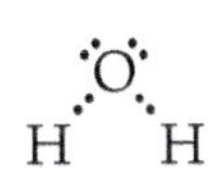

There are eight valence electrons (dots) in this diagram. Electron bookkeeping means asking the question "How many electrons does each atom in the molecule own?" Because the answer depends on how the bookkeeping is done (in other words, on how we define the word *own*), we need to agree on a few rules. The first rule states that:

Lone pairs of electrons are assigned completely to the atom on which they are drawn in the dot diagram.

This makes sense because lone pairs are not shared with another atom.

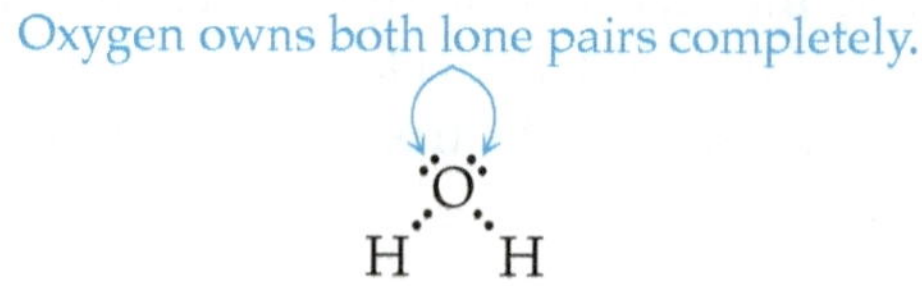

What about the bonding electrons? In our discussion of covalent bonding and the octet rule in Chapter 5, we double-counted the bonding electrons in a dot structure. For example, consider one of the O—H bonds in water. The oxygen atom was said to own both shared electrons, and the hydrogen atom was also said to own both shared electrons. This is how we satisfied the octet rule for each atom (or the duet rule in the case of hydrogen).

In the oxidation-state method of bookkeeping, we count electrons differently. The rule for shared electrons in this method is:

> **For a chemical bond, the more electronegative atom gets complete ownership of both shared electrons.**

Recall from Section 5.6 that electronegativity is a measure of an atom's ability to attract shared electrons to itself. It is often useful to remember that the four most electronegative elements in the periodic table are fluorine (EN = 4.0), oxygen (EN = 3.5), chlorine (EN = 3.0), and nitrogen (EN = 3.0), the electron "hogs." Hydrogen has only a moderate electronegativity value (EN = 2.1).

Let's now apply the oxidation-state method of bookkeeping to the water molecule:

Electron counting by the octet rule:
Shared electrons are double-counted.

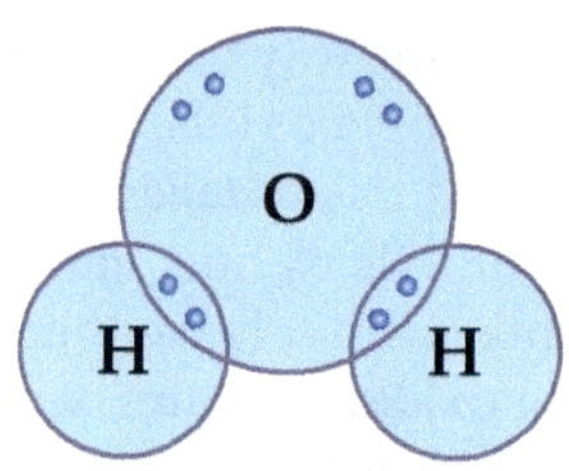

The O has **eight** electrons.
Each H has **two** electrons.

Electron counting by oxidation-state bookkeeping:
Shared electrons are awarded to the more electronegative atom.

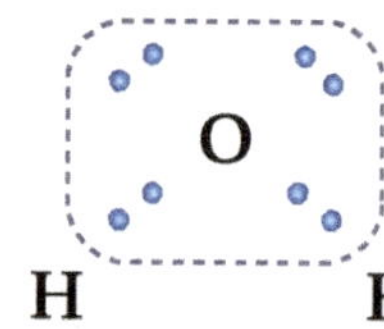

The O has **eight** electrons.
Each H has **zero** electrons.

Because oxygen is more electronegative than hydrogen, oxygen is assigned all the electrons in the O—H bonds. The hydrogen atoms get none of these electrons. Oxygen also gets its own lone pairs. So, by the oxidation-state method of electron bookkeeping, oxygen owns all eight valence electrons in the molecule, and the hydrogens own none.

At least for now, the only way we can do this type of bookkeeping is to have a dot diagram to look at. (In Chapter 6 we saw that it was necessary to have a dot diagram in order to use VSEPR theory to predict molecular shape. You can take it as a general truth that many times in chemistry, you will often find that the first step in understanding a problem is to draw a dot diagram.)

WORKPATCH 10.1 Using the oxidation-state method of electron bookkeeping, determine how many electrons each atom owns in NH_2F. (*Hint*: Start by drawing a dot diagram. Nitrogen is the central atom.) ●

You should have found that only one atom in NH_2F owns an octet by this method of bookkeeping. Was it the most electronegative atom? Check your answer against the one at the end of the chapter.

At this point, you might be puzzled by the existence of two systems for electron bookkeeping. First we saw the double-counting method, which we used to satisfy the octet rule. Now we've introduced the oxidation-state method, which does not double-count shared electrons but instead assigns them to the more electronegative atom. The two counting methods come up with different answers for how many electrons each atom owns. Which method is right? The answer is, both of them. Which one you use depends on what you want to know. If you want to know whether a particular dot diagram is valid with respect to the octet rule, then double-count. If you want to know whether electrons are transferred from one atom to another in a chemical reaction, use the oxidation-state method.

Now that we know how to use the oxidation-state method to assign electrons to atoms in a molecule or polyatomic ion, we are ready to calculate oxidation states for the atoms. We do this by comparing the number of electrons each atom owns (using oxidation-state bookkeeping) with the number of valence electrons a free atom of that element owns. For the representative elements, as we learned in Chapter 4, the number of valence electrons in a free atom is equal to the element's roman-numeral group number in the periodic table.

For the representative elements, the number of valence electrons in a free atom equals the roman-numeral group number.

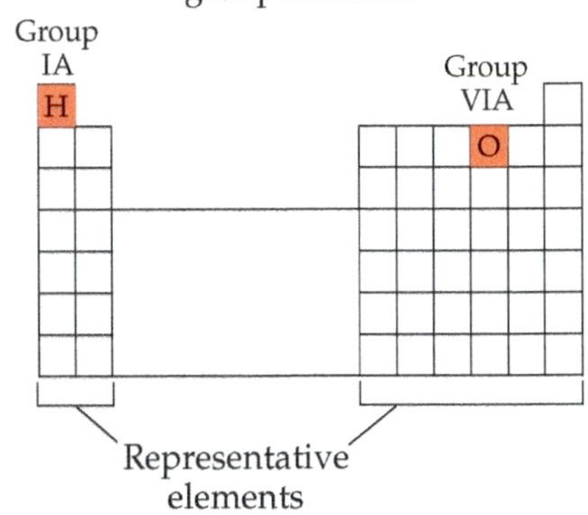

The **oxidation state**, or **oxidation number**,* of an atom in a molecule or polyatomic ion is the difference between the number of valence electrons in a free atom of that element and the number of electrons assigned to the atom by oxidation-state bookkeeping:

	Number of valence electrons in free atom of element
−	Number of electrons assigned to atom by oxidation-state bookkeeping
	Oxidation state

For water,

		Oxygen	Hydrogen
	Number of valence electrons in free atom of element	6	1
−	Number of electrons assigned to atom by oxidation-state bookkeeping	− 8	− 0
	Oxidation state	−2	+1

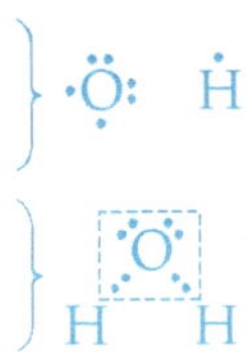

*These two terms are exact synonyms, and we use them interchangeably throughout this text.

Notice from the above illustration that the oxidation state can be either negative or positive. It is negative if the atom owns more electrons by oxidation-state bookkeeping than it would as a free atom; it is positive if the atom owns fewer electrons than it would as a free atom. An oxidation state of zero means that the number of electrons the atom owns by oxidation-state bookkeeping is the same as the number it owns as a free atom. To summarize:

If the oxidation state of an atom is $+n$, the atom owns n fewer electrons in a molecule or ion than it would as a free atom.

If the oxidation state of an atom is $-n$, the atom owns n more electrons in a molecule or ion than it would as a free atom.

The oxygen atom in water has a -2 oxidation state because, by oxidation-state bookkeeping, it owns two more electrons than it would as a free atom. The hydrogen atoms in water are in the $+1$ oxidation state because they each own one fewer electron than the free atom.

At this point, it is very important to realize that *oxidation states are artificial numbers that result from arbitrary bookkeeping rules.* They do not indicate real charges. For example, even though the oxygen atom in a water molecule is partially negative ($\delta-$) owing to the polar covalent O—H bond, it does *not* carry a full -2 charge.

Using this method of assigning oxidation states to atoms in a molecule or polyatomic ion, it is possible for the same type of atom to have different oxidation states in the same molecule. Let's examine this point by considering ozone, O_3.

WORKPATCH 10.2 Assign oxidation states for all the oxygen atoms in ozone, O_3, using this rule:

When two atoms of the same element (and therefore of identical electronegativity) share electrons, the oxidation-state method of electron bookkeeping divides the electrons equally between the atoms.

To do the WorkPatch, you'll need a dot diagram:

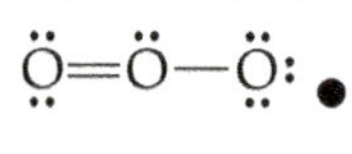

This WorkPatch includes our first mention of assigning oxidation states to molecules that contain electrons shared between atoms of identical electronegativity. Using the rule in the WorkPatch, you should have found that two of the oxygen atoms in ozone have nonzero oxidation states.

The chart at the top of the facing page sums up the rules for assigning oxidation states by the oxidation-state bookkeeping method.

Practice Problems

10.1 Consider the NH_2F molecule from WorkPatch 10.1. Assign an oxidation state to each atom in the molecule.

Answer: In WorkPatch 10.1, you should have assigned eight electrons to F, six to N, and none to either H atom, using the oxidation-state method of electron bookkeeping. Now we just subtract from each of these numbers the number of valence electrons in the free atoms:

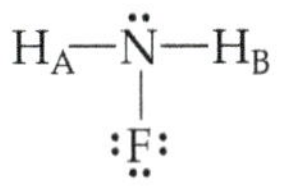

	Fluorine	Nitrogen	$Hydrogen_A$	$Hydrogen_B$
Number of valence electrons in free atom of element	7	5	1	1
− Number of electrons assigned to atom by oxidation-state bookkeeping	− 8	− 6	− 0	− 0
Oxidation state	−1	−1	+1	+1

Assigning oxidation states by the electron-bookkeeping method

Step 1: Draw a valid dot diagram for the molecule.

Step 2: Assign every valence electron to an atom using the following rules:

- Assign each lone pair to the atom on which it is drawn in the dot diagram.
- Assign all the electrons of each bond to the atom of higher electronegativity.
- If bonding electrons are shared by atoms of equal electronegativity, divide the electrons evenly between the atoms.

Step 3: Calculate the oxidation state for each atom by subtracting the electron count of the atom (as determined in step 2) from the number of valence electrons in a free atom of that element.

Step 4: Check your answer. The oxidation states for all the atoms should add up to the charge on the molecule or ion.

10.2 In carbon dioxide, CO_2,

(a) Assign an oxidation state to each atom in the molecule. (*Hint*: Draw a dot diagram first.)
(b) How many electrons does the C atom own by oxidation-state electron bookkeeping?
(c) How many more or fewer valence electrons are assigned to the C atom here than are present in a free C atom?
(d) Is it correct or incorrect to say that the C atom in CO_2 has a complete octet of valence electrons? Explain.

10.3 In methane, CH_4,

(a) Assign an oxidation state to each atom in the molecule. (*Hint*: Draw a dot diagram first.)
(b) How many electrons does the C atom own by oxidation-state electron bookkeeping?
(c) How many more or fewer valence electrons are assigned to the C atom here than are present in a free C atom?

10.4 In chloroform, $CHCl_3$,

(a) Assign an oxidation state to each atom in the molecule. (*Hint*: Draw a dot diagram first.)
(b) How many electrons does the C atom own by oxidation-state electron bookkeeping?
(c) How many more or fewer valence electrons are assigned to the C atom here than are present in a free C atom?
(d) Is it correct or incorrect to say that the C atom in $CHCl_3$ has a complete octet of valence electrons? Explain.

Shortcut Method for Assigning Oxidation States

The good news about the method we have been using to assign oxidation numbers is that it always works. The bad news is that it requires a dot diagram and drawing one takes time, if it can be done at all. For this reason, chemists have devised a shortcut for assigning oxidation states without the need for a dot diagram or even a table of electronegativities. However, what we gain in time we lose in accuracy because this shortcut method may not always produce a correct answer. Nevertheless, it works reliably for many of the compounds that chemists routinely encounter.

Shortcut rules

Rule 1. The oxidation state of an atom in the most abundant naturally occurring form of an element is assigned as zero.

The atoms in the substances shown below, including atoms in diatomic molecules, are assigned an oxidation state of zero.

Element	Naturally occurring elemental form	Oxidation state of atoms
Hydrogen	H_2	0
Oxygen	O_2	0
Nitrogen	N_2	0
Fluorine	F_2	0
Chlorine	Cl_2	0
Bromine	Br_2	0
Iodine	I_2	0
Helium	He	0
Neon	Ne	0
Phosphorus	P_4	0
Sulfur	S_8	0
Iron	Fe	0
Aluminum	Al	0

The most abundant elemental form for nonmetals is usually (but not always) a diatomic molecule, whereas the noble gases are monatomic. For all metals, the elemental form is represented by the symbol of the element.

Rule 2. The oxidation state of an oxygen atom in a compound is almost always −2.

For example, the oxidation state of the oxygen atoms in the compounds H_2O, CO, CO_2, $C_6H_{12}O_6$ (glucose), and $Fe_2O_3 \cdot H_2O$ (rust) is −2. Molecules called peroxides, which contain an O—O single bond, are exceptions to this rule. Ozone, which also has O—O single bonds, is also an exception (see WorkPatch 10.2). The only other time this rule fails is when oxygen is bound to an element more electronegative than itself. Only fluorine (F) meets this criterion. Except for these situations, it is a safe bet that an oxygen atom in a compound has an oxidation state of −2.

Rule 3. The oxidation state of a hydrogen atom in a compound or polyatomic ion is +1.

For example, the oxidation state of the hydrogen atoms in the compounds H_2O, $C_6H_{12}O_6$ (glucose), CH_4 (methane), and C_2H_2 (acetylene) is +1. The exceptions to this rule, where H is bonded to a less electronegative atom (such compounds are called hydrides), are not considered in this text.

Rule 4. The oxidation state of any group IA (1) atom (alkali metals such as lithium, Li, sodium, Na, and potassium, K) in a compound is always +1.

That's *in a compound*! This means that the oxidation state of the lithium atoms in lithium chloride, LiCl, or lithium oxide, Li_2O, is +1. The oxidation state of lithium metal, Li, its elemental form, is zero by Rule 1.

Rule 5. The oxidation state of any group IIA (2) atom (alkaline earth metals such as magnesium, Mg, barium, Ba, and calcium, Ca) in a compound is always +2.

Again, that's *in a compound*! This means that the oxidation state of the magnesium atom in magnesium chloride, $MgCl_2$, is +2. The oxidation state of magnesium metal, Mg, its elemental form, is zero.

Rule 6. The oxidation state of a monatomic (single-atom) ion is equal to the charge of the ion.

Some examples are:

Monatomic ion	Oxidation state
Al^{3+}	+3
O^{2-}	−2
H^+	+1
Cl^-	−1
N^{3-}	−3

Finally, we need a way to assign oxidation states to all the atoms in the periodic table not covered by Rules 1–6. For this, we use an extremely important rule that applies in all situations and to which there are no exceptions.

Rule 7. The sum of the oxidation states of the atoms in a chemical formula must add up to the overall charge of the molecule or ion indicated by the formula.

If the formula indicates a neutral molecule, such as H_2O, it has no charge and the oxidation states of the atoms must add up to zero. If you are dealing with a polyatomic ion, such as sulfate, SO_4^{2-}, the oxidation states of the atoms must add up to the charge on the ion (for sulfate, −2).

Which of the seven rules to apply and the order of their application can change from example to example, but we recommend that you always try applying Rules 2 and 3 first, and save Rule 7 for last. A few examples using the shortcut rules are worked out below. Follow them through, but realize that we would have arrived at the same result if we had started with a dot diagram.

	Atom	Oxidation state	
H_2O	O	−2	Rule 2: O is −2
	H	+1	Rule 3: H is +1
	H	+1	
		Sum 0	Oxidation states add up to 0, the charge on the water molecule (Rule 7).

	Atom	Oxidation state	
OH^-	O	−2	Rule 2: O is −2
	H	+1	Rule 3: H is +1
		Sum −1	Oxidation states add up to −1, the charge on the hydroxide ion (Rule 7).

These were straightforward examples. Now we'll make it a bit more difficult by including an atom for which there is no rule. The trick is to start with Rules 2 and 3, then apply Rules 1, 4, 5, and 6 as needed, remembering to always save Rule 7 for last.

	Atom	Oxidation state	
SO_4^{2-}	O	−2	Rule 2: O is −2
	O	−2	
	O	−2	
	O	−2	
	S	?	No rule for S
		Sum −2	The sum must equal −2, the charge on the sulfate ion (Rule 7).

Rule 7 says that all the oxidation numbers in a chemical formula must add up to give the overall charge on the compound, which in this case is −2. We have four oxygen atoms at −2 each, giving a total of −8. What do we have to add to −8 to get the −2 charge of the sulfate ion? The answer is +6. The oxidation number of the sulfur atom must therefore be +6.

We'll work through two more examples, each containing one element for which there is no shortcut rule.

	Atom	Oxidation state	
NH_4^+	H	+1	Rule 3: H is +1
	H	+1	
	H	+1	
	H	+1	
	N	?	No rule for N
		Sum +1	The sum must equal +1, the charge on the ammonium ion (Rule 7).

We have four hydrogen atoms with oxidation states of +1, giving a total of +4. In order to satisfy Rule 7, what do we have to add to +4 to get +1, the specific charge on the ammonium ion? The answer is −3. The oxidation number of the nitrogen atom must therefore be −3.

	Atom	Oxidation state	
$Cr_2O_7^{2-}$	O	−2	Rule 2: O is −2
	O	−2	
	O	−2	
	O	−2	
	O	−2	
	O	−2	
	O	−2	
	Cr	?	No rule for Cr
	Cr	?	
		Sum −2	The sum must equal −2, the charge on the dichromate ion (Rule 7).

We have seven oxygen atoms with oxidation states of −2 each, giving a total of −14. To get −2, the charge on the dichromate ion, so as to satisfy Rule 7, we must add +12. Now we have to be careful, however. The unknown oxidation number of Cr is not +12 because there are two Cr atoms in this compound. In cases like this, we can usually assume that the two Cr atoms have the same oxidation state. This means that each chromium atom must have an oxidation state of $+12/2 = +6$.

Before giving you some practice problems, we want to give you one more shortcut rule, which we call the halide rule:

> **Any halogen atom in a compound most likely has an oxidation state of −1.**

Recall that the halogens are the group VIIA (17) atoms (F, Cl, Br, I). So, for example, we can now assign the oxidation states to the atoms in bromoform, $CHBr_3$:

	Atom	Oxidation state	
$CHBr_3$	H	+1	Rule 3: H is +1
	Br	−1	Halide rule: Br is −1
	Br	−1	
	Br	−1	
	C	?	No rule for C
		Sum 0	The sum must equal 0 (Rule 7).

We have one hydrogen atom at +1 and three bromine atoms at −1 each, giving a total of −2. What do we have to add to −2 to get 0? The answer is +2. The oxidation number of the carbon atom must therefore be +2.

You have to be careful with the halide rule because it does have exceptions, just as Rule 2 for oxygen and Rule 3 for hydrogen do. Part (c) in WorkPatch 10.3 tests your knowledge of a basic concept that enables you to determine when you can use the halide rule and when you cannot.

WORKPATCH 10.3 Suppose a chlorine atom in a compound has an oxidation state of −1.

(a) Is the number of valence electrons assigned to this atom by oxidation-state bookkeeping more or fewer than the number of valence electrons in a free chlorine atom?

(b) How many more or fewer?

(c) In order for a chlorine atom in a compound to have a −1 oxidation state, what must be true about the electronegativity of the atom to which it is bonded? ●

Check your answers to the WorkPatch to make sure you understand these basic concepts. Now let's examine a case where the halide rule does not apply.

WORKPATCH 10.4 Consider the chlorate ion, ClO_3^-, in which all three oxygen atoms are bound to a central chlorine atom. The oxidation state of the Cl atom in this ion is not −1.

(a) Why doesn't the halide rule apply for the chlorate ion?

(b) What is the oxidation state of the Cl atom in chlorate? ●

These last two WorkPatches are very important. They are reminders that in oxidation-state electron bookkeeping, it is the more electronegative atom that gets the shared electrons. More important, though, the WorkPatches tell you when the halogen rule works and when it doesn't. The rule holds only when the atom to which the halogen atom is bonded is less electronegative than the halogen atom. Make sure you understand this principle, and then try the following practice problems using the shortcut rules. For your convenience, the rules are summarized at the top of the opposite page.

Practice Problems

Use the shortcut rules to assign oxidation states to all atoms.

10.5 $COCl_2$ (oxygen and chlorine bonded to the central carbon)

Answer: O is −2 (Rule 2). Cl is −1 (halide rule applies because Cl is more electronegative than C). Therefore C is +4 (the sum of the oxidation numbers for the O and two Cl's is −4).

10.6 $MgBr_2$

10.7 Fe_2O_3

10.8 HClO (Cl and O bonded to each other)

10.9 CO_3^{2-}

10.10 Cu^{2+}

Chemists frequently use the shortcut rules to assign oxidation states, but when the rules are inadequate (for example, the rules don't cover a molecule like CS_2) or when they present some ambiguity, a dot diagram will always work. A good example of the ambiguity we mentioned is demonstrated by

Assigning oxidation states by the shortcut method

Rule 1: The oxidation state of an atom in the most abundant naturally occurring form of an element is typically zero.

Rule 2: The oxidation state of an oxygen atom in a compound is almost always −2.

Rule 3: The oxidation state of a hydrogen atom in a compound is almost always +1.

Rule 4: The oxidation state of any group IA (1) atom in a compound is +1.

Rule 5: The oxidation state of any group IIA (2) atom in a compound is +2.

Rule 6: The oxidation state of a monatomic ion equals the ion's charge.

Halide rule: A halogen atom in a compound usually has an oxidation state of −1, except when bonded to a more electronegative atom.

Rule 7: The sum of the oxidation states of the atoms in a chemical formula must add up to the overall charge of the molecule or ion represented by the formula.

acetic acid, $C_2H_4O_2$. By the shortcut rules, we would assign the following oxidation states:

	Atom	**Oxidation state**	
$C_2H_4O_2$	O	−2	Rule 2: O is −2
	O	−2	
	H	+1	Rule 3: H is +1
	H	+1	
	H	+1	
	H	+1	
	C	?	No rule for C
	C	?	
		Sum 0	The sum must equal 0 (Rule 7).

The hydrogen and oxygen oxidation numbers sum up to 0. Therefore, if we make the oxidation number of both carbons, Rule 7 will be obeyed.

But we can also satisfy Rule 7 by making one carbon +1 and the other −1, or +2 and −2, or +3 and −3, and so forth. There is no clear answer here. The only way out is a dot diagram.

WORKPATCH 10.5 Using this dot diagram for acetic acid, assign each atom an oxidation number.

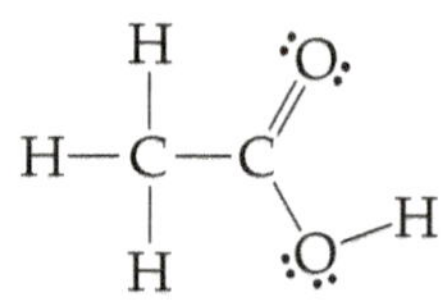

Using the dot diagram and applying the electronegativity rules for oxidation-state electron bookkeeping, you should have arrived at an unambiguous set of oxidation states for the carbon atoms in acetic acid. Check your answer against ours.

So far, we have simply been assigning oxidation states to the atoms in a chemical compound. Remember, however, that our original goal was to be able to determine whether, during a chemical reaction, any electrons are transferred from one atom to another. Oxidation numbers were designed to help us do just this, and we are now ready to use them for this purpose.

10.3 Recognizing Electron-Transfer Reactions

Oxidation numbers tell you how many electrons each atom in a compound owns relative to the free atom. Therefore *if the oxidation number of any atom changes during a chemical reaction, a transfer of electrons must have occurred.* Thus, looking for changes in oxidation numbers is a way to determine whether electron transfer has occurred during a chemical reaction. To see how this procedure works, look at this chemical reaction, where the oxidation number of each atom is printed above the atom:

$$\overset{+1}{\mathrm{Na}}\overset{-1}{\mathrm{Cl}} + \overset{+1}{\mathrm{Li}}\overset{-1}{\mathrm{Br}} \rightarrow \overset{+1}{\mathrm{Na}}\overset{-1}{\mathrm{Br}} + \overset{+1}{\mathrm{Li}}\overset{-1}{\mathrm{Cl}}$$

Consider each atom individually. Does the oxidation number of sodium change on going from reactant to product? Does the oxidation number of chlorine change on going from left to right? What about lithium and bromine? In this example, none of the oxidation numbers change during the reaction. This means that even though a chemical reaction takes place, there is no transfer of electrons from one atom to another. Each atom owns the same number of electrons before the reaction as after it.

The situation is quite different, however, in the reaction

$$\overset{-4}{\mathrm{C}}\overset{+1}{\mathrm{H_4}} + 2\,\overset{0}{\mathrm{O_2}} \rightarrow \overset{+4}{\mathrm{C}}\overset{-2}{\mathrm{O_2}} + 2\,\overset{+1}{\mathrm{H_2}}\overset{-2}{\mathrm{O}}$$

Carbon starts with an oxidation state of −4 and ends up as +4. Each oxygen begins as 0 and ends up as −2. This means that, according to the oxidation-state method of electron bookkeeping, a transfer of electrons must have taken place. In going from 0 to −2, each oxygen atom gained two electrons.

When an atom gains electrons, we say it has been *reduced* (or has *undergone reduction*); **reduction** is defined as a gain of electrons. This makes sense if you think about reduction not in terms of the number of electrons but in terms of the oxidation state of the atom, which is reduced to a lower value.

Reduction

- Gain of electrons (by oxidation-state bookkeeping)
- Reduction in oxidation number (from a higher to a lower value)

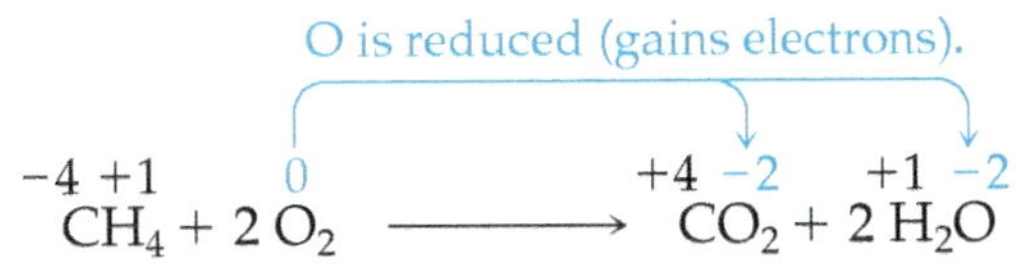

In this reaction, the oxygen atoms gain electrons, so we say the atoms are reduced. In general, an atom is reduced when its oxidation number *becomes smaller or more negative* as the atom goes from reactant to product.

If the oxygen atoms gain electrons in this electron-transfer reaction, where do the electrons come from? The key word here is *transfer*. It implies that, for oxygen to gain electrons, some other atom must lose them. Electron transfer implies *gain* and *loss*—you can never have one without the other. To find the atom that loses electrons, simply look for the one whose oxidation number increases (becomes more positive) during the reaction. In this reaction, it is carbon. To go from an oxidation state of −4 to +4, each carbon atom must lose eight electrons.

When an atom loses electrons, we say it has been *oxidized* (or has *undergone oxidation*); **oxidation** is defined as the loss of electrons. The carbon atoms are oxidized in this reaction.

Oxidation

- Loss of electrons (by oxidation-state bookkeeping)
- Increase in oxidation number (from a lower to a higher value)

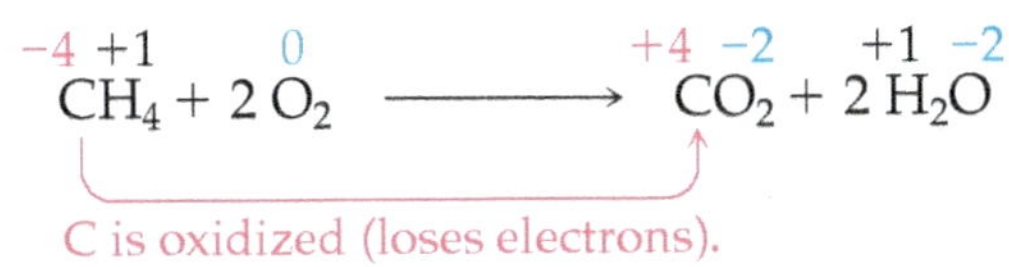

Now remember, by the octet method of electron bookkeeping, no atom loses or gains electrons. For example, the carbon atom in both methane and carbon dioxide has an octet:

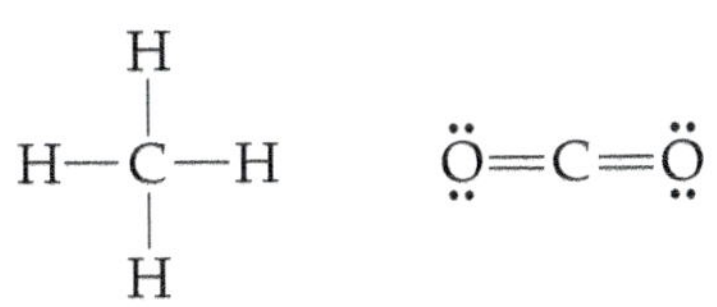

It is only when we switch to the oxidation-state method of electron bookkeeping that we can talk about the loss and gain of electrons.

Reduction (the gain of electrons) and oxidation (the loss of electrons) go together. One cannot occur in a chemical reaction without the other. We acknowledge this fact by referring to chemical reactions in which electrons are transferred as **redox** (pronounced ree-dox) **reactions**.

The reaction of methane, CH_4, with oxygen in the previous reaction is a redox reaction. The burning of any substance is a redox reaction, including the multistep "burning" of glucose in the cells of your body. Many chemical reactions are redox reactions, but not all of them. In order for a reaction to be classified as redox, electron transfer must take place during the reaction—that is, oxidation numbers must change.

Practice Problems

Indicate whether each reaction is a redox reaction. If it is, which atom gets oxidized and which atom gets reduced? Consult the shortcut rules.

10.11 $P_4 + 6\,Br_2 \rightarrow 4\,PBr_3$ (*Hint*: Br is more electronegative than P.)

Answer:

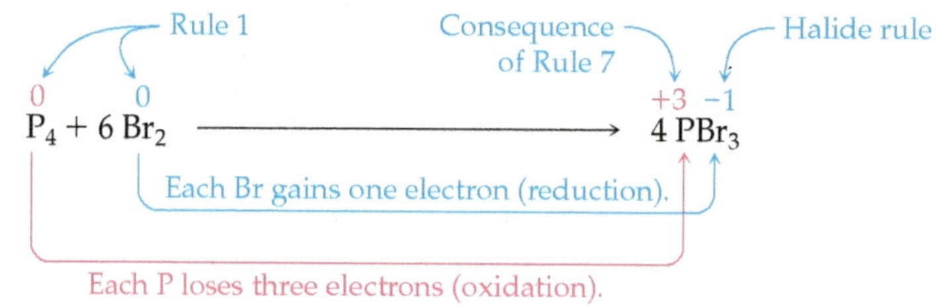

10.12 $4\,Fe + 3\,O_2 \rightarrow 2\,Fe_2O_3$

10.13 $Ca + 2\,H^+ \rightarrow Ca^{2+} + H_2$

10.14 $2\,NaBr + MgO \rightarrow MgBr_2 + Na_2O$

We are getting close to an explanation of how we can use electron-transfer (redox) reactions to produce electricity. But first we need to introduce a bit more terminology. To do this, let's go back to the combustion of methane:

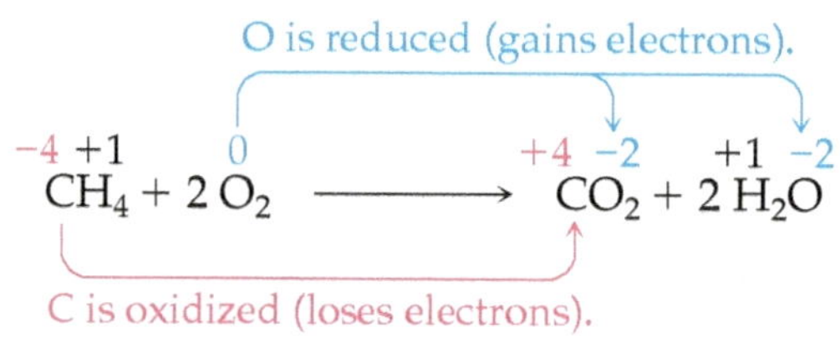

Up to now, we have been indicating the individual atoms that get oxidized or reduced. The C atom in CH_4 is oxidized. The O atoms in O_2 are reduced. While this way of looking at things is completely correct, chemists usually don't think in terms of the oxidation or reduction of individual atoms in a compound. Instead, they say that the entire CH_4 molecule is oxidized and the entire O_2 molecule is reduced. When the situation is described using whole molecules instead of individual atoms, we use the terms *oxidizing agent* and *reducing agent*. An **oxidizing agent** is a substance that oxidizes something else. A **reducing agent** is a substance that reduces something else. In our example, O_2 is an oxidizing agent because it oxidizes something else, CH_4. It does this by gaining the electrons that methane loses. This means that the oxidizing agent, O_2, must be reduced during the reaction.

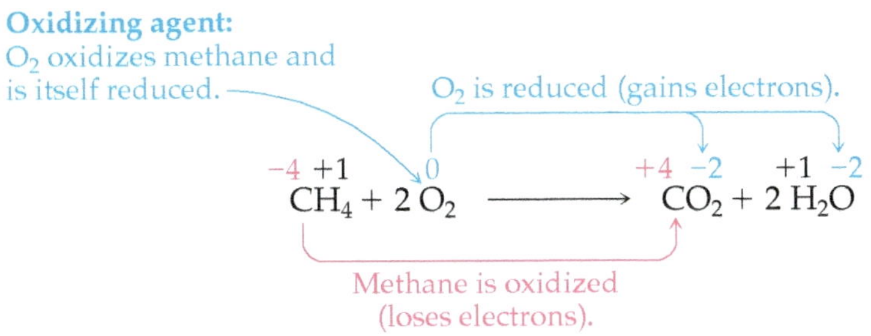

The same kind of logic works in identifying the reducing agent. In our example, CH_4 is a reducing agent because it reduces O_2. It does this by providing the electrons that O_2 gains. This means that the reducing agent, CH_4, must be oxidized during the reaction.

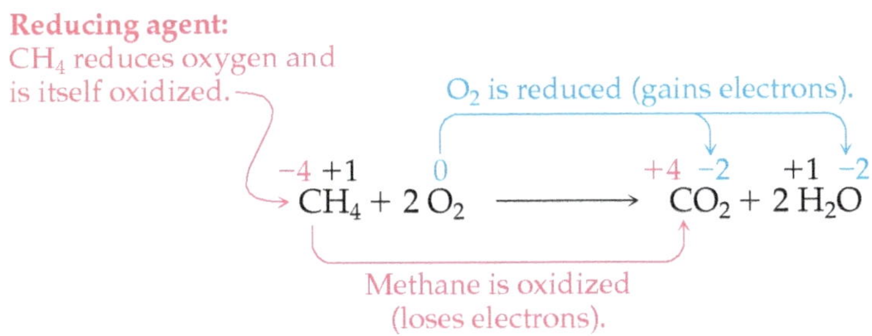

This terminology of *oxidizing agent* and *reducing agent* can be confusing, but it is something you will have to become comfortable with, and practice helps. Just think of it this way: Whichever compound (or ion) gets oxidized (whichever compound has an atom whose oxidation number increases) is the reducing agent, and whichever compound gets reduced (whichever compound has an atom whose oxidation number decreases) is the oxidizing agent.

The following are a repeat of Practice Problems 10.11–10.13, but this time you are asked to identify the oxidizing and reducing agents.

Practice Problems

For each redox reaction, indicate which substance is the oxidizing agent and which is the reducing agent.

10.15 $P_4 + 6\,Br_2 \rightarrow 4\,PBr_3$

Answer:

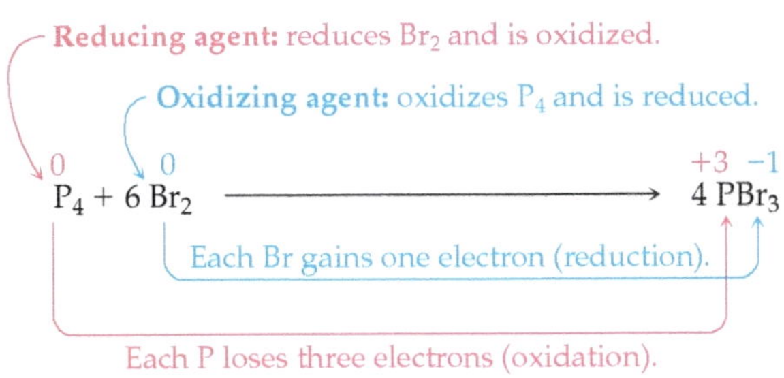

10.16 $4\,Fe + 3\,O_2 \rightarrow 2\,Fe_2O_3$

10.17 $Ca + 2\,H^+ \rightarrow Ca^{2+} + H_2$

Now let's carry out a real redox reaction. Put a strip of zinc metal into a beautiful blue aqueous solution of Cu^{2+} ions. Immediately, a chemical reaction begins. Within minutes, the blue color of the liquid begins to fade and the silvery zinc strip becomes plated with copper metal.

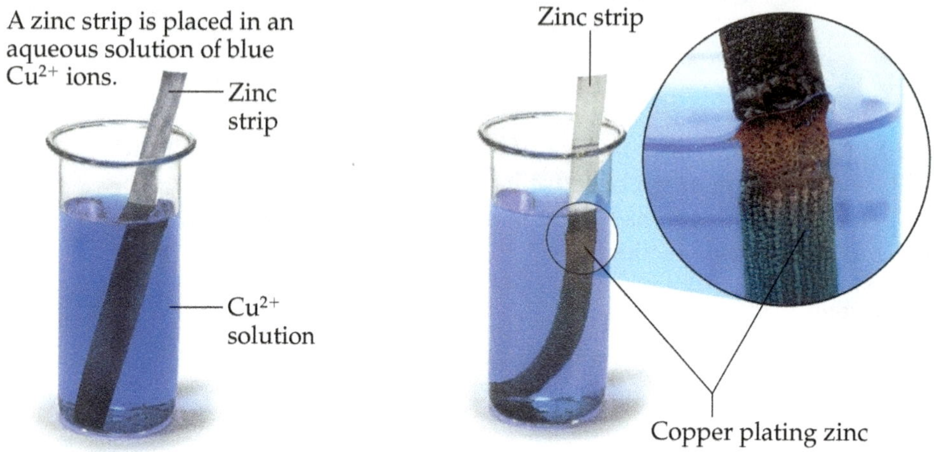

The reaction is $Zn + Cu^{2+} \longrightarrow Zn^{2+} + Cu$.

By now, you should be able to look at the balanced equation and quickly determine that this is indeed a redox reaction. Just to be sure, complete the following WorkPatch.

WORKPATCH 10.6

Step 1. **Use Rules 1–7 to assign oxidation states to the atoms and ions in the reaction**

$$Zn + Cu^{2+} \rightarrow Zn^{2+} + Cu$$

Step 2. **Now answer the following questions:**

(a) Is this a redox reaction? How can you tell?

(b) If so, what was oxidized?

(c) If so, what was reduced?

(d) If so, what was the oxidizing agent?

(e) If so, what was the reducing agent?

Without a doubt, electrons are transferred from Zn to Cu^{2+} in this reaction. Now let's harness some electricity from this redox reaction. ●

10.4 Electricity from Redox Reactions

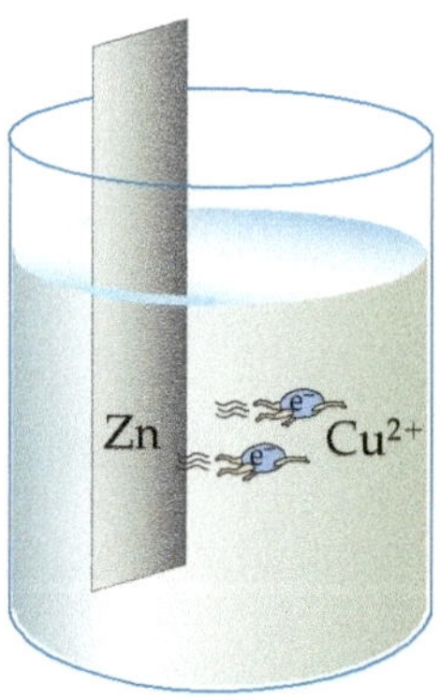

How do the electrons get from Zn to Cu^{2+}? **Not this way!**

Here again is the redox reaction we have chosen to harness electricity from:

$Zn + Cu^{2+} \rightarrow Zn^{2+} + Cu$

The zinc metal is oxidized, so the Zn atoms in the zinc strip each lose two electrons, while the copper ions (Cu^{2+}) are reduced, each picking up two electrons. But exactly how do the electrons get from the strip of Zn metal to the aqueous Cu^{2+} ions in solution? Because water is such a poor conductor of electricity, they can't just leap off the zinc strip and swim out to the Cu^{2+} ions. Instead, the Cu^{2+} ions must migrate to the zinc strip and touch it in order for electron transfer from Zn to Cu^{2+} to take place.

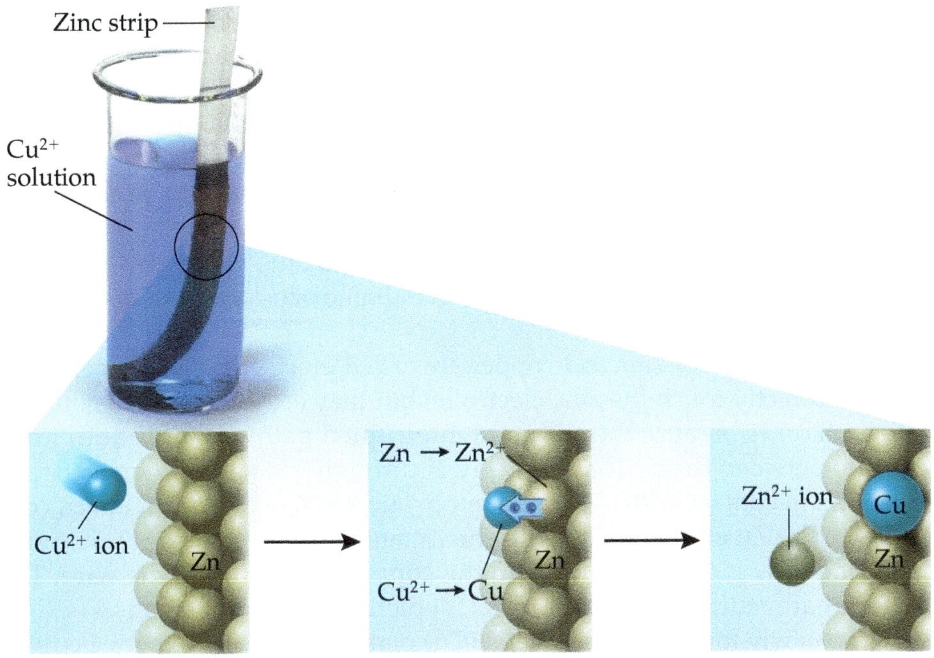

A Cu^{2+} ion touches a Zn atom.

Two electrons jump from the Zn atom to the Cu^{2+} ion.

This electron transfer reduces Cu^{2+} to Cu and oxidizes Zn to Zn^{2+}:

$$Cu^{2+} + Zn \longrightarrow Cu + Zn^{2+}$$

The Cu atom plates out on the Zn strip. The new Zn^{2+} ion goes into solution.

When a Cu^{2+} ion touches the zinc strip, the ion accepts two electrons and is reduced to elemental copper (Cu). This is where the copper metal that plates out on the Zn strip comes from. This also explains why the solution becomes less blue: blue Cu^{2+} ions are being removed from solution, and it is the presence of those ions in water that produces the blue color. At the same time, Zn atoms in the zinc strip are being oxidized to colorless Zn^{2+} ions. These newly formed Zn^{2+} ions enter the solution. If we could somehow get the electrons that transfer from the zinc to the copper ions to travel through a wire, we would be creating electricity (recall our description of electricity as a flow of electrons in a wire). This would be electricity produced by a chemical (redox) reaction, and it would last as long as there are reactants converting to products.

One way to accomplish this is to separate the Zn strip from the Cu^{2+} ions so that they are no longer in direct contact with each other, and then provide a wire path for the electrons to flow across. A common laboratory setup for this is shown in the figure at left. This setup, called a **galvanic cell** by chemists, is what you would

A Zinc–Copper Battery

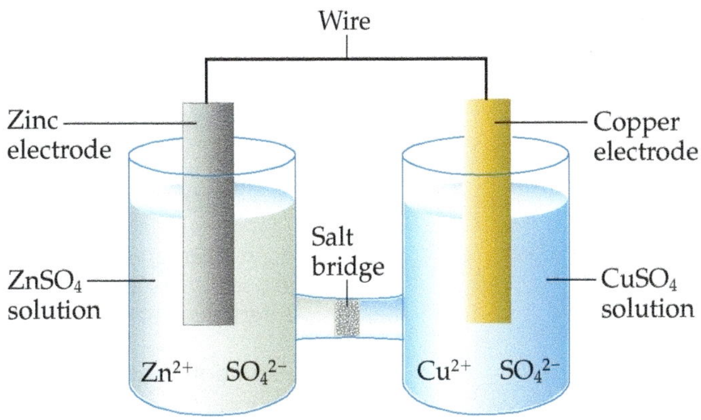

call a **battery**. In this example, a strip of zinc metal is placed in an aqueous solution of Zn^{2+} ions (created by dissolving zinc sulfate, $ZnSO_4$, in water), and a strip of copper metal is placed in a solution of Cu^{2+} ions (created by dissolving copper(II) sulfate, $CuSO_4$, in water). In other words, each strip of metal is placed in a solution of its own ions. Both solutions also contain sulfate, SO_4^{2-}, ions. The metal strips are then connected with a wire that can conduct electrons from one strip to the other. An additional component, called a *salt bridge*, is necessary to complete the circuit. This bridge can be as simple as a plug of absorbent cotton saturated with a solution of a salt, or it can be a tube like the one shown in the figure. The salt used in the salt bridge should be composed of *spectator ions*—that is, ions that are not directly involved in the redox reaction. A common bridge salt is sodium sulfate, Na_2SO_4, which provides Na^+ and SO_4^{2-} spectator ions to the system. We shall discuss the necessity for the salt bridge shortly.

The metal strips of zinc and copper are called **electrodes**. No longer can a Cu^{2+} ion directly touch the zinc electrode, but they are in "electrical contact" with each other because there is an uninterrupted pathway for the electrons to travel between them. That pathway is through the wire. When a copper ion comes in contact with the copper electrode, the Cu^{2+} ion is, in effect, connected to the Zn electrode (it's as if the Zn and Cu^{2+} are touching). The zinc metal can now give up electrons to the copper ions through the wire. The Cu^{2+} ions are reduced to copper metal, depleting the $CuSO_4$ solution of Cu^{2+} ions. The newly formed copper metal (Cu) plates out on the copper electrode, causing its mass to increase. The Zn atoms in the zinc electrode are oxidized to Zn^{2+}, and these newly formed ions enter the $ZnSO_4$ solution around the electrode, increasing the amount of Zn^{2+} ions in the solution. This reaction causes the zinc electrode to dissolve away over time. All of this is shown diagrammatically in the following figure. Study the sequence carefully to see what is happening.

How a Battery Generates Electrical Current

2. The electrons flow through the wire from the zinc electrode to the copper electrode.

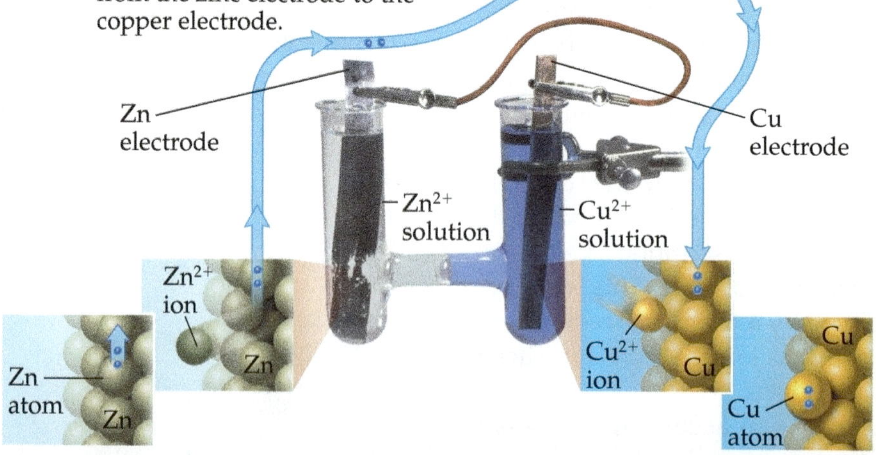

1. A Zn atom gives up two electrons to become a Zn^{2+} ion, which enters the $ZnSO_4$ solution.

Oxidation reaction: $Zn \rightarrow Zn^{2+} + 2\,e^-$

3. The electrons reduce a Cu^{2+} ion from the solution to a Cu atom. The Cu atom plates out on the electrode.

Reduction reaction: $Cu^{2+} + 2\,e^- \rightarrow Cu$

Now we can better appreciate the presence of the salt bridge. Without it, the zinc side of the galvanic cell becomes more and more positively charged as new Zn^{2+} ions are created. This buildup of positive charge eventually makes it impossible for negatively charged electrons to leave the zinc side. Likewise, the copper side of the cell becomes increasingly more negative as the Cu^{2+} ions are reduced and removed from solution, resulting in a relative excess of SO_4^{2-} ions. This excess negative charge repels electrons away from the copper side. This buildup of charge would cause the redox reaction to come to a halt, and electrons would no longer flow through the wire. In short, the cell (battery) would not function.

With the salt bridge in place, things are different. The cotton plug in the center of the bridge allows ions to move from one side to the other, but keeps the two solutions from mixing too quickly with each other. Now as the zinc side becomes more positively charged with Zn^{2+} ions, SO_4^{2-} ions from the salt bridge can move through the porous plug into the $ZnSO_4$ solution and neutralize the excess positive charge. Likewise, Na^+ ions from the salt bridge move into the $CuSO_4$ solution and neutralize the excess negative charge created on the copper side as Cu^{2+} ions plate out on the Cu electrode. In other words, a salt bridge is necessary for this type of reaction in order to maintain a charge balance in the two halves of the system.

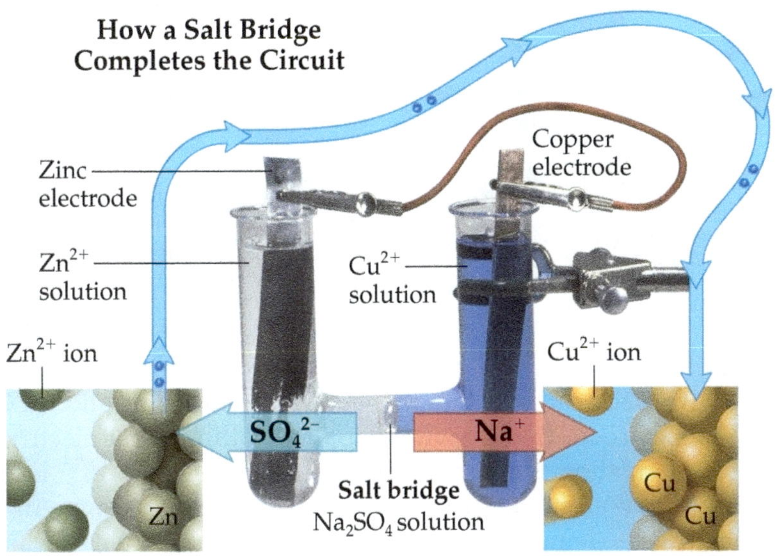

$Zn \longrightarrow Zn^{2+} + 2\,e^-$

As the oxidation reaction produces Zn^{2+} ions, SO_4^{2-} ions flow in from the salt bridge to prevent a buildup of positive charge.

$Cu^{2+} + 2\,e^- \longrightarrow Cu$

As the reduction reaction consumes Cu^{2+} ions, Na^+ ions flow in from the salt bridge to prevent a buildup of negative charge.

Because the electrons in our galvanic cell must flow through a wire to get from zinc atoms to copper ions, we are producing usable electricity. In addition, there is a measurable voltage pushing the electrons along. This zinc–copper battery produces about 0.9 V, enough to light a small bulb or run a motor to do some work. Until one of the reactants (Zn or Cu^{2+}) is used up, this battery will keep on going and going and going.

In principle, any redox reaction can be used to make a battery. All we need to do is separate the thing being oxidized from the thing being reduced and then connect them with a wire.

WORKPATCH 10.7 Design and label a battery based on the redox reaction between magnesium metal and silver ions:

$$Mg + 2\,Ag^{+} \rightarrow Mg^{2+} + 2\,Ag$$

Does your battery show electrons flowing from the Mg electrode to the Ag electrode? It should! •

Astounding as it may seem, you can make a battery using just electrical wires, strips of zinc and copper, and a lemon or other citrus fruit. You can even use a potato. Here is the lemon battery we built for this book:

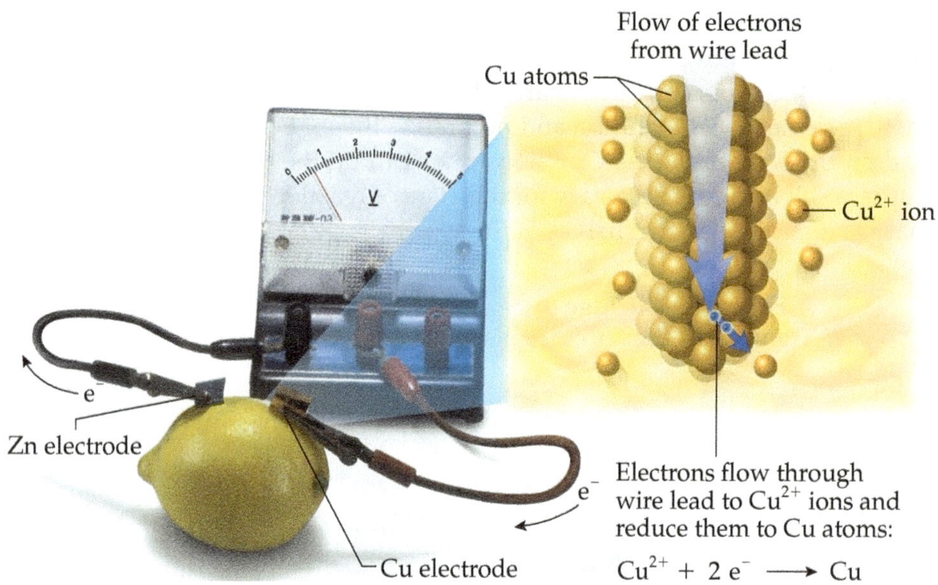

The only significant difference between this lemon battery and the zinc–copper battery we made earlier is that we haven't provided a solution of Cu^{2+} around the Cu electrode. Exactly what happens at the copper electrode is still an area of speculation. One reasonable explanation is that Cu^{2+} ions are already present on the electrode surface as a result of prior corrosion. Another is that the acids in the lemon cause some of the Cu atoms from the copper electrode to oxidize to Cu^{2+} ions. These ions remain in the vicinity of the copper electrode. As soon as the two electrodes are connected by a wire, zinc atoms begin to oxidize, releasing electrons. These electrons travel through the wire and reduce the Cu^{2+} ions back to Cu metal. The lemon itself serves as a salt bridge.

As we said, any redox reaction can potentially be used to make a battery if you can separate the substance that gets oxidized from the one that gets reduced, and then connect the two with a wire. Interestingly, you can simply place a zinc strip and a copper strip in the same beaker of salt water and run a low-current digital clock (a setup like this sits on the desk of one of the authors of this book). This and our lemon battery are a bit unconventional. Perhaps the ones in the photograph look more familiar to you.

The standard dry-cell battery is nothing more than a redox reaction in a can. A cross section of one type is shown at the top of the next page. The reactants are manganese(IV) oxide, MnO_2, and metallic zinc, Zn. As you can see from the diagram, these reactants are not in direct contact with each other.

A semimoist paste containing NH_4Cl (ammonium chloride) that serves as a salt bridge separates them. In this battery, the zinc metal is oxidized (loses two electrons) to become Zn^{2+} ions, so Zn is the reducing agent:

Zinc is oxidized (loses two electrons).
Each Zn atom becomes a Zn^{2+} ion.

The manganese(IV) oxide is changed to manganese(III) oxide, Mn_2O_3, during the reaction. The manganese atom in manganese dioxide is in the +4 oxidation state but becomes +3 in Mn_2O_3. So the manganese is reduced, and MnO_2 is the oxidizing agent:

Manganese is reduced (gains one electron).
Each Mn^{4+} ion in MnO_2 becomes a Mn^{3+} ion in Mn_2O_3.

Cross Section of a Dry-Cell Battery

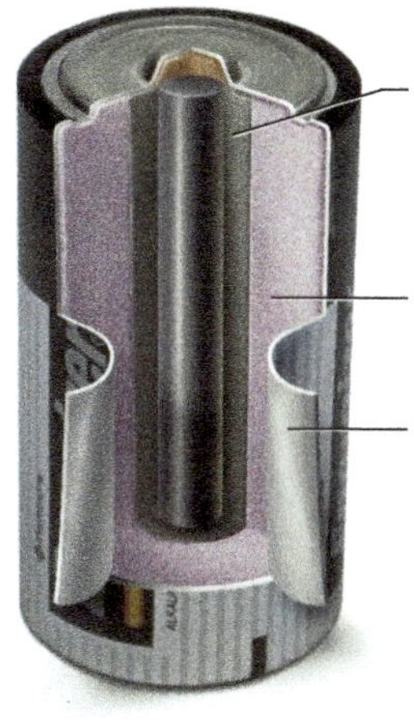

The overall redox reaction for the battery is:

Oxidizing agent

Reducing agent

Mn is reduced (gains one electron).

$$\overset{0}{Zn} + 2\,\overset{+4}{Mn}O_2 + 2\,NH_4^+ \longrightarrow \overset{+2}{Zn^{2+}} + \overset{+3}{Mn_2}O_3 + H_2O + 2\,NH_3$$

Zn is oxidized (loses two electrons).

Connect a wire between the two terminals and the reaction begins—electrons flow from Zn to MnO_2. Put a light bulb in the circuit and you've got a flashlight.

Redox reactions are at the heart of all batteries, from the battery in your car to the tiny battery in a hearing aid. All involve at least one reactant that loses electrons (oxidation) and another reactant that simultaneously gains them (reduction). You have probably noticed the + and − signs used to label the electrodes on a battery. In our zinc–manganese dry cell, which electrode would be marked with which sign? Recall that electrons are negatively charged and are therefore always attracted to anything with a positive charge. In the zinc–manganese dry cell, the electrons flow from the Zn electrode to the MnO_2 electrode. This means that the MnO_2 electrode is assigned the + sign.

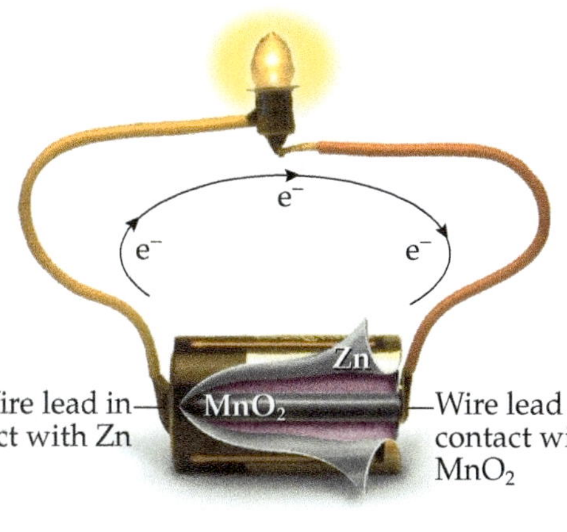

Not surprisingly, chemists have names for these two electrodes. The positive electrode is called the **cathode**, and the negative electrode is called the **anode**. To help you remember this, just recall the names of positive and negative ions. A positive ion is a *cat*ion, and a negative ion is an *an*ion. For both batteries and ions, the prefixes *cat-* and *an-* mean positive and negative, respectively.

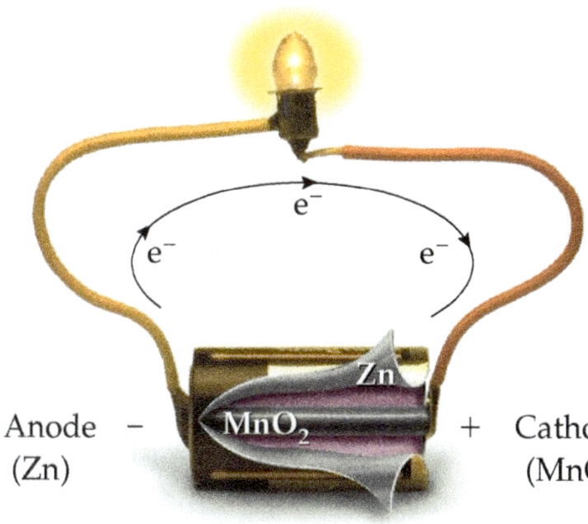

Electrons move from anode (−) to cathode (+).

WORKPATCH 10.8 Return to the drawing of the battery you made in WorkPatch 10.7 and add the labels, cathode, +, and anode, −. ●

It's important that you be able to do WorkPatch 10.8 because it leads us to a fundamental question. What ultimately decides which way the electrons flow? For the zinc–copper battery, you were told that zinc is oxidized and feeds electrons to the Cu^{2+} ions. That is, we started with the balanced redox reaction $Zn + Cu^{2+} \rightarrow Zn^{2+} + Cu$. But what if this were not the case? What if the diagram of the cell were put in front of you and you were asked, "Which way do the electrons flow? What gets oxidized? What gets reduced? Why is the reaction written with electrons flowing from zinc to copper and not the other way?"

WORKPATCH 10.9 Multiple choice: In general, which way do electrons flow in a battery?

(a) From what gets oxidized to what gets reduced.
(b) From what gets reduced to what gets oxidized.
Explain how you made your choice. ●

This WorkPatch should remind you that the ultimate answer to the question "Which way do the electrons flow?" requires knowing what is oxidized and what is reduced. In the next section, we shall see how you can determine this. For now, try the following practice problems, where the necessary information is given.

Practice Problems

10.18 Construct a battery in which the redox reaction is:

$Ni + Pb^{2+} \rightarrow Ni^{2+} + Pb$

To answer, draw a battery similar to the one you drew for WorkPatch 10.7, label which way the electrons flow, and label cathode, +, and anode, −.

Answer: The nickel electrode is the anode (−). The lead electrode is the cathode (+). Electrons flow from the nickel anode to the lead cathode.

10.19 For the nickel–lead battery of Practice Problem 10.18,

(a) What is the oxidizing agent?
(b) What is the reducing agent?
(c) Which electrode gets larger over time?
(d) Which electrode is eaten away over time?

10.20 A battery is constructed from iron, Fe, and silver, Ag, by dipping a strip of each metal into a solution of its ions (Fe^{3+} and Ag^{+}, respectively). As the battery operates, the Fe^{3+} concentration increases while the Ag^{+} concentration decreases.

(a) What is getting oxidized?
(b) What is getting reduced?
(c) Draw a battery similar to the one you drew for WorkPatch 10.7, label which way the electrons flow, and label cathode, +, and anode, −.

Let's finish this section by examining a battery that has become extremely popular. Its small size, its ability to be recharged over and over again, and its high energy density (it can power your devices for a relatively long time) make it the preferred power source for cell phones, laptop computers, iPods, and almost all of your favorite high-tech electronic gadgets. Once in a while, it also makes the news when one occasionally overheats and catches fire in someone's pocket (ouch!). This is the lithium ion battery, which comes in both cylindrical and rectangular shapes. Here is a diagram of one.

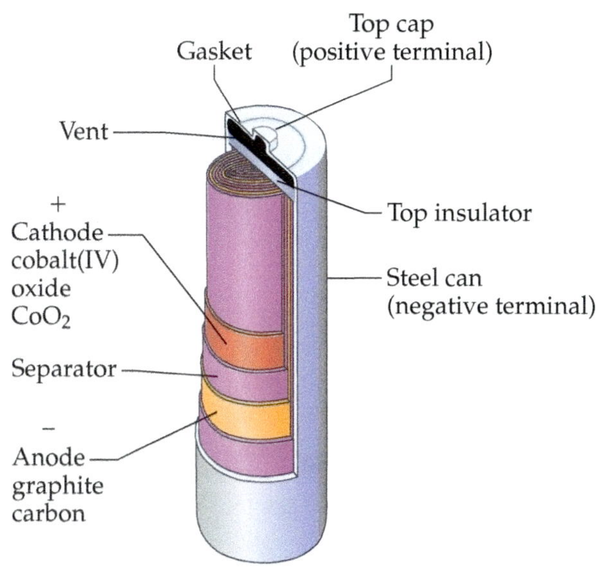

The three key components are the (+) cathode made of cobalt(IV) oxide [CoO_2] which contains Co^{+4} ions, the (−) anode made of neutral lithium (Li) atoms interspersed throughout graphite carbon powder, and a separator membrane that allows only Li^+ ions to pass through it. When the battery is in use, lithium atoms within the graphite anode give up their one valence electron (they are oxidized) and become Li^+ cations. The electrons pass through whatever device they are powering and travel into the cathode, where they reduce the Co^{+4} ions within cobalt(IV) oxide to Co^{+3}. This presents a problem because as Co^{+4} ions become Co^{+3} ions, the cathode becomes less and less positive (more and more negative), which would stop the flow of electrons. However, each time a Co^{+4} is reduced to a Co^{+3}, a Li^+ ion from the other side passes through the separator membrane to restore charge neutrality. In this way, Li^+ ions passing through the separator membrane act in the same fashion as a salt bridge. Ultimately, the CoO_2 cathode is converted into $LiCoO_2$, which contains Co^{+3} ions. Here is what the redox process looks like at the atomic level.

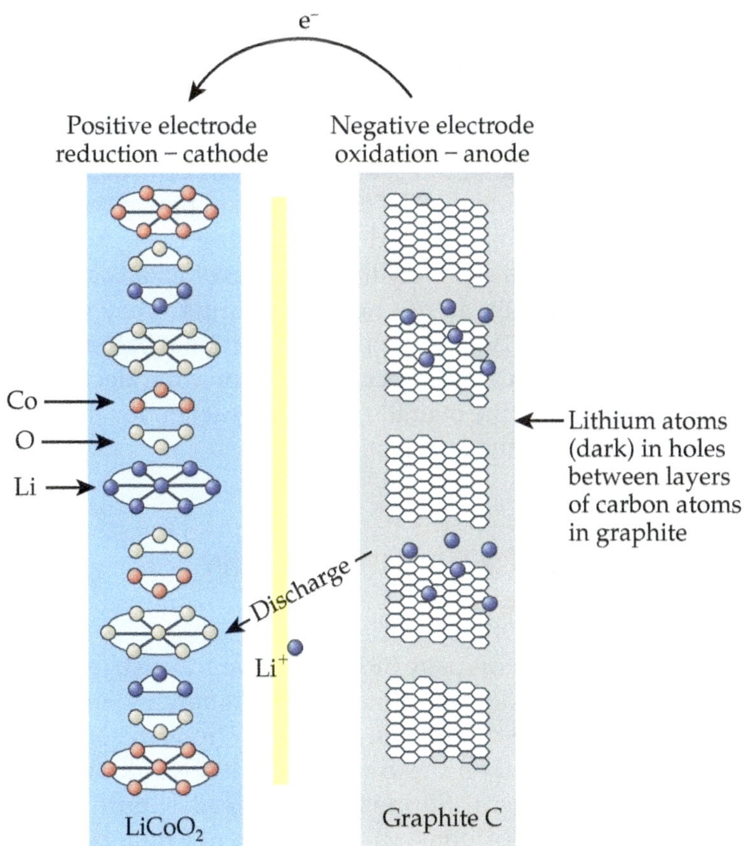

Source: Panasonic

An even more unique battery often in the news is the fuel cell. A fuel cell is a battery in which one of the reactants is a combustible fuel and the other is oxygen. For example, the fuel cells used by NASA to supply electricity and water for space vehicles employ the controlled, catalyzed combustion of hydrogen and oxygen (shown below).

Anode: $2\ H_2(g) + 4\ OH^-(aq) \longrightarrow 4\ H_2O(l) + 4e^-$

Cathode: $O_2(g) + 2\ H_2O(l) + 4\ e^- \longrightarrow 4\ OH^-(aq)$

Overall: $2\ H_2(g) + O_2(g) \longrightarrow 2\ H_2O(l)$

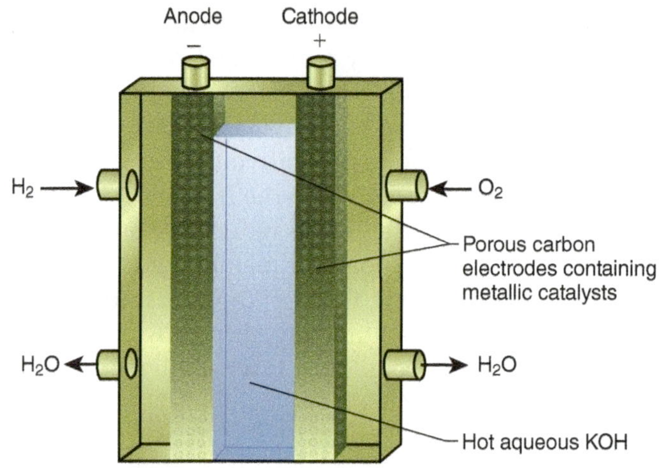

The hydrogen and oxygen gas never actually meet. Rather, a platinum catalyst in the anode causes the hydrogen molecules to split into positive hydrogen ions and electrons. The hydrogen ions can react with hydroxide ions from the hot aqueous KOH to convert into water, but the electrons must travel through the conductive graphite along an external circuit to the cathode. Today, automobile companies are experimenting with fuel cells to power cars. Such cars would fuel up on hydrogen gas instead of liquid gasoline, giving momentum to the so-called hydrogen economy that futurists often talk about. Unlike traditional batteries, fuel cells do not need recharging. They simply need a continuous source of fuel. As for any of the traditional batteries we have discussed, recharging involves using electricity to force the electrons to flow in the reverse directions and the overall net redox reaction for the cell to run in reverse, which bring us to the next section.

10.5 Which Way Do Electrons Flow?—The EMF Series

Look back at the illustration on page 381. When we put a strip of zinc into a solution of Cu^{2+} ions, a redox reaction occurred spontaneously. Neutral Zn atoms spontaneously lost electrons to the Cu^{2+} ions in solution, causing copper metal to plate out on the zinc strip. What would happen if we tried the opposite arrangement? Suppose that instead of putting a strip of Zn metal into a solution of Cu^{2+} ions, we put a strip of Cu metal into a solution of Zn^{2+} ions.

Would the Zn^{2+} ions be reduced (gain electrons) and the Cu atoms be oxidized (lose electrons)? In other words, would we see zinc metal plate out on the copper strip? The equation for this hypothetical reaction is:

$$Zn^{2+} + Cu \xrightarrow{?} Zn + Cu^{2+}$$

This is the opposite (reverse) of the reaction that occurs in a zinc–copper battery:

$$Zn + Cu^{2+} \rightarrow Zn^{2+} + Cu$$

In fact, our hypothetical reaction doesn't occur because when you immerse a copper strip in a solution of Zn^{2+} ions, nothing happens. By setting up these two experiments, we learn that:

What Happens Here?

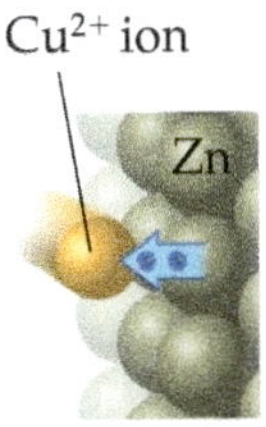

Zn metal gives up electrons to Cu^{2+} ions to form Cu atoms,

BUT

Cu metal does not give up electrons to Zn^{2+} ions.

Why is this so? Recall from Chapter 4 that the chemical definition of a metal is "an element that readily loses its valence electrons." But, given the results summarized above, it must be that all metals are not created equal with respect to this definition. *Some metals lose electrons more readily than others.* In the competition we set up between the two systems Zn/Cu^{2+} and Zn^{2+}/Cu, a spontaneous reaction occurred only in the first case, where Zn metal gives up electrons and copper accepts them. Because of this, we say that zinc is a more *active* metal than copper. An **active metal** is one that easily loses valence electrons. Zinc atoms have a greater tendency than copper atoms to lose valence electrons (in other words, zinc is more easily oxidized than copper), so zinc is the more active metal. Because this is so, zinc metal can "force" Cu^{2+} ions to take its electrons, but copper metal cannot "force" Zn^{2+} ions to take its electrons. The following diagram sums this up. In a zinc–copper system, the electron flow is always from zinc to copper because of zinc's greater tendency to lose electrons.

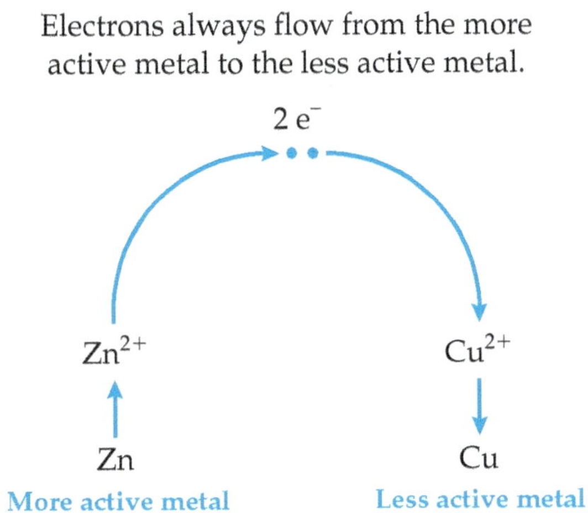

We should point out that the first time these metals were studied, there was no way to know whether it was the Zn or the Cu that was more active. The result had to be determined by setting up the two experiments described above and noting which one reacted spontaneously.

In principle, we could compare any two metals to determine which one is more active in this way. We could carry out this same experiment using zinc and magnesium, for example, by dipping each metal into a solution of the other's ions and noting which one reacts spontaneously. This experiment is shown below. We observe that nothing at all happens in the beaker with the zinc strip immersed in a solution of Mg^{2+} ions. However, in the beaker on the right, the situation is quite different. The magnesium strip quickly becomes coated with zinc metal.

Test to Find Out Which Metal Is More Active—Zn or Mg

From these experimental observations, we conclude the following about the relative activities of magnesium and zinc:

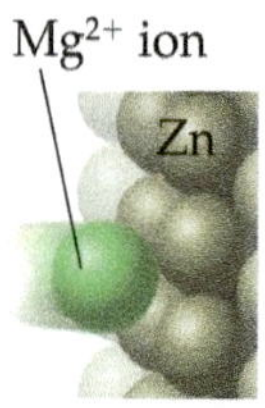

Zn metal does not give up electrons to Mg^{2+} ions,

BUT

Mg metal gives up electrons to Zn^{2+} ions to form Zn atoms.

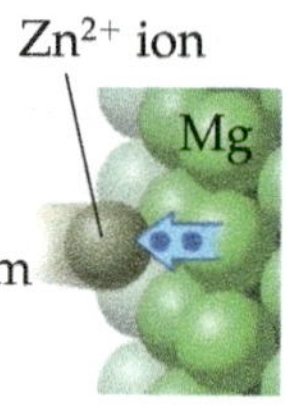

Electrons flow from Mg (more active) to Zn (less active).

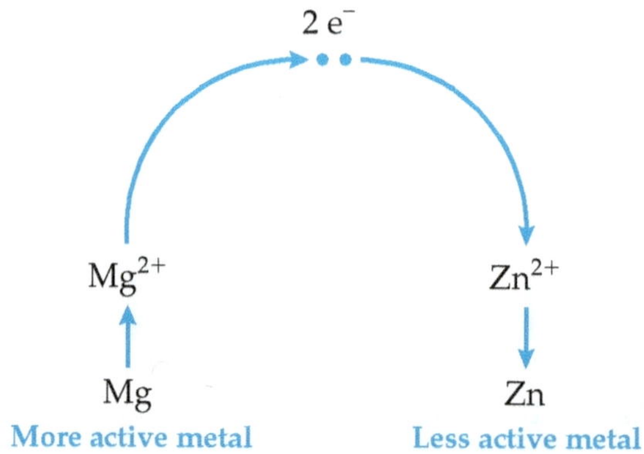

In other words, magnesium is a more active metal than zinc (magnesium's "desire" to lose electrons is greater than zinc's). And because zinc is more active than copper, magnesium must also be more active than copper. This relative ordering of activities is called the **electromotive force series** (or **EMF series**, for short):

An Electromotive Force Series

Mg Most active metal (loses electrons most easily)

↑

Zn Increasing activity

Cu Least active metal (loses electrons least easily)

We could extend this list by testing additional pairs of metals and metal ions, but chemists have already done this for us over the years. A portion of the resulting EMF series is shown in the margin.

Remember that a more active metal can always force electrons onto the ions of a less active metal. Using the EMF series, we can predict which redox reactions will be spontaneous and which way electrons will flow in a battery. For example, if we set up this experiment, in which beaker will a spontaneous redox reaction occur?

Copper is higher on the EMF list than silver (not by much, but it is higher). This means that copper is more active than silver, and so neutral Cu atoms give electrons to Ag^+ ions. The redox reaction

$$Cu + 2\,Ag^+ \rightarrow Cu^{2+} + 2\,Ag$$

occurs spontaneously in the beaker on the right. The Ag^+ ions are reduced to elemental silver and plate out on the copper strip (see next page). As for the beaker on the left, absolutely nothing happens in it. The reverse of the spontaneous redox reaction does not occur because silver is less active than copper. In a copper–silver battery, electrons flow from the copper electrode (more active metal) to the silver electrode (less active metal). This means that copper is the anode (−) and silver is the cathode (+). Copper metal is oxidized, and Ag^+ ions are reduced.

EMF Series

Most active (loses electrons most easily)

↑ Increasing activity

Li
K
Ba
Sr
Ca
Na
Mg
Al
Mn
Zn
Cr
Fe
Cd
Co
Ni
Sn
Pb
H
Sb
As
Bi
Cu
Ag
Pd
Hg
Pt
Au

Least active (loses electrons least easily)

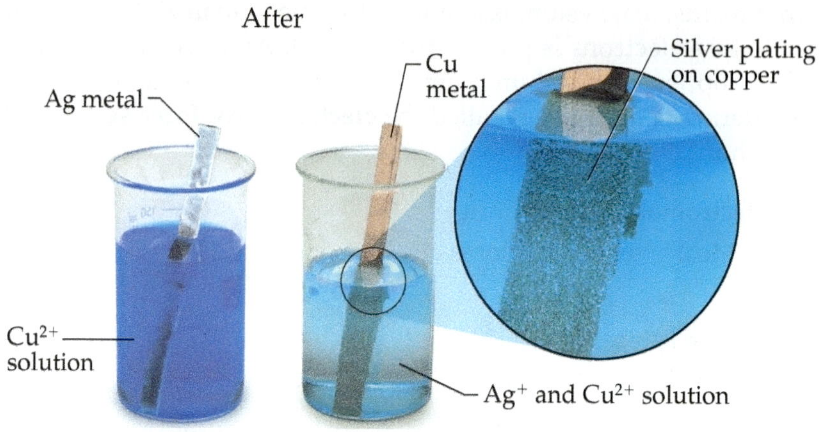

Nothing happens. Solution begins to turn blue as Cu atoms from the metal become (blue) Cu^{2+} ions in solution.

All these results can be predicted from a glance at the EMF series. We can even take advantage of this knowledge to make a beautiful Christmas decoration by bending a piece of copper wire into the shape of a Christmas tree and placing it in a solution of Ag^+ ions. The spontaneous redox reaction $Cu + 2\,Ag^+ \rightarrow Cu^{2+} + 2\,Ag$ occurs, and pure silver metal plates out on the copper tree. In addition, the Cu^{2+} ions formed color the solution a bright blue.

Practice Problems

10.21 What will happen if a piece of Zn metal is placed in a solution of Al^{3+} ions?

Answer: Absolutely nothing; Al is higher on the EMF series and is thus the more active metal. This means that Al could give its valence electrons to Zn^{2+} ions, but Zn does not spontaneously give its electrons to Al^{3+} ions.

10.22 Which of the following reactions is spontaneous?

(a) $3\,Zn^{2+} + 2\,Al \rightarrow 3\,Zn + 2\,Al^{3+}$
(b) $3\,Zn + 2\,Al^{3+} \rightarrow 3\,Zn^{2+} + 2\,Al$

10.23 Using a diagram, show how you would construct a zinc–aluminum battery. Label the cathode, the anode, the + electrode, and the − electrode. Indicate the direction of electron flow, and identify the beaker in which oxidation occurs and the beaker in which reduction occurs.

10.24 Consider a battery made from lead, Pb, and copper, Cu, along with Pb^{2+} ions and Cu^{2+} ions. In such a battery, what is the oxidizing agent, and what is the reducing agent? (*Hint*: Start by consulting the EMF series, and determine which metal oxidizes and which metal ion gets reduced.)

Understanding the EMF series allows us to predict which redox reactions are spontaneous and how to make batteries. It can also help us combat an enemy that costs us billions of dollars each year, an enemy that weakens our bridges, destroys our infrastructure, and even tried to tear down the Statue of Liberty! Who is this unpatriotic fiend? Read on and find out.

10.6 Another Look at Oxidation: The Corrosion of Metals

Oxidation is defined as the loss of electrons. We are not going to change this definition, but we do want to embellish it a bit. Earlier in this chapter, we talked about the burning of methane, CH_4:

$$\overset{-4+1}{CH_4} + 2\,\overset{0}{O_2} \rightarrow \overset{+4-2}{CO_2} + 2\,\overset{+1-2}{H_2O}$$

We say that methane is oxidized because the oxidation state of its carbon atom changes from -4 to $+4$, signifying a loss of eight electrons. However, there is an older definition of oxidation. One hundred years ago, before chemists knew of the existence of electrons, they still would have said that the methane was oxidized upon burning. At that time, oxidation literally meant "combining with oxygen," which is what happens when something burns. Many elements, including metals, react with oxygen ("burn"). To see an example of this, just go to any junkyard and find a '65 Chevy, or should we say, what's left of one. Mainly what you will see is rust! Rust is iron that has combined with oxygen, iron that has "burned":

$$\underset{\text{Iron metal}}{4\,Fe} + 3\,O_2 \rightarrow \underset{\text{Iron(III) oxide (rust)}}{2\,Fe_2O_3}$$

Of course, most cars don't burst into flames when they rust. This "burning" of metal, this combining with oxygen, occurs slowly. We call this slow burn **corrosion**.

Before we go on, let's get our definitions sorted out. We have the modern definition of oxidation—the loss of electrons. We also have the older definition of oxidation—combining with oxygen. Are these definitions compatible? Or, put another way, does an element lose electrons when it combines with oxygen ("burns")? The answer is yes, if the element combining (forming a new bond) with oxygen is less electronegative than oxygen. Because oxygen is more electronegative than every element in the periodic table except fluorine, the answer is nearly always yes. When a metal (or any other element except fluorine) forms new bonds with oxygen, the metal loses the bonding electrons according to the bookkeeping rules of the oxidation-state method.

Examine the oxidation numbers of the atoms in the reaction, and you will see that iron is indeed oxidized by the modern definition.

$$\overset{0}{4\,Fe} + \overset{0}{3\,O_2} \rightarrow 2\,\overset{+3\;-2}{Fe_2O_3}$$

Iron metal — Iron(III) oxide (rust)

Each iron atom loses three electrons in this reaction. So we see that the new and the old definitions of oxidation are truly one and the same.

Corrosion (rusting, oxidation) of iron or steel (steel is iron combined with small amounts of carbon) is a major economic problem in any modern society because so many things are built with steel. We try to prevent the corrosion of things like cars and bridges by covering the iron with paint or applying sealants, but this ultimately ends up as a losing battle. We live in an atmosphere full of oxygen, and sooner or later, nature will claim your car. Sometimes, however, we can unwittingly make the situation worse! Such was the case with the Statue of Liberty. She was built with a "time bomb" ticking away inside, an unintentional design flaw involving a redox reaction that almost brought her down.

The statue is made from sculpted sheets of copper metal hung on a steel skeleton that provides structural support. The statue sits on a small island in New York harbor and so is constantly exposed to the wind and spray of salt water. Needless to say, the exposed copper is subject to corrosion. Copper was chosen for the external skin of the statue because when it corrodes, it becomes coated with a green material called a patina. The composition of the patina is a mixture of a hydroxy carbonate $Cu_2(OH)_2CO_3$ and a hydroxy sulfate $Cu_2(OH)_2SO_4$. The copper in these compounds is in the +2 oxidation state. This means that, in forming the green patina, the exposed surface of the metallic copper has been oxidized to Cu^{2+}:

$$\overset{0}{Cu} \xrightarrow[\text{air and moisture}]{\text{In the presence of}} \overset{+2}{Cu^{2+}}$$

Ions in the form of copper hydroxy carbonate and copper hydroxy sulfate

Cu is oxidized (loses two electrons).

By itself, this is not a problem for the statue. Unlike Iron(III) oxide (rust), which flakes off and so continuously exposes fresh iron to further corrosion, oxidized copper (the patina) forms a tough, impenetrable coating over the underlying copper, protecting it from contact with oxygen and further corrosion. Given these properties of copper, the Statue of Liberty should not have corroded after the initial patina had formed. For about 100 years, this was true. Then, almost overnight, the statue started to fall apart. Things got so bad that the arm holding the torch was in serious danger of falling off!

The problem was inside the statue—the steel (iron) skeleton was reacting with the copper sheets. Consulting the EMF series on page 391, you can see that iron is a more active metal than copper. If iron metal comes in contact with copper in its oxidized state (Cu^{2+}), the iron will spontaneously oxidize and give electrons to the Cu^{2+} ions, reducing them back to elemental copper.

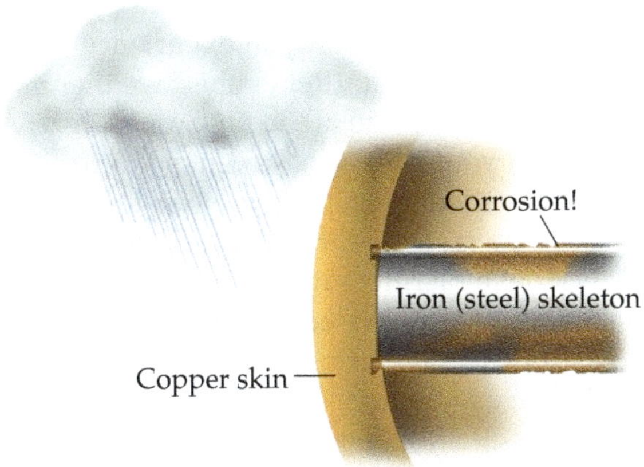

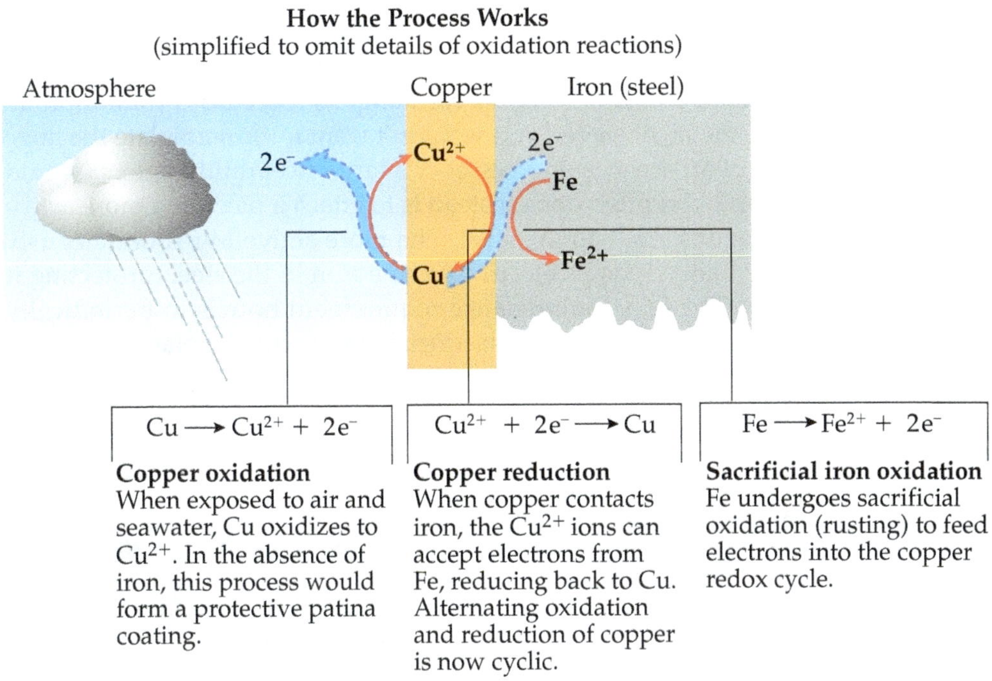

More active metals lose electrons to ions of less active metals. This is great for the copper because the iron "sacrificed" itself to save the copper. As air and moisture reoxidized the copper, the iron in the skeleton kept losing its electrons to Cu^{2+} ions, reducing them back to elemental Cu. The net result was that the inner steel skeleton of the statue oxidized and began to crumble very quickly.

Actually, the designers of the statue knew that this could happen. To prevent it, they put thick asbestos padding between the steel framework and the copper skin to insulate them from each other. Unfortunately, over the years, the padding rotted away, allowing the copper and iron to come into direct contact. At that point, the statue became one huge redox reaction, with her steel skeleton as one of the reactants.

After ignoring the problem for a number of years, authorities launched a stupendous effort to save the statue. Many millions of dollars later, she now has a reinforced, corrosion-resistant stainless steel skeleton incorporating high-tech plastics to insulate it from the outer copper skin.

Scaffolding surrounded the Statue of Liberty as corrosion damage was being repaired.

An active metal sacrificing itself for a less active one is also responsible for that rather painful sensation you experience when you bite down on a piece of aluminum foil. (Remember, we mentioned this at the beginning of this chapter.) Some of the tin, Sn, or silver, Ag, used in the mercury amalgam of your fillings has been oxidized to Sn^{2+} or Ag^{+} by continuous exposure to air and acids in your mouth. Aluminum, being a far more active metal than either tin or silver (it is far above Sn and Ag in the EMF series), sacrifices itself on contact with your filling. The Al oxidizes to Al^{3+} and feeds electrons to the Sn^{2+} and Ag^{+} ions, reducing them to their elemental forms. This electron transfer gives the nerves in your teeth the equivalent of an electric shock.

The phenomenon of a sacrificial metal can be used in the never-ending battle to prevent corrosion. One way to protect iron (or steel) from corrosion is to coat it with a more active metal, such as chromium, Cr. Many active metals (like Cr) form tough oxide coatings that strongly adhere to the metal underneath, protecting it from further corrosion. This works well for small pieces of iron like car bumpers. Because chromium is an expensive metal, however, you wouldn't want to chrome-plate the huge steel hull of an ocean-going vessel or a large steel utility pole. It would cost too much. What is often done instead is to attach a piece of a more active metal—say, magnesium,—to the steel. The more active Mg spontaneously oxidizes to Mg^{2+} and feeds its electrons to the iron in the steel, protecting it from oxidation. The only maintenance requirement now is to periodically replace the piece of magnesium, the sacrificial (more active) metal.

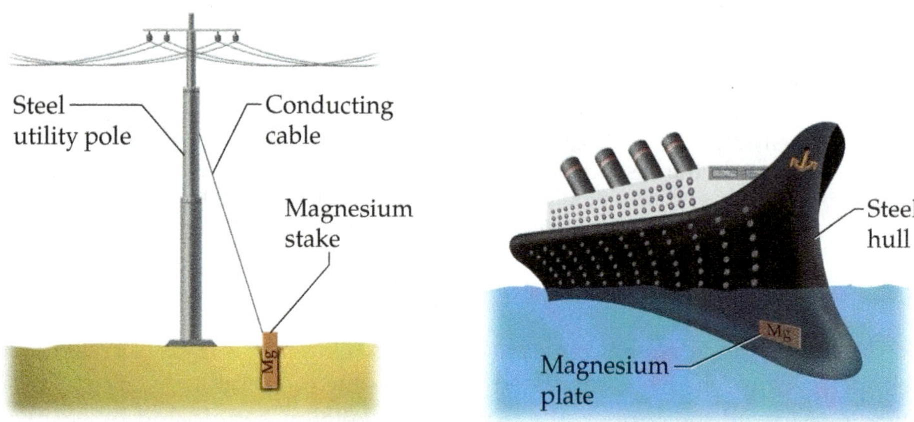

The magnesium sacrifices itself to prevent corrosion of the iron (steel).

Of course, there are some instances, often symbolic, where we want metals to last a lifetime without any signs of corrosion.

WORKPATCH 10.10 If marriage is truly supposed to be "till death do us part," what metal does the EMF series indicate should be used for wedding rings? Explain your answer. ●

Certainly, a ring made of iron, which corrodes rapidly, would not be appropriate in this case. Do you see why the metals on the bottom of the EMF list are usually the ones used for jewelry and coins?

From the "burning" of glucose in the cells of our bodies to the batteries in our electronic devices to the combustion of methane, chemical reactions that transfer electrons from one atom to another can be found almost everywhere. Not all chemical reactions involve electron transfer, but you now know how to recognize the ones that do.

+1 One More Thing

10.7 Less Common Oxidation States and Chemical Reactivity

In Section 10.2 you learned about the shortcut rules for assigning oxidation states to atoms in molecules, and they worked reasonably well. Not surprisingly though, there are times when the shortcut rules don't work. Often times the failure is based on electronegativity considerations. For example, the halide rule (halogen atoms have an oxidation state of −1 in a compound) works only if the halogen atom is attached to atoms less electronegative than itself. Rule 2 (oxygen has a −2 oxidation state in a compound) works only if the oxygen atom is bound to atoms less electronegative than itself. Rule 3 (hydrogen has a +1 oxidation state in a compound) works only if the hydrogen atom is bound to an atom more electronegative than itself. WorkPatch 10.2 (ozone) and 10.4 (chlorate ion) had you work examples where Rule 2 and the halide rule did not work. In both cases the electronegativity criterion was violated. A good way out was to draw a dot diagram and count the number of electrons each atom "owned" according to the oxidation-state method of electron bookkeeping and then assign oxidation states.

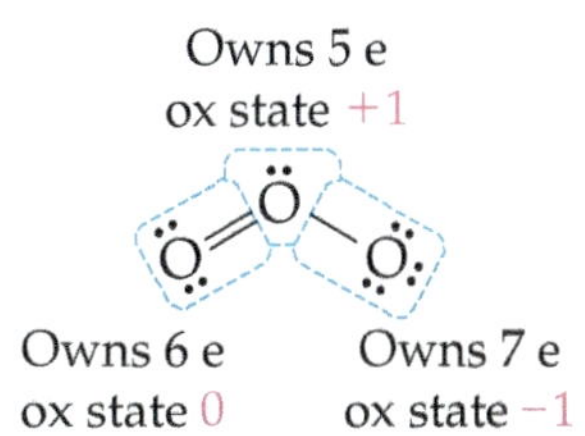

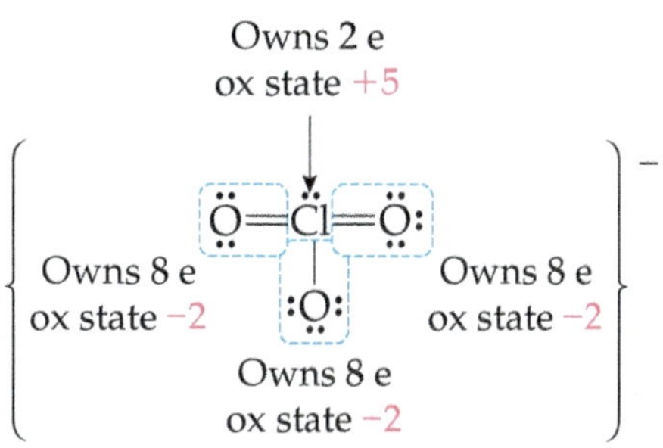

Oxidation states of +1 for oxygen and +5 for chlorine can be thought of as less common, the common values being those predicted by the shortcut rules. An example of when Rule 3 does not work is potassium hydride, KH. Here, hydrogen is bound to an atom less electronegative than itself, resulting in an ionic compound whose lattice consists of K^+ cations and $(H:)^-$ anions, called hydride ions. For monoatomic ions, the oxidation state is equal to the charge of the ion, so KH possesses H atoms in a compound with a less common −1 oxidation state instead of the much more common +1 value.

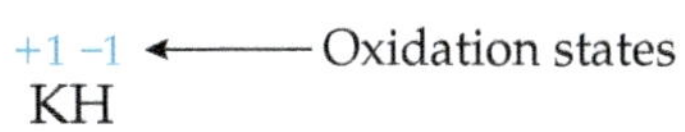

When atoms find themselves in less common oxidation states, such as +5 for the Cl in ClO_3^-, −1 for the H in KH, and +1 for one of the O's in O_3, it is often the case that these atoms are very reactive and can be thought of as having a tremendous drive to donate or accept electrons so as to return to more common values of their oxidation state. For example, KH undergoes a violent reaction with water, liberating hydrogen gas as shown below:

$$\overset{+1\,-1}{\text{KH}} + \overset{+1\,-2}{\text{H}_2\text{O}} \longrightarrow \overset{+1-2+1}{\text{KOH}} + \overset{0}{\text{H}_2}$$

This is a redox reaction in which the hydride ion gives up an electron to (reduces) one of the hydrogen atoms in water to give elemental hydrogen, H_2. KH is therefore known as a powerful reducing agent. The above reaction converts the hydride ion from the less common −1 oxidation state to an oxidation state of 0 that atoms have in their elemental state. Likewise, the +5 oxidation state chlorine atom in the chlorate (ClO_3^-) ion can be thought of as having a tremendous drive to accept electrons so it can attain its more common −1 oxidation state. The +5 oxidation state chlorine atom will therefore take electrons from many other molecules (i.e., it will oxidize other molecules). This makes the chlorate ion a very reactive oxidizing agent, so powerful it is used to bleach wood pulp for making paper.

What about carbon? There are literally millions of known molecules possessing one or more carbon atoms. There is no oxidation state shortcut rule for carbon atoms in a molecule, so you will often have to resort to a dot diagram to determine the oxidation states of the carbon atoms as you did in WorkPatch 10.5 for acetic acid ($C_2H_4O_2$). While we will not cite any particular oxidation state for carbon as being less common, we can say that knowledge of them can help us understand the chemistry of compounds that contain carbon atoms. For example consider acetone, the active ingredient in nail-polish remover and a common industrial solvent. Acetone has the formula C_3H_6O. If we use the shortcut rules and assign the O atom an oxidation state of −2 and each H atom an oxidation state of +1, then the sum of these oxidation states will be +4.

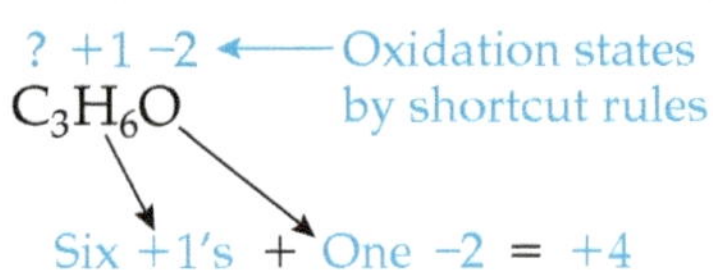

This means the oxidation states of the three carbon atoms must sum to −4 (this makes the sum of all the oxidation states equal the charge of the molecule, which is zero—shortcut Rule 7). How can we accomplish this? One way is to take the −4 charge and divide it equally among the three carbon atoms, giving each C atom an oxidation state of −4/3. However, we could also give two of the C atoms an oxidation state of −1 and the remaining C atom an oxidation state of −2. Or, we could give two of the C atoms an oxidation state of 0 and the remaining C atom an oxidation state of −4. It's a mess, and the shortcut rules give us no guidance. Once again, the way out is to draw the dot diagram and do explicit oxidation-state-method electron bookkeeping.

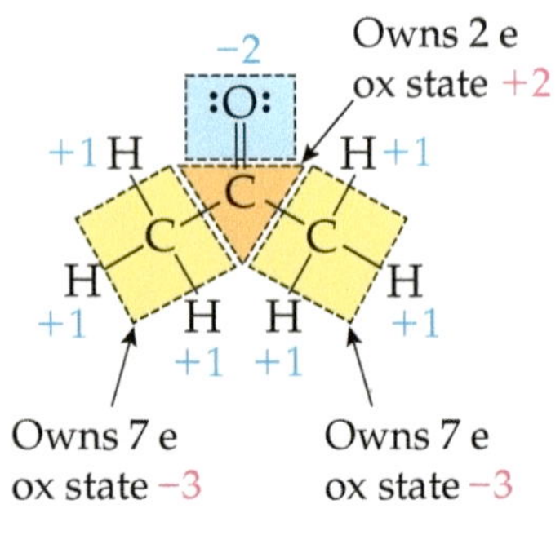

While oxidation states are not the actual charges on the atoms in a compound, except for simple ionic compounds, it is safe to say that the central carbon atom of acetone, with its +2 oxidation state, has some positive character to it. We might predict therefore that the nitrogen atom in ammonia, which possesses a lone pair of electrons and a negative −3 oxidation state, might be attracted to and react at the positive +2 oxidation-state carbon atom in acetone. Indeed, such a reaction occurs. The mechanism for this reaction has been shown to begin with an "attack" of ammonia's negative-oxidation-state N atom on the positive oxidation state C atom of acetone and results in the formation of new bonds between the amine N atom and the central C atom of acetone.

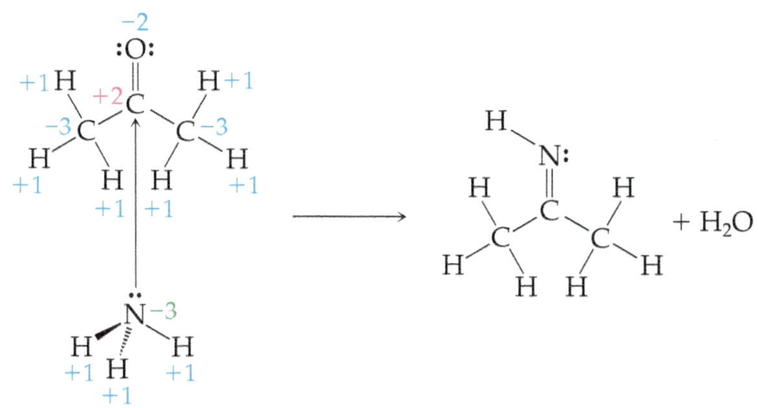

The lesson here is that knowledge of oxidation states can help us predict not only if electron transfer has occurred but also the type of chemical change that accompanies it and, if the oxidation states are less common, where the reactive sites on the molecules might be.

Have You Learned This?

Learning Outcomes

10.1	What is Electricity?	Describe what electricity is. [25, 26]
10.2	Electron Bookkeeping—Oxidation States	Assign oxidation states to atoms in compounds and molecules. [33–36, 43–48]
10.3	Recognizing Electron-Transfer Reactions	Use oxidation states to decide if electron transfer is occurring during a chemical reaction. [56, 58, 62] Identify the oxidizing agent and reducing agent in an electron-transfer reaction. [59, 60, 63]
10.4	Electricity from Redox Reactions	Construct (diagram) a battery and label its parts and direction of electron flow given a balanced redox reaction. [64–71]
10.5	Which Way Do the Electrons Flow—The EMF Series	Use the EMF series to construct (diagram) a battery and label its parts and direction of electron flow. [72–77, 80, 81, 83]
10.6	Another Look at Oxidation—The Corrosion of Metals	Describe the corrosion of metals in terms of electron-transfer reactions. [84–88]
10.7	Less Common Oxidation States and Chemical Reactivity	Describe the reactivity that might result from the presence of a less common oxidation state. [139, 140]

Skills to Know

Assigning oxidation states by the electron-bookkeeping method
This method always works.

Example: Assign an oxidation state to each atom in hydrogen peroxide, H_2O_2.

1. Draw a valid dot diagram for the ion or molecule.

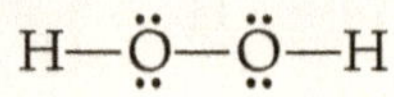

2. Assign every valance electron to an atom.
   - Assign each lone pair to the atom on which it is drawn in the dot diagram.
   - Assign the electrons of each bond to the bonding atom of higher electronegativity.
   - If bonding electrons are shared by atoms of equal electronegativity, divide the electrons evenly between the bonding atoms.

O has higher EN than H, so O gets both shared electrons.

Each O atom gets its own lone-pair electrons.

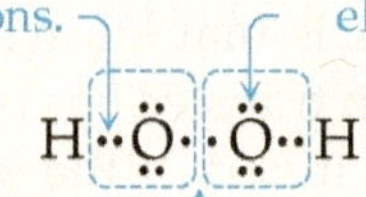

All O atoms have the same EN, so these electrons are split between the O atoms.

3. Calculate the oxidation state for each atom by subtracting the electron count of the atom (as determined in step 2) from the number of valence electrons on a free atom of that element.

	Each H atom	Each O atom
Number on free atom	1	6
Number assigned	−0	−7
Oxidation state	+1	−1

Check your answer
The oxidation states for all the atoms should add up to the charge on the molecule or ion.

$1 + 1 + (-1) + (-1) = 0$ OK!

Example: Assign an oxidation state to each atom in the bicarbonate ion, HCO_3^-.

Step 1: 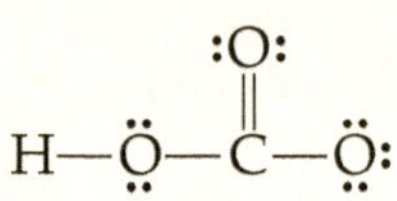 Valid dot diagram

Step 2: 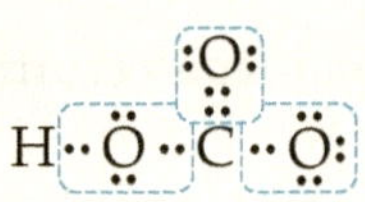 O has higher EN than H or C, so the O atoms get all the shared electrons (plus their own lone pairs).

Step 3:

	H atom	C atom	Each O atom
Number on free atom	1	4	6
Number assigned	−0	−0	−8
Oxidation state	+1	+4	−2

Check answer: $1 + 4 + (-2)+(-2) + (-2) = -1$ OK!

Assigning oxidation states by the shortcut method

This method is quick once you know the rules, although it sometimes fails to give a definitive answer.
Example: Determine the oxidation state of each atom in the bicarbonate ion, HCO_3^-.

Rule 1
The oxidation state of an atom in the most abundant naturally occurring form of an element is set to zero.

Rule 2
The oxidation state of an oxygen atom in a compound is almost always −2.

Rule 3
The oxidation state of a hydrogen atom in a compound is +1.

Step 1: List the elements and apply the element-specific rules:

Atom	Oxidation state	
H	+1	← Rule 3
C	?	← No rule
O	−2	Rule 2
O	−2	Rule 2
O	−2	Rule 2
	Sum: −5	

Rule 4
The oxidation state of any group IA atom in a compound is +1.

Rule 5
The oxidation state of any group IIA atom in a compound is +2.

Rule 6
The oxidation state of a monatomic ion equals the ion's charge.

Halide rule
A halogen atom in a compound usually has an oxidation state of −1, except when bonded to a more electronegative element.

Rule 7
The sum of the oxidation states for every atom in a formula must add up to the formula's overall charge.

Step 2: Use Rule 7 to assign an oxidation state to C:

Algebraic sum of all atoms except C → -5; C oxidation number → x; Charge on ion → -1

$$-5 + x = -1$$
$$x = -1 + 5 = +4$$

Atom	Oxidation state
H	+1
C	+4
O	−2
O	−2
O	−2
	Sum: −1

Rule 7 says that the oxidation states must add up to the overall charge (−1 in this case). Therefore, the oxidation state of C must be +4.

If this shortcut method fails to give a definitive answer for all atoms, you must draw a dot diagram and use the oxidation-state bookkeeping method (which always works). For instance, the shortcut rules do not give the correct answer for the H_2O_2 example starting on page 400.

Concept Questions

1. Voltage is:
   (a) The same as electricity.
   (b) The force or "push" behind the electrons flowing through a wire.
   (c) The flow of electrons in a wire.
   (d) The amount of electrons that flow through a wire per unit time.

2. Oxidation states are useful for:
   (a) Determining the charges of atoms in a molecule.
   (b) Confirming whether the octet rule is obeyed.
   (c) Deciding if electron transfer is occurring during a chemical reaction.
   (d) All of the above are correct.

3. When electron bookkeeping by the oxidation-state method, it is true that:
   (a) The most electronegative element in a bond "owns" the shared electrons.
   (b) Each atom "owns" half of the shared electrons in a bond between identical elements.
   (c) Lone pairs are completely owned by the atom they are on.
   (d) All of the above are correct.

4. To calculate an oxidation state for an atom in a molecule you need to know:
   (a) How many electrons on the atom are core electrons.
   (b) The number of valence electrons in a free atom of the element.
   (c) The number of protons the atom owns.
   (d) The total number of valence electrons in the molecule.

5. According to the shortcut rules for assigning oxidation states:
   (a) Halide atoms usually have a zero oxidation state.
   (b) Oxygen atoms usually have a +2 oxidation state.
   (c) The sum of all the oxidation states must be zero for molecules and ions.
   (d) The sum of all the oxidation states must equal the charge of the molecule or ion.

6. The way to recognize an electron transfer reaction from a balanced chemical equation is:
   (a) To draw a dot diagram of every reactant and product and see that the octet rule is not obeyed for some atoms.
   (b) To assign oxidation states to every atom and see if they always add to give the charge of the molecule.
   (c) To assign oxidation states to every atom and see if any change on going from reactants to products.
   (d) To assign oxidation states to every atom and see if any go from positive to negative or negative to positive.

7. Which of the following is true?
   (a) Oxidation is the gain of electrons.
   (b) The reducing agent loses electrons and is itself oxidized.
   (c) Reduction is the loss of electrons.
   (d) The oxidizing agent loses electrons and is itself reduced.
   (e) All the statements are true.

8. In a battery:
   (a) Oxidation occurs at the anode.
   (b) Reduction occurs at the anode.
   (c) Oxidation occurs at the cathode.
   (d) Electrons flow from what gets reduced to what gets oxidized.

9. The EMF series is useful for:
   (a) Determining oxidation states.
   (b) Determining which way the electrons will flow in a battery.
   (c) Balancing electron transfer reactions.
   (d) Determining if a salt bridge is required.

10. The corrosion of metals:
    (a) Is more likely the more active a metal is.
    (b) Involves a redox reaction in which the metal is oxidized.
    (c) Involves an electron transfer reaction in which the metal is oxidized.
    (d) All of the above statements are correct.

Answers: 1b; 2c; 3d; 4b; 5d; 6c; 7b; 8a; 9b; 10d

Electricity and Electron Bookkeeping

10.25 How is electricity similar to water flowing in a hose?

10.26 How is electricity mechanically produced in a power plant?

10.27 What is an electron-transfer reaction?

10.28 Why did chemists find it necessary to invent oxidation states?

10.29 What do we mean by electron bookkeeping?

10.30 Which method of assigning shared electrons is correct—the double-counting method of the octet rule or the oxidation-state method?

10.31 When assigning an oxidation state to an atom, why do we need to know how many valence electrons are present on a free atom of that element?

10.32 According to the oxidation-state method of electron bookkeeping, which atom is assigned the shared electrons in a bond between two atoms?

10.33 Draw a Lewis dot diagram for NF_2H, and use the oxidation-state method of electron bookkeeping to determine how many electrons each atom should be assigned.

10.34 Draw a Lewis dot diagram for H_2O_2 (hydrogen peroxide), and use the oxidation-state method of electron bookkeeping to determine how many electrons each atom should be assigned.

10.35 Draw a Lewis dot diagram for NO_3^-, and use the oxidation-state method of electron bookkeeping to determine how many electrons each atom should be assigned.

10.36 Assign an oxidation state to every atom in the molecules in Problems 10.33–10.35.

10.37 Consider the following molecules. Without using the shortcut rules, determine the oxidation state for each atom in the molecules and show how you calculated it.
(a) H_2
(b) O_2
(c) Cl_2
(d) IF

10.38 What does the answer to Problem 10.37 teach you?

10.39 What must always be true when you add up all the oxidation states for the atoms in a molecule?

10.40 Suppose an atom has an oxidation state of +3.
(a) Would more or fewer electrons be assigned to this atom by oxidation-state bookkeeping than are present on a free atom of that element?
(b) How many more or fewer?

10.41 Suppose an atom has an oxidation state of −2.
(a) Would more or fewer electrons be assigned to this atom by oxidation-state bookkeeping than are present on a free atom of that element?
(b) How many more or fewer?

10.42 Suppose the shortcut rules can determine the oxidation state of every atom in a compound except one. How can you find the oxidation state of the remaining atom?

10.43 Use the shortcut rules to assign an oxidation state to each atom in:
(a) PCl_3
(b) H_2S
(c) MnO_4^-
(d) HNO_3
(e) HCOOH
(f) $S_2O_3^{2-}$

10.44 Use the shortcut rules to assign an oxidation state to each atom in:
(a) CO
(b) CH_2Cl_2
(c) $HCOO^-$
(d) $PtCl_6^{2-}$

10.45 Use the shortcut rules to assign an oxidation state to each atom in:
(a) O^{2-}
(b) Li_3N
(c) $MgSO_4$
(d) MnO_2

10.46 Consider ClO^- and $AlCl_3$. For one of these substances, the halide shortcut rule works. For the other, it does not.
(a) Which one does it work for, and why?
(b) Why doesn't it work for the other?
(c) Assign oxidation states to all the atoms in both substances.

10.47 WorkPatch 10.2 had you assign oxidation states to the oxygen atoms in ozone. The answer was −1 and 0 for the terminal (end) oxygen atoms and +1 for the central oxygen atom. But now, a student claims the oxidation states are +1 for the central oxygen atom and −1/2 for the terminal oxygen atoms. He is also right. Explain why.

10.48 Consider the organic compound $C_3H_6O_2$, methyl acetate.
(a) What is the problem with using the shortcut method to assign oxidation numbers to the atoms in methyl acetate?
(b) The dot diagram of methyl acetate is shown below. Assign an oxidation state to every atom.

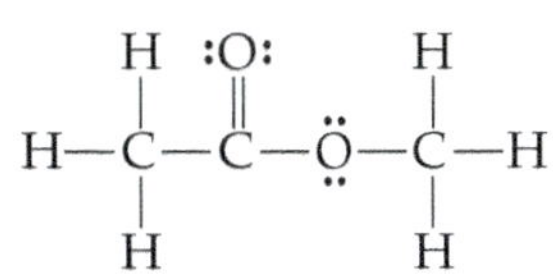

Recognizing Electron-Transfer Reactions

10.49 What do the terms *oxidation* and *reduction* mean with regard to valence electrons?

10.50 When is an oxidizing agent also the species that gets oxidized? Explain.

10.51 How are oxidation states useful in determining whether a reaction is a redox reaction?

10.52 What happens to an atom's oxidation state when the atom is reduced?

10.53 What happens to an atom's oxidation state when the atom is oxidized?

10.54 What does the word *transfer* imply about an electron-transfer reaction?

10.55 Why can we always call an electron-transfer reaction a redox reaction?

10.56 Which of the following are redox reactions?
(a) $2\,Na + 2\,H_2O \rightarrow 2\,NaOH + H_2$
(b) $MgBr_2 + 2\,NaF \rightarrow MgF_2 + 2\,NaBr$
(c) $2\,CO + O_2 \rightarrow 2\,CO_2$
(d) $SO_2 + H_2O \rightarrow H_2SO_3$

10.57 For those reactions in Problem 10.56 that are redox reactions:
(a) Indicate which atoms get oxidized and which atoms get reduced.
(b) Indicate which reactant is the oxidizing agent.
(c) Indicate which reactant is the reducing agent.

10.58 Which of the following are electron-transfer reactions?
(a) $2\,CrO_4^{2-} + 2\,H^+ \rightarrow Cr_2O_7^{2-} + H_2O$
(b) $Fe + NO_3^- + 4\,H^+ \rightarrow Fe^{3+} + NO + 2\,H_2O$
(c) $2\,C_2H_6 + 7\,O_2 \rightarrow 4\,CO_2 + 6\,H_2O$
(d) $2\,AgBr \rightarrow 2\,Ag + Br_2$

10.59 For those reactions in Problem 10.58 that are electron-transfer reactions:
(a) Indicate which atoms get oxidized and which atoms get reduced.
(b) Indicate which reactant is the oxidizing agent.
(c) Indicate which reactant is the reducing agent.

10.60 The following reaction is responsible for producing electricity in your car battery (often called a lead storage battery):

$$Pb + PbO_2 + 2\,H_2SO_4 \rightarrow 2\,PbSO_4 + 2\,H_2O$$

(a) Assign an oxidation state to each atom. (*Hint*: For $PbSO_4$, you can figure out the charge of the Pb if you remember that the charge of the sulfate ion is −2 (SO_4^{2-}). Then use shortcut Rule 7 to get the oxidation state of the Pb in $PbSO_4$.)
(b) Identify the atom that gets oxidized and the atom that gets reduced.
(c) Identify the reactant that is the oxidizing agent and the reactant that is the reducing agent.

10.61 Hydrogen gas burns very well in the presence of oxygen to give water:

$$2\,H_2 + O_2 \rightarrow 2\,H_2O$$

In principle, should it be possible to use this chemical reaction to produce electricity? Explain.

10.62 In the upper atmosphere, sunlight can convert oxygen to ozone:

$$2\,O_2 \rightarrow \underset{\text{Ozone}}{O_3} + O$$

Is this a redox reaction? Completely justify your answer.

10.63 The hydroquinone molecule can be converted to the quinone molecule as shown below:

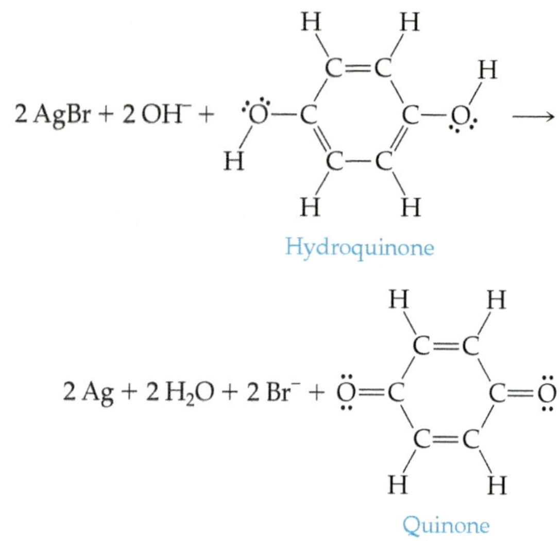

You may want to use a combination of shortcut rules and dot diagrams to assign oxidation numbers before answering the following questions:
(a) Which atom or atoms are oxidized?
(b) Which atom or atoms are reduced?
(c) What is the oxidizing agent?
(d) What is the reducing agent?

Electricity from Redox Reactions

10.64 If you put a piece of iron in an aqueous solution of blue Cu^{2+} ions, the spontaneous redox reaction

$$Fe + Cu^{2+} \rightarrow Fe^{2+} + Cu$$

will occur. An aqueous solution of Fe^{2+} ions is red-brown.
(a) What is oxidized?
(b) What is reduced?
(c) What is the oxidizing agent?
(d) What is the reducing agent?
(e) What will happen to the piece of iron over time?
(f) Do the electrons move from the oxidizing agent to the reducing agent or from the reducing agent to the oxidizing agent?
(g) How could you use this reaction to make a battery? (Explain and show your battery in a diagram, using a nail and a penny as your source of iron and copper, respectively.)

In Problems 10.65–10.69, you will be given the spontaneous redox reaction (or enough information regarding it) so that you can construct and label a battery.

10.65 Include a salt bridge filled with $NaCl(aq)$ and indicate which way the $Na^{+}(aq)$ and $Cl^{-}(aq)$ ions flow.

$$Cr + 3\,Ag^{+} \rightarrow Cr^{3+} + 3\,Ag$$

Draw a battery similar to the one you drew for WorkPatch 10.7. Make sure you label which way the electrons flow, and also label the cathode, +, and anode, −.

10.66 For the chromium–silver battery of Problem 10.65:
(a) What is the oxidizing agent?
(b) What is the reducing agent?
(c) Which electrode gains mass with time?
(d) Which electrode is eaten away with time?

10.67 Include a salt bridge filled with $NaCl(aq)$ and indicate which way the $Na^{+}(aq)$ and $Cl^{-}(aq)$ ions flow.

$$Mg + Ni^{2+} \rightarrow Mg^{2+} + Ni$$

Draw a battery similar to the one you drew for WorkPatch 10.7. Make sure you label which way the electrons flow, and also label the cathode, +, and anode, −.

10.68 For the magnesium–nickel battery of Problem 10.67:
(a) What is the oxidizing agent?
(b) What is the reducing agent?
(c) Which electrode gains mass with time?
(d) Which electrode is eaten away with time?

10.69 A battery is constructed from tin and copper by dipping strips of each metal into a solution of its ions (Sn^{2+} and Cu^{2+}, respectively). As the battery operates, the Sn^{2+} concentration increases and the Cu^{2+} concentration decreases.
(a) What is getting oxidized?
(b) What is getting reduced?
(c) Draw a battery similar to the one you drew for WorkPatch 10.7. Make sure you label which way the electrons flow, and also label the cathode, +, and anode, −.

10.70 A student claims that in his battery, the electrons flow from the positive to the negative electrode. In fact, this is not true for any battery. Explain why electrons would never flow in this direction and how the direction of electron flow tells you that the cathode is where reduction occurs.

10.71 How can the words *cation* and *anion* help you remember if + or − is associated with the cathode and anode of a battery?

Which Way Do Electrons Flow?—The EMF Series

10.72 Suppose you have two metals, A and B, and solutions of their ions, A^{+} and B^{+}. Metal B is more active than metal A.
(a) Write the spontaneous redox reaction for this situation.
(b) What is the oxidizing agent?
(c) What is the reducing agent?
(d) Describe which way the electrons flow.

10.73 A zinc electrode and a copper electrode are used to make a battery, as shown below:
(a) Replace the three question marks with appropriate labels.
(b) Indicate which way the electrons flow.
(c) Write the spontaneous redox reaction.
(d) Add the labels +, −, anode, and cathode to the drawing.

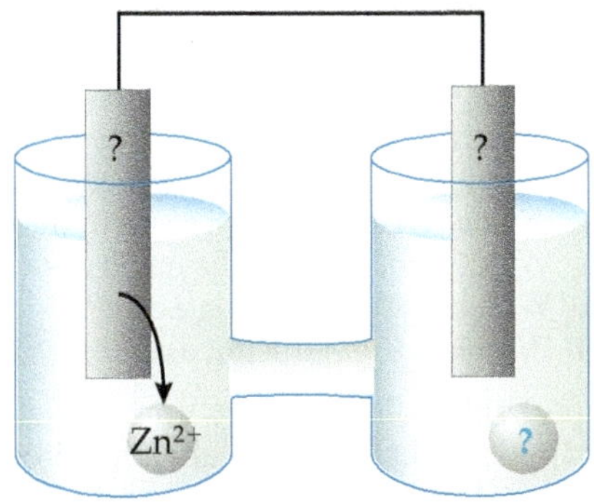

(e) Write "oxidation occurs at this electrode" under the appropriate beaker.
(f) Write "reduction occurs at this electrode" under the appropriate beaker.
(g) Indicate in which beaker the concentration of metal ions increases with time.
(h) Indicate in which beaker the concentration of metal ions decreases with time.
(i) Indicate in which beaker the electrode appears eaten away with time.
(j) Indicate in which beaker the electrode gains mass with time.

10.74 Suppose you have three different metals, X, Y, and Z, each of which also exists as a +2 ion. Design an experiment to place them properly in order of their activity (a mini-EMF series).

10.75 You are trapped on an island with lots of different metals, chemicals, and supplies but no EMF series table. How would you go about deriving an EMF series, and how would it be useful?

10.76 What happens when you place an active metal in a solution of ions of a less active metal?

10.77 What happens when you place a less active metal in a solution of ions of a more active metal?

10.78 Using the EMF series on page 391, decide which of the following redox reactions is spontaneous. Explain your answer.
(a) $3\,K + Al^{3+} \rightarrow Al + 3\,K^{+}$
(b) $Al + 3\,K^{+} \rightarrow 3\,K + Al^{3+}$

10.79 Using the EMF series on page 391, decide which of the following redox reactions is spontaneous. Explain your answer.
(a) $3\,Ag + Au^{3+} \rightarrow Au + 3\,Ag^+$
(b) $Au + 3\,Ag^+ \rightarrow 3\,Ag + Au^{3+}$

10.80 You are trapped on a desert island with plenty of water (both fresh and salt), a drinking glass, some wire, a radio, and no batteries. You do have a tin cup, a tube of toothpaste containing stannous fluoride (SnF_2, a source of Sn^{2+} ions), a silver pendant, and undeveloped black-and-white film (such film has silver bromide, AgBr, in it, a source of Ag^+ ions).
(a) How would you use the above materials to construct a battery? Show how with a diagram, including an arrow over the wire to show which way the electrons flow. (You can make a salt bridge by soaking a sock in salt water and then dipping one end of the sock in one cell and the other end in the other cell.)
(b) Which metal would be eaten away? Explain.
(c) Which is the oxidizing agent?
(d) Which is the reducing agent?

10.81 Show with a diagram how you would construct a copper–nickel battery. Label the direction of flow of electrons, cathode, +, and anode, −. Also, indicate in which beaker oxidation occurs and in which beaker reduction occurs.

10.82 Recharging a battery means forcing a spontaneous redox reaction to run backwards—in the nonspontaneous direction—once all the reactants have been used up. In Practice Problem 10.24, you considered a battery made from lead and copper.
(a) Write the spontaneous redox reaction for this battery.
(b) Write the recharging reaction for this battery.

10.83 Lithium is a very active metal, so active that it reacts with water to make flammable hydrogen gas.

$Li(s) + H_2O(l) \rightarrow LiOH(s) + 1/2H_2(g)$

(a) What metal would you use in combination with Li to make a battery having the highest voltage? (Consult the EMF series.)
(b) Too rapid a discharge of a lithium battery (which can occur if there is a short circuit) can overheat a lithium battery and cause it to explode and catch fire. Why is trying to put out such a fire with water a bad idea?
(c) Would lithium be the anode or cathode in a lithium ion battery? Explain.

Another Look at Oxidation—The Corrosion of Metals

10.84 There are two definitions of oxidation, one involving oxygen, the other not.
(a) State the two definitions of oxidation.
(b) Are they compatible? Explain.
(c) When we speak of corrosion of metals, what chemical reaction are we usually talking about? Use Fe as an example.

10.85 A piece of magnesium can be attached to the iron hull of a boat to prevent it from rusting.
(a) Why is the magnesium called a sacrificial metal?
(b) How does the Mg keep the iron hull from rusting?
(c) Which can be thought of as the positive cathode, the Mg or the steel hull? Explain.
(d) Putting a block of Mg on a steel boat hull does not create a battery. Why not?

10.86 Suppose gold were not available for a wedding ring, but you still wanted a ring that would last "forever" and not corrode. According to the EMF series on page 391, what would be a good alternative metal to use?

10.87 Explain what almost brought down the Statue of Liberty and what was done to keep the problem from happening again.

10.88 What problem might develop with the casing of the dry-cell battery shown on page 385 if it is not made thick enough? Explain.

Additional Problems

10.89 What is the oxidation state of the iodine atom in $H_2IO_6^{3-}$?

10.90 Identify the oxidizing and reducing agents in the reaction

$(NH_4)_2Cr_2O_7 \rightarrow N_2 + Cr_2O_3 + 4\,H_2O$

10.91 Burning octane, C_8H_{18}, in your car engine forms water and carbon dioxide:

$2\,C_8H_{18} + 25\,O_2 \rightarrow 18\,H_2O + 16\,CO_2$

What gets oxidized and what gets reduced?

10.92 Which metal is most easily oxidized: Pt, Hg, Fe, Mg, Zn?

10.93 Metal strips are immersed in aqueous solutions of various salts. In which combinations do you expect a spontaneous redox reaction:
(a) Silver strip in $CuBr_2(aq)$
(b) Copper strip in $ZnSO_4(aq)$
(c) Zinc strip in $AgNO_3(aq)$
(d) Gold strip in $FeCl_2(aq)$
(e) Copper strip in $Hg(NO_3)_2(aq)$

10.94 Can you protect a steel (predominantly iron) structure from corrosion by connecting it by wire to a small plate made of nickel? What elements (if any) would make better sacrificial metals?

10.95 Which ion is most difficult to reduce: Mn^{2+}, Hg^{2+}, Fe^{3+}, Mg^{2+}, Li^+?

10.96 The blood alcohol level of a person can be detected by reacting a sample of blood plasma with dichromate ion, $Cr_2O_7^{2-}$, which takes part in an electron-transfer reaction with ethanol, C_2H_5OH, in the blood:

$$2\,Cr_2O_7^{2-} + C_2H_5OH + 16\,H^+ \rightarrow 4\,Cr^{3+} + 2\,CO_2 + 11\,H_2O$$

Assign an oxidation state to each atom in this reaction and indicate the oxidizing agent and the reducing agent.

10.97 A battery was produced using copper metal in a solution of Cu^{2+} ions connected to rhodium metal in a solution of Rh^{3+} ions. Copper is the anode and rhodium is the cathode. Is rhodium higher or lower than copper in the EMF series?

10.98 Assign an oxidation state for each nitrogen atom in N_3^-. (*Hint*: Begin with a dot diagram for the ion.)

10.99 Which of the following are electron-transfer reactions? For those that are, indicate which reactant is the reducing agent and which reactant is the oxidizing agent.
(a) $SeO_3^{2-} + 4\,I^- + 6\,H^+ \rightarrow Se + 2\,I_2 + 3\,H_2O$
(b) $HI + KOH \rightarrow KI + H_2O$
(c) $4\,HCl + O_2 \rightarrow 2\,Cl_2 + 2\,H_2O$
(d) $SiO_2 + H_2O \rightarrow H_4SiO_4$

10.100 Tin cans are actually iron plated with tin. What is the advantage of the tin coating?

10.101 Fluorouracil is a compound administered to cancer patients as a part of chemotherapy. Assign an oxidation state to every atom:

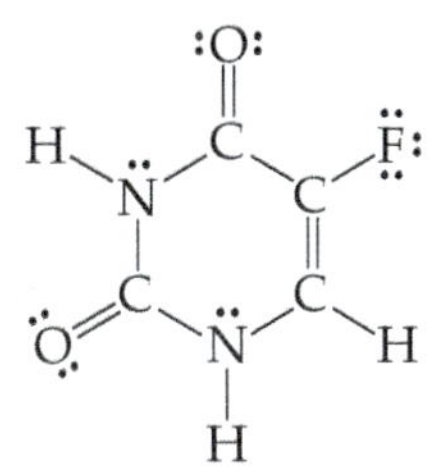

10.102 Which is a better reducing agent, Zn or Pb?

10.103 Assign an oxidation state to each carbon in:
(a) H_3CCH_3
(b) H_2CCH_2
(c) HCCH

10.104 The spontaneous redox reaction

$$Mn + Cd^{2+} \rightarrow Mn^{2+} + Cd$$

takes place in a battery.
(a) What is the oxidizing agent?
(b) What is the reducing agent?
(c) Which metal is the cathode?
(d) Which metal is the anode?

10.105 Assign an oxidation number to each atom and identify the oxidizing agent and reducing agent:

$$3\,Na_2SO_3 + 2\,KMnO_4 + H_2O \rightarrow 3\,Na_2SO_4 + 2\,MnO_2 + 2\,KOH$$

10.106 You are trying to determine the identity of an unknown metal. You place a strip of it in a solution of Mg^{2+}, and no reaction occurs. You then place a strip of it in a solution of Zn^{2+}, and zinc metal plates out on the strip. Name one possibility for the identity of the unknown metal.

10.107 Use the shortcut rules to assign an oxidation state to each atom:
(a) PF_6^-
(b) $Mo_2O_7^{2-}$
(c) $HPbO_2^-$
(d) $HC_2O_4^-$

10.108 When you turn on an electrical appliance, are you consuming electrons?

10.109 You create a battery using a zinc anode in a solution of Zn^{2+} ions connected to a nickel cathode in a solution of Ni^{2+} ions. The two beakers are also connected with a salt bridge containing K^+ ions and Cl^- ions.
(a) Toward which electrode do K^+ ions flow?
(b) Toward which electrode do Cl^- ions flow?

10.110 Assign an oxidation state to each atom in the amino acid glycine:

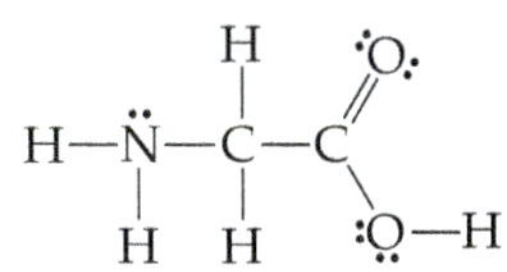

10.111 Identify the oxidizing agent and reducing agent in the reaction

$$IO_4^- + 7\,I^- + 8\,H^+ \rightarrow 4\,I_2 + 4\,H_2O$$

10.112 The black tarnish that forms on silver is silver sulfide, Ag_2S. The tarnish can be removed by wrapping the silver with a piece of aluminum foil and placing it in a solution of $NaHCO_3$ (baking soda) to allow for the flow of ions. Explain how the aluminum foil aids in removing the tarnish.

10.113 Under what circumstances does fluorine *not* have an oxidation state of -1?

10.114 Which of the following are electron-transfer reactions? For those that are, indicate which reactant is the reducing agent and which reactant is the oxidizing agent.
(a) $3\,H_2SO_3 + 2\,HNO_3 \rightarrow 3\,H_2SO_4 + 2\,NO + H_2O$
(b) $Mg + 2\,H_2O \rightarrow Mg(OH)_2 + H_2$
(c) $SO_3^{2-} + 2\,H^+ \rightarrow SO_2 + H_2O$
(d) $PbO + CO \rightarrow Pb + CO_2$

10.115 Assign an oxidation state to each atom in HCN. (*Hint*: Begin with a dot diagram.)

10.116 A battery is made by connecting strips of lead and palladium and dipping each metal into a solution of its ions (Pb^{2+} and Pd^{2+}, respectively). Over time, the mass of the lead strip decreases and the mass of the palladium strip increases.
(a) Which metal is the anode?
(b) Which metal is the cathode?
(c) Write the spontaneous redox reaction going on in this battery.

10.117 What is the range of oxidation states possible for carbon? (*Hint*: Consider that carbon bonds first to atoms that are all higher in electronegativity and then to atoms all lower in electronegativity.)

10.118 Identify the oxidizing agent and reducing agent in the reaction

$$NO_3^- + 4\,Zn + 7\,OH^- + 6\,H_2O \rightarrow 4\,Zn(OH)_4^{2-} + NH_3$$

10.119 Galvanized steel is steel coated with zinc. Why is galvanized steel less likely to rust than uncoated steel? (*Hint*: Remember, steel is mainly iron.)

10.120 Because of their small size and long life, zinc–mercury batteries were once used in hearing aids, pacemakers, and watches.
(a) Which metal is the anode and which is the cathode?
(b) Which electrode loses mass over time?

10.121 Which of the following are electron-transfer reactions? For those that are, indicate which reactant is the reducing agent and which reactant is the oxidizing agent.
(a) $PF_3 + 3\,H_2O \rightarrow H_3PO_3 + 3\,HF$
(b) $H_2 + Cl_2 \rightarrow 2\,HCl$
(c) $2\,Cr_2O_3 + 3\,Si \rightarrow 4\,Cr + 3\,SiO_2$
(d) $HCl + NaOH \rightarrow NaCl + H_2O$

10.122 A street vendor sells you a ring he claims is pure gold. However, when you place the ring in a solution of Cu^{2+} ions, copper plates out on the ring. How do you know the ring is a fake? Could it be made of silver?

10.123 Use the shortcut rules to assign an oxidation state to each atom:
(a) $NaPO_3$
(b) $B(OH)_3$
(c) V_2O_5
(d) K_2TiF_6

10.124 The reaction $Cu^{2+} + I_2 \rightarrow Cu^+ + 2\,I^-$ is not possible as written. Assign an oxidation number to each atom and explain what is wrong with this reaction.

10.125 A battery consists of a strip of titanium metal in a solution of Ti^{2+} ions connected to a strip of zinc metal in a solution of Zn^{2+} ions. Over time, the concentration of Zn^{2+} ions decreases and the concentration of Ti^{2+} ions increases.
(a) Which metal is the anode?
(b) Which metal is the cathode?
(c) Where must titanium be on the EMF scale relative to zinc?
(d) Write an equation that describes the spontaneous electron-transfer reaction occurring in this battery.

10.126 Consider the following galvanic cell:

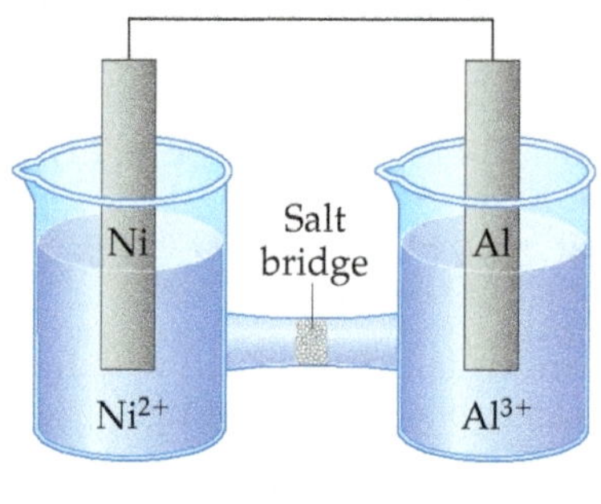

(a) Which way will the electrons flow in the wire?
(b) What is the oxidizing agent?
(c) What is the reducing agent?
(d) Which electrode is positive (cathode)?
(e) Which electrode is negative (anode)?

10.127 Hydrogen peroxide (H_2O_2) can be made to decompose into water and oxygen as shown below. (The hydrogen peroxide in the beaker is decomposing to give a cloud of water vapor. Oxygen gas is also being produced.)
Is this a redox reaction? If so, prove it.

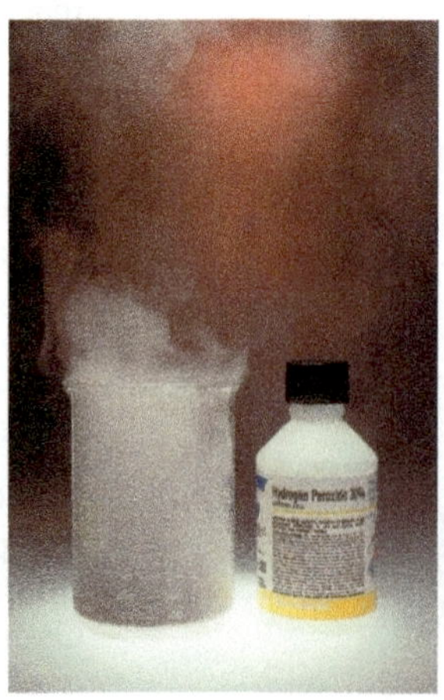

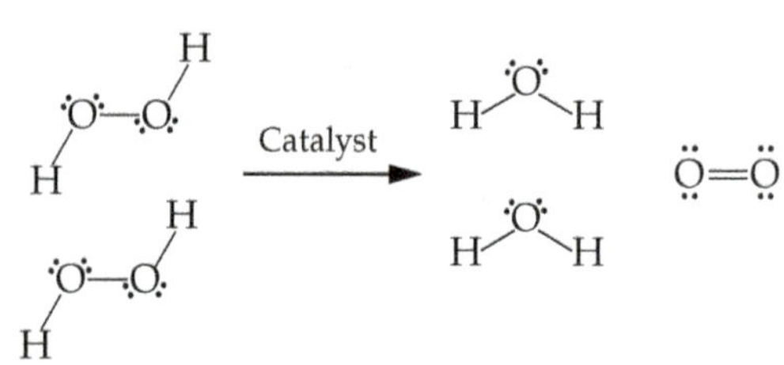

10.128 Consider the central carbon atom in each of the three molecules below.
(a) One molecule contains a central carbon in its highest possible oxidation state. Which molecule is it, what is its oxidation state, and why is this the highest oxidation state you will ever find a carbon atom in?
(b) One molecule contains a central carbon in its lowest possible oxidation state. Which molecule is it, what is its oxidation state, and why is this the lowest oxidation state you will ever find a carbon atom in?
(c) Which molecule(s) has/have a central carbon atom that satisfies the octet rule?

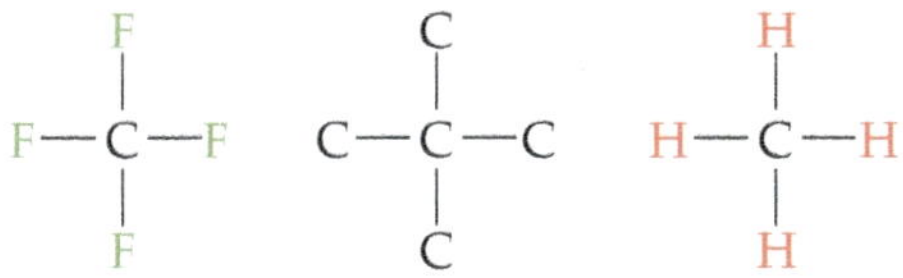

(1) (2) (3)

10.129 $F_6I_2O_4$ is an interesting compound whose structure is shown below. Complete the dot diagram (add lone pairs) and then determine the oxidation state for every atom in the molecule not using the shortcut rules. Then use the shortcut rules and see if they give the same result.

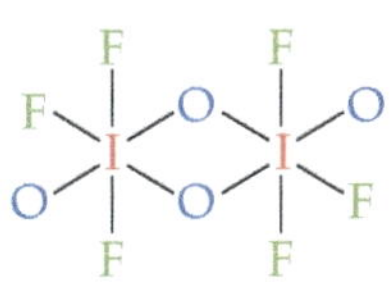

10.130 In the interconversion shown below, one carbon atom in the molecule can be thought of as being oxidized, while another carbon atom in the molecule can be thought of as being reduced. These two carbon atoms are highlighted in red and blue.
Which gets oxidized and which gets reduced? Supply proof.

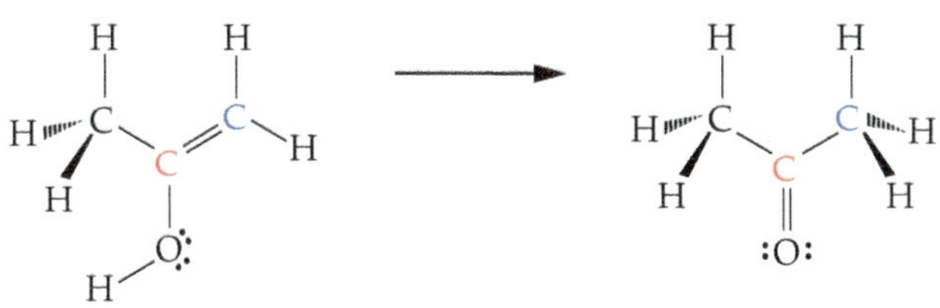

One Step Beyond

10.131 Most combustible materials will burn in chlorine (Cl_2) gas. For example, methane burns to yield CH_3Cl, CH_2Cl_2, $CHCl_3$, and CCl_4. Why is the word "burn" used in this case?

10.132 In a nickel–cadmium battery, the relevant redox reaction is:

$$2\,NiO(OH) + Cd + 2\,H_2O \rightarrow 2\,Ni(OH)_2 + Cd(OH)_2$$

Does this agree with the EMF series? What are the oxidation states of nickel before and after the reaction?

10.133 Ethane can be converted into an aldehyde and an alcohol (shown below):

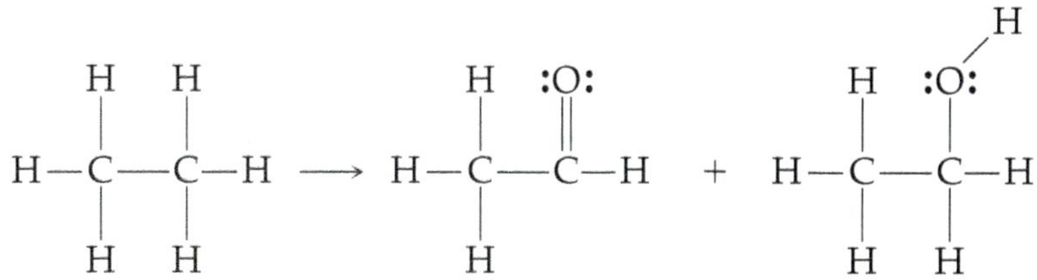

Ethane (a) an aldehyde (b) an alcohol

(a) Explain why this is an example of oxidation.
(b) Which is the more oxidized form of ethane, the aldehyde or the alcohol? (*Hint*: Assign oxidation states to all atoms).

10.134 Consider the following hypothetical reaction:

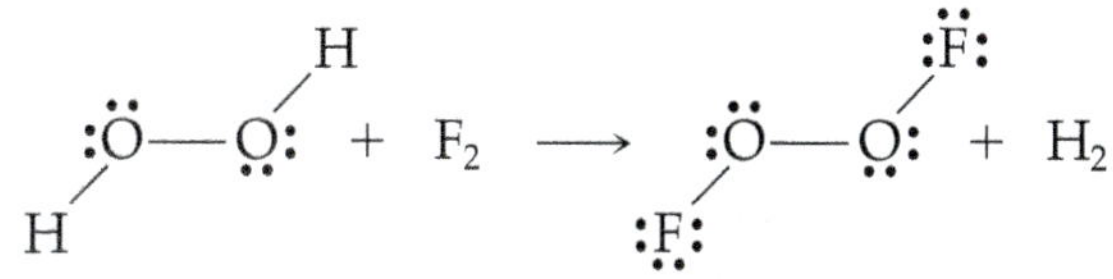

(a) Is this an electron transfer (redox) reaction? Justify your answer.
(b) Besides F versus H, what is a fundamental difference between H_2O_2 and F_2O_2?

10.135 What conclusion would you come to regarding electron transfer for the reaction in Problem 10.134 if you had used octet rule electron bookkeeping instead of oxidation state electron bookkeeping? Explain.

10.136 Acids like HCl ionize completely when dissolved in water to produce $H^+(aq)$ and $Cl^-(aq)$ ions. Metals (M) react with acids (H^+) as follows:

$$M(s) + H^+(aq) \rightarrow M^{+n}(aq) + H_2(g)$$

where n depends on the specific metal involved. Based on the activity series, which of the choices below would be the **best** for the lining of a tank intended for use in the storage of hydrochloric acid (HCl)?
(a) copper (b) zinc
(c) nickel (d) iron
(e) tin

10.137 (a) Use the following results to arrange the four metals, W, X, Y, Z in the vertical column on the right from the most active (at the top) to least active (on the bottom).

$W + X^+ \rightarrow W^+ + X$ most active ________
$X + Z^+ \rightarrow X^+ + Z$ ________
$Y^+ + Z \rightarrow$ no reaction ________
$X + Y^+ \rightarrow X^+ + Y$ least active ________

(b) Based on your series in part a, circle ALL the reactions **below** that you would expect to occur spontaneously as written. Explain.
$W^+ + Y \rightarrow W + Y^+$ $\quad$ $W^+ + Z \rightarrow W + Z^+$

10.138 Here is a diagram of a battery made from nickel and lead:

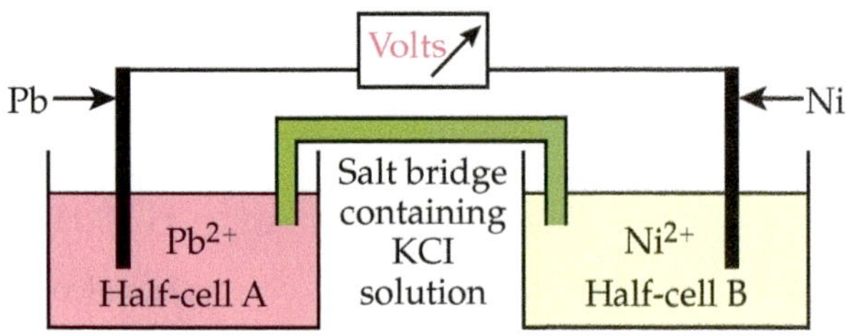

(a) In the diagram above, label the anode and the cathode **and** draw an arrow above the voltmeter to show the direction of the electron flow.
(b) Write the net ionic equation for the spontaneous reaction that occurs in this battery.
(c) In which half cell does reduction take place? (Circle one.)
Reduction occurs in the Pb half cell
Reduction occurs in the Ni half cell
(d) Into which half cell will the potassium (K^+) ions in the salt bridge flow? (Circle one.)
the Pb half cell $\quad$ the Ni half cell

10.139 Consider the perchlorate (ClO_4^-) ion.
(a) Assign oxidation states to all of the atoms in perchlorate.
(b) Explain why the chlorine atom in perchlorate does not follow the halide rule.
(c) Do you think perchlorate would be a powerful oxidizing agent or a powerful reducing agent? Explain.
(d) Would it be possible to have an even higher oxidation state for the Cl atom in some other compound? Explain. (*Hint*: Chlorine is in group VIIA).

10.140 Ozone (O_3) is a strong oxidizing agent. Explain why this is so.

Group Exercises (Work on these exercises together with someone else in the class.)

1. We can define movie agents by the action they perform. A movie agent is someone who gets you a job starring in a movie. Discuss how you would define an oxidizing agent and a reducing agent in terms of an action they perform. Then suppose you wanted to rob an electron bank. Would you hire a reducing agent or an oxidizing agent?

2. Two people have a joint savings account together. To simplify their bookkeeping, one person will be in charge and get to spend the money while the other watches. Are they abiding by the oxidation-state formalism or octet-rule formalism? How would they spend their money if they used the other formalism? Which would get them into trouble?

WORKPATCH SOLUTIONS

10.1 N is assigned six electrons because it is more electronegative than H but less electronegative than F.

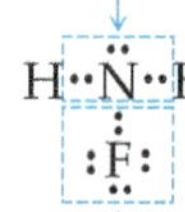

H is assigned zero electrons because it is less electronegative than N.

F is assigned eight electrons because it is more electronegative than N.

10.2 This O is assigned five electrons. Because a free O atom has six electrons, the oxidation state is 6 − 5 = +1.

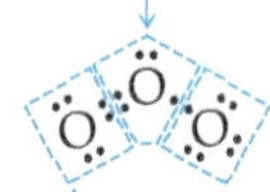

This O is assigned seven electrons. Because a free O atom has six electrons, the oxidation state is 6 − 7 = −1.

This O is assigned six electrons. Because a free O atom has six electrons, the oxidation state is 6 − 6 = 0.

10.3 (a) A negative oxidation state means that more valence electrons are assigned to the atom by oxidation-state bookkeeping than are present in a free atom of the element.
(b) An oxidation state of −1 means that oxidation-state bookkeeping assigns that chlorine atom one more valence electron than the number present in a free chlorine atom.
(c) Chlorine has seven valence electrons as a free atom (group VIIA). To have an oxidation state of −1 and therefore eight valence electrons, it must share its unpaired electron with an unpaired electron from an atom that is less electronegative than itself.

10.4 (a) The chlorine atom is bonded to atoms that are more electronegative than it is.

Oxidation-state bookkeeping always assigns shared bonding electrons entirely to the more electronegative atom. Therefore, all the bonding electrons in this ion are assigned to the oxygen atoms. The chlorine atom gets none of them, so the halide rule does not apply.

(b) If each O atom has an oxidation state of −2 (Rule 2), the oxidation states of the three O atoms add up to −6. Therefore, the Cl atom must have an oxidation state of +5 in order for the sum of the oxidation states of all the atoms to add up to the −1 charge on the ClO_3^- ion.

10.5

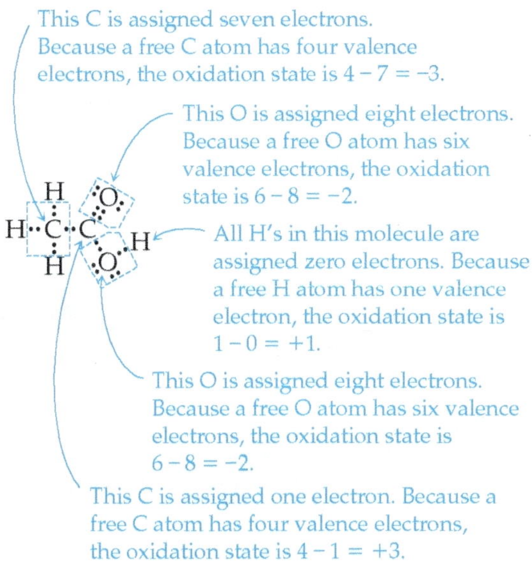

10.6 **Step 1:** The oxidation states are Zn 0, Cu^{2+} + 2, Zn^{2+} +2, and Cu 0.

Step 2: (a) Yes. You can tell by the fact that the oxidation state of each reactant changed.

(b) Zn was oxidized to Zn^{2+}.

(c) Cu^{2+} was reduced to Cu.

(d) Cu^{2+} was reduced (gained two electrons), so it was the oxidizing agent.

(e) Zn was oxidized (lost two electrons), so it was the reducing agent.

10.7 Electrons flow through wire from Mg to Ag^+.

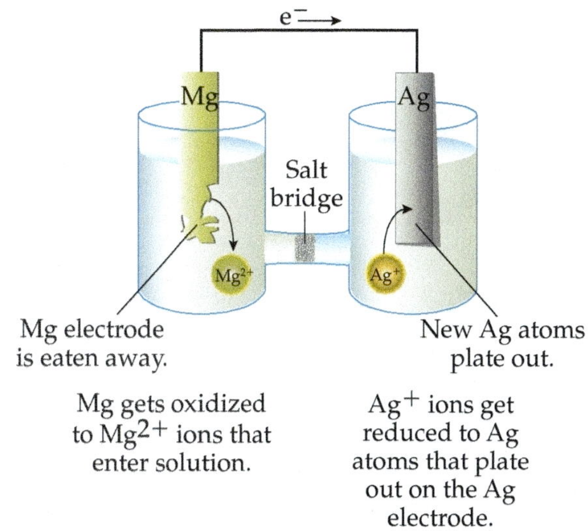

10.8 Electrons flow through wire from Mg to Ag^+.

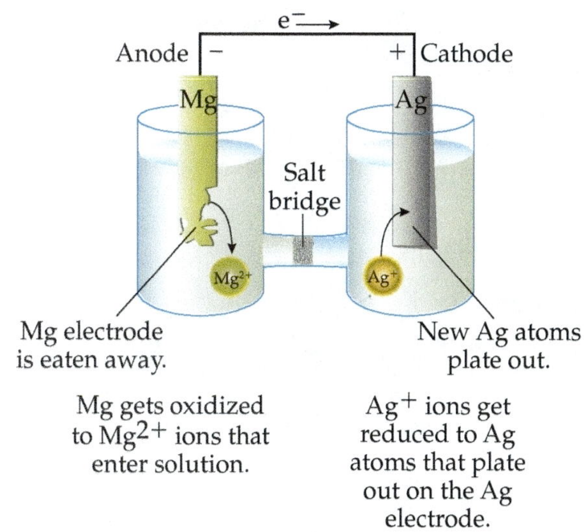

10.9 The correct answer is (a). Reduction is the gain of electrons. Oxidation is the loss of electrons. When something is oxidized, it loses electrons, which are then available to be gained by what is being reduced.

10.10 At the bottom of our abbreviated EMF series is gold, Au. It is the least active metal listed, and therefore very resistant to oxidation and thus corrosion.

PV=nRT

CHAPTER 11

What If There Were No Intermolecular Forces? The Ideal Gas

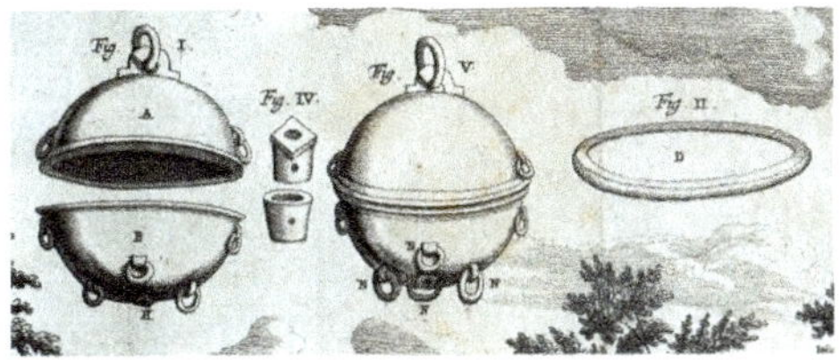

What is a vacuum? Philosophers have been pondering that question for thousands of years. More than 300 years before the birth of Christ, the Greek philosopher Aristotle declared that a vacuum could not exist. In medieval Europe, the Catholic Church regarded the idea of a vacuum as against nature and heretical since the absence of everything implied the absence of God. Then along came Otto von Guericke, who in 1654 was the mayor of the town of Magdeburg in Germany.

Von Guericke had just invented a hand-operated vacuum pump designed to remove the air from any closed vessel it was connected to. To demonstrate the existence of a vacuum, he placed two copper bowls together to form a hollow sphere of about 14 inches in diameter, and used his pump to remove the air from inside the sphere. Then he set up the following scientific experiment.

No matter how they were coaxed, the horses could not pull the sphere apart, thus demonstrating a vacuum and the tremendous force of air pressure outside the sphere that was responsible for holding it together. Using a barometer to measure atmospheric pressure (the barometer had been invented about a decade earlier), he could calculate that slightly over 4.5 tons of force would have to be applied to counteract the atmospheric pressure and pull the spheres apart. That much force is too much to overcome, even for two teams of horses as the illustration of the experiment shows. It is time for us to learn about the behavior of matter in the gas state.

11.1 Describing the Gas Phase—*P*, *V*, *n*, and *T*

Here is a 1000-year-old recipe discovered by the Chinese, a recipe that changed the world forever. Find some potassium nitrate, KNO_3, some carbon, and some sulfur. Potassium nitrate, also called saltpeter, can be found as a delicate crust on some rocks and stone walls. Charcoal is a source of carbon. Elemental sulfur can be found in various locations in the form of a yellow mineral sometimes called brimstone. Grind these ingredients *separately* into powders and then make a mixture that is, by mass, 75% potassium nitrate, 15% carbon, and 10% sulfur. Ignite the mixture with a flame or spark, and you'll see a chemical reaction that rapidly releases great amounts of heat and light—and a large amount of gas:

$$2KNO_3(s) + 3C(s) + S(s) \rightarrow N_2(g) + 3CO_2(g) + K_2S(s)$$

This reaction is not particularly dangerous unless you do it in a confined space, such as the space between the sealed end of a metal tube and a heavy metal ball:

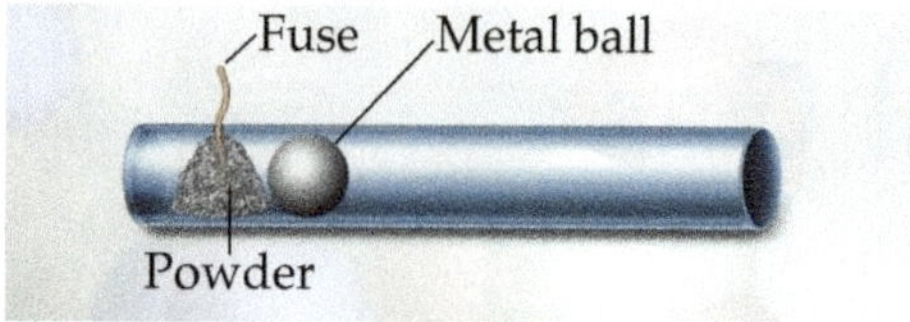

The powder, of course, is gunpowder. The ball becomes a lethal projectile because of all the gas produced by the combustion reaction. Trapped and heated, this gas develops a tremendous pressure that propels the ball out of the tube at high speed, with the gas expanding behind it. A similar phenomenon propels your car. In the engine cylinders, a fine spray of gasoline mixes with oxygen and combusts to produce hot gases at high pressure. The gases push down on a piston that is connected to a crankshaft that is in turn connected to the car's wheels. This combustion reaction is:

$$2\,C_8H_{18}(l) + 25\,O_2(g) \longrightarrow 16\,CO_2(g) + 18\,H_2O(g)$$

Octane, a component of gasoline

Given the usefulness of expanding gases, it's no surprise that, over the centuries, scientists have expended great effort in studying them. After all, if you want a speedier bullet, a more powerful locomotive, or a jetpack that lets you leap tall buildings in a single bound, you had better understand gases. One of the first scientists to contribute to our knowledge about gases was Robert Boyle, who in 1661 studied how gas pressure changes with volume. In 1787, Jacques Alexandre Charles studied how the volume of a gas changes with temperature. Shortly thereafter, Amedeo Avogadro (in whose honor Avogadro's number is named) studied the relationship between the volume of a gas and the number of gas molecules in that volume.

As a result of the work of these and other scientists, it soon became apparent that, under most conditions, all gases behave pretty much the same. For example, when it comes to producing pressure, it does not matter if the gas is O_2, N_2, CO_2, steam, or anything else. In fact, the only factors that typically matter are the number of moles of gas, its temperature, and the volume it occupies. This finding led scientists to develop a model gas, called an *ideal gas*, that captures the essence of gas behavior. An **ideal gas** can be thought of as consisting of pointlike particles that do not attract or repel one another at all. Thus, the "molecules" (the particles) in an ideal gas experience no intermolecular forces. The particles of an ideal gas fly around in straight-line paths, changing direction only when they collide with one another or with the walls of their container. When they collide, they rebound elastically.

An Ideal Gas

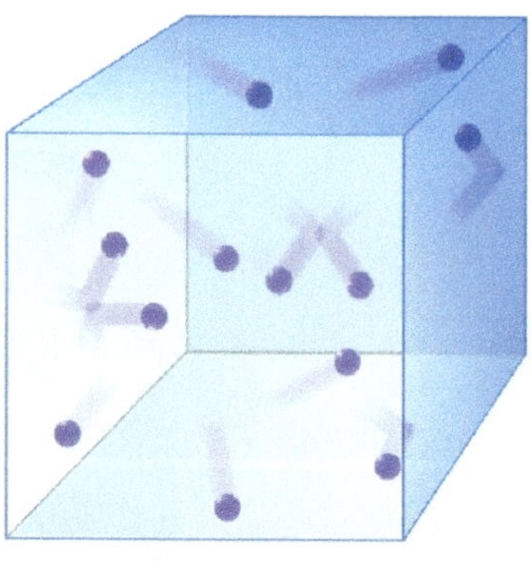

Because there are no attractive forces between ideal gas particles (no London forces, no dipolar forces, no hydrogen-bond forces—nothing), an ideal gas will never condense into a liquid or solid, no matter how much it is cooled. Clearly, no such gas exists, but under conditions far from those that cause condensation, real gases come close to behaving this way. That's because real gas molecules travel so fast and are so far apart that they easily overcome the weak attractive forces that *do* exist between them. The only time this isn't true is when a gas is either cooled to near the temperature at which it liquefies, or compressed to such a high density that the molecules are quite close to one another. In any other situation, the ideal gas model is appropriate for describing any real gas. Looking at a real gas like an ideal gas explains why any real gas always expands to completely fill its container.

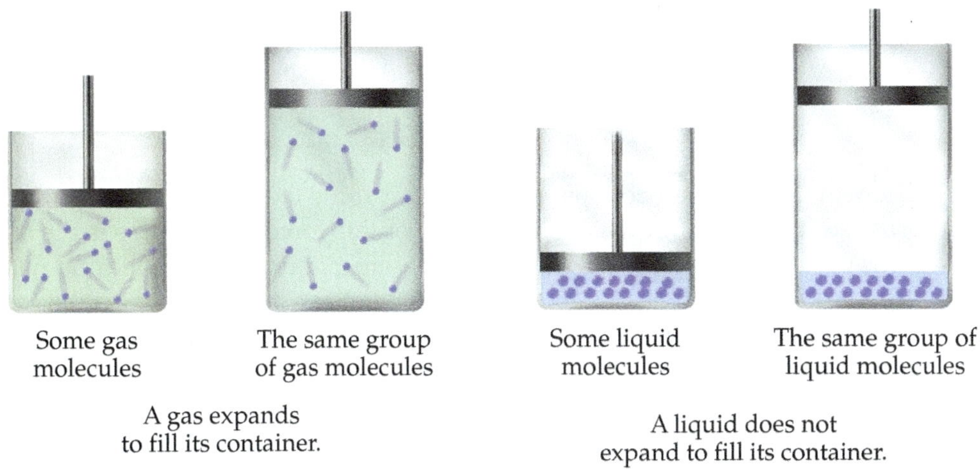

Because ideal gas molecules don't attract one another, they move freely until they hit something—such as the wall of their container. Thus, a gas expands to fill its container. In a condensed phase, the molecules are not free to wander all over the container because they are attracted to one another via various intermolecular forces.

WORKPATCH 11.1 Why is it never a reasonable approximation to ignore intermolecular forces in liquids and solids? ●

Your answer to this WorkPatch should remind you of why liquids and solids are referred to as *condensed phases*.

It turns out that we can *completely* describe the behavior of an ideal gas using just four variables:

P, the pressure of the gas
V, the volume the gas occupies
T, the temperature of the gas
n, the number of moles of gas in our sample

In Section 11.2, we'll develop a simple equation that connects these variables to one another. This equation allows us to make predictions about real gases that are reasonably accurate under normal circumstances. Before we get to the equation, however, you must become very familiar with the four variables that characterize an ideal gas. Let's start with pressure.

We live in an atmosphere of pressurized gas. Although you're usually not aware of it, the atmospheric pressure you experience is substantial. Perhaps you've seen the demonstration shown here, in which removing the air from a metal can causes the can to collapse.

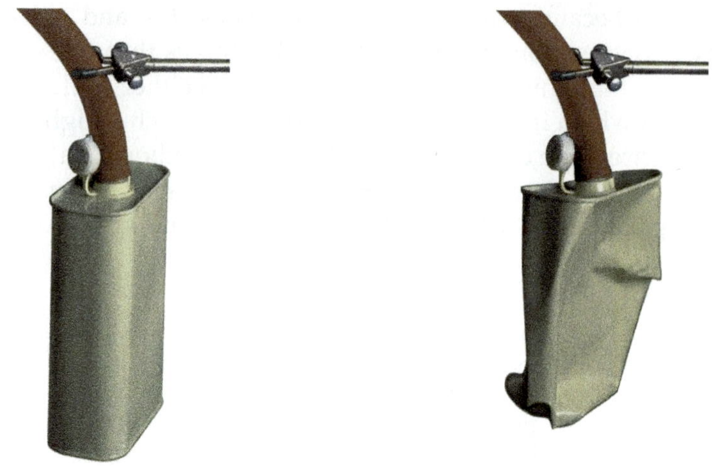

The way that the can collapses tells us that some external force must have crushed it inward. But what is the source of this force? The source is the billions of fast-moving air molecules in the atmosphere that are colliding with the outside walls of the can. Each collision exerts an inward force on the can. The force from these collisions is felt over the entire surface area of the can and is referred to as either *air pressure* or *atmospheric pressure*.

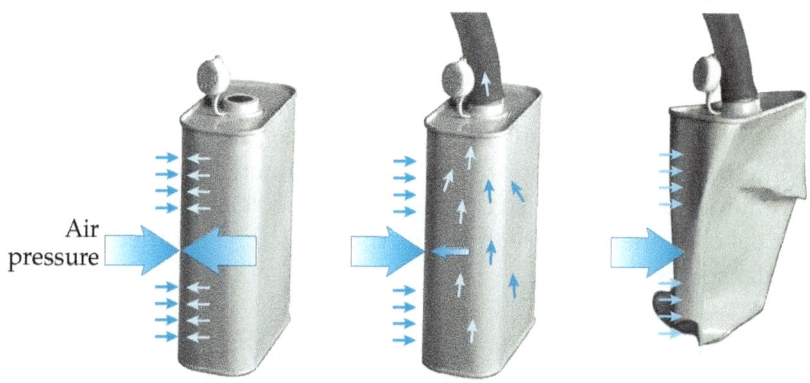

When the can is "filled" with air, molecules batter the outside and inside of the can equally, and as a result the can walls feel no net pressure.

Air is pumped out of the can.

With no opposition from inside, the pressure exerted by the outside air crushes the can.

Pressure (abbreviated P) is defined as force per unit area ($P = F/A$). The same atmospheric pressure that crushed the can is, at this very moment, attempting to crush you. So why don't you collapse? For the same reason the can didn't collapse until the air inside was pumped out. Before they were removed, the air molecules inside the can were colliding with the walls, pushing out with a force equal to the force with which the outside molecules were pushing in. The net result is that nothing happens until the inside air is removed.

This same idea can be used to construct a *mercury barometer*, a device used to measure atmospheric pressure. To make a barometer, we immerse one end of a glass tube into a dish of liquid mercury, Hg. Then we attach a vacuum pump to the other end of the tube and remove all the air from the tube. Once the tube is evacuated, the mercury rises into the tube to some specific height and then stops.

A Mercury Barometer

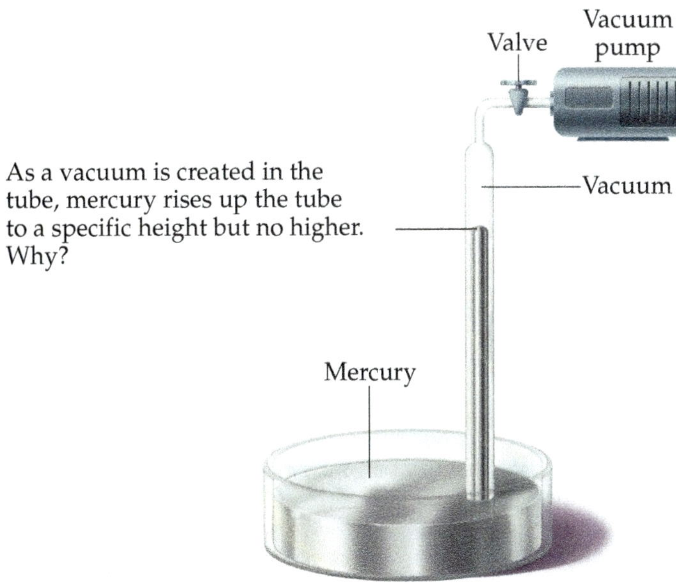

What makes the mercury rise up the tube is the same thing that crushed the can—air pressure. Before we removed the air from the tube, molecules of air both outside and inside the tube collided with the surface of the mercury in the dish, pushing down on the surface with the same force.

When we removed the air from the tube, we eliminated the air pressure inside the tube. Thus, the mercury was pushed up into the tube by the air pressure exerted on the mercury by the air molecules outside the tube.

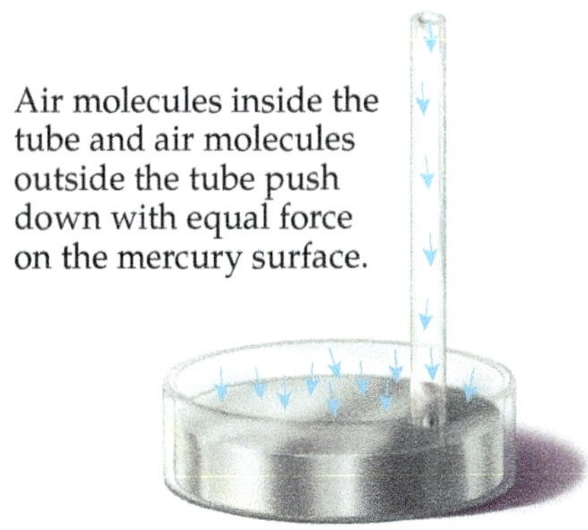

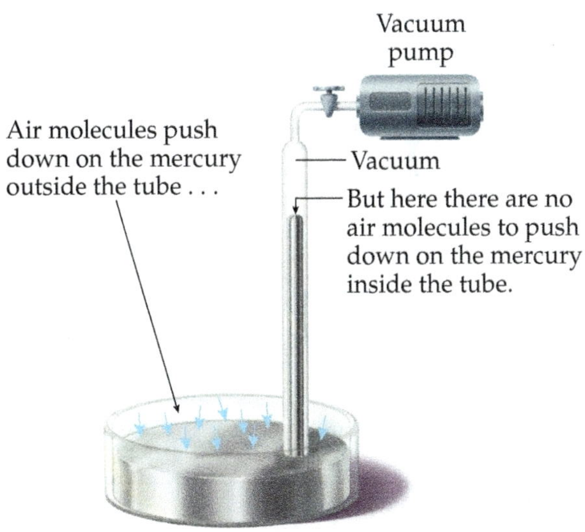

The mercury is pushed up into the tube by the outside force (atmospheric pressure) until the weight of the mercury column supplies enough opposing force to resist any more movement. On an average day at sea level, this stopping point occurs when the mercury column rises to a height of 760 mm. We therefore define **normal atmospheric pressure** as being equal to 760 mm of mercury (mm Hg) or 1 atmosphere (atm).

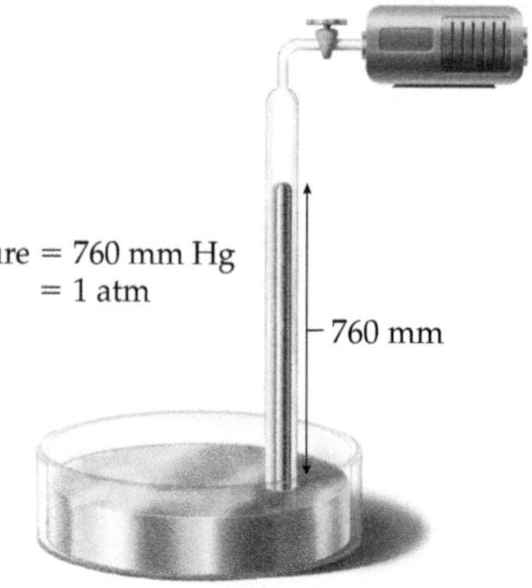

Mercury barometer on average day at sea level

Weather forecasters use barometers to measure atmospheric pressure, although in the United States they usually report the pressure in inches of mercury rather than millimeters (760 mm = 29.9 in. therefore, normal atmospheric pressure is 29.9 in. of mercury). When the atmospheric pressure drops, the column of mercury falls below its normal 760-mm height. Then it's time to grab your umbrella because low atmospheric pressure is often associated with rain.

As we saw earlier, pressure is just one of the variables we can measure to describe an ideal gas. There are three others. We can easily measure the volume, V, that a sample of gas occupies because the volume of the gas is always equal to the volume of the container confining the gas. We can also easily measure the temperature, T, of the gas sample. It is conventional to report the temperature of a gas using the Kelvin temperature scale discussed in Chapter 2. This is a bit of an inconvenience because most thermometers are calibrated in °F or °C. However, it is a simple matter to measure a temperature using one scale and then convert the reading to another scale. Recall that if you measure the temperature of a gas in °C, you convert to kelvins by adding 273.15:

$$\text{Temperature in K} = \text{Temperature in °C} + 273.15$$

Along with pressure, volume, and temperature, we also need to know the amount of gas we have. By convention, we specify the amount of gas in moles, and we use the symbol n to represent this quantity. When we say that $n = 1$ for a gas, we mean that we have 1 mole, or 6.02×10^{23} molecules, of the gas.

To summarize:

Measurable variables used to describe a gas
Pressure P, specified in millimeters of mercury or atmospheres
Volume V, specified in liters
Temperature T, specified in kelvins
Amount n, specified as number of moles

Practice Problems

11.1 An "empty" 0.500-gallon milk carton sits in a room where the temperature is 25.20 °C. The barometric pressure is exactly 1 atm.

(a) What is the Kelvin temperature of the air inside the carton?
(b) What is the pressure in millimeters of mercury of the air inside the carton?
(c) What is the volume in liters of the air inside the carton? (1 gallon = 3.78 L)

Answer:

(a) 25.20 °C + 273.15 = 298.35 K

(b) Because the pressure is 1 atm, $P = 1\ atm = 760\ mm\ Hg$.

(c) $0.500\ \cancel{\text{gallon}} \times \dfrac{3.78\ \text{L}}{1\ \cancel{\text{gallon}}} = 1.89\ \text{L}$

11.2 A container is evacuated with a vacuum pump and its mass is measured. Then it is filled with H_2 gas and its mass is measured again. If the mass increase is 10.50 g, how many moles of H_2 gas are in the container? (*Hint*: You will need the molar mass of H_2.)

11.3 The gas in a pressurized container is at five times normal atmospheric pressure.

(a) What is the gas pressure in millimeters of mercury?
(b) What is the gas pressure in inches of mercury? (2.54 cm = 1 in.)

11.4 A mercury barometer develops a leak, allowing some air to enter the glass tube. Will such a barometer read too high or too low a pressure? Explain your answer.

11.2 Describing a Gas Mathematically—The Ideal Gas Law

Using the quantities P, V, n, and T, we can characterize the state of any gas. If we phone some friends in another country and give them these values for a particular sample of gas, they can construct a sample identical to ours. In fact, we do not even need to specify all four values—any three will do. That is because P, V, n, and T are not independent of one another. If you change one of these variables, you can predict how the others will change, and if you know three of them, you can figure out the fourth.

In this section, we shall develop the simple mathematical equation that describes how the P, V, n, and T values for an ideal gas are related to one another. Because we use the ideal gas model to develop it, this equation is called either the *ideal gas equation* or the *ideal gas law*. This equation is very useful because we can use it to predict how real gases respond to changes in P, V, n, or T. (Remember that real gases behave almost ideally under most circumstances.) Suppose, for instance, that you leave a tank of air in the sun and notice that it has warmed by 10 °C. What has happened to the pressure in the tank? Could the tank be close to exploding? You can use the ideal gas equation to answer this question.

We can derive the ideal gas equation by thinking about our model for an ideal gas and asking three questions about pressure. Remember that a gas in any container exerts pressure because the fast-moving gas particles (atoms or molecules in the case of a real gas) strike the walls of the container, exerting

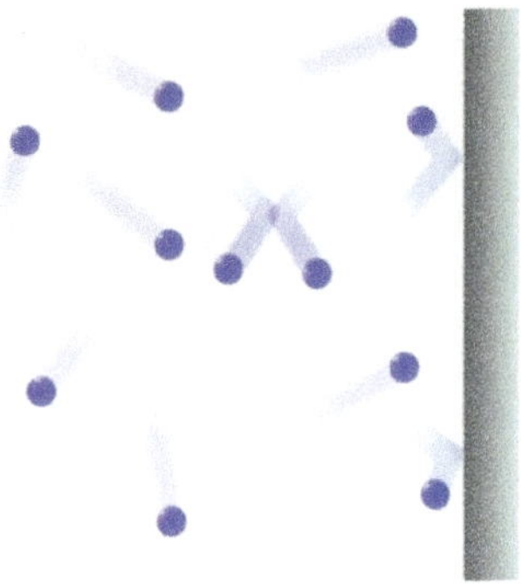

a force on them with each collision. Here then are our three questions about pressure:

1. What happens to the pressure exerted by a gas confined in a container when we increase or decrease the *volume* of the container?
2. What happens to the pressure exerted by a gas confined in a container when we increase or decrease the *temperature* of the gas?
3. What happens to the pressure exerted by a gas confined in a container when we increase or decrease the *amount* of gas in the container?

To predict the answers to these questions, we need nothing more than our simple model of an ideal gas. Let's begin with the first question. We've already seen that pressure comes about from gas molecules colliding with the container walls. What happens to the pressure when we decrease the volume of the container? You can answer this question graphically, with just two drawings, as the following WorkPatch shows.

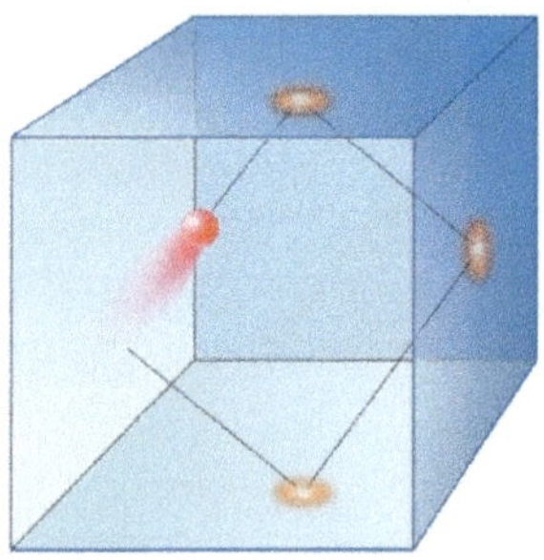

WORKPATCH 11.2 Draw two boxes like this one, making one larger than the other. Put a single fast-moving ideal gas particle inside each box (traveling at the same speed in both boxes). Show the path of the particle over the same time interval for both boxes, and make your interval long enough to allow for several collisions with the walls. Is the pressure in the smaller box the same as, less than, or greater than the pressure in the larger box? ●

What about the number of collisions with the walls in each box? There should be a difference. Do you see why? Check your answer against the one at the end of the chapter. This simple picture reveals why the pressure of an ideal gas always increases when the volume of its container decreases (and nothing else is changed). Your drawings should have revealed that the number of collisions per unit time increases as the volume of the box decreases. Thus, as the volume of the container gets *smaller*, the pressure exerted by the gas gets *larger*.

Rule 1: The pressure of a gas increases when the volume the gas occupies decreases.

Examine rule 1 very closely. It tells you that pressure and volume behave oppositely. As one increases, the other decreases. We can restate rule 1 mathematically:

Rule 1: *P* is inversely proportional to *V*.

These two statements of rule 1 for an ideal gas say the same thing. We can, however, simplify the mathematical version a bit by dropping the word *inversely*. Instead of saying P is inversely proportional to V, we can say P is directly proportional to $1/V$. This is a mathematical restatement you can always make for any two inversely related quantities.

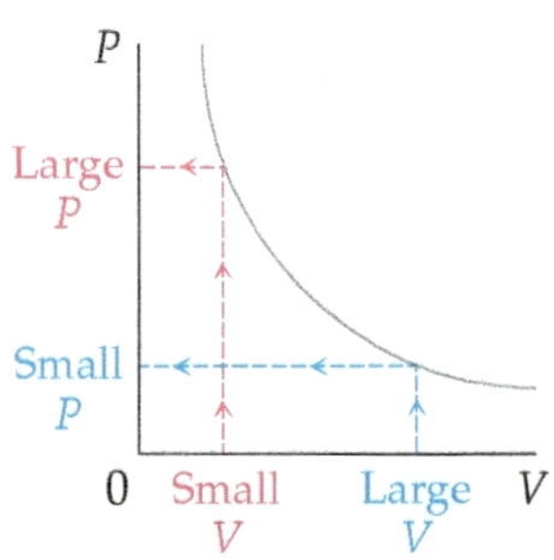

Rule 1 – Pressure-Volume Relationship: P is proportional to $\frac{1}{V}$.

This rewording puts the volume in the denominator of the expression. All three statements of rule 1 say the same thing—pressure increases as volume decreases (or pressure decreases as volume increases). We can even show this graphically. A plot of P versus V for a gas shows that a small volume generates a large pressure, whereas a large volume generates a small pressure (at constant n and T). They behave oppositely, and your answer to WorkPatch 11.2 tells you why.

Now let's consider the second question: "What happens to the pressure exerted by a gas confined in a container when we increase or decrease the temperature of the gas?" Once again, we use the ideal gas model. As we heat the gas, we are putting energy into it. This added energy causes the gas molecules to move faster, on average. Indeed, heating a gas is like stepping on the accelerator pedal in your car.

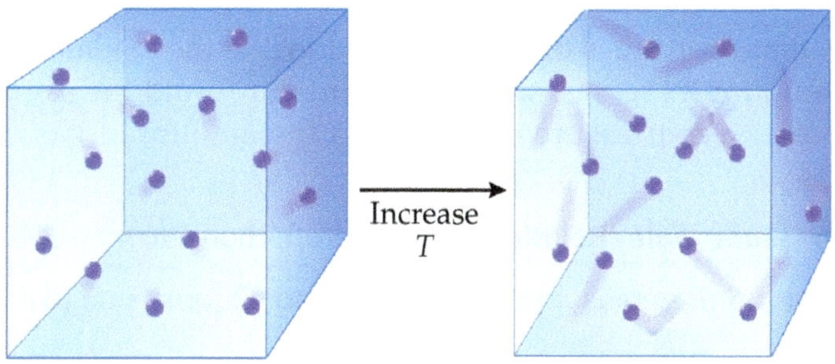

The higher temperature causes the molecules to move faster.

Because the molecules are moving faster, they collide with the walls of the container more often per unit of time, and since they are moving faster they hit the walls with greater force. Both effects result in more force being exerted on the walls per unit area, and so the pressure increases. Taking this argument in the reverse direction shows that the gas pressure decreases when you lower the temperature. Thus, our second rule is:

Rule 2 – Pressure-Temperature Relationship: The pressure of a gas increases when its temperature increases and decreases when its temperature decreases.

Look closely at rule 2. Unlike pressure and volume, which we found were inversely related, pressure and temperature are directly related. Whenever the temperature changes, the pressure changes in the same direction. We can once again say the same thing in a mathematical way:

Rule 2: *P* is proportional to *T*.

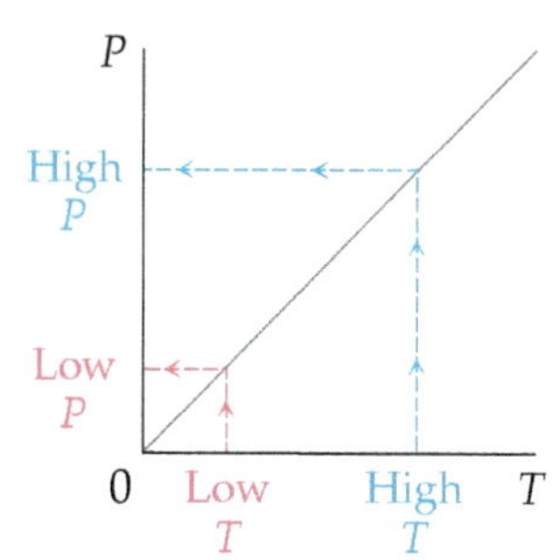

This is illustrated graphically in the plot of P versus T (at constant n and V) shown at left.

Our third question about pressure is "What happens to the pressure exerted by a gas confined in a container when we increase or decrease the amount of gas in the container?" In what is getting to be a theme, we once again use the model of an ideal gas to answer this question. In the illustration below, the left box has only a few gas molecules in it. The right box has lots more. In which box is the gas pressure higher?

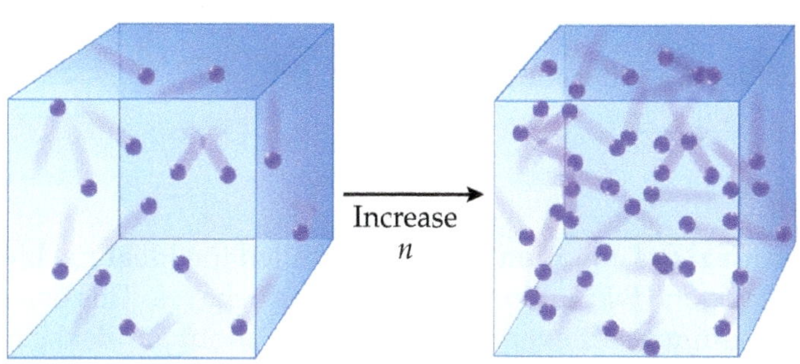

More molecules in the right box mean more collisions with the walls per unit time. More collisions mean more force pushing outward—that is, more pressure. And the reverse is true in the left box—fewer molecules in the same volume mean fewer collisions with the walls and therefore less pressure. This gives us a third rule about the behavior of gases:

Rule 3: The pressure of a gas increases when the number of moles of gas increases and decreases when the number of moles decreases.

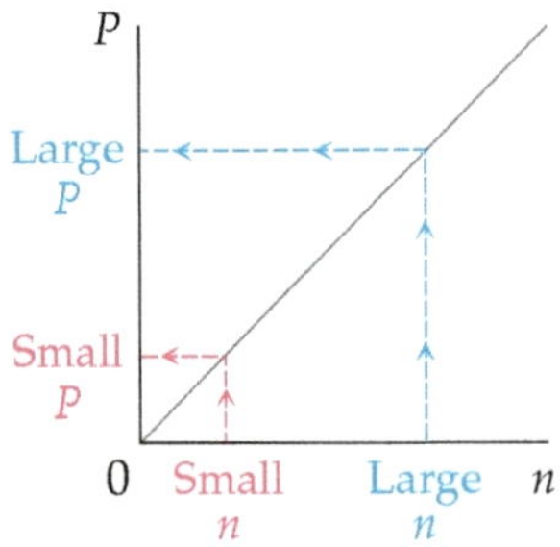

Rule 3 states that pressure is directly proportional to the amount of gas present. As the amount increases, so does the pressure. And the same goes for smaller samples: gas pressure decreases as the amount of gas decreases. Mathematically,

Rule 3 – Pressure-Moles Relationship: *P* is proportional to *n*.

Therefore, a plot of P versus n (at constant V and T) looks just like the previous plot.

We are done. Here is what we have concluded:

Rule 1: P is proportional to $\frac{1}{V}$.

Rule 2: P is proportional to T.

Rule 3: P is proportional to n.

We can combine these three rules into one summary rule:

P is proportional to $\frac{nT}{V}$.

Notice that n and T are in the numerator (they are directly proportional to P), and V is still in the denominator (it is inversely proportional to P).

We can turn this half-symbol, half-word statement into a full-fledged mathematical equation by replacing the words *proportional to* with an equal sign. However, to convert a proportionality to an equality requires introducing a constant of proportionality:

$$P = \text{Constant} \times \frac{nT}{V}$$

In principle, the value of this constant is not difficult to determine. All we need to do is take a container of gas (one whose behavior is very close to ideal) and experimentally determine the values of our four variables—P, n, T, and V. Then we can use the above equation to determine a value for the constant. Of course, we don't have to do this at all because others before us have already done it. This constant of proportionality is called the **ideal gas constant** and is represented by the letter R. Its value is:

$$\text{Ideal gas constant } R = 0.0821 \text{ L} \cdot \text{atm/K} \cdot \text{mol}$$

When we include R in the summary rule, we get:

$$P = R \times \frac{nT}{V} = \frac{nRT}{V}$$

This is the **ideal gas equation** (sometimes called the **ideal gas law**) we have been seeking, the one that can predict how a gas behaves. This one mathematical equation summarizes everything we know about an ideal gas. It can be

algebraically rearranged to be written $PV = nRT$, as it is on the bottle rocket being launched in the chapter-opening illustration.

Notice that R has units. *They are important!* When you use this value of R, you *must* express volume in liters, pressure in atmospheres, temperature in kelvins, and the amount of gas in moles in order to use the ideal gas equation. If you use any other units for these variables and attempt to use the ideal gas law with this value of R, you will get a wrong answer.

Now that we have derived the ideal gas law, let's apply it. Suppose we take a closed 2.0-L glass bottle filled with 0.10 mole of air initially at 25 °C and heat it while measuring the temperature of the air in the bottle.

At 600 °C, the bottle explodes!

Why did this happen? Because heating the air molecules made them move faster, causing them to collide with the bottle walls with more force and more often, creating more pressure. The pressure eventually increased to a point where the bottle could no longer contain it, and so the bottle exploded. Note that during the experiment, we never measured the pressure inside the bottle. The only thing we measured was the temperature, and all we know is that the bottle blew up when the air inside reached 600.00 °C. Even without a barometer, however, we can figure out what the pressure was when the bottle exploded. Knowing the values of n, T, and V, we can use the ideal gas law to calculate the pressure. Here is what we know about the experimental conditions at the moment the bottle blew up:

Amount of gas $n = 0.10$ mol
Volume of gas $V = 2.0$ L
Temperature of gas $T = 600.00$ °C

But remember! Everything must have the units of the gas constant R. The temperature must therefore be converted to kelvins:

$$K = {}^\circ C + 273.15$$
$$K = 600.00\ {}^\circ C + 273.15$$
$$K = 873.15\ K$$

Now we can plug the values for n, V, T, and R into our equation and solve for P:

$$P = \frac{nRT}{V}$$

$$P = \frac{0.0821 \frac{\cancel{L} \cdot atm}{\cancel{K} \cdot \cancel{mol}} \times 0.10\ \cancel{mol} \times 873.15\ \cancel{K}}{2.0\ \cancel{L}} = 3.6\ atm$$

All the units cancel except the units of pressure (atm), which means our answer is expressed in these units. The pressure when the bottle blew up was 3.6 atm, which is 3.6 times greater than normal atmospheric pressure. The ideal gas law has allowed us to determine the pressure of a gas without measuring it directly.

Here is another problem that makes use of the ideal gas law. Suppose you work in a factory filling balloons with helium. One day your boss walks in and accuses you of putting too much helium in each balloon, needlessly cutting into corporate profits. She demands that you tell her how many grams of helium you are putting in each balloon. This is not an easy problem to solve.

You can't simply determine the mass of a helium-filled balloon by weighing it because the balloon would float right off the balance!

But then you remember the ideal gas law, $P = nRT/V$. Your boss asked for the number of grams of helium in each balloon. The ideal gas law doesn't have grams in it, but it does have moles (n). You can therefore use the law to calculate the number of moles of helium in a balloon and then convert to grams. Rearranging the ideal gas equation to solve for n gives:

$$n = \frac{PV}{RT}$$

Because you know the value of R, the only data you need are the values of P, V, and T for one of the balloons. A thermometer hanging on the wall reads 22.00 °C. Because the balloon has been in the room for a while, you reason that the gas in the balloon must be at the same temperature. Converting this Celsius temperature to Kelvin, you get:

$$22.00\ °\text{C} + 273.15 = 295.15\ \text{K}$$

A barometer in the room contains a column of mercury 748 mm high. The pressure of the gas in the balloon must be the same as the air pressure in the room. (If this were not true, the balloon would expand or shrink until the gas pressure inside was equal to the air pressure outside. Actually, the pressure in the balloon is a little higher than atmospheric pressure because of the elastic force of the rubber balloon, but we'll ignore this slight difference.) To use the ideal gas equation, you need the pressure in atmospheres and therefore use the conversion factor 760 mm Hg = 1 atm:

$$748\ \cancel{\text{mm Hg}} \times \frac{1\ \text{atm}}{760\ \cancel{\text{mm Hg}}} = 0.984\ \text{atm}$$

You now have P and T. That leaves just the volume of the helium to be determined. Because the volume of a gas is always the same as the volume of its container, you realize that all you need to do is determine the volume of the balloon. Remembering that the volume of an object can be determined by water displacement (Section 2.7), you fill a bucket with water to the brim and submerge the balloon in the water. Collecting the overflowing water and pouring it into a graduated cylinder, you find that the volume is 240 mL. To be used in the ideal gas law, the volume must be in liters, not milliliters. There being 1000 mL in 1 L, 240 mL converts to 0.240 L.

You now have the P, V, and T values for the helium in the balloon, all in the appropriate units. Putting them in the ideal gas law and solving for n, the number of moles of helium in the balloon, you get:

$$n = \frac{PV}{RT}$$

$$n = \frac{0.984\ \cancel{\text{atm}} \times 0.240\ \cancel{\text{L}}}{0.0812\ \dfrac{\cancel{\text{L}} \cdot \cancel{\text{atm}}}{\cancel{\text{K}} \cdot \text{mol}} \times 295.15\ \cancel{\text{K}}} = 0.0100\ \text{mol He}$$

The balloon is filled with 0.010 mole of helium. You are not finished yet, though, because your boss asked for the amount in grams. The periodic table tells you that the atomic mass of helium is 4.003. One mole of helium atoms therefore has a mass of 4.003 g, and you use this information to calculate the grams of helium in each balloon:

$$0.0100\,\cancel{\text{mol He}} \times \frac{4.003\text{ g He}}{1\,\cancel{\text{mol He}}} = 0.0400\text{ g He}$$

With answer in hand, you proudly march into your boss's office and report that each balloon is being filled with 0.040 g of helium. Unfortunately, you are fired. The company limit is 0.030 g per balloon.

As we stated earlier, the answers we get when we use real gas data with the ideal gas equation are not exactly correct. In both the exploding-bottle problem, where we solved for pressure, and the helium-balloon problem, where we solved for amount, the calculated answers do not *exactly* match the answers we would get if we measured these quantities. This is because the ideal gas equation was developed using the model of an *ideal* gas, which assumes that the molecules experience absolutely no attractive intermolecular forces. All real gas molecules do interact somewhat, however, and the small interactions are enough to make the behavior of a real gas deviate from that predicted by the ideal gas law. Usually, though, if we avoid low temperatures and/or high pressures, the deviations are small enough to tolerate in most cases.

Practice Problems

11.5 According to the ideal gas law, what happens to the pressure of a gas in a container when:

(a) You double the absolute temperature of the gas?
(b) You double the number of liters the container can hold?
(c) You double the number of moles of the gas?
(d) You double both the absolute temperature of the gas and the number of liters the container can hold?

Answer: Solve the ideal gas law for P and use the result to reason it out. $P = \dfrac{nRT}{V}$

(a) Because T is in the numerator of the ideal gas equation, if it doubles, so does P.

(b) Because V is in the denominator of the ideal gas equation, if it doubles, P is halved.

(c) Because n is in the numerator of the ideal gas equation, if it doubles, so does P.

(d) This question asks what happens when you do (a) and (b) simultaneously. Because (a) doubled P and (b) halved it, the net result is that P does not change.

11.6 Suppose that 3.00 g of gaseous nitrogen, N_2, is placed in a 2.00 L container. The pressure is measured to be 450.5 mm Hg. What is the Celsius temperature of the gas? (*Hint*: Your first step should be solving the ideal gas equation for T.)

11.7 An automobile tire is filled with O_2 gas to a total pressure of 40.0 lb/in.^{2}. The temperature is 22.5 °C. The inside volume of the inflated tire is 10.5 gallons. How many grams of O_2 are in the tire? (760 mm Hg = 14.696 lb/in.^{2}; 1 gallon = 3.785 L. (*Hint*: Your first step should be solving the ideal gas equation for n.)

11.3 Getting the Most from the Ideal Gas Law

With just a little algebraic manipulation and some substitutions, we can get even more from the ideal gas law than P, V, n, and T values. For example, our assistant, Albert, has been assigned the task of analyzing an unknown gas leaking from an old, unmarked storage tank. Having learned about combustion analysis back in Chapter 7, he determines the mass of a sample of the gas to be 0.125 g, and then burns the sample. The products are water and carbon dioxide, and he measures the masses of these products. These masses tell him the gas is 14.37% by mass H and 85.63% by mass C. Using what he learned in Chapter 7 about combustion analysis, he concludes that the empirical formula must be CH_2, but of course this cannot be the molecular formula. (Review Chapter 7 if you've forgotten the relationship between empirical formulas and molecular formulas.) Albert then generates a table to help him determine the molecular formula:

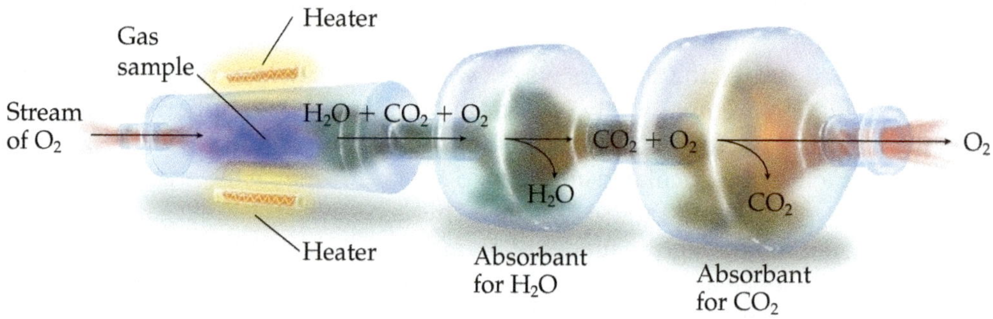

Molecular formula	Molar mass (g/mol)
C_2H_4	28
C_3H_6	42
C_4H_8	56
.	.
.	.
.	.
C_nH_{2n}	$14n$

If Albert knew the molar mass of the gas, he could easily pick the correct molecular formula. But how do you get the molar mass without knowing the molecular formula? From the ideal gas law of course! In Section 11.2, we solved the ideal gas equation for n, the number of moles of gas in a sample:

$$n = \frac{PV}{RT}$$

The key to getting molar mass from this equation is that Albert knew the mass of the sample before it was burned—0.125 g. He uses the symbol m_{sample} to represent the mass of the sample before combustion: $m_{\text{sample}} = 0.125$ g. Any time you have a known mass of some pure substance, you can convert to moles by dividing the mass by the molar mass MM (see Section 7.6 for review):

Number of moles of substance present → $$n = \frac{m_{\text{sample}}}{MM}$$

Measured mass of substance (m_{sample})

Molar mass of substance (MM)

This conversion formula is extremely important. *You should always remember it!*

Because both m_{sample}/MM and PV/RT equal n, these two terms must equal each other:

$$\frac{m_{sample}}{MM} = \frac{PV}{RT}$$

Now all Albert has to do is rearrange this equation to solve for the molar mass:

$$MM = \frac{m_{sample}RT}{PV}$$

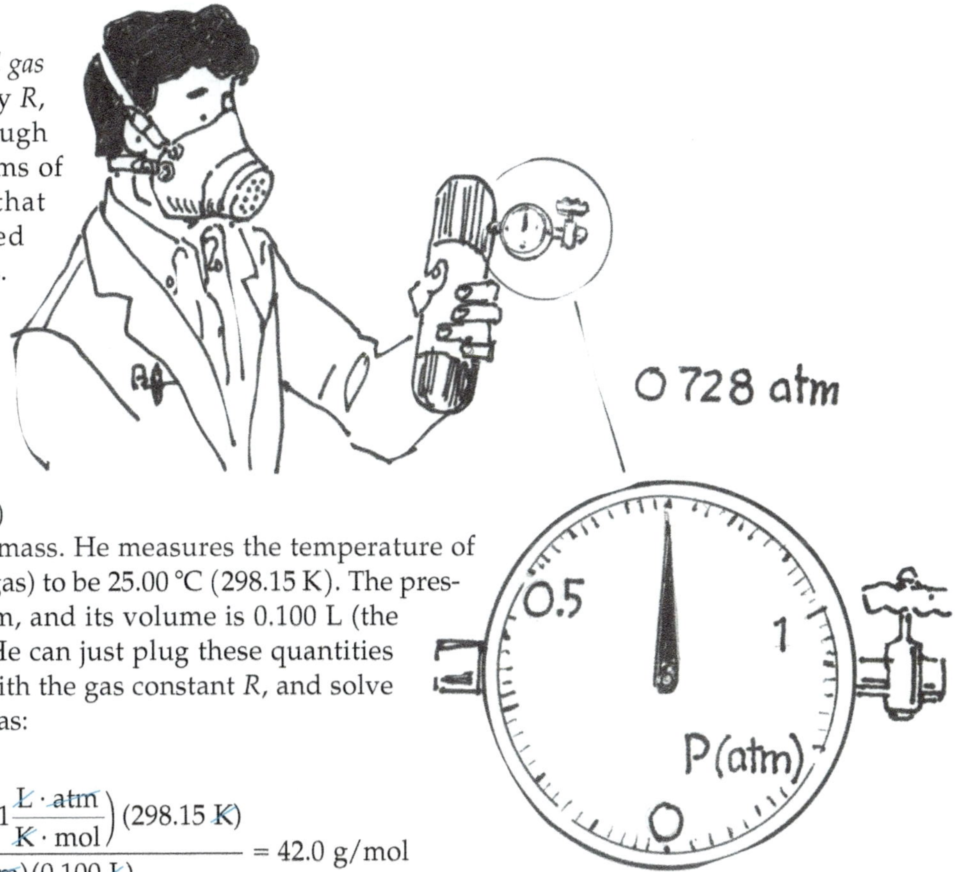

This is still a form of the ideal gas law because it still has only R, P, V, T, and n in it, although n is now expressed in terms of m_{sample} and MM. Recall that Albert initially collected $m_{sample} = 0.125$ g of the gas. He collected it in a 0.100-L steel cylinder equipped with a pressure gauge. He is pictured at right with his sample. He is able to measure all the variables (m_{sample}, T, P, V) in the equation for molar mass. He measures the temperature of the room (and thus of the gas) to be 25.00 °C (298.15 K). The pressure of the gas is 0.728 atm, and its volume is 0.100 L (the volume of the cylinder). He can just plug these quantities into the equation, along with the gas constant R, and solve for the molar mass of the gas:

$$MM = \frac{(0.125\text{ g})\left(0.0821\frac{\cancel{\text{L}}\cdot\cancel{\text{atm}}}{\cancel{\text{K}}\cdot\text{mol}}\right)(298.15\ \cancel{\text{K}})}{(0.728\ \cancel{\text{atm}})(0.100\ \cancel{\text{L}})} = 42.0\text{ g/mol}$$

From his table of possible formulas, Albert can now conclude that the molecular formula of the gas is C_3H_6.

This form of the ideal gas equation is a powerful tool because it allows us to determine the molar mass of a gas even when we don't know the identity of the gas! All we need to do is determine the mass of a sample of the gas and then measure P, V, and T for the sample. Remember, however, to use the correct units. All the variables must have units that agree with those of the gas constant R—L · atm/K · mol. Check out the solution to the following practice problem, where all the variables have been converted to the proper units.

Practice Problems

11.8 A steel cylinder contains 0.01378 kg of an unknown gas. Combustion analysis indicates that the gas has the empirical formula H_2S. The volume of the cylinder is 2.20×10^3 mL, and the pressure inside the cylinder is 3.42×10^3 mm Hg. The cylinder is stored in a closet at 25.0 °C. What is the molecular formula of the gas?

Answer:

$$MM = \frac{m_{\text{sample}}RT}{PV}$$

For gas law problems you must always convert T to kelvins—*always*

$$= \frac{\underbrace{(0.013\,78\ \text{kg})\left(\frac{1000\ \text{g}}{\text{kg}}\right)}_{m_{\text{sample}}}\underbrace{\left(0.0821\frac{\text{L}\cdot\text{atm}}{\text{K}\cdot\text{mol}}\right)}_{R}\underbrace{([25.0+273.15]\ \text{K})}_{T}}{\underbrace{(3.42\times10^3\ \text{mm Hg})\left(\frac{1\ \text{atm}}{760\ \text{mm Hg}}\right)}_{P}\underbrace{(2.20\times10^3\ \text{mL})\left(\frac{1\ \text{L}}{1000\ \text{mL}}\right)}_{V}}$$

$$= 34.1\ \text{g/mol}$$

If you calculate the molar mass of H_2S with atomic mass values from the periodic table, you get 34.082 g/mol. This is essentially identical to the molar mass we just calculated, telling us that the molecular formula in this case is the empirical formula, H_2S.

11.9 Suppose you have a sample of CO_2 gas and want to know its mass without bothering to use a balance. How could you do this?

Answer: Take the form of the ideal gas equation that has molar mass in it and solve for the mass of the sample:

$$MM = \frac{m_{\text{sample}}RT}{PV} \xrightarrow{\text{rearrange}} m_{\text{sample}} = \frac{PVMM}{RT}$$

To find the mass of the sample, measure P, V, and T for the gas and substitute the numeric values into the rearranged equation along with the value of R and the molar mass of CO_2, which is 44.01 g/mol. Make sure all the variables have units that agree with those in R, and solve for m_{sample}.

11.10 A sample of hydrogen gas, H_2, is in an outdoor 1000.0-gallon tank. It is winter, and the temperature is −4.50 °C. A pressure gauge indicates that the pressure inside the tank is 32.6 atm. How many pounds of hydrogen are left in the tank? (*Hint*: Examine the solution to Problem 11.9.)

11.11 What do you get when you divide the mass in grams of a sample of a pure substance by the molar mass of the substance? Prove your answer is correct.

Let's not stop here. We can rearrange the molar mass form of the ideal gas equation once again, this time solving for m_{sample}/V:

$$MM = \frac{m_{\text{sample}}RT}{PV} \xrightarrow[\text{and cross out all variables that cancel.}]{\text{Multiply both sides by } \frac{P}{RT}} \frac{MMP}{RT} = \frac{m_{\text{sample}}\cancel{RT}\cancel{P}}{\cancel{P}V\cancel{RT}}$$

Doing so gives

$$\frac{m_{\text{sample}}}{V} = \frac{MMP}{RT}$$

We arrive at an equation for m/V, which is the definition of density. This equation says that if you know the molar mass of a gas and measure its pressure and temperature, you can calculate its density. Let's try it. Below are two balloons, one filled with helium and the other with carbon dioxide. Both are in a room that is at 22.00 °C, and the pressure inside each balloon is 755.0 mm Hg (slightly less than 1 atm). The densities of both gases are calculated below (notice how we were careful to convert units to those of the gas constant R).

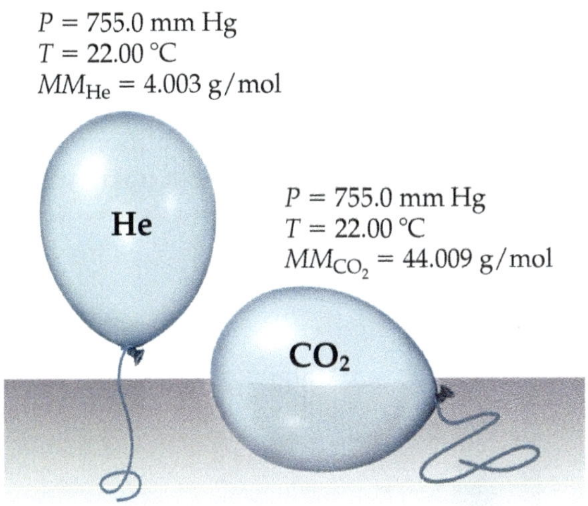

$$\text{Density} = \frac{m_{\text{sample}}}{V} = \frac{MMP}{RT}$$

For helium:

$$\frac{m_{\text{sample}}}{V} = \frac{\left(\frac{4.003\text{ g}}{\text{mol}}\right)(755.0\text{ mm Hg})\left(\frac{1\text{ atm}}{760\text{ mm Hg}}\right)}{\left(0.0821\frac{\text{L}\cdot\text{atm}}{\text{K}\cdot\text{mol}}\right)(295.15\text{ K})} = 0.164\text{ g/L}$$

For carbon dioxide:

$$\frac{m_{\text{sample}}}{V} = \frac{\left(\frac{44.009\text{ g}}{\text{mol}}\right)(755.0\text{ mm Hg})\left(\frac{1\text{ atm}}{760\text{ mm Hg}}\right)}{\left(0.0821\frac{\text{L}\cdot\text{atm}}{\text{K}\cdot\text{mol}}\right)(295.15\text{ K})} = 1.80\text{ g/L}$$

Look at the densities we calculated. Under the same conditions, carbon dioxide gas is more than ten times as dense as helium gas. Given that the density of air is 1.30 g/L, it should be no mystery why a helium balloon floats in air but a balloon filled with CO_2 would sink.

By now you should be catching on to the idea that rearranging the ideal gas equation can be quite a useful thing to do. For example, try the following WorkPatch, and then we'll discuss your answer.

STOP **WORKPATCH 11.3 A steel tank contains 12.992 kg of oxygen gas, O_2. Measurements indicate that the pressure of the gas is 100.0 atm and its temperature is 300.0 K. The internal volume of the tank is 100.0 L. Evaluate the numeric value of PV/nT (include units in your answer), and compare what you get with the accepted value of the gas constant R = 0.0821 L · atm/K · mol. ●**

Your WorkPatch answer is extremely important! For any gas that behaves ideally, knowing the number of moles n, the pressure P in atmospheres, the volume V in liters, and the temperature T in kelvins allows you to calculate the gas constant R from the relationship $R = PV/nT$.

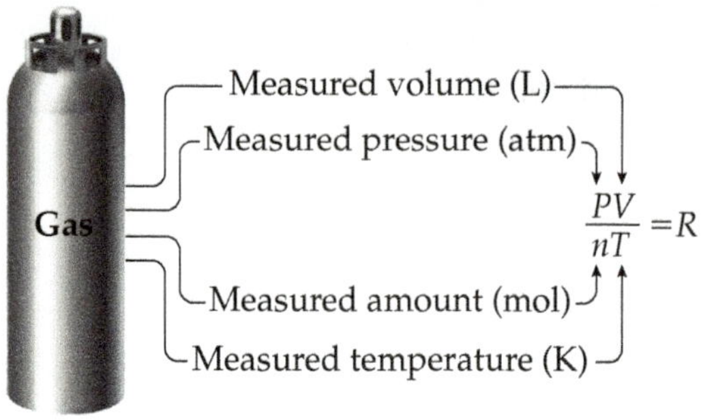

Knowing this makes it much easier to solve a type of problem called an initial-condition, final-condition problem. For example, suppose we have a steel cylinder filled with a gas. The pressure of this gas is 5.00 atm, and its temperature is 200.0 K. The volume of the cylinder is unknown, but it is fixed (unchanging) because the cylinder is made of solid steel. The number of moles of gas in the cylinder is also unknown. So these are our initial (i) conditions for the gas: $P_i = 5.00$ atm, $T_i = 200.0$ K, V_i and n_i both unknown:

Initial Conditions

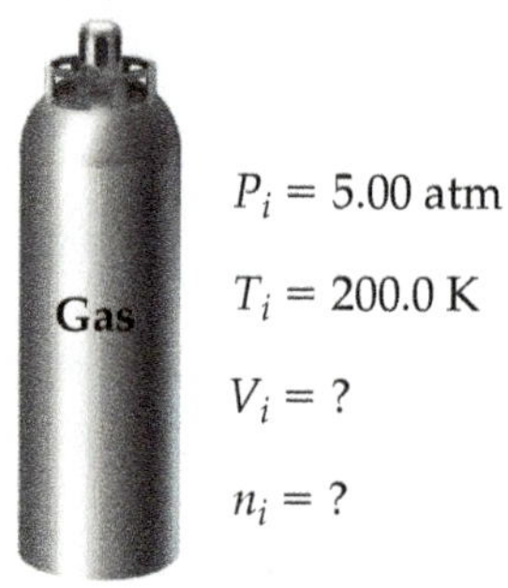

Now let's change the temperature by heating the cylinder to 400.0 K. Because this is the only change we make, our final (f) conditions are:

Final Conditions

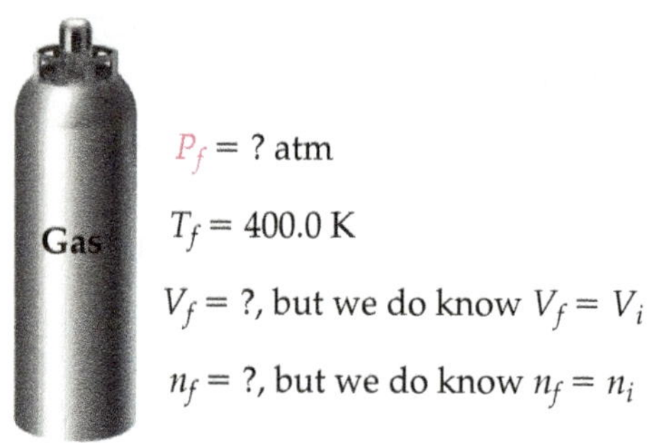

Notice that although we still do not know the volume of the gas or the number of moles of gas present, we *can* say that $V_f = V_i$ and $n_f = n_i$ because we did not change these quantities on going from the initial conditions to the final conditions. The pressure, however, has changed (it increased as the temperature increased because the volume could not change). What is the new pressure P_f inside the cylinder? This is an important question. If a storage cylinder of gas gets too hot, the increase in pressure can cause it to explode.

To determine P_f, just remember that PV/nT is always equal to R. This means that P_iV_i/n_iT_i is equal to R, and P_fV_f/n_fT_f is also equal to R. And, of course, because these expressions both equal R, they must be equal to each other:

$$\frac{P_iV_i}{n_iT_i} = \frac{P_fV_f}{n_fT_f}$$

This is an extremely useful result. It is the starting point for solving initial-condition, final-condition problems. The second step is to apply the rule of algebra that allows us to cancel *equal* quantities that appear in the same position on both sides of the equation. In our case, V_i on the left side and V_f on the right side are equal, and both appear in the numerator, which means they cancel each other. Likewise, n_i on the left and n_f on the right are equal and both are in the denominator, which means they also cancel. These cancellations simplify our equation to:

$$\frac{P_i\cancel{V_i}}{\cancel{n_i}T_i} = \frac{P_f\cancel{V_f}}{\cancel{n_f}T_f} \xrightarrow{\text{simplifies to}} \frac{P_i}{T_i} = \frac{P_f}{T_f}$$

All the variables with unknown values have disappeared! The final step is to rearrange the equation to solve for P_f and then plug in the numeric values for P_i, T_f, and T_i:

$$P_f = \frac{P_iT_f}{T_i} = \frac{(5.00\ \text{atm})(400.0\ \cancel{\text{K}})}{200.0\ \cancel{\text{K}}} = 10.0\ \text{atm}$$

This answer tells us that the pressure doubled when we doubled the temperature while holding the volume constant. Make sure you can do this algebraic rearrangement. (Review Chapter 2 if you can't!)

A summary of the method we just described appears below.

Solving initial-condition, final-condition gas problems

Step 1: List P, V, n, and T for the initial and final conditions.
- Write in numeric values for any variables you know.
- Where possible, express the final variable in terms of the initial one. For instance, if volume does not change, you can say $V_f = V_i$.

Step 2: Write the expression

$$\frac{P_iV_i}{n_iT_i} = \frac{P_fV_f}{n_fT_f}$$

- Whenever possible, rewrite this expression to show final variables in terms of initial ones (from step 1).
- Cancel factors that are identical on the two sides of the expression.

Step 3: Solve the equation algebraically for the desired variable. Then plug in numeric values for the quantities you know and do the calculation to get your answer.

Try the following initial-condition, final-condition practice problems. The first teaches you a valuable lesson about temperature. The second shows you how a seemingly difficult problem can be, in fact, quite solvable.

Practice Problems

11.12 A steel cylinder is filled with a gas. The initial pressure of this gas is 5.00 atm, and the initial temperature is 200.00 °C. The cylinder is heated to a final temperature of 400.00 °C. It appears that the temperature has doubled, but the final pressure is not 10.0 atm. Why doesn't the pressure double as it did in the previous example? What is the final pressure? What valuable lesson does this question teach you?

Answer:

Step 1: *List P, V, n, and T for the initial and final conditions.*

Initial conditions	Final conditions
$P_i = 5.00\ atm$	$P_f = ?$
$V_i = ?$	$V_f = V_i$
$n_i = ?$	$n_f = n_i$
$T_i = 200.00\ °C\ (473.15\ K)$	$T_f = 400.00\ °C\ (673.15\ K)$

Step 2: *Write the key expression and cancel identical terms:*

$$\frac{P_iV_i}{n_iT_i} = \frac{P_fV_f}{n_fT_f} \xrightarrow{\text{Wherever possible, write final variables in terms of initial variables.}} \frac{P_iV_i}{n_iT_i} = \frac{P_fV_i}{n_iT_f}$$

$$\frac{P_i\cancel{V_i}}{\cancel{n_i}T_i} = \frac{P_f\cancel{V_i}}{\cancel{n_i}T_f} \xrightarrow{\text{Cancel identical factors on the two sides of the equation.}} \frac{P_i}{T_i} = \frac{P_f}{T_f}$$

Step 3: *Solve algebraically for the desired variable (P_f in this case), and then plug in the known numeric values:*

$$\frac{P_i}{T_i} = \frac{P_f}{T_f} \longrightarrow P_f = \frac{T_fP_i}{T_i}$$

$$P_f = \frac{T_fP_i}{T_i} = \frac{(673.15\ K)(5.00\ atm)}{473.15\ K} = 7.11\ atm$$

The final pressure is only 7.11 atm, not 10.0 atm. The pressure did not double because the Kelvin temperature did not double! The valuable lesson to be learned is that you can use only the Kelvin scale for temperature when doing gas law problems. Never use any other scale. Never!

11.13 A steel cylinder initially contains some amount of O_2 gas at 500.0 K. Additional O_2 gas is added until the number of moles of gas has doubled. The cylinder is then cooled to 250.0 K. What happened to the pressure inside the cylinder?

Answer:

Step 1: *List P, V, n, and T for the initial and final conditions.*

Initial conditions	Final conditions
$P_i = ?$	$P_f = ?$
$V_i = ?$	$V_f = V_i$
$n_i = ?$	$n_f = 2n_i$ ← We don't know the number of moles of oxygen, but we do know the final number is twice the initial number.
$T_i = 500.0\ K$	$T_f = 250.0\ K$

Step 2: *Write the key expression and cancel identical terms.*

$$\frac{P_iV_i}{n_iT_i} = \frac{P_fV_f}{n_fT_f} \xrightarrow{\text{Wherever possible, write final variables in terms of initial variables.}} \frac{P_iV_i}{n_iT_i} = \frac{P_fV_i}{(2n_i)T_f}$$

$$\frac{P_i\cancel{V_i}}{\cancel{n_i}T_i} = \frac{P_f\cancel{V_i}}{(2\cancel{n_i})T_f} \xrightarrow{\text{Cancel identical terms on the two sides of the equation.}} \frac{P_i}{T_i} = \frac{P_f}{2T_f}$$

Step 3: *Solve algebraically for the desired variable (P_f in this case), and then plug in the known numeric values:*

$$\frac{P_i}{T_i} = \frac{P_f}{2T_f} \longrightarrow P_f = \frac{2T_fP_i}{T_i}$$

$$P_f = \frac{2T_fP_i}{T_i} = \frac{2(250.0\text{ K})P_i}{500.0\text{ K}} = \frac{500.0\text{ K}}{500.0\text{ K}} \times P_i = P_i$$

We find that $P_f = P_i$—meaning that the pressure did not change! There's no pressure change because the effect of doubling the amount of gas (a change that doubles the pressure) is counterbalanced by the effect of halving the temperature (a change that halves the pressure).

11.14 A gas initially at $P_i = 2.00$ atm, $T_i = 200.0$ K, and $V_i = 50.0$ mL is in a cylinder that has a movable piston. The cylinder is heated to $T_f = 220.0$ K and then the piston is pushed down so that the volume of the gas is decreased to $V_f = 25.0$ mL.

(a) Calculate the final pressure P_f of the gas, leaving all volumes in milliliters for your calculation.
(b) Calculate the final pressure P_f of the gas with all volumes expressed in liters.
(c) When you are doing initial-condition, final-condition problems, must volumes always be in liters?
(d) When you are using the ideal gas equation in the form $PV = nRT$, volumes must always be in liters (never milliliters or anything else). Why?

The solution to Practice Problem 11.14 reveals that in initial-condition, final-condition problems, the volumes can be in any units. This is true because R is missing from the equation. The same is true for pressures and amounts of gas. However, *temperature must always be expressed in kelvins.*

We can get other interesting information about gases from the ideal gas equation. For example, we know that real gases act ideally under most circumstances—that is, real gases follow the ideal gas law to a good approximation. We can therefore use our ideal gas model to answer the question "What volume does 1 mole of any gas occupy at standard temperature and pressure?" (**Standard temperature and pressure**, also written **STP**, is defined as 0 °C = 273.15 K and 1 atm pressure.) To find the answer, all we need to do is solve the ideal gas equation for volume and plug in the appropriate numeric values:

$$V = \frac{nRT}{P} = \frac{(1\text{ mol})\left(0.0821\frac{\text{L}\cdot\text{atm}}{\text{K}\cdot\text{mol}}\right)(273.15\text{ K})}{1\text{ atm}} = 22.4\text{ L}$$

This is, to a good approximation, the **molar volume** (the volume occupied by 1 mole) of *any* gas at STP. Thus 1 mole of H_2 gas at STP has a volume of 22.4 L. So does 1 mole of CO_2 gas at STP. So does 1 mole of O_2 gas at STP, and so on.

The upshot of all this is that *equal volumes of gas at the same temperature and pressure must contain equal numbers of moles*. For example, suppose we have one tank filled with a gas A and a second tank filled with a gas B. The tanks are the same size (same volume), in the same room (same temperature), and at the same pressure.

They must therefore contain the same number of moles of gas. (The first person to demonstrate this phenomenon was Avogadro.)

The two tanks must contain the same number of moles regardless of the identities of A and B.

Finally, we can even use the ideal gas equation for stoichiometry problems. This is useful when some of the reactants or products of a reaction are gases. For example, oxygen gas can be produced by heating solid potassium chlorate (a decomposition reaction, Chapter 7):

$$2KClO_3(s) \rightarrow 2KCl(s) + 3O_2(g)$$

Suppose you are in a spaceship that has an internal volume of 1000.0 m^3 (note that there are 1000 L per cubic meter). All your oxygen has leaked out, but luckily you have a stash of solid potassium chlorate on board. Your plan is to electrically heat some of it and thereby decompose it and generate enough oxygen gas to fill the ship at STP. How much $KClO_3$ do you need? This being a stoichiometry problem, you can be sure the first step is to convert to moles, and we do this with the ideal gas equation, converting the volume of 1000.0 m^3 of oxygen gas at STP to moles:

$$n = \frac{PV}{RT} = \frac{\overbrace{(1\ \text{atm})(1000.0\ \text{m}^3)\left(\frac{1000\ \text{L}}{\text{m}^3}\right)}^{V \text{ in liters}}}{\left(0.0821\frac{\text{L}\cdot\text{atm}}{\text{K}\cdot\text{mol}}\right)(273.15\ \text{K})} = 4.46 \times 10^4 \text{ mol } O_2 \text{ gas}$$

Now we can use the balanced chemical equation (which "speaks" in moles) to finish the problem:

$$(4.46 \times 10^4 \text{ mol } O_2)\overbrace{\left(\frac{2 \text{ mol } KClO_3}{3 \text{ mol } O_2}\right)}^{\text{From balanced reaction}}\overbrace{\left(\frac{122.548 \text{ g } KClO_3}{\text{mol } KClO_3}\right)}^{\text{Molar mass of } KClO_3} = 3.64 \times 10^6 \text{ g } KClO_3$$

That's over 4 tons of potassium chlorate! Maybe you should look for some spare oxygen tanks instead.

This chapter may seem very mathematical. We certainly gave the ideal gas equation, $PV = nRT$, a healthy algebraic workout. But remember, we derived this equation by visualizing a simple model of an ideal gas while asking three questions about pressure. Everything we did was based on that simple model. Chemists may forget equations and facts, but the underlying pictures and models stay with them forever, and this is where true understanding comes from.

+1 One More Thing

11.4 Deviations from Ideality

This chapter has been devoted to understanding and predicting the behavior of an ideal gas. The problem is, of course, that no such gas actually exists. To see why not, all we need to do is ask a question of the equation that describes how an ideal gas behaves, the ideal gas law $PV = nRT$. First, let's solve it for V.

$$V = \frac{nRT}{P}$$

Now we can ask the question, "What happens to the volume of a gas as we cool it all the way down to absolute zero ($T = 0$ K)?" Since T is in the numerator, when $T = 0$, V will also equal zero. Stated simply, the ideal gas law predicts that if we cool a gas to absolute zero, it will disappear from the universe. No such gas exists. For a real gas, cooling it to absolute zero will condense it to a liquid and then to a solid which, because its constituent molecules are made of matter and occupy space, will occupy a smaller but definitely nonzero volume. Another problem with an ideal gas is that by definition it is one in which there are absolutely no intermolecular forces of any kind operating between the molecules. Once again, no such gas exists. Be they London forces, dipolar forces, or other forces, all molecules have some sort of intermolecular force between them that increases in strength as the molecules get closer to one another and become more significant as the gas is cooled and the molecules slow down. If not, you could cool a gas to absolute zero and it would remain a gas, but there is no real gas that can't be sufficiently cooled so as to yield a condensed phase. This means that the ideal gas law might give answers that deviate from reality, and in fact it does under certain conditions. Below is a graph that demonstrates this for five real gases. The plot is the result of experiments that vary the pressure of a gas and then measure its volume at a fixed temperature and fixed number of moles and calculate the quantity PV/nRT at each pressure. Since $PV = nRT$ for an ideal gas, the quantity PV/nRT should always be equal to 1 for every pressure and volume measured. Here is what you actually get for five different real gases.

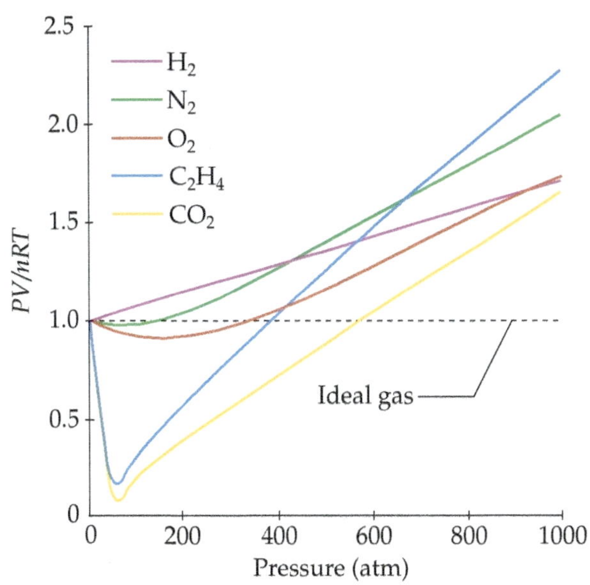

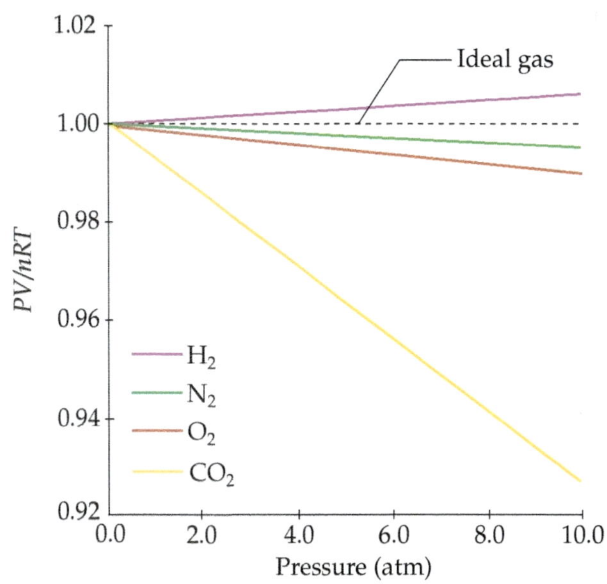

None of the gases produce PV/nRT values of 1 that always lie on the dotted (ideal gas) line. The plot on the right is just an expanded look at the beginning of the plot on the left. It reveals that the real gases come close to behaving ideally only at lower pressures of around 2 atm or less. These deviations from ideality are due to the fact that real gas molecules are made of matter and thus occupy space and have their own volume, and real gases do have intermolecular forces operating between their molecules. The way to fix this problem is to modify the ideal gas equation to take these facts into account, and because the type and strength of the intermolecular attractions and the size of the molecules will depend on the chemical identity of the molecules, the modified equation needs to be flexible enough so it can tailor itself to different gases. One such equation is the van der Waals equation.

$$\left\{P_{\text{real}} + \mathrm{a}\left(\frac{n}{V_{\text{real}}}\right)^2\right\}\left\{\left(\frac{V_{\text{real}}}{n}\right) - \mathrm{b}\right\} = RT$$

P, V, n, R, and T have the same meaning as they do in the ideal gas law, but now P_{real} is the real measured pressure and V_{real} is the real measured volume as opposed to the ideal gas values. The constant $\boldsymbol{a}$ provides a correction due to intermolecular forces, and the constant $\boldsymbol{b}$ corrects for the fact that molecules have their own size and volume that is nonzero. Below is a table of $\boldsymbol{a}$ and $\boldsymbol{b}$ constants for the gases in the above plot.

Gas	a (atm L^2/mol^2)	b (L/mol)
H_2	0.244	0.0266
N_2	1.39	0.0391
O_2	1.36	0.0318
CO_2	3.59	0.0427
C_2H_4 (ethylene)	4.612	0.0582

If we compare the ideal versus the van der Waals equation for 1 mole of a gas we see:

$$P_{\text{ideal}} \times V_{\text{ideal}} = RT$$

$$\left(P_{\text{real}} + \mathrm{a}/V_{\text{real}}^2\right) \times \left(V_{\text{real}} - \mathrm{b}\right) = RT$$

For these two equations to be identical, $P_{ideal} = (P_{real} + \mathbf{a}/V_{real}^2)$ and $V_{ideal} = V_{real} - \mathbf{b}$. Since real gas molecules are actually attracted to each other, they slow down a bit when they pass each other, and therefore the force with which they hit the walls of the container (the pressure) is actually somewhat less than it would be if there were no attractive forces (as in an ideal gas). So we have to add a bit to the "real" pressure to bring it up to what it would be if it behaved ideally. Likewise, a real gas molecule does take up a small amount of space; it has a volume, so the available volume of the container is actually somewhat smaller than it would be if it were an ideal gas, which is assumed to take up no space at all. Hence, we must subtract a small bit from the "real" volume to make it equal the ideal volume.

So the $\mathbf{a}/v^2$ term, the correction for the attractive forces between the molecules, should get larger as the attractive forces between the real molecules become stronger. In the same way, the $\boldsymbol{b}$ term, which corrects for the volume of a real molecule, should get bigger as the molecule gets bigger. If we look at the experimentally determined values for the ***a*** and ***b*** constants this is exactly what we see.

As an example, we can consider 1 mole of ethylene (C_2H_4) gas in a 1.00-L container at −160.00 °C (this is only about 10 °C above its boiling point of −169.2 °C, which means if we cooled it by another 10 °C, it would liquefy). Being barely above its boiling point and having the strongest intermolecular forces (largest ***a***) in the above table, we can expect some pretty significant deviation from ideal behavior. Indeed, if we solve both the ideal gas law and the van der Waals equation for *P* and plug in the given volume, moles, and temperature, we get $P_{ideal} = 9.29$ atm and $P_{vdw} = 5.25$ atm.

−160°C

$$P_{\text{ideal}} = \frac{nRT}{V} = \frac{(1 \text{ mole})\,(0.0821 \text{ L at/K mole})\,(113.15 \text{ K})}{(1 \text{ L})}$$

$$= 9.29 \text{ atm}$$

Finally, it is important to point out that real gases behave ideally and the ideal gas equation will give good results if the conditions allow the gas to approximate ideal behavior. These conditions are low pressure (see the above plot) and/or high temperature. One way to achieve low pressure is to expand the gas into a large volume. At large volumes, the volume occupied by the molecules themselves becomes negligible compared to the container volume. Also, at high temperatures (temperatures significantly higher than the boiling point of the substance), the molecules are moving so fast that they can easily overcome any intermolecular attractions, and so the attractions can be ignored.

Had we done the above pressure calculations for ethylene gas at a much higher temperature and larger volume, we would find that the two equations give much closer answers.

Have You Learned This?

Ideal gas (p. 414)
Pressure (p. 416)
Normal atmospheric pressure (p. 417)
Rule 1 Pressure-volume relationship (p. 420)
Rule 2 Pressure-temperature relationship (p. 421)
Rule 3 Pressure-moles relationship (p. 422)
Ideal gas constant (p. 422)
Ideal gas equation (p. 422)
Ideal gas law (p. 422)
Standard temperature and pressure (STP) (p. 433)
Molar volume (p. 434)

Learning Outcomes

11.1	**Describing the Gas Phase—*P, V, n,* and *T***	Describe the units of pressure and how pressure is created by a gas. [15, 17–24]
11.2	**Describing a Gas Mathematically—The Ideal Gas Law**	Describe the relationship between any two variables that characterize a gas when the remaining variables are held constant. [29–40] Perform calculations using the ideal gas law. [41–55]

11.3	**Getting the Most from the Ideal Gas Law**	Calculate the molar mass of a gas using the ideal gas law. [56–58] Calculate the density of a gas using the ideal gas law. [59–61] Calculate a result for an "initial condition" to "final condition" gas problem. [63–67] Calculate a result for a reaction stoichiometry problem involving gases. [68–71]
11.4	**Deviations from Ideality**	Calculate the pressure of a gas and the deviation from ideal behavior using the van der Waals equation. [131]

Skills to Know

Using the ideal gas law

1. Four variables characterize an ideal gas: pressure (P), volume (V), amount of gas in moles (n), and temperature (T).

2. The relationships among P, V, n, and T can be summarized by the three relations involving pressure:

P is proportional to $\frac{1}{V}$

If you double the volume, the pressure drops by half. (The reverse is also true: If you halve the volume, the pressure doubles.)

P is proportional to T

If you double the absolute temperature, the pressure also doubles. (The reverse is also true: If you halve the absolute temperature, the pressure is halved.)

P is proportional to n

If you double the number of particles, the pressure also doubles. (The reverse is also true: If you halve the number of particles, the pressure is halved.)

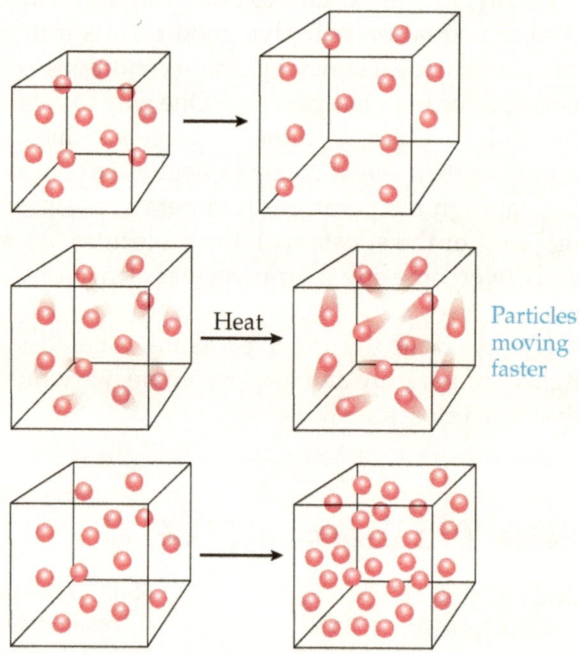

3. The **ideal gas equation** summarizes the above relationships:

$P = \frac{nRT}{V}$ where R is the ideal gas constant.

$$R = 0.0821 \frac{\text{L} \cdot \text{atm}}{\text{K} \cdot \text{mol}}$$

Always L and atm

Always K and moles

4. Any time you use a gas equation that contains R, the variables P, V, n, and T must have units that agree with those of R.

For the above value of R:
P must be in atmospheres
V must be in liters
n must be in moles
T must be in kelvins

5. You should feel comfortable with rearranging the ideal gas equation to solve for any of the variables.	$V = \frac{nRT}{P}$ Solved for V $n = \frac{PV}{RT}$ Solved for n and so on
6. Standard (sea-level) atmospheric pressure is 760 mm Hg = 1 atm exactly. This relationship provides conversion factors for changing between atmospheres and millimeters of mercury.	Useful conversion factors: $\frac{760 \text{ mm Hg}}{1 \text{ atm}}$ $\frac{1 \text{ atm}}{760 \text{ mm Hg}}$

Using the ideal gas equation to identify an unknown gas

You can use the ideal gas equation to identify a pure sample of an unknown gas. First rewrite the equation so you can solve it for molar mass. To do that, replace n in the equation with m/MM, where m is the sample mass and MM is the molar mass. Now measure P, V, and T for the gas under a given set of conditions, plus the mass m of the sample. Then solve for molar mass and use the molar mass value to figure out what the gas must be. Make sure you use units compatible with the units of R.	For any pure substance, $n = \frac{m}{MM}$ where m is the mass of the sample in grams. That is, if you divide the mass of the sample by the molar mass of the substance, you get the number of moles present. Substituting this expression for n in the ideal gas equation and solving for molar mass, you get: $MM = \frac{mRT}{PV}$

Solving initial-condition, final-condition gas problems

Example: An air bubble rises through the ocean from a depth of 1000 feet to the surface. As it rises, its pressure decreases from 31.3 atm to 1.00 atm and its volume increases. Its initial volume is 5.00 mL. The bubble's temperature remains constant, and it gains and loses no gas. What is its final volume?

1. List P, V, n, and T for the initial and final conditions.

- Write in numerical values for any variables you know.
- Where possible, express the final variable in terms of the initial one. For instance, if volume does not change, you can say $V_f = V_i$.

Initial conditions	Final conditions
$P_i = 31.3$ atm	$P_f = 1.00$ atm
$V_i = 5.00$ mL	$V_f = ?$
$n_i = ?$	$n_f = n_i$
$T_i = ?$	$T_f = T_i$

2. Write the expression $\frac{P_iV_i}{n_iT_i} = \frac{P_fV_f}{n_fT_f}$ • Whenever possible, rewrite this expression to show final variables in terms of initial ones (from step 1). • Cancel quantities that are identical on the two sides of the expression.	$\frac{P_iV_i}{n_iT_i} = \frac{P_fV_f}{n_iT_i}$ Because $n_f = n_i$ and $T_f = T_i$, we can replace variables on the right side of the equation with those from the left side. $\frac{P_iV_i}{\cancel{n_iT_i}} = \frac{P_fV_f}{\cancel{n_iT_i}} \longrightarrow P_iV_i = P_fV_f$
3. Solve the equation algebraically for the desired variable. Then plug in values for the quantities you know and solve numerically for an answer.	$P_iV_i = P_fV_f \rightarrow V_f = \frac{P_iV_i}{P_f}$ $V_f = \frac{P_iV_i}{P_f} = \frac{(31.3\text{ atm})(5.00\text{ mL})}{1.00\text{ atm}} = 157\text{ mL}$

Concept Questions

1. In a gas, which one of the following is true?
   (a) The molecules are in constant contact with one another.
   (b) The molecules move in straight-line motion in random directions.
   (c) The molecules create pressure by colliding with each other.
   (d) The molecules do not attract one another at all in the gas phase.

2. Which one of the following is true for an ideal gas?
   (a) It is called ideal because the molecules do not move.
   (b) It is called ideal because the molecules travel at ideal velocities.
   (c) It is called ideal because its pressure is independent of temperature.
   (d) It is called ideal because there are no intermolecular attractions between molecules.

3. A gas produces pressure because:
   (a) All gases have an urge to escape.
   (b) Its molecules are heavier than air.
   (c) Its molecules collide with the container walls.
   (d) Its molecules are constantly accelerating.

4. If a sample of gas in a steel container has its temperature increased:
   (a) Its pressure will increase because the number of moles will increase.
   (b) Its pressure will increase because the molecules move faster.
   (c) Its pressure will increase because the electrons in the gas molecules get excited.
   (d) Its pressure will decrease because its volume stays constant.

5. The statement that for a gas, $P = \text{constant}/V$ tells us that:
   (a) Volume is proportional to pressure.
   (b) Pressure is constant when V changes.
   (c) There is not enough information in the equation to say anything.
   (d) Pressure in inversely proportional to volume.

6. An inverse proportionality between variables means:
   (a) They are independent of one another but still related by a negative sign.
   (b) As one variable gets larger, so does the other.
   (c) As one variable gets smaller, so does the other.
   (d) As one variable gets larger, the other gets smaller.

7. The pressure of a gas is proportional to the number of moles of gas because:
   (a) More moles results in more forceful collisions with the container walls.
   (b) More moles results in more collisions per unit time with the container walls.
   (c) More moles causes repulsions between gas molecules, which gives more forceful collisions with the container walls.
   (d) The pressure of a gas is not proportional to the number of moles of gas; it is inversely proportional to it.

8. According to the ideal gas law, at STP, which one of the following is true?
   (a) All gases will have the same molar volume.
   (b) All gases will have molecules moving with the same average velocity.
   (c) All gases will have a volume equal to 1-L.
   (d) All gases will condense into a liquid.

9. If you measure *P*, *V*, *n*, and *T* for any sample of an ideal gas, the following will be true:
   (a) $PV/nRT = 1$
   (b) $PV/nRT = R$
   (c) $P/V = n/T$
   (d) $nT/PV = R$

10. We can always write $P_iV_i/n_iT_i = P_fV_f/n_fT_f$ for an ideal gas because:
   (a) Ideal gases are non compressible.
   (b) The quantity PV/nT will always equal R.
   (c) The quantity PV/nT will always equal 1.
   (d) Some of the variables are guaranteed to cancel.

Answers: 1b; 2d; 3c; 4b; 5d; 6d; 7b; 8a; 9a; 10b

Describing the Gas Phase—*P*, *V*, *n*, and *T*

11.15 Why is it proper to think of the gas phase of matter as being more chaotic than either of the condensed phases?

11.16 What assumption is made for an ideal gas, and what gives us the right to make that assumption?

11.17 Consider a container of gas with the pressure inside the container the same as the room pressure outside the container. If a tiny hole is punched in the side of the container, will the gas leak out? Explain your answer.

11.18 How does a gas create pressure?

11.19 Why does liquid rise up a straw when you suck on the liquid through the straw?

11.20 Describe how a mercury barometer works.

11.21 What is the value of normal atmospheric pressure in millimeters of mercury? In atmospheres?

11.22 True or false? 1 atm = 76 mm Hg = 760 cm Hg. If false, fix it.

11.23 A weather forecaster reports the barometric pressure as 29.7 inches of mercury. [1 in. = 2.54 cm]
   (a) How many millimeters of mercury is this?
   (b) How many atmospheres is this?

11.24 The pressure in a tank of oxygen is 2000.5 lb/in.^{2} [760.00 mm Hg = 14.696 lb/in.^{2}].
   (a) How many millimeters of mercury is this?
   (b) How many atmospheres is this?

11.25 How do you convert from °C to K? Convert room temperature (22.0 °C) to Kelvin.

11.26 Regarding temperatures in kelvins:
   (a) Convert −100.5 °C to kelvins.
   (b) What is wrong with someone telling you the temperature of something is −100.5 K?
   (c) What is the coldest possible temperature in °C?

11.27 A balloon of methane gas, CH_4, has a temperature of −2.0 °C and contains 2.35 g of the gas. What is the temperature of the gas in Kelvin? How many moles of the gas does the balloon contain?

11.28 A tank of acetylene gas (C_2H_2) contains 48.5 lb of the gas and is at a pressure of 600.2 lb/in.^{2}. Express the pressure of the gas in atmospheres and the amount of gas in moles. [760.0 mm Hg = 14.696 lb/in.^{2}, 453.6 g = 1 lb]

Describing a Gas Mathematically—The Ideal Gas Law

11.29 What happens to the pressure of a gas when the volume of its container is increased? Explain with pictures and words.

11.30 What happens to the pressure of a gas when its temperature is decreased? Explain with pictures and words.

11.31 What happens to the pressure of a gas when the amount of gas inside the container is increased? Explain with pictures and words.

11.32 Consider the following diagrams representing different gas samples all at the same temperature:

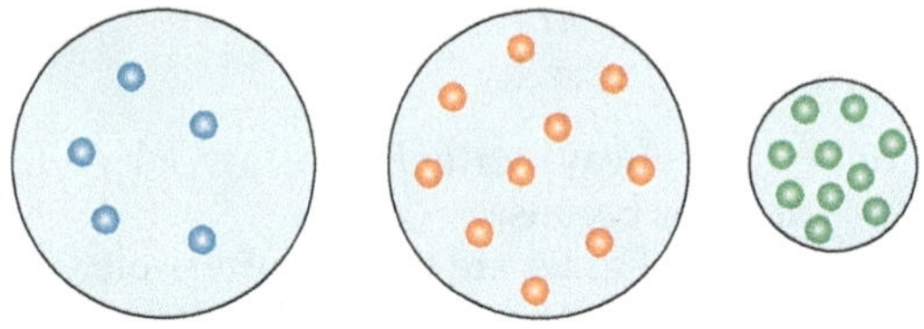

(1) (2) (3)

(a) Which gas is at the lowest pressure? Explain.
(b) Which gas is at the highest pressure? Explain.

11.33 Which of the two "gas" samples below is at the lower temperature? Explain, and also tell why we put the word ***gas*** in quotation marks.

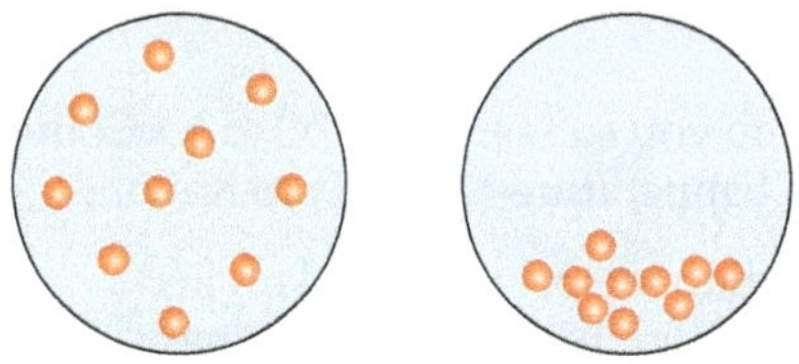

(1) (2)

11.34 State how the pressure of a gas depends on its volume at constant moles and temperature both with a mathematical relationship and an English statement. Also, make a graph that demonstrates the mathematical relationship.

11.35 State how the pressure of a gas at constant moles and volume depends on its temperature both with a mathematical relationship and an English statement. Also, make a graph that demonstrates the mathematical relationship.

11.36 State how the pressure of a gas at constant volume and temperature depends on the amount of gas present both with a mathematical relationship and an English statement. Also, make a graph that demonstrates the mathematical relationship.

11.37 What do we mean by inverse proportionality? By direct proportionality? Give an example of each using the way the pressure of a gas depends on something else.

11.38 Suppose the variable x is proportional to $1/y$. What does this tell you about how the numeric value of x changes as the numeric value of y changes?

11.39 "The older I get, the fewer hairs I have on my head." What kind of relationship (proportion or inverse proportion) exists between this gentleman's age and his hair? Explain your answer.

11.40 What kind of relationship (proportion or inverse proportion) exists between the strength of intermolecular forces and the distance between molecules? Explain your answer. (*Hint*: See Table 7.1.)

11.41 What are the units used in this book for the ideal gas constant R? Why are they important?

11.42 Rewrite the ideal gas law solving for V. Also show how all units cancel to leave you with just units of volume.

11.43 Rewrite the ideal gas law solving for T. Also show how all units cancel to leave you with just units of temperature.

11.44 Rewrite the ideal gas law solving for n. Also show how all units cancel to leave you with just units of moles.

11.45 According to the ideal gas law:
(a) If you measured P, V, n, and T for any gas sample and then calculated the quantity PV/nT, what would be the units and numerical value of the result?
(b) If you measured P, V, n, and T for any gas sample and then calculated PV/nRT, what would be the units and numerical value of the result?

11.46 According to the ideal gas law, what would happen to the pressure of a gas if you doubled the amount of gas in a container while also decreasing the volume of the container to one-half its initial volume? Explain.

11.47 According to the ideal gas law, what would happen to the pressure of a gas if you doubled the amount of gas in a container while also tripling the Kelvin temperature of the gas? Explain.

11.48 A student thinks he remembers reading that if you double the temperature of an ideal gas, its pressure doubles. He is given a problem where he has an ideal gas at 25.0 °C and 2.5 atm. He is asked what the temperature must be raised to in order to double the pressure to 5.0 atm. He answers, "50.0 °C, of course." Why is he wrong? What lesson should he learn about using the ideal gas law? What is the temperature increase in Celsius degrees that will double the pressure?

11.49 The gas inside a balloon is characterized by the following measurements: pressure = 745.5 mm Hg; volume = 250.0 mL; temperature = 25.5 °C. What is the number of moles of gas in the balloon?

11.50 A gas is in a container whose volume is variable. The container is in an ice bath at 0.00 °C, and there are 2.0 moles of gas in it. What must the volume in liters be if the gas has a pressure of 2.5 atm?

11.51 What must the Celsius temperature be if 2.0 moles of a gas in a 4.0-L steel container has a measured pressure of 100 atm?

11.52 According to the ideal gas law, what would be the volume of a gas at absolute zero (0 K)? Does the answer make sense? Describe what really happens to any gas as it is cooled toward this temperature.

11.53 An automobile tire at 22 °C with an internal volume of 20.0 L is filled with air to a total pressure of 30 psi (pounds per square inch). [1 atm = 14.696 lb/in.^{2}]
(a) What is the amount in moles of air in the tire?
(b) If the air were entirely nitrogen (N_2), how many grams of it would be in the tire? How many pounds of it would be in the tire? [453.6 g = 1 lb]

11.54 Why are the results that are calculated using the ideal gas law not exactly equal to the "true" results obtained by an experimental measurement?

11.55 Suppose you want to carry out the chemical reaction:

$H_2(g) + Cl_2(g) \rightarrow 2\,HCl(g)$

and you have 1 mole of Cl_2 gas.
(a) What volume in liters of H_2 gas would you need in order to have 1 mole of H_2, given that the H_2 pressure and temperature are 1.00 atm and 22.5 °C?
(b) What would the volume of the product be if it were collected at 1.00 atm and 22.5 °C?

Getting the Most from the Ideal Gas Law

11.56 Given some mass m of a known pure substance, what is the quickest way to determine the number of moles n you have of the substance?

11.57 A 7.24-g sample of gas is contained in a 4.00-L flask. Its pressure is 765.0 mm Hg, and its temperature is 25.0 °C. What is the molar mass of this gas?

11.58 A 1.56-g sample of gas is contained in a 250.0-mL cylinder. Its pressure is 1255.6 mm Hg, and its temperature is 22.7 °C.
(a) What is the molar mass of the gas?
(b) Combustion analysis reveals the empirical formula of this gas to be NO_2. What is the molecular formula?

11.59 A balloon is filled with H_2 gas to a volume of 1610.2 mL. The pressure of the gas in the balloon is 745.4 mm Hg, and the temperature is 22.7 °C. What is the mass in grams of the H_2 in the balloon?

11.60 In a room at 24.5 °C, a balloon is filled with helium gas until the pressure inside the balloon is 768.0 mm Hg. What is the density of the gas in grams per liter and in grams per milliliter?

11.61 A balloon filled with He gas and another balloon filled with H_2 gas have the same values for P and T.
(a) The density of the He gas is greater than the density of the H_2 gas. How can you prove this using the ideal gas law?
(b) How much more dense than the H_2 gas is the He gas?

11.62 Carbon dioxide and carbon monoxide are very different gases. For example, you exhale CO_2, but CO is extremely toxic. Suppose you have two balloons, one filled with 1.00 mole of CO and the other filled with 1 mole of CO_2. Both gases are at 1.00 atm and 25.0 °C.
(a) What is the volume in liters of each balloon?
(b) What did you learn about gases from doing this problem?

11.63 A tank of O_2 gas is stored in a cool basement (T = 10.0 °C). Its pressure gauge reads 50.5 lb/in.^{2}. Diver Dan then places the tank in his truck, where the sun warms it to 20.0 °C. What does the pressure gauge read at this elevated temperature?

11.64 In Problem 11.63, the temperature doubles but the pressure does not.
(a) Explain why the pressure does not double as well.
(b) To what temperature in °C would Diver Dan have to warm the tank in order to double the pressure?

11.65 On an atomic basis, and given that the average kinetic energy of motion of molecules in the gas phase is given by KE = $3RT/2$, explain why the pressure of a gas doubles when its temperature in Kelvin is doubled. (*Hint*: Pressure is due to collisions of molecules on the container walls and the force of those collisions.)

11.66 A tank initially at a pressure of 6.70 atm contains 20.0 moles of oxygen gas at 25.0 °C. Its temperature is then increased to 45.0 °C, and an additional 10.0 moles of gas are added. What is the final pressure in the tank?

11.67 The Kelvin temperature of a gas is doubled. At the same time, its volume is cut in half by compressing it with a piston. What happens to the pressure of the gas? (*Hint*: Set $T_f = 2\,T_i$ and $V_f = 0.5\,V_i$.)

11.68 Consider the reaction

$2\,H_2(g) + O_2(g) \rightarrow 2\,H_2O(g)$

If this reaction takes place at STP, what volume in liters of H_2 gas would be required to react completely with 22.4 L of O_2 gas? What volume in liters of $H_2O(g)$ would be produced? Explain how you arrived at your answers.

11.69 Consider the reaction

$2\,SO_2(g) + O_2(g) \rightarrow 2\,SO_3(g)$

(a) What volume in liters of oxygen gas at STP is required to produce 2 moles of sulfur trioxide gas?
(b) What volume in liters of oxygen gas at 25.0 °C and 1 atm pressure is required to produce 2 moles of sulfur trioxide gas?

11.70 To produce 500.0 g of gaseous H_2O, what volumes in liters of hydrogen gas and oxygen gas, both at STP, must combine via the reaction

$$2\,H_2(g) + O_2(g) \rightarrow 2\,H_2O(g)$$

11.71 Zinc metal reacts with hydrochloric acid to form hydrogen gas via the reaction

$$Zn(s) + 2\,HCl(aq) \rightarrow ZnCl_2(aq) + H_2(g)$$

How many grams of zinc would be required to produce enough hydrogen gas to fill a 50.0-L cylinder to a pressure of 10.0 atm at 25.0 °C?

Additional Problems

11.72 Consider two samples of water vapor, one at 101 °C and the other at 200 °C. Which behaves more ideally and why?

11.73 According to the ideal gas model, why does the chemical identity of a gas not matter when it comes to producing pressure?

11.74 A 5.00-L container is filled with 0.8004 g of helium gas at 107.0 °C. What is the pressure of the gas in atmospheres and in millimeters of mercury?

11.75 A chemical reaction produces 100.0 mL of hydrogen gas at 745.5 mm Hg and 24.0 °C. What is the amount of H_2 produced in moles and in grams?

11.76 What is the density in grams per liter of nitrogen gas at STP?

11.77 A gaseous compound of carbon and hydrogen contains 80.0% by mass carbon. A 2.00-L sample has a mass of 2.678 g at STP.
(a) Calculate the molar mass of the compound.
(b) What is the empirical formula of the compound?
(c) What is the molecular formula of the compound?

11.78 When it is heated, potassium chlorate decomposes to potassium chloride and oxygen:

$$2\,KClO_3(s) \rightarrow 2\,KCl(s) + 3\,O_2(g)$$

A sample of solid $KClO_3$ decomposes to produce 150.0 mL of oxygen at 22.0 °C and 780.5 mm Hg. How many grams of $KClO_3$ decomposed?

11.79 To what temperature must a gas initially at 20.0 °C be heated to double the volume and triple the pressure?

11.80 A sample of argon at 0.0 °C and 700.0 mm Hg occupies 15.0 mL. What is its volume at STP?

11.81 A sample of gas at 25.0 °C and 655 mm Hg has a density of 2.26×10^{-3} g/mL. What is the molar mass of the compound?

11.82 In the Haber process, nitrogen reacts with hydrogen to produce ammonia:

$$N_2(g) + 3\,H_2(g) \rightarrow 2\,NH_3(g)$$

(a) Suppose 2.0 L of N_2 gas at STP is combined with 6.0 L of H_2 gas, with the two gases being at the same temperature and pressure. Is this reaction being run in a balanced fashion or in a limiting fashion? Explain how you can tell without doing any calculations.
(b) If 50.0 L of N_2 gas at 200.0 lb/in.^{2} and 22.0 °C is combined with 100.0 L of H_2 gas at 240.0 lb/in.^{2} and 22.0 °C, what mass in grams of ammonia is produced? [14.70 lb/in.^{2} = 760.0 mm Hg]

11.83 How many atoms of nitrogen are there in a 230.0-mL sample of N_2O_4 gas that has a pressure of 745.0 mm Hg at 34.0 °C?

11.84 A sample of gas at 25 °C and 759.0 mm Hg has a volume of 1.58 L. If the temperature is raised to 35 °C but the pressure is held constant at 759.0 mm Hg, will the volume increase or decrease? Explain your answer.

11.85 Consider two identical 1-L containers, both at room temperature (300 K). One of them contains 1 mole of helium gas, and the other contains 1 mole of hydrogen gas. Is the pressure higher in the helium container, higher in the hydrogen container, or the same in the two containers?

11.86 Consider two identical 1-L containers, both at room temperature (300 K). One of them contains 1.0 g of helium gas, and the other contains 1.0 g of hydrogen gas. Is the pressure higher in the helium container, higher in the hydrogen container, or the same in the two containers?

11.87 What volume in milliliters of N_2O_4 gas at 0.996 atm and 25.0 °C contains 0.200 g of oxygen?

11.88 Carbon dioxide in a gas cylinder of unchangeable volume is at a pressure of 25.0 atm at 25 °C. When placed in the sunlight on a hot summer day, the temperature increases to 40 °C. What is the new pressure of the CO_2?

11.89 When a balloon is placed in a freezer, the size of the balloon decreases. Explain why.

11.90 A sample of Ne gas in a 10.0-L container at 25.0 °C exerts a pressure of 7.35 atm. What is the number of moles of Ne in the container? What is the mass in grams of the Ne?

11.91 Of the values for P, V, n, and T, which ones remain constant during the following changes?
(a) Air in a sealed glass bulb is heated from 25 °C to 32 °C.
(b) An He-filled balloon is cooled from 23 °C to 10 °C.
(c) The amount of CO_2 in a sealed steel cylinder is increased by adding 50.0 g of CO_2.

11.92 Under what two conditions are gas molecules unable to overcome the weak attractive forces that exist between them?

11.93 Arrange these gases, all at STP, in order of increasing density: CO_2, H_2, O_2, CH_4, He.

11.94 If a gas sample occupies 1.80 L at 250 °C and 792 mm Hg, what is the number of moles in the sample?

11.95 Which variable is not needed to describe the behavior of an ideal gas: volume, number of moles, temperature, molar mass, or pressure?

11.96 (a) How many liters of O_2 are there in 5.38 moles at STP?
(b) How many moles of NH_3 are there in 859 mL at STP?
(c) How many moles of H_2 are there in 0.518 L at STP?

11.97 Fill in the missing values:

	P	V	n	T
Gas A	757 mm Hg	952 mL	0.300	? °C
Gas B	1.20 atm	? L	5.00	27.0 °C
Gas C	800.0 mm Hg	1.20 L	?	298 K
Gas D	? atm	750 mL	0.0875	0.0 °C

11.98 Are gases denser or less dense than liquids and solids? Explain.

11.99 If a gas occupies 2.40 L at 1 atm and 22 °C, at what Celsius temperature will it occupy 7.20 L at 1 atm?

11.100 A chemistry student realizes she has forgotten the value of R and needs to determine it experimentally. She decides to measure the mass, volume, pressure, and temperature of a sample of carbon dioxide gas. She finds that 4.505 g of the gas occupies 2.50 L at 23 °C and 0.9960 atm. Calculate the value of R she determines from these data.

11.101 A gas at 25.0 °C occupies a volume of 5.00 gallons and exerts a pressure of 755 mm Hg. Express (a) the temperature of the gas in Kelvin, (b) the volume of the gas in liters, and (c) the pressure of the gas in atmospheres.

11.102 In the ideal gas law, pressure must be expressed in units of ______, volume must be expressed in units of ______, temperature must be expressed in ______, and n represents the number of ______.

11.103 A radial tire has an air pressure of 2.20 atm at 32 °F. After hours of high-speed driving, the temperature of the air in the tire reaches 80 °F. What is the pressure in the tire at this temperature? Assume that the volume of the tire does not change.

11.104 An 84.0-g sample of an unknown gas occupies 60.8 L at 23 °C and 1.2 atm. What is the molar mass of the gas?

11.105 True or false? In order to determine the volume occupied by 3.0 moles of an ideal gas at 28 °C and 1.4 atm, the identity of the gas must be known. If the statement is false, explain why.

11.106 What is the density of He gas in a cylinder if the pressure of the gas is 15.2 atm and its temperature is 22 °C?

11.107 In the laboratory preparation of oxygen gas, the following reaction is carried out:

$$2\,KClO_3(s) \rightarrow 2\,KCl(s) + 3\,O_2(g)$$

(a) How many moles of oxygen gas can be produced from the decomposition of 500.0 g of $KClO_3$?
(b) What volume in liters would the oxygen gas produced in part (a) occupy at 24.0 °C and 750.0 mm Hg?
(c) What volume in liters would this gas occupy at STP?

11.108 In an ideal gas, the molecules are considered to have no ______ forces.

11.109 (a) If the temperature of a gas is doubled while the pressure is kept constant, the volume of the gas ______.
(b) If the pressure of a gas is halved while the temperature is kept constant, the volume of the gas ______.

11.110 What is the Celsius temperature of a gas if 3.200 moles of it occupy 12.00 L at 8.500 atm?

11.111 A 1.25-g sample of a gas of unknown identity occupies 2.50 L and exerts a pressure of 0.400 atm at 0 °C.
(a) What is the molar mass of the gas?
(b) Combustion analysis determines that the empirical formula of the gas is CH_2. What is its molecular formula?

11.112 If a quantity of gas occupies 850 mL at 300 K and 750 mm Hg, what volume in milliliters will the gas occupy at 200 K and 1200 mm Hg?

11.113 Perform the following conversions:
(a) 30.2 in. Hg to millimeters of mercury
(b) 890.0 mm Hg to atmospheres
(c) 300.0 lb/in.^{2} to atmospheres

11.114 If 8.50 moles of He gas occupy a volume of 25.0 L at 28 °C, what pressure in atmospheres does the gas exert?

11.115 If 48.3 g of an unknown gas occupies 10.0 L at 40 °C and 3.10 atm, what is the molar mass of the gas?

11.116 A gas tank contains CO_2 at a pressure of 6.80 atm. What would the CO_2 pressure be if the container were (a) twice as large and (b) one-fourth as large?

11.117 What is the molar mass of a gas whose density is 1.52 g/L at 0 °C and 1 atm pressure?

11.118 Explain why gases always occupy the entire container they are in, but solids and liquids do not.

11.119 What happens to the volume of a gas in a cylinder with a movable piston if:
(a) The pressure is doubled while the temperature is held constant?
(b) The temperature is doubled while the pressure is held constant?
(c) The pressure and the temperature are both doubled?
(The movable piston means the volume of the cylinder, and therefore of the gas, can change.)

11.120 Consider the following two plots:

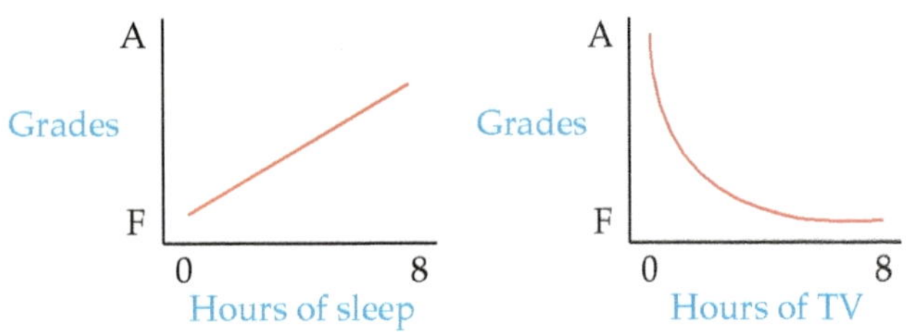

(a) Which of the two plots represents an inverse relationship? Express the inverse relationship in an English statement.
(b) A student looks at the plot and declares that grades are proportional to 1/hours of TV. Is he correct? Explain why.
(c) A student looks at the plot and declares that grades are proportional to 1/hours of sleep. Is this correct? Explain why.

11.121 Something is done to this tank of gas to make the pressure change as shown.

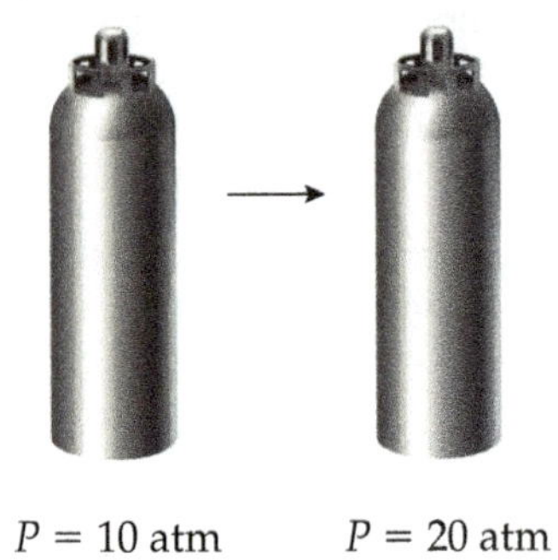

Give at least two different ways this could have been accomplished.

11.122 Consider two balloons, both in the same room and both exactly the same volume. Consider also the balanced reaction between these gases.

$$2H_2 + O_2 \longrightarrow 2H_2O$$

If we allow the gases from these balloons to mix and react, will there be a limiting reactant? If so, which is it, and how did you know?

11.123 If we continuously compress a real gas in a cylinder with a movable piston while keeping the container at a constant temperature in an ice bath:
(a) What will eventually happen to the gas and why?
(b) How well will the ideal gas law apply to the real gas shown when it is close to doing what you said it will do in part (a)?

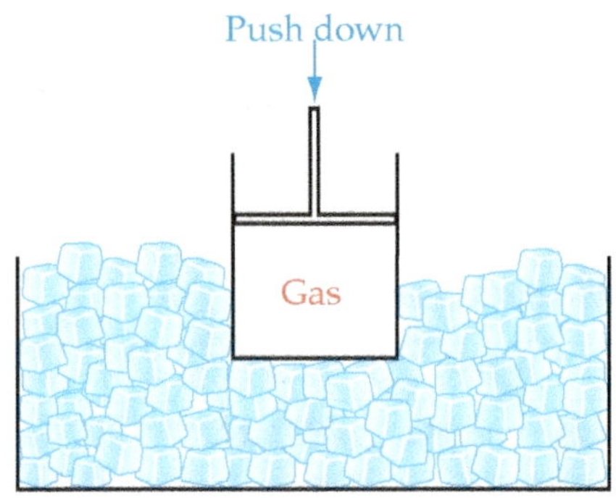

11.124 You are trapped on a desert island without any reference books, and you need the value of the ideal gas constant R. Luckily, you stumble upon the following tank of helium (see below). Use what you found to calculate the value of R. Make sure your units are L · atm/K · mol. Oh, by the way, while you forgot the value of R, you do remember that helium has a molar mass of 4.003 g/mol and also how to convert to Kelvin.

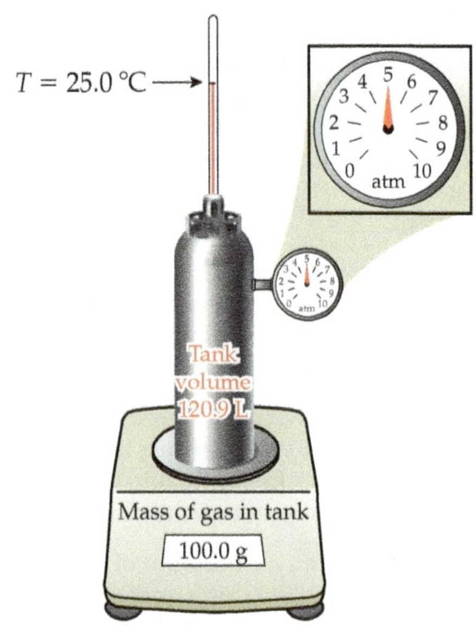

One Step Beyond

11.125 Real gases are not ideal gases. Explain why this is so in terms of intermolecular forces, and then describe (and justify) what conditions with regard to temperature and pressure will bring a real gas close to behaving like an ideal gas.

11.126 Consider three identical-sized cylinders, all at the same temperature. Cylinder A contains 2 moles of nitrogen (N_2) gas. Cylinder B contains 1 mole of nitrogen gas and 1 mole of methane (CH_4) gas. Cylinder C contains 2 moles of methane gas. Which cylinder has the higher pressure?

11.127 Suppose the pressure in cylinder A in the previous problem was 2.50 atm. What do you think the pressure due to methane is in cylinder B? Justify your answer with some sort of explanation. Then Google "Dalton's Law" and explain how it relates to this problem.

11.128 Normal atmospheric pressure will push a column of mercury up an evacuated glass tube (a barometer) to a height of 76 cm (760 mm Hg), which we call one atmosphere. Suppose water were used as the liquid in a barometer instead of mercury. How high up would the atmosphere push a column of water in centimeters and in feet? Also, why would mercury give a more accurate indication of the atmospheric pressure then water? Some data you may need: Density of mercury, 13.6 g/mL; Density of water, 1.00 g/mL; Boiling point of mercury, 357 °C; Boiling point of water, 100 °C.

11.129 Normal atmospheric pressure is 1 atm = 760 mm Hg. However, it is also 14.7 lbs/in.^{2} (14.7 pounds per square inch). Given that liquid mercury has a density of 13.6 g/mL, explain (show a calculation) why 1 atm also equals 14.7 lbs/in.^{2}. Then, go back to the introduction of this chapter and calculate that the atmospheric force holding the Magdeburg spheres together is indeed close to the stated 4.5 tons (the surface area of a sphere is given by $4\pi r^2$).

11.130 You are on the balcony of the upper floor of a high-rise building in Chicago. Way down below on the street, your friend is engaged in conversation with someone and is drinking a particularly flavorful mixed drink. You decide to take a sip from his glass without his knowing, so you collect together every drinking straw in the apartment and start taping them together to create a very long drinking straw. If the atmospheric pressure is currently 760 mm Hg, what is the longest straw you can use to get a sip of your friend's drink? (Assume the density of the drink is the same as pure water, 1.0 g/mL.)

11.131 Consider a container that contains 1.00 mole of $CO_2(g)$ at 298 K.

(a) What does the ideal gas law predict the pressure to be in atm?
(b) What does the van der Waals equation predict the pressure to be?
(c) What is the percent difference of the van der Waals pressure from the ideal pressure?
(d) Suppose you increased the temperature to 1000 K. Would you expect the percent difference to increase or decrease compared to your answer in (c)? Explain.

Group Exercises (Work on these exercises together with someone else in the class.)

1. Design and then draw a device that allows you to increase the temperature of a gas without increasing its pressure.
2. List everything you need to know about a gas to determine its molar mass without having knowledge of the molecular formula. Then develop a set of instructions that tells someone how to calculate the molar mass.

WORKPATCH SOLUTIONS

11.1 In the condensed phases, the molecules are very close to one another, and as a result, the intermolecular forces (no matter what type) are relatively strong and thus can't be ignored. Remember, attractive forces between molecules are stronger when the molecules are closer together.

11.2 The pressure in the smaller box is greater because, in any given time interval, there are more particle– wall collisions in the smaller box than in the larger box.

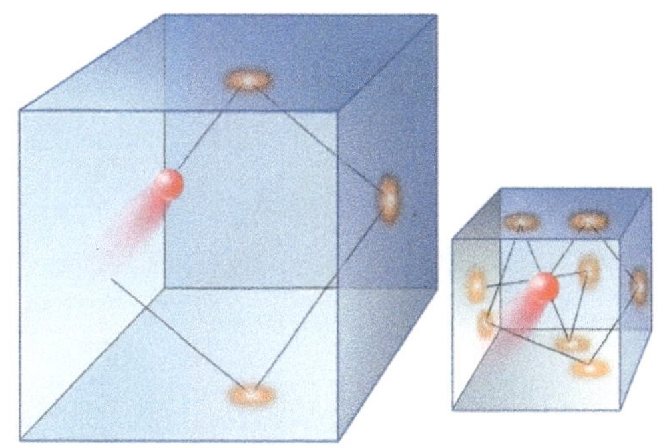

11.3

$$\frac{(100.0\ \text{atm})(100.0\ \text{L})}{(12.992 \times 10^3\ \text{g}\ O_2)\left(\dfrac{1\ \text{mol}\ O_2}{31.998\ \text{g}\ O_2}\right)(300.0\ \text{K})}$$

$$= \underbrace{0.08210\ \frac{\text{L} \cdot \text{atm}}{\text{K} \cdot \text{mol}}}_{\text{This equals } R!}$$

Glossary

absolute temperature scale Another term for Kelvin temperature scale.

accuracy The degree to which a measured value is close to the true value. Compare precision.

acid A substance that increases the concentration of H^+ ions in an aqueous solution.

acid–base neutralization reaction The reaction between an H_3O^+ ion supplied by an acid and an OH^- ion supplied by a base. A hydronium ion and a hydroxide ion react with each other to give two molecules of water.

acidic solution An aqueous solution in which the concentration of hydronium ions (H_3O^+) is less than 10^{-7} M, giving the solution a pH value below 7. Compare basic solution, neutral solution.

actinides The elements with atomic numbers 90 through 103.

activation energy (E_a) The minimum amount of energy, specific to each chemical reaction, that the reactants must absorb from their surroundings in order for the reaction to occur.

active metal In any pair of metals, the one that more easily gives up its electrons.

active pharmaceutical ingredient Abbreviated API, is a pharmaceutically active compound or drug.

active site A precisely shaped location (on an enzyme molecule, for example) at which substrate molecules having a complementary shape may bind.

activity series See electromotive force series.

actual yield The experimental amount of product obtained in a chemical reaction. Compare theoretical yield.

air pressure The force exerted on a surface by air molecules colliding with the surface.

alcohol A functionalized hydrocarbon containing one or more OH functional groups. General formula: R—OH.

aldehyde A functionalized hydrocarbon containing a carbonyl (C—O) bonded to at least one hydrogen. General formula: R—CHO.

algebraic manipulation Rearrangement of an equation in order to solve for a desired quantity.

aliphatic hydrocarbon An early general name for long-chain hydrocarbons, stemming from the Greek aleiphas meaning "fat"; applied to these compounds because many fats and oils contain long-chain hydrocarbon molecules. A general term for all alkanes, alkenes, and alkynes.

alkali metal A general name for any element in group IA (except H).

alkaline earth metal A general name for any element in group IIA.

alkane A hydrocarbon containing only carbon–carbon single bonds; a saturated hydrocarbon. General formula: C_nH_{2n+2}.

alkene A hydrocarbon containing at least one carbon–carbon double bond; an unsaturated hydrocarbon. General formula: C_nH_{2n}.

alkyne A hydrocarbon containing at least one carbon–carbon triple bond; an unsaturated hydrocarbon. General formula: C_nH_{2n-2}.

allotrope One of a set of different structural forms of an element, such as the graphite, fullerene, and diamond forms of carbon.

alloy A solid solution in which both solute (or solutes in some cases) and solvent are metals.

α-amino acid Any amino acid whose amine functional group is separated from the acid (carboxyl) functional group by only a single carbon atom.

alpha decay A nuclear reaction in which a radioactive nucleus ejects an alpha particle (^4_2He) and thereby becomes a daughter isotope containing two fewer protons and two fewer neutrons.

alpha (α) particle A helium nucleus (^4_2He) ejected from some radioactive nuclei during a process known as alpha decay (or alpha emission).

amide A functionalized hydrocarbon that contains a derivative of the carboxyl group (COOH) in which the —OH has been replaced by an amine. General formulas: primary, R—$CONH_2$; secondary, R—CONHR; tertiary, R—$CONR_2$.

amide bond The covalent bond formed between the carbon atom of the carbonyl and the nitrogen atom in an amide.

amine An organic derivative of ammonia (NH_3) in which one, two, or all three of its hydrogen atoms are replaced by a hydrocarbon chain. See primary amine; secondary amine; tertiary amine.

amino acid An organic molecule that contains both a carboxylic acid group and an amine group.

amino group See amine.

amorphous solid A solid in which there is no overall order among the particles that make it up.

angstrom (Å) A non-SI unit of length, equal to 1×10^{-10} meter. Often used to express bond lengths in molecules.

anion An ion that has a negative charge. Compare cation.

anode The negative electrode in a galvanic cell; the electrode at which oxidation takes place. Compare cathode.

antimatter The oppositely charged version of matter. For example, a positron is the antimatter version of an electron.

aqueous solution A solution in which the solvent is water.

Arrhenius definition of acid The earliest identification of acids as a particular class of compounds, with "acid" defined as any electrolyte that ionizes to give H^+ ions when dissolved in water.

Arrhenius definition of base The earliest identification of bases as a particular class of compounds, with "base" defined as any electrolyte that ionizes to give OH^- ions when dissolved in water.

atmosphere The collection of gases that surround Earth. Also, a non-SI unit of pressure; 1 atm = 760 mm Hg = 29.9 in Hg at sea level.

atmospheric pressure See air pressure.

atom The smallest unit of an element that has the properties of that element. An atom consists of a central nucleus plus one or more electrons outside the nucleus; the number of electrons is equal to the number of protons, making an atom electrically neutral. Compare ion.

atomic mass A unitless number expressing the mass of a particular atom relative to the mass of an atom of carbon-12, which is arbitrarily assigned a mass of exactly 12.

atomic mass unit (amu) Exactly one-twelfth the mass of a $^{12}_{6}C$ atom and equal to $1.660\,54 \times 10^{-24}$ g; also known as a dalton.

atomic number (Z) The number of protons in the nucleus of an atom. All atoms of a particular element have the same atomic number, which is indicated by a subscript to the left of the element symbol.

atomic radius The distance from the center of the nucleus to the outermost electron shell.

autodissociation The process by which a small number of the water molecules in any given volume of water dissociate into hydronium and hydroxide ions. See K_w.

autoionization A synonym for autodissociation.

Avogadro's number 6.02×10^{23}, the number of units in 1 mole.

balanced equation A chemical equation written such that matter is conserved, which means the total number of atoms of each element on the right side of the equation is exactly equal to the number of atoms of those elements on the left side.

balancing coefficient A number placed in front of a chemical formula in a chemical equation to indicate the number of moles (or molecules) of the substance that take part in the chemical reaction represented by the equation.

band of stability The area on a band of stability plot representing all existing nuclei, both nonradioactive isotopes and all radioactive isotopes having a measurable half-life.

band of stability plot A plot of number of neutrons in a nucleus (represented on the vertical axis of the graph) as a function of number of protons in the nucleus (represented on the horizontal axis), with some indication as to whether an atom exists with that combination.

barometer A device used to measure atmospheric pressure.

barometric pressure Another term for air pressure or atmospheric pressure.

base A substance that increases the concentration of OH^- ions in an aqueous solution.

base unit Any one of seven fundamental SI units: meter, kilogram, second, kelvin, mole, ampere, candela. Compare derived unit.

basic solution An aqueous solution in which the concentration of hydroxide ions (OH^-) is greater than 10^{-7} M, giving the solution a pH value above 7. Compare acidic solution; neutral solution.

battery Common term for galvanic cell.

beta decay Another term for beta emission.

beta (β^-) emission A nuclear reaction in which a neutron changes to a proton plus a beta particle (electron); the proton remains in the nucleus, and the beta particle is ejected from the atom.

beta (β^-) particle An electron created during a nuclear reaction when a neutron changes to a proton; the particle is ejected from the nucleus.

bias A strong preference or inclination that inhibits impartial judgment.

bimolecular Involving two molecules.

binary compound A compound containing atoms of only two different elements. The compound may contain more than two atoms, but only two different elements.

binary covalent compound A covalent compound containing atoms of only two elements, both usually nonmetals.

binary ionic compound An ionic compound containing only one type of cation and one type of anion. The compound may contain more than two ions, but all the cations (usually a metal) are identical and all the anions (usually a nonmetal) are identical.

binding energy The energy released when a nucleus is created from protons and neutrons. Also, the energy required to break up a nucleus into its constituent protons and neutrons.

binding energy per nucleon The binding energy divided by the number of nucleons (protons and neutrons) in the nucleus.

biopolymer A polymer made by and used in living organisms.

Bohr diagram A diagram showing the electron shells of an atom as a series of concentric circles centered on the nucleus.

Bohr model of the atom A model proposed by Niels Bohr, in which the atom is pictured as a central nucleus surrounded by orbiting electrons. Because the energy of the electrons is quantized, the electrons can orbit the nucleus only at certain allowed distances. Compare Rutherford model; Thomson model.

boiling point (bp) The temperature at which a pure substance changes from a liquid to a gas at standard atmospheric pressure. Because condensation is the reverse of evaporation (boiling), the boiling point is numerically equal to the condensation point.

bomb A device that can be used for the combustion of a substance so as to determine its calorie content.

bond See chemical bond.

bonding pair Two electrons forming a covalent bond between two atoms, usually with one electron of the pair coming from one atom and the other electron coming from the other atom.

branched hydrocarbon A hydrocarbon in which at least one carbon atom is bonded to more than two other carbon atoms. Compare linear hydrocarbon.

break-even point For a nuclear reaction, the point at which the amount of energy recovered from the reaction equals the amount of energy input to start the reaction.

Brønsted–Lowry definition of acid A substance that is a proton (H^+ ion) donor in aqueous solution.

Brønsted–Lowry definition of base A substance capable of accepting a proton (H^+ ion) from a proton donor.

buffer A shorthand name for buffered solution.

buffered solution A solution that is able to maintain an essentially constant pH when acids or bases are added to it. The solution is made up of either a weak acid and a salt of its conjugate base or a weak base and a salt of its conjugate acid.

buffering ability The ability of a solution to resist changes to its pH.

c The symbol representing the speed of light, 3.00×10^8 meters per second $=$ 186,000 miles per second.

calorie A non-SI unit of energy; 1 cal $=$ 4.184 J. The amount of heat necessary to raise 1 g of water 1 °C.

Calorie (Note capital C.) The energy unit used to measure the amount of energy contained in food; equal to 1000 calories.

calorimeter An insulated vessel that measures the amount of heat released when a substance is burned or a chemical reaction takes place.

carbocation A cation containing a C^+ ion.

carbohydrate A general term for molecules made up of carbon, hydrogen, and oxygen, with many of the hydrogen and oxygen atoms forming alcohol (—OH) groups.

carbonyl A functional group consisting of a carbon atom and an oxygen atom joined by a double bond: $>C=O$.

carboxyl A functional group consisting of a carbon atom bonded to an —OH group by a single bond and to an oxygen atom by a double bond: —COOH.

carboxylic acid A functionalized hydrocarbon containing one or more carboxyl groups. General formula: R—COOH.

catalyst A substance that, by lowering the activation energy of a chemical reaction, changes the rate at which the reaction proceeds but is not itself consumed in the reaction.

catenation The ability of an atom to link together with like atoms.

cathode The positive electrode in a galvanic cell; the electrode at which reduction takes place. Compare anode.

cation An ion with a positive charge. Compare anion.

Celsius temperature scale A metric (but non-SI) temperature scale used for everyday purposes in all countries except the United States and often used for scientific work in the United States. On this scale, the freezing point of water is set at 0 °C and the boiling point of water is set at 100 °C. The unit of temperature on this scale is the degree Celsius.

centi- The SI prefix meaning one-hundredth (10^{-2}); uses the symbol c.

chain length The number of carbon atoms in the main chain of an organic molecule.

chain reaction (nuclear) A sequence of nuclear reactions in which the first reaction produces sufficient energy and daughter particles to enable multiple repeat reactions in the next step, and so on.

chalcogen A general name for any group VIA element.

chemical bond Any attractive electrostatic force between two atoms that binds them together. Ionic bonds, covalent bonds, and metallic bonds are all types of chemical bonds.

chemical equation A notation for showing what takes place in a chemical reaction. The chemical formulas of all reactants are written on the left side of the arrow, and the chemical formulas of the products are written on the right side.

chemical equilibrium The stage in a chemical reaction at which the rate of the forward reaction equals the rate of the reverse reaction.

chemical formula A combination of element symbols and numbers that gives the exact composition and makeup of a pure substance.

chemical kinetics The study of the rates at which chemical reactions take place.

chemical periodicity The fact that, if you arrange the elements in order of increasing atomic number, elements spaced at periodic (regular) intervals have similar chemical properties. For the first 18 elements, an element's properties are similar to those of the elements eight positions ahead of it and eight positions behind it. Beyond element 18, the interval (periodicity) increases to 18 or 32 elements. This periodicity is the basis for the periodic table.

chemical property A property of a substance that can be studied only by having the substance undergo a chemical change; a property that determines the chemical reactions the substance can undergo. Compare physical property.

chemical reaction A process in which one or more reactants are transformed into one or more products through the breaking of chemical bonds in the reactant molecules and the formation of new chemical bonds in the product molecules.

chemical synthesis The work of planning, performing, and perfecting the preparation of a chemical compound.

chemical transformation Another term for chemical reaction.

chemistry The branch of science that studies matter and the changes it undergoes.

classical physics Physics as described by Newton, and as it was understood until the end of the nineteenth century, before the discovery that energy can be quantized. The laws of classical physics work well when applied to everyday objects because at that level the fact that energy can be quantized can be ignored, but the laws fail when applied to objects that are the size of atoms and electrons. Compare quantum physics.

cleaving The breaking of a solid along a perfectly straight plane.

colligative property A property that depends on the number of solute particles in a solution but not on their identity.

combination reaction A reaction in which two or more substances combine to give another substance. Compare decomposition reaction.

combustion A synonym for burning. In any combustion reaction, the substance being burned combines with oxygen from the air.

combustion analysis A procedure for determining the chemical formula of an unknown substance. The substance is burned in a closed container, and the products of the combustion are then isolated and their masses are determined. The masses are then used to calculate the empirical formula of the unknown substance.

combustion reaction The reaction of any substance with oxygen (O_2).

common name A traditional name for a substance, usually one that gives no information about its composition.

complete ionic equation An equation that shows all the ions, both those that take part in the reaction and the spectator ions.

compound A pure substance composed of two or more atoms of different elements. Compare elemental substance.

concentration A measure of the amount of solute dissolved in a given amount of solvent or solution.

condensation A change from the gas phase to the liquid phase. Compare vaporization.

condensed phase Either the liquid phase or the solid phase of matter.

conjugate acid For a given base, the substance that contains all the atoms of the base plus one additional H^+ per molecule. The part of a basic molecule that exists after chemical reaction with an acid. Example: base CH_3COO^-; conjugate acid CH_3COOH. Compare conjugate base.

conjugate base For a given acid, the substance that contains all the atoms of the acid minus one H^+ per molecule. The part of an acid molecule that exists after chemical reaction with a base. Example: acid HF; conjugate base F^-. Compare conjugate acid.

conjugate pair An acid and its conjugate base, or a base and its conjugate acid.

contact process An industrial process for producing sulfuric acid from elemental sulfur and oxygen gas.

continuous spectrum An electromagnetic spectrum in which the wavelengths of visible light blend smoothly from one color to the next. When you pass sunlight through a prism, the rainbow you see is a continuous spectrum. Compare line spectrum.

conversion factor A ratio used in unit analysis to change given units to the units wanted in the answer.

core electron An electron in an atom that occupies any shell other than the valence shell.

corrosion The oxidation of metal at a very slow pace, with the oxidizing agent being O_2 from the air. Rust (iron oxide) is the most common evidence of this type of reaction.

Coulomb's law A statement of the fact that the attractive force between two objects of opposite electric charge is directly proportional to the magnitude of the charges and inversely proportional to the distance between them.

covalent bond An attractive force between two atoms in a molecule, resulting from the sharing of valence electrons. Compare ionic bond; metallic bond.

covalent compound A compound containing all covalent or polar covalent bonds.

critical mass In a nuclear fission reaction, the amount of radioactive material needed to sustain a chain reaction.

crystal lattice The regular, three-dimensional, geometric arrangement of atoms in a crystalline solid.

crystalline solid A solid in which the particles that make it up, be they atoms, ions, or molecules, are arranged in a highly ordered fashion.

cubic meter The SI unit of volume.

D The element symbol for deuterium.

Δ The Greek capital letter delta; used to represent either a change in a quantity or a difference between two values.

ΔE The symbol used to represent a change in the energy of a system.

ΔE_{rxn} The symbol representing the net energy change for a chemical reaction, defined to be the difference between the energy of the products and the energy of the reactants. This amount of energy is released during an exothermic reaction and is absorbed during an endothermic reaction. Exothermic reactions have a negative value of ΔE_{rxn}, and endothermic reactions have a positive value of ΔE_{rxn}.

$\delta +, \delta -$ The Greek lowercase letter delta, used with a plus or minus sign to indicate fractional or partial charge.

Dalton's atomic theory The theory of matter formulated by John Dalton: (1) All matter is made of atoms; (2) atoms can be neither created nor destroyed; (3) all atoms of a given element are identical; (4) atoms of different elements are different from one another; (5) all chemical reactions involve the combining or separating of atoms in the reactant compounds.

daughter isotope An isotope created in a nuclear reaction.

daughter nucleus Another name for daughter isotope.

deci- The SI prefix meaning one-tenth (10^{-1}); uses the symbol d.

decomposition reaction A reaction in which a compound breaks down (decomposes) into two or more simpler compounds or substances. Compare combination reaction.

density A measure of the compactness of matter. For any substance, it is the amount of mass per unit volume; D = Mass/Volume.

deoxyribonucleic acid (DNA) The biopolymer present in every living cell that carries the hereditary information the cell uses to synthesize all the protein molecules it needs to function and to reproduce itself.

derived SI unit Any SI unit of measurement obtained from a combination of two or more SI base units.

deuterium (D) The isotope of hydrogen containing one proton and one neutron; ^2_1H.

diamond The allotrope of carbon in which every carbon atom is bonded to four other carbon atoms, all arranged in a three-dimensional crystal lattice. Compare fullerene; graphite.

diatomic Consisting of two atoms. Compare monatomic; polyatomic.

dilution equation An equation for solving a problem that involves diluting a stock solution to a desired concentration: $M_{\text{stock solution}} \times V_{\text{stock solution}} = M_{\text{diluted solution}} \times V_{\text{diluted solution}}$.

dipeptide A small protein molecule containing only two amino acids joined by a peptide linkage. Compare polypeptide; protein.

dipolar force A synonym for dipole–dipole force.

dipole–dipole force The intermediate-strength intermolecular attractive force exerted by polar molecules on each other. Compare hydrogen bond; London force.

dipole moment An experimental measure of the polarity of a molecule. A measure of both the strength of the dipole charges and the distance between them.

dipole vector A synonym for dipole moment; a form of notation used to represent the size and direction of the dipole moment; $\mapsto$.

diprotic acid An acid that, when it dissociates, can yield a maximum of two H_3O^+ ions per molecule of acid.

disaccharide A sugar molecule consisting of two monosaccharide (simple sugar) units joined together.

dispersion force Another name for London force.

dissociation The process by which either the molecules of a covalent compound or the cations and anions bound in the lattice of an ionic compound break up into individual cations and anions.

DNA The shortened name for deoxyribonucleic acid.

dot diagram A notation used to show valence electrons and bonding in covalent compounds.

double bond A multiple covalent bond between two atoms that is formed by two bonding electron pairs.

double helix The structural arrangement of deoxyribonucleic acid (DNA), which consists of two single strands of DNA polymer that wind around each other and are held together by hydrogen bonds.

double-replacement reaction A reaction in which two reactants AB and CD react to form products AD and CB.

dynamic equilibrium Another name for chemical equilibrium.

E_a The symbol for the activation energy of a chemical reaction.

effective collision In a chemical reaction, a collision between reactant molecules that leads to the formation of a product molecule.

electrical charge The property of any particle that determines how it behaves in an electric (or magnetic) field. Particles carrying like electric charges repel each other, those carrying opposite electric charges attract each other, and those carrying no electric charge are neutral.

electricity The flow of electrons through a conducting medium, such as a wire.

electrochemical cell See galvanic cell.

electrode The anode or cathode of a galvanic cell.

electrolyte A water-soluble substance that dissociates into anions and cations when dissolved in water, creating a solution that can conduct an electric current. Compare nonelectrolyte.

electromagnetic radiation All wavelengths of radiant energy.

electromagnetic spectrum The continuum comprising all wavelengths of electromagnetic radiation, from low-energy forms (radio waves and microwaves) to high-energy forms (X-rays and gamma rays), with visible light midway between these extremes.

electromotive force (EMF) series A ranking scheme indicating the relative (oxidizing) activities (tendency to give up valence electrons) of metals.

electron A subatomic particle that has a mass 1/1836 that of a hydrogen atom, carries an electric charge of −1, and moves about at high velocity around the nucleus of an atom.

electron capture A nuclear reaction in which a proton in a nucleus changes to a neutron by absorbing (capturing) an electron.

electron configuration The way the electrons of an atom are distributed among the atom's shells and subshells.

electron configuration notation A way of indicating the distribution of electrons in an atom, consisting of the principal quantum number (indicating the main shell) followed by the subshell letter followed by a superscript indicating the number of electrons in that subshell. For example, the electron configuration of lithium is $1s^22s^1$.

electron-deficient atom Also called octet-deficient atom, is generally a period 2 atom in a stable molecule that possesses less than an octet of electrons.

electron group Some number of electrons that share the same region of space in a molecule.

electron pair With respect to an electron dot diagram, two valence electrons in close proximity to one another.

electron shell See shell.

electron-transfer reaction A type of reaction that can be used to push electrons through a wire and thereby produce electricity.

electron tunneling The ability of electrons to pass through barriers that supposedly should stop them.

electronegativity (EN) An atom's ability to attract shared electrons in a chemical bond to itself.

electrostatic force Forces of attraction between unlike charges and repulsion between like charges.

element A pure substance that cannot be broken down into simpler substances. Atoms of an element get their identity from the unique number of protons in the nucleus.

elemental substance A pure substance composed of atoms of a single element. Compare compound.

elementary step In a multistep chemical reaction, any one of the steps in the mechanism that occurs exactly as written in the equation.

EMF series A shortened name for electromotive force series.

empirical formula A chemical formula in which the numeric subscripts indicate the ratio of elements in a substance expressed as the lowest possible numbers. One molecule of the substance may contain the number of atoms shown in the empirical formula or a multiple of those numbers. Compare molecular formula.

endothermic reaction A chemical reaction that absorbs heat from its surroundings; ΔE_{rxn} is positive. Compare exothermic reaction.

energy (E) A measure of the capacity to do work.

energy barrier A graphical representation of the activation energy of a chemical reaction.

energy density The amount of energy in mega joules (MJ) that can be released per kilogram of reactant for an electricity-producing chemical (redox) reaction.

energy-level diagram A diagram that shows the energies of the shells and subshells that electrons can occupy in an atom.

energy-neutral reaction A reaction that neither absorbs energy from the surroundings nor releases energy to the surroundings.

entropy The entropy of a system is a function of the probability for it. Increased entropy can be equated with a more probable arrangement of a system.

enzyme A class of large protein molecules that act as catalysts in biochemical reactions.

equilibrium A shorthand name for dynamic equilibrium.

equilibrium constant (K_{eq}) A number that expresses the position of equilibrium for a particular chemical reaction at a particular temperature. See also K_a; K_b; K_w.

essential amino acids The ten amino acids that are required in the human diet because they cannot be synthesized in the body.

ester A functionalized hydrocarbon containing a carboxyl group in which the hydrogen atom of the —OH group has been replaced by another hydrocarbon group. General formula: R—COOR.

ether A functionalized hydrocarbon consisting of a central oxygen atom attached to two hydrocarbon groups. General formula: R—O—R.

evaporation A change from the liquid state to the gas state, usually occurring below the boiling point.

exact number A number that has no uncertainty. Compare measured number.

excess reactant A reactant that is present in excess relative to the limiting reactant. When the reaction is complete, the limiting reactant will be completely used up, replaced by product and unused excess reactant(s).

excited state For an atom, a state in which one or more electrons have absorbed energy and jumped from their ground state position to a higher-energy shell or subshell, leaving behind an electron vacancy. Compare ground state.

exothermic reaction A chemical reaction that, as it proceeds, releases energy into the surroundings; ΔE_{rxn} is negative. Compare endothermic reaction.

expanded octet atom An atom of period 3 or beyond that can possess more than eight electrons in its valence shell.

experiment A procedure carried out to study a phenomenon or to test a theory.

experimental rate law The rate law found by doing kinetics experiments. Example: Rate $= k[A_2]^2$.

extensive property Any property of a substance that depends on the amount present. Compare intensive property.

Fahrenheit temperature scale The (non-SI) temperature scale used for nonscientific work in the United States, in which the freezing point of water is set at 32 °F and the boiling point of water is set at 212 °F. The unit of temperature on this scale is the degree Fahrenheit.

fatty acid Long carbon-chain carboxylic acids, so called because they can be derived from fats.

first ionization energy The amount of energy needed to remove an electron from the outermost shell of a neutral (uncharged) atom.

fission A shorthand name for nuclear fission.

formula Another term for chemical formula.

forward reaction In a reversible chemical equation, the reaction reading from left to right (by convention).

freezing The process in which a substance changes from the liquid state to the solid state.

freezing point The temperature at which a substance changes from a liquid to a solid at standard atmospheric pressure. Because melting is the reverse of freezing, the freezing point is numerically equal to the melting point.

fuel cell A galvanic cell in which the chemical reaction is the direct combination of oxygen and hydrogen to form water and create electricity, with both the water and the electricity being usable products.

fullerene An allotrope of carbon in which every carbon atom is bonded to three other carbon atoms, creating a series of hexagons and pentagons forming a sphere. Compare diamond; graphite.

functional group A commonly occurring atom or group of atoms (other than carbon or hydrogen), attached to a hydrocarbon. The properties of the functionalized hydrocarbon are different from the properties of the parent hydrocarbon.

functionalized hydrocarbon A hydrocarbon to which one or more functional groups have been added.

fused ring A cyclic hydrocarbon consisting of two or more rings of carbon atoms that share one or more carbon atoms.

fusion A shorthand name for nuclear fusion.

galvanic cell The name used by scientists for what is commonly called a battery, consisting of two compartments connected by a salt bridge. One compartment contains a cathode (where reduction takes place) and the other an anode (where oxidation takes place). The oxidation–reduction reactions create electricity.

gamma decay Another name for gamma emission.

gamma emission A nuclear reaction in which a nucleus emits energy in the form of gamma radiation; no particles are emitted, so the identity of the decaying nucleus is not changed.

gamma (γ) ray The form of very high–energy electromagnetic radiation most frequently emitted by radioactive nuclei during decay.

gas A state of matter in which fast-moving atoms or molecules behave relatively independently of one another, only interacting upon collision. A gas expands or contracts as necessary to assume the shape and volume of any vessel used to contain it. Compare liquid; solid.

gas-phase reaction A chemical reaction in which all reactants are gases.

giga- The SI prefix meaning 1 billion (10^9); uses the symbol G.

graphite An allotrope of carbon in which every carbon atom is bonded to three other carbon atoms, creating a series of hexagons arranged in two-dimensional sheets stacked one on top of another. Compare diamond; fullerene.

ground state For an atom, the state in which all of its electrons are arranged so as to have the lowest possible total energy. The arrangement in which each electron occupies the lowest-energy subshell available. Compare excited state.

group A column of the periodic table. All the elements in each column have similar valence electron configurations.

h The symbol representing Planck's constant, 6.626×10^{-34} J · s.

half-life The time needed for one-half of the radioactive nuclei in a sample to decay.

halide rule A statement of the fact that any halogen atom in a covalent compound most typically has an oxidation state of −1.

halogen A general name for any group VIIA element.

heavy hydrogen Another name for deuterium.

heavy water Water in which each molecule contains two deuterium ^{2}H atoms instead of two hydrogen ^{1}H atoms; D_2O.

Henry's law A statement of the fact that the solubility of a gas in a liquid increases with increasing pressure.

heteroatom In the language of organic chemistry, any atom other than a carbon or hydrogen.

heterogeneous chemical reaction A chemical reaction that takes place between reactants that are in different phases.

heterogeneous mixture A mixture whose composition is not uniform throughout. Compare homogeneous mixture.

heterogeneous reaction A reaction that occurs at the interface between two or more phases, that is, at the surface where different phases are in contact with one another. Example: the rusting of a nail exposed to water and air.

homogeneous chemical reaction A reaction in which reactants and products are in the same phase. Examples: reactants and products are all gases or reactants and products are all dissolved.

homogeneous mixture A mixture whose composition is uniform throughout; a solution. Compare heterogeneous mixture.

homologous series A group of similar compounds in which one member differs from the next by a constant number of atoms. Example: alkanes.

hydration Solvation with water as the solvent.

hydration energy As a substance dissolves, the energy given off as solute ions are solvated by water molecules.

hydride A binary covalent compound consisting of one or more hydrogen atoms bonded to atoms that are less electronegative than hydrogen. This gives the H atom a −1 oxidation number. Example: LiH.

hydrocarbon A compound containing only carbon and hydrogen atoms.

hydrocarbon tail The nonpolar section of a soap or detergent molecule, consisting of nonpolar C and H atoms.

hydrogen bond An intermolecular dipole–dipole attraction between a δ + H atom covalently bonded to either an O, N, or F atom in one molecule and an O, N, or F atom in another molecule. Compare dipole–dipole force; London force.

hydronium ion (H_3O^+) The cation created when an H^+ ion is hydrated with one water molecule.

hydrophilic Referring to the section of a soap or detergent molecule (the polar head) that is soluble in water ("water-loving"). Compare hydrophobic.

hydrophobic Referring to the section of a soap or detergent molecule (the nonpolar hydrocarbon tail) that is insoluble in water ("water-fearing"). Compare hydrophilic.

hydroxide ion An anion with formula OH^-.

hyperbaric High-pressure.

hypothesis What a theory is called the first time it is postulated, and before it is well enough tested to be accepted as a theory.

ideal gas A gas in which absolutely no intermolecular forces are present between molecules and whose molecules occupy no volume whatsoever. Such a gas, which is only an abstraction and does not exist in the real world, remains a gas at all temperatures and all pressures and obeys the ideal gas law.

ideal gas constant (R) The constant of proportionality in the ideal gas law; $R = 0.0821\ \text{L}\cdot\text{atm/K}\cdot\text{mol}$.

ideal gas equation A mathematical expression relating the pressure P, volume V, temperature T (in kelvins), and number (n) of moles of a sample of an ideal gas: $PV = nRT$, where R is the ideal gas constant. Also called the ideal gas law.

ideal gas law Another name for the ideal gas equation.

indicator Any substance that changes color with the strength (or concentration) of acid (or base) present.

inert gas Outdated name for a group VIIIA element. A synonym for noble gas.

inherent rate factor In a reaction rate equation, any term that does not depend on the concentrations of the reactants. All inherent factors are collected in the rate constant k for a given reaction.

inner-shell electron Another term for core electron.

intensive property Any property of a substance that does not depend on the amount present. Examples: density; boiling point. Compare extensive property.

interface The region of contact between two different phases.

intermolecular attractive force A force exerted by one molecule on another molecule, arising from the separation of positive and negative electrical charges in the molecules. The positive part of one molecule is attracted to the negative part of the other molecule. The three types of intermolecular forces are London forces (momentary charge separation), dipole–dipole forces (permanent charge separation), and hydrogen bonds (permanent charge separation). Compare intramolecular force.

intramolecular force A force exerted by one atom in a molecule on another atom in the same molecule. Ionic bonds and covalent bonds are two types of intramolecular forces. Compare intermolecular force.

ion Any atom or group of atoms carrying either a net negative or a net positive electrical charge. See anion; cation.

ion–dipole attraction Another name for ion–dipole force.

ion–dipole force The force of attraction exerted by an ion and a polar molecule on each other.

ionic bond A bond between two ions, formed when the atoms of one give up electrons (form cations) and the atoms of the other accept electrons (form anions). Compare covalent bond; metallic bond.

ionic compound A compound containing all ionic bonds.

ionic lattice The three-dimensional ordered arrangement of the ions making up an ionic solid, consisting of alternating cations and anions.

ionic solid A solid made up of individual ions arranged in an ionic lattice held together by ionic bonds. Compare molecular solid; network solid.

ionization energy The amount of energy needed to remove an electron from an atom or ion. Each electron in any atom or ion has a specific ionization energy. Compare first ionization energy.

ionizing radiation Radiation capable of ionizing atoms, especially those in the cells of biological organisms.

isomer Any one of a set of molecules that have the same chemical formula but different structures.

isotopes Different forms of an element having the same number of protons but different numbers of neutrons (and therefore different atomic masses).

IUPAC The acronym for International Union of Pure and Applied Chemistry.

IUPAC nomenclature system A set of rules for naming chemical compounds uniquely and unambiguously.

joule (J) The SI unit of energy.

K_a The equilibrium constant for the reaction in which a weak acid partially dissociates in aqueous solution.

K_b The equilibrium constant for the reaction in which a weak base partially dissociates in aqueous solution.

K_{eq} The equilibrium constant of a chemical reaction.

K_w A modified equilibrium constant for the autodissociation of water; $K_w = [OH^-] \times [H_3O^+] = 10^{-14}$.

kelvin (K) The SI base unit of temperature.

Kelvin temperature scale The SI temperature scale, where the freezing point of water is set at 273.15 K and the boiling point of water is set at 373.15 K. The unit of temperature on this scale is called the kelvin. Sometimes called the absolute temperature scale.

ketone A functionalized hydrocarbon in which the functional group is a carbonyl attached to two hydrocarbon groups. General formula: R—CO—R.

kilo- The SI prefix meaning 1000 (10^3); uses the symbol K.

kilogram The SI base unit of mass; equal to about 2.2 pounds.

kinetic energy The energy an object has whenever it is moving; the faster the object is moving, the higher its kinetic energy.

kinetics experiment An experiment run to determine the rate of a chemical reaction.

lanthanides Elements 58–71 on the periodic table. Also known as the rare earth elements.

lattice energy The energy it would take to break the ionic lattice up into individual, isolated ions.

law A statement that describes the way things are consistently observed to behave under a given set of circumstances. Compare theory.

law of conservation of energy A statement of the fact that energy can be neither created nor destroyed; it just goes from one form to another.

law of conservation of matter A statement of the fact that no matter is ever created or destroyed in any chemical reaction.

law of constant composition A statement of the fact that the amounts of the elements in a compound are determined by the identity of the compound.

law of definite proportions Another name for the law of constant composition.

law of Mendeleev The pattern discovered by Mendeleev in which each element in his version of the periodic table has properties similar to those of the elements eight positions to the left and eight positions to the right. Sometimes called the law of octaves.

Le Châtelier's principle A statement of the fact that any time the equilibrium of a chemical reaction is disturbed, the forward and reverse reaction rates shift in the direction that acts to partially remove the disturbance.

leading zero Any zero to the left of the first nonzero digit in a number.

Lewis dot diagram Another name for dot diagram.

limiting reactant In a chemical reaction, a reactant that, because there is less of it than is needed to react with the total amount of the other reactants, determines how much product is produced. All of the limiting reactant is used up in the reaction.

line drawing A shorthand way of representing hydrocarbon molecules using line segments.

line spectrum An electromagnetic spectrum consisting of discrete lines separated by dark spaces of various widths. The lines indicate radiation (energy) emitted by an atom when excited electrons in the atom lose energy and thereby fall to a lower-energy state. Compare continuous spectrum.

linear hydrocarbon A hydrocarbon in which each carbon atom, except the two end carbons, is bonded to two other carbon atoms. Compare branched hydrocarbon.

liquid A state of matter in which molecules or atoms are in contact (touch) and are also in motion, jostling by one another. A liquid changes shape to match the shape of the vessel used to contain it but maintains a constant volume. Compare gas; solid.

liter A non-SI unit of volume, equal to 1.057 quarts.

lithium ion battery A relatively high energy–density battery in which lithium atoms are oxidized at the anode to Li^+ ions, which then cross a membrane and travel to the cathode.

litmus A plant-derived dye that is red in the presence of an acid and blue in the presence of a base.

lock-and-key model A model describing enzyme action as the binding of a substrate molecule (the key) at an active site (the lock) on an enzyme molecule.

logarithm For any numeric value, the power to which 10 must be raised to get that value.

London force The relatively weak intermolecular attractive forces exerted by molecules on each other, caused by brief, constantly shifting electron imbalances in the molecules that cause them to act as momentary dipoles. Compare dipole–dipole force; hydrogen bond.

lone pair Two valence electrons that usually are not used for sharing with another atom in a covalent bond. Compare unpaired electrons.

macromolecule A general term for any very large organic or biochemical molecule, where molecular weights can be as high as 50,000 g/mol and more.

magnetic resonance imaging (MRI) A technique for creating images of the interior of a body.

main chain In a branched hydrocarbon, the longest continuous chain of carbon atoms in the molecule. When a multiple bond occurs in the molecule, the longest continuous chain that includes the multiple bond.

main-group element Another name for representative element.

mass A measure of the quantity of matter in an object. Mass is independent of the location of the object, with a given object having the same mass no matter where in the universe it is located. Compare weight.

mass defect The difference between the actual mass of an atom and the theoretical mass calculated by adding up the masses of all the protons, neutrons, and electrons in the atom.

mass number The number of protons plus neutrons in the nucleus of an atom.

mass percent The mass of any component of a sample divided by the total mass of the sample, with this quotient multiplied by 100.

mass spectrometer An instrument that can determine the mass of single atoms or molecules by ionizing them, accelerating them toward a detector, and bending their path with an electromagnet.

matter Anything that has mass and takes up space.

maximum Also called maximum electron capacity; with respect to an atom's Bohr shells, the most number of electrons a particular shell of quantum number n can hold.

measured number A number obtained by using a measuring device. Measured numbers always have some uncertainty because no measuring device is perfect. Compare exact number.

measurement The determination of a numeric value via a device such as a ruler or a balance.

mega- The SI prefix meaning 1 million (10^6); uses the symbol M.

melting The process in which a substance changes from the solid state to the liquid state.

melting point (mp) The temperature at which a substance changes from a solid to a liquid at standard atmospheric pressure. Because freezing is the reverse of melting, the melting point is numerically equal to the freezing point.

mercury barometer A device that uses mercury as the liquid to measure the pressure exerted by the atmosphere.

metal An element that tends to lose one or more valence electrons in chemical reactions. In their elemental form, metals typically are shiny and bendable and conduct heat and electricity well.

metal hydroxide A compound consisting of a metal cation and one or more hydroxide anions, having the general formula $M(OH)_n$. Those that are soluble produce basic solutions when dissolved in water.

metallic bond Bonds between metal atoms that are the result of electrons being shared by all the atoms. Compare covalent bond; ionic bond.

metalloid An element whose properties are intermediate between those of metals and those of nonmetals.

meter The SI base unit of length, equal to 39.37 inches.

metric system A system of units in which the basic length unit is the meter.

micelle The spherical structure formed by a group of soap or detergent molecules in water. The polar heads orient relative to each other so that they create the surface of the sphere, and the hydrocarbon tails all point to the interior. The polarity of the surface causes the micelle to dissolve in water, while (nonpolar) grease and oil molecules dissolve in the nonpolar interior.

micro- The SI prefix meaning one-millionth (10^{-6}); uses the symbol μ.

micron A synonym for micrometer.

milli- The SI prefix meaning one-thousandth (10^{-3}); uses the symbol m.

millimeters of mercury A non-SI unit of pressure; 760 mm Hg = 1 atm.

mixture Matter that consists of two or more substances in variable amounts.

model A verbal description or physical construction used to visualize something a theory describes.

molar mass The mass of 1 mole of any substance, expressed in grams.

molar volume The volume occupied by 1 mole of a substance.

molarity (M) A unit of concentration for solutions, indicating the number of moles of solute contained in each liter of a solution; M = moles/liter.

mole The SI base unit of amount of a substance, defined as the amount of any substance that contains 6.02×10^{23} atoms or molecules of the substance. For elements, this amount is equal to the atomic mass expressed in grams; for molecules, this amount is equal to the molar mass expressed in grams.

molecular compound A compound made up of individual molecules.

molecular dipole moment The dipole moment for an entire molecule. It points from the $\delta +$ portion of the entire molecule to the $\delta -$ portion.

molecular electrolyte A substance that, despite containing only covalent bonds, either partially or completely dissociates into ions when dissolved in water.

molecular formula A chemical formula in which the numeric subscripts indicate the actual composition of one molecule of a substance. Compare empirical formula.

molecular solid A solid made up of discrete molecules attracted to each other by intermolecular forces. Compare ionic solid; network solid.

molecular weight A synonym for molar mass.

molecule A stable collection of two or more atoms bound together by covalent bonds.

molten In the liquid state.

monatomic Consisting of one atom. Compare diatomic; polyatomic.

monoalkene An alkene containing only one carbon–carbon double bond.

monoalkyne An alkyne containing only one carbon–carbon triple bond.

monomer A small molecule that forms chains to create a polymer.

monomer unit A monomer once it has been incorporated into a polymer.

monoprotic acid An acid that, when it dissociates in water, can yield only one H_3O^+ ion per molecule of acid.

monosaccharide A sugar molecule consisting of one simple sugar.

MRI An acronym for magnetic resonance imaging.

multiple bond Another name for multiple covalent bond.

multiple covalent bond More than one bond between two atoms, formed by the sharing of more than one electron from each atom. Can be either a double bond (two bonding pairs) or a triple bond (three bonding pairs).

n/p ratio Shorthand for neutron-to-proton ratio.

nano- The SI prefix meaning one-billionth (10^{-9}); uses the symbol n.

net energy change (ΔE_{rxn}) The difference between the energy of the products of a chemical reaction and the energy of the reactants. This amount of energy is released during an exothermic reaction (ΔE_{rxn} is negative) and is absorbed during an endothermic reaction (ΔE_{rxn} is positive).

net ionic equation An equation that shows the reactants and products of the reaction, but not the spectator ions.

network covalent substance Another name for network solid.

network solid A nonmolecular solid made up of atoms held together by covalent bonds.

neutral solution A solution in which the concentration of hydronium ions is the same as the concentration of hydroxide ions, giving the solution a pH value of exactly 7. Compare acidic solution; basic solution.

neutral substance A substance that gives no signs of being either acidic or basic.

neutralization reaction The reaction of an acid with a base to form a salt and water. The combining of one hydronium ion (H_3O^+) and one hydroxide ion (OH^-) to form two molecules of water.

neutron A subatomic particle that has a mass approximately equal to that of the hydrogen atom, carries zero electrical charge, and is located in the nucleus of an atom.

neutron-to-proton ratio The ratio of neutrons to protons in an atomic nucleus.

noble gas Any element in group VIIIA of the periodic table.

nonelectrolyte A water-soluble substance that does not dissociate into ions when dissolved in water, with the result that the solution cannot conduct an electric current. Compare electrolyte.

nonmetal An element that tends to gain one or more valence electrons in a chemical reaction. In their elemental form, nonmetals do not conduct heat or electricity well, meaning they are thermal and electrical insulators.

nonmolecular solid Any solid that does not consist of discrete molecules.

nonpolar molecule A covalent molecule in which there is no net separation of electrical charge in the molecule. Compare polar molecule.

nonstoichiometric compound A compound where one or more of the subscripts in the formula is a nonintegral number.

normal atmospheric pressure The air pressure that raises the liquid column in a mercury barometer to a height of 760 mm = 29.9 in = 1 atm.

normal boiling point The boiling point of a substance at standard pressure (760 mm Hg = 1 atm).

normal freezing point The freezing point of a substance at standard pressure (760 mm Hg = 1 atm).

normal melting point The melting point of a substance at normal atmospheric pressure (1 atm).

NR An abbreviation for no reaction.

nuclear decay Another name for radioactive decay.

nuclear fission The splitting apart of a large parent nucleus into two smaller daughter nuclei.

nuclear fusion The fusing (joining together) of two small nuclei into one larger nucleus.

nuclear radiation A general term for the particles and electromagnetic radiation emitted by a nucleus undergoing radioactive decay.

nuclear reaction A process in which change occurs to the nuclei of the atoms involved. Compare chemical reaction.

nucleon A particle found in the nucleus of an atom; a general term for protons and neutrons.

nucleus The tiny, positively charged core of an atom, made up of protons plus neutrons (except in the case of the hydrogen nucleus, which contains no neutrons). The electric charge of the nucleus binds the electrons to the atom, and the number of protons determines which element the atom is.

ocean of instability All areas on a band of stability plot except the band of stability and the small "island" representing predicted stable isotopes having atomic numbers 114 and above.

octahedron An eight-sided polygon and the guiding shape for molecules possessing an expanded octet central atom with six electron groups around it.

octet rule A statement of the fact that when two or more elements react to form a compound, the atoms usually combine in such a way that, once the compound is formed, each atom has eight electrons in its valence shell.

orbital A region of space surrounding the nucleus of an atom inside which one is most likely to find an electron.

order An experimentally determined number, shown as an exponent in the rate law for a chemical reaction, that expresses how the concentration of a reactant affects the reaction rate. Example: Rate $= k[A]^1[B]^2$ says that the reaction is first order with respect to A and second order with respect to B.

organic chemistry The branch of chemistry that deals with carbon and the compounds it forms.

orientation factor A number between 0 and 1 indicating how important proper orientation between reactants is in a chemical reaction, with a value of 1 indicating orientation does not matter.

outer-shell electron Another name for valence electron.

overall order of the reaction A number obtained by adding together all the exponents in the rate expression for a chemical reaction.

oxidation The loss of electrons by an atom, resulting in an increase in oxidation state. Compare reduction.

oxidation number Another term for oxidation state.

oxidation–reduction reaction An electron transfer reaction in which one reactant loses electron(s) (is oxidized) and another reactant gains electron(s) (is reduced).

oxidation state A numeric value indicating the change in the number of valence electrons in a covalently bonded atom, more (−) or less (+) than in the atom in its free (unbonded) state.

oxidation state bookkeeping method A technique for determining the number of valence electrons "owned by" a covalently bonded atom.

oxidizing agent The reactant in an electron-transfer reaction that accepts electrons. The oxidizing agent is reduced during the reaction. Compare reducing agent.

paired electrons See electron pair.

parent isotope In a nuclear reaction, the radioactive isotope that undergoes decay.

pascal The SI unit of pressure.

peptide linkage An amide bond when it occurs in a protein molecule.

percent by mass (mass %) A unit of concentration for solutions, used most frequently for solutions made up of a solid solute dissolved in a liquid solvent and defined as

$$\frac{\text{Grams of solute}}{\text{Grams of solution}} \times 100$$

percent by volume (vol %) A unit of concentration for solutions, used most frequently for solutions made up of a liquid solute dissolved in a liquid solvent and defined as

$$\frac{\text{Volume of solute}}{\text{Volume of solution}} \times 100$$

percent composition A measure of solution concentration defined in terms of mass (percent by mass), volume (percent by volume), or mass and volume (percent by mass/volume).

percent mass/volume (mass/vol %) A unit of concentration for solutions, defined as

$$\frac{\text{Grams of solute}}{\text{Volume of solution}} \times 100$$

percent yield The actual yield of product from a chemical reaction divided by the theoretical yield, with this quotient multiplied by 100 to give a value in percent.

period A horizontal row of the periodic table.

periodic behavior Another name for chemical periodicity.

periodic table A table of the elements arranged in order of increasing atomic number. The elements are arranged in rows (periods) in such a way that all the elements in a given column (group) have similar chemical properties.

pH The negative of the logarithm of the hydronium ion concentration in an aqueous solution.

physical change A change that alters a physical property of a substance without changing its identity.

physical property A property of a substance that characterizes the substance without the substance undergoing a chemical reaction; examples are density, boiling point, and color. Compare chemical property.

physical transformation A change in a pure substance that leaves it as the same substance but in a different physical state.

physics See classical physics; quantum physics.

pico- The SI prefix meaning one-trillionth (10^{-12}); uses the symbol p.

Planck's constant (h) An unchanging numeric value used in calculating the energy of electromagnetic radiation; $6.626 \times 10^{-34}\ \text{J} \cdot \text{s}$.

polar covalent bond A covalent bond in which the electrons are shared unequally.

polar molecule A covalent molecule containing atoms of different electronegativities arranged in such a way that there is a net (unbalanced) separation of electrical charge in the molecule. Compare nonpolar molecule.

polyatomic Containing three or more atoms. Compare diatomic; monatomic.

polyatomic ions Ions made from many atoms.

polymer A very large molecule made from repeating units called monomers.

polypeptide A protein molecule containing between three and fifty amino acids joined by peptide linkages. Compare dipeptide; protein.

polysaccharide A sugar molecule consisting of many monosaccharide (simple sugar) units.

positron (β^+) A subatomic particle having the same mass as an electron but carrying an electrical charge of +1; the antimatter of the electron.

positron (β^+) emission A nuclear reaction in which a proton changes to a neutron plus a positron, with the neutron staying in the nucleus and the positron being ejected from the atom.

precipitate An insoluble solid that forms in a precipitation reaction when two or more aqueous solutions are mixed.

precipitation reaction A reaction in which an insoluble solid (the precipitate) forms when two or more aqueous solutions are mixed.

precision The degree to which a set of measured values of the same quantity agree with each other. Compare accuracy.

predicted rate law The rate law for a reaction predicted by a postulated reaction mechanism.

pressure The force per unit area exerted on a surface.

pressure–moles relationship The pressure of a gas increases when the number of moles of gas increases and decreases when the number of moles decreases (assumes temperature and volume held constant).

pressure–temperature relationship The pressure of a gas increases when its temperature increases and decreases when its temperature decreases (assumes moles and volume held constant).

pressure–volume relationship The pressure of a gas increases when the volume the gas occupies decreases (assumes moles and temperature held constant).

primary amine An amine in which the amine functional group is bonded to one hydrocarbon chain and two hydrogen atoms. General formula: $R{-}NH_2$.

principal quantum number (n) The number that indicates which shell an electron occupies in the atom. Electrons with $n = 1$ occupy the first (lowest-energy) shell; electrons with $n = 2$ occupy the second shell counting outward from the nucleus, and so forth.

product A substance produced in the course of a chemical reaction by the transformation of one or more reactants. Compare reactant.

proof A unit used to indicate the concentration of ethanol in alcoholic beverages; equal to twice the percent by volume of ethanol in the beverage.

protein A biopolymer containing 50 or more amino acid monomer units linked by peptide linkages. Compare dipeptide; polypeptide.

proton A subatomic particle that has a mass approximately equal to that of the hydrogen atom, carries an electrical charge of +1, and is located in the nucleus of an atom.

pure substance Matter with constant composition that consists of just one element or compound in just one chemical form.

pyramidal vs. tetrahedral A pyramidal geometry is derived from a tetrahedral geometry via removal of the top part of the tetrahedron.

quantized energy Energy that occurs only in particular discrete amounts.

quantum jump The movement of an electron in an atom from one allowed energy level to another.

quantum mechanics See quantum physics.

quantum physics Physics that takes into account the fact that energy is quantized and therefore can be used to describe the behavior of atoms and subatomic particles. Compare classical physics.

R group notation A shorthand method for drawing functionalized hydrocarbon molecules in which the hydrocarbon chain is represented by the symbol R.

radioactive atom An atom whose nucleus is unstable and therefore tends to undergo a nuclear reaction in which it changes spontaneously (without the addition of energy from an external source) to the nucleus of an atom of some other element.

radioactive dating The process of determining an object's age by measuring the amount of some radioactive isotope contained in the object and then using that information plus the half-life of the isotope to calculate age.

radioactive decay The spontaneous change that an unstable nucleus undergoes.

rare earth elements The elements with atomic numbers 57 through 71.

rare gas A synonym for noble gas.

rate constant (k) A numeric constant, specific to each chemical reaction, that represents the product of all the inherent rate factors for the reaction.

rate-determining step The slowest step in any chemical reaction.

rate law An equation that expresses the rate of a chemical reaction as the product of the rate constant for the reaction times the concentration of each reactant raised to an exponent called an order.

rate of reaction Another term for reaction rate.

reactant A starting material for a chemical reaction; a substance that enters into the reaction and is transformed into one or more product substances. Compare product.

reaction See chemical reaction.

reaction coordinate The horizontal axis of a reaction energy profile; represents the progress of the reaction.

reaction-energy profile A graph showing the relative energies of the reactants and products in a chemical reaction, with the horizontal axis representing the progress of the reaction and the vertical axis representing the energy of the reaction.

reaction intermediate A product of one step of a multistep chemical reaction that serves as a reactant in a subsequent step.

reaction mechanism The sequence of elementary steps in a chemical reaction.

reaction rate The change in concentration of a reactant or product per unit time during a chemical reaction.

reagent See reactant.

redox reaction A shorthand name for an oxidation–reduction (electron-transfer) reaction.

reducing agent The reactant in an electron-transfer reaction that loses electrons. The reducing agent is oxidized during the reaction. Compare oxidizing agent.

reduction The gain of electrons by an atom, resulting in a decrease in oxidation state. Compare oxidation.

relaxation The process by which an excited electron falls to a lower energy level. The difference in energy between the two levels is often emitted as electromagnetic radiation and creates a line spectrum.

representative element Any element in groups IA, IIA, IIIA, IVA, VA, VIA, VIIA, and VIIIA of the periodic table.

resonance forms Two or more equally correct dot representations of the same molecule; the average of these representations is the actual structure of the molecule.

reversible reaction A chemical reaction in which the rate at which products are converted back to reactants is significant.

Rutherford model of the atom A model proposed by Ernest Rutherford in which the atom is pictured as being mostly empty space surrounding a tiny, but massive, positively charged nucleus. The atom's electrons occupy the "empty" space and fly around the nucleus. Compare Bohr model; Thomson model.

sacrificial metal A metal used in a structure to protect other metals in the structure from corrosion.

salt Another name for ionic compound.

salt bridge A connector between the two compartments of a galvanic cell. The salt bridge contains an electrolyte that provides anions to the compartment containing the metal being oxidized and cations to the compartment containing the metal ion being reduced.

saturated hydrocarbon A hydrocarbon in which all the carbon–carbon bonds are single bonds, which means that the maximum number of hydrogen atoms are present.

saturated solution A solution containing the maximum possible amount of a given solute.

science The experimental investigation and explanation of natural phenomena.

scientific method The method of making observations, proposing theories, and testing those theories through experimentation.

scientific notation A convention in which a numeric quantity is written as a number between 1 and 10 multiplied by 10 raised to the appropriate power.

secondary amine An amine in which the amine functional group is bonded to two hydrocarbon chains (which may or may not be identical) and one hydrogen atom. General formula: $R_2—NH$.

semimetal A synonym for metalloid.

shell One of the distances at which electrons orbit the nucleus (corresponding to allowed electron energy levels). Each shell is assigned a principal quantum number $n = 1, 2, 3, \ldots$, with $n = 1$ being the shell of lowest energy.

SI units (SI stands for Système Internationale.) The system of units used in science, including seven base units, with the five most important to chemistry being the meter for length, the kilogram for mass, the second for time, the kelvin for temperature, and the mole for amount of substance.

side reaction A chemical reaction that takes place along with a main reaction, usually yielding unwanted products.

significant digits A synonym for significant figures.

significant figures The digits in a measured number (or a number calculated from measured numbers) that are known with certainty and/or within the limits of estimation.

single bond A covalent bond between two atoms that is formed by one bonding pair of electrons.

single-replacement reaction A reaction in which one element replaces another element in a compound.

solid A state of matter in which molecules or atoms are in contact with one another and remain in fixed positions. A solid maintains a constant shape and volume. Compare gas; liquid.

solid solution A solution in which both solute and solvent are solids at room temperature. The solution is prepared from molten solute and molten solvent.

solubility The maximum amount of a solute that dissolves in a given amount of solvent at a given temperature and, for a gas solute, at a given pressure.

solubility product (K_{sp}) The equilibrium constant for sparingly soluble salts, which is simply the product of the ion concentrations in an aqueous solution at equilibrium each raised to the appropriate power.

solute In a solution, the substance dissolved in the solvent. Compare solvent.

solution A homogeneous mixture of a solute or solutes dissolved in a solvent.

solvation The process whereby solvent molecules in a solution completely surround a solute ion.

solvent In a solution, the substance present in the greatest amount. Compare solute.

solvent cage A hollow sphere of water molecules completely surrounding an ion. The water molecules are attracted to the ion by ion–dipole attractions.

specific heat The amount of heat energy needed to warm 1 g of a substance by 1 °C.

spectator ion An ion present in the vessel in which a chemical reaction takes place that does not take part in the reaction.

stable isotope An isotope that is not radioactive, or, for radioactive elements located in the band of stability, a radioactive isotope that has a measurable half-life.

standard temperature and pressure (STP) Defined as 0 °C = 273.15 K and 1 atm pressure.

states of matter The physical states in which matter can occur: solid, liquid, and gas.

steric number The number of electron groups about an atom in a molecule, equal to the sum of the number of lone pairs of electrons and the number of attached atoms.

stock solution A solution of a known concentration that can be used to make more dilute solutions of the same solute.

stoichiometry The branch of chemistry that deals with measuring the quantities of reactants and products involved in chemical reactions.

strong acid An acid that dissociates completely when dissolved in water.

strong base A base that dissociates completely when dissolved in water.

strong electrolyte An electrolyte that dissociates completely when dissolved in water. Compare weak electrolyte.

strong force The very strong attractive force that holds an atomic nucleus together.

structural formula A drawing showing all the atoms and bonds in a compound arranged in their correct relative positions, as they actually appear in the molecule. Compare chemical formula.

subatomic particle Any particle that is more fundamental than the atom, such as protons, neutrons, and electrons.

sublimation A change from the solid state directly to the gas state.

subshell A subdivision of an electron shell, designated by letters *s, p, d, f, …* , with the s subshell being the one of lowest energy.

substitution reaction A chemical reaction in which one atom or group of atoms on a reactant molecule is replaced by an atom or group of atoms from another reactant molecule.

substrate For a given enzyme, a molecule that reacts at the active site of the enzyme.

superconductor Any material that offers zero resistance to the passage of electricity.

technology The application to practical problems of knowledge gained by doing science.

termolecular Involving three molecules.

tertiary amine An amine in which the amine functional group is bonded to three hydrocarbon chains (which may or may not be identical). General formula: $R_3—N$.

tetrahedron A four-sided polygon in which all sides are identical equilateral triangles.

theoretical yield For a chemical reaction, the maximum amount of product possible for a given amount of reactants. Compare actual yield.

theory A tentative explanation for a set of observations. A theory must be consistent with every one of the observations and is tested through experiments. Compare law.

Thomson model of the atom A model proposed by J. J. Thomson in which the atom is pictured as a cloud of positive electricity with electrons embedded in the cloud. Compare Bohr model; Rutherford model.

titration A procedure for determining solution concentrations that entails using a known amount of a solution of known concentration to determine the concentration of some other solution.

trailing zero Any zero to the right of the final non-zero digit in a number.

transition metal Any element in groups IB through VIIIB of the periodic table.

transition state A transient state of the reactants in a chemical reaction; the reactants have absorbed the required activation energy and are therefore in a high-energy, unstable condition.

triglyceride Also known as fats and oils, compounds derived from the chemical combination of three fatty acid molecules with glycerol.

trigonal bipyramid A six-sided polygon and the guiding shape for molecules possessing an expanded octet central atom with five electron groups around it.

triple bond A multiple covalent bond between two atoms that is formed by three bonding pairs of electrons.

triprotic acid An acid that, when it dissociates in water, can yield a maximum of three H_3O^+ ions per molecule of acid.

tritium The isotope of hydrogen containing one proton and two neutrons; ${}^3_1H$.

tunneling See electron tunneling.

uncertainty The amount by which a measured number may vary from the true value of the quantity being measured.

uncertainty principle A statement of the fact that it is impossible to know, at a given instant, both the velocity and the position of any subatomic particle.

unimolecular Involving one molecule.

unit analysis A problem-solving strategy in which data given in the problem statement are multiplied by appropriate conversion factors in order to obtain the correct units for the answer.

unpaired electrons Single valence electrons that are used for sharing with another atom in a covalent bond. Compare lone pair.

unsaturated hydrocarbon A hydrocarbon containing at least one carbon–carbon double or triple bond, which means that the carbon atoms in the multiple bond(s) have fewer than their full complement of hydrogen atoms.

valence electron An electron in the outermost shell of an atom.

valence shell The occupied shell of an atom that has the highest principal quantum number.

valence shell electron pair repulsion (VSEPR) theory A model used to predict molecular shape, the basic tenet being that the valence electrons in a molecule repel each other because they carry negative charges.

van der Waals attraction Intermolecular forces as a whole, including London forces.

van der Waals equation A corrected version of the ideal gas law that corrects for nonideal behavior of gases.

vapor A synonym for gas.

vapor pressure The pressure exerted by the gas molecules when the liquid and gas states of a substance are in dynamic equilibrium.

vaporization A change from the liquid phase to the gas phase. Compare condensation.

vector A quantity having both magnitude (size) and direction.

virus A nonliving parasite that consists of a protein shell surrounding strands of genetic material, often DNA.

visible light The part of the electromagnetic spectrum between approximately 380 and 750 nm.

voltage The intensity of the "pressure" or "push" behind electrons that causes them to flow in an electric current.

voltaic cell See galvanic cell.

volume The amount of space occupied by an object.

VSEPR An acronym for valence shell electron pair repulsion.

wavelength The distance between any two adjacent comparable points on a wave, such as the distance from one crest to the next or the distance from one trough to the next.

weak acid An acid that dissociates only partially when dissolved in water.

weak base A base that yields a very low concentration of OH^- ions when dissolved in water.

weak electrolyte An electrolyte that dissociates only partially when dissolved in water. Compare strong electrolyte.

weight A measure of how strongly Earth's gravitational force (or the gravitational force exerted by any other body) pulls on an object. As the gravitational force changes, the object's mass stays constant but its weight changes. Compare mass.

weighted average atomic mass The atomic mass of an element listed on the periodic table. It is obtained by multiplying the atomic mass of each isotope by the fraction of its naturally occurring abundance and then summing all the products.

white light Light containing all the wavelengths of the visible part of the electromagnetic spectrum, from 380 to 750 nm.

Z The symbol used to represent atomic number.

Selected Answers

Note: Blue numbers in the text indicate practice problems and end-of-chapter problems for which answers are provided here.

Chapter 1

1.2 Heterogeneous mixture.
1.3 Heterogeneous mixture.
1.5 (d) and (e)
1.6 True.
1.8 (c)
1.9 False.
1.10 Heat the metal until it just begins to melt; then measure the temperature of the liquid to see if it is the same as the melting point of gold.
1.12 Methane and oxygen are the reactants; water and carbon dioxide are the products.
1.13 (b)
1.15 A law summarizes experimental data.
1.16 If the theory is used to predict the results of proposed experiments and then the data from those experiments agree with the prediction, this is good evidence the theory is correct.
1.17 Science is the experimental investigation and explanation of natural phenomena. Technology is the application of scientific knowledge. Here, the discovery of electromagnetic induction is an example of science and the use of magnets to produce electricity is an example of technology.
1.18 Chemistry is the study of matter and the transformations it undergoes.
1.22 Matter is anything that has mass and occupies space.
1.23 Yes.
1.25 Heterogeneous mixture.
1.26 (a) Heterogeneous mixture. (b) Solution. (c) Solution.
1.28 An element is one of the basic building blocks of matter. There are 113 known elements today.
1.30 An atom.
1.31 Lead, Pb; molybdenum, Mo; tungsten, W; chromium, Cr; mercury, Hg.
1.33 Ti, titanium; Zn, zinc; Sn, tin; He, helium; Xe, xenon; Li, lithium.
1.35 An elemental substance contains only one type of atom, and a compound contains two or more different types of atoms. Here, both S_8 and S_6 are elemental substances, while SO_2 is a compound.
1.36 The chemical formula tells how many of each type of atom are present in the smallest possible piece of a pure substance.
1.38 F_2, P_4, Ar, and Al are elemental substances; $BrCl_3$, C_2H_2, HCl, and Al_2O_3 are compounds.
1.40 H_2O_2
1.42 $C_6H_{12}O_6$
1.44 Solid, liquid, and gas.
1.46 Condensation.
1.48 $-117.3\ °C$
1.50 Melting point $0\ °C = 32\ °F$; boiling point $100\ °C = 212\ °F$.
1.53 (a) Sugar has undergone a chemical change because its chemical composition has been changed. New compounds, CO_2 and H_2O, have been formed.
(b) Condensation: $H_2O(g)$ condenses into $H_2O(l)$.
(c) The fog cannot be due to condensation of $CO_2(g)$ as the temperature of the window is well above the condensation point of CO_2 of $-78\ °C$.
1.55 Gasoline (C_8H_{18}) reacts with oxygen (O_2) in a car's engine, forming new compounds: $CO_2(g)$ and $H_2O(g)$. When water vapor condenses at the end of the exhaust pipe, it becomes $H_2O(l)$ and drips out.
1.57 Chemical change.
1.58 The properties of any substance formed in a chemical reaction are usually very different from the properties of the reactants used.
1.59 A law is a statement that summarizes experimental data. A theory proposes an explanation of why a law is true.
1.64 $C(s)$, $N_2(g)$, and $N_2(l)$ are elemental substances; $H_2O(l)$ and $HNO_3(l)$ are compounds.
1.66 Sulfuric acid.
1.68 (a) Mixture. (b) Mixture. (c) Compound. (d) Compound. (e) Elemental substance.
1.72 Chemical.
1.74 Table salt and table sugar.
1.78 Numerous answers possible. Example: brass is a homogeneous mixture of solid copper and solid zinc. You would melt these two metals in a single container and then cool the mixture.
1.80 Sublimation.
1.82 Chemical.
1.86 N_2
1.88 Reactants: $C_6H_{12}O_6$, O_2; products: CO_2, H_2O. A chemical change.
1.90 Fog.
1.92 Ozone.
1.94 True.
1.96 Sample (b) would be considered homogeneous because the gold and silver atoms are distributed evenly on an atomic scale.

Chapter 2

2.2 Ike is more accurate. Mike is more precise.
2.3 Jack.
2.5 ±0.05 gallon
2.6 (c)
2.8 By ending it with a decimal point: 600.

2.9

	Number of significant figures	Uncertainty
10.0	3	0.1
0.004 60	3	0.000 01
123	3	1

2.11 0.473
2.12 47,325
2.14 0.002 35
2.15 6000
2.17 $4.710\,000\,0 \times 10^{13}$
2.18 $4.710\,000 \times 10^{13}$
2.20 44 miles^2
2.21 660. hours
2.24 (a) 6.1×10^2 lb/in (b) 6.11×10^2 lb/in or 611 lb/in (c) 86.88 cm
2.26 1556 cm, rounded from 1555.801 cm
2.27 142 cm
2.29 4.736 km
2.30 25 mm
2.32 2500 mL
2.33 246.7 mL
2.34 K = °C + 273.15; therefore

°C = K − 273.15 = 263.5 K − 273.15 = −9.7 °C.

$$°\text{F} = 32 + \frac{9}{5}°\text{C} = 14.5\ °\text{F}$$

2.36 $$\frac{4.70\text{ g}}{1000\text{ mm}^3} = \frac{470\text{ g}}{1.00\text{ cm}^3} = 4.70\text{ g/cm}^3 = 4.70\text{ g/mL}$$

2.37 $$\frac{500.0\text{ g}}{150.5\text{ mL}} = 3.322\text{ g/mL}$$

2.39 $$\frac{1\text{ day}}{24\text{ h}} \qquad \frac{24\text{ h}}{1\text{ day}}$$

2.40 $$50.0\text{ miles} \times \frac{1\text{ h}}{600.0\text{ miles}} = 0.0833\text{ h}$$

2.41 $$\frac{600.0\text{ miles}}{1\text{ h}} \times 50.0\text{ h} = 3.00 \times 10^4\text{ miles}$$

2.43 $$500.0\text{ L} \times \frac{1000\text{ mL}}{1\text{ L}} \times \frac{0.001\,30\text{ g}}{1\text{ mL}} = 650.\text{ g} = 6.50 \times 10^2\text{ g}$$

$$650.\text{ g} \times \frac{1\text{ kg}}{1000\text{ g}} = 0.650\text{ kg}$$

2.44 $$1.50\text{ lb} \times \frac{453.6\text{ g}}{1\text{ lb}} \times \frac{1\text{ mL}}{11.4\text{ g}} = 59.7\text{ mL}$$

2.45 $$6955\text{ g flour} \times \frac{1\text{ cup flour}}{120.0\text{ g flour}} \times \frac{1\text{ cake}}{6\text{ cups flour}} = 9.660\text{ cakes}$$

You can bake nine cakes (it's not possible to bake a partial cake).

2.46 $$250.0\frac{\text{m}^2}{\text{h}} \times \frac{1\text{ h}}{60\text{ min}} \times \left(\frac{1\text{ ft}}{60\text{ in}}\right)^2 \times \left(\frac{1\text{ in}}{2.54\text{ cm}}\right)^2 \times \left(\frac{100\text{ cm}}{1\text{ m}}\right)^2 = 44.85\text{ ft}^2/\text{min}$$

(Four significant digits allowed because 1 in is *exactly* 2.54 cm.)

2.52 $$50.0\text{ mL} \times \frac{0.785\text{ g}}{\text{mL}} = 39.3\text{ g ethanol}$$

$$2.43\frac{\text{J}}{\text{g}\cdot°\text{C}} \times 39.3\text{ g} \times 38.0\ °\text{C} \times \frac{\text{kJ}}{1000\text{ J}} = 3.63\text{ kJ}$$

2.53 3.63 kJ = 3630 J;

3.63 kJ × 1 Cal/4.184 kJ = 0.868 Cal = 868 cal

2.54 $$1.000\frac{\text{cal}}{\text{g water}\cdot°\text{C}} \times 1000\text{ g water} \times 13.5\ °\text{C} = 1.35 \times 10^4\text{ cal}$$

$$\frac{1.35 \times 10^4\text{ cal}}{0.100\text{ g candy}} \times \frac{1\text{ Cal}}{1000\text{ cal}} = 1.35 \times 10^2\text{ Cal/g candy}$$

2.55 The 3 in "3 feet in a yard" is an exact number. The 3 in "a certain piece of wood is 3 feet long" comes from a measurement and therefore has some uncertainty associated with it.

2.57 You should choose the accurate result because a precise result that is not accurate is useless. An average of accurate results that were not precise usually gets you closer to the true value. On the other hand, an average of inaccurate but precise results may be far off of the true value. 1 inch above the average height measured accurately will be safer than 1 inch above the average height measured precisely, but inaccurately.

2.59 The person with the tape measure.

2.61 The uncertainty lies in the last digit written in the number. We often assume an uncertainty of ±1 in the position of the uncertain digit. Some typical examples: 15.2 cm is 15.2 ± 0.1 cm (if measured using a rule marked with centimeters only) 1534 cm^3 is 1534 ± 1 cm^3 (if measured using a cylinder marked every 10 cm^3) 0.00987 g is 0.00987 ± 0.00001 g (if measured using a so-called analytical balance)

2.63 (a) 12.60 ± 0.01 cm (b) 12.6 ± 0.1 cm (c) 0.000 000 03 ± 0.000 000 001 inch (d) 125 ± 1 ft

2.64 (a) Four. (b) Three. (c) One. (d) Three.

2.67 (a) $12.2\underline{0}2$ (b) No significant zeros. (c) $2\underline{0}5$ (d) $0.01\underline{0}$

2.69 It is not clear whether 30 has one or two significant figures. The decimal point with no digits following indicates that the zero in 30. is significant, meaning 30. has two significant figures.

2.71 (a) 56.0 kg (three significant figures).
(b) 0.000 25 m (two significant figures).
(c) 5,600,000 miles (four significant figures, but you cannot tell by looking at the normal notation).
(d) 2 ft (one significant figure).

2.73 (a) 3×10^1 ft (b) 3.0×10^1 ft (c) 3.00×10^1 ft

2.75 (a) 2.26×10^2 (b) 2.260×10^2 (c) 5.0×10^{-10} (d) 3×10^{-1} (e) 3.0×10^{-1} (f) 9.00×10^8 (g) $9.000\,006 \times 10^8$

2.77 102 in.; ±1 in.

2.79 3.873 14 miles (six significant figures because 5280 is an exact number).

2.80 (a) 2.55×10^5 km
(b) 1.000×10^{18} J
(c) 2.11×10^2 m
(d) 4.00×10^4 L

2.82 Liter and milliliter. They are volumes more commonly encountered in everyday situations and in the laboratory.

2.84 To eliminate confusion.

2.85 (a) 2.31×10^9 m (b) 5.00×10^{-6} m (c) 1.004×10^0 m (d) 5.00×10^{-12} m (e) 2.5×10^2 m

2.87 The Celsius and Fahrenheit scales can have negative temperature values. The Kelvin scale cannot because the zero point on the Kelvin scale is absolute zero, the lowest possible temperature.

2.88 (a) 72.5 °F; 295.6 K (b) −19.4 °C; 253.7 K (c) 373.2 K; 212.0 °F (d) 149 °F; 338.2 K

2.91 (a) Left cylinder: ±0.01 mL
Right cylinder: ±1 mL
(b) 98 mL + 1.18 mL = 99 ±1 mL

2.93 The student who reports 1.5 cm used the ruler incorrectly by not reporting enough digits.

2.96 997 g

2.98 9×10^2 g

2.100 The two students measure the same density because density does not depend on the mass of a sample.

2.102 5.14×10^5 s

2.104 0.007 083 dollars/s

2.106 0.264 gallon

2.109 (a) 101.4 cm (b) 1.043×10^6 cm^3 (c) 0.106 g/mL

2.110 5.51 L

2.111 179 miles/h

2.112 In any equation, the two sides are equal to each other, by definition. In order for the sides to remain equal, whatever is done to one side must be done to the other.

2.114 $x = z - y$

2.116 57.5 g

2.118 The capacity for doing work.

2.120 (a) 4500 cal = 4.50×10^3 cal
(b) 2510. kJ
(c) 0.2390 cal
(d) 2.09×10^5 J

2.122 The aluminum block requires more heat energy because aluminum has the higher specific heat (aluminum 0.215 cal/g°C, iron 0.107 cal/g°C). How much more is determined from the ratio of the specific heats: (0.215 cal/g°C)/(0.107 cal/g°C) = 2.01 times as much, or if you use SI, (0.901 J/g°C)/(0.449 J/g°C) = 2.01 times as much.

2.123 1.51×10^5 J = 151 kJ

2.125 5.5×10^3 J, 2.2×10^3 J/g

2.126 2.44×10^4 J

2.128 The temperature of the block increases to a blistering 4916 °C!

2.130 (a) 7.98×10^{17} L (b) 3.00×10^{-6} g (c) 1.11×10^5 gallons

2.132 9.3 g/cm^3

2.134 26.8 m/s

2.136 (a) 22 °C, 295 K (b) 10.4 °F, 261 K (c) −95 °C, −139 °F

2.138 1.400 g/mL

2.139 (a) $°C = \frac{5}{9}(°F - 32)$

(b) $T = \dfrac{PV}{nR}$

(c) $\lambda = \dfrac{hc}{E}$

2.141 (a) 1.34×10^{-3} g/mL
(b) 1.34×10^{-3} kg/L
(c) 1.34×10^{-6} kg/mL

2.142 2.7×10^2 g

2.145 (a) Neither. (b) Both.
(c) Accurate but not precise.
(d) Precise but not accurate.

2.146 8.0×10^3 cal; 8.0 Cal

2.149 2.69 g/mL

2.150 6.12 kJ

2.154 ±0.001 mL

2.157 (a) ±0.1 m
(b) ±0.0001 g
(c) $\pm 0.001 \times 10^3$ L = ±1 L
(d) ±1 cm (e) No uncertainty.

2.160 Because precision refers to how close to one another a series of measurements are, this term can never be used to describe a single measurement.

2.162 2.2×10^{-4} m^3

2.163 (a) 404 atomic mass units
(b) 6.71×10^{-22} g (c) 20.2 g

2.165 (a) 15.0 mL (b) 16.4 mL (c) 2×10^1 mL (d) 1.32 mL (e) 1.10 mL (f) 8.3×10^4 mL

2.167 (a) Six. (b) Three. (c) Two. (d) Four. (e) Four.
2.170 (a) 0.249 cal/g°C (b) 0.20 cal/g°C
2.173 (a) 5.02×10^5 (b) 3.8402×10^{-5} (c) 4.36×10^8 (d) 8.47×10^3 (e) 5.91×10^{-3} (f) 6.58×10^{-1}
2.175 227 U.S. dollars
2.177 (a) Two. (b) Two. (c) None. (d) Cannot tell. (e) Three.
2.180 The two units are the same size.
2.182 (a) 0.0179 (b) 0.000 000 008 76 (c) 48,800,000,000 (d) 75.2 (e) 8.37 (f) 41,840
2.183 (a) 7.80 g/cm^3 (b) 64.1 cm^3 (c) Floats.
2.185 (a) 1.3×10^{-3} kg/L (b) 1.1×10^{-2} lb/gallon
2.188 (a) 5.70×10^2 g (b) 3.93×10^1 g (c) 7.0×10^{-2} g (d) 4.2×10^{-3} g (e) 6.80×10^2 g (f) 9.65×10^2 g
2.190 8.50 atm
2.192 Incorrect because 1 ft equals exactly 12 in. by definition, with no uncertainty.
2.194 507 U.S. dollars
2.197 (a) Four. (b) Two. (c) Three. (d) Eight. (e) Four.
2.199 Compressing means squeezing a given mass of the gas into a smaller volume. Because the relationship is density = mass/volume and the mass is constant, decreasing the volume causes the density to increase.
2.200 (a) 269 J (b) 541 J (c) 84 J (d) 2.51×10^3 J
2.202 Shooter A is more precise (there is little spread of data—all the arrows hit the same mark). Shooter B is more accurate (all three arrows come closer to a bull's-eye) than Shooter A.

Chapter 3

3.5 The law of conservation of matter requires that the total mass of the substances produced equal the mass of coal plus the mass of oxygen reacting. The coal seems to disappear because it is being converted to carbon dioxide, a colorless, odorless gas. If the carbon dioxide were captured and weighed, the mass would equal the mass of the coal and oxygen reacting to produce it.

3.6

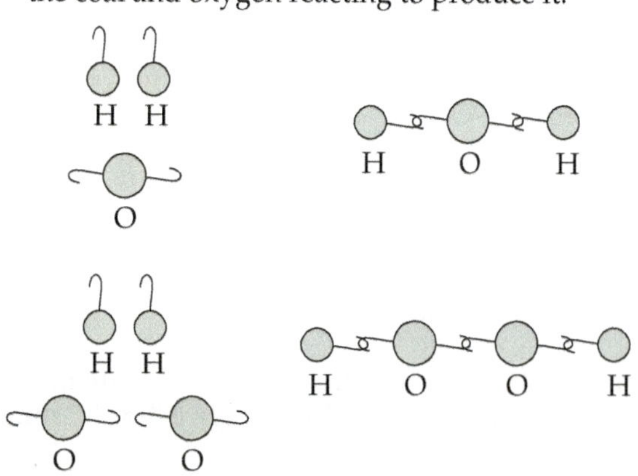

3.8 $^{79}_{35}Br$, $^{81}_{35}Br$
3.9 The atoms of both isotopes have 35 electrons.

3.10

	$^{14}_{7}N$	$^{24}_{12}Mg$	$^{23}_{11}Na$	$^{59}_{26}Fe$
Mass number	14	24	23	59
Atomic number	7	12	11	26
Number of protons	7	12	11	26
Number of neutrons	7	12	12	33
Number of electrons	7	12	11	26

3.12 $3.982\,81 \times 10^{-5}$ g
3.13 (a) 100% − 75.77% = 24.23% (the 100 is taken to be an exact number).

(b) $$\left(34.969\ \text{amu} \times \frac{75.77}{100}\right) + \left(36.966\ \text{amu} \times \frac{24.23}{100}\right)$$
$$= 35.45\ \text{amu}$$

(c) $^{37}_{17}Cl$ is $\frac{36.966}{34.969} = 1.0571$ times more massive than $^{35}_{17}Cl$.

3.15 No.
3.16 Similar physical and chemical properties.
3.18 Iodine (I) and astatine (At); periodicity for I is 18 and for At is 32.
3.19 The periodicity would be 2, 8, 8, 8, 8, 8, 8 instead of 2, 8, 8, 18, 18, 32, 32.
3.21 Rb < Zr < Ag < I
3.22 $^{56}_{26}Fe^{3+}$; group VIIIB (8).
3.23 $^{16}_{8}O^{2-}$; chalcogen group.
3.25 (a) %S: $$\frac{94.08\ \text{g S}}{100.00\ \text{g sample}} \times 100\% = 94.08\%\ \text{S}$$

%H: $$\frac{(100.00\ \text{g} - 94.08\ \text{g})}{100.00\ \text{g sample}} \times 100\% = 5.92\%\ \text{H}$$

(b) The law of constant composition.

3.27 (a) %I: $$\frac{126.9\ \text{g I}}{162.4\ \text{g compound A}} \times 100\% = 78.14\%\ \text{I}$$

%Cl: 100% − 78.14% I = 21.86% Cl

(b) %I: $$\frac{126.9\ \text{g I}}{233.3\ \text{g compound A}} \times 100\% = 54.39\%\ \text{I}$$

%Cl: 100% − 54.39% I = 45.61% Cl

3.30 (a)

Compound	Mass of Na present
A	11.50 g
B	45.98 g

3.32 mass of proton/mass of electron = 1.007 *amu*/5.486×10^{-4} *amu* = 1836
3.34 It would imply that any positive charges in the atom were evenly spread throughout the atom and not concentrated in a tiny area in the center. This would mean that the structure of the atom would be more consistent with Thomson's "plum-pudding" model.
3.35 The mass number alone is not sufficient in determining elemental identity as there can be atoms of different elements with the same mass number, for instance: $^{15}_{8}O$ and $^{15}_{7}N$. Here, the atomic numbers of 8 and 7 identify atoms of

oxygen and nitrogen, respectively, yet both have the same mass number of 17.

3.36 Yes. In a neutral atom the number of electrons is equal to the number of protons, and any atom's identity is determined by its proton number.

3.38

	$^{15}_{8}O$	$^{16}_{8}O$	$^{37}_{17}Cl$	$^{23}_{11}Na$
Mass number	15	16	37	23
Atomic number	8	8	17	11
Number of protons	8	8	17	11
Number of neutrons	7	8	20	12
Number of electrons	8	8	17	11

3.40 (a) The atomic number and the symbol do not agree. The element with the atomic number of 7 is nitrogen, not carbon.
(b) To make the symbol correct by adjusting the atomic number, the atomic number must be changed to 6: $^{14}_{6}C$. It symbolizes an atom of a carbon-14 isotope.

3.42 $^{1}_{1}H$

3.44 The atomic mass of an element is the weighted average of the masses of all its naturally occurring isotopes. The mass number is the sum of the protons and neutrons in an atom of the element.

3.46 $^{12}_{6}C$ is the only isotope whose atomic mass and mass number are equal. This is true because chemists have assigned $^{12}_{6}C$ a mass of exactly 12 amu.

3.48 An "average" titanium atom is 3.990 times more massive than the $^{12}_{6}C$ atom: $\frac{47.88}{12} = 3.990$

3.51 (a) Exactly 235 amu.
(b) Because with uranium-235 as the standard reference, the atomic masses of all the lighter elements would be a very small fraction of 235.

3.52 (a) 49.31%
(b) Atomic mass of Br

$$= \left(78.918\,336 \text{ amu} \times \frac{50.69}{100}\right) +$$
$$\left(80.916\,289 \text{ amu} \times \frac{49.31}{100}\right) = 79.90 \text{ amu}$$

3.53 The naturally occurring hydrogen on the other planet must contain more of the more massive $^{2}_{1}H$ isotope than on Earth.

3.55 (a) According to atomic masses.
(b) He discovered repeating behavior every 8 elements.
(c) Mendeleev's ordering was by atomic mass; the modern ordering is by atomic number.

3.57 Periodicity refers to the fact that the chemical properties repeat themselves every 8, 18, or 32 elements.

3.58 8

3.60 IA (1), alkali metals; IIA (2), alkaline earth metals; VIA (16), chalcogens; VIIA (17), halogens; VIIIA (18), noble gases.

3.62 A group is a vertical column of elements with similar chemical properties. A period is a horizontal row of elements.

3.64 (a) $BeCl_2$, $CaCl_2$, $SrCl_2$, $BaCl_2$, and $RaCl_2$
(b) $BeBr_2$, $MgBr_2$, $CaBr_2$, $SrBr_2$, $BaBr_2$, and $RaBr_2$
(c) The chemical behavior of elements in the same group will be similar.

3.65 They are among the most unreactive substances known, and they are the only group in the periodic table that contains only gases.

3.67 The transition metal portion is 10 elements wide; the lanthanide/actinide portion is 14 elements wide.

3.68 The change from 8 to 18 is due to the transition metals; the change from 18 to 32 is due to the lanthanides and actinides.

3.70 Opposites attract! A negatively charged electron in an atom is attracted to the positively charged nucleus via coulombic forces. To remove an electron from an atom means to "fight" against these attractive forces, which required energy. The first ionization energy is the minimum amount of energy needed to remove an electron from a neutral atom.

3.72 (a) Adding three electrons to a neutral atom would convert it into an anion of charge −3.
(b) Removing four electrons from a neutral atom would convert it into a cation of charge +4.
(c) Adding two neutrons to the nucleus of a neutral atom would have no effect on its charge.

3.74

	$^{15}_{8}O^{+}$	$^{27}_{13}Al^{3+}$	$^{31}_{15}P^{3-}$	$^{58}_{28}Ni^{+}$
Mass number	15	27	31	58
Atomic number	8	13	15	28
Number of protons	8	13	15	28
Number of neutrons	7	14	16	30
Number of electrons	7	10	18	27
Charge on ion	+1	+3	−3	+1

3.76 Wrong. After the first electron has been removed from an atom, the number of protons in the nucleus remains the same, resulting in the atom being converted into a +1 cation. The second electron now must be pulled away from this +1 cation whereas the first electron had to be pulled away from a neutral atom. Since opposite charges attract, it is harder to remove the second electron, so the second ionization energy is always greater than the first ionization energy.

3.78 $^{41}_{20}Ca^{2+}$. Creating this ion from the neutral atom requires the removal of two electrons. This will always take more energy than the first ionization energy, which is the minimum energy required to remove just one electron from the atom.

3.80 Mg is the most difficult to ionize; K has the smallest first ionization energy.

3.82 (a) Sodium easily loses an electron to form the cation Na^+, and each Cl atom in Cl_2 easily gains an electron to form the anion Cl^-.
(b) Being a group IA (1) metal, lithium has a low first ionization energy and easily loses one electron to become a Li^+ cation (just like sodium). Being a group VIIA (17) nonmetal, each bromine atom in Br_2 tends to gain one electron to become a Br^- anion (just like chlorine). The result is the formation of lithium bromide, LiBr, similar to sodium chloride, NaCl.

3.85 A line spectrum is made up of specific colors separated by dark regions. A continuous spectrum is made up of all colors, blending smoothly from one to the next.

3.87 Oxide A, 21.23% O; oxide B, 11.88% O.

3.88 You would have to show that the sum

Mass of wood before burning +
Mass of oxygen consumed

was equal to the sum

Mass of ash + Mass of carbon dioxide +
Mass of water vapor

3.90 2 g of B; the law of conservation of matter.

3.92 Incorrect. Isotopes of a given element differ in numbers of neutrons but have the same numbers of protons.

3.94 Adding a proton doesn't create a cation, even though protons carry a positive charge. Adding a proton changes the atom's identity. In this case, sodium (11 protons) becomes magnesium (12 protons).

3.96 Protons 26, electrons $26 - 2 = 24$, neutrons $56 - 26 = 30$.

3.100 All elements in a given group had to have similar chemical properties.

3.102 Cs < Na < S < Cl

3.106 Magnesium, Mg

3.108

Name of group or classification	Period	Group	Elemental symbol	Atomic number	Atomic mass (amu)	Metal, metalloid, or nonmetal?
Transition metal	4	VIIIB (8)	Fe	26	55.845	Metal
Noble gas	1	VIIIA (18)	He	2	4.003	Nonmetal
Halogen	5	VIIA (17)	I	53	126.905	Nonmetal
Alkali metal	4	IA (1)	K	19	39.098	Metal
Halogen	3	VIIA (17)	Cl	17	35.453	Nonmetal
Noble gas	2	VIIIA (18)	Ne	10	20.180	Nonmetal
Chalcogen	2	VIA (16)	O	8	15.999	Nonmetal
—	3	IVA (14)	Si	14	28.086	Metalloid
Actinide	7	—	U	92	238.029	Metal

3.109 (a) 70.01 g Cl (b) 36.92% Ca (c) 63.08% Cl

3.111 The first ionization energy decreases as you go down any group (it becomes easier to remove an electron).

3.114

	Atomic number	Mass number	Protons	Neutrons	Electrons
$^{27}_{13}Al$	13	27	13	14	13
$^{60}_{27}Co$	27	60	27	33	27
$^{200}_{79}Au$	79	200	79	121	79
$^{238}_{92}U$	92	238	92	146	92
$^{127}_{53}I$	53	127	53	74	53

3.117 (a) Alkali metals. (b) Alkaline earth metals. (c) Halogens. (d) Noble gases or rare gases.

3.119 (a) $^{17}_{8}O$ (b) $^{119}_{50}Sn$ (c) $^{23}_{11}Na$ (d) $^{58}_{28}Ni$ (e) $^{137}_{56}Ba$

3.122 (a) Fluorine. (b) Calcium. (c) 1− charge. (d) 2+ charge. (e) $^{40}_{20}Ca^{2+}$, $^{19}_{9}F^-$.

3.125 Ne < Be < Li < Na < Cs

3.128 (a) 30.83% (b) 63.55 amu

3.132 Chart (b) is correct, since the naturally occurring mass is much closer to that of Cl-35 than it is to Cl-37.

3.134

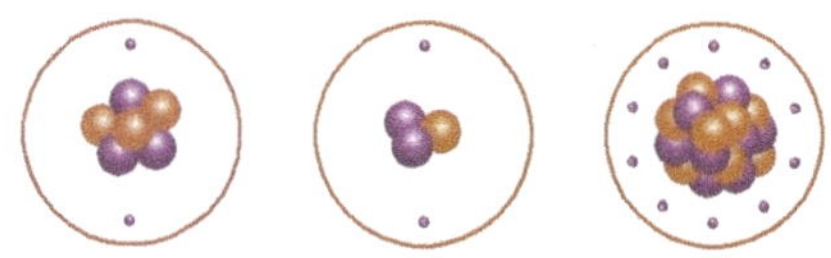

Element	Li	H	O
Ion (yes or no)	yes	yes	yes
Charge of atom	+1	−1	−2
Atomic number	3	1	8
Mass number	6	3	16

3.135 $$100{,}000 \times 24.25\ \text{mm}\left(\frac{\text{cm}}{10\ \text{mm}}\right) \times \left(\frac{\text{m}}{100\ \text{cm}}\right) \times \left(\frac{\text{mile}}{1609.344\ \text{m}}\right) = 1.507\ \text{miles}$$

Chapter 4

4.4 Radio and television waves; this is very low energy electromagnetic radiation.

4.6

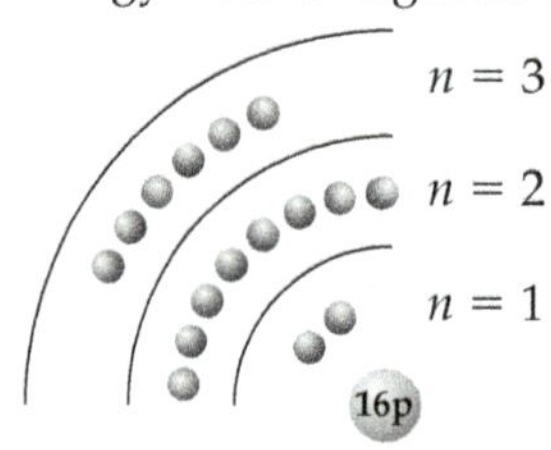

Two more electrons.

4.7 An electron in a shell with a low value of n is closer to the nucleus and feels a stronger attraction than one that is farther from the nucleus. The stronger attraction makes the closer electron more stabilized.

4.8 $2 \times 5^2 = 50$ electrons

4.10

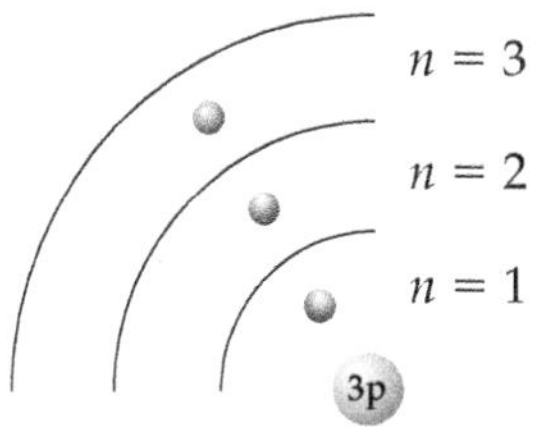

4.11

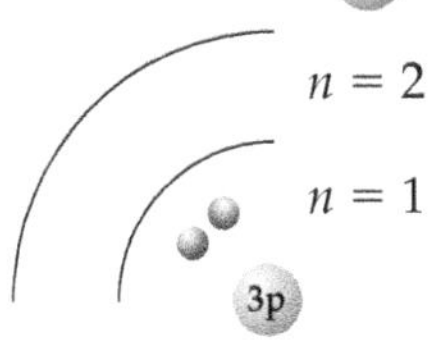

4.12 (a) Carbon. (b) Anion with −2 charge.
(c)

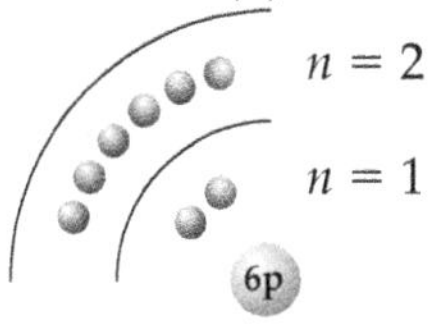

4.13 The $n = 1$ shell has 3 electrons in this diagram; the maximum is 2 electrons for this shell. The $n = 2$ shell has only 6 electrons; the maximum is 8 for this shell, and it should be filled before an electron is placed in the $n = 3$ shell.

4.15 As has a total of 33 electrons: $1s^22s^22p^63s^23p^63d^{10}4s^24p^3$; yes, there are 5 valence electrons, and the group number is VA.

4.16 Sc has 21 electrons: $1s^22s^22p^63s^23p^63d^14s^2$.

4.18 $1s^22s^22p^63s^23p^64s^23d^{10}4p^6$; it is proper for Kr to be in group VIIIA because it has 8 valence electrons.

4.19 Pd has 46 electrons: $1s^22s^22p^63s^23p^64s^23d^{10}4p^65s^24d^8$.

4.21 $1s^22s^22p^63s^23p^64s^23d^{10}4p^65s^24d^{10}5p^66s^24f^{14}5d^{10}6p^67s^2$;[Rn]$7s^2$

4.22 $1s^22s^22p^63s^23p^64s^23d^{10}4p^65s^24d^{10}5p^66s^24f^{14}5d^{10}6p^67s^26d^15f^3$; [Rn]$7s^26d^15f^3$

4.24 The highest value of n is 4, so it is in the fourth period; the element is vanadium (V).

4.25 Period 6; group IA; the element is cesium (Cs).

4.31

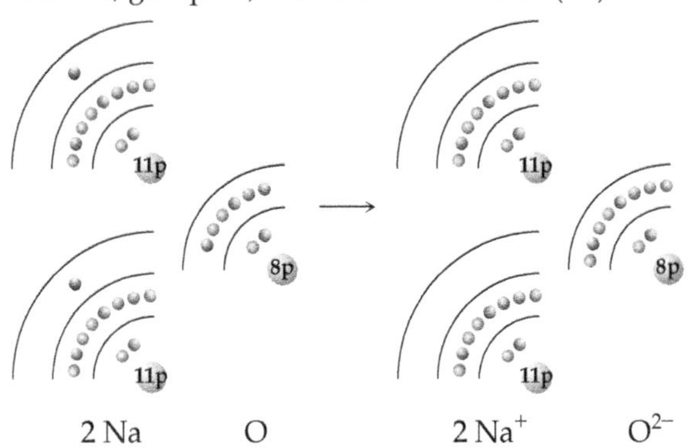

4.32 Ba is in group IIA and will lose 2 electrons to become Ba^{2+}.
F is in group VIIA and will gain 1 electron to become F^-.
The formula of the compound is BaF_2.

4.33 Al is in group IIIA and will lose 3 electrons to become Al^{3+}.
O is in group VIA and will gain 2 electrons to become O^{2-}.
The formula of the compound is Al_2O_3.

4.35 $E = hc/\lambda$, so $E\lambda/h = c = 3.00 \times 10^8$ m/s. Since $E\lambda/h$ is a constant, its value does not depend on the wavelength of light. In other words, since the energy and the wavelength are inversely proportional, when λ goes up, E goes down and $E\lambda$ remains unchanged.

4.36 (a) According to the calculation below, X-rays are 1000 times more energetic than green light:

For X-rays: $E = (6.626 \times 10^{-34}\,\text{J}\cdot\text{s}) \times (3.00 \times 10^8\,\text{m/s})/(1 \times 10^{-10}\,\text{m}) = 1.99 \times 10^{-15}\,\text{J}$

For green light: $E = (6.626 \times 10^{-34}\,\text{J}\cdot\text{S}) \times (3.00 \times 10^8\,\text{m/s})/(1 \times 10^{-7}\,\text{m}) = 1.99 \times 10^{-18}\,\text{J}$

The ratio $= 1.99 \times 10^{-15}\,\text{J}/1.99 \times 10^{-18}\,\text{J} = 10^3 = 1000$

(b) X-rays are at the high-energy end of the electromagnetic spectrum and can cause damage to cells and tissues. Green light carries less energy and is, in fact, part of the visible light, which surrounds us all.

4.38 8.3 minutes for both.

4.39 (2) is true because in the relationship equation, wavelength is in the denominator: $E = hc/\lambda$.

4.42 $$E = \frac{(6.63 \times 10^{-34}\,\text{J}\cdot\text{s})(3.00 \times 10^8\,\text{m/s})}{10\,\text{m}}$$
$$= 1.99 \times 10^{-26}\,\text{J}$$

4.45 You would appear at the starting line, then a certain time later you would instantaneously appear some distance ahead, and later still some farther distance ahead, until finally you would appear instantaneously at the finish line.

4.47 "Quantized energy" means that the energy can have only certain allowable values and no values in-between.

4.49 It tells us that something inside the atom is allowed to possess only certain energies, as opposed to any energy.

4.50 Energy and stability are inversely related to each other. The higher the energy of an object, the less stable it is; the lower the energy, the more stable.

4.51 (a)

4.53 As n increases, an electron's energy and distance from the nucleus increase.

4.55 Electron shell.

4.56 (a) Energy must be added to an atom to remove an electron from the nucleus because the attractive forces holding the negative electron near the positive nucleus must be overcome.
(b) The higher the value of n, the further away from the nucleus the electron tends to be and the weaker the attraction between the nucleus and the electron. The weaker this attraction to the nucleus is, the less energy is needed to remove the electron from an atom.

4.59 The lower shells are of lower energy; the electrons will go into lower-energy positions before beginning to go into higher-energy (less stable) positions.

4.60 Excited state, because the $n = 2$ energy shell is not filled, yet there is an electron in the $n = 3$ shell.

4.63 The energy of an electron depends on distance from the nucleus. The Bohr model assigns each allowed distance from the nucleus an energy value. Therefore, saying an electron can be only at certain distances from the nucleus means the electron can have only certain energy values.

4.65 Because they have identical valence-shell configurations.

4.66 A line spectrum is a record of something giving off discrete amounts of energy. The Bohr model says electrons in an atom give off discrete amounts of energy as they change orbits, which means the atomic spectra they produce are line spectra.

4.68 True; the H^+ cation has no electrons.

4.70 (a) 12.1 eV (b) 12.1 eV (c) Ultraviolet.

4.72 The energy of the shells would have to be continuous, not quantized.

4.73 (a) 1: Nothing wrong. The fact that it has 7 protons but only 6 electrons makes it a cation. No shell overfilled.
2: Nothing wrong. Same number of protons and electrons. No shell overfilled.
3: Too many electrons in $n = 1$ shell, which holds a maximum of 2 electrons.
(b) The full atomic symbol for (1) is ${}^{14}_{7}N^+$; that for (2) is ${}^{14}_{7}N$.
(c) Diagram (1) is a ground state; diagram (2) is an excited state.

4.74 (a) The (roman-numeral) group number signifies the number of valence-shell electrons in an atom. E.g., Na is in group 1 and has one valence-shell electron.
(b) The period number signifies the maximum value of the principal quantum number, n, which in turn determines the valence shell of a ground state atom, E.g., Na is in the third period, and its valence-shell electrons are found in the $n = 3$ shell.

4.76 1.24×10^{-6} m; 1.24×10^{3} nm

4.77 $E = 0.700$ eV; electron relaxes to $n = 3$ shell.

4.78 The simple Bohr model predicts that there should be 9 electrons in the $n = 3$ shell. However, the chemical behavior of K suggests that it has 1 valence electron in the $n = 4$ shell.

4.80 (a) $n = 1, 2, 3, 4$, etc.
(b) s, p, d, f
(c) The number of subshells in a given shell is equal to n; for example, $n = 1$ has one subshell (s); $n = 2$ has two subshells (s and p); $n = 3$ has three subshells (s, p, and d); etc.

4.82 In the 4s subshell instead of in the 3d subshell.

4.84 (a) $1s^22s^22p^1$
(b) $1s^22s^22p^63s^23p^64s^23d^1$
(c) $1s^22s^22p^63s^23p^64s^23d^7$
(d) $1s^22s^22p^63s^23p^64s^23d^{10}4p^4$
(e) $1s^22s^22p^63s^23p^64s^23d^{10}4p^65s^24d^6$

4.87 (a) $[Xe]6s^2$ (b) $[Xe]6s^24f^{14}5d^4$
(c) $[Xe]6s^24f^{14}5d^{10}6p^2$ (d) $[Xe]6s^25d^14f^2$
(e) $[Rn]7s^26d^15f^2$

4.88 (a) Group VA; period 2.
(b) Group IA; period 3.
(c) Group VIIA; period 4.

4.89 (a), (b), (c), (d), and (e) are all incorrect; (f) is correct.
(a) is incorrect because 2s should come after 1s.
(b) is incorrect because the 2s subshell should be followed by the 2p subshell.
(c) is incorrect because the 1s subshell must come first.
(d) is incorrect because the 2p subshell holds a maximum of 6 electrons.
(e) is incorrect because the 2p subshell would accommodate 6 electrons before the 3s subshell would be used.

4.90 (a) 5 (b) 1 (c) 2 (d) 8 (e) 2

4.92 O: $1s^22s^22p^4$; O^{2+}: $1s^22s^22p^2$; O^{2-}: $1s^22s^22p^6$; O^{2-} should be found in most compounds of oxygen because the octet rule is satisfied.

4.94 The s block is 2 elements wide because each s subshell holds 2 electrons.
The p block is 6 elements wide because each p subshell holds 6 electrons.
The d block is 10 elements wide because each d subshell holds 10 electrons.
The f block is 14 elements wide because each f subshell holds 14 electrons.

4.95 (a) Since the electrons are being placed in the same valence shell, one might expect the size to stay the same.
(b) The atomic size decreases because 1 electron is added to the valence shell and 1 proton is added to the nucleus. The added proton makes the nucleus more positive, which causes it to more strongly attract the surrounding electrons. The increased pull on the electrons shrinks the atom.

4.97 Be; the greater nuclear charge for Be (+4) compared to Li (+3) pulls the electrons in closer to the nucleus, shrinking the subshells.

4.100 Na; atoms get larger going down a group.

4.101 Si is the smallest, next is Mg, then Na, and then Rb is the largest.

4.102 (a) As you move from left to right one step across the period, one proton is added to the nucleus, and one more electron is added outside the nucleus.
(b) The added positive nuclear charge pulls the electrons in closer to the nucleus, causing the size of the atoms to decrease.

4.104 F $<$ S $<$ Se $<$ Ca

4.105 As you go down a group, the size of the atom increases and an outermost electron is farther away from the nucleus. It thus feels less pull from the nucleus, is easier to remove, and its separation requires less energy. As you go across a period, the atomic size decreases, and an outermost electron is closer to the positively charged nucleus. More energy is thus needed to remove the electron.

4.106 The octet rule states that elements will react in such a way as to put 8 electrons in the valence shell. Having 8 valence electrons gives them a noble-gas valence electron configuration and makes them exceptionally stable.

4.108 By gaining enough electrons to have 8 in their outermost occupied shell.

4.110

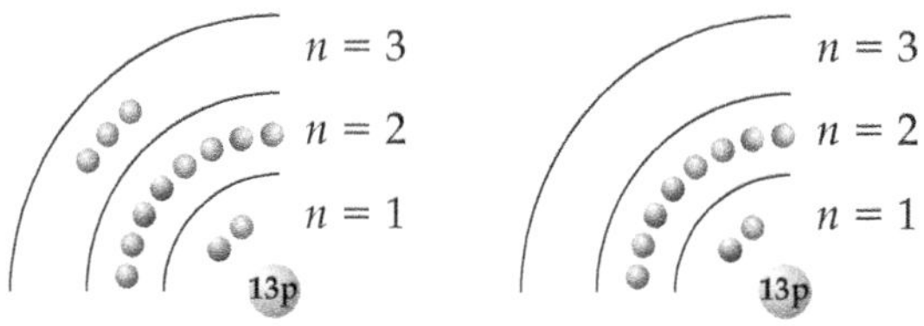

Al, 3 electrons in valence shell — Al^{3+}, 8 electrons in valence shell

4.112 Each lithium atom has 1 valence electron, which it will lose to obtain a full outer shell (because Li^+ has only two electrons). Each nitrogen atom has 5 valence electrons and will gain 3 to have an octet. Thus, each nitrogen atom will need 3 lithium atoms, and the formula of the compound will be Li_3N.

4.114 (a) $^{23}_{11}Na$ and $^{31}_{15}P$

(b) Na is the metal, and P is the nonmetal.

(c) Na_3P

(d)

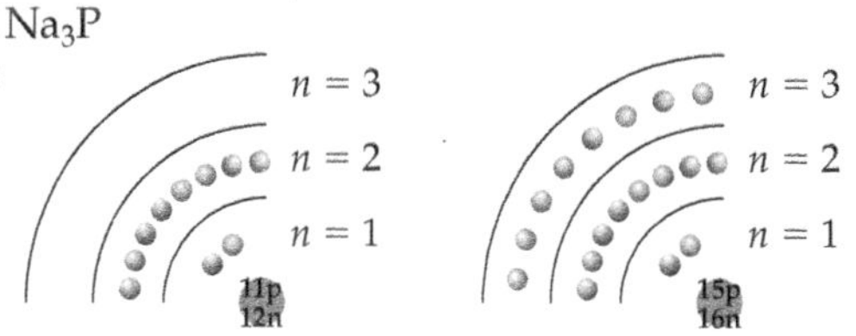

Na^+ ion — P^{3-} ion

4.116 Hydrogen will sometimes lose its valence electron, similar to what the other elements in group IA do to obtain an octet. However, when hydrogen loses its electron, it does not obtain an octet; it is left with no electrons.

4.117 Aluminum (Al) has 3 valence electrons, which it will lose to form a +3 cation. Both oxygen and sulfur have 6 valence electrons and will gain 2 electrons to form −2 anions. Because Al needs to lose 3 electrons and oxygen and sulfur both need to gain 2 electrons, each compound will contain 2 Al^{3+} cations and 3 anions. The formulas are similar because both oxygen and sulfur have the same number of valence electrons.

4.119 It is true that Mg^{2+} and Na^+ have identical configurations, but they do not have identical properties. The ions have different charges, different sizes, and different chemical reactivities.

4.120 The Roman-numeral group number is equal to the number of valence electrons the metal will lose. The cation will have a positive charge equal to the Roman-numeral group number.

4.122 (a) The charges on the cation and the anion tell how many electrons are lost and gained, respectively, to form the ions. The formula of the compound is simply the minimum number of each ion needed to have an equal number of electrons transferred.

(b) To form Ca^{2+} and N^{3-} ions, Ca must have lost two electrons and N must have gained three electrons. To form a compound, the number of electrons lost and gained must be the same. Therefore, three Ca^{2+} ions will combine with two N^{3-} ions for form Ca_3N_2 and the number of electrons transferred in this reaction is six.

4.123 Bohr's model says we can know precisely the present and future positions of an electron; quantum mechanical theory says we cannot.

4.125 An orbital is a region of space around the nucleus of an atom in which it is most probable that an electron will be found.

4.126

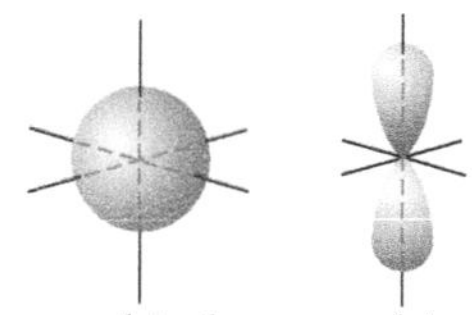

s orbital — *p* orbital

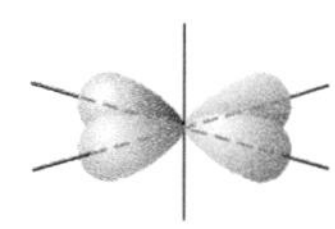

d orbital

4.127 You would see a smaller, fuzzy, spherical cloud convert into a still fuzzy, but larger, dumbbell shaped cloud.

4.129 1.61×10^{-2} s

4.131 2.50×10^4 nm; 33.3 times longer.

4.133

Wavelength (m)	Wavelength (nm)	Energy (J)
3.52×10^{-8} m	35.2 nm	5.65×10^{-18} J
2.6020×10^{-6} m	2602.0 nm	7.64×10^{-20} J
7.85×10^{-12} m	7.85×10^{-3} nm	2.53×10^{-14} J

4.135 Released.

4.137 (a) Lower-energy excited state:

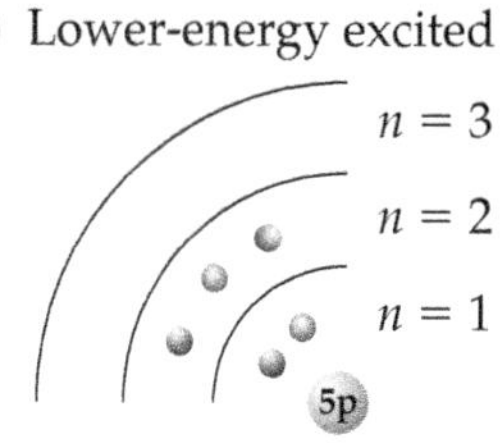

(b) Ground state:

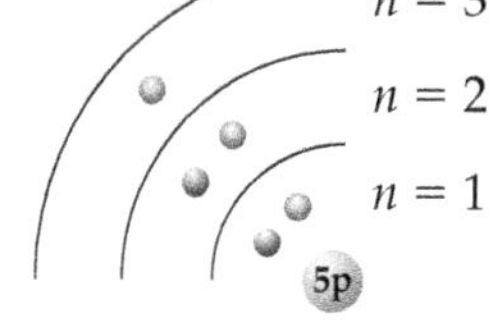

4.139 2.11×10^{-18} J, 94.2 nm
4.141 (a) Ground state. (b) Incorrect (electrons go into 4*s* before 3*d*). (c) Incorrect (2*p* overfilled).
4.143 *d* subshell. It gets filled *after* the next-higher *s* subshell.
4.145 They all have the same electron configuration.
4.147 (a) N (b) Ba (c) Ca
4.151 Rb < Ca < Se < S < F
4.152 Li_3N
4.154 0.199 nm
4.156 (c)
4.158 (a)
4.160 (e)
4.162 3.98×10^{-19} J
4.164 (a) $CaBr_2$ (b) K_3N (c) Al_2S_3 (d) NaI (e) MgO
4.166 Decreasing wavelength, increasing energy.
4.170 Mg $n = 3$ shell; Ge $n = 4$ shell; W $n = 6$ shell; Cl $n = 3$ shell; Cs $n = 6$ shell.
4.171 1.99×10^{-13} J
4.173 Germanium, atomic number 32; group IVA (14), period 4.
4.176 5.36×10^{-3} s
4.178 Phosphorus, 9. Magnesium, 6. Selenium, 16. Zinc, 12.
4.182 (a) Bromine. (b) Boron. (c) Rubidium. (d) Uranium. (e) Calcium.
4.184 (a) 4 (b) 4 (c) 5 (d) 6
4.185 (a) Relaxation occurs from the $n = 4$ energy level to a point between allowed energy levels $n = 1$ and $n = 2$. This is strictly not allowed. Relaxation must be to the allowed energy levels only.
(b) 12.8 eV
(c)

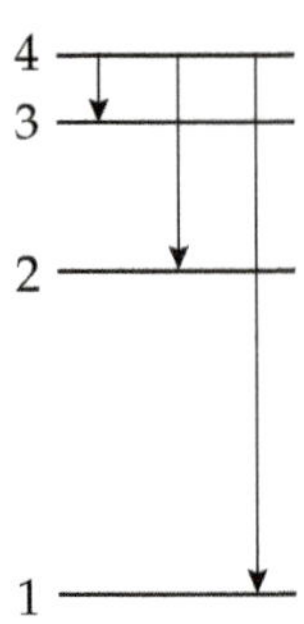

4.187 b, d, and e are correct statements.
4.189 Lithium's valence electron is most likely to be found in the volume described by a spherical shaped cloud about its nucleus.

Chapter 5

5.2 PH_3
5.3 $SiBr_4$
5.5 H:F̤̈:
5.6 Oxygen has 6 valence electrons. To achieve an octet, it will form two bonds with hydrogen to give H:Ö̤:H.
5.8 H—Ö̤—C̤̈l:
5.9

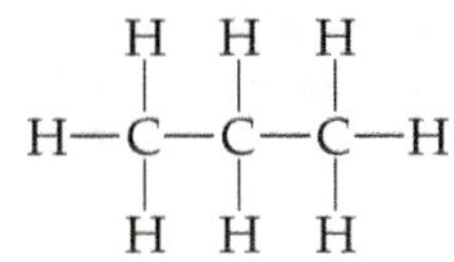

5.11

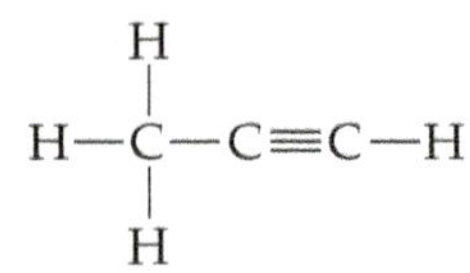

5.12 H—C≡N:
5.13

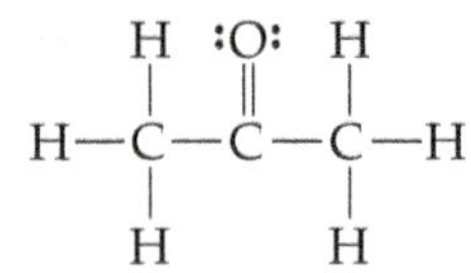

5.15

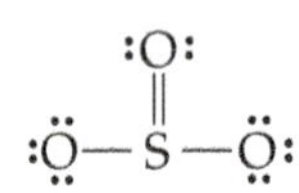

(plus two additional resonance forms)

5.16

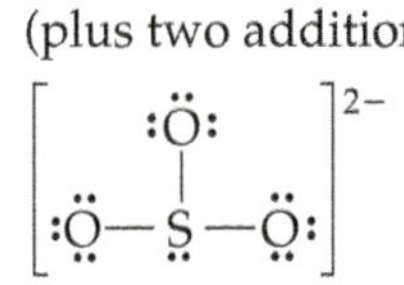

5.17 $[:N{\equiv}O:]^+$
5.18 No, because the atoms are arranged differently.
5.20 Identical atoms mean $\Delta EN = 0$ and therefore both bonds are pure covalent.
5.21 EN Si = 1.8, EN O = 3.5; $\Delta EN = 1.7$; a polar covalent.
5.22 Because Be is at the top of group IIA, she knows that Be is more electronegative than Ba. Therefore, ΔEN for Ba/Cl is greater than ΔEN for Be/Cl, making $BaCl_2$ more ionic.
5.24

Formula	Cation	Anion	Name
CaF_2	Ca^{2+}	F^-	Calcium fluoride
CsBr	Cs^+	Br^-	Cesium bromide
Al_2S_3	Al^{3+}	S^{2-}	Aluminum sulfide
K_2O	K^+	O^{2-}	Potassium oxide

5.25 All the group IIA bromides contain two bromide ions.
5.27 Titanium(III) chloride.
5.28 SnF_2
5.30 Nitrogen monoxide.
5.31 Phosphorus pentachloride.
5.32 P_4S_{10}
5.34 Na_2CO_3
5.35 Calcium hypochlorite.

5.36 $Mg(HCO_3)_2$

5.40 Each potassium atom would lose its one and only valence electron to become a K+ cation. The oxygen atom would gain these two electrons to become an O^{2-} anion.

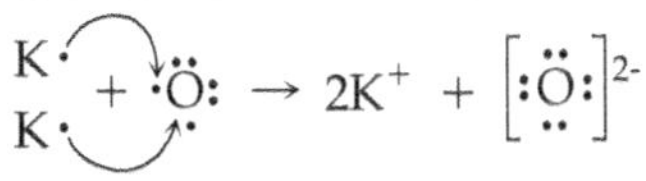

5.41 (a) $MgBr_2$ (b) BeO (c) NaI

5.50 No; molecules that contain atoms of the same element are elemental substances (for example, O_2).

5.58 Because the electrons are attracted to both nuclei; this positive–negative attractive force is what bonds the atoms to each other.

5.60 The added attractive force created when the covalent bond forms lowers the energy of the H—H system and makes the molecule more stable than the separated atoms.

5.63 Only the valence electrons are used for bonding because any inner electrons are too close to and too strongly attracted to the atom's nucleus to be shared with other nuclei.

5.66 We mean that the 2 electrons shared by 2 atoms are both counted as valence electrons for each atom sharing them. Hence, these bonding electrons are counted twice.

5.68 1 electron

5.69 The nitrogen atom is correct. The oxygen atom should be ·Ö:. Fluorine should have 7 valence electrons and should be ·F̤̈:.

5.74 Oxygen has 6 valence electrons; hence, it combines with two hydrogens to form H_2O and complete its valence-shell octet. HO would leave the oxygen 1 electron short of an octet.

5.76 ·P̤̈· and ·B̤̈r: combine to form

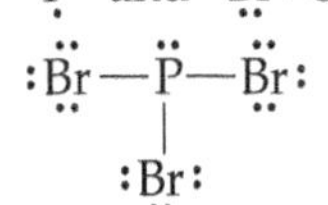

or PBr_3

5.78

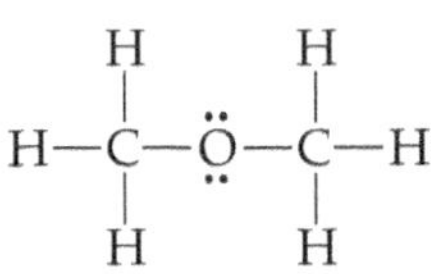

There are 8 bonding pairs and 2 lone pairs.

5.83 Three bonds; the atom has 5 valence electrons and needs 3 to achieve an octet.

5.84

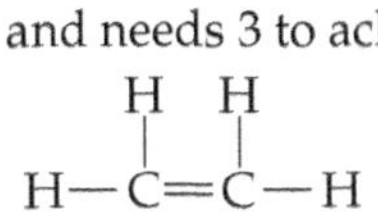

5.88 There are two resonance forms:
:Ö—Ö=Ö: and :Ö=Ö—Ö:

5.90 The O_3 molecule is an average of the two resonance forms shown in Problem 5.88. Therefore, each O—O bond is an average of a single bond and a double bond, which is a 1.5 bond.

5.92 Yes, the structure shown has too many electrons. The correct structure is :C≡O:.

5.96 :O≡O:

5.98

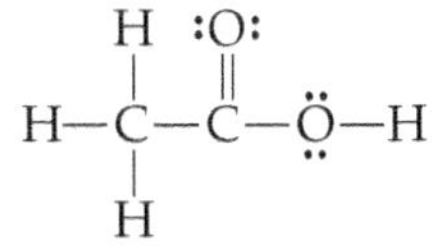

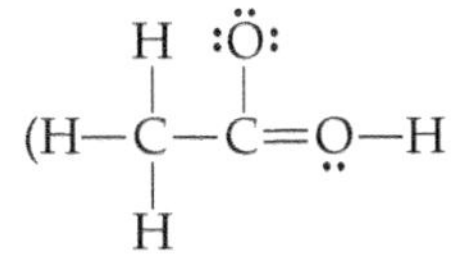

(allowed but not likely to occur)

5.103 8 valence electrons on each carbon atom and 2 valence electrons on each hydrogen atom.

5.107 Fluorine (F) is the most electronegative element; francium (Fr) is the least electronegative element. They would form the ionic compound FrF.

5.112 The electronegativities are C, 2.5, and H, 2.1. Because $\Delta EN > 0$ but < 1.7, the C—H bonds are classified as polar covalent.

5.114 H_2O

5.117 A is more electronegative because it carries the partial negative charge, indicated by $\delta-$.

5.120 If the two atoms are of the same element.

5.123 The suffix *-ide* indicates the negative part in a binary ionic compound; the negative ion gets this suffix.

5.127 (a) Calcium nitride. (b) Aluminum fluoride. (c) Sodium oxide. (d) Calcium sulfide.

5.130 (a) CuCl is copper(I) chloride or cuprous chloride; $CuCl_2$ is copper(II)chloride or cupric chloride.
(b) $Fe(OH)_2$ is iron(II) hydroxide or ferrous hydroxide; $Fe(OH)_3$ is iron(III) hydroxide or ferric hydroxide.

5.134 An ionic bond is the attractive force between oppositely charged ions that were formed by a transfer of electrons. It is similar to a covalent bond in that both have approximately the same strength and both are due to the attraction between relatively large positive and negative charges. An ionic bond differs from a covalent bond in that an ionic bond is due to a transfer of electrons, whereas a covalent bond is due to a sharing of electrons.

5.137 (a) Ammonia is the electrically neutral molecule, NH_3. Ammonium is the cation, NH_4^+.
(b) By analogy, phosphine is the electrically neutral molecule, PH_3, while phosphonium is the cation, PH_4^+.

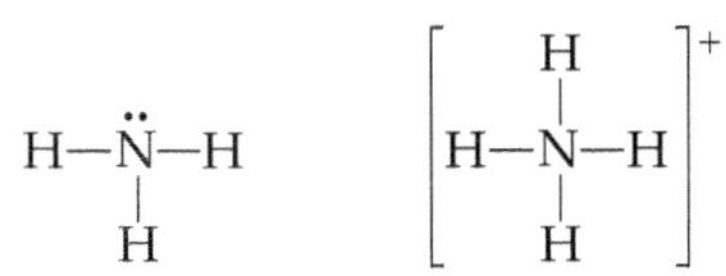

Ammonia   Ammonia ion

5.138 IO_4^-, $Mg(IO_4)_2$

5.140 The group IIA metal Mg forms a 2+ cation and therefore combines with two OH^- ions. In the formula $MgOH_2$, the subscript 2 applies only to the H. To indicate two OH^- ions, parentheses must be used: $Mg(OH)_2$.

5.142 Na_2O is a binary *ionic* compound; Greek number prefixes are used only in naming binary *covalent* compounds. Because Na always forms a 1+ cation and the oxide ion always carries a 2− charge, it is obvious without any *di-* in the name that this compound contains two Na^+ for every one O^{2-}.

5.143

Anion	Anion name	Acid formula	Acid name
F^-	Fluoride	HF	Hydrofluoric acid
NO_3^-	Nitrate	HNO_3	Nitric acid
Cl^-	Chloride	HCl	Hydrochloric acid
$C_2H_3O_2^-$	Acetate	$HC_2H_3O_2$	Acetic acid
NO_2^-	Nitrite	HNO_2	Nitrous acid

5.145 NO_3^-, nitrate ion; HNO_3, nitric acid; HNO_2, nitrous acid.

5.147 Hypochlorous acid, HClO; perchloric acid, $HClO_4$.

5.149 Subtract the Roman-numeral group number of the element from 8.

5.151 Because each noble gas has a filled valence shell.

5.153 $H{-}C{\equiv}N$

5.154 (b) is correct. (a) has only three bonds on C. (c) has only three bonds on O and violates the guideline that the least electronegative atom (not counting H) be the central atom.

5.157 (a) Two.
(b) Six.

5.159 Two.

5.160 (a) is correct. The second (b) diagram has too few electrons on S. The second (c) diagram has too few electrons on one O and too many on the other O.

5.162 (c)

5.164 (a)

5.166 $PBr_3 < MgBr_2 < KBr < CsBr$

5.168

Name	Formula
Sodium hydrogen carbonate or sodium bicarbonate	$NaHCO_3$
Magnesium acetate	$Mg(C_2H_3O_2)_2$
Barium hypochlorite	$Ba(ClO)_2$
Iron(III) nitrate or ferric nitrate	$Fe(NO_3)_3$
Ammonium sulfate	$(NH_4)_2SO_4$
Calcium phosphate	$Ca_3(PO_4)_2$
Cobalt(III) chromate or cobaltic chromate	$Co_2(CrO_4)_3$

5.170

Name	Formula
Xenon tetrafluoride	XeF_4
Xenon difluoride	XeF_2
Iodine monochloride	ICl
Bromine trichloride	$BrCl_3$
Diboron hexahydride	B_2H_6
Dinitrogen monoxide	N_2O
Tetrasulfur dioxide	S_4O_2

5.173 Because the less electronegative atom is listed first in the molecular formulas of binary covalent compounds.

5.176 (a) $Ca_3(PO_4)_2$ (b) K_2HPO_4
(c) $Mg(CN)_2$ (d) $Ba(ClO_3)_2$

5.177 (a) Iron(III) sulfite, ferric sulfite.
(b) Gold(III) nitrate.
(c) Sodium dihydrogen phosphate.
(d) Lead acetate or lead(II) acetate.

5.179 (a) $K < Li < Be$ (b) $Si < S < O$
(c) $Te < I < Br$

5.182 $N{-}Cl < \overset{\delta-}{O}{-}\overset{\delta+}{Cl} < \overset{\delta+}{S}{-}\overset{\delta-}{O} < \overset{\delta-}{O}{-}\overset{\delta+}{H} < \overset{\delta+}{C}{-}\overset{\delta-}{F}$

5.185 Manganate ion.

5.186 There are two more positive–negative attractions relative to the isolated atoms. We call this extra attraction a covalent bond.

5.188 The part that is true is "each oxygen atom has an octet." The part that is false is "the diagram is correct." It is false because oxygen is a group VIA atom, so the dot diagram should have a total of 12 dots. The one she drew has only 10 dots.

5.190 (a) There are the four most-electronegative elements in the periodic table.
(b) On the bottom left-hand side.
(c) The trends in electronegativity. Electronegativity increases on going from left to right across a period, or from the bottom to the top of a group.

Chapter 6

6.2 Electron groups all 109.5° apart, molecular shape tetrahedral:

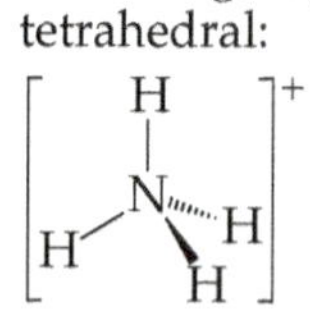

6.3 Electron groups 180° apart, molecular shape linear: $Cl{-}C{\equiv}C{-}Cl$.

6.4 Electron groups all in one plane at corners of triangle; molecular shape trigonal planar:

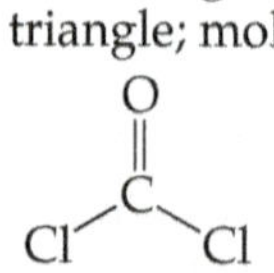

6.6 Both electron-group geometry and molecular shape tetrahedral:

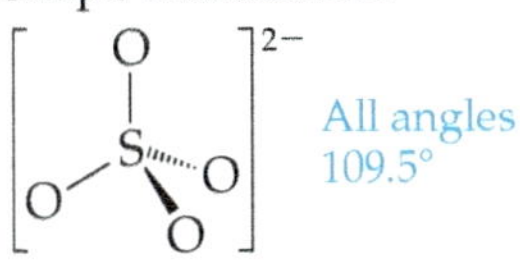

6.7 Both electron-group geometry and molecular shape linear:

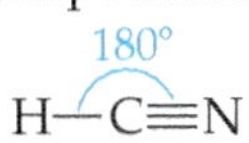

6.8 Electron-group geometry, trigonal planar; molecular shape, bent:

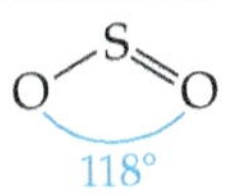

6.9 Both electron-group geometry and molecular shape trigonal planar:

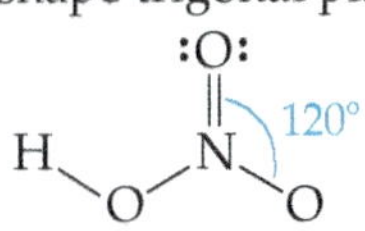

6.11 Yes, it is polar.

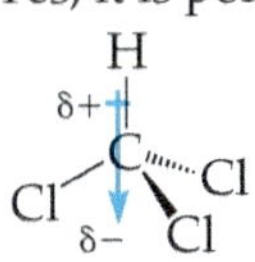

6.12 PCl_3:

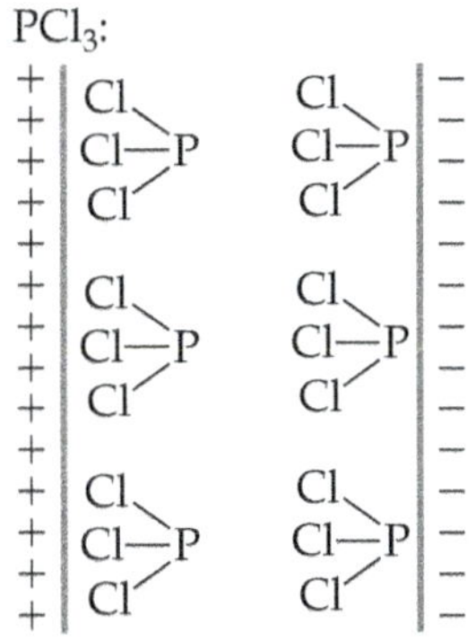

$CHCl_3$:

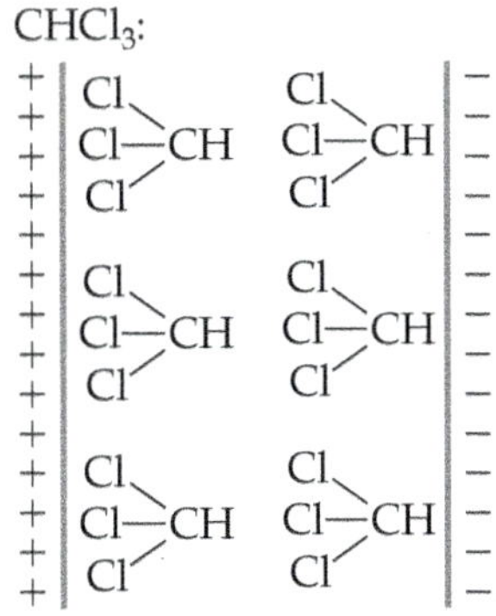

6.13 Because the bonding pair electrons can be farther apart with 109.5° angles.

6.16 (a)

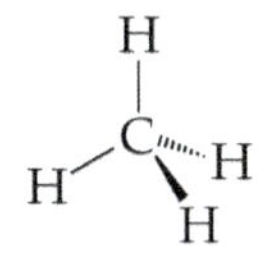

(b)

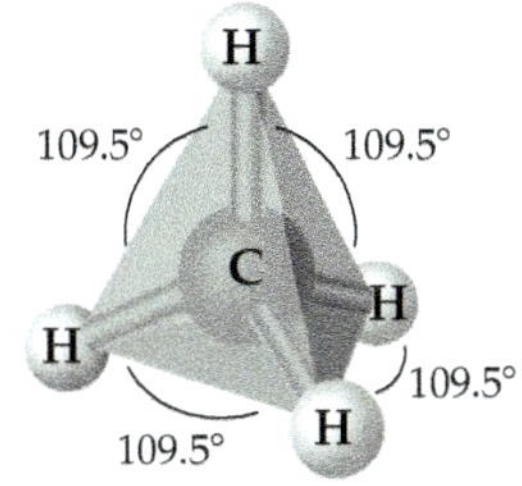

6.18 (a) (1)

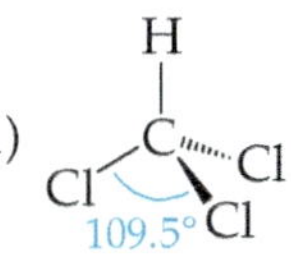

(2)

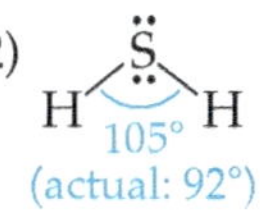

(3)

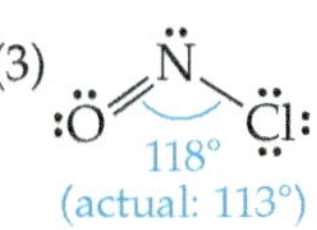

(4)

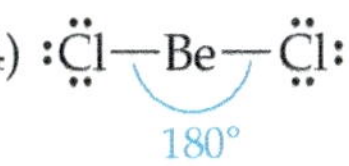

(b) (1) Electron-group geometry and molecular shape both tetrahedral.
(2) Electron-group geometry, tetrahedral; molecular shape, bent.
(3) Electron-group geometry, trigonal planar; molecular shape, bent.
(4) Electron-group geometry and molecular shape both linear.

6.20 All the electrons in a multiple bond occupy roughly the same region of space.

6.21 In the case where the bond angle is approximately 118°, the central atom has two bonding pairs and one lone pair. In the case where the bond angle is approximately 105°, the central atom has two bonding pairs and two lone pairs.

6.23 The shape of a molecule is determined by the arrangement of only the *atoms* in the molecule. One of the electron groups on the N in NH_3 is a lone pair and thus ignored in determining molecule shape. Three bonding groups and one lone pair make the NH_3 molecule pyramidal.

6.25 (a)

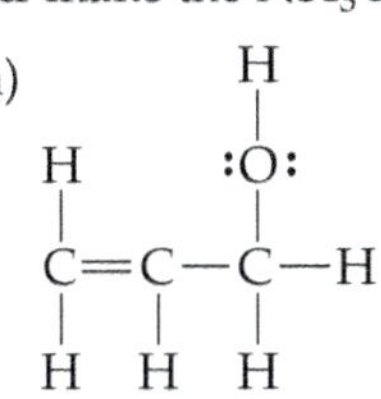

(b)

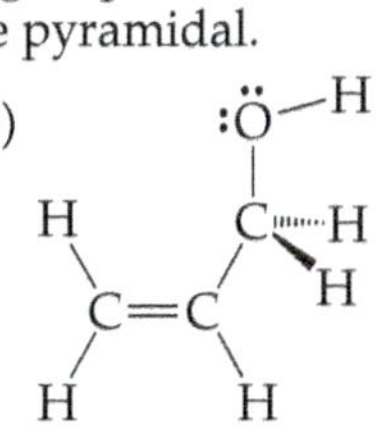

(c)

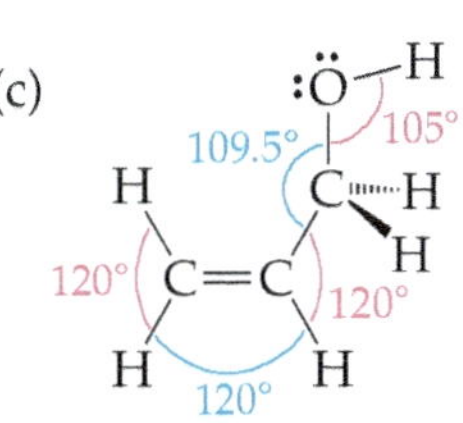

6.26 The shape of a polyatomic ion depends on the number of electron groups around the central atom. That number can be determined only by using a correct Lewis dot diagram. Lewis structures reveal CH_3^+ has a steric number of 3 (electron geometry is trigonal planar), whereas CH_3^- has a steric number of 4 (electron geometry is tetrahedral).

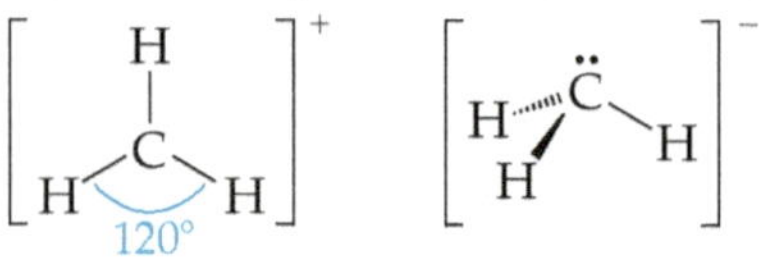

6.28

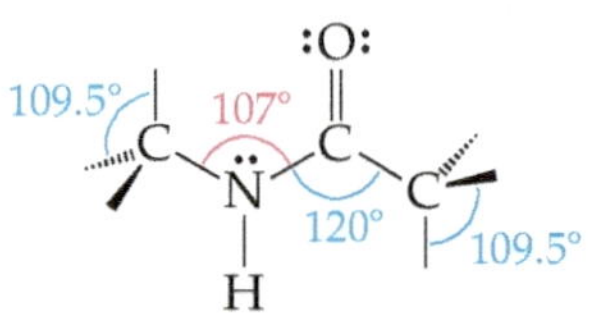

Leftmost C, tetrahedral; N, pyramidal; middle C, trigonal planar; rightmost C, tetrahedral.

6.30 Yes; you must know which, if any, bonds in the molecule are polar (based on the relative electronegativities of the connected atoms) and the shape of the molecule.

6.32 (a) HCl, because the electronegativity difference is greater in HCl than in HBr.
(b) HCl, because the molecular dipole moment is greater.

6.34 Because a dipole moment has a direction associated with it.

6.37 SO_2 is polar, while CO_2 is nonpolar. Because the molecule of SO_2 is bent (two bonding groups, one lone pair), the two bond dipole moments do not cancel each other. Instead they add together to give a nonzero molecular dipole moment:

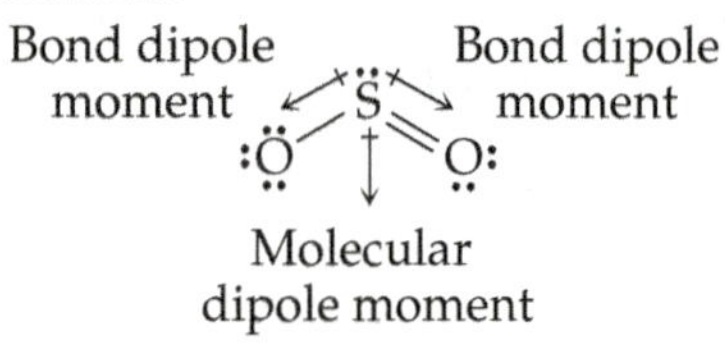

On the other hand, the molecule of CO_2 is linear and the two bond dipoles moments cancel each other.

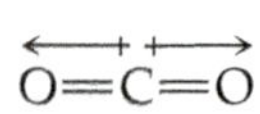

Net dipole = 0

6.38 (a) phosphorus trichloride; polar P—Cl bonds and their bond dipole moments add in the pyramidal shape. The C—Cl bond in a Cl_2 molecule is nonpolar.
(b) sulfur difluoride; S—F bond dipoles add in the bent molecule; C═S bond dipoles cancel due to molecule's linear shape.
(c) nitrogen trifluoride; both are polar, but electronegativity difference greatest between N and F.
(d) hydrogen cyanide; polar carbon-nitrogen bond, while Si—Br bond dipoles cancel due to tetrahedral shape of molecule.
(e) oxygen dichloride; both molecules are polar. but O—Cl bonds are more polar.

6.41 (a)

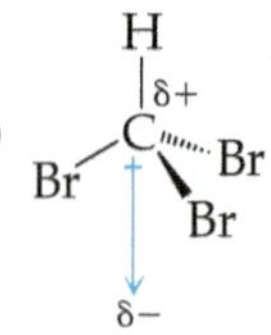

(b)

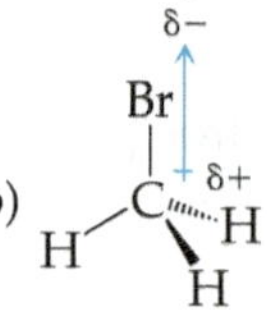

(c)

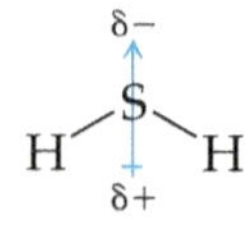

(d)

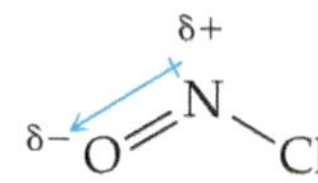

(e) C_2Cl_2 is nonpolar: Cl—C≡C—Cl

6.43 No, dipole–dipole forces have only a fraction of the strength of the forces between ions because ions carry full charges and dipole charges are partial.

6.45

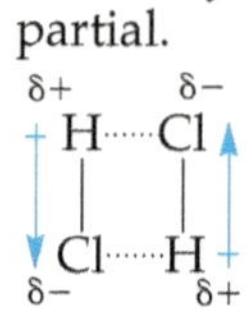

6.47 The lone pairs are important in determining the electron-group geometry around an atom and thus in determining the shape of the molecule.

6.48 (a)

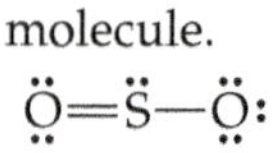

(b) The correct diagram predicts a bent shape and polar molecule. The incorrect diagram predicts a linear shape and nonpolar molecule.

6.49 (a)

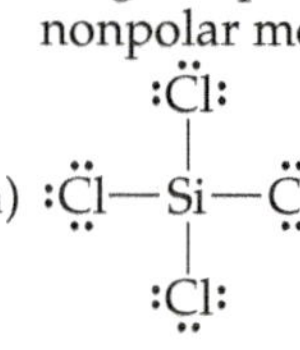

(b)

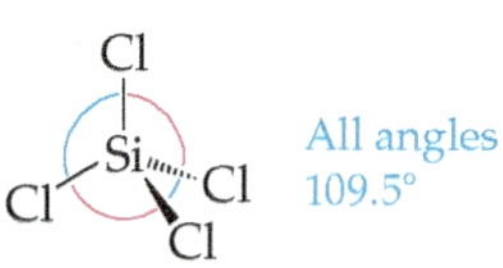

(c) Tetrahedral

(d)

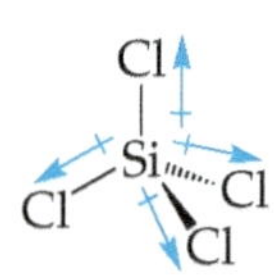

(e) No, the molecule is nonpolar.

6.51 (a)

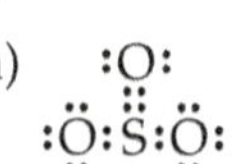

(b)

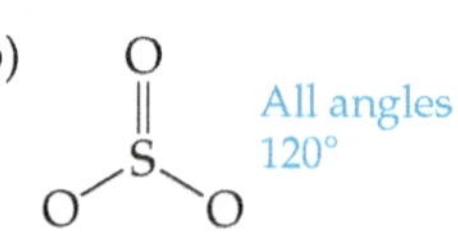

(c) Trigonal planar

(d)

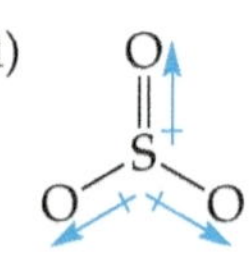

(e) No, the molecule is nonpolar.

6.53 (a) 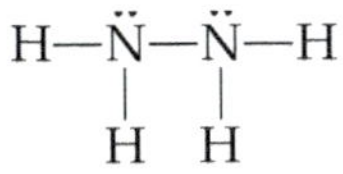

(b) 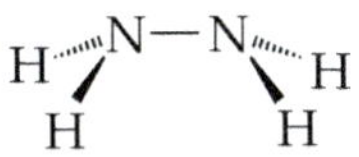
All angles 107° (2% compression rule for period 2 atoms)

(c) Pyramidal around each N

(d) 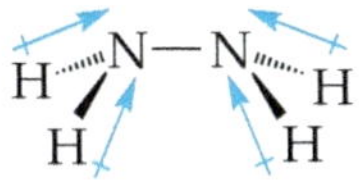

(e) Yes, the molecule is polar.

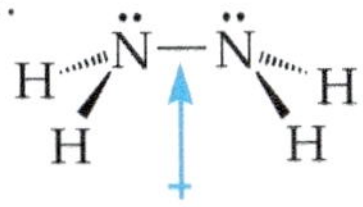

6.55 (a) :N≡N—Ö: (b) N≡N—O (180°) (c) Linear.

(d) N≡N—O (e) Yes, the molecule is polar. N≡N—O

6.56 (a) NO_2 has an odd number of valence electrons: $5 + 6(2) = 17$. This makes it impossible to show only electron pairs in the dot diagram.
(b) :Ö—N̈=Ö:
(c) Bent; the molecule is polar.

6.58 As the molecular dipole moment was increased, the intermolecular attractive forces would become stronger and the boiling point would increase.

6.61 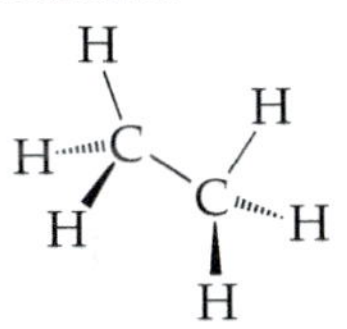
All angles 109.5°

6.63 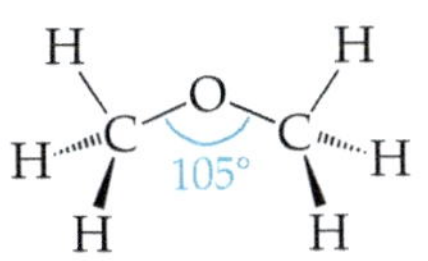
All other angles 109.5°

6.67 (a) 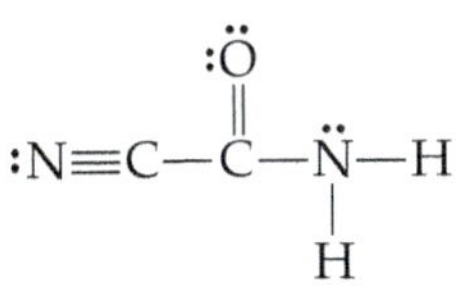

(b) 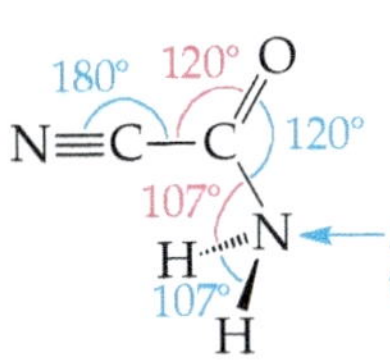

All angles around this N 107°

6.69 (a) [:C≡N:]$^-$ Linear molecular shape

(b) 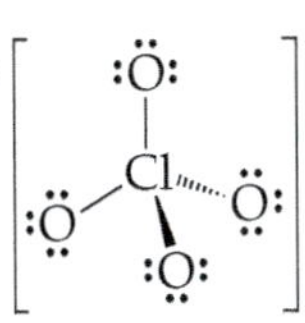
Tetrahedral molecular shape, no dipole moment

(c) 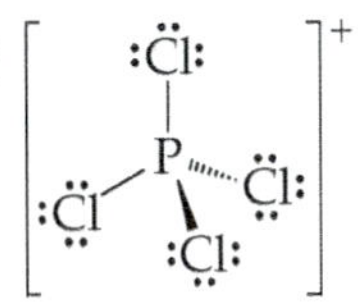
Tetrahedral molecular shape, no dipole moment

(d) 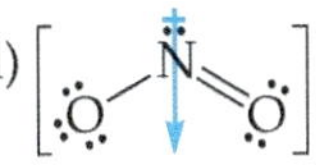
Bent molecular shape

6.74 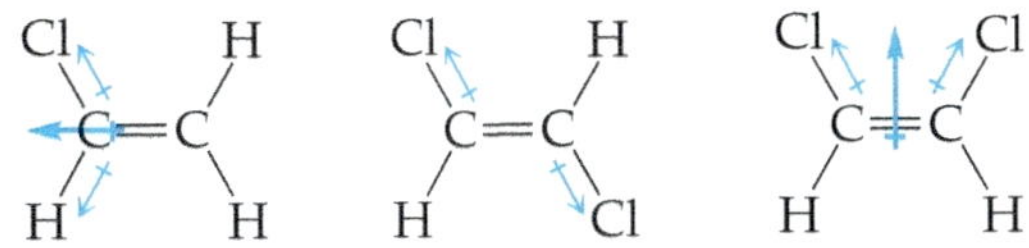
Polar Nonpolar Polar

6.75 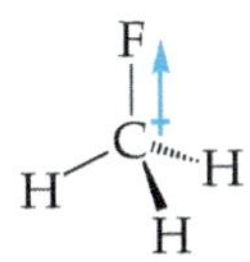
Tetrahedral shape, all angles 109.5°, polar

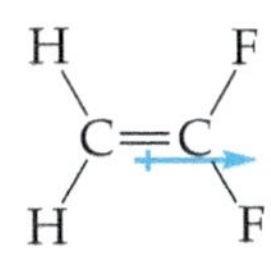
Trigonal planar shape, all angles 120°, polar

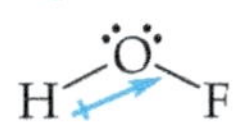
Bent shape, 105° angle, polar

6.77 CH_3F

6.79 The answer considers not the three-dimensional configuration of the atoms but only the two-dimensional configurations she drew. The N has two bonding groups and two lone pairs, and Table 6.2 tells her that this combination gives a bent molecular shape.

6.81 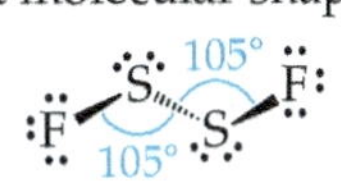
Tetrahedral electron-group geometry around each S.

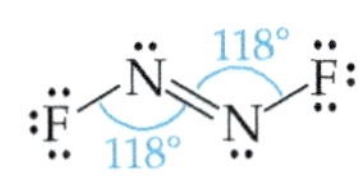
Trigonal planar electron-group geometry around each N.

6.84 (a) Ö=N̈—Ö—Ö—H

(b) 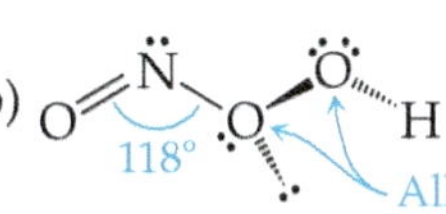

All angles 105° around both these O atoms

(c) N—electron-group geometry, trigonal planar; molecular shape, bent to 118°. Both O—electron-group geometry, tetrahedral; molecular shape, bent to 105°.

6.86 (a) 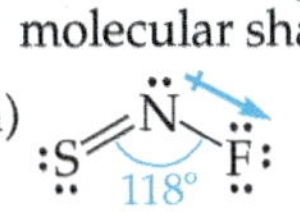
Trigonal planar electron-group geometry around N, bent molecular shape

(b) [:N≡C—Ö:]$^-$ Linear electron-group geometry around C, linear molecular shape

(c) 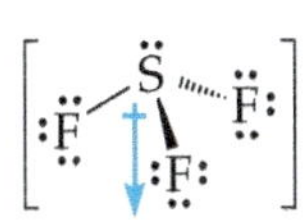
Tetrahedral electron-group geometry around S, pyramidal molecular shape

6.89 The molecular shape is linear, which means the bond angle is 180°:

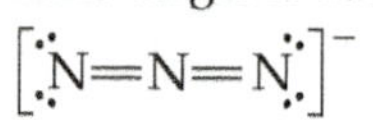

6.92 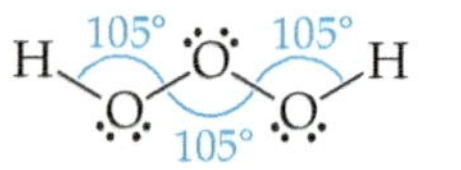

6.95 BH_3—electron-group geometry around B trigonal planar, molecular shape trigonal planar. PH_3—electron-group geometry around P tetrahedral, molecular shape pyramidal. BH_3 is nonpolar.

6.96 (a) 4
(b) Molecule (a).
(c) Molecule (a) is nonpolar as its flat shape leads to canceling of individual bond dipole moments. Molecule (b) is polar since the tetrahedral shape does not lead to canceling of individual bond dipole moments.

6.98 Against. He has shown an approach such that the $\delta+$ faces of both molecules get close. Both being $\delta+$, these faces would repel. More than likely, the $\delta-$ N end of one molecule approaches the $\delta+$ face of another.

6.100 (a) Given that ΔEN for S−F equals 1.5, the bonds would be classified as polar covalent.
(b) Although each bond is polar, the individual bond dipole moments for such a highly symmetric molecule would all cancel, rendering the entire molecule nonpolar. (Put yourself at sulfur's position and do the tug-of-war test. You would not move.)

Chapter 7

7.2 HF

7.3 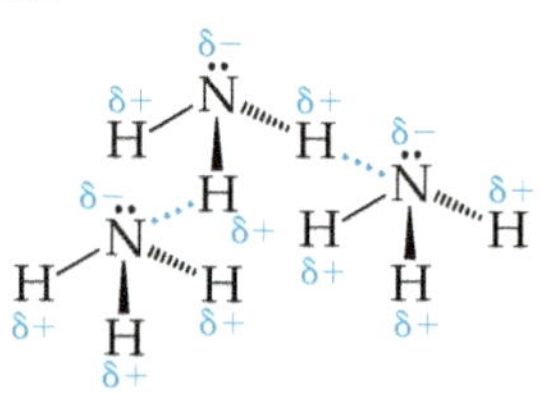

7.4 H_2O

7.6 Both ethylene and polyethylene are nonpolar molecules. Therefore, the only intermolecular forces involved are London forces. Polyethylene is a very large molecule containing many electrons, which means substantial London forces between molecules and consequently a high melting point, making this substance a solid at room temperature. The much smaller ethylene molecule contains a relatively small number of electrons. This means weak London forces between molecules and low melting and boiling points, making this substance a gas at room temperature.

7.7 (a) 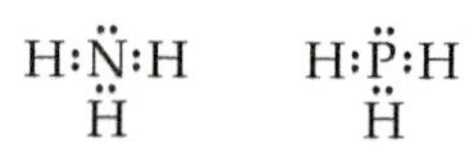
(b) NH_3 is polar; PH_3 is nonpolar.
(c) PH_3 has the stronger London forces because it has more electrons.
(d) PH_3 because it has stronger London forces. However, this is not the case. See Practice Problem 7.8 for more information.

7.9 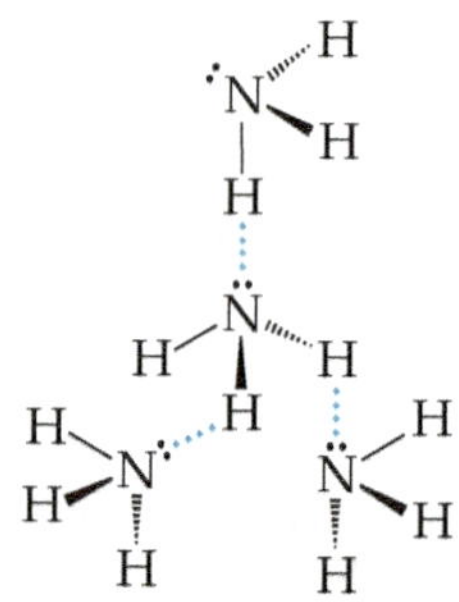

7.10 (a) Because HF is the only one of the four where the molecules in the liquid form hydrogen bonds.
(b) Because HI has more electrons than HBr and HCl and therefore stronger London forces.

7.12 The forces holding these solids together—metallic bonds in iron and ionic bonds in NaCl—are both true chemical bonds, and a large amount of heat energy is therefore required to melt either substance.

7.13 Intermolecular London forces.

7.14 (a) Metallic. (b) Metallic. (c) Ionic. (d) Network covalent. (e) Molecular.

7.16 (a) Cooling a gas slows the molecules down to a speed at which they are not moving fast enough to overcome the attractive forces that will make them condense to a liquid.
(b) The substance would remain gaseous as there would be no attractive forces causing the gas particles to aggregate.

7.18 Yes. Molecules in the liquid phase are still very much in motion. They are jostling past one another constantly but remain in close proximity to one another.

7.20 Water has stronger intermolecular attractive forces because it will condense into the liquid phase with less cooling than nitrogen requires.

7.22 Gas molecules are not bound to one another as are the molecules of solids and liquids. Gas molecules are essentially independent of one another and can travel around in the entire volume of their container.

7.24 (a) C—D. The larger dipole moment means stronger intermolecular attractive forces, which in turn means less energy needs to be removed from the liquid in order to form a solid.
(b) C—D; the stronger intermolecular attractive forces mean more energy must be added to the liquid to allow the molecules to overcome the attractive forces and escape into the gas phase.

7.25 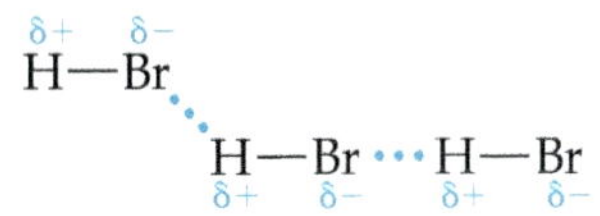

Dipole–dipole interactions.

7.27 CBr_4; because it is the larger molecule, it has more electrons and therefore stronger London forces.

7.28 (1) Chloromethane is a polar molecule and therefore subject to dipolar interactions.
(2) Chloromethane is a larger molecule and therefore has stronger London forces.

7.30 The strength of London forces depends on the size of the molecule, so that the London forces in some large nonpolar molecules can be stronger than the dipole–dipole forces in some small polar molecules.

7.31 (a) HF, because the electronegativity difference is greater.
(b) HF, because the partial charges on the H and F atoms are greater than in HCl.
(c) HF. The larger dipole moment means stronger intermolecular attractive forces, which in turn mean less energy needs to be removed from the gas in order to form a liquid. The stronger intermolecular forces also mean more energy must be added to HF to allow liquid molecules to escape to the gas phase. Thus HF has the higher boiling point.
(d) Yes, both molecules experience London forces. HCl has greater London forces because it is a larger molecule with more electrons.

7.34 Because waxes are very large molecules, they have many electrons and therefore quite strong London forces—strong enough to make them solid at room temperature.

7.36 The hydrogen atom has no electrons other than those it is sharing and resembles a naked proton when bonded to a very electronegative atom such as O, N, or F.

7.38 (a) Because the size of the molecules is increasing as you go from S to Se to Te and London forces are stronger.
(b) H_2O has a much higher boiling point than expected because of the relatively strong hydrogen-bonding present.

7.40 Because hydrogen bonds are not covalent bonds within a molecule (which we represent by solid lines). Hydrogen bonds are unusually strong intermolecular attractions.

7.42 DNA molecules must be very stable but still able to break into two strands when the DNA is involved in protein building. Covalent bonds would be too strong (the strands would not be able to break apart), and London forces would be too weak (the DNA could not be kept intact).

7.43 If a nonpolar molecule is very large with a large number of electrons, its London forces will be quite substantial and can be stronger than the dipole–dipole interactions present in small polar molecules.

7.45 Sometimes.

7.47 Because the nonmolecular solids are held together by much stronger forces—namely, covalent, ionic, or metallic bonds.

7.49 The high melting points of some metals, such as gold and iron, are evidence that metallic bonds can be as strong as ionic or covalent bonds.

7.51 (a) Metallic solid. (b) Ionic solid.
(c) Network solid. (d) Molecular solid.

7.53 Because the attraction is the result of opposite charges attracting each other. When two molecules are oriented with a $\delta+$ H near a $\delta-$ O, the opposite charges create a hydrogen bond:

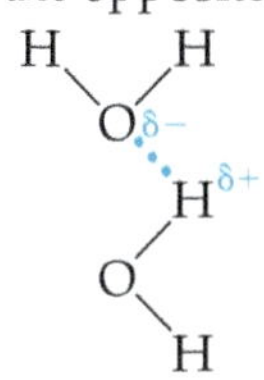

When two molecules are oriented with two $\delta+$ H near each other or two $\delta-$ O near each other, there are no hydrogen bonds formed:

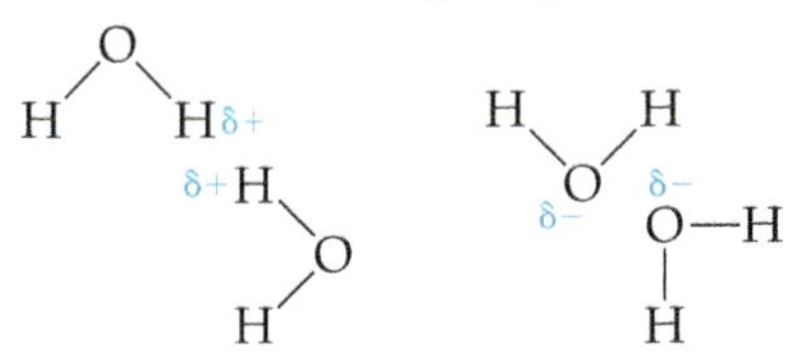

No attraction　　No attraction

7.54 The molecules move faster and faster.

7.56 (d)

7.58 (b) because this molecule is linear and nonpolar.

7.60 (d) because it contains no H—O, H—N, or H—F bonds.

7.62 (b). Choice (a) is ionic, and (c) is a metal; both therefore have high melting points and very high boiling points. (b) and (d) are both small covalent molecules, but only (b) is nonpolar.

7.64 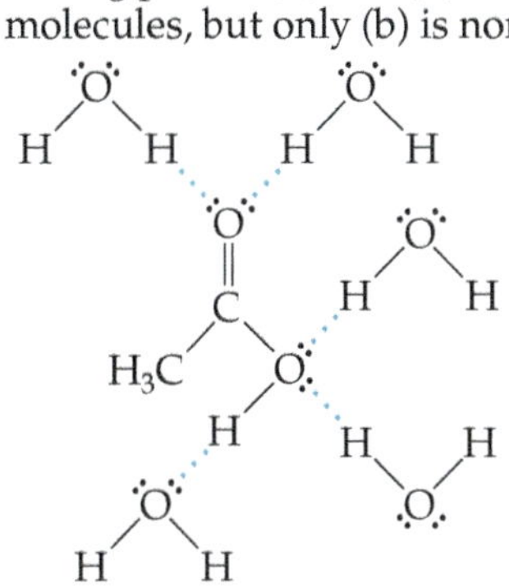

7.66 (a) 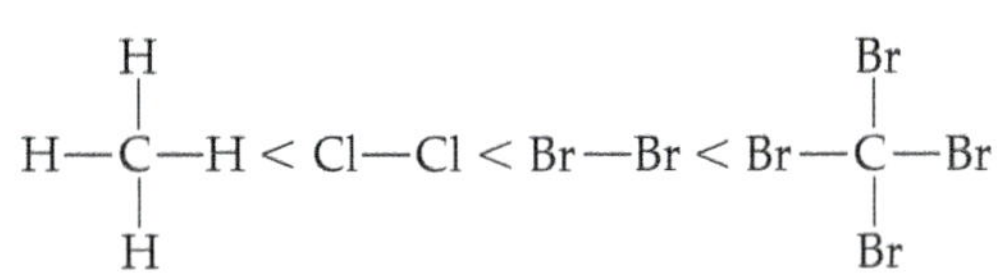

(b) All four molecules are nonpolar, meaning the only intermolecular forces in the liquids are London forces. Because London forces depend on numbers of electrons, the boiling point ranking goes from smallest molecule to largest.

7.68 (a) and (b) because both are nonpolar molecules.
7.70 (a) and (c)
7.72 $C_2H_6 < CH_3OH < NaCl < SiO_2$
7.73 Ethylene glycol boils at the higher temperature because of all the hydrogen-bonding resulting from the —OH parts of the molecules.
7.75 (a) London forces.
(b) The strong covalent bonds in the diamond network are difficult to break, making diamond hard. In graphite, the London forces between sheets are very weak, allowing the sheets to slide easily over one another and making the substance soft and slippery.
7.77 In a piece of metal, the valence electrons are free to "swim" throughout the entire piece.
7.79 The great number of electrons in the very large eicosane molecules create London forces that are stronger than the hydrogen bonds in water.
7.80 Nothing, because melting involves breaking bonds between molecules (*inter*molecular bonds); it does not involve the covalent bonds inside the molecules (*intra*molecular bonds).
7.82 (a) goes with $TiCl_4$ and (b) goes with $ZrCl_4$ (even though each Zr is bound to 6 Cl atoms, the bridging ones are only 1/2 owned by each Zr atom). This makes $TiCl_4$ a molecular compound, with only weak London forces between the nonpolar molecules, giving rise to a low melting point. $ZrCl_4$, on the other hand, is nonmolecular (it is network covalent). To melt it requires breaking Zr–Cl polar covalent bonds, which is quite difficult. Thus it has a very high melting point.
7.84 The molecules are aligned such that the $\delta-$ side of one is close to the $\delta-$ side of another, and the $\delta+$ side of one is close to the $\delta+$ side of another. Both are repulsive situations.
7.86 Going from a linear to a bent geometry would convert CO_2 from nonpolar to polar, giving it another (dipolar) intermolecular attractive force. This would increase its melting point.

Chapter 8

8.2 $CaH_2 + 2\,H_2O \rightarrow Ca(OH)_2 + 2\,H_2$
8.3 $C_6H_{12}O_6 + 6\,O_2 \rightarrow 6\,CO_2 + 6\,H_2O$
8.4 $2\,HCl + Na_2CO_3 \rightarrow 2\,NaCl + CO_2 + H_2O$
8.6 $2\,C_6H_{12}(l) + 5\,O_2(g) \rightarrow 2\,H_2C_6H_8O_4(l) + 2\,H_2O(l)$
8.7 $4\,CH_3NH_2(g) + 9\,O_2(g) \rightarrow 4\,CO_2(g) + 10\,H_2O(g) + 2\,N_2(g)$
8.9 $2\,Pb(NO_3)_2(s) \xrightarrow{\text{Heat}} 2\,PbO(s) + 4\,NO(g) + 3\,O_2(g)$; decomposition
8.10 $C_2H_4 + C_2Cl_4 \xrightarrow{\text{Catalyst}} 2\,C_2H_2Cl_2$; combination.
8.12 (a) Step 1: $Pb^{2+}, NO_3^-, Na^+, SO_4^{2-}$
Step 2: Pb^{2+} with SO_4^{2-}, Na^+ with NO_3^-
Step 3: Table 8.1 indicates $PbSO_4$ is insoluble in water. All Na^+ salts are water-soluble, meaning Na^+ and NO_3^- do not form a precipitate. Na^+ and NO_3^- are spectator ions.
(b) Step 4: $Pb^{2+}(aq) + SO_4^{2-}(aq) \rightarrow PbSO_4(s)$
8.13 (a) Step 1: $Ni^{2+}, NO_3^-, NH_4^+, PO_4^{3-}$
Step 2: Ni^{2+} with PO_4^{3-}, NH_4^+ with NO_3^-
Step 3: Table 8.1 says most phosphate salts are insoluble in water and does not list $Ni_3(PO_4)_2$ as an exception. All ammonium salts and all nitrate salts are water-soluble, meaning NH_4^+ and NO_3^- do not form a precipitate. They are spectator ions.
(b) Step 4: $3\,Ni^{2+} + 2\,PO_4^{3-} \rightarrow Ni_3(PO_4)_2$
8.15 Each mole of H_2SO_4 dissociates into 2 moles of H^+ and 1 mole of SO_4^{2-}, but each mole of NaOH dissociates into 1 mole of Na^+ and 1 mole of OH^-. Therefore, it takes twice as much NaOH to provide the 2 moles of OH^- needed to neutralize all the H^+ from H_2SO_4.
8.16 Intact-molecule: $2\,HF(aq) + Ca(OH)_2(aq) \rightarrow CaF_2(s) + 2\,H_2O(l)$
Net ionic: $2\,H^+(aq) + 2\,F^-(aq) + 2\,Ca^{2+}(aq) + 2\,OH^-(aq) \rightarrow CaF_2(s) + 2\,H_2O(l)$
The salt is calcium fluoride.
8.17 Only (a) represents a chemical reaction because the substances on the right (water and oxygen) are different from the substance on the left (hydrogen peroxide, H_2O_2). In (b), all substances remain chemically unchanged.
8.20 Under *no* circumstances; it *always* takes energy to break a bond.
8.21 Two Cl—Cl bonds and two C—H bonds must be broken. Two H—Cl bonds and two C—Cl bonds are formed.
8.23 (a) Cooling the reaction mixture slows down the molecules, meaning they have less energy available to break bonds when they collide.
(b) Increasing the number of reactant molecules yields more collisions per time period, which means more changes for reactant bonds to break in that time period, which yields a faster reaction.
8.25 No, because it violates the law of conservation of matter, meaning there must be the same number of atoms of each type on both sides of the equation. This one has iron atoms on one side and gold atoms on the other.
8.27 Changing the subscripts in a chemical formula changes the chemical identity of a substance.
8.28 $C_2H_4 + 3\,O_2 \rightarrow 2\,CO_2 + 2\,H_2O$
8.29 1 mole of methane and 2 moles of oxygen react to give 1 mole of carbon dioxide and 2 moles of water. Or 1 molecule of methane and 2 molecules of oxygen react to give 1 molecule of carbon dioxide and 2 molecules of water.
8.31 $Al_2Cl_6 + 6\,H_2O \rightarrow 2\,Al(OH)_3 + 6\,HCl$
8.33 (a) $2\,C_2H_2 + 5\,O_2 \rightarrow 2\,H_2O + 4\,CO_2$
(b) Two moles of acetylene and five moles of oxygen react to give two moles of water and 4 moles of carbon dioxide. Can also say molecules instead of moles.
8.35 $2\,N_2O_5(g) \rightarrow 4\,NO_2(g) + O_2(g)$; decomposition.
8.36 $2\,CO(g) + 2\,NO(g) \rightarrow 2\,CO_2(g) + N_2(g)$; single replacement (O in NO replaced by N to form NN).

8.38 $2\,SO_2(g) + O_2(g) \rightarrow 2\,SO_3(g)$; combination.

8.39 $3\,Li(s) + \frac{1}{2}N_2(g) \rightarrow Li_3N(s)$ or $6\,Li(s) + N_2(g) \rightarrow 2\,Li_3N(s)$; combination.

8.40 (a) $C_5H_{10}O_2(l) + \frac{13}{2}O_2(g) \rightarrow 5CO_2(g) + 5H_2O(g)$ or $2C_5H_{10}O_2(l) + 13O_2(g) \rightarrow 10CO_2(g) + 10H_2O(g)$
(b) While a bit of a stretch, you can call it a double replacement reaction since the carbon in the $C_5H_{10}O_2$ gives its hydrogens to oxygen to form water, while O_2 gives some O atoms to carbon to form carbon dioxide.
(c) Combustion reaction.

8.42 A combination reaction brings two or more substances together to form one new substance: A + B → AB. A decomposition reaction breaks one substance up into two or more new substances: AB → A + B.

8.44 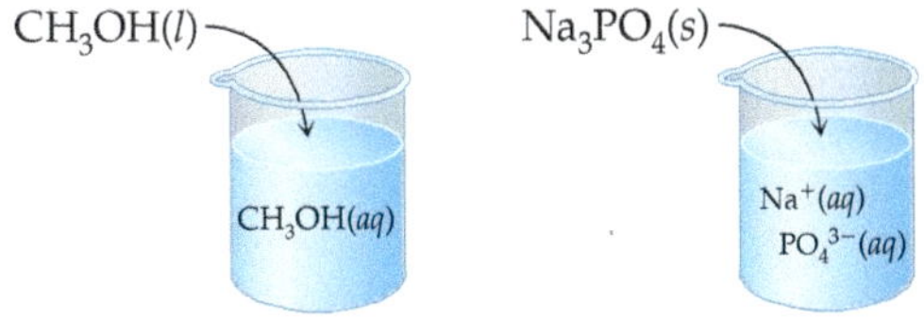

Methanol does not dissociate into ions in aqueous solution. Sodium phosphate does.

8.46 (a) $Pb^{2+}(aq) + 2\,Cl^-(aq) \rightarrow PbCl_2(s)$
(b) $Ba^{2+}(aq) + SO_4^{2-}(aq) \rightarrow BaSO_4(s)$
(c) No precipitation.

8.48 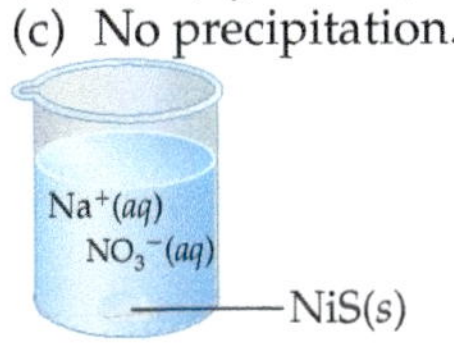

$Ni^{2+}(aq) + S^{2-}(aq) \rightarrow NiS(s)$

8.50 Calcium carbonate precipitates: $Ca^{2+}(aq) + CO_3^{2-}(aq) \rightarrow CaCO_3(s)$.

8.52 Table 7.1 tells you hydroxides are insoluble in water except for NaOH, KOH, $Ca(OH)_2$, and $Ba(OH)_2$. Mix an aqueous solution of any one of these with an aqueous solution of any water-soluble Ni^{2+} salt, and then filter off the $Ni(OH)_2$ that forms.

8.54 (a) $Ca^{2+}(aq) + SO_4^{2-}(aq) \rightarrow CaSO_4(s)$
(b) $Ca(NO_3)_2(aq) + (NH_4)_2SO_4(aq) \rightarrow CaSO_4(s) + 2\,NH_4NO_3(aq)$

8.55 (a) $H_2S(aq) + 2\,Na^+(aq) + 2\,OH^-(aq) \rightarrow 2\,Na^+(aq) + S^{2-}(aq) + 2\,H_2O(l)$
(b) Sodium sulfide, Na_2S.

8.56 (a) $H_2S(aq) \rightarrow 2\,H^+(aq) + S^{2-}(aq)$
(b) It produces H^+ ions in water.
(c) $Cu^{2+}(aq) + S^{2-}(aq) \rightarrow CuS(s)$

8.58 $H_2SO_4(aq) + Ca(OH)_2(aq) \rightarrow 2\,H_2O(l) + CaSO_4(s)$
The salt is calcium sulfate.

8.59 (a) Lead(II) sulfate.
(b) $Pb^{2+}(aq) + SO_4^{2-}(aq) \rightarrow PbSO_4(s)$
(c) No, because $Pb(NO_3)_2$ is not a base, meaning no OH^- ions were present to neutralize the $H^+(aq)$ ions created when the H_2SO_4 dissociated.

8.60 (a) $NH_4^+(aq) \rightarrow NH_3(aq) + H^+(aq)$
(b) $NH_3(aq)$, $NaCl(aq)$, and $H_2O(l)$

8.62 (a) $H^+(aq) + Br^-(aq) + Na^+(aq) + OH^-(aq) \rightarrow Br^-(aq) + Na^+(aq) + H_2O(l)$
$H^+(aq) + OH^-(aq) \rightarrow H_2O(l)$
Sodium bromide.
(b) $Ca^{2+}(aq) + 2OH^-(aq) + 2H^+(aq) + 2NO_3^-(aq) \rightarrow Ca^{2+}(aq) + 2\,NO_3^-(aq) + 2\,H_2O(l)$
$OH^-(aq) + H^+(aq) \rightarrow H_2O(l)$
Calcium nitrate.
(c) $Ca^{2+}(aq) + 2\,OH^-(aq) + 2\,H^+(aq) + 2\,Br^-(aq) \rightarrow Ca^{2+}(aq) + 2\,Br^-(aq) + 2\,H_2O(l)$
$OH^-(aq) + H^+(aq) \rightarrow H_2O(l)$
Calcium bromide.
(d) $Mg(OH)_2(s) + 2\,H^+(aq) + 2\,Cl^-(aq) \rightarrow Mg^{2+}(aq) + 2\,Cl^-(aq) + 2\,H_2O(l)$
$Mg(OH)_2(s) + 2\,H^+(aq) \rightarrow Mg^{2+}(aq) + 2\,H_2O(l)$
Magnesium chloride.

8.64 The source is the formation of bonds as the precipitate molecules are created. Bond formation always releases energy, usually in the form of heat.

8.65 Energy is needed to pull each electron in the covalent bond away from the two positively charged nuclei it is attracted to.

8.68 (a) Yes, because iron(III) sulfide is insoluble in water.
(b) $2\,Fe^{3+}(aq) + 3\,S^{2-}(aq) \rightarrow Fe_2S_3(s)$

8.70 Both salts dissolve; no precipitate forms.

8.71 Any sodium salt whose anion forms an insoluble precipitate with Cu^{2+}: Na_3PO_4, Na_2CO_3, Na_2S, or NaOH. Add the right amount, and all the anion you add combines with Cu^{2+}, leaving only added Na^+ and meaning no new types of ions are in the solution.

8.73 (a) 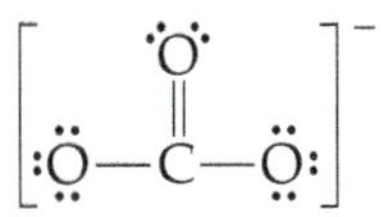

(b) 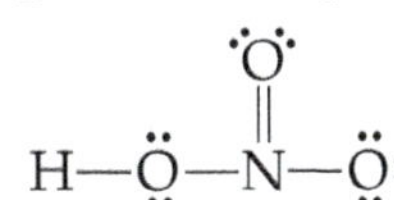

(c) Nitrate ion, because HNO_3 dissociates in water.

8.74 $HClO_4(aq) + NaOH(aq) \rightarrow NaClO_4(aq) + H_2O(l)$
$H^+(aq) + OH^-(aq) \rightarrow H_2O(l)$
Sodium perchlorate.

8.75 Add $NaCl(aq)$, $NaBr(aq)$, or $NaI(aq)$ to precipitate Pb^{2+} as the insoluble halide salt. Filter off the solid halide, and then add any sodium salt whose anion forms an insoluble salt with Ba^{2+}: Na_2SO_4, Na_3PO_4, Na_2CO_3, or Na_2S. Filter off the precipitate that forms, and your filtered solution contains $Na^+(aq)$ but no $Pb^{2+}(aq)$ and no $Ba^{2+}(aq)$.

8.77 $C_7H_{16}(l) + 11\,O_2(g) \rightarrow 7\,CO_2(g) + 8\,H_2O(g)$

8.80 (a) Broken 1 N—N, 3 Br—Br; formed 6 N—Br.
(b) Broken 4 P—P, 6 H—H; formed 12 P—H.
(c) Broken 1 K—F, 1 Na—I; formed 1 K—I, 1 Na—F.
(d) Broken 1 C—H, 1 Cl—Cl; formed 1 C—Cl, 1 H—Cl.

8.82 $BaCO_3$ insoluble; K_2S soluble; $CaSO_4$ insoluble; $MgNO_3$ soluble; $(NH_4)_3PO_4$ soluble.

8.84 (a)

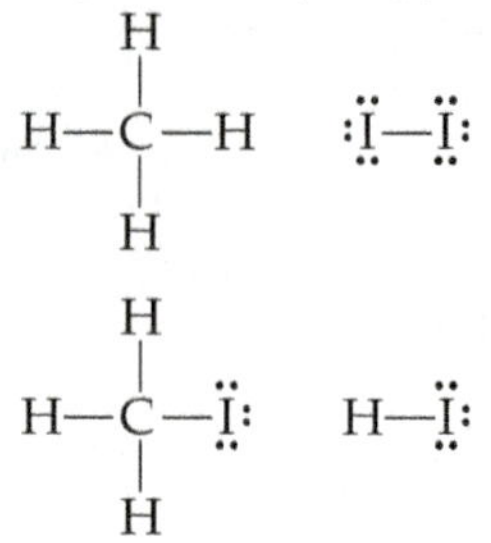

Broken: 1 C—H, 1 I—I; formed 1 C—I, 1 H—I.

(b)

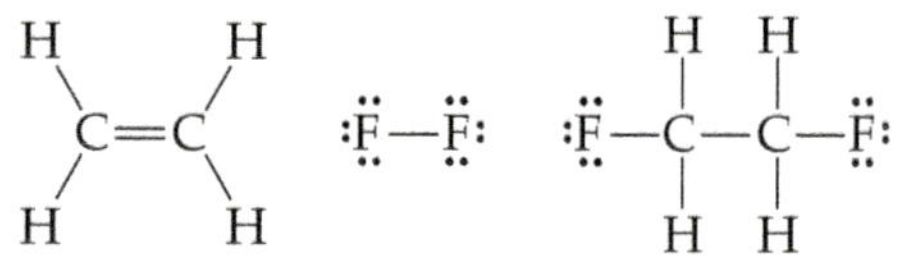

Broken: one-half of C═C, 1 F—F; formed 2 C—F.

8.87 KCl salt; CH_3COOH acid; $Al(OH)_3$ base; H_3BO_3 acid; $LiC_2H_3O_2$ salt.

8.89 Balanced:
$2\,HNO_3 + Ca(OH)_2 \rightarrow Ca(NO_3)_2 + 2\,H_2O$
Complete ionic:
$2\,H^+ + 2\,NO_3^- + Ca^{2+} + 2\,OH^- \rightarrow Ca^{2+} + 2\,NO_3^- + 2\,H_2O$
Net ionic: $H^+ + OH^- \rightarrow H_2O$

8.91 (a) $Al(OH)_3$
(b) $Ca_3(PO_4)_2$
(c) $MgCO_3$

8.94 $4\,NH_3 + 3\,Cl_2 \rightarrow 3\,NH_4Cl + NCl_3$

8.97 $2\,KClO_3 \rightarrow 2\,KCl + 3\,O_2$

8.100 $CaCO_3 \xrightarrow{\text{Heat}} CO_2 + CaO$

8.102 (a) Potassium hydroxide and nitric acid.
(b) Calcium hydroxide and phosphoric acid.
(c) Lithium hydroxide and sulfuric acid.
(d) Sodium hydroxide and hydroiodic acid.

8.105 $C_3H_8(g) + 5\,O_2(g) \rightarrow 3\,CO_2(g) + 4\,H_2O(l)$

8.107 $CO(g) + H_2O(g) \rightarrow CO_2(g) + H_2(g)$

8.111 $2\,H_2S + 3\,O_2 \rightarrow 2\,SO_2 + 2\,H_2O$

8.113 (a) $Cl^-, Br^-, I^-, PO_4^{3-}, CO_3^{2-}, S^{2-}$, or OH^-.
(b) SO_4^{2-}, PO_4^{3-}, or CO_3^{2-}.

8.115 (a) $Al^{3+}(aq) + 3\,OH^-(aq) \rightarrow Al(OH)_3(s)$
(b) $3\,Ca^{2+}(aq) + 2\,PO_4^{3-}(aq) \rightarrow Ca_3(PO_4)_2(s)$
(c) $Mg^{2+}(aq) + CO_3^{2-}(aq) \rightarrow MgCO_3(s)$

8.118 (a) AgI(*s*) (b) None. (c) None.

8.121 Balanced:
$$2\,KCl(aq) + Pb(NO_3)_2(aq) \rightarrow PbCl_2(s) + 2\,KNO_3(aq)$$
Complete ionic:
$$2\,K^+ + 2\,Cl^- + Pb^{2+} + 2\,NO_3^- \rightarrow PbCl_2 + 2\,K^+ + 2\,NO_3^-$$
Net ionic: $Pb^{2+} + 2\,Cl^- \rightarrow PbCl_2$

8.124 $H_2CO_3(aq) + 2\,LiOH(aq) \rightarrow Li_2CO_3(s) + 2\,H_2O(l)$

8.127 Slower, because the increased space the reactant molecules have to move around in means there will be fewer collisions between molecules.

8.130 (a) $6\,Ca(s) + P_4(s) \rightarrow 2\,Ca_3P_2(s)$
(b) $2\,Na(s) + 2\,H_2O(l) \rightarrow 2\,NaOH(aq) + H_2(g)$
(c) $2\,NH_4NO_3(s) \rightarrow 2\,N_2(g) + O_2(g) + 4\,H_2O(g)$

8.132 Complete ionic:
$$2\,H^+(aq) + 2\,Br^-(aq) + Ba^{2+}(aq) + 2\,OH^-(aq) \rightarrow 2\,Br^-(aq) + Ba^{2+}(aq) + 2\,H_2O(l)$$
Net ionic:
$$H^+(aq) + OH^-(aq) \rightarrow H_2O(l)$$

8.133 Beaker (a) is incorrect. CH_2Cl_2 is not an ionic compound, and as such, it does not dissociate into ions upon dissolving.

8.135 $CH_4(g) + 2Cl_2(g) \rightarrow CH_2Cl_2(l) + 2HCl(g)$
1 mole of methane and 2 moles of chlorine react to give 1 mole of dichloromethane and 2 moles of hydrogen chloride.

8.136 $H_2O(g) + CO(g) \rightarrow CO_2(g) + H_2(g)$. A single replacement reaction.

Chapter 9

9.2 1 cup of sugar requires 5 eggs; therefore 3 cups of sugar requires 15 eggs.

9.2 **Second time**
Conversion factors:
$$\frac{1\text{ cup sugar}}{5\text{ eggs}} \text{ and } \frac{5\text{ eggs}}{1\text{ cup sugar}}$$
Use the second one because you want your answer in eggs:
$$3\text{ cups sugar} \times \frac{5\text{ eggs}}{1\text{ cup sugar}} = 15\text{ eggs}$$

9.3 **Second time**
$$21\text{ blocks cream cheese} \times \frac{5\text{ eggs}}{3\text{ blocks cream cheese}} = 35\text{ eggs}$$

9.5 (a) $0.10\text{ mol U} \times \dfrac{6.022\times10^{23}\text{ atoms U}}{1\text{ mol U}} = 6.0\times10^{22}\text{ atoms U}$

(b) $0.10\text{ mol U} \times \dfrac{238.029\text{ g U}}{1\text{ mol U}} = 23.803\text{ g U}$,

which you must report as 24 g because 0.10 has only two significant digits.

9.6 (a) $120.11\text{ g C} \times \dfrac{1\text{ mol C}}{12.011\text{ g C}} = 10.000\text{ mol C}$

(b) $10.000\text{ mol C} \times \dfrac{6.022\times10^{23}\text{ atoms C}}{1\text{ mol C}} = 6.022\times10^{24}\text{ atoms C}$

9.8 From Practice Problem 9.7, 1 mole of propane has a mass of 44.096 g; 2 moles of propane has a mass of 88.192 g.

9.9 6 moles of carbon atoms

9.10 1 mole of propane and 5 moles of oxygen react to give 3 moles of carbon dioxide and 4 moles of water.

9.12 (a) 72.3 g (b) 3.40×10^{23} molecules
9.13 (a) 8.30 moles (b) 133 g
9.15 (a) $C_6H_6 + 3\,H_2 \rightarrow C_6H_{12}$
(b) 1 mole of benzene and 3 moles of hydrogen react to give 1 mole of cyclohexane.
(c) 78.113 g C_6H_6; 6.0474 g H_2
(d) 84.161 g (e) 28.5%
9.16 (a) $C_6H_{12}O_6 + 6\,O_2 \rightarrow 6\,CO_2 + 6\,H_2O$
(b) 1 mole of glucose and 6 moles of oxygen react to give 6 moles of carbon dioxide and 6 moles of water.
(c) 180.155 g glucose; 191.99 g O_2
(d) 264.05 g
(e) 74.23%

9.18 $$10.0\text{ g glucose} \times \frac{1\text{ mol glucose}}{180.155\text{ g glucose}} \times \frac{6\text{ mol }H_2O}{1\text{ mol glucose}} \times \frac{18.015\text{ g }H_2O}{1\text{ mol }H_2O} = 6.00\text{ g }H_2O$$

9.19 $$10.0\text{ g glucose} \times \frac{1\text{ mol glucose}}{180.155\text{ g glucose}} \times \frac{6\text{ mol }CO_2}{1\text{ mol glucose}} \times \frac{44.009\text{ g }CO_2}{1\text{ mol }CO_2} = 14.7\text{ g }CO_2$$

9.20 $$10.0\text{ g }CO_2 \times \frac{1\text{ mol }CO_2}{44.009\text{ g }CO_2} \times \frac{1\text{ mol glucose}}{6\text{ mol }CO_2} \times \frac{180.155\text{ g glucose}}{1\text{ mol glucose}} = 6.82\text{ g glucose}$$

9.22 $$10.0\text{ g }Al_2O_3 \times \frac{1\text{ mol }Al_2O_3}{101.96\text{ g }Al_2O_3} \times \frac{2\text{ mol Al}}{1\text{ mol }Al_2O_3} \times \frac{6.022\times10^{23}\text{ Al atoms}}{1\text{ mol Al}} = 1.18\times10^{23}\text{ Al atoms}$$

9.23 $$10.0\text{ g }H_2O \times \frac{1\text{ mol }H_2O}{18.015\text{ g }H_2O} \times \frac{6.022\times10^{23}\ H_2O\text{ molecules}}{1\text{ mol }H_2O} = 3.34\times10^{23}\ H_2O\text{ molecules}$$

9.25 (a) 362.9 g O_2 (b) 163.4 g H_2O
9.26 (a) $2\,ZnS + 3\,O_2 \rightarrow 2\,ZnO + 2\,SO_2$
(b) ZnS is the limiting reactant. Using it to calculate theoretical yield gives 8.39 g ZnO and 6.60 g SO_2.
(c) 5.06 g O_2 left over.
(d) 89.4% yield.
9.28 (a) C_2H_6O
(b) 52.4% C; 13.1% H; 34.5% O
(c) $C_2H_6O + 3\,O_2 \rightarrow 2\,CO_2 + 3\,H_2O$

9.30 (a) $$(85.62\text{ g C})\left(\frac{\text{mole C}}{12.011\text{ g C}}\right) = \frac{7.13\text{ mole C}}{7.13} = 1$$

$$(14.37\text{ g H})\left(\frac{\text{mole H}}{1.0079\text{ g H}}\right) = \frac{14.26\text{ mol H}}{7.13} = 2$$

No oxygen since % C + % H = 100
The empirical formula is CH_2
(b) C_2H_4
(c) $C_2H_4 + 3\,O_2 \rightarrow 2\,CO_2 + 2\,H_2O$
9.32 37.02% C; 2.22% H; 18.50% N; 42.27% O
9.33 The empirical formula and molecular formula are both $CHCl_3$.
9.34 The empirical formula is C_2H_4O; the molecular formula is $C_4H_8O_2$.
9.36 (a) $$\frac{1\text{ egg}}{2\text{ cups flour}}, \frac{1\text{ egg}}{3\text{ cups sugar}}, \frac{2\text{ cups flour}}{3\text{ cups sugar}}, \frac{\frac{1}{4}\text{ pound butter}}{3\text{ cups sugar}}, \frac{1\text{ cup milk}}{1\text{ dozen sugar cookies}}, \frac{1\text{ dozen sugar cookies}}{\frac{1}{4}\text{ pound butter}}, \text{ etc.}$$

(b) $$30\text{ cookies} \times \frac{1\text{ dozen cookies}}{12\text{ cookies}} \times \frac{2\text{ cups flour}}{1\text{ dozen cookies}} = 5\text{ cups flour}$$

(c) You don't know how many cups of milk are in the container.
(d) $$1\text{ container of milk} \times \frac{4\text{ cups milk}}{1\text{ container of milk}} \times \frac{1\text{ egg}}{1\text{ cup milk}} = 4\text{ eggs}$$

(e) 1 egg + 2 cups flour + 3 cups sugar + $\frac{1}{4}$ pound butter + 1 cup milk → 1 dozen sugar cookies

9.37 (a) $$3\text{ cakes} \times \frac{5\text{ eggs}}{1\text{ cake}} = 15\text{ eggs}$$

(b) $$63\text{ blocks of cream cheese} \times \frac{5\text{ eggs}}{3\text{ blocks of cream cheese}} = 105\text{ eggs}$$

(c) $$5\text{ cakes} \times \frac{5\text{ eggs}}{1\text{ cake}} = 25\text{ eggs};$$

Then add 2 eggs: 25 + 2 = 27.

(d) $$3.011 \times 10^{24}\text{ eggs} \times \frac{1\text{ cake}}{5\text{ eggs}} \times \frac{1\text{ mol of cakes}}{6.022 \times 10^{23}\text{ cakes}} = 1\text{ mole of cakes!}$$

9.39 Always.
9.41 (a) There are 6.022×10^{23} bicycles in 1 mole of bicycles.
(b) Because there are two tires on every bicycle, the number of bicycle tires in 1 mole of bicycles is 2 times 6.022×10^{23}, or 1.204×10^{24}.
(c) $$1\text{ mole of bicycles} \times \frac{6.022 \times 10^{23}\text{ bicycles}}{1\text{ mole of bicycles}} \times \frac{2\text{ tires}}{1\text{ bicycle}} \times \frac{24\text{ spokes}}{1\text{ tire}} = 2.9 \times 10^{25}\text{ spokes}$$

(d) $1 \text{ mole of spokes} \times \frac{6.022 \times 10^{23} \text{ spokes}}{1 \text{ mole of spokes}}$

$\times \frac{1 \text{ tire}}{24 \text{ spokes}} \times \frac{1 \text{ bicycle}}{2 \text{ tires}} = 1.3 \times 10^{22} \text{ bicycles}$

9.43 $2.5 \text{ moles of pennies} \times \frac{6.022 \times 10^{23} \text{ pennies}}{1 \text{ mole of pennies}}$

$= 1.5 \times 10^{24} \text{ pennies}$

$1.5 \times 10^{24} \text{ pennies} \times \frac{1 \text{ dollar}}{100 \text{ pennies}}$

$= 1.5 \times 10^{22} \text{ dollars}$

9.45 2 molecules of sufur dioxide react with 1 molecule of oxygen to give 2 molecules of sulfur trioxide. 2 moles of sulfur dioxide react with 1 mole of oxygen to give 2 moles of sulfur trioxide.

9.46 There is no such thing as a fractional molecule, but we *can* have a fraction of a mole.

9.48 The mass of 1 mole of any molecule is the molecule's mass expressed in grams.

9.50 Molar mass allows us to form conversion factors between moles and grams.

9.51 6.022×10^{23}

9.52 (a) 6.022×10^{23}

(b) $1 \text{ mol C} \times \frac{6.022 \times 10^{23} \text{ atoms of C}}{1 \text{ mol C}}$

$\times \frac{1 \text{ dozen}}{12 \text{ atoms of C}} = 5.018 \times 10^{22} \text{ dozens of C atoms}$

(c) 1 mole

$1 \text{ mol C} \times \frac{6.022 \times 10^{23} \text{ atoms of C}}{1 \text{ mol C}}$

$\times \frac{1 \text{ dozen}}{12 \text{ atoms of C}} = 5.018 \times 10^{23} \text{ dozens of C atoms}$

9.54 (a) 3 moles
(b) 6 moles
(c) 2 times $6.022 \times 10^{23} = 1.204 \times 10^{24}$

9.56 The actual yield is the amount of product that is actually recovered after a chemical reaction.

9.58 $\% \text{ yield} = \frac{\text{Actual yield}}{\text{Theoretical yield}} \times 100$

9.61 (a) $\frac{2.00 \text{ g}}{2.0158 \text{ g}} \times 100 = 99.2\%$

(b) 2 moles of hydrogen chloride and 1 mole of zinc react to give 1 mole of molecular hydrogen and 1 mole of zinc chloride.

(c) Molar mass of HCl is (1.0079 g/mol H) + (35.453 g/mol Cl) = 36.461 g/mol HCl. Because the reaction calls for 2 moles of HCl, the mass of HCl needed is 72.922 g HCl. Molar mass of Zn is 65.39 g. This is the mass of 1 mole of Zn, which is what the equation calls for.

(d) Molar mass of H_2 is 2.0158 g, which is the theoretical yield.

(e) $\frac{2.00 \text{ g}}{2.0158 \text{ g}} \times 100 = 99.2\%$

9.63 (a) $\frac{6 \text{ mol C atoms}}{1 \text{ mol } C_6H_{12}O_6}, \frac{12 \text{ mol H atoms}}{1 \text{ mol } C_6H_{12}O_6}, \frac{1 \text{ mol } C_6H_{12}O_6}{6 \text{ mol O atoms}}, \frac{6 \text{ mol C atoms}}{12 \text{ mol H atoms}}$

(b) Half of 72,000, or 36,000 carbon atoms.

(c) $72{,}000 \text{ H atoms} \times \frac{1 \text{ molecule of } C_6H_{12}O_6}{12 \text{ H atoms}}$

$\times \frac{1 \text{ mole of } C_6H_{12}O_6}{6.022 \times 10^{23} \text{ molecules of } C_6H_{12}O_6}$

$\times \frac{180.157 \text{ g}}{1 \text{ mole of } C_6H_{12}O_6} = 1.8 \times 10^{-18} \text{ g}$

9.65 (a) $I_2 + 3\,Cl_2 \rightarrow 2\,ICl_3$

(b) $\frac{1 \text{ mol } I_2}{3 \text{ mol } Cl_2}, \frac{1 \text{ mol } I_2}{2 \text{ mol } ICl_3}, \frac{3 \text{ mol } Cl_2}{2 \text{ mol } ICl_3}, \frac{3 \text{ mol } Cl_2}{1 \text{ mol } I_2}, \frac{2 \text{ mol } ICl_3}{1 \text{ mol } I_2}, \frac{2 \text{ mol } ICl_3}{3 \text{ mol } Cl_2}$

(c) $\frac{2 \text{ mol I atoms}}{1 \text{ mol } I_2 \text{ molecules}}, \frac{2 \text{ mol Cl atoms}}{1 \text{ mol } Cl_2 \text{ molecules}}, \frac{1 \text{ mol I atoms}}{1 \text{ mol } ICl_3 \text{ molecules}}, \frac{3 \text{ mol Cl atoms}}{1 \text{ mol } ICl_3 \text{ molecules}},$

plus four more that are the reciprocals of these four.

9.67 (a) $24.0 \text{ g } O_2 \times \frac{1 \text{ mol } O_2}{31.998 \text{ g } O_2} = 0.750 \text{ mol } O_2$

(b) $0.750 \text{ mol } O_2 \times \frac{2 \text{ moles O atoms}}{1 \text{ mol } O_2}$

$= 1.50 \text{ moles O atoms}$

9.68 (a) $92.5 \text{ g } O_2 \times \frac{1 \text{ mol } O_2}{31.998 \text{ g } O_2} = 2.89 \text{ mol } O_2$

(b) $2.89 \text{ mol } O_2 \times \frac{2 \text{ moles O atoms}}{1 \text{ mol } O_2}$

$= 5.78 \text{ moles O atoms}$

(c) $5.78 \text{ moles O atoms} \times \frac{6.022 \times 10^{23} \text{ O atoms}}{1 \text{ mol O atoms}}$

$= 3.48 \times 10^{24} \text{ O atoms}$

This number is twice as large as the number of O_2 molecules present.

9.71 $1.00 \times 10^{9} \text{ molecules } H_2O$

$\times \frac{1 \text{ mol } H_2O}{6.022 \times 10^{23} \text{ molecules } H_2O}$

$\times \frac{18.015 \text{ g } H_2O}{1 \text{ mol } H_2O} = 2.99 \times 10^{-14} \text{ g } H_2O$

9.73 (a) $20.0\ \text{g H}_2\text{O}_2 \times \frac{1\ \text{mol H}_2\text{O}_2}{34.014\ \text{g H}_2\text{O}_2} \times \frac{2\ \text{mol H}_2\text{O}}{2\ \text{mol H}_2\text{O}_2} \times \frac{18.015\ \text{g H}_2\text{O}}{1\ \text{mol H}_2\text{O}} = 10.6\ \text{g H}_2\text{O}$

(b) $20.0\ \text{g H}_2\text{O} \times \frac{1\ \text{mol H}_2\text{O}}{18.015\ \text{g H}_2\text{O}} \times \frac{2\ \text{mol H}_2\text{O}_2}{2\ \text{mol H}_2\text{O}} \times \frac{34.014\ \text{g H}_2\text{O}_2}{1\ \text{mol H}_2\text{O}_2} = 37.8\ \text{g H}_2\text{O}_2$

(c) $20.0\ \text{g O}_2 \times \frac{1\ \text{mol O}_2}{31.998\ \text{g O}_2} \times \frac{2\ \text{mol H}_2\text{O}_2}{1\ \text{mol O}_2} \times \frac{34.014\ \text{g H}_2\text{O}_2}{1\ \text{mol H}_2\text{O}_2} = 42.5\ \text{g H}_2\text{O}_2$

9.75 Since B was the limiting reactant, some of the excess reactant A will contaminate the final product C. Since A and C are both solids, it will be difficult and expensive to separate them.

9.77 To determine which is the limiting ingredient, multiply each given amount by the appropriate conversion factor:

$25\ \text{eggs} \times \frac{1\ \text{cheesecake}}{5\ \text{eggs}} = 5\ \text{cheesecakes}$

$9\ \text{blocks cream cheese} \times \frac{1\ \text{cheesecake}}{3\ \text{blocks cream cheese}} = 3\ \text{cheesecakes}$

$4\ \text{cups sugar} \times \frac{1\ \text{cheesecake}}{1\ \text{cup sugar}} = 5\ \text{cheesecakes}$

Because it gives the smallest number, the cream cheese is the limiting ingredient. You can make 3 cheesecakes.

9.78 (a) $5.00\ \text{g H}_2 \times \frac{1\ \text{mol H}_2}{2.0158\ \text{g H}_2} \times \frac{2\ \text{mol H}_2\text{O}}{2\ \text{mol H}_2} \times \frac{18.015\ \text{g H}_2\text{O}}{1\ \text{mol H}_2\text{O}} = 44.7\ \text{g H}_2\text{O}$

(b) $5.00\ \text{g H}_2\text{O} \times \frac{1\ \text{mol H}_2\text{O}}{18.015\ \text{g H}_2\text{O}} \times \frac{1\ \text{mol O}_2}{2\ \text{mol H}_2\text{O}} \times \frac{31.998\ \text{g O}_2}{1\ \text{mol O}_2} = 4.44\ \text{g O}_2$

(c) $100.0\ \text{g H}_2 \times \frac{1\ \text{mol H}_2}{2.0158\ \text{g H}_2} \times \frac{1\ \text{mol O}_2}{2\ \text{mol H}_2} \times \frac{31.998\ \text{g O}_2}{1\ \text{mol O}_2} = 793.7\ \text{g O}_2$

(d) $50.0\ \text{g O}_2 \times \frac{1\ \text{mol O}_2}{31.998\ \text{g O}_2} \times \frac{2\ \text{mol H}_2\text{O}}{1\ \text{mol O}_2} \times \frac{18.015\ \text{g H}_2\text{O}}{1\ \text{mol H}_2\text{O}} = 56.3\ \text{g H}_2\text{O}$

(e) $56.3\ \text{g H}_2\text{O} \times \frac{1\ \text{mol H}_2\text{O}}{18.015\ \text{g H}_2\text{O}} \times \frac{6.022 \times 10^{23}\ \text{molecules H}_2\text{O}}{1\ \text{mol H}_2\text{O}} = 1.88 \times 10^{24}\ \text{molecules H}_2\text{O}$

9.80 The balanced equation is: $3\ H_2 + N_2 \rightarrow 2\ NH_3$

(a) $10.0\ \text{g H}_2 \times \frac{1\ \text{mol H}_2}{2.0158\ \text{g H}_2} \times \frac{1\ \text{mol N}_2}{3\ \text{mol H}_2} \times \frac{28.014\ \text{g N}_2}{1\ \text{mol N}_2} = 46.3\ \text{g N}_2$

(b) $10.0\ \text{g H}_2 \times \frac{1\ \text{mol H}_2}{2.0158\ \text{g H}_2} \times \frac{2\ \text{mol NH}_3}{3\ \text{mol H}_2} \times \frac{17.031\ \text{g NH}_3}{1\ \text{mol NH}_3} = 56.3\ \text{g NH}_3$

or

$46.3\ \text{g N}_2 \times \frac{1\ \text{mol N}_2}{28.014\ \text{g N}_2} \times \frac{2\ \text{mol NH}_3}{1\ \text{mol N}_2} \times \frac{17.031\ \text{g NH}_3}{1\ \text{mol NH}_3} = 56.3\ \text{g NH}_3$

(c) $56.3\ \text{g NH}_3 \times \frac{1\ \text{mol NH}_3}{17.031\ \text{g NH}_3} \times \frac{6.022 \times 10^{23}\ \text{molecules NH}_3}{1\ \text{mol NH}_3} = 1.99 \times 10^{24}\ \text{molecules NH}_3$

9.82 (a) $Cl_2 + 3\ F_2 \rightarrow 2\ ClF_3$
(b) F_2
(c) 4.10 mol ClF_3
(d) 379 g ClF_3
(e) 0.45 mol Cl_2
(f) 32 g Cl_2

9.83 (a) $2\ Na(s) + Br_2(l) \rightarrow 2\ NaBr(s)$
(b) Na
(c) 22.3 g NaBr
(d) 12.7 g Br_2
(e) 65.9%

9.86 (a) $2\ C_4H_{10} + 13\ O_2 \rightarrow 8\ CO_2 + 10\ H_2O$
(b) O_2
(c) 8.463 g CO_2, 4.331 g H_2O
(d) 7.21 g C_4H_{10}
(e) 25.78 g O_2

9.88 (a) C_2H_2 (b) C_4H_4 (c) C_6H_6

9.90 (a) $(76.59\ \text{g C})\left(\frac{\text{mole C}}{12.011\ \text{g C}}\right) = \frac{6.38\ \text{mole C}}{1.06} = 6$

$(6.39\ \text{g H})\left(\frac{\text{mole H}}{1.0079\ \text{g H}}\right) = \frac{6.34\ \text{mole H}}{1.06} = 6$

$(17.02\ \text{g O})\left(\frac{\text{mole O}}{15.999\ \text{g O}}\right) = \frac{1.06\ \text{mole O}}{1.06} = 1$

Note that the difference between 100% and (%C + %H) = 17.02 g O.
The empirical formula is C_6H_6O

(b) C_6H_6O (c) C_6H_6O

9.92 (a) The % by mass of H in the sample is:

$$\frac{0.2874 \text{ g H}}{2.000 \text{ g sample}} \times 100\% = 14.37\% \text{ H}$$

(b) Since the information was given in grams, not % by mass composition, you need to somewhat modify your approach. Since the compound is said to contain only C and H, you can obtain the mass of C from the difference between the mass of sample and the mass of H:

mass of C = 2.000 g − 0.2874 g = 1.713 g C

Convert grams of each element to moles of each element. This gives you the subscript in the formula for that element.

$$\text{Moles of C} = 2.713 \text{ g C} \times \frac{1 \text{ mol C}}{12.011 \text{ g C}}$$

$$= 0.1426 \text{ mol C}$$

$$\text{Moles of H} = 0.2874 \text{ g H} \times \frac{1 \text{ mol H}}{1.0079 \text{ g H}}$$

$$= 0.2851 \text{ moles H}$$

Divide by the smallest number to obtain the subscripts in the empirical formula. The smallest number is 0.1426.

$$\text{Subscript for C: } \frac{0.1426}{0.1426} = 1$$

$$\text{Subscript for H: } \frac{0.2851}{0.1426} = 2$$

The empirical formula is therefore CH_2.

(c) To find the molecular (actual) formula of the compound, divide the given molar mass of the compound by the molar mass of the empirical formula $(1 \times 12.011 + 2 \times 1.0079 = 14.027)$

$$\frac{71}{14.027} = 5$$

The molecular formula is therefore $C_{1\times5}H_{2\times5} \rightarrow C_5H_{10}$

9.94 (a) $(24.78 \text{ g C})\left(\frac{1 \text{ mole C}}{12.011 \text{ g C}}\right) = 2.06 \text{ mole C}$

$(2.08 \text{ g H})\left(\frac{1 \text{ mole H}}{1.0079 \text{ g H}}\right) = 2.06 \text{ mole H}$

$(73.14 \text{ g Cl})\left(\frac{1 \text{ mole Cl}}{35.453 \text{ g Cl}}\right) = 2.06 \text{ mole Cl}$

Note that this differnce between 100% and (%C + %H) = 73.14 g Cl.
The empirical formula is CHCl

(b) The actual formula would be four times the empirical formula, or $C_4H_4Cl_4$

9.96 C_4H_4ON

9.98 KClO

9.100 38.7% C; 9.74% H; 51.6% O

9.102 23.19% C; 1.43% H; 1.80% N; 8.24% O; 65.34% I

9.104 (a) S atom; (b) 0.670 mol O_3

9.106 (a) $2\,Na(s) + 2\,H_2O(l) \rightarrow 2\,NaOH(aq) + H_2(g)$
(b) Water.

9.109 (a) 342.29 g/mol
(b) 428 g sucrose
(c) 7.53×10^{23} sucrose molecules
(d) 1.66×10^{25} H atoms
(e) 27.8 g hydrogen

9.111 (a) $2\,Cu_2O(s) + C(s) \rightarrow 4\,Cu(s) + CO_2(g)$
(b) 40.07 kg coke

9.113 (a) 4.78 g CO
(b) 3.4×10^2 g CO
(c) 0.171 mol CO
(d) Theoretical yield = 17.1 g CO; percent yield = 78.9%.

9.115 (a) 0.0294 mol $AgNO_3$
(b) 0.0882 mol O
(c) 0.412 g N
(d) 1.77×10^{22} Ag atoms

9.117 (a) 0.33 mol $N_2(g)$
(b) 5.3 mol $N_2(g)$
(c) None left over.
(d) 200 molecules $N_2(g)$.
(e) 1.31 g $N_2(g)$
(f) 2.8 mg $N_2(g)$

9.118 \$23.25

9.119 (a) $Ca_3P_2 + 6\,H_2O \rightarrow 3\,Ca(OH)_2 + 2\,PH_3$
(b) 35.6 g H_2O
(c) 22.4 g PH_3

9.122 39.2 g

9.124 6.98×10^{23} atoms

9.127 $PbCrO_4$

9.129 17.0 g

9.130 $(40.93 \text{ g C})\left(\frac{\text{mole C}}{12.011 \text{ g C}}\right) = \frac{3.41 \text{ mole C}}{3.40}$

$= 1 \xrightarrow{X3} = 3$

$(4.58 \text{ g H})\left(\frac{\text{mole H}}{1.0079 \text{ g H}}\right) = \frac{4.54 \text{ mole H}}{3.40}$

$= 1.33 \xrightarrow{X3} = 4$

$(54.49 \text{ g O})\left(\frac{\text{mole O}}{15.999 \text{ g O}}\right) = \frac{3.40 \text{ mole O}}{3.40}$

$= 1 \xrightarrow{X3} = 3$

Difference between 100 and C + H
The emperical formula is $C_3H_4O_3$, which has a molar mass of 88.06 g/mol

$$\frac{\text{Actual molar mass}}{\text{Empirical molar mass}} = \frac{176 \text{ g/mol}}{88.06 \text{ g/mol}} = 2$$

So the actual formula is $C_6H_8O_6$

9.133 (a) $SiCl_4 + 2\,H_2O \rightarrow SiO_2 + 4\,HCl$
(b) 139.8 g $SiCl_4$, 29.64 g H_2O
(c) 49.43 g SiO_2

9.136 0.693 g CO_2, 0.196 g H_2O

9.138 1.38×10^{23} molecules

9.140 15.77% Al, 28.11% S, 56.11% O

9.142 $(93.66\text{ g C})\left(\dfrac{1\text{ mole C}}{12.011\text{ g C}}\right) = \dfrac{7.80\text{ mole C}}{6.26}$

$= 1.25 \xrightarrow{x4} = 5$

$(6.31\text{ g H})\left(\dfrac{1\text{ mole H}}{1.0079\text{ g H}}\right) = \dfrac{6.26\text{ mole H}}{6.26}$

$= 1 \xrightarrow{x4} = 4$

No oxygen as %C + %H = 100
So the empirical formula is C_5H_4, which has a molar mass of 64.0 g/mol.
Since 128 g/mol ÷ 64.0 g/mol = 2, the actual formula is $C_{10}H_8$.

9.144 212 g
9.146 $C_8H_{10}N_4O_2$
9.149 113.7 g CO, 125.2 g CaC_2
9.151 (a) $S + 2\,H_2SO_4 \rightarrow 3\,SO_2 + 2\,H_2O$ (b) 15.87 g
9.155 18.2 g CaF_2, 20.5 g N_2O, 16.8 g H_2O
9.157 (a) PbS (b) 12.1 g
9.159 2.06 g
9.161 3.98×10^{-11} g
9.165 12.6 g of Cu_2S
9.167 3.53 g
9.169 8.35×10^{20} aspirin molecules, 7.52×10^{21} C atoms.

9.171 $(63.08\text{ g C})\left(\dfrac{1\text{ mole C}}{12.011\text{ g C}}\right) = \dfrac{5.25\text{ mole C}}{1.98} = 2.65$

$(5.30\text{ g H})\left(\dfrac{1\text{ mole H}}{1.0079\text{ g H}}\right) = \dfrac{5.26\text{ mole H}}{1.98} = 2.66$

$(31.62\text{ g O})\left(\dfrac{1\text{ mole O}}{15.999\text{ g O}}\right) = \dfrac{1.98\text{ mole O}}{1.98} = 1$

Difference between 100 and C + H
Final a multiplier that takes C and H subscripts to whole numbers

$3 \times 2.65 = 7.95 \sim 8$

So empirical formula is $C_8H_8O_3$, which has a molar mass of 152 g/mole, so the empirical formula is also the actual molecular formula.

9.174 (a) H_2O (b) 7.9 g
9.175 (a)

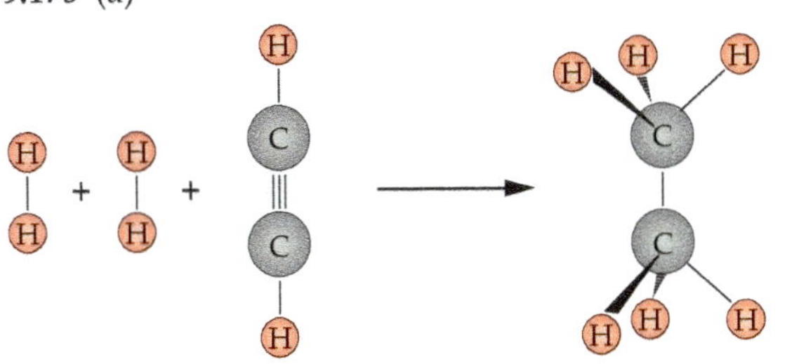

(b) $2H_2(g) + C_2H_2(g) \rightarrow C_2H_6(g)$
(c) 2 moles of hydrogen gas and 1 mole of acetylene (C_2H_2) gas react to give 1 mole of ethane (C_2H_6) gas.
(d) $(10.0\text{ g } C_2H_6)\left(\dfrac{\text{mole } C_2H_6}{30.0694\text{ g } C_2H_6}\right)\times\left(\dfrac{2\text{ mole } H_2}{\text{mole } C_2H_6}\right)\times$

$\left(\dfrac{2.0158\text{ g } H_2}{\text{mole } H_2}\right) = 1.34\text{ g } H_2$

9.177 (a) 1 cup sugar + 3 pkgs. cream cheese + 4 eggs → 12 dessert squares
(b) Still only 12 dessert squares.
(c) The sugar.
(d) 637 cups sugar/1 = 637
2498 pkgs. cream cheese/3 = 832.7
4024 eggs/4 = 1006
The limiting reagent is sugar. 637 cups sugar × (12 dessert squares/1 cup sugar) = 7644 dessert squares.
637 cups sugar × (3 pkgs. cream cheese/1 cup sugar) = 1911 pkgs. cream cheese used, leaving 2498 − 1911 = 587 pkgs. cream cheese left over
637 cups sugar × (4 eggs/1 cup sugar) = 2548 eggs used, leaving 4024 − 2548 = 1476 eggs left over

9.179 (a) $CH_4(g) + 2O_2(g) \rightarrow 2H_2O(g) + CO_2(g)$
(b) 5 moles O_2 × (1 mole CO_2/2 moles O_2) = 2.5 moles CO_2
(c) 5 g O_2 × (1 mole O_2/31.998 g O_2) × (1 mole CO_2/2 moles O_2) × (6.022×10^{23} molecules CO_2/mole CO_2) = 4.70×10^{22} molecules CO_2

Chapter 10

10.2 (a) :Ö=C=Ö:
C oxidation state is +4; each O oxidation state is −2.
(b) None.
(c) 4 valence electrons fewer.
(d) It is correct to say that the C atom has a complete octet because assigning it no electrons in part (b) is only a technique for determining oxidation number and does *not* reflect the number of electrons actually around the C in the CO_2 molecule.

10.3 (a)
```
   H
H:C:H
   H
```
C oxidation state is −4; each H oxidation state is +1.
(b) 8 electrons.
(c) 4 valence electrons more.

10.4 (a)
```
     H
:Cl:C:Cl:
    :Cl:
```
C oxidation state is +2; H oxidation state is +1; each Cl oxidation state is −1.
(b) 2 electrons.
(c) 2 valence electrons fewer.
(d) It is correct to say that the C atom has a complete octet because assigning it only two electrons in part (b) is only a technique for determining oxidation number and does *not* reflect the number of electrons actually around the C in the molecule.

10.6 Mg is +2 (rule 5); each Br is −1 (halide rule).

10.7 Each O is −2 (rule 2); each Fe is +3 (rule 7).
10.8 H is +1 (rule 3); O is −2 (rule 2); Cl is +1 (rule 7).
10.9 Each O is −2 (rule 2); C is +4 (rule 7).
10.10 Cu is +2 (rule 6).
10.12 This is a redox reaction. The oxidation number of Fe changes from 0 to +3 (each Fe loses 3 electrons), and the oxidation number of O changes from 0 to −2 (each O gains 2 electrons). The Fe is oxidized, and the O is reduced.
10.13 This is a redox reaction. The oxidation number of Ca changes from 0 to +2 (Ca loses 2 electrons), and the oxidation number of H changes from +1 to 0 (each H gains 1 electron). The Ca is oxidized, and the H is reduced.
10.14 This is not a redox reaction.
10.16 Fe is the reducing agent, and O_2 is the oxidizing agent.
10.17 Ca is the reducing agent, and H^+ is the oxidizing agent.
10.19 (a) Pb^{2+} (b) Ni
(c) Pb cathode. (d) Ni anode.
10.20 (a) Fe (b) Ag^+
(c)

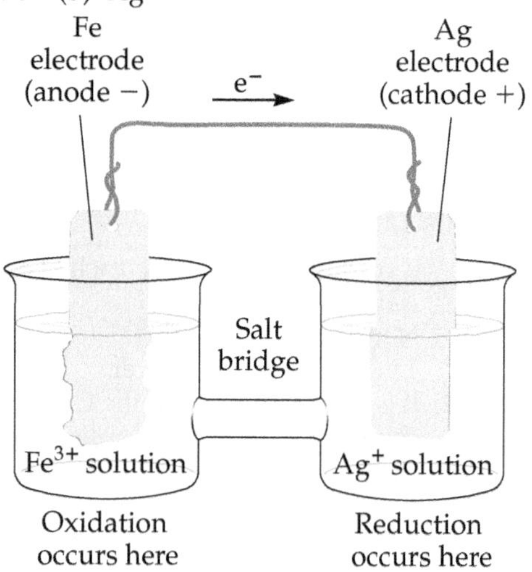

10.22 (a)
10.23

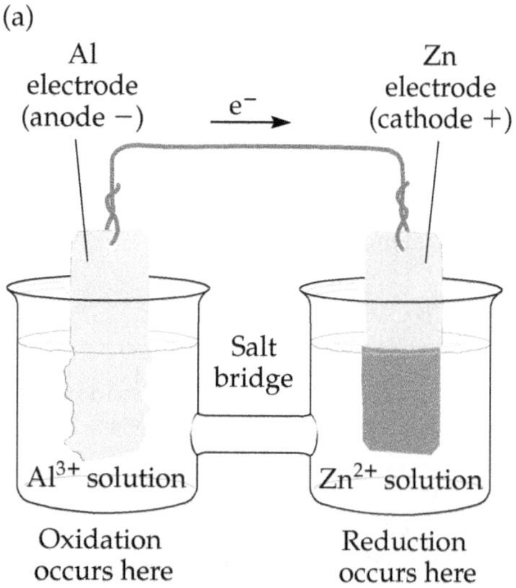

10.24 Lead is higher than copper in the EMF series. Therefore, the oxidizing agent is Cu^{2+}, and the reducing agent is Pb.
10.26 By moving copper wire through a magnetic field. Coils of copper wire are wound around a rotor and spun in a magnetic field by a turbine powered by steam or moving water (hydroelectric).
10.28 To keep track of how many electrons an atom owns, so we can determine whether electron transfer has occurred during a chemical reaction.
10.30 Both methods are correct.
10.32 The more electronegative atom owns the shared electrons.
10.34 H:Ö:Ö:H

Each O owns 7 electrons (both of its lone pairs, both electrons from the O—H bonds because H is less electronegative than O, and 1 electron from the O—O bond). Each H owns 0 electrons because both electrons in the O—H bond go to O.
10.35

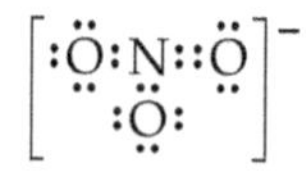

Each O owns 8 electrons (all of its lone pairs and the electrons from the N—O bond because N is less electronegative than O). The N therefore owns 0 electrons.
10.37 (a)–(c) All have oxidation state 0. (d) F has −1 oxidation state; I has +1 oxidation state.
10.39 The sum must always equal the overall charge on the molecule.
10.41 (a) More. (b) 2 more electrons.
10.42 Add up all the other oxidation numbers of the atoms in the compound. Subtract this sum from the total charge on the species. The difference is the oxidation number of the remaining atom.
10.44 (a) O is −2; C is +2.
(b) Each H is +1; each Cl is −1; C is 0.
(c) H is +1; each O is −2; C is +2.
(d) Each Cl is −1; Pt is +4.
10.46 (a) The halide rule works for $AlCl_3$ because Cl is more electronegative than Al. This means both electrons from an Al—Cl bond go to Cl, giving each Cl 8 electrons and therefore an oxidation state of −1.
(b) The halide rule doesn't work for ClO^- because Cl is less electronegative than O. The O therefore owns the electrons in the bond, and the Cl has a positive oxidation state instead of the −1 predicted by the halide rule.
(c) ClO^-: O is −2; Cl is +1.
$AlCl_3$: each Cl is −1; Al is +3.
10.48 (a) 2 oxygens at −2 each and 6 hydrogens at +1 each (using shortcut rules) sums to +2. This means the sum of the oxidation numbers of the 3 carbons must be −2 to get neutrality

(the molecule is neutral). This is not enough information to assign the carbons a unique set of oxidation numbers.

(b) 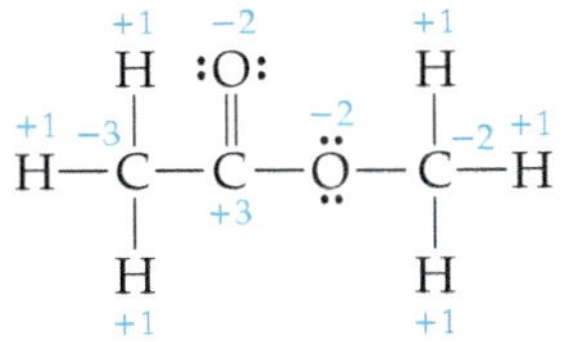

10.51 If the oxidation state of any atom changes during a reaction, a redox reaction has occurred.

10.53 The oxidation state increases.

10.55 Whenever electron transfer occurs, there must be both reduction and oxidation occurring simultaneously; hence the name redox.

10.58 (b), (c), and (d)

10.59 (a) For reaction (b): Fe is oxidized; N in NO_3^- is reduced.
For reaction (c): C in C_2H_6 is oxidized; O in O_2 is reduced.
For reaction (d): Br in AgBr is oxidized; Ag in AgBr is reduced.
(b) For reaction (b): NO_3^- is the oxidizing agent.
For reaction (c): O_2 is the oxidizing agent.
For reaction (d): AgBr is the oxidizing agent.
(c) For reaction (b): Fe is the reducing agent.
For reaction (c): C_2H_6 is the reducing agent.
For reaction (d): AgBr is the reducing agent.

10.61 Yes, because it is an electron-transfer (redox) reaction.

10.63 (a) In the hydroquinone the two carbons bonded to the oxygens are oxidized from +1 to +2.
(b) Ag is reduced.
(c) The oxidizing agent is AgBr.
(d) The reducing agent is hydroquinone.

10.65

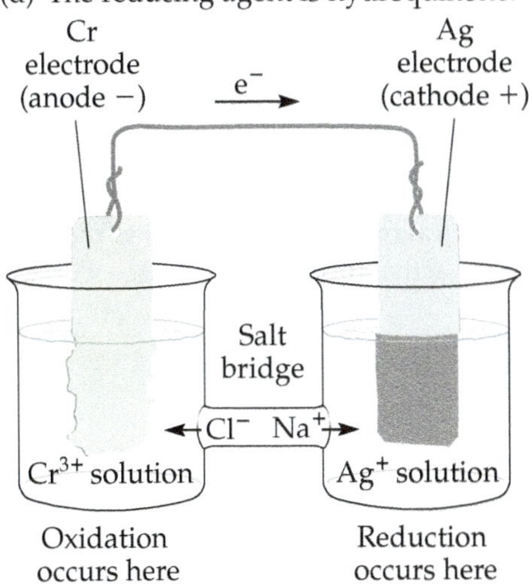

10.66 (a) Ag^+ is the oxidizing agent.
(b) Cr is the reducing agent.
(c) Ag electrode.
(d) Cr electrode.

10.69 (a) The tin (Sn) is getting oxidized (from 0 to +2).
(b) The Cu^{2+} is getting reduced (from +2 to 0).
(c)

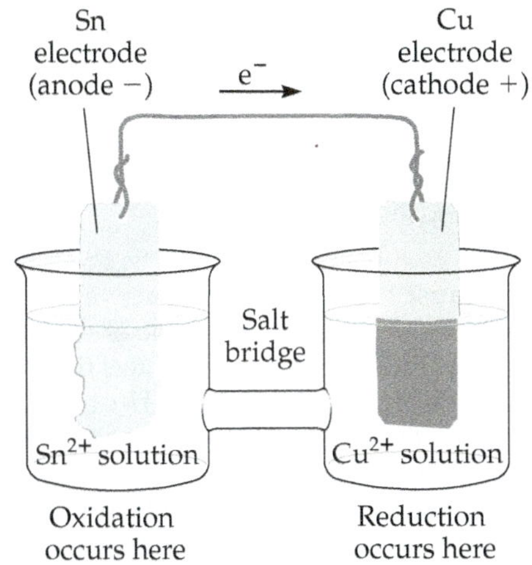

10.71 Electrons are negative and flow from the negative electrode (like charges repel) to the positive electrode (unlike charges attract). Because electrons always flow from the anode to the cathode, the anode must be negative and the cathode positive. Or, you can remember that a *cat*ion is positive (so is a *cat*hode) and an *an*ion is negative (so is an *an*ode).

10.72 (a) $B + A^+ \rightarrow B^+ + A$
(b) The oxidizing agent is A^+.
(c) The reducing agent is B.
(d) Electrons flow from B to A^+.

10.74 You would have to do competitive redox experiments, two metals at a time. For example, you could first test metals X and Y, and see which electrode gets eaten away (and which one gains mass). The metal that gets eaten away (let's say it's X) would be the more active metal. Next, you would repeat the experiment for Y and Z; then again for X and Z. By comparing the results, you could properly order X, Y, and Z according to their relative activities.

10.77 Nothing happens.

10.79 Reaction (a) is spontaneous because Ag is more active than Au.

10.80 (a)

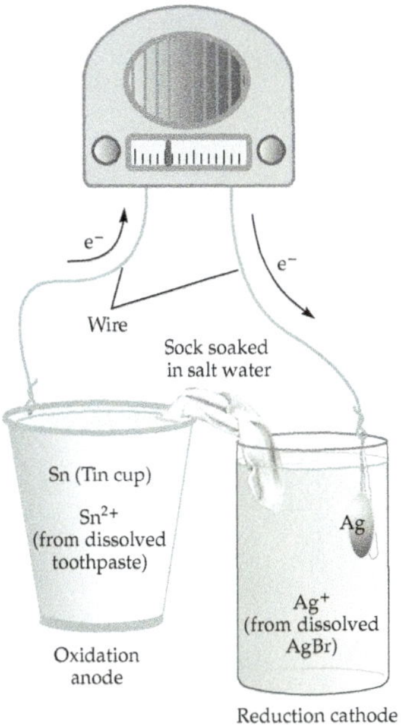

(b) The tin metal would be eaten away as Sn was oxidized to Sn^{2+}.
(c) Ag^+ is reduced, meaning it is the oxidizing agent.
(d) Sn is oxidized, making it the reducing agent.

10.82 (a) $Pb + Cu^{2+} \rightarrow Pb^{2+} + Cu$
(b) $Pb^{2+} + Cu \rightarrow Pb + Cu^{2+}$

10.86 The next-best metal would be Pt (platinum) because it is the next least-active metal.

10.87 The Statue of Liberty is made of sculpted sheets of copper metal placed over a steel (iron) skeleton. Iron is higher on the EMF series than copper. When the copper sheets began to oxidize to Cu^{2+}, the iron skeleton gave up its electrons to reduce the copper ions. Therefore, the iron skeleton was being eaten away and converted to ions. (The asbestos placed in the statue to separate the copper and the iron had rotted away.) The solution to the problem was to replace the dissolving steel skeleton with a reinforced, corrosion-resistant, stainless steel skeleton incorporating high-tech plastics to insulate it from copper.

10.88 If the casing is not made thick enough, the conversion of Zn to Zn^{2+} could result in weak spots along the surface of the battery.

10.89 +7

10.91 C in C_8H_{18} oxidized (from either −3 to +4 or −2 to +4); O in O_2 reduced (from 0 to −2).

10.93 (c) and (e)

10.95 Li^+

10.97 Lower.

10.99 (a) and (c). In (a), I^- is the reducing agent and SeO_3^{2-} is the oxidizing agent. In (c), HCl is the reducing agent and O_2 is the oxidizing agent.

10.101

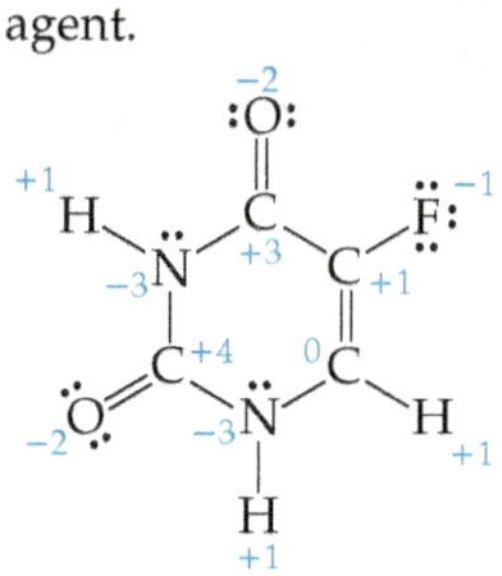

10.103 (a) −3 for both C.
(b) −2 for both C.
(c) −1 for both C.

10.104 (a) Cd^{2+} (b) Mn (c) Cd (d) Mn

10.107 (a) P +5, each F −1.
(b) Each Mo +6, each O −2.
(c) H +1, Pb +2, each O −2.
(d) H +1, each C +3, each O −2.

10.109 (a) Cathode. (b) Anode.

10.110

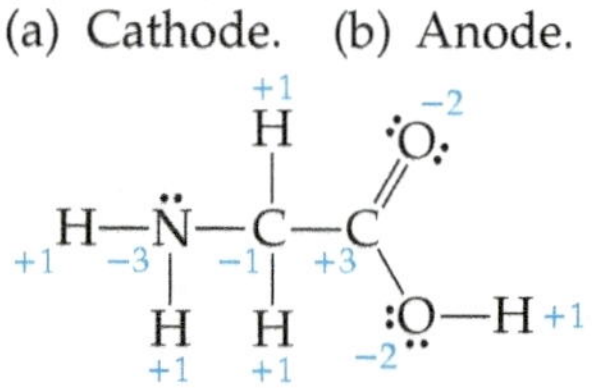

10.112 The Al metal gives up electrons to Ag^+ in Ag_2S, reducing Ag^+ to Ag metal and destroying the tarnish.

10.115 H—C≡N:
+1 +2 −3

10.116 (a) Pb (b) Pd (c) $Pb + Pd^{2+} \rightarrow Pb^{2+} + Pd$

10.118 Oxidizing agent NO_3^-; reducing agent Zn.

10.120 (a) Zinc is the anode; mercury is the cathode.
(b) The zinc anode.

10.121 (b) and (c). In (b), H_2 is the reducing agent and Cl_2 is the oxidizing agent. In (c), Si is the reducing agent and Cr_2O_3 is the oxidizing agent.

10.123 (a) Na +1, P +5, each O −2.
(b) B +3, each O −2, each H +1.
(c) Each V +5, each O −2.
(d) Each K +1, Ti +4, each F −1.

10.125 (a) Ti (b) Zn (c) Above.
(d) $Ti + Zn^{2+} \rightarrow Ti^{2+} + Zn$

10.126 (a) From Al to Ni.
(b) Ni^{2+}
(c) Al
(d) Ni
(e) Al

10.128 (a) Molecule (a) has carbon in its highest oxidation state, +4.
(b) Molecule has carbon in its lowest oxidation state, −4.
(c) All three molecules have an octet-satisfied central carbon atom.

10.130 $(40.93\text{ g C})\left(\dfrac{\text{mole C}}{12.011\text{ g C}}\right) = \dfrac{3.41\text{ mole C}}{3.40} = 1 \xrightarrow{x3} = 3$

$(4.58\text{ g H})\left(\dfrac{\text{mole H}}{1.0079\text{ g H}}\right) = \dfrac{4.54\text{ mole H}}{3.40} = 1.33 \xrightarrow{x3} = 4$

$(54.49\text{ g O})\left(\dfrac{\text{mole O}}{15.999\text{ g O}}\right) = \dfrac{3.40\text{ mole O}}{3.40} = 1 \xrightarrow{x3} = 3$

Difference between 100 and C + H
The empirical formula is $C_3H_4O_3$, which has molar mass of 88.06 g/mol

$\dfrac{\text{Actual molar mass}}{\text{Empirical molar mass}} = \dfrac{176\text{ g/mol}}{88.06\text{ g/mol}} = 2$

So the actual formula is $C_6H_8O_6$

Chapter 11

11.2 $10.50\text{ g }H_2 \times \dfrac{1\text{ mol }H_2}{2.0158\text{ g }H_2} = 5.209\text{ mol }H_2$

11.3 (a) $5 \times 760\text{ mm Hg} = 3800\text{ mm Hg} = 3.80 \times 10^3\text{ mm Hg}$

(b) $3.80 \times 10^3\text{ mm Hg} \times \dfrac{1\text{ cm}}{10\text{ mm}} \times \dfrac{1\text{ inch}}{2.54\text{ cm}} = 150\text{ inches Hg}$

11.4 Too low, because the air in the tube will push down on the mercury.

11.6 Use $P = \dfrac{nRT}{V}$ and solve for T: $T = \dfrac{PV}{nR}$.

$$\text{Mol } N_2 = 3.00 \text{ g } N_2 \times \frac{1 \text{ mol } N_2}{28.014 \text{ g } N_2} = 0.107 \text{ mol } N_2$$

$$\text{Pressure in atm} = 450.5 \text{ mm Hg} \times \frac{1 \text{ atm}}{760.0 \text{ mm Hg}} = 0.5928 \text{ atm}$$

$$T = \frac{0.5928 \text{ atm} \times 2.00 \text{ L}}{0.107 \text{ mol } N_2 \times \dfrac{0.0821 \text{ L} \cdot \text{atm}}{\text{K} \cdot \text{mol}}} = 135 \text{ K}$$

$$°\text{C} = 135 \text{ K} - 273.15 = -138 \text{ °C}$$

11.7 Use $P = \dfrac{nRT}{V}$ and solve for n: $n = \dfrac{PV}{RT}$.

$$\text{Pressure in atm} = 40.0 \text{ lb/in.}^2 \times \frac{760 \text{ mm Hg}}{14.696 \text{ lb/in.}^2} \times \frac{1 \text{ atm}}{760 \text{ mm Hg}} = 2.72 \text{ atm}$$

$$\text{Volume in liters} = 10.5 \text{ gal} \times \frac{3.785 \text{ L}}{1 \text{ gal}} = 39.7 \text{ L}$$

$$T = 22.5 \text{ °C} + 273.15 = 295.7 \text{ K}$$

$$n = \frac{2.72 \text{ atm} \times 39.7 \text{ L}}{\dfrac{0.0821 \text{ L} \cdot \text{atm}}{\text{K} \cdot \text{mol}} \times 295.7 \text{ K}} = 4.45 \text{ mol } O_2$$

$$\text{grams } O_2 = 4.45 \text{ mol } O_2 \times \frac{31.998 \text{ g } O_2}{1 \text{ mol } O_2} = 142 \text{ g } O_2$$

11.10 $1000.0 \text{ gal} \times \dfrac{3.785 \text{ L}}{1 \text{ gal}} = 3785 \text{ L}$

$$\text{K} = °\text{C} + 273.15 = -4.50 \text{ °C} + 273.15 = 268.65 \text{ K}$$

Molar mass of H_2 = 2.0158 g/mol

$$m = \frac{PV(MM)}{RT} = \frac{32.6 \text{ atm} \times 3785 \text{ L} \times 2.0158 \text{ g/mol}}{0.0821 \dfrac{\text{L} \cdot \text{atm}}{\text{K} \cdot \text{mol}} \times 268.65 \text{ K}}$$

$$= 1.13 \times 10^4 \text{ g } H_2 \times \frac{1 \text{ lb}}{453.6 \text{ g}} = 24.9 \text{ lb } H_2$$

11.11 You get the number of moles in the sample:

$$\frac{\text{Mass}}{\text{Molar mass}} = \frac{\text{Grams}}{\text{Grams/mol}} = \text{Gram} \div \frac{\text{Grams}}{\text{Mol}} = \text{Grams} \times \frac{\text{Mol}}{\text{Grams}} = \text{Mol}$$

11.14 Step 1: $P_i = 2.00$ atm, $P_f = ?$
$T_i = 200.0$ K, $T_f = 220.0$ K
$V_i = 50.0$ mL, $V_f = 25.0$ mL
$n_i = ?$, $n_f = ? = n_i$

Step 2: $\dfrac{P_iV_i}{n_iT_i} = \dfrac{P_fV_f}{n_fT_f} \longrightarrow \dfrac{P_iV_i}{n_iT_i} = \dfrac{P_fV_f}{n_iT_f} \longrightarrow \dfrac{P_iV_i}{T_i} = \dfrac{P_fV_f}{T_f}$

Step 3:

(a) $P_f = \dfrac{T_fP_iV_i}{V_fT_i} = \dfrac{220.0 \text{ K} \times 2.00 \text{ atm} \times 50.0 \text{ mL}}{25.0 \text{ mL} \times 200.0 \text{ K}} = 4.40 \text{ atm}$

(b) $50.0 \text{ mL} \times \dfrac{1 \text{ L}}{1000 \text{ mL}} = 0.0500 \text{ L}$

$25.0 \text{ mL} \times \dfrac{1 \text{ L}}{1000 \text{ mL}} = 0.0250 \text{ L}$

$P_f = \dfrac{220.0 \text{ K} \times 2.00 \text{ atm} \times 0.0500 \text{ L}}{0.0250 \text{ L} \times 200 \text{ K}} = 4.40 \text{ atm}$

(c) No.

(d) Because one of the units in R is liters.

11.16 The assumption is that the molecules of an ideal gas are moving independently of one another and feel no attractive forces from one another. We are justified in making this assumption because the molecules are moving very fast in straight-line paths and are relatively far from one another. If they attracted one another, they would not move in straight-line paths.

11.18 By colliding with the walls of its container.

11.21 760 mm Hg, 1 atm

11.23 (a) $29.7 \text{ inches Hg} \times \dfrac{2.54 \text{ cm}}{1 \text{ inch}} \times \dfrac{10 \text{ mm}}{1 \text{ cm}} = 754 \text{ mm Hg}$

(b) $754 \text{ mm Hg} \times \dfrac{1 \text{ atm}}{760 \text{ mm Hg}} = 0.992 \text{ atm}$

11.25 K = °C + 273.15
K = 22.0 °C + 273.15 = 295.2 K

11.28 $600.2 \text{ lb/in.}^2 \times \dfrac{760.0 \text{ mm Hg}}{14.696 \text{ lb/in.}^2} \times \dfrac{1 \text{ atm}}{760.0 \text{ mm Hg}} = 40.84 \text{ atm}$

$$48.5 \text{ lb} \times \frac{453.6 \text{ g}}{1 \text{ lb}} \times \frac{1 \text{ mol } C_2H_2}{26.038 \text{ g}} = 845 \text{ mol } C_2H_2$$

11.30 The pressure decreases because gas pressure is directly proportional to temperature.

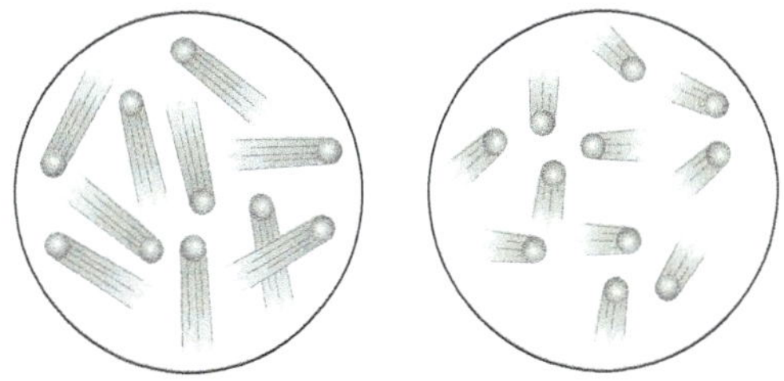

High-temperature, fast-moving molecules collide more frequently and more forcefully with container walls, creating higher pressure.

Low-temperature, slow-moving molecules collide less frequently and less forcefully with container walls, creating lower pressure.

11.32 (a) Sample 1 because it has the fewest molecules in the available space and therefore the molecules collide less frequently with the walls of the container.
(b) Sample 3 because it has the most molecules in the available space and therefore the molecules collide most frequently with the walls of the container.

11.33 Sample 2. In fact, the temperature of sample 2 has been lowered to the point where the gas has condensed to the liquid phase, which is why the molecules are clumped together at the bottom of the container and which is why we had to use quotation marks when referring to this liquid as a "gas."

11.34 The pressure of a gas is inversely proportional to its volume. P is proportional to $1/V$.

$$P = \frac{nRT}{V} = \underbrace{nRT}_{\text{constant}}\frac{1}{V} \Rightarrow P \propto \frac{1}{V}$$

11.35 The pressure of a gas is directly proportional to its temperature. P is proportional to T.

$$P = \frac{nRT}{V} = \underbrace{\frac{nR}{V}}_{\text{constant}}T \Rightarrow P \propto T$$

11.37 Inverse proportionality means that as one value *increases* the other one *decreases*. Direct proportionality means that as one value *increases* the other one *also increases*.

11.39 Inverse proportion; as his age increases, the number of hairs on his head decreases.

11.41 L·atm/K·mole; the units are important because they indicate the units in which the volume, pressure, temperature, and quantity of gas must be expressed when using the ideal gas law.

11.43 $T = PV/nR$;

$$\text{units of } T = \frac{\text{atm}\cdot\text{L}}{\text{mol}\cdot\frac{\text{L atm}}{\text{mol K}}} = \frac{1}{\frac{1}{\text{K}}} = \text{K}$$

11.45 (a) You should get the value of R, which is 0.0821 L·atm/K·mol.
(b) No units, and a numerical value of 1.

11.48 The temperatures must be expressed in kelvins when using the ideal gas law. To get the correct answer, he must convert 25.0 °C to kelvins, double that number, then reconvert to °C:

K = 25.0 °C + 273.15 = 298.2 K

The pressure doubles when this absolute temperature is doubled to 596.4 K, which is equivalent to 323.3 °C.

11.50 K = 0.00 °C + 273.15 = 273.15 K

$$V = \frac{nRT}{P}$$

$$= \frac{(2.0\ \text{mol})\left(\frac{0.0821\ \text{L}\bullet\text{atm}}{\text{K}\bullet\text{mol}}\right)(273.15\ \text{K})}{2.5\ \text{atm}} = 18\ \text{L}$$

11.52 Because $V = nRT/P$, if $T = 0$, $V = 0$. The volume of the gas would be 0 L. That does not make sense. What really happens is that the gas becomes a liquid as it approaches this temperature, and the ideal gas law no longer governs its behavior.

11.55 (a) 22.5 °C + 273.15 = 295.7 K

$$V = \frac{nRT}{P}$$

$$= \frac{(1\ \text{mol})\left(\frac{0.0821\ \text{L}\bullet\text{atm}}{\text{K}\bullet\text{mol}}\right)(295.7\ \text{K})}{1.00\ \text{atm}}$$

$= 24.3\ \text{L}\ H_2$

(b) $1\ \text{mol}\ H_2 \times \frac{2\ \text{mol HCl}}{1\ \text{mol}\ H_2} = 2\ \text{mol HCl}$

$$V = \frac{(2\ \text{mol})\left(\frac{0.0821\ \text{L}\cdot\text{atm}}{\text{K}\cdot\text{mol}}\right)(295.7\ \text{K})}{1.00\ \text{atm}}$$

$= 48.6$ L HCl

11.57 44.0 g/mol

11.59 0.131 g

11.61 (a) By using the ideal gas equation in the form that gives density:

$$\text{Density} = \frac{m}{V} = \frac{P(MM)}{RT}$$

The values for P, R, and T are the same whether you are using the equation for the H_2 sample or the He sample. Because the molar mass of He is greater than the molar mass of H_2, the equation tells you that the density of the He must be greater than the density of the H_2.

(b) Because molar mass is in the numerator of the equation, the densities are directly proportional to the molar masses:

$$\frac{4.003\ \text{g/mol}}{2.0158\ \text{g/mol}} = 1.986$$

The density in the He balloon is 1.986 times the density in the H_2 balloon.

11.63 52.3 lb/in.^{2}

11.65 When the temperature in *kelvins* doubles, so does the average kinetic energy, leading to more collisions with the wall per unit time and more forceful collisions, both effects leading to the increase in pressure.

11.67 The pressure quadruples.

11.69 (a) 22.4 L (b) 24.5 L

11.71 1.34×10^3 g

11.73 Because the model assumes that the gas particles being studied do not interact with one another. If they don't interact, their identity cannot matter.

11.75 4.02×10^{-3} mol, 8.10×10^{-3} g

11.76 1.249 g/L

11.78 0.5191 g

11.81 64.2 g/mol

11.83 1.077×10^{22} atoms
11.85 The same in the two containers.
11.87 76.9 mL
11.88 26.3 atm
11.90 3.00 mol, 60.5 g
11.91 (a) V, n (b) n (c) V, T
11.94 0.0437 mol
11.96 (a) 121 L (b) 0.0383 mol (c) 0.0231 mol
11.99 612 °C
11.101 (a) 298.2 K (b) 18.9 L (c) 0.993 atm
11.103 2.42 atm
11.104 28.0 g/mol
11.106 2.51 g/mol
11.107 (a) 6.123 mol (b) 151.4 L (c) 137.3 L
11.110 115.1 °C
11.112 354 mL
11.114 8.40 atm
11.117 34.1 g/mol
11.120 (a) Grades versus hours of TV is an inverse relationship. As the number of hours of TV increases, the grade decreases.
(b) He is correct. An inverse relationship between variables x and y is expressed as x is proportional to $1/y$.
(c) He is incorrect. The relationship between grades and hours of sleep is direct, not inverse. A direct relationship between variables x and y is expressed as x is proportional to y.
11.122 Yes, there is a limiting reactant, and it is hydrogen. Since Avogadro taught us that at a given temperature, equal volumes of gases contain equal moles, the number of moles of hydrogen is equal to the number of moles of oxygen. However, the balanced equation tells us that we need twice as many moles of hydrogen as moles of oxygen, so hydrogen is in short supply.
11.124

$$PV = nRT, \text{ so } R = \frac{PV}{nT}$$

$$n = (100.0 \text{ g He})\left(\frac{\text{mol He}}{4.003 \text{ g He}}\right) = 24.98 \text{ mol He}$$

$T = 298.15$ K
$P = 5.00$ atm
$V = 120.9$ L

$$R = \frac{(5.00 \text{ atm})(120.9 \text{ L})}{(24.98 \text{ mol})(298.15 \text{ K})} = 0.0812 \text{ L atm/K mol}$$

Chapter 12

12.2 (a) Water is the solvent; acetone is the solute.
(b) Nitrogen is the solvent; oxygen, water vapor, and other gases are the solutes.
(c) Iron is the solvent; chromium, nickel, and carbon are the solutes.
(d) Water is the solvent; aspirin is the solute.
12.3 Because 135-proof vodka is 67.5% alcohol and 32.5% water, alcohol is the major component and therefore the solvent.
12.5 The energy released in dissolving NaCl will be less in carbon tetrachloride than in water. Because carbon tetrachloride is a nonpolar molecule, it has no dipoles that will attract ions. With little or no attraction between the Na^+ and Cl^- ions and the carbon tetrachloride molecules, there will be little or no solvation energy released.
12.6 The energy required for pulling apart the lattice would be greater for $MgCl_2$. The attraction of the +2 magnesium ions for the negative chloride ions would be stronger than the attraction of the +1 sodium ions because the attractive force between oppositely charged particles increases as the magnitude of the charge increases.
12.8 $\Delta E_{\text{solute separation}}$ would be larger for dissolving the solid ionic solute. Because gas molecules have very little attraction for one another and are already separated, $\Delta E_{\text{solute separation}}$ is negligible for a gaseous solute.
12.9 The more negative ΔE_{total}, the more negative $\Delta E_{\text{solvation}}$ is because solvation is the only negative step in the process. Because $\Delta E_{\text{solvation}}$ is a measure of how strongly the solute and solvent are attracted to each other, the more negative ΔE_{total}, the stronger the solute–solvent attraction, and the more likely the solute will dissolve. Also, a very negative ΔE_{total} means that the solvation step releases more than enough energy to balance the energy absorbed in the solute-separation and solvent-separation steps.
12.10 (a) Because Mg is in group IIA of the periodic table and has 2 valence electrons that are lost when Mg forms an octet.
(b) $\Delta E_{\text{solute separation}}$ is a measure of the energy needed to break up the lattice of the solute into individual ions. Because the Mg^{2+} ions have a stronger attraction for the Cl^- ions than do the Na^+ ions, more energy is needed to break the $MgCl_2$ lattice, and $\Delta E_{\text{solute separation}}$ is greater for $MgCl_2$.
(c) $MgCl_2$ is more soluble in water than NaCl because even though $\Delta E_{\text{solute separation}}$ is greater for $MgCl_2$ than for NaCl, $\Delta E_{\text{solvation}}$ is much larger for Mg^{2+} than for Na^+ because the strength of the ion–dipole attractions with water increases as the magnitude of the charge increases.
12.12

$$500.0 \text{ mL water} \times \frac{1.000 \text{ g water}}{1 \text{ mL water}} \times \frac{56.7 \text{ g KCl}}{100 \text{ g water}} = 2.84 \times 10^2 \text{ g KCl}$$

12.13 The biggest difference is in the solute-separation step. Gases do not exist in a lattice, and the particles are already separated. Hence, $\Delta E_{\text{solute separation}}$ is essentially zero for gases.
12.15

$$25.0 \text{ g glucose} \times \frac{1 \text{ mol glucose}}{180.155 \text{ g glucose}} \times \frac{1 \text{ L solution}}{2.55 \text{ mol glucose}} \times \frac{1000 \text{ mL solution}}{1 \text{ L solution}} = 54.4 \text{ mL solution}$$

12.16 $200.0\ \text{mL solution} \times \frac{1\ \text{L solution}}{1000\ \text{mL solution}}$
$\times \frac{2.00\ \text{mol ethanol}}{1\ \text{L solution}} \times \frac{46.068\ \text{g ethanol}}{1\ \text{mol ethanol}}$
$= 18.4\ \text{g ethanol}$

12.17 $400.0\ \text{mL solution} \times \frac{1\ \text{L solution}}{1000\ \text{mL solution}}$
$\times \frac{2.00\ \text{mol NaCl}}{1\ \text{L solution}} \times \frac{58.443\ \text{g NaCl}}{1\ \text{mol NaCl}}$
$= 46.8\ \text{g NaCl required;}$

add water until a volume of 400.0 mL is reached.

12.18 First get moles of NaCl needed:

$400.0\ \text{mL solution} \times \frac{1\ \text{L solution}}{1000\ \text{mL solution}}$
$\times \frac{2.00\ \text{mol NaCl}}{1\ \text{L solution}} = 0.800\ \text{mol NaCl}$

Next get volume of stock solution that contains this many moles of NaCl:

$0.800\ \text{mol NaCl} \times \frac{1\ \text{L solution}}{3.00\ \text{mol}}$
$\times \frac{1000\ \text{mL solution}}{1\ \text{L solution}}$
$= 2.67 \times 10^2\ \text{mL stock solution}$

267 mL of stock solution is needed; place this volume of the stock solution in a 400-mL volumetric flask and add water until a volume of 400.0 mL is reached.
You can also solve this problem with the dilution equation:

$V_{\text{stock}} = \frac{2.00\ \text{M} \times 400.0\ \text{mL}}{3.00\ \text{M}} = 267\ \text{mL}$

12.20 $\text{Percent by volume} = \frac{90.0\ \text{mL ethanol}}{1000\ \text{mL solution}} \times 100$
$= 9.00\ \text{vol \%}$

12.21 $65.0\ \text{g solution} \times \frac{12.5\ \text{g sucrose}}{100\ \text{g solution}}$
$= 8.12\ \text{g sucrose}$

12.24 $255.0\ \text{mL solution} \times \frac{1\ \text{L solution}}{1000\ \text{mL solution}}$
$\times \frac{0.998\ \text{mol glucose}}{1\ \text{L solution}} = 0.254\ \text{mol glucose}$

12.25 $0.500\ \text{mol BaCl}_2 \times \frac{1\ \text{L solution}}{0.350\ \text{mol BaCl}_2}$
$= 1.43\ \text{L solution}$

12.26 $0.500\ \text{mol Cl}^- \times \frac{1\ \text{mol BaCl}_2}{2\ \text{mol Cl}^-}$
$\times \frac{1\ \text{L solution}}{0.350\ \text{mol BaCl}_2} = 0.714\ \text{L solution}$

12.28 $\frac{8.24\ \text{g}}{9.70\ \text{g}} \times 100 = 84.9\%\ \text{yield}$

12.29 Step 1: $Fe^{3+}(aq) + 3\,OH^-(aq) \rightarrow Fe(OH)_3(s)$

Step 2: $20.0\ \text{g Fe(OH)}_3 \times \frac{1\ \text{mol Fe(OH)}_3}{106.9\ \text{g Fe(OH)}_3}$
$= 0.187\ \text{mol Fe(OH)}_3$

Step 3: $0.187\ \text{mol Fe(OH)}_3 \times \frac{1\ \text{mol Fe}^{3+}}{1\ \text{mol Fe(OH)}_3}$
$= 0.187\ \text{mol Fe}^{3+}$

$0.187\ \text{mol Fe(OH)}_3 \times \frac{3\ \text{mol OH}^-}{1\ \text{mol Fe(OH)}_3}$
$= 0.561\ \text{mol OH}^-$

Step 4: $0.187\ \text{mol Fe}^{3+} \times \frac{1\ \text{mol Fe(NO}_3)_3}{1\ \text{mol Fe}^{3+}}$
$\times \frac{1\ \text{L Fe(NO}_3)_3\ \text{solution}}{0.250\ \text{mol Fe(NO}_3)_3}$
$= 0.748\ \text{L Fe(NO}_3)_3\ \text{solution}$

$0.561\ \text{mol OH}^- \times \frac{1\ \text{mol Ba(OH)}_2}{2\ \text{mol OH}^-}$
$\times \frac{1\ \text{L Ba(OH)}_2\ \text{solution}}{0.150\ \text{mol Ba(OH)}_2}$
$= 1.87\ \text{L Ba(OH)}_2\ \text{solution}$

Combine 0.748 L of the ferric nitrate solution with 1.87 L of the barium hydroxide solution, and then filter off the precipitated $Fe(OH)_3$. The amount to be collected for a 67.5% yield is

$\text{Actual yield} = \frac{\%\ \text{yield} \times \text{Theoretical yield}}{100}$
$= \frac{67.5\% \times 20.0\ \text{g}}{100} = 13.5\ \text{g}$

12.31 Step 1: $H^+ + OH^- \rightarrow H_2O$

Step 2: $0.03860\ \text{L NaOH solution}$
$\times \frac{0.100\ \text{mol OH}^-}{1\ \text{L NaOH solution}}$
$= 0.00386\ \text{mol OH}^-$

Step 3: $0.00386\ \text{mol OH}^- \times \frac{1\ \text{mol H}^+}{1\ \text{mol OH}^-}$
$\times \frac{1\ \text{mol H}_3\text{PO}_4}{3\ \text{mol H}^+}$
$= 1.29 \times 10^{-3}\ \text{mol H}_3\text{PO}_4$

Step 4: $\frac{1.29 \times 10^{-3}\ \text{mol H}_3\text{PO}_4}{0.05000\ \text{L}}$
$= 0.0258\ \text{M H}_3\text{PO}_4$

12.32 (a) Step 1: $H^+ + OH^- \rightarrow H_2O$

Step 2: $0.02748\ \text{L NaOH solution}$
$\times \frac{1.000\ \text{mol OH}^-}{1\ \text{L NaOH solution}}$
$= 0.02748\ \text{mol OH}^-$

Step 3: $0.02748 \text{ mol OH}^- \times \dfrac{1 \text{ mol H}^+}{1 \text{ mol OH}^-} \times \dfrac{1 \text{ mol acid}}{1 \text{ mol H}^+}$

$= 0.02748 \text{ mol acid}$

Step 4: $\dfrac{0.02748 \text{ mol acid}}{0.02500 \text{ L}} = 1.099 \text{ M acid}$

(b) Molar mass $= \dfrac{1.65 \text{ g acid}}{0.02748 \text{ mol acid}}$

$= 60.0 \text{ g/mol}$

(c) The molar mass of CH_2O is 30.0 g/mol, making the molecular formula twice the empirical formula, or $C_2H_4O_2$ (Section 9.5). Writing the one acidic H separately gives the formula $HC_2H_3O_2$. Because $C_2H_3O_2^-$ is the acetate ion, the acid is acetic acid.

12.35 Each mole of calcium nitrate dissociates in water to produce 3 moles of solute particles: $Ca(NO_3)_2 \rightarrow Ca^{2+} + 2\,NO_3^-$.

$$100 \text{ g Ca(NO}_3)_2 \times \frac{1 \text{ mol Ca(NO}_3)_2}{164.1 \text{ g Ca(NO}_3)_2} \times \frac{3 \text{ mol solute particles}}{1 \text{ mol Ca(NO}_3)_2}$$

$$= 1.828 \text{ mol solute particles}$$

The temperature at which this solution has a vapor pressure of 760 mm Hg is the temperature at which the solution boils, which you find from the expression

$$\Delta T_b = \frac{0.52 \dfrac{°\text{C} \cdot \text{kg solvent}}{\text{mol solute particles}} \times 1.828 \text{ mol solute particles}}{0.4500 \text{ kg solvent}}$$

$$= 2.1 \text{ °C}$$

The elevated boiling point is therefore 102.1 °C, and it is at this temperature that the vapor pressure is 760 mm Hg.

12.36 Table 12.2 gives 6.47 °C as the normal freezing point of cyclohexane. The freezing point change is thus 6.47 °C − 2.70 °C = 3.77 °C = ΔT_f. Rearranging the expression for ΔT_f gives

$$\text{Mol solute} = \frac{\Delta T_f \times \text{kg solvent}}{K_f}$$

$$= \frac{3.77 \text{ °C} \times 0.225 \text{ kg solvent}}{20.0 \dfrac{°\text{C} \cdot \text{kg solvent}}{\text{mol solute}}}$$

$$= 0.0424 \text{ mol solute}$$

The molar mass is

$$\frac{45.0 \text{ g}}{0.0424 \text{ mol}} = 1.06 \times 10^3 \text{ g/mol}$$

12.38 No, because the sample still consists of flour particles and sugar particles in a nonuniform mixture.

12.40 A dust filled atmosphere would be a heterogeneous mixture and not a solution since the dust particles (solute) and atmosphere (solvent) are in different phases.

12.41 The solute is acetic acid; the solvent is water.

12.43 (a) Solution.
(b) Solution.
(c) Heterogeneous mixture.
(d) Solution.
(e) Heterogeneous mixture.

12.45 A heterogeneous mixture because each solid exists in its own lattice and is not mixed into the lattice of the other solid.

12.46 The + end of the dipole vector is incorrectly drawn near the more electronegative oxygen atom. The head of the arrow should be near the oxygen, and the + end should be near the hydrogens.

12.48 Water would always exist in the gaseous phase if there were no attractive forces.

12.50 In melting there are no solute–solvent interactions to provide energy for the lattice of ions to break apart, whereas in dissolving in water, the attractions between the ions and the water molecules aid in the solution process, therefore requiring less external energy.

12.52 Making room for the solute particles absorbs energy because the attractive forces between solvent molecules must be overcome.

12.55 (1) Solute–separation: overcoming the solute–solute attractive forces.
(2) Solvent–separation: overcoming the solvent–solvent attractive forces.
(3) Solvation: forming solute–solvent attractive forces.

12.57 The associations formed between the ions and water molecules (hydration) help keep the ions from coming back together.

12.59 A water molecule could form hydrogen bonds with the alcohol:

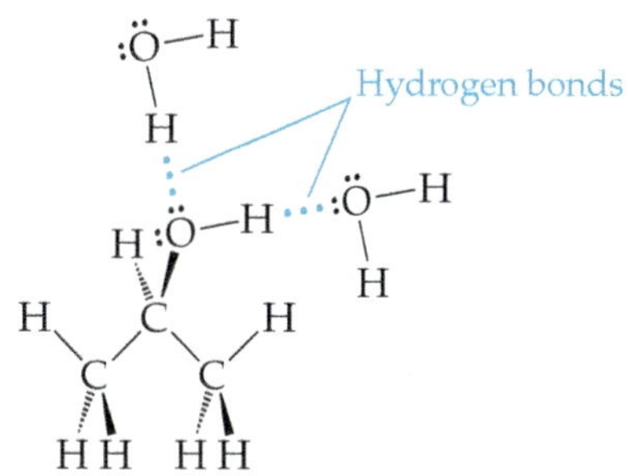

Yes, this interaction is hydration.

12.61 The solid would probably not be soluble in the liquid because the relatively low solvation energy means ΔE_{total} is a positive number.

12.63 Much less energy is required for the solute-separation and solvent-separation steps compared to the energy being released in the solvation (hydration) step.

12.65 Because hexane is a nonpolar compound, there will be little interaction between the ions and the solvent. This means little or no energy released in solvation, with the result that NaCl will not form a solution with C_6H_{14}.

12.67 $AlCl_3$, because the Al^{3+} ions have a much stronger interaction with water than the Na^+ ions.

12.69 "Like dissolves like" means that polar solutes dissolve in polar solvents and nonpolar solutes dissolve in nonpolar solvents.

12.71 (a) Because attractive forces between solute particles are being broken.
(b) Because attractive forces between solvent particles are being broken.
(c) Because attractive forces between solute particles and solvent particles are being formed.
(d) Soluble, because the total energy change is negative.
(e) Warmer, because the solvation step releases more energy than is absorbed by the solute-separation and solvent-separation steps.

12.72 The (*aq*) means that the ion is surrounded by and attracted to water molecules.

12.74 The solute-separation step because there is no lattice to be broken and the solute particles are already separated.

12.76 CCl_4 would be least soluble in water because it is nonpolar and will not form attractive forces with the polar water solvent molecules. CH_3OH would be most soluble because it can form hydrogen bonds with water molecules.

12.79 The natural tendency is for entropy to increase.

12.81 Because the state of higher entropy is the one in which the dye is dispersed.

12.82 Entropy is important when the total energy change, ΔE_{total}, is slightly positive. When ΔE_{total} is either negative or very positive, entropy will not be the deciding factor.

12.84 Because NaCl is a polar compound and hexane is a nonpolar compound, they form no significant solute–solvent interactions. The total energy change is too positive for entropy to be the deciding factor.

12.86 (a) Solubility increases;
(b) solubility decreases.

12.88 The bends occurs when dissolved nitrogen in the blood is released too rapidly if a diver ascends too quickly. The decrease in pressure when the diver rises to the surface causes a decrease in the solubility of nitrogen, and the gas escapes rapidly.

12.90 NaCl shows very little increase in solubility as the temperature increases. Cerium sulfate becomes less soluble as the temperature increases.

12.91 Aquatic life suffers from oxygen depletion due to a decreased concentration of dissolved oxygen as the water is heated.

12.93 $$100.0\ \text{lb water} \times \frac{453.6\ \text{g water}}{1\ \text{lb water}} \times \frac{420.0\ \text{g sucrose}}{100.0\ \text{g water}} = 1.905 \times 10^5\ \text{g sucrose}$$

12.94 (a)

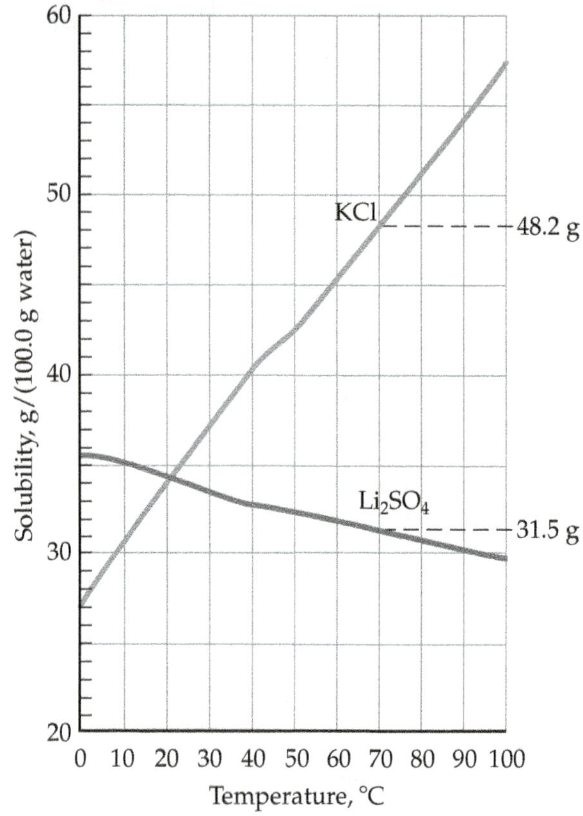

(b) KCl: 48.2 g/100.0 g water; Li_2SO_4: 31.5 g/100.0 g water

(c) $$75\ \text{g water} \times \frac{48.2\ \text{g KCl}}{100.0\ \text{g water}} = 36\ \text{g KCl}$$

$$75\ \text{g water} \times \frac{31.5\ \text{g Li}_2\text{SO}_4}{100.0\ \text{g water}} = 24\ \text{g Li}_2\text{SO}_4$$

12.96 $$5.00\ \text{lb water} \times \frac{453.6\ \text{g water}}{1\ \text{lb water}} \times \frac{40.0\ \text{g KCl}}{100.0\ \text{g water}} = 907\ \text{g KCl}$$

12.98 Molarity is the number of moles of solute present in 1 L of solution: molarity = moles of solute/L of solution.

12.100 $$1.00\ \text{L solution} \times \frac{1.00\ \text{mol C}_{12}\text{H}_{22}\text{O}_{11}}{1\ \text{L solution}} \times \frac{342.295\ \text{g C}_{12}\text{H}_{22}\text{O}_{11}}{1\ \text{mol C}_{12}\text{H}_{22}\text{O}_{11}} = 342\ \text{g C}_{12}\text{H}_{22}\text{O}_{11}$$

Place 342 g of $C_{12}H_{22}O_{11}$ in the volumetric flask, dissolve in about 300 mL of water, and then add enough water to bring the total volume to 1.00 L.

(c) No. To prepare this solution means adding sufficient water to the sucrose to get 45.0 mL of solution, and not adding 45.0 mL of water.

12.102 $0.500\ \text{L solution} \times \frac{1.50\ \text{mol } C_{12}H_{22}O_{11}}{1\ \text{L solution}} \times \frac{342.295\ \text{g } C_{12}H_{22}O_{11}}{1\ \text{mol } C_{12}H_{22}O_{11}} = 257\ \text{g } C_{12}H_{22}O_{11}$

Place 257 g of sucrose in the volumetric flask, dissolve in about 250 mL of water, and then add enough water to bring the volume to 0.50 L.

12.104 You would fire him. To prepare a 2.00 M solution of NaCl, enough water should be added to 116.886 g of NaCl to bring the total volume of solution to 1.00 L. Adding exactly 1.00 L of water will not result in a solution volume of 1.00 L, which means the NaCl concentration will not be 2.00 M.

12.106 (a) $45.0\ \text{mL} \times \frac{1\ \text{L}}{1000\ \text{mL}} \times \frac{0.250\ \text{mol sucrose}}{1\ \text{L}} = 0.0113\ \text{mol sucrose}$

(b) $0.0113\ \text{mol sucrose} \times \frac{342.295\ \text{g sucrose}}{1\ \text{mol sucrose}} = 3.85\ \text{g sucrose}$

(c) No. To prepare 45.0 mL of this solution requires adding sufficient water to the sucrose to get 45.0 mL of solution, not adding 45.0 mL of water.

12.108 $100.0\ \text{g glucose} \times \frac{1\ \text{mol glucose}}{180.1548\ \text{g glucose}} \times \frac{1\ \text{L solution}}{0.250\ \text{mol glucose}} \times \frac{1000\ \text{mL solution}}{1\ \text{L solution}} = 2.22 \times 10^3\ \text{mL}$

12.110 $V_{\text{stock solution}} = \frac{4.00\ \text{M} \times 100.0\ \text{mL}}{4.50\ \text{M}} = 88.9\ \text{mL}$

Instructions: Put 88.9 mL of stock solution in a 100-mL volumetric flask, and add enough water to bring the total volume to 100.0 mL.

12.112 (a) Percent by mass, percent by volume, and percent by mass/volume.
(b) Solvent.
(c) (1) $\text{Percent by mass} = \frac{\text{Grams of solute}}{\text{Grams of solution}} \times 100$

(2) $\text{Percent by volume} = \frac{\text{Volume of solute}}{\text{Volume of solution}} \times 100$

(3) $\text{Percent by mass/volume} = \frac{\text{Grams of solute}}{\text{Milliliters of solution}} \times 100$

12.114 $\frac{22.5}{22.5 + 49.6} \times 100 = 31.2\ \text{mass \%}$

12.116 $2.00\ \text{kg solution} \times \frac{1000\ \text{g solution}}{1\ \text{kg solution}} \times \frac{30.0\ \text{g NaCl}}{100\ \text{g solution}} = 600\ \text{g NaCl}$

Place 600 g of NaCl in a flask. Add 1.40 kg of water to give a total of 2.00 kg of solution.

12.118 $200.0\ \text{mL alcohol} \times \frac{100.0\ \text{mL solution}}{35.00\ \text{mL alcohol}} = 571.4\ \text{mL solution}$

12.120 A sphere of water molecules surrounding a dissolved ion.

12.122 (a) $Pb^{2+}(aq) + S^{2-}(aq) \rightarrow PbS(s)$
(b) 41.8 mL $Pb(NO_3)_2$ solution, 27.9 mL Na_2S solution.
(c) 83.6 mL $Pb(NO_3)_2$ solution, 55.8 mL Na_2S solution.

12.124 (a) $2\,Fe^{3+}(aq) + 3\,CO_3^{2-}(aq) \rightarrow Fe_2(CO_3)_3(s)$
(b) 19.0 g
(c) 0.263 M CO_3^{2-}

12.126 (a) $H^+(aq) + OH^-(aq) \longrightarrow H_2O\,(l)$

(b) The number of moles of OH^- used to neutralize the acid is:

$0.04328\ \text{L NaOH} \times \frac{0.1001\ \text{mol NaOH}}{1\ \text{L NaOH}} \times \frac{1\ \text{mol OH}^-}{1\ \text{mol NaOH}} = 0.004332\ \text{mol OH}^-$

(c) From the stoichiometry of the equation in part (a) and the fact that there is one H^+ in HBr:

$0.004332\ \text{mol OH}^- \times \frac{1\ \text{mol H}^+}{1\ \text{mol OH}^-} \times \frac{1\ \text{mol HBr}}{1\ \text{mol H}^+} = 0.004332\ \text{mol HBr}$

(d) This number of moles in the 25.00 mL acid sample means the molar concentration of HBr is

$\frac{0.004332\ \text{mol}}{0.02500\ \text{L}} = 0.1733\ \text{M}$

12.128 (a) 0.03686 M (b) 74.1 g/mol

12.129 (a) −115.0 °C (b) −115.0 °C
(c) 78.4 °C (d) 78.4 °C
(e) Because the temperature of a substance stays constant while the substance is changing phase at its freezing or boiling point.
(f) Ethanol.

12.131 To break the hydrogen bonds holding the H_2O molecules together in the liquid phase.

12.133 The pressure exerted by the gas phase of a substance when the gas phase is in dynamic equilibrium with the liquid phase of the substance.

12.135 Reduce the air pressure above the water surface to 17.54 mm Hg.

12.137 760 mm Hg

12.139 336 g

12.141 28.1 g

12.142 (a) 102.5 °C (b) −8.76 °C

12.144 (a) 194 g/mol (b) $C_8H_{10}N_4O_2$

12.146 (a)

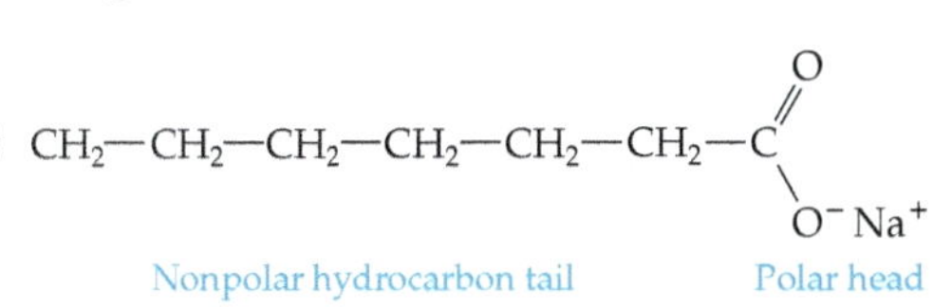

(b) The nonpolar tail is derived from fatty acids found in animal fats. Animal fats are esters of glycerol and fatty acids. When animal fat is cooked with NaOH or KOH, the long chained anions of fatty acids are released from the fat and form soluble salts with Na^+ or K^+ ions from NaOH or KOH. In their pure forms, the soaps containing Na^+ are liquid and soaps containing K^+ are solid.

(c) The nonpolar tail has an affinity for nonpolar molecules, such as grease and oil, while the polar head has an affinity for polar molecules, such as water.

12.148 Because the nonpolar hydrophobic tail of each soap molecule needs to be isolated from water, the soap molecules form a sphere to allow the tails to reside inside and away from the water. If the micelle were flat and two-dimensional, the water would be in direct contact with the hydrophobic tails.

12.151 The hydrophobic portion is the hydrocarbon tail. The hydrophilic portion is the

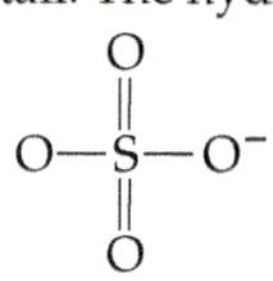
head group.

12.152 $Mg(OH)_2(s)$ releases the most, $CO_2(g)$ the least.

12.154 The lattice in $Mg(OH)_2$.

12.156 (c)

12.157 1.50×10^{-4} mol

12.159 Pour 14.6 mL of the stock solution into a 250-mL volumetric flask and dilute to the mark with water.

12.161 (a) 0.0400 mol Na^+, 0.0400 mol Cl^-
(b) 0.210 mol K^+, 0.0700 mol PO_4^{3-}
(c) 0.288 mol Al^{3+}, 0.864 mol NO_3^-

12.163 (a) 0.196 M (b) 1.96×10^{-4} mol
(c) 3.09 mass %

12.165 $C_6H_8O_6$

12.167 0.83 gallon

12.169 (a) 0.0714 M (b) 0.214 M (c) 0.0714 M

12.171 300 mL

12.173 245 mL; Na^+ and $C_2H_3O_2^-$

12.174 (a) 9.192×10^{-3} M (b) 193 g/mol

12.176 (a) Solvent copper, solute zinc.
(b) Solvent water, solute ammonia.
(c) Solvent sucrose, solute water.

12.179 More soluble in CCl_4 than in H_2O because I_2 and CCl_4 are both nonpolar, H_2O is polar, and like dissolves like.

12.181 The $Al^{3+}—PO_4^{3-}$ bonds in the $AlPO_4$ lattice are stronger than the $Na^+—PO_4^{3-}$ bonds in the Na_3PO_4 lattice.

12.182 The ethanol solvation step releases a lot more energy than does the diethyl ether solvation step because more hydrogen-bonding is possible in ethanol:

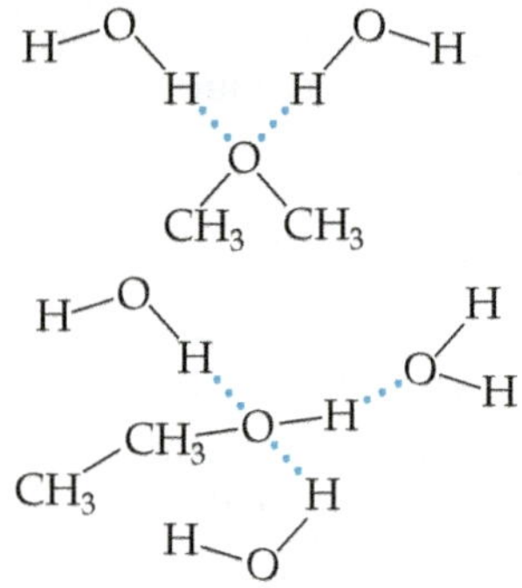

12.184 A solute can dissolve even when the energy *released* by the solute–solvent interactions is less than the energy absorbed by the solute–solute and solvent–solvent interactions because the entropy of the system *increases*.

12.186 (a) is correct because it says s is directly proportional to P—s increases as P increases and decreases as P decreases.
(b) wrongly says s is inversely proportional to P—s increases as P decreases and decreases as P increases.
(c) wrongly says s does not depend at all on P because the two P cancel, leaving you with the incorrect $s = k$.

12.188 Yes.

12.190 The higher temperature makes the H_2O molecules move faster, on average, giving them more energy. The more energetic molecules collide more forcefully with the solid and thus break more solute–solute bonds.

12.192 The molecules migrate to the surface and orient as they do so that the hydrophobic tails are as far away from the water as possible.

12.194 (a)
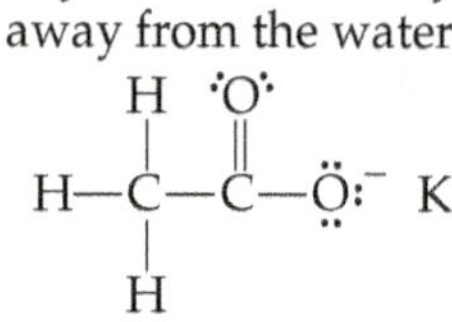

(b) Both consist of a cation plus an anion containing a $—C(=O)—O^-$ unit. The potassium acetate does not have the hydrocarbon tail found in the soap molecule.

(c)
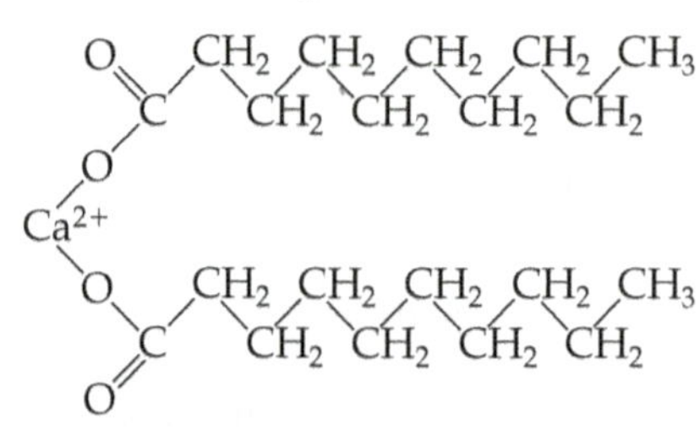

$Ca(C_{10}H_{19}O_2)_2$

12.196 (a) 0.0750 M (b) 0.0750 M (c) 0.225 M
12.198 0.188 M
12.200 2.80 g
12.202 0.68 M
12.204 33.3 mL
12.207 0.498 mass %
12.209 55 g
12.211 1.30×10^4 g
12.213 (a) 0.194 g (b) 0.00210 M
12.214 (a) $Fe^{3+}(aq) + 3\,OH^-(aq) \rightarrow Fe(OH)_3(s)$
(b) 16.0 g
(c) 68.4%
(d) 0.17 M
12.216 (a) The solvent separation step whereby holes or spaces within the solvent are created for the solute.
(b) This step always requires energy.
(c) Less energy, since the intermolecular forces between CCl_4 molecules (boiling point 77 °C) are weaker than the intermolecular forces between H_2O molecules (boiling point 100 °C).
12.218 You could either heat the water such that the vapor pressure grows to match the external pressure *P*, or you could decrease the external pressure *P* using a vacuum pump so that it shrinks to match the vapor pressure.
12.220 Because the act of titrating only tells you the moles of acid (H^+) that were present in the flask. To determine the molar concentration of H^+ in the flask, you must divide the moles of H^+ by the volume of solution that it was in (molarity of H^+ = moles H^+/volume it is in).

Credits

Chapter 1

p. 3 (l): Lunamarina/Fotolia; p. 3(b): Adrian Neal/Stone/Getty Images; p. 3 (t): National Library of Medicine; p. 4 (cl): Corbin17/Alamy; p. 4 (cr): Rupert Robertson/Fotolia; p. 4 (l): BanksPhotos/E+/Getty Images; p. 4 (r): Photodisc/Getty Images; p. 5: NASA/Jet Propulsion Laboratory; p. 6 (l): Pearson Education/Pearson Science; p. 6 (r): Pearson Education/Pearson Science; p. 7 (cl): Photodisc/Getty Images; p. 7 (l): Pearson Education/Pearson Science; p. 7 (c): Pearson Education/Pearson Science; p. 7 (cr): Pearson Education/Pearson Science; p. 7 (r): Pearson Education/Pearson Science; p. 8: Andrew Dunn/Alamy; p. 9 (l): Photodisc/Getty Images; p. 9 (tl): Steve Wisbauer/Photodisc/Getty Images; p. 9 (tr): Pearson Education/Pearson Science; p. 9 (bl): Pearson Education/Pearson Science; p. 9 (bc): Pearson Education/Pearson Science; p.9 (br): Pearson Education/Pearson Science; p. 9 (r): Eky Studio/Shutterstock; p. 10 (l); p. 10 (r): Pearson Education/Pearson Science; p. 11 (l): Pearson Education/Pearson Science; p. 11 (r): Pearson Education/Pearson Science; p. 12 (t): Pearson Education/Pearson Science; p. 12 (b): Pearson Education/Pearson Science; p. 13 (l): Pearson Education/Pearson Science; p. 13 (c): Pearson Education/Pearson Science; p. 13 (r): Pearson Education/Pearson Science; p. 14 (t): Pearson Education/Pearson Science; p. 14 (b): Pearson Education/Pearson Science; p. 19: Pearson Education/Pearson Science; p. 27: Iaroslav Lazunov/Fotolia;

Chapter 2

p. 29: NASA/Jet Propulsion Laboratory; p. 33: Pearson Education/Pearson Science; p. 40: Pearson Education/Pearson Science; p. 46: Pearson Education/Pearson Science; p. 48: Pearson Education/Pearson Science; p. 51: Pearson Education/Pearson Science;

Chapter 3

p. 78: Science Source; p. 80 (tl): Pearson Education/Pearson Science; p. 80 (tl): Pearson Education/Pearson Science; p. 80 (tr): Pearson Education/Pearson Science; p. 80 (tr): Pearson Education/Pearson Science; p. 80 (b): Pearson Education/Pearson Science; p. 81: Pearson Education/Pearson Science; p. 87: Tom Pennington/Getty Images; p. 90: Pearson Education/Pearson Science; p. 94 (cl): Pearson Education/Pearson Science; p. 94 (bl): Pearson Education/Pearson Science; p. 94 (bc): Pearson Education/Pearson Science; p. 94 (br): Pearson Education/Pearson Science; p. 95 (tl): Pearson Education/Pearson Science; p. 95 (tc): Pearson Education/Pearson Science; p. 95 (tr): Pearson Education/Pearson Science; p. 95 (bl): Pearson Education/Pearson Science; p. 95 (bc): Pearson Education/Pearson Science; p. 95 (br): Pearson Education/Pearson Science; p. 100 (inset): Lawrence Stepanowicz/Alamy; p. 100 (cl): Bogdan Vasilescu/Fotolia; p. 100 (b): Aleksandar Todorovic/Fotolia; p. 107 (tl): Emmanuel Faure/Photodisc/Getty Images; p. 107 (tr): Science Source; p. 107 (bl): Richard Treptow/Science Source; p. 107 (br): Richard Treptow/Science Source; p. 117 (cr): David Mager/Pearson Learning Photo Studio;

Chapter 4

p. 121: Richard Treptow/Science Source; p. 122 (t): NASA/Jet Propulsion Laboratory; p. 122 (b): Science Source;

Chapter 5

p. 168: WimL/Fotolia; p. 169: SCIMAT/Science Source; p. 170: Anthony Berenyi/Shutterstock; p. 171: Dustin Dennis/Shutterstock; p. 175: Pearson Education/Pearson Science; p. 195: Bluerain/Shutterstock;

Chapter 6

p. 216: Pearson Education/Pearson Science; p. 217: Pearson Education/Pearson Science; p. 236: Pearson Education/Pearson Science;

Chapter 7

p. 252: Pearson Education/Pearson Science; p. 253: (cl): imageegami/Fotolia; p. 253 (cr): PT1/WENN/Newscom; p. 254 (cl): Pearson Education/Pearson Science; p. 254 (c): Pearson Education/Pearson Science; p. 254 (cr): Pearson Education/Pearson Science; p. 255: Pearson Education/Pearson Science; p. 257: Pearson Education/Pearson Science; p. 258: Pearson Education/Pearson Science; p. 267 (l): Steve Cole/Photodisc/Getty Images; p. 267 (cr): Pearson Education/Pearson Science; p. 268 (b): Photodisc/Getty Images; p. 268 (cl): Pearson Education/Pearson Science; p. 268 (cr): Pearson Education/Pearson Science; p. 269 (cr): Photodisc/Getty Images; p. 269 (cl): Pearson Education/Pearson Science; p. 269 (cl): Pearson Education/Pearson Science; p. 269 (cr): Pearson Education/Pearson Science; p. 271: Richard T. Nowitz/Science Source;

Chapter 8

p. 283 (br): Stonemeadow Photography/Alamy; p. 283 (tr): U.S. Navy News Photo; p. 283 (cl): U.S. Navy News Photo; p. 283 (cr): U.S. Navy News Photo; p. 284 (tl): Junko Kimura/Getty Images; p. 296 (cl): Richard Megna/Fundamental Photographs; p. 299: Alexander Dashewsky/Shutterstock;

Chapter 9

p. 315 (cr): abimages/Shutterstock; p. 315 (l): Luisa Puccini/Shutterstock; p. 316 (tl): Luisa Puccini/Shutterstock; p. 316 (tr): abimages/Shutterstock; p. 317: Dustin Dennis/Fotolia; p. 321: Pearson Education/Pearson Science; p. 346: Jeff Weissman/Photographic Guide to Mineral Species; p. 359 (bc): Viktor/Fotolia; p. 359 (r): Luisa Puccini/Shutterstock; p. 359 (l): FoodCollection/SuperStock;

Chapter 10

p. 380 (tl): Pearson Education/Pearson Science; p. 380 (tc): Pearson Education/Pearson Science; p. 380 (tr): Pearson Education/Pearson Science; p. 381 (tl): Pearson Education/Pearson Science; p. 381 (br): Pearson Education/Pearson Science; p. 382: Pearson Education/Pearson Science; p. 383 (l): Pearson Education/Pearson Science; p. 383 (r): Pearson Education/Pearson Science; p. 384 (c): Pearson Education/Pearson Science; p. 384 (bl): Pearson Education; p. 390 (l): Pearson Science/Pearson Education; p. 390 (cr): Pearson Education/Pearson Science; p. 391: Pearson

Education/Pearson Science; p. 392 (tl): Pearson Education/Pearson Science; p. 392 (tc): Pearson Education/Pearson Science; p. 392 (bl): Pearson Education/Pearson Science; p. 392 (br): Pearson Education/Pearson Science; p. 394: MasterLu/Fotolia; p. 396: Marty Honig/Jupiter Images; p. 408: Charles D. Winters/Science Source;

Chapter 11

p. 412: Pearson Education/Pearson Science; p. 413: Science and Society/SuperStock;

Chapter 12

p. 448: Pearson Education/Pearson Science;

Index

Useful Conversion Factors

Length

SI unit: meter (m)

1 km = 0.621 37 mi

1 mi = 5280 ft (exactly)

= 1.6093 km

1 m = 3.28 ft

= 39.37 in

= 1.0936 yd

1 in = 2.54 cm (exactly)

1 cm = 0.393 70 in

1 Å = 10^{-10} m

1 nm = 10^{-9} m

Mass/Weight

SI unit: kilogram (kg)

1 kg = 10^3 g = 2.2046 lb

1 lb = 16 oz = 453.6 g

1 ton = 2000 lb

1 atomic mass unit = 1.66×10^{-24} g

Temperature

SI unit: Kelvin (K)

0 K = −273.15 °C

= −459.67 °F

K = °C + 273.15

$°C = \frac{5}{9}(°F - 32)$

$°F = \frac{9}{5}(°C) + 32$

Energy

SI unit: Joule (J)

1 J = 0.239 01 cal

= 1 C × 1 V

1 cal = 4.184 J

1 eV = 1.602×10^{-19} J

Pressure

SI unit: Pascal (Pa)

1 Pa = 1 kg/(m · s^2)

1 atm = 101,325 Pa

= 760 mm Hg (torr)

= 29.9 in Hg

= 14.696 lb/in^2

Volume

SI unit: cubic meter (m^3)

1 L = 10^{-3} m^3

= 1 dm^3

= 10^3 cm^3

= 1.057 qt

1 L = 0.264 gal

1 m^3 = 264 gal

1 gal = 4 qt

= 3.7854 L

1 cm^3 = 1 mL

= 10^{-6} m^3

1 in^3 = 16.4 cm^3

Fundamental Constants

Avogadro's number = 6.022×10^{23}/mole

Gas constant, R = 0.0821 L · atm/K · mole

Mass of electron, m_e = $9.109\,390 \times 10^{-31}$ kg
= 1/1836 of mass of H

Mass of neutron, m_n = $1.674\,929 \times 10^{-27}$ kg
≈ mass of H

Mass of proton, m_p = $1.672\,623 \times 10^{-27}$ kg
≈ mass of H

Planck's constant, h = 6.626×10^{-34} J · s

Speed of light, c = 3.00×10^8 m/s